Contents

*Numbers in brackets refer to Bibliography on pp. 635–642.

Biographical Note on Jacob Wolfowitz

This volume of selected papers of Jacob Wolfowitz is published to honor the 70th birthday of this leader in research in statistical inference and information theory. He was born on March 19, 1910, in Warsaw, Poland, and came to the United States in 1920. He was educated in New York City schools, and graduated from the College of the City of New York in 1931. The depression was then at its depth, and he supported himself until 1942 by teaching in various high schools. During part of this period he was also working toward his Ph.D. at New York University, where he received his degree in 1942.

In 1934 Wolfowitz married Lillian Dundes, who has been his cherished companion for over 45 years. Their children, Laura Mary and Paul Dundes, were born in 1941 and 1943, respectively. Laura is a biologist, and Paul a political scientist with a special interest in defense problems. Laura lives in Israel with her Israeli husband and three children, while Paul is a senior Defense Department official in Washington, where he lives with his wife and child. The grandchildren have received citations in the work of a proud grandfather.

In 1938 Wolfowitz met Abraham Wald, and almost at once began the collaboration which lasted until Wald's tragic death in an airplane crash in India in 1950. They were the closest of friends, and Wolfowitz regarded Wald as his teacher as well as his co-worker. Their work together produced some of the most important and striking results in theoretical statistics. During World War II both worked for the war effort in the Statistical Research Group at Columbia. In 1945 Wolfowitz went to the University of North Carolina as an associate professor, returning to Columbia the following year. During the late 1940's Columbia attracted other distinguished faculty members and a large body of graduate students. In 1951, after Wald's death, Wolfowitz moved to Cornell University and remained there until 1970, when he moved to the University of Illinois at Urbana. Upon his retirement from Illinois in 1978 he became Distinguished Professor at the University of South Florida in Tampa, where he now lives.

Wolfowitz was a visiting professor at UCLA in 1952, at the University of Illinois in 1953, at the Israel Technion in 1957 and 1967, at the University of Paris in 1967, and at the University of Heidelberg in 1969. The Technion awarded him an honorary doctorate in 1975. His other honors include election to the National Academy of Sciences of the U.S.A. and the American Academy of Arts and Sciences, and election as Fellow of the Econometric Society, the International Statistics Institute, and the Institute of Mathematical Statistics; he served a term as president of the latter. He has given invited addresses at an International Congress of Mathematicians and at numerous scientific meetings in the U.S. and in Germany, Japan, the USSR, Czechoslovakia, and Hungary. He is one of the few statisticians to have been both the Rietz Lecturer and also the Wald Lecturer of the IMS. In 1979 he was the Shannon Lecturer of the Institute of Electrical and Electronic Engineers.

Wolfowitz is a superb teacher. His lectures reflect his own insistence on understanding the essential features of a proof. "Let's see what makes things tick," his classes hear, and his students and audiences at scientific meetings have the privilege of receiving a lively and lucid exposition that enables them to appreciate the crucial ideas of a subject much more than does the customary formal lecture or line-by-line proof. A large number of co-workers—Wald, Dvoretzky, Hoeffding, Chung, Kiefer, Weiss, and Teicher—attest to the stimulating experience of doing joint research with him. In research discussions he is energetic, probing, critical, humorous, and very inventive. His students, some of whom became his colleagues, always found generosity, patience, and deep personal concern along with his helpful criticism. He is a voracious reader, and his knowledge of, and intense interest in, all facets of the state of the world, make him an interesting person with whom to discuss almost anything.

As this volume demonstrates, the breadth of his research interests is remarkable. The work in asymptotic statistical theory alone, or the work in information theory alone, puts him at the top of his profession; and there is much more work outside these two areas. He has always exhibited the gift of being able to step back and see his subject from a vantage point that allows a broad perspective, thus enabling him to select the most significant areas for exploration. We may hope that he enjoys many more fruitful years in this journey.

Introduction to the Research of Jacob Wolfowitz

It is difficult to describe in a few pages the stream of 40 years of research by one whose interests have been as broad as Wolfowitz's. Three of us have shared this task, partly to improve the chance that we do justice to the material, partly because it would be inappropriate for a paper to be reviewed only by one of its co-authors.

We have made an attempt to organize the material into sections based on subject matter. Since Wolfowitz's many interests have persisted over time, the sections do not represent chronological divisions. Moreover, there is necessarily some arbitrariness; some papers that prove asymptotic results are listed under "Nonparametric inference," and some results in nonparametric inference under "Asymptotic statistical theory." The assignments reflect an attempt to keep related papers together, and perhaps a judgment about the question that engendered a series of papers.

Our style is informal, uneven in detail (as we emphasize some ideas more than others or find that some take longer to explain), and is deficient in reference to other authors' related work or in mention of the other authors of each co-authored paper. The latter are listed in the publication list, while one can look at the papers themselves to find their background, and leafing through obvious titles in such publications as the *Current Index to Statistics* will yield a list of papers that depend on, or are related to, those of Wolfowitz. It did not seem worth doubling or tripling the length of this introduction to include and explain these hundreds of related papers, only a few of which might interest the reader.

Most of Wolfowitz's papers are described to some degree in this introduction. The selection of those papers that have been reprinted in this volume has been extremely difficult. We have tried to choose the papers we regarded as most important among Wolfowitz's work in terms of their further influence, or sometimes a paper that contains what we found a striking idea of his. A further consideration was the lucidity of his explanations that we mentioned in describing his teaching; some of his papers contain intuition or heuristics that no statistician should miss. The obvious limitations of space made us exclude many worthy candidates.

The informality of our discussion of Wolfowitz's research is prompted by the desire to make it easy and quick to read, thereby also giving a feeling for the scope of the work without having the reader bogged down by technical details. We use few symbols and usually make no attempt to list regularity conditions. The distribution function of independent, identically distributed random variables is the df of iidrv's, and all other abbreviations are defined as they are introduced. No attempt has been made to define all the terms we use; they will be familiar to readers with even a slight statistical background.

We thank Walter Kaufmann-Bühler for his indispensable help in making this book materialize.

1. Nonparametric Inference.

Settings in which the class of possible probability laws of the observations cannot be represented smoothly as a finite dimensional set are termed nonparametric. Increasingly, interest in the well-developed parametric theory of statistics has given way to nonparametric developments as practitioners realized that an assumption such as normality of measurement errors was sometimes unrealistic, and that methodologies based on such false models could yield disastrously inaccurate results. Nonparametric inference was the earliest of Wolfowitz's research interests, one that he still occasionally pursues. His first paper [1], written with Wald, gives methods for computing confidence bands, not necessarily of fixed width, on a df F, based on the empiric df from n iid observations on F. It remains a starting point for current research. In [5] the limiting "permutation distribution" of the serial correlation coefficient is obtained. In [9] a general treatment of the distribution of a linear function of a random permutation is given, with applications of the limiting normal law to Fisher–Pitman tests for such applications as the two-sample problem and tests of independence. This is a pioneering paper, whose influence on asymptotic nonparametric developments is still seen.

The empiric df is shown to be asymptotically minimax for estimating an unknown continuous df, for a wide variety of loss functions, in [49]. This conclusion is extended to the multivariate case in [61], using results of [55], although the empiric df no longer has distribution-free deviations, and its difficult exact distribution theory is still unknown in the multivariate case. In [113] and [115] estimation of a df under an order restriction (e.g., decreasing density) is considered, and the first optimality results in that area are obtained.

The titles of [101, 104, 110] tell the story of interesting nonparametric problems whose solutions, for large samples, are almost as efficient as if the actual families of distributions were known.

2. Sequential Analysis

This concerns the important class of procedures for which the number of observations is not specified in advance of the experiment, but is determined by the chance actual outcome of the observations. Sequential methods are important both because they yield solutions to some problems that cannot be treated adequately with a predetermined sample size, and also because, even when fixed sample size procedures are adequate, sequential procedures are often more efficient in requiring, on the average, fewer observations. The basic theory was developed during World War II by Wald, working in the Statistical Research Group at Columbia, where Wolfowitz was a co-worker. The latter's first papers in this area [13, 14] were concerned with a sequential estimator of a Bernoulli parameter, given by Girshick, Mosteller, and Savage, and shown by them to be unbiased under certain conditions on the stopping region. Wolfowitz generalized these conditions, gave criteria on the region for the unbiased estimator to be unique, and gave conditions that insure the consistency of a sequence of such estimators. In [15] he gave his celebrated "sequential Cramér-Rao inequality," which has been used by many subsequent authors to obtain bounds on the efficiency, or prove such properties as minimaxity, of various sequential estimators.

His proof with Wald [16, 85] of the optimum character of the sequential probability ratio test (SPRT) for testing between two simple hypotheses, is one of the strikingly beautiful results of theoretical statistics. It asserts that a SPRT with error probabilities α, β minimizes the expected sample size, under *both* hypotheses, among all sequential tests whose error probabilities are $\leq \alpha,\beta$. The surprise is that one procedure minimizes both of these quantities, but the proof and those of [18, 23] concerning the Bayes character of the SPRT make the reason clear. The optimum character results have seemed to Wolfowitz's associates the part of his distinguished collaboration with Wald of which he was justifiably most proud.

Stochastic approximation refers to estimation of some characteristic of a regression function M, assumed to belong to a nonparametric class, by sequential choice of the values of the independent variable for which observations are to be taken. Robbins and Monro introduced this important framework, and treated the problem of estimating the intercept of M; in [35] Wolfowitz proved stochastic convergence of their method under conditions they had conjectured to suffice and which were more general than those used in their work. In [36] a method for estimating the maximum of M was given and shown to converge. In [51] he greatly simplified a convergence result of Dvoretzky's that implies the two stochastic approximation convergence results previously mentioned and many others.

In [107] a sequential method is given for testing that a normal law, with unknown nonnegative mean and variance, has mean 0. The test is efficient in the sense that, for alternatives with mean near 0, it does almost as well as the corresponding SPRT for the problem of testing between simple hypotheses with those parameter values.

In [106], fixed length, sequential confidence limits for a translation parameter are given in the nonparametric case. These are shown to be asymptotically efficient.

3. Statistical Decision Theory

The ideas of decision theory are present in the formulation of many of the problems Wolfowitz worked on. Here we mention the papers that contain general decision-theoretic results and methods.

In [18, 23], he and Wald characterized Bayes solutions for various sequential decision problems. In particular, this fundamental work develops the recurrence formula that relates the problem with the number of observations bounded above by n, to that with upper bound $n+1$; in the limit one obtains the integral equation that characterizes the stationary Bayes procedure for infinite horizon.

In characterizing a complete class of procedures (a class containing, for each procedure outside the class, one that is better in a strong sense), one must in general include procedures which reach a final decision by randomizing according to some chance device external to the experiment. One can either choose in advance at random among nonrandomized procedures or let the randomization probabilities depend on the outcome of the experiment. The latter appears more general, but in [28] it is shown that the two methods achieve equivalent results. At the same time, the practical shortcomings of randomization lead one to search for conditions under which any randomized procedure can be replaced by a nonrandomized one with the same performance. The development of [24, 26, 27] implies that randomization can be dispensed with in reasonable statistical settings involving only absolutely continuous probability laws; the main tool is a generalization of Liapounoff's theorem on the range of a vector measure.

Although the paper [25] is concerned with estimating the mean of a normal law, it has broader impact; it was the first paper in which the Bayes method was used to find explicitly a minimax procedure (which turned out to be a fixed sample size procedure), and at the same time employed a sequence of prior laws because the minimax procedure is not genuinely Bayes. In [29] a method is given for considering in turn the performance of each procedure for every one of an ordered set of prior laws, to eliminate the inadmissible Bayes procedures (those dominated by other procedures).

Processes with continuous time parameter are considered in [39, 40, 52], and many of the results of ordinary discrete-time sequential analysis are carried over. Additionally, explicit exact expressions for some of the operating characteristics of sequential procedures can be obtained for some distributions, in contrast with the approximations of the discrete-time setting.

4. Asymptotic Statistical Theory

This heading, which refers to behavior of statistical procedures for large sample sizes, also describes work contained in some of the other sections; for example, much of nonparametric theory has that flavor. Certain ideas, though, have become especially associated in the literature with the phrase "asymptotic theory," and this section discusses Wolfowitz's work on those ideas. One of Wolfowitz's major contributions, the maximum probability method, is reserved for the next section.

R.A. Fisher instituted the pattern of justifying a method for producing a sequence of estimators, one for each sample size, in terms of the asymptotic properties of consistency and efficiency. Consistency of a sequence of estimators means convergence in probability to the true parameter value, whatever it may be. Efficiency refers to the rate of convergence compared with that of other sequences of procedures. Fisher was of course interested primarily in maximum likelihood (ML) estimators.

Wolfowitz has had a long and lasting interest in the ML method, its virtues and shortcomings, and in other methods that may work better than ML for certain models. Wald's approach was extended in [50] to yield consistency for a wide class of settings including nonparametric ones.

Efficiency is a more delicate concept. A number of leading statisticians have presented possible definitions of efficiency that made precise Fisher's somewhat faulty notion, or have given other (best asymptotic normal, including regular Bayes) procedures that shared ML's asymptotic properties and were often easier to compute; these authors, in addition to Wolfowitz, include Bahadur, Le Cam, Neyman, and Rao. The early paper [38] gives a lucid explanation of the good properties of ML, and a precise sense of efficiency: under usual smoothness conditions on f_θ, the probability density function of X_1 when the true parameter value is θ, and for any smooth positive prior density on θ, with probability approaching 1, the posterior law of θ is approximately normal, centered about the ML estimator. Hence, for symmetric monotone loss functions the ML estimators are asymptotically Bayes relative to every smooth prior density. In effect, Wald's complete class almost collapses to a single procedure whose risk function dominates that of all other regular estimator sequences. In [80] Wolfowitz returned to the notion of efficiency, dissatisfied with frequent presentations that had made the notion depend on restricting competing estimator sequences so that they were asymptotically normal, or making other assumptions on them that eliminated the possibility of "superefficiency" compared with ML, on a necessarily small set of parameter points. He assumed only that the suitable normalized dfs of the competing estimators approached some limit uniformly, arguing that without that the asymptotic behavior was meaningless for real-life applicability. The ML estimators are shown, under the usual conditions, to give asymptotically largest probability of being in any appropriately normalized interval containing the true θ value.

The derivation of this result uses a Bayesian device somewhat finer than that of [38], which shows, roughly, that the ML estimator is almost Bayes with respect to a prior law concentrated in a very small interval about the true parameter value. This was the beginning of the development of maximum probability (first called generalized ML) estimators by Wolfowitz, partly in collaboration with Weiss. The basic ideas are developed further in [83, 88, 91, 95, 100]; the last of these also clarifies Fisher's idea that the ML estimators are approximately sufficient, i.e., contain almost all the information (in a precise sense) in the sample, concerning θ. More details will be found in Section 5 below.

A different approach was used earlier, beginning with [37], on a number of nonparametric problems in which ML was often known not even to be consistent. This "minimum distance" technique, while not yielding efficient estimators, is powerful in the generality of its applicability. Typically, if F_θ is the df of iid rvs $X_1, X_2, \ldots, X_n$, where θ can be infinite dimensional, one estimates θ by the value that minimizes the L_∞ (or some other) distance between F_θ and the empiric df of the X_i. The motivation is that the latter converges strongly, at a rapid rate, to the true F_θ, as $n \to \infty$. This method is developed and applied in a variety of settings in [42, 44, 45, 53]. In [47], a similar idea is used to develop a test of whether the X_is have a law that belongs to a specified parametric family. Further work on tests of this type, by a number of authors, has continued to the present, and the power of minimum distance estimation makes it a useful part of other developments, such as Robbins' empiric Bayes theory.

5. Maximum Probability Estimators

Let $X_1, \ldots, X_n$ be independent observations (chance variables) with the same frequency function $f(\cdot \mid \theta)$ which depends on an unknown (to the statistician) value of a parameter. Let $k(n)$ be a normalizing factor for a "good" estimator of θ. An estimator $\Psi_n(X_1, \ldots, X_n)$ is now defined to be asymptotically efficient with respect to a set R in the space of θ if, among all estimators which satisfy a natural regularity condition, the limit of $P_\theta\{k(n)(\Psi_n - \theta) \in R\}$ is greatest, for all admissible values θ. An estimator $Z_n(X_1, \ldots, X_n)$ is called a maximum probability (MP) estimator if it maximizes the integral

$$(*) \qquad \int \prod_{i=1}^{n} f(X_i \mid \theta) d\theta$$

over the set $\{\theta \mid \theta \in d - [k(n)]^{-1}R\}$, with respect to d. It is shown in [88] that the MP estimator is asymptotically efficient with respect to R. Actually, this is shown under conditions much more general than those in which the observations are independent and identically distributed. The parameter θ need not be one-dimensional. Certain modifications of the solution of (*) (of the MP estimator) are shown to be asymptotically equivalent and hence also asymptotically efficient (with respect to R). R itself may depend on θ.

It is also shown in [88] that the asymptotic efficiency properties of the ML estimator are due to its being an MP estimator when the regularity conditions are those of the usual "regular" case, and R is a convex set symmetric about the origin. This happens to be the most important case where the MP estimator does not depend on R. (Such estimators are relatively rare.) It explains why the MP estimator was found relatively late (because efficiency was not defined relative to R) and hence why this explanation of the efficiency of the ML estimator was also so late in coming.

In [108, 111, 114] Wolfowitz considered classes of problems which do not fall into the "regular" case. To these, therefore, standard ML theory does not apply. The asymptotic distributions are not always normal, and even when normal the normalizing factor is not always the standard $n^{1/2}$. By suitable modification of (*) above the asymptotically efficient MP estimator Z_n with respect to $R = (-r,r)$, r arbitrary, is obtained. Sometimes it is the ML estimator.

In [95] the theory of the MP estimator is extended to apply for a general loss function. The loss function which corresponds to a set R is the $0-1$ lost function. In [100] a statistic is defined to be asymptotically sufficient if (asymptotically) the MP estimator is a function only of this statistic. Such a statistic is found for several important classes of problems. Again, the ML theory turns out to be a special case.

The address [102] and the monograph [109] summarize much of the material.

6. Design of Experiments

The modern theory developed by Kiefer is concerned with the choice of values in a given space $\mathscr{X}$ of the independent (controllable) variable in a regression setting, that will make the least squares estimator (LSE) or regression coefficients approximately optimum in a specified sense, for large sample sizes. In [60] the solution for optimum estimation a single one of the regression coefficients is given in terms of a related Chebyshev approximation problem. Also, the minimization of the determinant of the covariance matrix (generalized variance) of estimators of any number of coefficients is considered. When all coefficients are to be estimated, this criterion is shown in [64] to be equivalent to the criterion of choosing the design that minimizes the maximum over $\mathscr{X}$ of the variance at any point therein of the LSE of regression. This first of a number of "equivalence" results for other criteria, which have appeared subsequently in the design area, yields useful methods for computing optimum designs, both analytically and by iterative computer routines.

Problems of extrapolating the regression to a region outside $\mathscr{X}$, or of interpolating to a region which is a subset of $\mathscr{X}$, are considered in [77, 78] when the region is a set symmetric about 0 and the model is polynomial regression with $\mathscr{X} = [-1, 1]$. Hoel and Levine showed, somewhat surprisingly, that an asymmetric extrapolation region such as $[1, b]$ with $b > 1$ yields a simpler and more elegant solution; their work is generalized to models other than polynomials in [81, 79].

7. Other Topics in Statistical Inference

Wolfowitz's interests have extended to a wide variety of statistical topics not listed in earlier sections.

In [11] he and Wald treated the computation of tolerance limits for the normal distribution with unknown mean and variance. In [12] he gave approximations for the surprisingly difficult problem of determining a confidence interval for the probability such a normal law assigns to a specified interval.

The paper [21] is a classic. It derives optimum properties of various classical normal theory tests, anticipating later publications in which similar results were obtained using invariance (Hunt – Stein) theory. The ideas of [59] have influenced the research of several subsequent authors, concerned with conditions on several sets of distributions that permit one to decide correctly, with high probability for large enough sample size, in which set the true law falls.

His concern with the directions in which statistics has moved, and with the systems of foundations that have inspired such movements, is shown in [73, 90, 94]. In particular, his skepticism of the claims of some early Bayesian authors, that their approach eliminated so many statistical difficulties, led him not only to comment cogently on various practical aspects of those developments, but also to study in detail the axioms of the Bayesian approach, and to formulate striking criticism of these axioms.

8. Information Theory

Wolfowitz's book *Coding Theorems of Information Theory* ([71], 1961, 1964, and [116], 1978) is a classic, and the only book which concentrates on statistical and probabilistic aspects of noisy channel communication theory. It is also a handy introductory text because of its brief and simple formulations of problems and estimates. Yet it is comprehensive and at the limits of present research. The completely revised third edition [116] is indispensable for specialists, as the other two editions were before. It contains the core of the ideas of Wolfowitz's papers and of research influenced by him, which already means that the main stream of present research in this theory is covered. "Coding Theorems of Information Theory" would be the proper headline for Wolfowitz's research in information theory, too.

The ambiguous, albeit natural, basic concept of Shannon (1948) for mathematical channel theory had two very different aspects which did not fit well together: an emphasis on probabilistic channel models (stochastic transition matrices) on the one hand, and an emphasis on a treatment of abstract coding machines for the optimization of practical transmission (regularity conditions for coding) on the other hand. Although very important results and methods had been obtained and the importance of entropy parameters and of the weak law of large numbers was noticed (mainly by Shannon), the focus on heuristics was actually aimed between those two issues.

Some of the main methods deserve mention: Shannon's random coding, Feinstein's lower estimate on lengths of codes, and Fano's upper convexity estimate on lengths of codes. Shannon's lower estimate on lengths of codes for some randomized version of codes (by random coding) and Fano's lemma together yielded the first asymptotic stability statement for the exponential growth rate of codes with time when the error probability λ of these particular codes goes to zero slowly (a "coding theorem" together with a "weak converse", implying a notion of time asymptotic "transmission capacity" C). The weak law of large numbers is applied for the latter statements from one side only, e.g., for the lower estimate of code lengths. Estimates like these shifted focus more to the probabilistic framework. There was no precise frame for the coding problem, though. The gap between the two aspects of Shannon's concept is rather large, and there are too many practical aspects which allow one to see the theory in very different ways.

This was the situation when Wolfowitz started writing on channels. The gap between coding and framework narrows as a result of the way he concentrated on a discussion of coding decisions for probabilistic channels. His ideas may be summarized like this:

Channel problems and coding problems are basically problems for finite time (first). Channels approximating more realistic ones are primarily product channels (discrete memoryless channels (d.m.c.), channels with independent letter transmissions of words being elements of a Cartesian product) at times $\{1,2,\ldots,n\}$. The first main point is this: The statistical decision problems for finite time and the approximation problems require one to treat a "minimax situation" for coding. Sender and receiver know the channel and agree upon a sequence $\{(U_i, A_i)\}_{1\leqslant i\leqslant N}$ (code) of input words U_i of length n for the transmission of messages and of disjoint sets A_i of output words of length n. The code has error probability at most λ ($0 < \lambda < 1$) if $\min_{1\leqslant i\leqslant N} P(A_i|U_i) > 1-\lambda$, where $P(A_i|U_i)$ is the probability of receiving a word in A_i when U_i is sent. (Thus, this means best coding against the "worst" of N messages if one tries to minimize λ. If one characterizes the channel only, then there is no a priori probability to message sets.) This is the proper coding situation, which should have priority. The second main point is now that tight estimates on lengths of codes should be proved for finite time if one is really interested in a probabilistic setup of the theory. A label fitting this program might perhaps be "geometric channel characterization for finite time by means of the dependences of different channel parameters on each other, with concentration always on decision systems."

Wolfowitz shows the following for the d.m.c. in his first papers. Let $N(n, \lambda)$ denote the maximal length of a code for the time $\{1,2,\ldots,n\}$ with error probability at most λ. Then there is a constant $C \geqslant 0$ and a constant $K(\lambda) \geqslant 0$ (both depending on the channel) such that $n^{-1}\log N(n,\lambda) \leqslant C-K(\lambda)n^{-1/2}$ (coding theorem) and $n^{-1}\log N(n,\lambda) \leqslant C+K(\lambda)n^{-1/2}$ (strong converse). Now the transmission capacity C is a principal channel parameter for finite time. Moreover, C can actually be computed with the knowledge of the channel at the first instant of time. (It is supposed that the channels are the same for all instants of time $1,2,\ldots$.)

If only $|\log N(n,\lambda) - nC| \leqslant o(n)$ is stated, this is called a weak form of the coding theorem and of the strong converse, respectively. Shannon's coding theorem states that $\log N(n,\lambda) \geqslant nC - o(n)$ and Fano's inequality states that $\log N(n,\lambda) \leqslant [1/(1-\lambda)] (nC + 1)$. Together they imply $\lim_{\lambda\to 0} \lim_{n\to\infty} \inf n^{-1}\log N(n,\lambda) = \lim_{\lambda\to 0} \lim_{n\to\infty} \sup n^{-1}\log N(n,\lambda) = C$, a "coding theorem with weak converse."

Both inequalities of Wolfowitz are derived with stronger forms of the weak law of large numbers (Čebyšev's inequality, combinatorial probabilistic tools). The strong converse is a completely new type of theorem.

In his later papers, Wolfowitz considers various changes of the probabilistic framework in connection with different coding disciplines (partly together with other authors). Channels still sufficiently related to product channels (with memory, feedback, constraints, in particular with unknown states, compositions of channels) are discussed with computations showing the major discrepancies between randomized versions of codes (correlated coding), average error codes, maximum error codes (as above), and the theorems mentioned or weaker ones are proved. (Differences as just indicated do not come up for the d.m.c.) There are important and very realistic channels for which knowledge about good coding theorems is still very incomplete.

Of course, this is only a glance at his main theme in information theory. But it may simplify the orientation for a statistician who is a layman in this field.

We now survey the papers on information theory. In [54] a strict formulation of the coding problem for the d.m.c. is given, as well as a new proof of the coding theorem to within $0(\sqrt{n})$ and the first statement and proof of the strong converse, to within $0(\sqrt{n})$. (See above.) In this paper, π-sequences (now popularly called typical sequences) and generated sequences (now called jointly typical sequences) are introduced. Körner bases his approach on these ideas (page 179 of paper of J. Körner, in *Information Theory: New Trends and Open Problems*, edited by G. Longo; Springer, New York, 1977). The paper [55] also contains combinatorial lemmas which point out the significance of various entropies, and which are of fundamental use in subsequent work.

Extensions of these results to finite input channels with infinite output (semicontinuous channels) are in [62], with a weaker form of the strong converse. Simplifications and stronger versions of the strong converse have been obtained since then by Wolfowitz and Kemperman, separately. Elegant formulations are contained in [71], [93], and [116], section 7.8. Paper [57] (and [56]) gives the capacity of a channel with finite memory in a computable way (with coding theorem and strong converse), the first such result in the literature. This is continued in [58], with a general version in [66]. In [70] a particular infinite memory channel is treated (dependences increasing with time). Reduction to [66] is achieved, with coding theorem and strong converse and computable capacity. The constructions in these papers are now main parts of many constructions in the probabilistic theory. They are in his book.

From here to [99] below, the papers are concerned with realistic versions of channel approximation. The paper [75] is based on [74] (see also [71], section 6.6), and on constructions as in [66] for reductions. Blackwell, Breiman, and Thomasian (*Ann. Math. Stat.*, Vol. 29) prove the coding theorem for a finite state indecomposable channel, and a weak converse. In [75] a strong converse (weaker form) is proved, thus determining capacity. The latter is done in a computable way by proving an approximation theorem for the accuracy of its determination. In [65] (and [67] and [69]) the concern is with "simultaneous" channels, now called "compound" channels (uniform coding for different states of transition matrix). The strong converse is obtained to within $0(\sqrt{n})$. Moreover, the new theme that sender or receiver knows the channel is treated. In [76] and [116], section 7.7, an example (by means of a particular randomization argument for finite state channels) is constructed where the weak converse is valid but the strong converse is not. In [89], a comparison between a channel with memory and a memoryless channel linked to it is made, for a particular situation. The channel with memory has a larger capacity. In [97] and [98] (with Ahlswede), papers linked to [65], capacity functions for average error with randomized coding permitted are obtained, as are various changes of frames and coding disciplines. By a counterexample, it is shown that a result announced by Dobrushin is not valid. Results are very numerous. A simple characterization of their classification program is impossible. Channels without synchronization are studied in [105] (with Ahlswede). For finite alphabet channels, the receiver receives k_j letters for the jth letter of the sender. Thus, the received sequence has length $k_1 + \ldots k_n$. It is supposed that there are no gaps between the received blocks of letters and that the lengths k_j are determined at random. A strong converse of the coding theorem with capacity to within desired accuracy is obtained. This is also achieved when feedback is present, where capacity is obtained by computation with an algorithm. Also, these results are obtained with an arbitrary set of channel probability functions. Finally, a result of Dobrushin on continuous transmission with a fidelity criterion is supplemented.

The paper [72] (with Kiefer) is the first in the literature with results on arbitrarily varying channels (a.v.ch., channel coding simultaneously with respect to all states when the set of states is a Cartesian product with time). Necessary and sufficient conditions for the existence of positive code rates for any probability of error $\lambda(0<\lambda<1)$ are proved with a group coding argument. This is also done for the cases where only the sender or only the receiver knows the states of transmission. For the case that both know the state for transmission, the capacity is determined (with coding theorem and strong converse). In [99] (with Ahlswede), a coding theorem and strong converse with capacity are proved in a computable way for the a.v.ch. with binary output. (There are "worst" states for this a.v.ch. such that good maximum likelihood codes for them are good codes for the other states.)

A simple solution to the problem of optimality of the Schalkwijk–Kailath scheme (1966) is given in [93] by proving a strong converse for the Gaussian channel with feedback and power constraint. The paper [112] is a study of first order regression for the Gaussian channel with power constraint (results extend immediately to general regression). An optimal linear signalling scheme is given.

The basic situation treated in the next papers is sketched in the last part of [116]. In [82] a new approach to a theorem of Shannon's about approximation with a fidelity criterion is given. Further, a converse to this theorem is proved. (A more explicit treatment of the latter has since been given by Gallager.) In [118] an upper bound on the rate distortion function for partial side information for source coding is proved. This is more difficult than the one for complete side information of Wyner and Ziv in their prize-winning paper in IEEE *Trans. Info. Th.*, IT-22, 1-10, 1976. In [120] Wolfowitz obtains a lower bound showing that the upper bound is tight, and thus obtains the rate decoding function itself. The chance variables $\{Z^t\}$ introduced in this paper will play a fundamental role in future problems on source coding, for solving the crucial problem of a tight lower bound on the possible rate. In Section 7, a new method of deriving the result of [118] is given, which is so expeditious that it almost makes the problem trivial. In [117] a coding theorem and strong converse (both to within $0(\sqrt{n})$) are proved for list codes with a suitable definition of error. (List codes were first introduced by Elias.) The coding theorem and strong converse for the d.m.c. are special cases of these results. In [119] the study of [117] is deepened. Application of the results obtained is made to obtain expeditiously coding theorems for degraded channels and for the Wyner-Ziv channel generalized in [118] and [120].

It is fitting that the latest of Wolfowitz's papers [120] records the contents of the Shannon Lecture he gave at the IEEE Information Theory Group International Meeting at Grignano, Italy (near Trieste) in June of 1979. A promising new method is introduced for obtaining bounds, and it is applied to obtain the (more difficult) lower bound of Wyner and Ziv, as well as to give a streamlined proof of their upper bound and that of [118].

9. Other Optimization Problems

The spirit of characterizing efficient procedures, which has permeated Wolfowitz's work in statistics and information theory, extends to other stochastic problems he has attacked. In [33] the inventory problem suggests a development which applies to the broader class of settings now often referred to as discounted dynamic programming, in which an optimal policy is sought for adjusting a chance process so that the sum of discounted rewards over many time periods is maximized. A particularly simple form of policy, suggested in earlier work of Arrow, Harris, and Marshak, is shown in [41] to be optimum under certain conditions. The case in which the chance law that governs the process is unknown and thus, in effect, has to be estimated as time periods pass so as to help determine what actions to take, is treated in [34].

In [84] work of Chow and Robbins and of Dvoretzky is generalized, to demonstrate the existence of optimal stopping rules for rewards of the form c_nS_n or $c_nS_n^2$ for certain sequences $\{c_n\}$, where S_n is a sum of iid rvs with zero mean and finite variance.

The early paper [10] gives a procedure for quality control inspection that insures a specified average outgoing quality while minimizing the amount of inspection needed when the inspected process is in control.

10. Probability Theory

Probabilistic developments are, of course, central to much of the research described in other sections. Many of these developments are of interest beyond that of the statistical results that motivated them. Here we list those papers in which the motivation usually does not stem from a statistical problem.

In [20] Wolfowitz gave an exceedingly simple proof of Poincaré's recurrence theorem (which refers to finite, not necessarily probability, measures) and of Kac's formula for the mean recurrence time. In [86] higher moments were treated.

The research of [19], carried out by Wolfowitz during World War II, deals with the geometric configuration of a moving convex set upon its contact with a randomly moving particle.

In [30] he studied, with Dvoretzky, conditions under which the sums $S_n = X_1 + \ldots + X_n$, where the X_i are independent but not identically distributed, are asymptotically uniformly distributed mod m, and the rate of convergence to uniformity. This has applications to computer construction of uniform rvs. In [32] he and Chung generalized a celebrated result of Erdös, Feller, and Pollard, which is also a consequence of the work of Kolmogorov, by showing that for integer valued iid summands X_i of positive (possibly infinite) expectation m, the expected number of times n that $S_n = j$ $(1 \leq n < \infty)$ approaches t/m as $j \to \infty$; here t is the gcd of support values of the probability law of X_1.

In [43] it is shown that, for iid multivariate rvs $\mathbf{X}_1, \mathbf{X}_2, \ldots \mathbf{X}_n$ with independent coordinates, the empiric frequency of (proportion of observations falling in) any halfspace $H_{\mathbf{a},b} = \{\mathbf{x}: \mathbf{a}'\mathbf{x} < b\}$ converges strongly to the probability that $\mathbf{X}_1 \in H_{\mathbf{a},b}$ uniformly in ${}_{\mathbf{a},b}$, as $n \to \infty$. While this work was motivated by application to the minimum distance method described in Section 4, it led to further work in [63] in which the conclusion was proved without restriction on the df of $\mathbf{X}_1$.

In [46], general results are obtained on the convergence in probability of waiting times and other quantities of interest in queueing systems. The methods yield results on random walks [48], such as characterization of random walks S_n based on iid summands, for which $E(\max S_n)^k < \infty$.

Bounds on the probability of large deviations of the distance between the true and empiric df of vector chance variables are obtained in [55], and it is shown that the limiting behavior of such deviations can be deduced from that of a limiting Gaussian process with multidimensional time — a generalization to this setting where the probabilities are no longer independent of the true df, of the classical one-dimensional invariance principle of Doob and Donsker.

In [74] it is shown that, if a system of stochastic matrices satisfies a certain condition, the product of n matrices has its rows "almost" identical, uniformly in n, in a precisely defined sense. This result has applications in semi-group theory and in information theory.

Permissions

Springer-Verlag would like to thank the original publishers of J. Wolfowitz's scientific papers for granting permissions to reprint a selection of his papers in this volume. The following credit lines were specifically requested:

[1] Reprinted from Ann. Math. Stat. **10,** 1939, © 1939 by the Inst. Math. Stat.

[9] Reprinted from Ann. of Math. Stat. **15,** © 1944 by the Inst. Math. Stat.

[13] Reprinted from Ann. of Math. Stat. **17,** © 1946 by the Inst. of Math. Stat.

[15] Reprinted from Ann. of Math. Stat. **18,** © 1947 by the Inst. Math. Stat.

[16] Reprinted from Ann. of Math. Stat. **19,** © 1948 by the Inst. of Math. Stat.

[20] Reprinted from Bulletin of the Amer. Math. Soc. **55,** © 1949 by the Amer. Math. Soc.

[21] Reprinted from Ann. Math. Stat. **20,** © 1949 by the Inst. Math. Stat.

[23] Reprinted from Ann. Math. Stat. **21,** © 1950 by the Inst. Math. Stat.

[25] Reprinted from Ann. Math. Stat. **21,** © 1950 by the Inst. Math. Stat.

[27] Reprinted from Ann. Math. Stat. **22,** © 1951 by the Inst. Math. Stat.

[30] Reprinted from Duke Math. J. **18,** © 1951 by the Duke U. Press

[36] Reprinted from Ann. Math. Stat. **23,** © 1952 by the Inst. Math. Stat.

[49] Reprinted from Ann. Math. Stat. **27,** © 1956 by the Inst. Math. Stat.

[50] Reprinted from Ann. Math. Stat. **27,** © 1956 by the Inst. Math. Stat.

[53] Reprinted from Ann. of Math. Stat. **28,** © by the Inst. Math. Stat.

[60] Reprinted from Ann. Math. Stat. **30,** © 1959 by the Inst. Math. Stat.

[61] Reprinted from Ann. Math. Stat. **30,** © 1959 by the Inst. Math. Stat.

[73] Reprinted from Proc. Amer. Math. Soc. **14,** © 1963 by the Amer. Math. Soc.

[81] Reprinted from Ann. Math. Stat. **36,** © 1965 by the Inst. Math. Stat.

[85] Reprinted from Essays in Probability and Statistics, edited by R. C. Bose and others, © 1969 The University of North Carolina Press.

[86] Reprinted from Ann. Math. Stat. **37,** © 1966 by the Inst. Math. Stat.

[87] Reprinted from Proc. Amer. Math. Soc. **18,** © 1967 by The Amer. Math. Soc.

[90] Reprinted from the New York Statistician **18,** © 1967 by the Inst. Math. Stat.

[108] Reprinted from Ann. of Stat. **1,** © 1973 by the Inst. Math. Stat.

Reprinted from THE ANNALS OF MATHEMATICAL STATISTICS
Vol. X, No. 2, June, 1939

CONFIDENCE LIMITS FOR CONTINUOUS DISTRIBUTION FUNCTIONS[1]

BY A. WALD[2] AND J. WOLFOWITZ

1. **Introduction.** The theory of confidence limits for unknown parameters of distribution functions has been considerably developed in recent years. This theory assumes that there is given a family F of systems of n stochastic variables $X_1(\theta_1, \cdots, \theta_k), \cdots, X_n(\theta_1, \cdots, \theta_k)$ depending upon k parameters $\theta_1, \cdots, \theta_k$ and such that the distribution function of every element of F is known.

For the case $k = 1$, for example, this theory proceeds as follows:

Denote by E an n-tuple $x_1, \cdots, x_n$ of observed values of the stochastic variables $X_1(\theta), \cdots, X_n(\theta)$ of which we know only that they constitute a system which is an element of F. E can be represented as the point $x_1, \cdots, x_n$ in an n-dimensional Euclidean space. Let there be given a positive number α, $0 < \alpha < 1$. Then to each pair E, α there is constructed a θ-interval, $[\underline{\theta}(E, \alpha), \bar{\theta}(E, \alpha)]$ with the following property: If we were to draw a sample from the system $X_1(\theta), \cdots, X_n(\theta)$, the probability is exactly α that we shall get a system of observations $E = x_1, \cdots, x_n$ such that the interval corresponding to E, α will include θ (i.e., that $\underline{\theta}(E, \alpha) \leq \theta \leq \bar{\theta}(E, \alpha)$).

In this paper we do not limit ourselves to a family of systems of n stochastic variables depending upon a finite number of parameters, but consider the family G of all systems of n stochastic variables $X_1, \cdots, X_n$ subject only to the condition that $X_1, \cdots, X_n$ are independently distributed with the same continuous distribution function.

Let E be the point in an n-dimensional Euclidean space which corresponds to the observed values $x_1, \cdots, x_n$ of the n stochastic variables $X_1, \cdots, X_n$ of which we know only that they constitute an element of the family G, i.e., that they are independently distributed with the same continuous distribution function. Let us denote their distribution function by $f(x)$; the probability that $X_i < x$ is $f(x)$, $i = 1, \cdots, n$. Let α be a number such that $0 < \alpha < 1$. To each pair E, α we shall construct two functions, $\bar{l}_{E,\alpha}(x)$ and $\underline{l}_{E,\alpha}(x)$, with the following property: The probability is α that, if we were to draw a sample from the system $X_1, \cdots, X_n$, we would get a system of observations $E = x_1, \cdots, x_n$ such that $f(x)$ lies entirely between $\bar{l}_{E,\alpha}(x)$ and $\underline{l}_{E,\alpha}(x)$ (i.e., that $\underline{l}_{E,\alpha}(x) \leq f(x) \leq \bar{l}_{E,\alpha}(x)$ for all x). We shall call $\bar{l}_{E,\alpha}(x)$ and $\underline{l}_{E,\alpha}(x)$ the upper and lower confidence limits, respectively, corresponding to the confidence coefficient α.

---

[1] Presented to the American Mathematical Society at New York, February 25, 1939.

[2] Research under a grant-in-aid from the Carnegie Corporation of New York.

All the stochastic variables considered hereafter in this paper are to have continuous distribution functions.

2. A theorem on continuous distribution functions. Let $f(x)$ be the continuous distribution function of a stochastic variable X whose range is from $-\infty$ to $+\infty$. Let $\delta_1(x)$ and $\delta_2(x)$ be two functions defined for $0 \leq x \leq 1$ and satisfying the following requirements:

(a) $\delta_1(x)$ and $\delta_2(x)$ are non-negative and continuous for $0 \leq x \leq 1$.

(b) $l_1(x)$ and $l_2(x)$ are monotonically non-decreasing for all x, where

$$l_1(x) \equiv f(x) + \delta_1(f(x))$$

$$l_2(x) \equiv f(x) - \delta_2(f(x)).$$

(c) There exists a number h, such that $f(h) < 1$ and $l_1(h) = 1$.

(d) There exists a number h', such that $f(h') > 0$ and $l_2(h') = 0$.

(e) $l_1(x) \leq 1$ for all x
$l_2(x) \geq 0$ for all x

(f) $\delta_1(x) + \delta_2(x) \geq \frac{1}{n}$ for all x, where n is the number of random, independent observations of the stochastic variable X.

Let $\varphi(x)$ be the distribution function of such a system of observations, i.e., the ratio, to n, of the number of observations $<x$ is $\phi(x)$. $\varphi(x)$ is, of course, a multiple of $\frac{1}{n}$ for all x.

We shall consider the following problem:

What is the probability P that

$$l_2(x) \leq \varphi(x) \leq l_1(x) \tag{1}$$

for all x?

The reasons for restrictions (b), (c), (d), (e), and (f) on $\delta_1(x)$ and $\delta_2(x)$ are now apparent. If there exist two numbers $q_1 < q_2$, such that, for $q_1 < x < q_2$, $l_1(x) > l_1(q_2)$ and $l_1(q_1) = l_1(q_2)$, then, if we change $l_1(x)$ so that $l_1(x) = l_1(q_2)$ for $q_1 \leq x \leq q_2$, P will remain unchanged. An analogous process leads to a similar conclusion for $l_2(x)$. Hence $l_1(x)$ and $l_2(x)$ are to be monotonically non-decreasing. If there did not exist a number h or h', P would be 0. Hence requirements (c) and (d). Since $0 \leq \varphi(x) \leq 1$, there is no point to considering functions which do not satisfy (e). $\varphi(x)$ is a step-function whose saltuses are $\geq \frac{1}{n}$. If, for all x.

$$\delta_1(x) + \delta_2(x) < \frac{1}{n}$$

then $P = 0$. If there is an interval $[\beta, \gamma]$ within which $\delta_1(x) + \delta_2(x) < \frac{1}{n}$, then all samples in which one of the observed values lies in this interval are such that (1) does not hold for all x. For the sake of simplicity and because the situation described in (*f*) is the one of importance, we make the latter requirement.

It would appear that P depends upon $f(x)$, $\delta_1(x)$, $\delta_2(x)$, and n.

THEOREM: *P is independent of $f(x)$ and depends only upon $\delta_1(x)$, $\delta_2(x)$, and n.*

PROOF: Let $Y = f(X)$. Then Y is a stochastic variable distributed in the range 0 to 1 with a distribution function $\equiv x$. By this transformation $l_1(x)$ and $l_2(x)$ become respectively

$$(2) \qquad \left.\begin{aligned} l_1'(x) &= x + \delta_1(x) \\ l_2'(x) &= x - \delta_2(x) \end{aligned}\right\} \; 0 \leq x \leq 1.$$

Then P is the probability that the distribution function $\varphi(x)$ of a random sample of n of the stochastic variable Y shall be such that $l_2'(x) \leq \varphi(x) \leq l_1'(x)$ and is therefore independent of $f(x)$.

3. Computation of P. From the previous section it follows that, in computing P, we may confine ourselves to a stochastic variable X whose range is from 0 to 1 and whose distribution function $\equiv x$. Let $l_1(x)$ and $l_2(x)$ be the upper and lower limits, respectively, which are set for $\varphi(x)$. $l_1(x)$ and $l_2(x)$ are defined in (2), if the accents are omitted.

Consider the equations:

$$(3) \qquad l_1(x) = \frac{i}{n} \qquad (i = 1, 2, \cdots, n; 0 \leq x \leq 1).$$

If, for a certain i, the corresponding equation possesses one or more solutions in x, let a_i be the minimum of these solutions. If the first r of these equations (3) have no solutions, let

$$a_i = 0 \qquad (i = 1, \cdots, r).$$

If the ith, say, of the equations

$$(4) \qquad l_2(x) = \frac{i-1}{n} \qquad (i = 1, \cdots, n; 0 \leq x \leq 1)$$

possesses one or more solutions in x, let b_i be the maximum of these. If the last $n - s$ of the equations (4) have no solutions, let

$$b_i = 1 \qquad (i = s + 1, \cdots, n).$$

Obviously

$$a_i \leq a_{i+1}, \qquad b_i \leq b_{i+1}, \qquad a_i \leq b_i .$$

From restrictions, (e) and (f) on $l_1(x)$ and $l_2(x)$, it follows that $a_1 = 0$, $b_n = 1$.

Suppose the sample $E = x_1, \cdots, x_n$ has been obtained. Arrange the x's in ascending order, thus: $x_{p_1}, x_{p_2}, \cdots, x_{p_n}$ where $x_{p_1} \le x_{p_2} \le \cdots \le x_{p_n}$. Then necessary and sufficient conditions that (1) hold are:

$$a_i \le x_{p_i} \le b_i \qquad (i = 1, \cdots, n). \tag{5}$$

Let $P_k(t, \Delta t)$, $(k = 0, 1, \cdots, (n-1);\ a_{k+1} \le t \le b_{k+1})$ be the probability that a sample $E = x_1, \cdots, x_n$ shall fulfill the following conditions:

(a) $x_1 \le x_2 \le \cdots \le x_{k+1}$,

(b) $x_1, \cdots, x_k$ satisfy the first k inequalities (5),

(c) $t \le x_{k+1} \le t + \Delta t$.

Let

$$P_k(t) = \lim_{\Delta t \to 0} \frac{P_k(t, \Delta t)}{\Delta t}.$$

Since $f(x) \equiv x$, we get easily

$$P_0(t) \equiv 1. \tag{6}$$

We shall now develop a recursion formula for $P_{k+1}(t)$. For this purpose let us consider the following composite event: The observations $x_1, \cdots, x_n$ satisfy the conditions (a), (b), and

$$t' \le x_{k+1} \le t' + \Delta t'$$

and

$$t \le x_{k+2} \le t + \Delta t.$$

If $a_{k+1} \le t' \le b_{k+1}$, the probability of this event is $P_k(t', \Delta t')\Delta t$. Now

$$\lim_{\substack{\Delta t' \to 0 \\ \Delta t \to 0}} \frac{P_k(t', \Delta t')\Delta t}{\Delta t' \cdot \Delta t} = P_k(t').$$

$P_k(t')$ is obviously the probability density of the bivariate distribution of t' and t. In order to obtain $P_{k+1}(t)$ we have to integrate $P_k(t')\,dt'$ over the region defined by the two inequalities

$$t' \le t$$

$$a_{k+1} \le t' \le b_{k+1}.$$

Hence, omitting the now unnecessary accent, if

$$t \le b_{k+1} \tag{7}$$

then

$$P_{k+1}(t) = \int_{a_{k+1}}^{t} P_k(t)\,dt \qquad (k = 0, 1, \cdots, (n-2)), \tag{8}$$

and if

$$t > b_{k+1} \tag{9}$$

then

$$P_{k+1}(t) = \int_{a_{k+1}}^{b_{k+1}} P_k(t)\,dt \quad (k = 0, 1, 2, \cdots (n-2)). \tag{10}$$

Now, to obtain P, we cannot confine ourselves only to cases where $x_1 \leq x_2 \leq \cdots \leq x_n$, but have to consider all the $n!$ permutations of the n x's. Hence

$$P = n! \int_{a_n}^{b_n} P_{n-1}(t)\,dt. \tag{11}$$

The fact that there are two forms of the recursion formula corresponding to the two possible cases (7) and (9) makes actual calculation very cumbersome for n of any considerable size. We shall therefore give an approximation formula which is considerably easier to apply to practical calculations.

4. **Computation of $\overline{P}$ and $\underline{P}$.** Let $\overline{P}$ be the probability that, for a sample of n, $l_1(x) \geq \varphi(x)$ for all x. Let $\underline{P}$ be the probability that, for a sample of n, $\varphi(x) \geq l_2(x)$ for all x.

Consider the inequalities

$$\left.\begin{aligned} x_i &\geq a_i \quad (12)\\ x_i &\leq b_i \quad (13)\end{aligned}\right\} \quad (i = 1, 2, \cdots, n)$$

Let

$$\overline{P}_k(t, \Delta t), \qquad (k = 0, 1, \cdots, (n-1);\ t \geq a_{k+1})$$

be the probability that a sample $E = x_1, \cdots, x_n$ of the stochastic variable X should fulfill the following conditions:

(a) $x_1 \leq x_2 \leq \cdots \leq x_{k+1}$

(b) $x_1, \cdots, x_k$ satisfy the first k inequalities (12)

(c) $t \leq x_{k+1} \leq t + \Delta t$.

Let

$$\overline{P}_k(t) = \lim_{\Delta t \to 0} \frac{\overline{P}_k(t, \Delta t)}{\Delta t}.$$

Then, by an argument like that employed in the preceding section, we obtain

$$\overline{P}_0(t) \equiv 1, \tag{14}$$

and the recursion formula

$$\overline{P}_{k+1}(t) = \int_{a_{k+1}}^{t} \overline{P}_k(t)\,dt. \tag{15}$$

Let $\overline{P}_n(t)$ be defined formally by (15). Then, in the same way in which we obtained (11), we get

$$\overline{P} = n!\,\overline{P}_n(1). \tag{16}$$

In the same manner we shall obtain an expression for $\underline{P}$.

Let $\underline{P}_k(t, \Delta t)$, $(k = 0, 1, \cdots (n - 1);\ t \leq b_{n-k})$ be the probability that a sample $E = x_1, \cdots, x_n$ of the stochastic variable X should fulfill the following conditions:

(a) $x_{n-k} \leq x_{n-k+1} \leq \cdots \leq x_n$,

(b) $x_{n-k+1}, \cdots, x_n$ satisfy the last k inequalities (13),

(c) $t \leq x_{n-k} \leq t + \Delta t$.

Let

$$\underline{P}_k(t) = \lim_{\Delta t \to 0} \frac{\underline{P}_k(t, \Delta t)}{\Delta t}.$$

Then

$$\underline{P}_0(t) \equiv 1 \tag{17}$$

and by an argument very similar to that employed above,

$$\underline{P}_{k+1}(t) = \int_t^{b_{n-k}} \underline{P}_k(t)\, dt. \tag{18}$$

Let $\underline{P}_n(t)$ be defined formally by (18). Then

$$\underline{P} = n!\, \underline{P}_n(0). \tag{19}$$

The $\overline{P}_i(t)$ and $\underline{P}_i(t)$ are polynomials in t. Denote by c_j the constant term of $\overline{P}_j(t)$ and by d_j the constant term of $(-1)^j \underline{P}_j(t)$. Obviously

$$c_0 = 1 \tag{20}$$

$$d_0 = 1 \tag{21}$$

and

$$\overline{P}_i(t) = \frac{c_0}{i!} t^i + \frac{c_1}{(i-1)!} t^{i-1} + \cdots + c_{i-1} t + c_i \tag{22}$$

$$\underline{P}_i(t) = (-1)^i \left(\frac{d_0}{i!} t^i + \frac{d_1}{(i-1)!} t^{i-1} + \cdots + d_{i-1} t + d_i \right). \tag{23}$$

Since

$$\overline{P}_i(a_i) = 0, \qquad \underline{P}_i(b_{n-i+1}) = 0$$

we obtain

$$c_0 \frac{a_i^i}{i!} + c_1 \frac{a_i^{i-1}}{(i-1)!} + \cdots + c_{i-1} a_i + c_i = 0 \qquad (i = 1, 2, \cdots, n) \tag{24}$$

and

$$\frac{d_0}{i!} b_{n-i+1}^i + \frac{d_1}{(i-1)!} b_{n-i+1}^{i-1} + \cdots + d_{i-1} b_{n-i+1} + d_i = 0 \tag{25}$$

$$(i = 1. 2. \cdots, n)$$

The determinant of (20) and the first j equations (24) ($j = 1, \cdots, n$) considered as equations in $c_0, c_1, \cdots, c_j$ equals 1, since all the elements of the principal diagonal are 1 and all the elements above the principal diagonal are 0. Then

$$c_i = \begin{vmatrix} 1 & 0 & 0 & \cdots & 0 & 1 \\ a_1 & 1 & 0 & \cdots & 0 & 0 \\ \frac{a_2^2}{2!} & a_2 & 1 & \cdots & 0 & 0 \\ \cdots & \cdots & \cdots & \cdots & \cdots & \cdots \\ \frac{a_i^i}{i!} & \frac{a_i^{i-1}}{(i-1)!} & \frac{a_i^{i-2}}{(i-2)!} & \cdots & a_i & 0 \end{vmatrix} \tag{26}$$

$$= (-1)^i \begin{vmatrix} a_1 & 1 & 0 & \cdots & 0 \\ \frac{a_2^2}{2!} & a_2 & 1 & \cdots & 0 \\ \cdots & \cdots & \cdots & \cdots & \cdots \\ \frac{a_i^i}{i!} & \frac{a_i^{i-1}}{(i-1)!} & \frac{a_i^{i-2}}{(i-2)!} & \cdots & a_i \end{vmatrix}$$

From (16) and (22) for $i = n$, we get

$$\bar{P} = c_0 + nc_1 + n(n-1)c_2 + \cdots + n(n-1)\cdots(3)(2)c_{n-1} + n!\,c_n$$

$$= \begin{vmatrix} \frac{n!}{n!} & \frac{n!}{(n-1)!} & \frac{n!}{(n-2)!} & \cdots & \frac{n!}{1!} & \frac{n!}{0!} \\ a_1 & 1 & 0 & \cdots & 0 & 0 \\ \frac{a_2^2}{2!} & a_2 & 1 & \cdots & 0 & 0 \\ \cdots & \cdots & \cdots & \cdots & \cdots & \cdots \\ \frac{a_n^n}{n!} & \frac{a_n^{n-1}}{(n-1)!} & \frac{a_n^{n-2}}{(n-2)!} & \cdots & a_n & 1 \end{vmatrix} \tag{27}$$

In the same way, we obtain

$$d_i = \begin{vmatrix} 1 & 0 & 0 & \cdots & 0 & 1 \\ b_n & 1 & 0 & \cdots & 0 & 0 \\ \frac{b_{n-1}^2}{2!} & b_{n-1} & 1 & \cdots & 0 & 0 \\ \cdots & \cdots & \cdots & \cdots & \cdots & \cdots \\ \frac{b_{n-i+1}^i}{i!} & \frac{b_{n-i+1}^{i-1}}{(i-1)!} & \frac{b_{n-i+1}^{i-2}}{(i-2)!} & \cdots & b_{n-i+1} & 0 \end{vmatrix} \tag{28}$$

$$= (-1)^i \begin{vmatrix} b_n & 1 & 0 & \cdots & 0 \\ \frac{b_{n-1}^2}{2!} & b_{n-1} & 1 & \cdots & 0 \\ \cdots & \cdots & \cdots & \cdots & \cdots \\ \frac{b_{n-i+1}^i}{i!} & \frac{b_{n-i+1}^{i-1}}{(i-1)!} & \frac{b_{n-i+1}^{i-2}}{(i-2)!} & \cdots & b_{n-i+1} \end{vmatrix}$$

and from (19) and (23) for $i = n$,

$$\underline{P} = (-1)^n n!\, d_n . \tag{29}$$

Perhaps if the determinants in (27) and (28) were to be simplified it might be easier to calculate $\overline{P}$ and $\underline{P}$ that way than by the recursion formulas.

5. **The approximation of P.** Let J be the probability that, for a sample of n, there exists at least one pair of numbers ω_1, ω_2, such that

$$0 \le \omega_i \le 1 \qquad (i = 1, 2)$$

$$\varphi(\omega_1) > l_1(\omega_1)$$
$$\varphi(\omega_2) < l_2(\omega_2).$$

Recalling the definitions of P, $\overline{P}$, and $\underline{P}$, it is obvious that

$$1 - P = (1 - \overline{P}) + (1 - \underline{P}) - J. \tag{30}$$

Now if

$$J \le (1 - \overline{P})(1 - \underline{P}) \tag{31}$$

and $(1 - P)$ is small, the right member of (30) with J omitted furnishes an excellent approximation to $(1 - P)$. Suppose, for example, that it were desired to give upper and lower limits $l_1(x)$ and $l_2(x)$ such that $P = .95$. Choose $l_1(x)$ and $l_2(x)$ so that, for example, $\overline{P} = \underline{P} = .975$. Then P cannot differ from .95 by more than .000625. Even if

$$J \le K(1 - \overline{P})(1 - \underline{P}) \tag{32}$$

where K is a small factor, say 10, the approximation would still be excellent. It seems very plausible that even (31) holds. However, we have not yet succeeded in obtaining a rigorous proof.

6. **The construction of confidence limits.** We now proceed to the construction of $l_{E,\alpha}(x)$ and $\underline{l}_{E,\alpha}(x)$ which were defined in Section I of this paper.

A confidence coefficient $\alpha (0 < \alpha < 1)$ is selected to which it is desired that the confidence limits correspond. Functions $\delta_1(x)$ and $\delta_2(x)$ are chosen to be as defined in Section 2 and also to be such as to make $P = \alpha$. This can be done by application of the formulas for the evaluation of P.

The functions $l_{E,\alpha}(x)$ and $\underline{l}_{E,\alpha}(x)$ are to be known when E and α are known.

Since α is given, $l_{E,\alpha}(x)$ and $\underline{l}_{E,\alpha}(x)$ depend upon the outcome of the experiment which yields observed values of the stochastic variable X. Let $E = x_1, \cdots, x_n$ be this system of values and let $\varphi(x)$ be its distribution function. Consider the equations

$$\delta_2(\varphi(x) + \Delta_1(x)) = \Delta_1(x) \tag{33}$$

$$\delta_1(\varphi(x) - \Delta_2(x)) = \Delta_2(x). \tag{34}$$

For a fixed but arbitrary x, $-\infty < x < +\infty$, $\varphi(x)$ is known and (33) and (34) are equations in $\Delta_1(x)$ and $\Delta_2(x)$. If, for a certain x, (33) has one or more solutions, let $\epsilon_1(x)$ be the maximum of the set of solutions (for this x, of course). Similarly, if for a certain x, (34) has one or more solutions, let $\epsilon_2(x)$ be the maximum of the set of solutions.

We can now give $l_{E,\alpha}(x)$ and $\underline{l}_{E,\alpha}(x)$ as follows:

For an x such that (33) has at least one solution,

$$l_{E,\alpha}(x) = \varphi(x) + \epsilon_1(x). \tag{35}$$

For an x such that (33) has no solutions,

$$l_{E,\alpha}(x) = 1. \tag{36}$$

For an x such that (34) has at least one solution,

$$\underline{l}_{E,\alpha}(x) = \varphi(x) - \epsilon_2(x). \tag{37}$$

For an x such that (34) has no solution,

$$\underline{l}_{E,\alpha}(x) = 0. \tag{38}$$

We recapitulate briefly the meaning of $l_{E,\alpha}(x)$ and $\underline{l}_{E,\alpha}(x)$ which were defined in Section 1. These are two functions defined for $-\infty < x < +\infty$ which may be constructed as above after a confidence coefficient α has been assigned and after the outcome of the physical experiment which determines the stochastic point E is known. These functions have the following property: No matter what the distribution function $f(x)$ of each of n stochastic independent variables $X_1, \cdots, X_n$ may be, provided only that $f(x)$ is continuous and the same for each $X_1, \cdots, X_n$, the probability is exactly α that, if we were to perform the physical experiment which gives a set of values E of the stochastic system $X_1, \cdots, X_n$ and were then to construct $l_{E,\alpha}(x)$ and $\underline{l}_{E,\alpha}(x)$, the inequality

$$\underline{l}_{E,\alpha}(x) \leq f(x) \leq l_{E,\alpha}(x) \tag{39}$$

would hold for all x.

A less precise but more intuitive statement of the above result is as follows: If, in many experiments we were to proceed as above to construct $l_{E,\alpha}(x)$ and $\underline{l}_{E,\alpha}(x)$ and then, in each instance, we were to predict that the unknown $f(x)$ (which need not be the same in all experiments) satisfies (39), the relative frequency of correct predictions would be α.

The formal proof of this result is exceedingly simple. For any continuous $f(x)$, the probability is α that

$$l_2(x) \leq \varphi(x) \leq l_1(x) \tag{40}$$

will hold for all x. This is so because of the way in which $\delta_1(x)$ and $\delta_2(x)$ were chosen. To prove the required result it would therefore be sufficient to show that, if (39) holds for all x, (40) holds for all x and conversely.

Let x be fixed but arbitrary. We shall show that

$$f(x) \leq l_{E,\alpha}(x) \tag{41}$$

implies

$$l_2(x) \leq \varphi(x) \tag{42}$$

and conversely.

If (33) has no solution, $\varphi(x) > l_2(1) \geq l_2(x)$, $l_{E,\alpha}(x) = 1$, and (41) and (42) are trivial. Assume therefore that (33) has at least one solution. For this situation, then, we have to show that

$$f(x) \leq \varphi(x) + \epsilon_1(x) \tag{43}$$

implies

$$l_2(x) \leq \varphi(x) \tag{44}$$

and conversely.

With x and hence $\varphi(x)$ and $\epsilon_1(x)$ fixed, consider the equation in x':

$$l_2(x') = \varphi(x). \tag{45}$$

Since $\varphi(x) \leq l_2(1)$, (45) has at least one solution. Let x'_m be the maximum of these solutions for a fixed x. Then from the definition of $\epsilon_1(x)$ it follows that

$$f(x'_m) - l_2(x'_m) = \epsilon_1(x), \tag{46}$$

or, on account of the definition of x'_m,

$$f(x'_m) = \varphi(x) + \epsilon_1(x). \tag{47}$$

Now, if (43) holds, $x \leq x'_m$ because of (47). Then, from the definition of x'_m and the fact that $l_2(x')$ is monotonically non-decreasing (44) follows.

If (44) holds, then $x \leq x'_m$ (by the definition of x'_m and the monotonic character of $l_2(x')$). Hence, because of (47), (43) is true. This shows the equivalence of (43) and (44).

In a similar manner, it may be shown that

$$l_{E,\alpha}(x) \leq f(x) \tag{48}$$

implies

$$\varphi(x) \leq l_1(x) \tag{49}$$

and conversely. This completes the proof.

7. **Miscellaneous remarks.** An expedient way of choosing $\delta_1(x)$ and $\delta_2(x)$ is such that, with c a constant,

$$(50) \qquad \begin{aligned} x + \delta_1(x) &\equiv \min [x + c, 1] \\ x - \delta_2(x) &\equiv \max [x - c, 0]. \end{aligned} \qquad 0 \le x \le 1$$

Tables of double entry could be constructed giving the c corresponding to specified α and n. With such tables available the construction of confidence limits would be quick and simple in practice. In this case, $\epsilon_1(x) = \epsilon_2(x) = c$.

Another expedient and plausible way of choosing $\delta_1(x)$ and $\delta_2(x)$ might be to choose them so that

$$(51) \qquad \begin{aligned} x + \delta_1(x) &\equiv \min [px + q, 1] \\ x - \delta_2(x) &\equiv \max [p'x + q', 0] \end{aligned} \qquad 0 \le x \le 1$$

where p, p', q, and q' are constants. The actual construction of confidence limits could then be handled with dispatch if similar tables were constructed.

$$\bar{l}_{E,\alpha}(x) \quad \text{and} \quad \underline{l}_{E,\alpha}(x)$$

are, like $\varphi(x)$, step-functions. The situation may occur where, for $x = e$,

$$\lim_{(x<e),x\to e} \bar{l}_{E,\alpha}(x) < \lim_{(x>e),x\to e} \underline{l}_{E,\alpha}(x).$$

This would give a prediction, corresponding to the confidence coefficient α, that $f(x)$ is not continuous. If $f(x)$ is continuous the probability of such a situation is 0.

8. **Further problems.** Even with α fixed, the functions $\delta_1(x)$ and $\delta_2(x)$ may be chosen in many ways. Each different choice gives, in general, different confidence limits. Which is to be preferred? This very problem also arose in the theory of parameter estimation and the testing of hypotheses and gave rise to the Neyman-Pearson theory. It would be desirable to develop such a theory for the confidence limits discussed in this paper.

We have treated here only the case where $f(x)$ is continuous. A similar theory is needed for the case where $f(x)$ is not continuous.

It would be of practical value to construct tables such as those described in Section 7. The construction of tables could be greatly facilitated if the formulas for P or $\overline{P}$ and $\underline{P}$ could be simplified so as to render them more practical for calculation or else if they were to be replaced by asymptotic expansions.

9. **An example.** To illustrate the method we shall consider an example for the case of samples of size 6, i.e. $n = 6$.

Let $\delta_1(x)$ and $\delta_2(x)$ be given as follows:

$$\delta_1(x) = d \quad \text{for} \quad 0 \leq x \leq 1 - d,$$

$$\delta_1(x) = 1 - x \quad \text{for} \quad 1 - d < x \leq 1,$$

$$\delta_2(x) = x \quad \text{for} \quad 0 \leq x \leq d,$$

and

$$\delta_2(x) = d \quad \text{for} \quad d < x \leq 1.$$

Denote by $\bar{P}$ the probability that

$$\varphi(x) \leq f(x) + \delta_1[f(x)],$$

by $\underline{P}$ the probability that

$$\varphi(x) \geq f(x) - \delta_2[f(x)]$$

and by P the probability that

$$f(x) - \delta_2[f(x)] \leq \varphi(x) \leq f(x) + \delta_1[f(x)].$$

$\varphi(x)$ denotes the sample distribution and $f(x)$ denotes the population distribution.

Since $\delta_2(x) = \delta_1(1 - x)$, we obviously have

$$\bar{P} = \underline{P}.$$

Let us calculate $\bar{P} = \underline{P}$ in case $d = \frac{1}{2}$. According to (3) we have

$$a_1 = a_2 = a_3 = 0, \qquad a_4 = \tfrac{1}{6}, \qquad a_5 = \tfrac{1}{3}, \qquad a_6 = \tfrac{1}{2}.$$

According to (16)

$$\bar{P} = 6!\bar{P}_6(1)$$

where

$$\bar{P}_0(t) \equiv 1,$$

$$\bar{P}_k(t) \equiv \int_{a_k}^{t} \bar{P}_{k-1}(t)\,dt \qquad (k = 1, \cdots, 6).$$

Applying this recursion formula we get

$$\bar{P}_1(t) = t; \qquad \bar{P}_2(t) = \frac{t^2}{2}, \qquad \bar{P}_3(t) = \frac{t^3}{6},$$

$$\bar{P}_4(t) = \frac{t^4}{24} - \frac{1}{2^7 \cdot 3^5}$$

$$\bar{P}_5(t) = \frac{t^5}{120} - \frac{t}{2^7 \cdot 3^5} - \frac{11}{3^6 \cdot 2^7 \cdot 5}$$

$$\bar{P}_6(t) = \frac{t^6}{720} - \frac{t^2}{2^8 \cdot 3^5} - \frac{11t}{3^6 \cdot 2^7 \cdot 5} - \frac{11}{2^9 \cdot 3^6 \cdot 5}.$$

Hence

$$\bar{P} = \underline{P} = 6!\bar{P}_6(1) = 1 - \frac{85}{2592} = 0.967.$$

Let us now calculate $\bar{P} = \underline{P}$ in case $d = \frac{1}{3}$. We have

$$a_1 = a_2 = 0, \quad a_3 = \tfrac{1}{6}, \quad a_4 = \tfrac{1}{3}, \quad a_5 = \tfrac{1}{2} \text{ and } a_6 = \tfrac{2}{3}.$$

Applying the recursion formula we get

$$\bar{P}_0(t) = 1, \quad \bar{P}_1(t) = t, \quad \bar{P}_2(t) = \frac{t^2}{2}, \quad \bar{P}_3(t) = \frac{t^3}{6} - \frac{1}{2^4 \cdot 3^4},$$

$$\bar{P}_4(t) = \frac{t^4}{24} - \frac{t}{2^4 \cdot 3^4} - \frac{1}{2^4 \cdot 3^5},$$

$$\bar{P}_5(t) = \frac{t^5}{120} - \frac{t^2}{2^5 \cdot 3^4} - \frac{t}{2^4 \cdot 3^5} - \frac{11}{2^8 \cdot 3^5 \cdot 5},$$

$$\bar{P}_6(t) = \frac{t^6}{720} - \frac{t^3}{2^5 \cdot 3^5} - \frac{t^2}{2^5 \cdot 3^5} - \frac{11t}{2^8 \cdot 3^5 \cdot 5} - \frac{13}{2^7 \cdot 3^8 \cdot 5}.$$

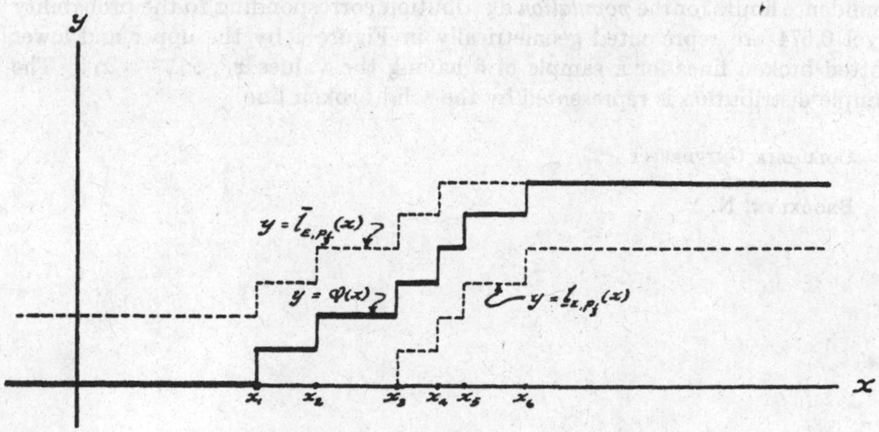

FIG. 1

Hence

$$\bar{P} = \underline{P} = 6!\bar{P}_6(1) = 1 - \frac{2483}{11664} = 0.787.$$

It is obvious that

$$1 - P = (1 - \bar{P}) + (1 - \underline{P}) - J,$$

where J denotes the probability that $\varphi(x)$ violates both limits. In case $d = \frac{1}{2}$ no $\varphi(x)$ exists which violates both limits, and therefore $J = 0$. If $d = \frac{1}{3}$,

J is not zero but so small that it can be neglected. Hence

$$P = 0.934 \quad \text{if} \quad d = \tfrac{1}{2}$$

and

$$P = 0.574 \quad \text{if} \quad d = \tfrac{1}{3}$$

P increases monotonically from 0.574 to 0.934 if d increases from $\frac{1}{3}$ to $\frac{1}{2}$. Denote by P_d the probability corresponding to d. According to (33)–(38), the confidence limits corresponding to the probability level P_d are given as follows:

$$\begin{aligned} l_{E,P_d}(x) &= \varphi(x) + d \text{ if } \varphi(x) + d \leq 1, \\ l_{E,P_d}(x) &= 1 \text{ if } \varphi(x) + d > 1, \\ \underline{l}_{E,P_d}(x) &= \varphi(x) - d \text{ if } \varphi(x) - d \geq 0 \end{aligned}$$

and

$$\underline{l}_{E,P_d}(x) = 0 \text{ if } \varphi(x) - d < 0.$$

Substituting for d the numbers $\frac{1}{2}$ and $\frac{1}{3}$, we get the confidence limits corresponding to the probability levels 0.934 and 0.574 respectively. The upper and lower confidence limits for the *population* distribution corresponding to the probability level 0.574 are represented geometrically in Figure 1 by the upper and lower dotted broken lines for a sample of 6 having the values $x_1, x_2, \cdots x_6$. The sample distribution is represented by the solid broken line.

Columbia University
and
Brooklyn, N. Y.

Made in the United States of America

Reprinted from THE ANNALS OF MATHEMATICAL STATISTICS
Vol. XII, No. 1, March, 1941

NOTE ON CONFIDENCE LIMITS FOR CONTINUOUS DISTRIBUTION FUNCTIONS

BY A. WALD* AND J. WOLFOWITZ

In a recent paper [1] we discussed the following problem: Let X be a stochastic variable with the cumulative distribution function $f(x)$, about which nothing is known except that it is continuous. Let $x_1, \cdots, x_n$ be n independent, random observations on X. The question is to give confidence limits for $f(x)$. We gave a theoretical solution when the confidence set is a particularly simple and important one, a "belt."

A particularly simple and expedient way from the practical point of view is to construct these belts of uniform thickness ([1], p. 115, equation 50). If the appropriate tables, as mentioned in our paper, were available, the construction of confidence limits, no matter how large the size of the sample, would be immediate.

Our formulas (11), (16), (19), (27) and (29) are not very practical for computation, particularly when the samples are large. We have recently learned that there exists a result by Kolmogoroff [2], generalized by Smirnoff [3],[1] which for large samples gives an easy method for constructing tables, i.e. of finding α when c and n are given (all notations as in [1]). The result of Kolmogoroff-Smirnoff is:

Let $c = \lambda/\sqrt{n}$. Then for any fixed $\lambda > 0$,

$$\lim_{n=\infty} \overline{P} = \lim_{n=\infty} \underline{P} = 1 - e^{-2\lambda^2}$$

$$\lim_{n=\infty} P = 1 - 2 \sum_{m=1}^{\infty} (-1)^{(m-1)} e^{-2m^2\lambda^2}$$

This series converges very rapidly.

REFERENCES

[1] WALD AND WOLFOWITZ, "Confidence limits for continuous distribution functions," *Annals of Math. Stat.*, Vol. 10(1939), pp. 105–118.

[2] A. KOLMOGOROFF, "Sulla determinazione empirica di una leggi di distribuzione," *Giornale dell'Instituto Italiana degli Attuari*, Vol. 11(1933).

[3] N. SMIRNOFF, "Sur les ecarts de la courbe de distribution empirique," *Recueil Mathematique (Mathematicheski Sbornik)*, New series, Vol. 6(48)(1939), pp. 3–26.

* Columbia University, New York City.

[1] In the French résumé of Smirnoff's article, on page 26, due to a typographical error this formula is given with a factor $(-1)^m$ instead of the correct factor $(-1)^{m-1}$. The correct result follows from equation (112), page 23, of the Russian text when t is set equal to zero.

Made in United States of America

Reprinted from THE ANNALS OF MATHEMATICAL STATISTICS
Vol. XV, No. 4, December, 1944

STATISTICAL TESTS BASED ON PERMUTATIONS OF THE OBSERVATIONS

A. WALD AND J. WOLFOWITZ
Columbia University

1. Introduction. One of the problems of statistical inference is to devise exact tests of significance when the form of the underlying probability distribution is unknown. The idea of a general method of dealing with this problem originated with R. A. Fisher [13, 14]. The essential feature of this method is that a certain set of permutations of the observations is considered, having the property that each permutation is equally likely under the hypothesis to be tested. Thus, an exact test on the level of significance α can be constructed by choosing a proportion α of the permutations as critical region. In an interesting paper H. Scheffé [2] has shown that for a general class of problems this is the only possible method of constructing exact tests of significance.

Tests based on permutations of the observations have been proposed and studied by R. A. Fisher, E. J. G. Pitman, B. L. Welch, the present authors, and others. Pitman and Welch derived the first few moments of the statistics used in their test procedures. However, it is desirable to derive at least the limiting distributions of these statistics and make it practicable to carry out tests of significance when the sample is large. Such a large sample distribution was derived for a statistic considered elsewhere [1] by the present authors.

In this paper a general theorem on the limiting distribution of linear forms in the universe of permutations of the observations is derived. As an application of this theorem, the limiting distributions of the rank correlation coefficient and that of several statistics considered by Pitman and Welch, are obtained. In the last section the limiting distribution of Hotelling's generalized T in the universe of permutations of the observations is derived.

2. A theorem on linear forms. Let $H_N = (h_1, h_2 \ldots, h_N)$ ($N = 1, 2, \ldots$, ad inf.) be sequences of real numbers and let

$$\mu_r(H_N) = N^{-1} \sum_{\alpha=1}^{N} \left((h_\alpha - N^{-1} \sum_{\beta=1}^{N} h_\beta \right)^r$$

for all integral values of r. We define the following symbols in the customary manner: For any function $f(N)$ and any positive function $\varphi(N)$ let $f(N) = O(\varphi(N))$ mean that $|f(N)/\varphi(N)|$ is bounded from above for all N and let

$$f(N) = \Omega(\varphi(N))$$

mean that

$$f(N) = O(\varphi(N))$$

and that

$$\lim_N \inf |f(N)/\varphi(N)| > 0.$$

Also let

$$f(N) = o(\varphi(N))$$

mean that

$$\lim_{N\to\infty} f(N)/\varphi(N) = 0.$$

Let $[\rho]$ denote the largest integer $\leq \rho$.

We shall say that the sequences $H_N (N = 1, 2, \cdots, \text{ad inf.})$ satisfy the condition W if, for all integral $r > 2$,

$$\frac{\mu_r(H_N)}{[\mu_2(H_N)]^{r/2}} = O(1). \tag{2.1}$$

For any value of N let

$$X = (x_1, x_2, \cdots, x_N)$$

be a chance variable whose domain of definition is made up of the $N!$ permutations of the elements of the sequence $A_N = (a_1, a_2, \cdots, a_N)$. (If two of the $a_i (i = 1, 2, \cdots, N)$ are identical we assume that some distinguishing index is attached to each so that they can then be regarded as distinct and so that there still are $N!$ permutations of the elements $a_1, \cdots, a_N$). Let each permutation of A_N have the same probability $(N!)^{-1}$. Let $E(Y)$ and $\sigma^2(Y)$ denote, respectively, the expectation and variance of any chance variable Y.

We now prove the following:

Theorem. *Let the sequences $A_N = (a_1, a_2, \cdots, a_N)$ and $D_N = (d_1, d_2, \cdots, d_N)$ $(N = 1, 2, \cdots, \text{ad inf.})$ satisfy the condition W. Let the chance variable L_N be defined as*

$$L_N = \sum_{i=1}^{N} d_i x_i.$$

Then as $N \to \infty$, the probability of the inequality

$$L_N - E(L_N) < t\,\sigma(L_N)$$

for any real t, approaches

$$\frac{1}{\sqrt{2\pi}} \int_{-\infty}^{t} e^{-\frac{1}{2}x^2}\, dx.$$

For convenience the proof will be divided into several lemmas.

Since

$$L_N^* = \frac{L_N - E(L_N)}{\sigma(L_N)}$$

remains invariant if a constant is added to all the elements of D_N or of A_N, or if the elements of either of the latter are multiplied by any constant other than zero, we may, in the formation of L_N^*, replace A_N and D_N by the sequences A_N'

and D'_N, respectively, whose ith elements a'_i and $d'_i (i = 1, 2, \cdots, N)$ are, respectively

$$a'_i = [\mu_2(A_n)]^{-\frac{1}{2}} \left(a_i - N^{-1} \sum_{j=1}^{N} a_j \right) \tag{2.2}$$

and

$$d'_i = [\mu_2(D_N)]^{-\frac{1}{2}} \left(d_i - N^{-1} \sum_{j=1}^{N} d_j \right). \tag{2.3}$$

The sequences A'_N and D'_N satisfy the condition W. Furthermore,

$$\mu_1(A'_N) \equiv \mu_1(D'_N) \equiv 0 \tag{2.4}$$

and

$$\mu_2(A'_N) \equiv \mu_2(D'_N) \equiv 1. \tag{2.5}$$

LEMMA 1.

$$\sum_{\alpha_1 < \alpha_2 < \cdots < \alpha_k \le N} \cdots \sum a'_{\alpha_1} a'_{\alpha_2} \cdots a'_{\alpha_k} = O(N^{[k/2]}) \tag{2.6}$$

$$\sum_{\alpha_1 < \alpha_2 < \cdots < \alpha_k \le N} \cdots \sum d'_{\alpha_1} d'_{\alpha_2} \cdots d'_{\alpha_k} = O(N^{[k/2]}). \tag{2.7}$$

From (2.4), (2.5), and the fact that the A'_N and D'_N satisfy condition W, it follows that the A'_N and D'_N satisfy conditions *a*), *b*), and *c*) of the theorem on page 383 of [1]. Our lemma 1 is the same as lemma 1 of [1].

LEMMA 2. *Let*

$$V = (v_1, v_2, \cdots, v_N)$$

be the same permutation of the elements of A'_N *that* X *is of the elements of* A_N. *Let* $y = v_1 \cdots v_k z$ *where* $z = v_{(k+1)}^{i_2} \cdots v_{(k+r)}^{i_r}$, $i_j > 1$ $(j = 1, 2, \cdots, r)$, *and* $k, r, i_1, \cdots, i_r$ *are fixed values independent of* N.
Then

$$E(y) = O(N^{[k/2]-k}). \tag{2.8}$$

This is Lemma 2 of [1].

In a similar manner we obtain that

$$\sum_{\alpha_1, \alpha_2, \cdots, \alpha_{(k+r)}} \cdots \sum d'_{\alpha_1} \cdots d'_{\alpha_k} d'^{i_1}_{\alpha(k+1)} \cdots d'^{i_r}_{\alpha(k+r)} = O(N^{[k/2]-k}) \cdot O(N^{k+r}) = O(N^{[k/2]+r}). \tag{2.9}$$

The summation in the above formula is to be taken over all possible sets of $k + r$ distinct positive integers $\le N$.

LEMMA 3. *Let* $\alpha_1, \cdots, \alpha_{(k+r)}$ *be* $(k + r)$ *distinct positive integers* $\le N$. *Then*

$$E(v_1 v_2 \cdots v_k v_{(k+1)}^{i_1} \cdots v_{(k+r)}^{i_r}) = E(v_{\alpha_1} v_{\alpha_2} \cdots v_{\alpha_k} v_{\alpha(k+1)}^{i_1} \cdots v_{\alpha(k+r)}^{i_r}). \tag{2.10}$$

This follows from the fact that all permutations of A'_N have the same probability.

LEMMA 4. *Let*

$$L'_N = \sum_{i=1}^{N} d'_i v_i .$$

Then

$$E(L'^p_N) = O(N)^{[p/2]}). \tag{2.11}$$

PROOF: Expand L'^p_N and take the expected value of the individual terms. The contribution to $E(L'^p_N)$ of all the terms which are multiples of the type appearing in the right member of (2.10) with fixed $k, r, i_1, \cdots, i_r (k + i_1 + \cdots + i_r = p)$, is, by Lemmas 2 and 3

$$O(N^{[k/2]-k}) \cdot \underset{\substack{\alpha_1, \cdots, \alpha_{(k+r)} \\ \text{all different}}}{\sum \cdots \sum} d'_{\alpha_1} \cdots d'_{\alpha_k} d'^{i_1}_{\alpha_{(k+1)}} \cdots d'^{i_r}_{\alpha_{(k+r)}} = O(N^{[k/2]-k}) O(N^{[k/2]+r})$$

$$= O(N^{2[k/2]-k+r}).$$

Since $i_j > 1 (j = 1, \cdots, r)$, it follows from the fact that $k + i_1 + \cdots + i_r = p$ that $2r \leq p - k$ and that $2r = p - k$ only if $i_1 = \cdots = i_r = 2$. Now

$$2\left[\frac{k}{2}\right] - k + r \leq r \leq \frac{p-k}{2} \leq \frac{p}{2}. \tag{2.12}$$

Hence the maximum value of $2\left[\frac{k}{2}\right] - k + r$ is reached when $r = \left[\frac{p}{2}\right]$ and $k = 0$. This proves the lemma.

From the last remarks of the preceding paragraph we obtain

LEMMA 5.

$$E(L'^{2j}_N) - \frac{(2j)!}{j!2^j} \big(\underset{\substack{\alpha_1, \cdots, \alpha_j \\ \text{all different}}}{\sum \cdots \sum} d'^2_{\alpha_1} \cdots d'^2_{\alpha_j}\big) E(v_1^2 \cdots v_j^2) = o(N^j). \tag{2.13}$$

We now prove

LEMMA 6.

$$E(L'_N) = 0 \tag{2.14}$$

$$E(L'^2_N) = NE(v_1^2) + o(N) = N + o(N). \tag{2.15}$$

Equation (2.14) follows from (2.2). Consider the expectations of the various terms in the expansion of L'^2_N. The sum of all the terms of the type

$$d'_i d'_j E(v_i v_j)$$

is

$$\big(\sum_{i \neq j} d'_i d'_j\big) E(v_1 v_2) = O(N) O(N^{-1}) = O(1),$$

by Lemmas 1 and 2. The sum of all the terms of the type

$$d'^2_i E(v_i^2)$$

is

$$\left(\sum_{i=1}^{N} d_i'^2\right) E(v_1^2) = NE(v_1^2) = N,$$

by (2.2) and (2.3). This proves the lemma.

LEMMA 7.

$$E(v_1^2 \cdots v_j^2) = 1 + o\,(1) \tag{2.16}$$

$$\sum_{\substack{\alpha_1 \cdots \alpha_j \\ \text{all different}}} \cdots \sum d_{\alpha_1}'^2 \cdots d_{\alpha_j}'^2 = N^j + o(N^j). \tag{2.17}$$

From (2.2) and (2.3), and Lemma 3, it follows that it will be sufficient to prove (2.17), because (2.16) follows in the same manner. Consider the relation

$$N^j = \left(\sum_{i=1}^{N} d_i'^2\right)^j = \sum_{\substack{\alpha_1, \cdots, \alpha_j \\ \text{all different}}} \cdots \sum d_{\alpha_1}'^2 \cdots d_{\alpha_j}'^2 + \text{other terms}.$$

By (2.9) the sum of these other terms must be not larger than $O(N^{j-1})$. From this follows the lemma.

PROOF of the theorem: Since

$$L_N^* = \frac{L_N'}{\sigma(L_N')} = \frac{L_N - E(L_N)}{\sigma(L_N)},$$

it will be sufficient to show that the moments of L_N^* approach those of the normal distribution as $N \to \infty$. From (2.14), (2.15), and (2.11) we see that, when p is odd, the pth moment of L_N^* is $O(N^{-\frac{1}{2}})$ and hence approaches zero as $N \to \infty$. When p is even and $= 2s$ (say), it follows from Lemma 5 that

$$E(L_N'^{2s}) - \frac{(2s)!}{s!2^s} \Big(\sum_{\substack{\alpha_1, \cdots, \alpha_s \\ \text{all different}}} \cdots \sum d_{\alpha_1}'^2 \cdots d_{\alpha_s}'^2\Big) E(v_1^2 \cdots v_s^2) = o(N^s).$$

Hence from (2.16) and (2.17)

$$E(L_N'^{2s}) = \frac{(2s)!}{s!2^s} N^s + o(N^s). \tag{2.18}$$

From (2.18) and (2.15) we obtain that

$$\lim_{N \to \infty} E(L_N^{*2s}) = \frac{(2s)!}{s!2^s}.$$

This completes the proof of the theorem.

It will be noticed that nothing in the foregoing proof requires that, when $N < N'$, the sequences A_N and D_N be subsequences of $A_{N'}$ and $D_{N'}$. Indeed, the sequences were written as they were simply for typographic brevity. We have therefore

COROLLARY 1. *The theorem is valid for sequences*

$$A_N = (a_{N1}, \cdots, a_{NN})$$
$$D_N = (d_{N1}, \cdots, d_{NN})$$
$$(N = 1, 2, \cdots \text{ ad inf.})$$

provided they fulfill condition W.

COROLLARY 2. *If the elements* $a_i(i = 1, 2, \cdots$ ad inf.) *are all independent observations on the same chance variable, all of whose moments are finite and whose variance is positive, the sequences* $A_N(N = 1, 2, \cdots$, ad inf.) *will fulfill condition W with probability one.*

3. The rank correlation coefficient. For this well known statistic (see [3])

$$A_N \equiv D_N \equiv (1, 2, 3, \cdots, N).$$

The sequences A_N and D_N satisfy the condition W. For

$$\sum_{i=1}^{N} i^r = O(N^{r+1})$$

and hence, for $r \geq 3$

$$\mu_r(A_N) = \mu_r(D_N) = O(N^r).$$

Also

$$\mu_2(A_N) = \mu_2(D_N) = \Omega(N^2).$$

Hence the distribution of the rank correlation coefficient is asymptotically normal in the case of statistical independence. This result was first proved by Hotelling and Pabst [3].

4. Pitman's test for dependence between two variates. The distribution of the correlation coefficient in the population of permutations of the observations was used by Pitman [4] in a test for dependence between two variates which "involves no assumptions" about the distributions of these variates. In our notation, let

$$(a_i, d_i)(i = 1, 2, \cdots, N)$$

be N observations on the pair of variates A and D whose dependence it is desired to test. Then the value of the correlation coefficient is

$$N^{-1} \sum_{i=1}^{N} d_i' a_i'.$$

At the level β the observations are considered to be significant if the probability that $N^{-1} \mid L_N' \mid$ be equal to or greater than the absolute value of the actually observed correlation coefficient is $\leq \beta$.

In his paper ([4], page 227] Pitman points out that if the ratios of certain sample cumulants are "not too large," then, as $N \to \infty$, the first four moments of $N^{-\frac{1}{2}} L'_N$ will approach 0, 1, 0, and 3, respectively (the first moment is always zero). Our theorem and the relation (2.15) make clear that under proper circumstances all the moments will approach those of the normal distribution.

5. Pitman's procedure for testing the hypothesis that two samples are from the same population. For testing the hypothesis that two samples came from the same population Pitman [5] proposed the following procedure:
Let one sample be

$$a_1, a_2, \cdots, a_m$$

and the other

$$a_{m+1}, a_{m+2}, \cdots, a_{m+n}.$$

Write $m + n = N$, and construct the sequences A_N and A'_N as before defined. Let

$$d_i = 1 \qquad (i = 1, \cdots, m)$$

$$d_i = 0 \qquad (i = m + 1, \cdots, N)$$

and construct the sequences D_N and D'_N. Then the value of the statistic considered by Pitman is, except for a constant factor,

$$N^{-\frac{1}{2}} \left(\sum_{i=1}^{N} d'_i a'_i \right). \tag{5.1}$$

At the level β the observations are considered significant if the probability that $N^{-\frac{1}{2}} \mid L'_N \mid$ be equal to or greater than the observed absolute value of the expression (5.1) is $\leq \beta$.

Let $N \to \infty$, while $\frac{m}{n}$ is constant. Then the sequences D_N are seen to satisfy condition W. If then the sequences A_N satisfy condition W we may, for large N, employ the result of our theorem and expeditiously determine the critical value of Pitman's statistic.

6. Analysis of variance in randomized blocks. Welch [7] and Pitman [6] consider the following problem: Each of n different "varieties of a plant" is planted in one of the n cells which constitute a "block." It is desired to test, on the basis of results from m blocks, the null hypothesis that there is no difference among the varieties. In order to eliminate a possible bias caused by variations in fertility among the cells of a block, the varieties are assigned at random to the cells of a block. If the cells of the jth block are designated by $(j1)$, $(j2)$, $\cdots$, (jn), a permutation of the integers $1, 2, \cdots, n$ is allocated to the jth block by a chance process, each permutation having the same probability $(n!)^{-1}$.

Let x_{ijk} be the yield of the ith variety in the kth cell of the jth block to which it was assigned by the randomization process. It is assumed that

$$x_{ijk} = y_{jk} + \delta_i + \epsilon_{jk},$$

where y_{jk} is the "effect" of the kth cell in the jth block, δ_i is the "effect" of the ith variety, and ϵ_{jk} are chance variables about whose distribution we assume nothing. The null hypothesis states that

$$\delta_1 = \delta_2 = \cdots = \delta_n = 0.$$

Let a_{jk} be the yield in the kth cell of the jth block and x_{ij} the yield of the ith variety in the jth block. If the null hypothesis is true then, because of the randomization within each block described above, the conditional probability that, given the set $\{a_{jk}\}$ $(k = 1, 2, \cdots, n)$, the sequence $x_{1j}, x_{2j} \cdots, x_{nj}$, be any given permutation of the elements of $\{a_{jk}\}$ is $(n!)^{-1}$. Permuting in all the blocks simultaneously we have that, under the null hypothesis, given the set of mn values $\{a_{jk}\}$ $(j = 1, 2, \cdots m;\ k = 1, 2, \cdots, n)$, the conditional probability of any of the permutations is the same, $(n!)^{-m}$. This permits an exact test of the null hypothesis.

The classical analysis of variance statistic that would be employed in the conventional two-way classification with independent normally distributed observations is

$$F = \frac{(m-1)m \sum (x_{i\cdot} - x)^2}{\sum \sum (x_{ij} - x_{i\cdot} - x_{\cdot j} + x)^2}$$

where

$$x_{i\cdot} = m^{-1} \sum_j x_{ij} \qquad x = (mn)^{-1} \sum \sum x_{ij}.$$
$$x_{\cdot j} = n^{-1} \sum_i x_{ij}$$

The statistic W used by Welch and Pitman is

$$W = F(m - 1 + F)^{-1}.$$

Since W is a monotonic function of F and the critical regions are the upper tails, the two tests are equivalent. The distribution of F or W is to be determined in the same manner as that of the other statistics discussed in this paper, i.e., over the equally probable permutations of the values actually observed. The critical region is, as usual, the upper tail.

Since x_{ij} takes any of the values $a_{j1}, \cdots, a_{jn}$ with probability $1/n$, we have

$$E(x_{ij}) = n^{-1} \sum_k a_{jk} = a_j \quad \text{(say)}. \tag{6.1}$$

$$\sigma^2(x_{ij}) = n^{-1} \sum_k (a_{jk} - a_j)^2 = b_j \quad \text{(say)}. \tag{6.2}$$

$$\begin{aligned} \sigma(x_{i_1j} x_{i_2j}) &= [n(n-1)]^{-1} \sum_{k_1 \neq k_2} a_{jk_1} a_{jk_2} - a_j^2 \\ &= [n(n-1)]^{-1} [(\sum_k a_{jk})^2 - \sum_k a_{jk}^2] - a_j^2 \\ &= [n^2 a_j^2 - \sum_k a_{jk}^2][n(n-1)]^{-1} - a_j^2 \\ &= (n-1)^{-1} [a_j^2 - n^{-1} \sum_k a_{jk}^2] = -(n-1)^{-1} b_j . \end{aligned} \tag{6.3}$$

Hence

$$E(x_{i\cdot}) = m^{-1} \sum a_j . \tag{6.4}$$

$$\sigma^2(x_{i\cdot}) = m^{-2} \sum b_j = b \quad \text{(say)}. \tag{6.5}$$

$$\sigma(x_{i_1\cdot}\, x_{i_2\cdot}) = -[m^2(n-1)]^{-1} \sum b_j = c \quad \text{(say)}. \qquad i_1 \neq i_2 \tag{6.6}$$

Let

$$x^*_{ij} = \sum_v \lambda_{iv} x_{vj} \qquad (i, v = 1, \cdots, n)$$

where $\| \lambda_{iv} \|$ is an orthogonal matrix and

$$\lambda_{n1} = \lambda_{n2} = \cdots = \lambda_{nn} = n^{-\frac{1}{2}}.$$

Then it follows that

$$\begin{aligned} E(x^*_{i\cdot}) &= 0 \\ \sigma^2(x^*_{i\cdot}) &= b - c \qquad (i = 1, 2, \cdots, n-1) \\ \sigma(x^*_{i_1\cdot}\, x^*_{i_2\cdot}) &= 0 \qquad (i_1 \neq i_2 ; i_1, i_2 = 1, \cdots, n-1). \end{aligned} \tag{6.7}$$

Furthermore, we have

$$\sum_{i=1}^{n-1} x^{*2}_{i\cdot} = \sum_{i=1}^{n} (x_{i\cdot} - x)^2. \tag{6.8}$$

Applying the well known identity

$$\Sigma\Sigma(x_{ij} - x_{i\cdot} - x_{\cdot j} + x)^2 = \Sigma\Sigma(x_{ij} - x_{\cdot j})^2 - m\Sigma(x_{i\cdot} - x)^2$$

to the definitions of F and W we obtain

$$W = \frac{m \sum_i (x_{i\cdot} - x)^2}{\sum_i \sum_j (x_{ij} - x_{\cdot j})^2}. \tag{6.9}$$

The denominator of the right member of (6.9) is invariant under permutations *within* each block and equals

$$\sum_j \sum_k (a_{jk} - a_j)^2 = (n-1)m^2(b - c).$$

Hence

$$\begin{aligned} W &= [m(n-1)(b-c)]^{-1} \sum_{i=1}^{n} (x_{i\cdot} - x)^2 \\ &= [m(n-1)(b-c)]^{-1} \sum_{i=1}^{n-1} x^{*2}_{i\cdot}. \end{aligned} \tag{6.10}$$

If the joint distribution of the $x^*_{i\cdot}$ $(i = 1, 2, \cdots, n-1)$ over the set of admissible permutations approaches a normal distribution with non-singular correlation matrix as m, the number of blocks, becomes large, it follows from (6.7) and (6.10) that the distribution of $m(n-1)W$ approaches the x^2 distribution with $n-1$ degrees of freedom. Hence it remains to indicate conditions on the set $\{a_{jk}\}$ which would make the distribution of the $x_{i.}^*$ approach normality. Each $x_{i.}^*$ is the mean of independent variables, so these conditions need not be very restrictive.

According to Cramér [8], Theorem 21a, page 113, if the variances and covariances fulfill certain requirements (the limiting correlation matrix should also be non-singular) and if a generalized Lindeberg condition holds, normality in the limit will follow. Somewhat more restrictive conditions which are simpler to state and which will be satisfied in most statistical applications are that $o < c' < b_j < c''$ for all j, where c' and c'' are positive constants. Since the variance of x_{ij}^* is $(n-1)^{-1}nb_j$, it can be seen that the above inequalities imply the fulfillment of the conditions of the Laplace-Liapounoff theorem (see, for example, Uspensky [9], page 318). By [6.7] the correlation matrix is non-singular.

7. Hotelling's generalized T for permutation of the observations. In this section we shall restrict ourselves to bivariate populations, the extension to more than two variables being straightforward. Let $(u_{11}, u_{21}), \cdots, (u_{1m}, u_{2m})$ be m pairs of observations on the chance variables U_1, U_2, and $(u_{1(m+1)}, u_{2(m+1)}), \cdots, (u_{1N}, u_{2N})$, be n pairs of observations on the chance variables U_1', U_2', where $m + n = N$. If each of the pairs U_1, U_2, and U_1', U_2' is jointly normally distributed with the same convariance matrix, the Hotelling generalized T for testing the null hypothesis that

$$E(U_1) = E(U_1') \tag{7.1}$$

and

$$E(U_2) = E(U_2'), \tag{7.2}$$

is given (Hotelling [10]) by

$$T^2 = N^{-1}(mn) \sum_{j=1}^{2} \sum_{i=1}^{2} q_{ij}(\bar{u}_i - \bar{u}_i')(\bar{u}_j - \bar{u}_j')$$

where

$$m\bar{u}_i = \sum_{l=1}^{m} u_{il} \quad n\bar{u}_i' = \sum_{l=m+1}^{N} u_{il}$$

and the matrix $\| q_{ij} \|$ is the inverse of the matrix $\| b_{ij} \|$ with b_{ij} given by

$$(N-2)b_{ij} = \sum_{l=1}^{m} (u_{il} - \bar{u}_i)(u_{jl} - \bar{u}_j) + \sum_{l=m+1}^{N} (u_{il} - \bar{u}_i')(u_{jl} - \bar{u}_j').$$

In Hotelling's procedure the b_{ij} are sample estimates of the population covariances whose distribution is independent of that of the sample means. A constant multiple of the statistic T^2 has the analysis of variance distribution under the null hypothesis. If the population covariances were known and used in place of the b_{ij}, T^2 would have the χ^2 distribution with two degrees of freedom.

Let us now apply the generalized T over the permutations of the actually observed values, as was done with other statistics in previous sections. If we do this literally we will find that the b_{ij} are no longer independent of the sample means. To avoid this complication we shall use a slightly different statistic T' which, as will be shown later, is a monotonic function of T, so that the test based on T' is identical with that based on T. The statistic T' is defined as follows: Let

$$\bar{U}_i = N^{-1} \sum_{k=1}^{N} u_{ik}$$

$$c'_{ij} = N[(N-1)mn]^{-1} \sum_{k=1}^{N} (u_{ik} - \bar{U}_i)(u_{jk} - \bar{U}_j) \qquad (i, j, = 1, 2)$$

and

$$\| q'_{ij} \| = \| c'_{ij} \|^{-1}.$$

Then

$$T'^2 = \sum_{i=1}^{2} \sum_{j=1}^{2} q'_{ij}(\bar{u}_i - \bar{u}'_i)(\bar{u}_j - \bar{u}'_j). \tag{7.3}$$

The expression T'^2 is much simpler than T^2, since the coefficients q'_{ij} are constants in the population of permutations of the observations. We shall show that T'^2 is a monotonic function of T^2. Let

$$Q_{ij} = \sum_{k=1}^{m} (u_{ik} - \bar{u}_i)(u_{jk} - \bar{u}_j) + \sum_{k=m+1}^{N} (u_{ik} - \bar{u}'_i)(u_{jk} - \bar{u}'_j)$$

$$Q'_{ij} = \sum_{k=1}^{N} (u_{ik} - \bar{U}_i)(u_{jk} - \bar{U}_j)$$

$$\| Q^{ij} \| = \| Q_{ij} \|^{-1}$$

$$\| Q'^{ij} \| = \| Q'_{ij} \|^{-1}.$$

Then the expressions

$$T_1^2 = \sum_{i=1}^{2} \sum_{j=1}^{2} Q^{ij}(\bar{u}_i - \bar{u}'_i)(\bar{u}_j - \bar{u}'_j) \tag{7.4}$$

and

$$T_2^2 = \sum_{i=1}^{2} \sum_{j=1}^{2} Q'^{ij}(\bar{u}_i - \bar{u}'_i)(\bar{u}_j - \bar{u}'_j), \tag{7.5}$$

are constant multiples of T^2 and T'^2, respectively. Hence it is sufficient to show that T_2^2 is a monotonic function of T_1^2. We have

$$Q'_{ij} = Q_{ij} + m(\bar{u}_i - \bar{U}_i)(\bar{u}_j - \bar{U}_j) + n(\bar{u}'_i - \bar{U}_i)(\bar{u}'_j - \bar{U}_j). \tag{7.6}$$

Furthermore, we have

$$\bar{u}_i - \bar{U}_i = \bar{u}_i - \frac{m\bar{u}_i + n\bar{u}'_i}{m+n} = \frac{n(\bar{u}_i - \bar{u}'_i)}{m+n}. \tag{7.7}$$

Similarly

$$\bar{u}'_i - \bar{U}_i = \bar{u}'_i - \frac{m\bar{u}_i + n\bar{u}'_i}{m + n} = \frac{m(\bar{u}'_i - \bar{u}_i)}{m + n}. \tag{7.8}$$

From (7.6), (7.7) and (7.8) it follows that

$$\begin{aligned} Q'_{ij} &= Q_{ij} + \frac{mn^2}{(m+n)^2}(\bar{u}_i - \bar{u}'_i)(\bar{u}_j - \bar{u}'_j) + \frac{nm^2}{(m+n)^2}(\bar{u}_i - \bar{u}'_i)(\bar{u}_j - \bar{u}'_j) \\ &= Q_{ij} + \frac{mn}{m+n}(\bar{u}_i - \bar{u}'_i)(\bar{u}_j - \bar{u}'_j). \end{aligned} \tag{7.9}$$

Denote $\dfrac{mn}{m+n}$ by λ and $\bar{u}_i - \bar{u}'_i$ by h_i. Then we have

$$Q'_{ij} = Q_{ij} + \lambda h_i h_j. \tag{7.10}$$

Denote the cofactor of Q_{ij} in $\| Q_{ij} \|$ by R_{ij} and the cofactor of Q'_{ij} in $\| Q'_{ij} \|$ by R'_{ij}. Then

$$\frac{| Q_{ij} |}{| Q'_{ij} |} = \frac{| Q_{ij} |}{| Q_{ij} + \lambda h_i h_j |} = \frac{| Q_{ij} |}{| Q_{ij} | + \lambda \Sigma\Sigma R_{ij} h_i h_j} = \frac{1}{1 + \lambda T_1^2}. \tag{7.11}$$

Furthermore, we have

$$\frac{| Q_{ij} |}{| Q'_{ij} |} = \frac{| Q'_{ij} - \lambda h_i h_j |}{| Q'_{ij} |} = \frac{| Q'_{ij} | - \lambda \Sigma\Sigma R'_{ij} h_i h_j}{| Q'_{ij} |} = 1 - \lambda T_2^2. \tag{7.12}$$

From (7.11) and (7.12) it follows that T_2^2 is a monotonic function of T_1^2. Hence also T'^2 is a monotonic function of T^2 and, therefore, we do not change our test procedure by using T'^2 instead of T^2.

Let the sequence of pairs

$$(x_{11} . x_{21}), \cdots, (x_{1N}, x_{2N})$$

be a permutation of the actually observed pairs

$$(u_{11}, u_{21}), \cdots, (u_{1N}, u_{2N})$$

where to each permutation is ascribed the same probability $(N!)^{-1}$. Then one obtains for $i = 1, 2$,

$$E(\bar{x}_i - \bar{x}'_i) = 0 \tag{7.13}$$

$$\sigma^2(\bar{x}_i - \bar{x}'_i) = N[(N-1)mn]^{-1} \sum_{j=1}^{N} (u_{ij} - \bar{U}_i)^2 = c'_{ii} \tag{7.14}$$

$$E(\bar{x}_1 - \bar{x}'_1)(\bar{x}_2 - \bar{x}'_2) = N[(N-1)mn]^{-1} \sum_{j=1}^{N} (u_{1j} - \bar{U}_1)(u_{2j} - \bar{U}_2) = c'_{12}. \tag{7.15}$$

Hence $\| c'_{ij} \|$ is the covariance matrix of the variates

$$(\bar{x}_1 - \bar{x}'_1) \quad \text{and} \quad (\bar{x}_2 - \bar{x}'_2).$$

Now we shall show that the limiting distribution of T'^2, as $N \to \infty$, is the χ^2 distribution with 2 degrees of freedom, provided that the observation u_{ik} $(i = 1, 2; k = 1, \cdots, N)$ satisfy some slight restrictions. Since $\| q'_{ij} \|$ is the inverse of the covariance matrix $\| c'_{ij} \|$ our statement about the limiting distribution of T'^2 is obviously proved if we can show that $\bar{x}_1 - \bar{x}'$ and $\bar{x}_2 - \bar{x}'_2$ have a joint normal distribution in the limit.

Let $N \to \infty$ while m/n remains constant. Let the sequences A_N and D_N of Section II be defined as follows:
There are two sequences A_N, denoted respectively by A_{1N} and A_{2N}, such that

$$a_{ij} = u_{ij} \qquad (i = 1, 2; j = 1, \cdots, N).$$

Also

$$d_j = \frac{1}{m} \qquad (j = 1, \cdots, m)$$

$$d_j = -\frac{1}{n} \qquad (j = m + 1, \cdots, N).$$

Then the sequences D_N satisfy the condition W. If also the sequences A_{iN} satisfy the condition W, the distribution of $\bar{x}_i - \bar{x}'_i$ approaches the normal distribution as N increases, by the theorem of Section 2. If the joint distribution of $\bar{x}_1 - \bar{x}'_i$ and $\bar{x}_2 - \bar{x}'_2$ approaches a normal distribution with non-singular correlation matrix, the distribution of T'^2 approaches that of χ^2 with two degrees of freedom.

The correlation matrix of $(\bar{x}_1 - \bar{x}'_1)$ and $(\bar{x}_2 - \bar{x}'_2)$ will be of rank two in the limit if the correlation coefficient between $(\bar{x}_1 - \bar{x}'_1)$ and $(\bar{x}_2 - \bar{x}'_2)$ approaches a limit ρ, where $|\rho| < 1$. By (7.14) and (7.15) this is equivalent to saying that the absolute value of the angle between the vectors A'_{1N} and A'_{2N} is eventually greater than a positive lower bound. We shall show that, if the correlation coefficient approaches, as $N \to \infty$, a limit ρ whose absolute value is less than one, and if A_{1N} and A_{2N} satisfy the condition W, then $(\bar{x}_1 - \bar{x}'_1)$ and $(\bar{x}_2 - \bar{x}'_2)$ are *jointly* normally distributed in the limit.

Let δ_1 and δ_2 be any two real numbers not both zero. Then the sequence

$$A_N^* = (a_1^*, \cdots, a_N^*)$$

where

$$a_j^* = \delta_1 a_{1j} + \delta_2 a_{2j}$$

will be shown to satisfy the condition W. If either δ_1 or δ_2 is zero this is trivial; assume therefore that neither is zero. Without loss of generality we may assume that $\sum_{j=1}^{N} a_{ij} = 0$, for if this were not so we could replace the original a_{ij} by $a'_{ij} = a_{ij} - N^{-1} \sum_j a_{ij}$ as was done in Section 2. Let ρ' be such that $1 > \rho' > |\rho|$.

For N sufficiently large we have

$$\mu_2(A_N^*) \geq N^{-1}(\delta_1^2 \sum_j a_{1j}^2 - 2 \mid \delta_1 \delta_2 \sum a_{1j}a_{2j} \mid + \delta_2^2 \sum_j a_{2j}^2)$$

$$\geq N^{-1}(\delta_1^2 \sum_j a_{1j}^2 - 2\rho' \mid \delta_1 \delta_2 \mid \sqrt{(\sum_j a_{1j}^2)(\sum_j a_{2j}^2)} + \delta_2^2 \sum_j a_{2j}^2)$$

$$= N^{-1}[(\mid \delta_1 \mid \sqrt{\sum_j a_{1j}^2} - \mid \delta_2 \mid \sqrt{\sum_j a_{2j}^2})^2$$

$$+ 2(1 - \rho') \mid \delta_1 \delta_2 \mid \sqrt{(\sum_j a_{1j}^2)(\sum_j a_{2j}^2)}]$$

and

$$\mu_2(A_N^*) \leq 2(\delta_1^2\mu_2(A_{1N}) + \delta_2^2\mu_2(A_{2N})).$$

Hence

$$\mu_2(A_N^*) = \Omega[\max \{\mu_2(A_{1N}), \mu_2(A_{2N})\}]. \tag{7.16}$$

Also $\mu_r(A_N^*)$ is a sum of constant multiples of terms of the type

$$N^{-1} \sum_j a_{1j}^i a_{2j}^{r-i}.$$

By Schwarz' inequality

$$N^{-1} \sum_j a_{1j}^i a_{2j}^{r-i} \leq N^{-1}(\sum_j a_{1j}^{2i})^{\frac{1}{2}}(\sum_j a_{2j}^{2(r-i)})^{\frac{1}{2}} = (\mu_{2i}(A_{1N})\mu_{2(r-i)}(A_{2N}))^{\frac{1}{2}}. \tag{7.17}$$

The required result follows from (7.16) and (7.17).

Since the sequences A_N^* satisfy the condition W, the limiting distribution of

$$\delta_1(\bar{x}_1 - \bar{x}_1') + \delta_2(\bar{x}_2 - \bar{x}_2'),$$

for any pair δ_1, δ_2 not both zero, is normal. From this and a theorem of Cramér and Wold ([11] Theorem 1; see also [8], Theorem 31) it follows that if the joint distribution of $(\bar{x}_1 - \bar{x}_1')$ and $(\bar{x}_2 - \bar{x}_2')$ approaches a limit, this limit must be the normal distribution. From a theorem of Radon ([12]; see also Cramér [8], page 101) it follows that if the joint distribution of $(\bar{x}_1 - \bar{x}_1')$ and $(\bar{x}_2 - \bar{x}_2')$ does not approach a limit as $N \to \infty$ it is possible to find two subsequences of the sequence $(1, 2, \cdots, N, \cdots$ ad inf.) for each of which the joint distribution approaches a different limit. This contradicts the previous result. Hence the limit exists and is the normal distribution. This proves our statement that the limiting distribution of T'^2 is the χ^2 distribution with two degrees of freedom.

The statistic T'^2 seems to be appropriate for testing the null hypothesis that two bivariate distributions Π_1 and Π_2 are identical if the alternatives are restricted to the case where Π_2 differs from Π_1 only in the mean values, i.e., the distribution Π_2 can be obtained from Π_1 by a translation. This is no restriction as compared with Hotelling's T-test since also the T-test is based on the assumption that the two normal populations differ at most in their mean values, i.e., the covariance matrices in the two populations are assumed to be equal.

REFERENCES

[1] A. Wald and J. Wolfowitz, *Annals of Math. Stat.*, Vol. 14 (1943), p. 378.
[2] Henry Scheffé, *Annals of Math. Stat.*, Vol. 14 (1943), p. 305.
[3] H. Hotelling and M. R. Pabst, *Annals of Math. Stat.*, Vol. 7 (1936), p. 29.
[4] E. J. G. Pitman, *Supp. Jour. Roy. Stat. Soc.*, Vol. 4 (1937), p. 225.
[5] E. J. G. Pitman, *Supp. Jour. Roy. Stat. Soc.*, Vol. 4 (1937), p. 119.
[6] E. J. G. Pitman, *Biometrika*, Vol. 29 (1938), p. 322.
[7] B. L. Welch, *Biometrika*, Vol. 29 (1937), p. 21.
[8] Harald Cramér, *Random Variables and Probability Distributions*, Cambridge, 1937.
[9] J. V. Uspensky, *Introduction to Mathematical Probability*, New York, 1937.
[10] H. Hotelling, *Annals Math. Stat.*, Vol. 2 (1931), p. 359.
[11] H. Cramér and H. Wold, *Journal London Math. Soc.*, Vol. 11 (1936), pp. 290–4.
[12] J. Radon, *Sitzungsberichte Akademie Wien*, Vol. 122 (1913), pp. 1295–1438.
[13] R. A. Fisher, *The Design of Experiments*, Edinburgh, 1937. (Especially Section 21.)
[14] R. A. Fisher, *Statistical Methods for Research Workers*, Edinburgh, 1936. (Especially Section 21.02.)

Made in United States of America

Reprinted from The Annals of Mathematical Statistics
Vol. XVII, No. 4, December, 1946

ON SEQUENTIAL BINOMIAL ESTIMATION

By J. Wolfowitz
Columbia University

The present note, written after a reading of the very interesting paper by Girshick, Mosteller, and Savage [1], is for the purpose of adding a few remarks in the nature of a supplement. For the sake of brevity the notation and terminology of [1] are adopted in toto.

Theorem 1 below generalizes Theorem 1 of [1]. In Theorem 2′ we formulate explicitly the fact which lies at the basis of the GSM method of estimation. Parts of the proofs of Theorems 3 and 4 of [1] are simply proofs of special cases of this (e.g., equation (2) of [1]). We then use this fact repeatedly in proving Theorem 3, which states that the Girshick-Mosteller-Savage estimate is the only proper unbiased estimate for sequential tests defined by regions which we shall call doubly simple.

A doubly simple region is defined precisely below. Intuitively we may describe such a region as the one between two curves $y = f_1(x)$ and $x = f_2(y)$, where $f_1(x)$ is defined and monotonically non-decreasing for all non-negative x, $f_2(y)$ is defined and monotonically non-decreasing for all non-negative y, $f_1(0) > 0$, $f_2(0) > 0$. If the two curves intersect, the region is finite, and the values of the functions f_1 and f_2 beyond the point of intersection are of no interest. This description is of course purely heuristic, because in actual fact only integral values of the variables come into play, and intersection of the curves, for example, is not needed to make the region finite. Since the question of finite regions is completely settled by [1], Theorem 7, only non-finite regions remain to be discussed, and the precise definition given below is such as to imply that the region is not finite. It seems to the present writer that at least many of the non-finite sequential tests which may be developed for meaningful statistical problems will require doubly simple regions. The Wald sequential binomial test [2] defines such a region, which also falls within the scope of Theorem 6 of [1]. It is easy to see that there exist closed regions which are doubly simple and do not satisfy the conditions of this theorem.

By a "proper" estimate $p(\alpha)$ we shall mean an estimate such that $0 \leq p(\alpha) \leq 1$ for every α. It is difficult to see how any estimate which is not proper can make much sense.

Theorem 1. *A sufficient condition that a region R be closed is that* $\liminf_{n\to\infty} \frac{A(n)}{\sqrt{n}} < \infty$, *where* $A(n)$ *is the number of accessible points of index n.*

Proof: The hypothesis of the theorem implies that there exist a positive number H and an increasing sequence of positive integers $n_1, n_2, n_3, \cdots$, with the following properties:

a) $n_{i+1} > 2n_i$ ($i = 1, 2, \cdots$ ad inf.)

b) $A(n_i) < H\sqrt{n_i}$.

For n_i sufficiently large, the conditional probability of reaching the accessible points on $x + y = n_{i+1}$, when an accessible point on $x + y = n_i$ has been reached, is $< K < 1$ by the normal approximation to the binomial distibution, where K is constant (and depends on H). Hence the probability of passing through accessible points on all members of the set $x + y = n_i$ ($i = 1, 2, \cdots, L$) approaches zero as $L \to \infty$, so that the region is closed.

Theorem 2. *Let R be any region, B its boundary, and $t = (a, b)$; any accessible point in R. Let $l_t(\alpha)$ be the number of paths from t to $(x, y) = \alpha \in B$. Let $Q(t)$ be the conditional probability that a path, which has reached t, will reach the boundary B. Then*

$$\sum_{\alpha \in B} l_t(\alpha)p^y q^x = Q(t)p^b q^a.$$

Theorem 2′. (Corollary to Theorem 2)

If R is closed, then

$$\sum_{\alpha \in B} l_t(\alpha)p^y q^x = p^b q^a. \tag{1}$$

Proof: Let $k(t)$ be the number of paths in R from the origin to t. The probability of reaching $\alpha \in B$ by a path which passes through t is $k(t)l_t(\alpha)p^y q^x$. The probability of reaching t from the origin is $k(t)p^b q^a$, and hence the probability of reaching the boundary via t is $Q(t)k(t)p^b q^a$. From this the desired result follows.

We now define a doubly simple region. The boundary of the region consists of the two infinite sequences of points

$$(0, a_0), (1, a_1), (2, a_2), \cdots$$

and

$$(b_0, 0), (b_1, 1), (b_2, 2), \cdots$$

where $a_0, a_1, a_2, \cdots$ and $b_0, b_1, b_2, \cdots$ are two infinite non-decreasing sequences of positive integers. The accessible points of the region are all points which can be reached by a path from the origin which does not contain a boundary point. (It is to be noted that since a boundary point is, by definition, a point not in the region which can be reached by a path in the region, the above definition implies that a doubly simple region is not finite. The reason for making this so has been given above.)

Theorem 3. Let R be a closed doubly simple region. Then $\hat{p}(\alpha)$ is the unique proper unbiased estimate of p.

PROOF: Suppose there were two proper unbiased estimates $p_1(\alpha)$ and $p_2(\alpha)$. Writing $m(\alpha) = p_1(\alpha) - p_2(\alpha)$, we would have

$$\sum_{\alpha \in B} m(\alpha)k(\alpha)p^y q^x = 0 \tag{2}$$

with

$$| m(\alpha) | \leq 1 \tag{3}$$

First we prove

LEMMA 1. *If* $a_0 > 1$, *then* $m(b_0, 0) = 0$.

PROOF: Let $k^*(\alpha)$ denote the number of paths in R from the point $(0, 1)$ to the boundary point α. For all points $\alpha \in B$ except $(b_0, 0)$ we have

$$b_0 k^*(\alpha) \geq k(\alpha). \tag{4}$$

From (1), (2), (3), and (4) we have, since $k(b_0, 0) = 1$,

$$\begin{aligned} | m(b_0, 0) | \, q^{b_0} &= \Big| \sum_{\alpha \in B, \alpha \neq (b_0, 0)} m(\alpha)k(\alpha)p^y q^x \Big| \\ &\leq \sum_{\alpha \in B, \alpha \neq (b_0, 0)} k(\alpha)p^y q^x \leq b_0 \sum_{\alpha \in B} k^*(\alpha)p^y q^x = b_0 p. \end{aligned} \tag{5}$$

Now as $p \to 0$, the left member of the inequality (5) approaches $| m(b_0, 0) |$, and the right member approaches zero. This proves Lemma 1.

LEMMA 2. *For every* $z < a_0 - 1$, $m(b_z, z) = 0$.

PROOF: In view of Lemma 1 it is sufficient to prove the following:

If $Z \leq a_0 - 2$, and if $m(b_z, z) = 0$ for $z = 0, 1, \cdots, Z - 1$, then $m(b_Z, Z) = 0$. Let $k_{Z+1}(\alpha)$ denote the number of paths in R from $(0, Z + 1)$ to the boundary point α. For any point $\alpha \in B$ whose ordinate is $\geq Z + 1$ we have

$$b_0 b_1 \cdots b_Z k_{Z+1}(\alpha) \geq k(\alpha). \tag{6}$$

From (1), (2), (3), and (6) we have

$$\begin{aligned} | m(b_Z, Z) | \, k(b_Z, Z)p^Z q^{b_Z} &= | \Sigma m(\alpha)k(\alpha)p^y q^x | \leq \Sigma k(\alpha)p^y q^x \\ &\leq b_0 b_1 \cdots b_Z \Sigma k_{Z+1}(\alpha)p^y q^x = b_0 b_1 \cdots b_Z p^{Z+1} \end{aligned} \tag{7}$$

where the summations take place over all boundary points whose ordinates are $\geq Z + 1$. Hence

$$| m(b_Z, Z) | \, k(b_Z, Z)q^{b_Z} \leq b_0 b_1 \cdots b_Z p.$$

and letting $p \to 0$ we obtain the desired result.

LEMMA 3. $m(b_{a_0-1}, a_0 - 1) = 0$.

PROOF: Let s be the smallest integer such that (s, a_0) is an accessible point.

We proceed as in Lemma 2, with (s, a_0) playing the role of $(0, Z + 1)$, and eventually obtain the following inequality:

$$\begin{aligned} | m(b_{a_0-1}, a_0 - 1) | \, k(b_{a_0-1}, a_0 - 1)p^{a_0-1} q^{b_{a_0-1}} &= | \textstyle\sum_0 m(\alpha)k(\alpha)p^y q^x | \\ &\leq p^{a_0} \Big(\sum_{i=0}^{s-1} k(i, a_0)q^i + b_0 b_1 \cdots b_{a_0-1} q^s \Big), \end{aligned} \tag{8}$$

where Σ_0 denotes summation over all boundary points with ordinate $\geq a_0$. The desired result follows.

LEMMA 4. *Let $h(\geq a_0)$ be the smallest ordinate for which at least one boundary point (w^*, h) exists such that $m(w^*, h) \neq 0$ (If no such h exists the theorem is proved). Of all such points let w be the one with the smallest abscissa. Then the point (w, h) is a member of the sequence*

$$(0, a_0)\ (1, a_1),\ (2, a_2),\ \cdots$$

PROOF: If the lemma is not true, then for all boundary points α with ordinate h, $m(\alpha) = 0$, except that $m(b_h, h) \neq 0$. Let W be that accessible point of R whose ordinate is $h + 1$ and whose abscissa v is a minimum. Let $k_w(\alpha)$ be the number of paths in R from W to the boundary point α. For boundary points α accessible from W we have

$$b_0 b_1 \cdots b_h k_W(\alpha) \geq k(\alpha). \tag{9}$$

From (1), (2), (3), and (9) we have

$$| m(b_h, h) | k(b_h, h) p^h q^{b_h} = | \Sigma_1(m(\alpha)k(\alpha)p^y q^x | \leq \Sigma_2 k(\alpha) p^{h+1} q^x + b_0 b_1 \cdots b_h p^{h+1} q^v = K^* p^{h+1}, \tag{10}$$

where:

a) Σ_1 denotes summation over all $\alpha \in B$ for which $y > h$

b) Σ_2 denotes summation over all boundary points α of ordinate $h + 1$ and abscissa $< v$.

c) K^* denotes a constant.

From this it easily follows that $m(b_h, h) = 0$, in contradiction to the definition of h. This proves Lemma 4.

PROOF OF THEOREM 3: Let (w, h) be as defined in the statement of Lemma 4. From Lemma 4 it follows that, if any other boundary points with abscissa w exist, they must be members of the sequence $(b_0, 0)$, $(b_1, 1)$, $(b_2, 2)$, $\cdots$ and hence their ordinates are $< h$. From the definition of (w, h) and from Lemma 4 it follows that for any $\alpha \in B$ whose abscissa is $< w$, $m(\alpha) = 0$.

Now in the proofs of Lemmas 1–4 the roles of x and y are not symmetrical. However, symmetry of course exists, and analogous lemmas follow. In particular, the analogue to Lemma 4 has as a consequence that, since w is the smallest abscissa such that $m(\alpha) = 0$ when abscissa of $\alpha < w$, and $m(w, h) \neq 0$, there exists a boundary point (w, h'), such that $m(w, h') \neq 0$ and (w, h') is a member of $(b_0, 0)$, $(b_1, 1)$, $(b_2, 2)$, ... Then $h' < h$. But this contradicts the definition of h and proves the theorem.

It is easy to see that, if the boundary points of a closed region constitute a single "curve" instead of two "curves" as in a doubly simple region, the estimate $\hat{p}(\alpha)$ will be the only proper unbiased estimate of p.

It is interesting to consider some of the consequences of Theorem 3 for all unbiased estimates (not necessarily proper) for doubly simple regions. An examination of the proof of Theorem 3 shows that it would go through with little change if equation (3) were replaced by the requirement that $|m(\alpha)|$ be bounded. We therefore obtain the following result: If for a doubly simple region there exists an unbiased estimate $p(\alpha)$ of p, not identically equal to $\hat{p}(\alpha)$, then not only is $p(\alpha)$ not proper, but also, no matter how large M, there exists a boundary point α such that $|p(\alpha)| > M$. The uselessness of such an estimate is manifest.

The author is of the opinion that freedom from bias is not necessarily an indispensable characteristic of an optimum estimate. In general there is no reason for requiring the first moment of the estimate rather than any other moment to be the unknown parameter. The justification in any particular case must be based on special conditions of the problem.

The author is indebted to Mr. Howard Levene for reading the present paper and making valuable suggestions.

REFERENCE

[1] M. A. Girshick, Frederick Mosteller, and L. J. Savage, "Unbiased estimates for certain binomial sampling problems with applications," *Annals of Math. Stat.*, Vol. 17 (1946). pp. 13–23.

[2] A. Wald, "Sequential tests of statistical hypotheses," *Annals of Math. Stat.*, Vol. 16 (1945), pp. 117–186.

Reprinted from The Annals of Mathematical Statistics
Vol. XVIII, No. 2, June, 1947

THE EFFICIENCY OF SEQUENTIAL ESTIMATES AND WALD'S EQUATION FOR SEQUENTIAL PROCESSES

By J. Wolfowitz

Columbia University

1. Summary. Let n successive independent observations be made on the same chance variable whose distribution function $f(x, \theta)$ depends on a single parameter θ. The number n is a chance variable which depends upon the outcomes of successive observations; it is precisely defined in the text below. Let $\theta^*(x_1, \cdots, x_n)$ be an estimate of θ whose bias is $b(\theta)$. Subject to certain regularity conditions stated below, it is proved that

$$\sigma^2(\theta^*) \geq \left(1 + \frac{db}{d\theta}\right)^2 \left[EnE\left(\frac{\partial \log f}{\partial \theta}\right)^2\right]^{-1}.$$

When $f(x, \theta)$ is the binomial distribution and θ^* is unbiased the lower bound given here specializes to one first announced by Girshick [3], obtained under no doubt different conditions of regularity. When the chance variable n is a constant the lower bound given above is the same as that obtained in [2], page 480, under different conditions of regularity.[1]

Let the parameter θ consist of l components $\theta_1, \cdots, \theta_l$ for which there are given the respective unbiased estimates $\theta_1^*(x_1, \cdots, x_n), \cdots, \theta_l^*(x_1, \cdots, x_n)$. Let $\| \lambda_{ij} \|$ be the non-singular covariance matrix of the latter, and $\| \lambda^{ij} \|$ its inverse. The concentration ellipsoid in the space of $(k_1, \cdots, k_l)$ is defined as

$$\sum_{i,j} \lambda^{ij}(k_i - \theta_i)(k_j - \theta_j) = l + 2.$$

(This valuable concept is due to Cramér). If a unit mass be uniformly distributed over the concentration ellipsoid, the matrix of its products of inertia will coincide with the covariance matrix $\| \lambda_{ij} \|$. In [4] Cramér proves that no matter what the unbiased estimates $\theta_1^*, \cdots, \theta_l^*$, (provided that certain regularity conditions are fulfilled), when n is constant their concentration ellipsoid always contains within itself the ellipsoid

$$\sum_{i,j} \mu_{ij}(k_i - \theta_i)(k_j - \theta_j) = l + 2$$

where

$$\mu_{ij} = nE\left(\frac{\partial \log f}{\partial \theta_i} \frac{\partial \log f}{\partial \theta_j}\right).$$

---

[1] To whom this result is to be ascribed is not clear from the context in which Professor Cramér describes it (in [2]). After the present paper was completed the author learned of the papers by Rao [8] and Aitken and Silverstone [9], both of which deal with this question. The author is indebted to Prof. M. S. Bartlett for drawing his attention to these papers.

Consider now the sequential procedure of this paper. Let $\theta_1^*, \cdots, \theta_l^*$ be, as before, unbiased estimates of $\theta_1, \cdots, \theta_l$, respectively, recalling, however, that the number n of observations is a chance variable. It is proved that the concentration ellipsoid of $\theta_1^*, \cdots, \theta_l^*$ always contains within itself the ellipsoid

$$\sum_{i,j} \mu'_{ij}(k_i - \theta_i)(k_j - \theta_j) = l + 2$$

where

$$\mu'_{ij} = EnE\left(\frac{\partial \log f \, \partial \log f}{\partial \theta_i \quad \partial \theta_j}\right).$$

When n is a constant this becomes Cramér's result (under different conditions of regularity).

In section 7 is presented a number of results related to the equation $EZ_n = EnEX$, which is due to Wald [6] and is fundamental for sequential analysis.

2. Introduction. Let X be a chance variable whose distribution function $f(x, \theta)$ depends on the parameter θ. It is assumed that X either has a probability density function (which we then denote by $f(x, \theta)$) or that it can take only an at most denumerable number of discrete values (in the latter case $f(x, \theta) = P\{X = x\}$, where the latter symbol denotes the probability of the relation in braces). Let $\omega = x_1, x_2, \cdots$ be an infinite sequence of observations on X, and let Ω be the space of "points" ω. Let there be given an infinite sequence of Borel measurable functions $\varphi_1(x_1), \varphi_2(x_1, x_2), \cdots, \varphi_j(x_1, \cdots, x_j), \cdots$ defined for all ω in Ω, such that each takes only the values zero and one. It is well known that the function $f(x, \theta)$ defines a measure (probability) on a Borel field in Ω. We assume that everywhere in Ω, except possibly on a set whose probability is zero for all θ under consideration, at least one of the functions $\varphi_1, \varphi_2, \cdots$ takes the value one. Let $n(\omega)$ be the smallest integer at which this occurs. Thus $n(\omega)$ is a chance variable.

In statistical applications the chance variable $n(\omega)$ may be interpreted as a rule for terminating a sequence of observations on the chance variable X, the probability of termination being one, and the decision to terminate depending only upon the observations obtained. A sequential test is an example of this procedure. The converse is, however, not true, because the process described above does not require that any statistical decision should be reached when the process of drawing observations is terminated.

An "estimate" of θ is a function $\theta^*(x_1, \cdots, x_n)$ of the observations $x_1, \cdots, x_n$ (those obtained prior to the "termination" of the process of drawing observations). In the sequel we shall limit ourselves to estimates whose second moments are finite. The estimate is "unbiased" if $E\theta^*$, the expected value of θ^*, is θ. When this is not so $E\theta^* - \theta$ is called the bias, $b(\theta)$, of θ^*. In general the bias is a function of θ. It is obvious that the function θ^* may be undefined on a set of points $(x_1, \cdots, x_n)$ whose probability is zero for all θ under consideration.

In the present paper we shall be concerned with an upper bound on the efficiency of a sequential estimate, or, more precisely, with a lower bound on its variance. This lower bound is intimately related to certain results on the efficiency of the maximum likelihood estimate from a sample of fixed size. This is not surprising since fixed-size sampling is a special instance of sequential sampling. The results obtained in this paper are also obviously and intimately related to those due to Cramér [4] and those described by him in [2], pp. 477–488. Naturally the conditions of regularity (restrictions on $f(x, \theta)$, θ^*, etc.) under which the results are proved are different. For example, no restrictions on the sequential sampling procedure need appear in the statement of a theorem which deals only with samples of fixed size.

The argument below proceeds as if $f(x, \theta)$ were a probability density function. The results apply equally well to the case where $f(x, \theta)$ is the probability function of a discrete chance variable provided:

1). Integration is replaced by summation wherever this is obviously required.

2). The phrase "almost all points" in a Euclidean space of any finite dimensionality is understood

a). as all points in the space with the possible exception of a set of Lebesgue measure zero, when $f(x, \theta)$ is a probability density function

b). as all points in the space with the possible exception of points one of whose coordinates is a member of the set Z, when $f(x, \theta)$ is the probability function of a discrete chance variable. The set Z consists of all points z such that $f(z, \theta) = 0$ identically for all θ under consideration.

3. Conditions of regularity. In this section we shall formulate the restrictions which we impose on f, the estimates, and the sequential process. They are intended to be such as will be satisfied in most cases of statistical interest. No doubt they can be weakened, but the author has decided against attempting to do so here. The list may seem long for two reasons. Seldom in the literature are the assumptions which, for example, lead to validation of differentiation under the integral sign etc., formulated explicitly. The presence of a sequential procedure means that additional restrictions must be imposed.

In this section we assume that θ is a single parameter. The case where θ has more than one component is treated later.

(3.1). *The parameter θ lies in an open interval D of the real line. D may consist of the entire line or of an entire half-line.*

(3.2). *The derivative $\frac{\partial f}{\partial \theta}$ exists for all θ in D and almost all x. We define $\frac{\partial \log f(x, \theta)}{\partial \theta}$ as zero whenever $f(x, \theta) = 0$; thus $\frac{\partial \log f}{\partial \theta}$ is defined for all θ in D and almost all x. We postulate that $E \frac{\partial \log f(x, \theta)}{\partial \theta} = 0$ and that $E \left(\frac{\partial \log f(x, \theta)}{\partial \theta}\right)^2$ be not zero for all θ in D.*

(3.3).
$$E\left(\sum_{i=1}^{n}\left|\frac{\partial \log f(x_i, \theta)}{\partial\theta}\right|\right)^2$$

exists for all θ in D.

(3.4). *Let R_j, ($j = 1, 2, \cdots$), be the set of points $(x_1, \cdots, x_j)$ in the j-dimensional Euclidean space such that*

$$\varphi_i(x_1, \cdots, x_i) = 0 \qquad i = 1, 2, \cdots, j-1$$
$$\varphi_j(x_1, \cdots, x_j) = 1.$$

For any integral j there exists a non-negative L-measurable function $T_j(x_1, \cdots, x_j)$ such that

a).
$$\left|\theta^*(x_1, \cdots, x_j)\frac{\partial}{\partial\theta}\prod_{i=1}^{j} f(x_i, \theta)\right| < T_j(x_1, \cdots, x_j)$$

for all θ in D and almost all $(x_1, \cdots, x_j)$ in R_j

b).
$$\int_{R_j} T_j(x_1, \cdots, x_j)\, dx_1 \cdots dx_j$$

is finite.

(3.5). *Let*

$$t_j(\theta) = \int_{R_j}\theta^*(x_1, \cdots, x_j)\prod_{i=1}^{j} f(x_i, \theta)\, dx_i, \qquad (j = 1, 2, \cdots).$$

We postulate the uniform convergence of the series

$$\sum_j \frac{dt_j(\theta)}{d\theta}$$

(the existence of $\frac{dt_j(\theta)}{d\theta}$ is a consequence of Assumption (3.4)) for all θ in D.

4. The case of one parameter. In this section we assume that $f(x, \theta)$ depends on a single parameter θ. In sections 5 and 6 we shall discuss the case when θ is a vector with more than one component.

We have $E\,\frac{\partial \log f(x, \theta)}{\partial\theta} = 0$

by (3.2). Define the chance variable

$$Y_n = \sum_{i=1}^{n}\frac{\partial \log f(x_i, \theta)}{\partial\theta}.$$

By an argument almost identical with that of [1], Theorem 1, or of Theorem 7.1 below, we have

$$EY_n = 0. \tag{4.1}$$

From Theorem 7.2 below we obtain

$$\sigma^2(Y_n) = EnE\left(\frac{\partial \log f(x, \theta)}{\partial \theta}\right)^2. \tag{4.2}$$

Let θ^* $(x_1, \cdots, x_n)$ be an estimate of θ such that

$$E\theta^* = \theta + b(\theta).$$

Then

$$\sum_{j=1}^{\infty} \int_{R_j} \theta^*(x_1, \cdots, x_j) \prod_{i=1}^{j} f(x_i, \theta)\, dx_i = \theta + b(\theta). \tag{4.3}$$

Differentiation of both members of (4.3) with respect to θ (Assumptions (3.4) and (3.5)) gives

$$E\theta^* Y_n = 1 + \frac{db}{d\theta}. \tag{4.4}$$

From (4.1) it follows that (4.4) gives the covariance between θ^* and Y_n. Hence from (4.2)

$$\sigma^2(\theta^*) \geq \left(1 + \frac{db}{d\theta}\right)^2 \left[EnE\left(\frac{\partial \log f(x, \theta)}{\partial \theta}\right)^2\right]^{-1}. \tag{4.5}$$

When the bias $b(\theta)$ is constant, for example when $b(\theta) \equiv 0$ in case θ^* is an unbiased estimate, we have from (4.5)

$$\sigma^2(\theta^*) \geq \left[EnE\left(\frac{\partial \log f(x, \theta)}{\partial \theta}\right)^2\right]^{-1}. \tag{4.6}$$

The equality sign in (4.6) will hold if θ^* may be written as $Z'(\theta)Y_n + Z''(\theta)$, where Z' and Z'' are functions of θ. However, θ^* itself should not be a function of θ if our argument is to remain valid. The subject is connected with the question of the existence of a sufficient estimate.

Let $f(x, \theta)$ be defined as follows:

$$f(x, \theta) = \theta^x(1 - \theta)^{1-x}, \qquad (x = 0 \text{ or } 1; 0 < \theta < 1).$$

Then

$$\frac{\partial \log f(x, \theta)}{\partial \theta} = \frac{x}{\theta} - \frac{(1 - x)}{(1 - \theta)}, \qquad E\left(\frac{\partial \log f}{\partial \theta}\right)^2 = \frac{1}{\theta(1 - \theta)}.$$

Suppose θ^* is unbiased. Then $\sigma^2(\theta^*) \geq \theta(1 - \theta)(En)^{-1}$, a result first given by Girshick [3] under unspecified regularity conditions.

Let the functions $\varphi_1, \varphi_2, \cdots$ be such that $n(\omega)$ is a constant. We are then dealing with samples of fixed size. The result (4.5) is then given in [2], p. 480, under different conditions of regularity.

5. Regularity conditions for the case when θ has more than one component. We suppose that $\theta = (\theta_1, \cdots, \theta_l)$ and that simultaneous estimates

$\theta_1^*(x_1, \cdots, x_n), \cdots, \theta_l^*(x_1, \cdots, x_n)$ of the components of θ are under discussion. In the sequel we shall limit ourselves to the case when these estimates are all unbiased.

We postulate the following regularity conditions which are sufficient to validate section 6:

(5.1). *The covariance matrix of the estimates* $\theta_1^*, \cdots, \theta_l^*$ *is non-singular for all* θ *in* D *(this time* D *is an open interval of the l-dimensional parameter space).*

(5.2). *The conditions of section 3 are satisfied for each* θ_i *and* θ_i^* $(i = 1, \cdots, l)$.

6. The ellipsoid of concentration when θ has more than one component. Let

$$\theta = (\theta_1, \cdots, \theta_l).$$

We shall first describe briefly the result of Cramér [4] which refers to samples of fixed size $n > l$. Let $\theta_i^*(x_1, \cdots, x_n)$ be an unbiased estimate of θ_i, $(i = 1, \cdots, l)$. Let $\| \lambda_{ij} \|$ be the non-singular covariance matrix of the θ_i^*, and let $\| \lambda^{ij} \|$ be its inverse. The "ellipsoid of concentration" in the space of points $(k_1, \cdots, k_l)$ is defined as

$$\sum_{i,j=1}^{l} \lambda^{ij}(k_i - \theta_i)(k_j - \theta_j) = l + 2. \tag{6.1}$$

If a unit mass be distributed uniformly over this ellipsoid it will have the point $(\theta_1, \cdots, \theta_l)$ as its center of gravity and λ_{ij} as its product of inertia about the corresponding axes. Cramér proves that, subject to certain regularity conditions, there is a fixed ellipsoid

$$\sum_{i,j=1}^{l} \mu_{ij}(k_i - \theta_i)(k_j - \theta_j) = l + 2 \tag{6.2}$$

where

$$\mu_{ij} = nE\left(\frac{\partial \log f}{\partial \theta_i} \frac{\partial \log f}{\partial \theta_j}\right)$$

which is always contained entirely within the concentration ellipsoid of any set of unbiased estimates. The two ellipsoids coincide only under certain conditions, among which is that the θ_i^* be jointly sufficient estimates of the θ_i.

Let us now consider the sequential procedure of this paper and postulate the regularity conditions of section 5. Let

$$K = \| k_{ij} \|$$

be a matrix with real elements such that $| K | = 1$ and let

$$K^{-1} = \| k^{ij} \|$$

be its inverse. Let

$$\| \theta \| = \begin{Vmatrix} \theta_1 \\ \cdot \\ \cdot \\ \cdot \\ \theta_l \end{Vmatrix}, \qquad \| \theta^* \| = \begin{Vmatrix} \theta_1^* \\ \cdot \\ \cdot \\ \cdot \\ \theta_l^* \end{Vmatrix}, \qquad \| \psi \| = \begin{Vmatrix} \psi_1 \\ \cdot \\ \cdot \\ \cdot \\ \psi_l \end{Vmatrix}$$

be column matrices. Suppose

(6.3) $$\| \psi \| = K \| \theta \|.$$

Then

(6.4) $$\| \theta \| = K^{-1} \| \psi \|.$$

Define

$$\| \psi^* \| = \begin{Vmatrix} \psi_1^* \\ \cdot \\ \cdot \\ \cdot \\ \psi_l^* \end{Vmatrix} = K \| \theta^* \|.$$

From section 4 we have

(6.5) $$EnE \left(\frac{\partial \log f(x, \theta)}{\partial \psi_1} \right)^2 \geq [\sigma^2(\psi_1^*)]^{-1}$$

where the differentiation by which $\frac{\partial \log f}{\partial \psi_1}$ is obtained is performed with $\psi_2, \cdots, \psi_l$ held constant. Consider the last $(l - 1)$ rows of K as fixed and $(k_{11}, k_{12}, \cdots, k_{1l})$ as free to vary subject only to the restriction that $| K | = 1$. The left member of (6.5) is then a fixed quantity, while the right member is a function of the first row of K. The inequality (6.5) must remain valid for all admissible $(k_{11}, \cdots, k_{1l})$. Hence (6.5) will remain valid if the right member of (6.5) is replaced by its maximum with respect to $(k_{11}, \cdots, k_{1l})$. We shall obtain this maximum and find that (6.5) then implies a result about the minimal ellipsoid of concentration.

The problem is therefore to minimize $\sigma^2(\psi_1^*)$. Now·

(6.6) $$\sigma^2(\psi_1^*) = \sum_{i,j} \lambda_{ij} k_{1i} k_{1j}.$$

The family of ellipsoids in the space of $(k_{11}, \cdots, k_{1l})$

(6.7) $$\sum_{i,j} \lambda_{ij} k_{1i} k_{1j} = c,$$

where c is a running parameter, has all centers located at the origin. Let

$$(k_{11}^0, \cdots, k_{1l}^0)$$

be the sought-for maximizing values of $(k_{11}, \cdots, k_{1l})$. From the definitions of K and K^{-1} we have

(6.8) $$\sum_i k^{i1} k_{1i} = 1$$

where $(k^{11}, k^{21}, \cdots, k^{l1})$ are constants. It follows that the minimum value c_0 of $\sigma^2(\psi_1^*)$ is such that the ellipsoid

(6.9) $$\sum_{i,j} \lambda_{ij} k_{1i} k_{1j} = c_0$$

is tangent to the hyperplane (6.8) at the point $(k_{11}^0, \cdots, k_{1l}^0)$. Now the tangent plane to (6.9) at this point is given by

(6.10) $$\sum_{i,j} \lambda_{ij} k_{1i}^0 k_{1j} = c_0.$$

From (6.8) and (6.10) we obtain

$$c_0 k^{j1} = \sum_i k^0_{1i}\lambda_{ij}, \qquad (j = 1, \cdots, l). \tag{6.11}$$

Hence

$$c_0 \sum_i \lambda^{ij} k^{i1} = k^0_{1j}, \qquad (j = 1, \cdots, l) \tag{6.12}$$

from which

$$c_0 \sum_{i,j} \lambda^{ij} k^{i1} k^{j1} = 1. \tag{6.13}$$

We have

$$\begin{aligned} \frac{\partial \log f}{\partial \psi_1} &= \sum_i k^{i1} \frac{\partial \log f}{\partial \theta_i} \\ \left(\frac{\partial \log f}{\partial \psi_1}\right)^2 &= \sum_{i,j} k^{i1} k^{j1} \frac{\partial \log f}{\partial \theta_i} \frac{\partial \log f}{\partial \theta_j}. \end{aligned} \tag{6.14}$$

From (6.5), (6.13), (6.14), and the definition of c_0 we conclude that

$$\sum_{i,j} \mu'_{ij} k^{i1} k^{j1} \geq \sum_{i,j} \lambda^{ij} k^{i1} k^{j1} \tag{6.15}$$

where

$$\mu'_{ij} = EnE\left(\frac{\partial \log f}{\partial \theta_i} \frac{\partial \log f}{\partial \theta_j}\right). \tag{6.16}$$

We may restate (6.15) as follows: The concentration ellipsoid

$$\sum_{i,j} \lambda^{ij}(k_i - \theta_i)(k_j - \theta_j) = l + 2 \tag{6.17}$$

of the unbiased estimates $\theta_1^*, \cdots, \theta_l^*$ always contains within itself the ellipsoid

$$\sum_{i,j} \mu'_{ij}(k_i - \theta_i)(k_j - \theta_j) = l + 2 \tag{6.18}$$

where the μ'_{ij} are defined by (6.16).

The question of the coincidence of the two ellipsoids is connected with the question of the existence of sufficient estimates. It may be difficult to state any general results about the concentration ellipsoid of biased estimates without postulating some relationships among the biases and/or their derivatives.

7. On Wald's equation and related results in sequential analysis. In section 4 we referred to a proof by Blackwell [1] of an equation due to Wald [5] which is fundamental in the Wald theory of sequential tests of statistical hypotheses. Here we shall give a perhaps simpler proof of this equation, and then prove several new and related results of general interest for sequential analysis.

The results of Theorems 7.2 and 7.3 below can be obtained by differentiation of Wald's fundamental identity of sequential analysis ([6], [7]). However, the conditions under which we obtain these results are less stringent than any so far found sufficient to establish the identity and the validity of differentiating it. Theorem 7.4 and its corollaries refer to sequential processes where the chance variables may have different distributions or even be dependent. In the future we hope to return to the question of finding all central moments of Z_n, the problem of generalizing the fundamental identity, and related questions.

For Theorems 7.1, 7.2, and 7.3 we shall assume a chance variable X whose cumulative distribution function $F(x)$ is subject only to whatever restrictions may be explicitly imposed on it in each theorem. We assume the existence of a general sequential process such as is described above, which is subject only to such restrictions as may be explicitly formulated in each theorem. The sequential process of course defines the chance variable n. Let $x_1, x_2, \cdots$ be successive independent observations on X. We define $Z_n = \sum_{i=1}^{n} x_i$. If $E(X)$ and $\sigma^2(X)$ exist we shall denote them by w and σ^2, respectively.

THEOREM 7.1 (Wald [5], Blackwell [1]). *Suppose w and En exist. Then*

$$E(Z_n - nw) = 0. \tag{7.1}$$

The following theorem, which is a sort of partial converse of Theorem 7.1, is proved concomitantly with Theorem 7.1:

THEOREM 7.1.1. *If EZ_n exists, and if either $P\{X > 0\} = 0$ or $P\{X < 0\} = 0$, then w and En both exist, and*

$$EZ_n = wEn.$$

Actually the same proof suffices for a somewhat stronger form of Theorem 7.1.1:

THEOREM 7.1.2. *If EZ_n exists, and if*

$$E(X_i \mid n = j) \geq 0 \qquad (\text{or } \leq 0)$$

for all positive integral j such that $P\{n = j\} \neq 0$, and all $i \leq j$, then w and En both exist, and

$$EZ_n = wEn .$$

THEOREM 7.2. *If $E\left(\sum_{i=1}^{n} | x_i - w |\right)^2$ exists, then σ^2 and En both exist, and*

$$E(Z_n - nw)^2 = \sigma^2 En . \tag{7.2}$$

We have

$$\begin{aligned} E(Z_n - nw) = E\left(\sum_{i=1}^{n} (x_i - w)\right) &= \sum_{j=1}^{\infty} \int_{R_j} \left(\sum_{i=1}^{j} (x_i - w)\right) \prod_{i=1}^{j} dF(x_i) \\ &= \sum_{j=1}^{\infty} \sum_{i=j}^{\infty} \int_{R_i} (x_j - w) \prod_{m=1}^{m=i} dF(x_m). \end{aligned} \tag{7.3}$$

Also

$$\sum_{i=j}^{\infty} \int_{R_i} (x_j - w) \prod_{m=1}^{m=i} dF(x_m) = P\{n \geq j\} E(x_j - w) = 0. \tag{7.4}$$

Hence

$$\sum_{j=1}^{\infty} \sum_{i=j}^{\infty} \int_{R_i} (x_j - w) \prod_{m=1}^{m=i} dF(x_m) = 0. \tag{7.5}$$

From this (7.1) follows.

Suppose now that the conditions of Theorem 7.2 are fulfilled. We have

$$\begin{aligned} E(Z_n - nw)^2 &= \sum_{j=1}^{\infty} \int_{R_j} \left(\sum_{i=1}^{j} (x_i - w) \right)^2 \prod_{m=1}^{m=j} dF(x_m) \\ &= \sum_{j=1}^{\infty} \sum_{i=j}^{\infty} \int_{R_i} (x_j - w)^2 \prod_{m=1}^{m=i} dF(x_m) \\ &\quad + 2 \sum_{j=2}^{\infty} \sum_{s=1}^{j-1} \sum_{i=j}^{\infty} \int_{R_i} (x_s - w)(x_j - w) \prod_{m=1}^{m=i} dF(x_m). \end{aligned} \tag{7.6}$$

Let $s < j$ be any two positive integers. Then

$$\sum_{i=j}^{\infty} \int_{R_i} (x_s - w)(x_j - w) \prod_{m=1}^{m=i} dF(x_m) = 0. \tag{7.7}$$

Hence

$$\sum_{j=2}^{\infty} \sum_{s=1}^{j-1} \sum_{i=j}^{\infty} \int_{R_i} (x_s - w)(x_j - w) \prod_{m=1}^{m=i} dF(x_m) = 0. \tag{7.8}$$

In a similar manner we obtain

$$\sum_{i=j}^{\infty} \int_{R_i} (x_j - w)^2 \prod_{m=1}^{m=i} dF(x_m) = \sigma^2 P\{n \geq j\}. \tag{7.9}$$

From (7.6), (7.8), and (7.9) it therefore follows that

$$E(Z_n - nw)^2 = \sigma^2 \sum_{j=1}^{\infty} P\{n \geq j\} = \sigma^2 \sum_{j=1}^{\infty} jP\{n = j\} = \sigma^2 En \tag{7.10}$$

which is the desired result.

It remains to prove the validity of rearranging the series in (7.3) and (7.6). First, we have

$$\sum_{i=j}^{\infty} \int_{R_i} | x_j - w | \prod_{m=1}^{m=i} dF(x_m) = P\{n \geq j\} E | X - w |. \tag{7.11}$$

Hence it follows that

$$(7.12)\qquad \begin{aligned}\sum_{j=1}^{\infty}\sum_{i=j}^{\infty}\int_{R_i} | x_j - w | \prod_{m=1}^{m=i} dF(x_m) &= \sum_{j=1}^{\infty} P\{ n \geq j\} E | X - w | \\ &= E | X - w | \sum_{j=1}^{\infty} jP\{n = j\} = E | X - w | En.\end{aligned}$$

This justifies the rearrangement of terms in the series in (7.3). Second, the series (7.6) is dominated by the series

$$(7.13)\qquad \begin{aligned}&\sum_{j=1}^{\infty}\sum_{i=j}^{\infty}\int_{R_i} (x_j - w)^2 \prod_{m=1}^{m=i} dF(x_m) \\ &\quad + 2\sum_{j=2}^{\infty}\sum_{s=1}^{j-1}\sum_{i=j}^{\infty}\int_{R_i} | x_s - w |\cdot| x_j - w | \prod_{m=1}^{m=i} dF(x_m)\end{aligned}$$

all of whose terms are positive. The series (7.13) converges because

$$(7.14)\qquad E\left(\sum_{i=1}^{n} | x_i - w |\right)^2 < +\infty.$$

Hence the rearrangement of the series (7.6) is valid.

In the sequel we require certain sets $R'_j(j = 1, 2, \cdots)$ which we shall define now. Let R^*_{ij}, $i \leq j$, be the totality of all points $(x_1, \cdots, x_j)$ such that

$$(7.15)\qquad (x_1, \cdots, x_i) \in R_i.$$

Let R^j be the j-dimensional Euclidean space. Then

$$(7.16)\qquad R'_j = R^j - \sum_{i=1}^{j} R^*_{ij}.$$

We shall now prove:

THEOREM 7.3. *Suppose that* $E\left[\sum_{i=1}^{n} | x_i - w |\right]^3$ *and* $En\left[\sum_{i=1}^{n} | x_i - w |\right]$ *exist.*[2] *Then*

$$(7.17)\qquad E(Z_n - nw)^3 = w_3 En + 3\sigma^2 En(Z_n - nw)$$

where

$$w_3 = E(X - w)^3$$

exists.

---

[2] The author has succeeded in proving that the existence of $E\left[\sum_{i=1}^{n} | x_i - w |\right]^3$ implies the existence of $E\left[n\sum_{i=1}^{n} | x_i - w |\right]$ The proof will be published subsequently in connection with other results.

PROOF: We have

$$
\begin{aligned}
E(Z_n - nw)^3 &= \sum_{j=1}^{\infty} \int_{R_j} \left[\sum_{i=1}^{j} (x_i - w) \right]^3 \prod_{m=1}^{j} dF(x_m) \\
&= \sum_{j=1}^{\infty} \int_{R_j} \sum_{i=1}^{j} (x_i - w)^3 \prod_{m=1}^{j} dF(x_m) \\
&\quad + 3 \sum_{j=2}^{\infty} \int_{R_j} \sum_{i=2}^{j} \sum_{s=1}^{i-1} (x_s - w)(x_i - w)^2 \prod_{m=1}^{j} dF(x_m) \\
&\quad + 3 \sum_{j=2}^{\infty} \int_{R_j} \sum_{i=2}^{j} \sum_{s=1}^{i-1} (x_s - w)^2 (x_i - w) \prod_{m=1}^{j} dF(x_m) \\
&\quad + 6 \sum_{j=3}^{\infty} \int_{R_j} \sum_{i=3}^{j} \sum_{s=2}^{i-1} \sum_{t=1}^{s-1} (x_t - w)(x_s - w)(x_i - w) \prod_{m=1}^{j} dF(x_m).
\end{aligned} \tag{7.18}
$$

Considering the first term in the right member of (7.18), it follows that

$$
\begin{aligned}
&\sum_{j=1}^{\infty} \int_{R_j} \left[\sum_{i=1}^{j} (x_i - w)^3 \right] \prod_{m=1}^{j} dF(x_m) \\
&= \sum_{i=1}^{\infty} \sum_{j=i}^{\infty} \int_{R_j} (x_i - w)^3 \prod_{m=1}^{j} dF(x_m) \\
&= \sum_{i=1}^{\infty} w_3 P\{n \geq i\} \\
&= \sum_{i=1}^{\infty} i w_3 P\{n = i\} = w_3 En.
\end{aligned} \tag{7.19}
$$

All the rearrangements of terms in the operations involved in the proof of Theorem 7.3 are legitimate because the various series are absolutely convergent.

As for the second term in the right member of (7.18), we have

$$
\begin{aligned}
&\sum_{j=2}^{\infty} \int_{R_j} \sum_{i=2}^{j} \sum_{s=1}^{i-1} (x_s - w)(x_i - w)^2 \prod_{m=1}^{j} dF(x_m) \\
&= \sum_{s=1}^{\infty} \sum_{i=s+1}^{\infty} \sum_{j=i}^{\infty} \int_{R_j} (x_s - w)(x_i - w)^2 \prod_{m=1}^{j} dF(x_m) \\
&= \sigma^2 \sum_{s=1}^{\infty} \sum_{i=s+1}^{\infty} \int_{R'_{i-1}} (x_s - w) \prod_{m=1}^{i-1} dF(x_m) \\
&= \sigma^2 \sum_{s=1}^{\infty} \sum_{i=s}^{\infty} \int_{R'_i} (x_s - w) \prod_{m=1}^{i} dF(x_m).
\end{aligned} \tag{7.20}
$$

We now operate on $En(Z_n - nw)$, and obtain

$$
\begin{aligned}
En(Z_n - nw) &= \sum_{j=1}^{\infty} \int_{R_j} j \sum_{i=1}^{j} (x_i - w) \prod_{m=1}^{j} dF(x_m) \\
&= \sum_{j=1}^{\infty} \sum_{i=j}^{\infty} \int_{R_i} i(x_j - w) \prod_{m=1}^{i} dF(x_m).
\end{aligned} \tag{7.21}
$$

We observe that

$$\sum_{i=j}^{\infty}\int_{R_i} i(x_j - w)\prod_{m=1}^{i} dF(x_m)$$

$$= j\sum_{i=j}^{\infty}\int_{R_i}(x_j - w)\prod_{m=1}^{i} dF(x_m) + \sum_{s=j+1}^{\infty}\sum_{i=s}^{\infty}\int_{R_i}(x_j - w)\prod_{m=1}^{i} dF(x_m). \tag{7.22}$$

To evaluate the left member of (7.22), we proceed as follows: It is easy to see that

$$\sum_{i=j}^{\infty}\int_{R_i}(x_j - w)\prod_{=1}^{i} dF(x_m) = 0. \tag{7.23}$$

Moreover, when $s > j$,

$$\sum_{i=s}^{\infty}\int_{R_i}(x_j - w)\prod_{m=1}^{i} dF(x_m) = \int_{R'_{s-1}}(x_j - w)\prod_{m=1}^{s-1} dF(x_m). \tag{7.24}$$

Hence

$$\sum_{i=j}^{\infty}\int_{R_i} i\,(x_j - w)\prod_{m=1}^{i} dF(x_m) = \sum_{s=j}^{\infty}\int_{R'_s}(x_j - w)\prod_{m=1}^{s} dF(x_m). \tag{7.25}$$

Therefore

$$En(Z_n - nw) = \sum_{j=1}^{\infty}\sum_{s=j}^{\infty}\int_{R'_s}(x_j - w)\prod_{m=1}^{s} dF(x_m). \tag{7.26}$$

It remains now to consider the third term of the right member of (7.18). We have

$$\sum_{j=2}^{\infty}\int_{R_j}\sum_{i=2}^{j}\sum_{s=1}^{i-1}(x_s - w)^2(x_i - w)\prod_{m=1}^{j} dF(x_m)$$

$$= \sum_{s=1}^{\infty}\sum_{i=s+1}^{\infty}\sum_{j=i}^{\infty}\int_{R_j}(x_s - w)^2(x_i - w)\prod_{m=1}^{j} dF(x_m). \tag{7.27}$$

Now, suppose that in the expression

$$V_{sij} = \int_{R_j}(x_s - w)^2(x_i - w)\prod_{m=1}^{j} dF(x_m) \tag{7.28}$$

where $j \geq i > s$, we integrate with respect to all x_m for which $m \geq i$. Then it is not difficult to see that

$$\sum_{j=i}^{\infty} V_{sij} = 0 \tag{7.29}$$

for all s and i such that $1 \leq s < i$. Hence from (7.27)

$$\sum_{j=2}^{\infty}\int_{R_j}\sum_{i=2}^{j}\sum_{s=1}^{i-1}(x_s - w)^2(x_i - w)\prod_{m=1}^{j} dF(x_m) = 0. \tag{7.30}$$

In a similar way it is shown that the fourth term of the right member of (7.18) is zero.

The desired result (7.17) is a direct consequence of (7.18), (7.19), (7.20), (7.26), and (7.30).

Consider now an infinite sequence of chance variables $x_1, x_2, \cdots$, which need not have the same distribution and which may be dependent (in which case they must satisfy the obvious consistency relationships). We take successive observations on these chance variables and define a sequential process as above, which is subject only to such restrictions as we shall explicitly state. Let Z_n maintain its previous definition.

THEOREM 7.4. *Suppose that*

$$\nu_i = E(X_i \mid n \geq i) \tag{7.31}$$

exists for all positive integral i for which $P\{n \geq i\} \neq 0$. In those cases write

$$\nu_i' = E(|X_i - \nu_i| \mid n \geq i). \tag{7.32}$$

Suppose also that the series

$$\sum_{i=1}^{\infty} (\nu_1' + \cdots + \nu_i')P\{n = i\} \tag{7.33}$$

converges. Then

$$E\left[Z_n - \sum_{i=1}^{n} \nu_i\right] = 0. \tag{7.34}$$

It is regrettable but unavoidable that the mean values ν_i and ν_i' entering into (7.33) and (7.34) be conditional. The fundamental reason is that the sequential process may drastically modify the distribution of dependent chance variables, so that their distribution for our purposes can only be considered in conjunction with the sequential process itself. Consider the following example:

$$P\{X_1 = -1\} = \tfrac{1}{2}, \qquad P\{X_1 = 1\} = \tfrac{1}{2}$$

$$P\{X_2 = -2 \mid X_1 = -1\} = \tfrac{1}{2}$$

$$P\{X_2 = -1 \mid X_1 = -1\} = \tfrac{1}{2}$$

$$P\{X_2 = 1 \mid X_1 = 1\} = \tfrac{1}{2}$$

$$P\{X_2 = 2 \mid X_1 = 1\} = \tfrac{1}{2}.$$

We have $E(X_2) = 0$. Suppose we define the following sequential process: If $X_1 = -1, n = 1$, and if $X_1 = 1, n = 2$. It is then clear that for our purposes X_2 can take no negative values and the fact that $E(X_2) = 0$ is of no use to us.

If, however, the chance variables $X_1, X_2, \cdots$ are independent, this difficulty disappears, and we have the following.

COROLLARY 1 TO THEOREM 7.4. *If the chance variables $X_1, X_2, \cdots$ are independent, we have Theorem 7.4 with $\nu_i = E(X_i)$, and $\nu_i' = E \mid X_i - \nu_i \mid$.*

If further all the X_i have the same distribution, we see that Theorem 7.1 is a special case of Theorem 7.4, since the convergence of the series (7.33) is then a consequence of the existence of w and En. From this argument we see, however, that it is not necessary that all the X_i have the same distribution, and we may write the following generalization of Theorem 7.1:

COROLLARY 2 TO THEOREM 7.4. *Let the X_i be independent with, in general, different distributions. Suppose, however, that all ν_i are equal, and all ν_i' are equal, except perhaps for those i such that $P\{n \geq i\} = 0$. Suppose further that En exists. Then* (7.1) *holds.*

Among possible fields of application of Theorem 7.4 are sequential tests of composite statistical hypotheses, and the random walk of a particle governed by probability distributions which are functions of time and the position of the particle. The extension of this theorem to vector chance variables is straightforward. The extension to higher moments may present difficulties. We hope to return to some of these questions in the future.

PROOF OF THEOREM 7.4. This is very elementary. We have

$$E\Big(Z_n - \sum_{i=1}^{n} \nu_i\Big) = \sum_{j=1}^{\infty} \int_{R_j} \Big[\sum_{i=1}^{j} (x_i - \nu_i)\Big] dF(x_1, \cdots, x_j)$$

$$(7.35) \qquad = \sum_{j=1}^{\infty} \sum_{i=j}^{\infty} \int_{R_i} (x_j - \nu_j)\, dF(x_1, \cdots, x_i).$$

$$= \sum_{j=1}^{\infty} P\{n \geq j\} E(X_j - \nu_j \mid n \geq j) = 0.$$

The rearrangement of the series is valid because

$$(7.36) \qquad \begin{aligned} \sum_{j=1}^{\infty} \sum_{i=j}^{\infty} \int_{R_i} \mid x_j - \nu_j \mid dF(x_1, \cdots, x_i) &= \sum_{j=1}^{\infty} \nu_j' P\{n \geq j\} \\ &= \sum_{j=1}^{\infty} (\nu_1' + \cdots + \nu_j') P\{n = j\} \end{aligned}$$

which converges by (7.33).

REFERENCES

[1] DAVID BLACKWELL, "On an equation of Wald," *Annals of Math. Stat.*, Vol. 17 (1946), pp. 84–87.

[2] H. CRAMÉR, *Mathematical Methods of Statistics*, Princeton Univ. Press, 1946.

[3] M. A. GIRSHICK, FREDERICK MOSTELLER, AND L. J. SAVAGE, "Unbiased estimates for certain binomial sampling problems," *Annals of Math. Stat.*, Vol. 17 (1946), pp. 13–23.

[4] H. Cramér, "A contribution to the theory of statistical estimation," *Skandinavisk Aktuarietidskrift*, Vol. 29 (1946), pp. 85–94.

[5] A. Wald, "Sequential tests of statistical hypotheses," *Annals of Math. Stat.*, Vol. 16 (1945), pp. 117–186.

[6] A. Wald, "On cumulative sums of random variables," *Annals of Math. Stat.*, Vol. 15 (1944), pp. 283–296.

[7] A. Wald, "Differentiation under the expectation sign of the fundamental identity in sequential analysis," *Annals of Math. Stat.*, Vol. 17 (1946).

[8] C. R. Rao, "Information and the accuracy attainable in the estimation of statistical parameters," *Bull. Calcutta Math. Soc.*, Vol. 37, No. 3 (Sept., 1945), pp. 81-91.

[9] A. C. Aitken and H. Silverstone "On the estimation of statistical parameters," *Proc. Roy. Soc. Edinburgh*, Vol. 61 (1941), pp. 56–62.

Made in United States of America

Reprinted from THE ANNALS OF MATHEMATICAL STATISTICS
Vol. XIX, No. 3, September, 1948

OPTIMUM CHARACTER OF THE SEQUENTIAL PROBABILITY RATIO TEST

A. WALD AND J. WOLFOWITZ

Columbia University

1. Summary. Let S_0 be any sequential probability ratio test for deciding between two simple alternatives H_0 and H_1, and S_1 another test for the same purpose. We define $(i, j = 0, 1)$:

$\alpha_i(S_j)$ = probability, under S_j, of rejecting H_i when it is true;

$E_i^j(n)$ = expected number of observations to reach a decision under test S_j when the hypothesis H_i is true. (It is assumed that $E_i^1(n)$ exists.)

In this paper it is proved that, if

$$\alpha_i(S_1) \leq \alpha_i(S_0) \qquad (i = 0,1),$$

it follows that

$$E_i^0(n) \leq E_i^1(n) \qquad (i = 0, 1).$$

This means that of all tests with the same power the sequential probability ratio test requires on the average fewest observations. This result had been conjectured earlier ([1], [2]).

2. Introduction. Let $p_i(x)$, $i = 0, 1$, denote two different probability density functions or (discrete) probability functions. (Throughout this paper the index i will always take the values 0, 1). Let X be a chance variable whose distribution can only be either $p_0(x)$ or $p_1(x)$, but is otherwise unknown. It is required to decide between the hypotheses H_0, H_1, where H_i states that $p_i(x)$ is the distribution of X, on the basis of n independent observations $x_1, \cdots, x_n$ on X, where n is a chance variable defined (finite) on almost every infinite sequence

$$\omega = x_1, x_2, \cdots$$

i.e., n is finite with probability one according to both $p_0(x)$ and $p_1(x)$. The definition of $n(\omega)$ together with the rule for deciding on H_0 or H_1 constitute a sequential test.

A sequential probability ratio test is defined with the aid of two positive numbers, $A^* > 1$, $B^* < 1$, as follows: Write for brevity

$$p_{ij} = \prod_{k=1}^{j} p_i(x_k).$$

Then $n = j$ if

$$\frac{p_{1j}}{p_{0j}} \geq A^* \qquad \text{or} \qquad \leq B^*$$

and

$$B^* < \frac{p_{1k}}{p_{0k}} < A^*, \qquad k < j.$$

If

$$\frac{p_{1n}}{p_{0n}} \geq A^*, \qquad \text{the hypothesis } H_1 \text{ is accepted,}$$

if

$$\frac{p_{1n}}{p_{0n}} \leq B^* \text{ the hypothesis } H_0 \text{ is accepted.}$$

In this paper we limit consideration to sequential tests for which $E_i(n)$ exists, where $E_i(n)$ is the expected value of n when H_i is true (i.e., when $p_i(x)$ is the distribution of X). It has been proved in [3] that all sequential probability ratio tests belong to this class. The purpose of the paper is to prove the result stated in the first section. Throughout the proof we shall find it convenient to assume that there is an a priori probability g_i that H_i is true ($g_0 + g_1 = 1$; we shall write $g = (g_0, g_1)$). We are aware of the fact that many statisticians believe that in most problems of practical importance either no a priori probability distribution exists, or that even where it exists the statistical decision must be made in ignorance of it; in fact we share this view. Our introduction of the a priori probability distribution is a purely technical device for achieving the proof which has no bearing on statistical methodology, and the reader will verify that this is so. We shall always assume below that $g_0 \neq 0, 1$.

Let W_0, W_1, c be given positive numbers. We define

$$R = g_0(W_0\alpha_0 + cE_0(n)) + g_1(W_1\alpha_1 + cE_1(n)),$$

and call R the average risk associated with a test S and a given g (obviously R is a function of both). We shall say that H_i is accepted when the decision is made that $p_i(x)$ is the distribution of X. We shall say that H_0 is rejected when H_1 is accepted, and vice versa. The reader may find it helpful to regard W_i as a weight which measures the loss caused by rejecting H_i when it is true, c as the cost of a single observation, and R as the average loss associated with a given g and a test S. For mathematical purposes these are simply quantities which we manipulate in the course of the proof.

3. Role of the probability ratio. Let g, $W = (W_0, W_1)$, and c be fixed. Let S be a given sequential test, with $R(S)$ the associated risk and $n(\omega, S)$ the associated "sample size" function. Let $\psi(x_1, \cdots, x_n)$ be the "decision" function; this is a function which takes only the values 0 and 1, and such that, when $x_1, \cdots, x_n$ is the sample point, the hypothesis with index $\psi(x_1, \cdots, x_n)$ is *rejected*. Define the following decision function $\varphi(x_1, \cdots, x_n)$: $\varphi = 0$ when

$$\lambda = \frac{W_1 g_1 p_{1n}}{W_0 g_0 p_{0n}}$$

is greater than 1, and $\varphi = 1$ when $\lambda < 1$. When $\lambda = 1$, φ may be 0 or 1 at pleasure.

It must be remembered that an actual decision function is a single-valued function of $(x_1, \cdots, x_n)$. We note, however, that

a) the relevant properties of a test are not affected by changing the test on a set T of points ω whose probability is zero according to both H_0 and H_1, i.e., changing the definition on T of n and/or of the decision function, leaves α_0, α_1, $E_0(n)$ and $E_1(n)$ unaltered. In particular, the average risk R remains unchanged.

b) the set of points for which $p_{0n} = p_{1n} = 0$ and λ is indeterminate, has probability zero according to both H_0 and H_1.

In view of the above we decide arbitrarily, in all sequential tests which we shall henceforth consider, to define $n = j$, and $\psi = 0$, whenever $p_{0j} = p_{1j} = 0$, and $n \neq 1, \cdots, (j - 1)$. By this arbitrary action $R(S)$ will not be changed.

Let now

$$L_{in} = \frac{W_i g_i p_{in}}{g_0 p_{0n} + g_1 p_{1n}};$$

$$L_n = cn + \min (L_{0n}, L_{1n}).$$

We have

$$EL_{\psi n} = \Sigma g_i W_i \alpha_i$$

where the operator E denotes the expected value with respect to the joint distribution of H_i and $(x_1, \cdots, x_n)$, i.e., E is the operator $g_0 E_0 + g_1 E_1$. If now the event $\{\psi(S) \neq \varphi \text{ and } \lambda \neq 1\}$ has positive probability according to either H_0 or H_1, we would have, for $n = n(\omega, S)$,

$$EL_{\varphi n} < EL_{\psi n}.$$

Hence, if the decision function ψ connected with the test S were replaced by the decision function φ, R would be decreased. Since our object throughout this proof will be to make R as small as possible, we shall confine ourselves henceforth, except when the contrary is explicitly stated, to tests for which φ is the decision function. This will be assumed even if not explicitly stated.

The function φ has not yet been uniquely defined when $\lambda = 1$. A definition convenient for later purposes will be given in the next section. R is the same for all definitions.

We thus have that φ is a function only of λ, or, what comes to the same thing when W is fixed, of $r_n = \dfrac{p_{1n}}{p_{0n}}$. Define

$$r_j = \frac{p_{1j}}{p_{0j}}, \qquad j = 1, 2, \cdots.$$

We shall now prove

LEMMA 1. *Let g, W, and c be fixed. There exists a sequential test S^* for which the average risk is a minimum. Its sample size function $n(\omega, S^*)$ can be defined by means of a properly chosen subset K of the non-negative half-line as follows: For any ω consider the associated sequence*

$$r_1, r_2, \cdots$$

and let j be the smallest integer for which $r_j \in K$. Then $n = j$. The function n may be undefined on a set of points ω whose probability according to H_0 and H_1 is zero.

Let $a = (a_1, \cdots, a_d)$ be any point in some finite d-dimensional Euclidean space, provided only that $p_{0d}(a)$ and $p_{1d}(a)$ are not both zero. Let $b = \frac{p_{1d}(a)}{p_{0d}(a)}$ and let $l(a) = cd + \min(L_{0d}, L_{1d})$. Let D be any sequential test whatever for which $n(\omega, D) > d$ for any ω whose first d coördinates are the same as those of a, and for which $E(n \mid a, D) < \infty$, where $E(n \mid a, D)$ is the conditional expected value of n according to the test D under the condition that the first d coördinates of ω are the same as those of a. For brevity let G represent the set of points ω which fulfill this last condition, i.e., that the first d coördinates of ω are the same as those of a. Finally, let $E(L_n \mid a, D)$ be the conditional expected value of L_n according to D under the condition that ω is in the set G. We know that $\min(L_{0d}, L_{1d})$ depends only on $r_d(a) = b$.

Write

$$\nu(a) = \sup_D [l(a) - E(L_n \mid a, D)].$$

Let $a_0 = (a_{01}, \cdots, a_{0k})$ be any point such that

$$\frac{p_{1d}(a)}{p_{0d}(a)} = \frac{p_{1k}(a_0)}{p_{0k}(a_0)}.$$

Let D_0 be any sequential test whatever for which $n(\omega, D_0) > k$ for any ω whose first k coordinates are the same as those of a_0, and for which $E(n \mid a_0, D_0) < \infty$ Let

$$\nu(a_0) = \sup_{D_0} [l(a_0) - E(L_n \mid a_0, D_0)].$$

We shall prove that $\nu(a) = \nu(a_0)$. Thus we shall be justified in writing

$$\gamma(b) = \nu(a) = \nu(a_0).$$

Suppose, therefore that $\nu(a) > \nu(a_0)$. Let D_1 be a test of the type D such that

$$l(a) - E(L_n \mid a, D_1) > \frac{\nu(a) + \nu(a_0)}{2}.$$

We now partially define another sequential test D_{10} of the type D_0 as follows: Let

$$\bar{a} = a_1, \cdots, a_d, y_1, \cdots, y_t,$$

be any sequence such that $n(\bar{a}, D_1) = d + t$. Then for the sequence

$$\bar{a}_0 = a_{01}, \cdots, a_{0k}, y_1, \cdots, y_t,$$

let $n(\bar{a}_0, D_{10}) = k + t$. The decision function ψ_0 associated with D_{10} will be partially defined as follows:

$$\psi_0(\bar{a}_0) = \varphi(\bar{a}).$$

(The reader will observe that it may happen that $\psi_0(\bar{a}_0) \neq \varphi(\bar{a}_0)$). Since $r_d(a) = r_k(a_0)$ it follows that

$$l(a) - E(L_n \mid a, D_1) = l(a_0) - E(L_n \mid a_0, D_{10}) > \frac{\nu(a) + \nu(a_0)}{2} > \nu(a_0),$$

in violation of the definition of $\nu(a_0)$. A similar contradiction is obtained if $\nu(a) < \nu(a_0)$. Hence $\nu(a) = \nu(a_0)$ as was stated above.

We define K to consist of all numbers b which are such that there exist points a with $r_d(a) = b$, and for which $\gamma(b) \leq 0$. We shall now prove that the test S^* defined in the statement of the lemma is such that $R(S^*)$ is a minimum. Recall that the average risk is the expected value of L_n. Let S be any other test. Let $a^* = (a_1^*, \cdots, a_{d^*}^*)$ be any sequence such that either $n(a^*, S^*) = d^*$, or $n(a^*, S) = d^*$, but $n(a^*, S^*) \neq n(a^*, S)$. We exclude the trivial case that the probability of the occurrence of such a sequence, under both H_0 and H_1, is zero. Let $r_{d^*}(a^*) = b^*$. The sequence a^* may be one of three types:

1) $\gamma(b^*) < 0$. Hence $b^* \in K$, $n(a^*, S) > d^*$. It is more advantageous, from the point of view of diminishing the average risk, to terminate the sequential process at once, since $E(L_n \mid a^*, S) > l(a^*)$.

2) $\gamma(b^*) = 0$. Hence $b^* \in K$, $n(a^*, S) > d^*$. If $l(a^*) - E(L_n \mid a^*, S) = 0$, i.e., the supremum is actually attained by S, then, as far as the average risk is concerned, it makes no difference whether the sequential process is terminated with a^* or continued according to S. If, however, $l(a^*) - E(L_n \mid a^*, S) < 0$, it is clearly disadvantageous to proceed according to S. It is impossible that $l(a^*) - E(L_n \mid a^*, S) > 0$, since $\gamma(b^*) = 0$.

3) $\gamma(b^*) > 0$. Hence $b^* \notin K$, $n(a^*, S) = d^*$. Clearly it is more advantageous from the point of view of diminishing the average risk not to terminate the sequential process, but to continue with at least one more observation. After one more observation we are either in case 1 or 2, where it is advantageous to terminate the sequential process, or again in case 3, where it is advantageous to take yet another observation.

We conclude that $R(S^*)$ is a minimum, as was to be proved.

4. A fundamental lemma. Consider the complement of K with respect to the non-negative half-line, and from it delete all points b' for which there exists no point a in some d-dimensional Euclidean space such that $r_d(a) = b'$. The point 1 is never to be considered as of the type of b', i.e., 1 is never to be deleted. Designate the resulting set by $\bar{K}$.

Our proof of the theorem to which this paper is devoted hinges on the following lemma:

LEMMA 2. *Let W, g, c be fixed, and $\bar{K}$ be as defined above. There exist two positive numbers A and B, with $B \le \frac{W_0 g_0}{W_1 g_1} \le A$, such that*

a) if $b \in K$, then either $b \ge A$ or $b \le B$

b) if $b \in \bar{K}$, $B \le b \le A$.

Two remarks may be made before proceeding with the proof:

1) We may now complete the definition of φ for tests of the type of S^*. The reader will recall that φ was not uniquely defined when $\lambda = 1$, i.e., when $r_n = \frac{W_0 g_0}{W_1 g_1}$. Lemma 2 shows that it is necessary to define $\varphi(\lambda)$ only when $\lambda = \frac{W_0 g_0}{W_1 g_1} \in K$ and λ is therefore either A or B. We will define $\varphi\left(\frac{W_0 g_0}{W_1 g_1}\right)$ as 0 or 1, according as $\frac{W_0 g_0}{W_1 g_1}$ is A or B, and $A \ne B$. This is simply a convenient definition which will give uniqueness. When $A = B = \frac{W_0 g_0}{W_1 g_1} \in K$, the situation is completely trivial, and we may take $\varphi = 0$ arbitrarily.

2) If $1 \in K$ the above lemma shows that the average risk is minimized (for fixed W, g, c, of course) by taking no observations at all. We have $\varphi = 0$ or 1 according as $1 \ge A$ or $1 \le B$.

PROOF OF THE LEMMA: Let $h > \frac{W_0 g_0}{W_1 g_1}$ be a point in $\bar{K}$. We will prove that any point h' such that $\frac{W_0 g_0}{W_1 g_1} \le h' < h$, and such that there exists a point a' in some d'-dimensional Euclidean space for which $r_{d'}(a') = h'$, is also in $\bar{K}$. In a similar way it can be shown that, if $h_0 < \frac{W_0 g_0}{W_1 g_1}$ is any point in $\bar{K}$, any point h_0' such that $h_0 < h_0' \le \frac{W_0 g_0}{W_1 g_1}$, and such that there exists a point a_0' in some d''-dimensional Euclidean space for which $r_{d''}(a_0') = h_0'$, is also in $\bar{K}$. This will prove the lemma.

Let therefore h and h' be as above. Let S^* be the sequential test based on K, with the decision function φ. Let a be a point in d-space such that $r_d(a) = h$. Since $h \in \bar{K}$ we have $\gamma(h) > 0$.

We now wish to define partially another sequential test $\bar{S}$, with a decision function which may be different from φ, as follows: Let a' be defined as above. Write

$$a = (a_1, \cdots, a_d)$$

$$a' = (a_1', \cdots, a_{d'}').$$

Let

$$\bar{a} = a_1, \cdots, a_d, y_1, \cdots, y_t$$

be any sequence such that $n(\bar{a}, S^*) = d + t$. Then for the sequence

$$\bar{a}' = a_1', \cdots, a_{d'}', y_1, \cdots, y_t$$

let $n(\bar{a}', \bar{S}) = d' + t$. The decision function ψ associated with $\bar{S}$ will be partially defined as follows:

$$\psi(\bar{a}') = \varphi(\bar{a}).$$

Clearly

$$E_i(n \mid a, S^*) - d = E_i(n \mid a', \bar{S}) - d' \qquad (i = 0, 1) \tag{4.1}$$

and

$$E_i(\varphi \mid a, S^*) = E_i(\psi \mid a', \bar{S}) \qquad (i = 0, 1). \tag{4.2}$$

Furthermore, we have

$$\begin{aligned} l(a) &- E(L_n \mid a, S^*) \\ &= \frac{g_0}{g_0 + g_1 h} \{W_0 + cd - cE_0(n \mid a, S^*) - W_0[1 - E_0(\varphi \mid a, S^*)]\} \\ &\quad + \frac{g_1 h}{g_0 + g_1 h} \{cd - cE_1(n \mid a, S^*) - W_1 E_1(\varphi \mid a, S^*)\}. \end{aligned} \tag{4.3}$$

Since $\gamma(h) > 0$, and since

$$cd - cE_1(n \mid a, S^*) - W_1 E_1(\varphi \mid a, S^*) < 0, \tag{4.4}$$

we must have

$$W_0 + cd - cE_0(n \mid a, S^*) - W_0[1 - E_0(\varphi \mid a, S^*)] > 0. \tag{4.5}$$

From $h' < h$ it follows that

$$\frac{g_0}{g_0 + g_1 h'} > \frac{g_0}{g_0 + g_1 h}, \qquad \text{and} \qquad \frac{g_1 h'}{g_0 + g_1 h'} < \frac{g_1 h}{g_0 + g_1 h}. \tag{4.6}$$

Relations (4.1), (4.2), (4.4), (4.5) and (4.6) imply that the value of the right hand member of (4.3) is increased by replacing φ, h, a, S^* and d by ψ, h', a', $\bar{S}$, and d', respectively. This proves our lemma.

If there are values which r_j cannot assume the pair B, A might not be unique. For convenience we shall define A and B uniquely in the manner described below. We will always adhere to this definition thereafter.

We shall first define $\gamma(h)$ for all positive h in a manner consistent with the previous definition, which defined $\gamma(h)$ only for those values of h which could be assumed by r_j. Let h be any positive number and $D(h)$ be any sequential test with the following properties:

(4.7) there exists a set $Q(h)$ of positive numbers such that $n = j$ if and only if the j-th member of the sequence

$$hr_1, hr_2, hr_3, \cdots$$

is the first element of the sequence to be in $Q(h)$

$$E_i(n \mid D(h)) < \infty \qquad (i = 0, 1). \tag{4.8}$$

We define, for $h \geq \frac{W_0 g_0}{W_1 g_1}$,

$$\gamma(h \mid D(h)) = \frac{g_0}{g_0 + g_1 h} \{W_0 E_0(\varphi \mid D(h)) - cE_0(n \mid D(h))\} \tag{4.9}$$

$$+ \frac{g_1 h}{g_0 + g_1 h} \{-W_1 E_1(\varphi \mid D(h)) - cE_1(n \mid D(h))\},$$

$$\gamma(h) = \sup_{D(h)} \gamma(h \mid D(h)) \tag{4.10}$$

with a corresponding definition for $h \leq \frac{W_0 g_0}{W_1 g_1}$. Thus $\gamma(h)$ is defined for all positive h. This definition coincides with the previous definition whenever the latter is applicable. It is true that the supremum operation in (4.10) is limited to tests which depend only on the probability ratio, as (4.7) implies, but the argument of Lemma 1 shows that this limitation does not diminish the supremum. (It might appear that, for $h = \frac{W_0 g_0}{W_1 g_1}$, $\gamma(h)$ is not uniquely defined. We shall shortly see that this is not the case.)

The quantity $\gamma(h)$ depends, of course, on g_0 and g_1. To put this in evidence, we shall also write $\gamma(h, g_0, g_1)$. One can easily verify that

$$\gamma(h, g_0, g_1) = \gamma\left(1, \frac{g_0}{g_0 + g_1 h}, \frac{g_1 h}{g_0 + g_1 h}\right).$$

More generally, for any positive values h and h', we have $\gamma(h, g_0, g_1) = \gamma(h', \bar{g}_0, \bar{g}_1)$, where $\bar{g}_0$ and $\bar{g}_1$ are suitable functions of g_0, g_1, h, and h'. Thus, if h is not an admissible value of the probability ratio and h' is any admissible value, we can interpret the value of $\gamma(h, g_0, g_1)$ as the value of γ corresponding to h' and some properly chosen a priori probabilities $\bar{g}_0$ and $\bar{g}_1$.

We now define A as the greatest lower bound of all points $h \geq \frac{W_0 g_0}{W_1 g_1}$ for which $\gamma(h) \leq 0$. We define B as the least upper bound of all points $h \leq \frac{W_0 g_0}{W_1 g_1}$ for which $\gamma(h) \leq 0$. If $\gamma(h) \leq 0$ for all h the above definition implies $A = B = \frac{W_0 g_0}{W_1 g_1}$.

The argument of Lemma 2 shows that $\gamma(h)$ is monotonically increasing in the interval $\left(B, \frac{W_0 g_0}{W_1 g_1}\right)$, and that $\gamma(h)$ is monotonically decreasing in the interval $\left(\frac{W_0 g_0}{W_1 g_1}, A\right)$

We shall now define a sequential test $S^*(h)$ for every positive h. The decision function of $S^*(h)$ will be φ, and $n = j$ if and only if the j-th member of the sequence

$$\gamma(hr_1), \quad \gamma(hr_2), \quad \gamma(hr_3), \cdots$$

is the first element to be ≤ 0. We see that

$$\gamma(h) = \gamma(h \mid S^*(h)) \tag{4.11}$$

for *all* h. Incidentally, this proves that $\gamma(h)$ was uniquely defined at $h = \dfrac{W_0 g_0}{W_1 g_1}$.

We shall now prove

LEMMA 3. *The function* $\gamma(h)$ *has the following properties:*

a) *It is continuous for all* h.

b) $\gamma(A) = \gamma(B) = 0$

c) $\gamma(h) < 0$ *for* $h > A$ *or* $< B$.

Only a) and c) require proof, since b) is a trivial consequence of a) and the definition of A and B.

Let h be any point except $\dfrac{W_0 g_0}{W_1 g_1}$, and let z be any point in a neighborhood of h. Within a neighborhood of h both $E_0(n \mid S^*(z))$ and $E_1(n \mid S^*(z))$ are bounded. Let Δ be an arbitrarily given, positive number. Let h' and h'' be any two points in a sufficiently small neighborhood of h, to be described shortly. We proceed as in the argument of Lemma 2, with the present h' corresponding to h of Lemma 2, the present h'' corresponding to h' of Lemma 2, and with $S^*(h')$ corresponding to S^* of Lemma 2. Since $\dfrac{g_0}{g_0 + g_1 z}$ and $\dfrac{g_1 z}{g_0 + g_1 z}$ are continuous functions of z, and since $E_0(n \mid S^*(z))$ and $E_1(n \mid S^*(z))$ are bounded functions of z, we conclude that, when the neighborhood of h is sufficiently small,

$$\gamma(h'') \geq \gamma(h') - \Delta.$$

Reversing the roles of h' and h'' we obtain that in this neighborhood

$$\gamma(h') \geq \gamma(h'') - \Delta,$$

and conclude that

$$|\gamma(h') - \gamma(h'')| \leq \Delta.$$

Since Δ was arbitrary, this implies the continuity of $\gamma(h)$ everywhere, except perhaps at $h = \dfrac{W_0 g_0}{W_1 g_1}$.

To deal with the point $h = \dfrac{W_0 g_0}{W_1 g_1}$, proceed as follows: Using the above argument and the definition (4.9), (4.10), we prove that $\gamma(h)$ is continuous on the right at $h = \frac{W_0 g_0}{W_1 g_1}$. Using, at the point $h = \frac{W_0 g_0}{W_1 g_1}$, the definition of $\gamma(h \mid D(h))$ for $h \leq \frac{W_0 g_0}{W_1 g_1}$ i.e.,

$$\gamma(h \mid D(h)) = \frac{g_0}{g_0 + g_1 h} \{-W_0 E_0(1 - \varphi \mid D(h)) - cE_0(n \mid D(h))\}$$
$$+ \frac{g_1 h}{g_0 + g_1 h} \{W_1 E_1(1 - \varphi \mid D(h)) - cE_1(n \mid D(h))\}, \tag{4.12}$$

(4.10) and (4.11), we prove that $\gamma(h)$ is continuous on the left at $h = \frac{W_0 g_0}{W_1 g_1}$. This proves a).

To prove c), we proceed as follows: Suppose for $h_0 > A$ we had $\gamma(h_0) = 0$. Since

$$\{ -W_1 E_1(\varphi \mid S^*(h_0)) - cE_1(n \mid S^*(h_0))\} < 0,$$

we would have that

$$\{W_0 E_0(\varphi \mid S^*(h_0)) - cE_0(n \mid S^*(h_0))\} > 0.$$

An argument like that of Lemma 2 would then show that $\gamma(h) > 0$ for $\frac{W_0 g_0}{W_1 g_1} < h < h_0$. This, however, is impossible, because it is a violation of the definition of A.

In a similar way we prove that if $h < B$, $\gamma(h) < 0$. This proves c) and with it the lemma.

5. The behavior of A and B. LEMMA 4. *Let g and c be fixed. Then A and B are continuous functions of W_0 and W_1.*

PROOF: It will be sufficient to prove that A is continuous, the proof for B being similar. Suppose $A > B$. Let h_1 and h_2 be such that

a) $B < h_1 < A < h_2$;

b) $h_2 - h_1 < \Delta$ for an arbitrary positive Δ.

We write $\gamma(h)$ temporarily as $\gamma(h, W_0, W_1)$ in order to exhibit the dependence on W_0 and W_1. Then

$$\gamma(h_1, W_0, W_1) > 0;$$
$$\gamma(h_2, W_0, W_1) < 0.$$

It follows from (4.9) that $\gamma(h \mid D(h))$ is continuous in W_0, W_1, uniformly in $D(h)$. Hence $\gamma(h, W_0, W_1) = \sup_{D(h)} \gamma(h \mid D(h))$ is also continuous in W_0, W_1. Hence, for ΔW_0 and ΔW_1 sufficiently small,

$$\gamma(h_1, W_0 + \Delta W_0, W_1 + \Delta W_1) > 0;$$
$$\gamma(h_2, W_0 + \Delta W_0, W_1 + \Delta W_1) < 0.$$

Therefore

$$h_1 \leq A(W_0 + \Delta W_0, W_1 + \Delta W_1) \leq h_2,$$

which proves continuity, since Δ was arbitrary.

If $\frac{W_0 g_0}{W_1 g_1} = A = B$, we take $h_1 < \frac{W_0 g_0}{W_1 g_1} < h_2$, $h_2 - h_1 < \Delta$, and by a similar argument show that

$$\gamma(h_1, W_0 + \Delta W_0, W_1 + \Delta W_1) < 0;$$
$$\gamma(h_2, W_0 + \Delta W_0, W_1 + \Delta W_1) < 0.$$

Thus

$$h_1 \leq B(W_0 + \Delta W_0, W_1 + \Delta W_1) \leq A(W_0 + \Delta W_0, W_1 + \Delta W_1) \leq h_2.$$

This proves the lemma.

LEMMA 5. *Let g, c, and W_1 be fixed. A is strictly monotonic in W_0. As W_0 approaches* 0, *A approaches* 0; *as W_0 approaches $+\infty$, A also approaches $+\infty$.*

PROOF: Since $A \geq \frac{W_0 g_0}{W_1 g_1}$, $A \to +\infty$ as $W_0 \to +\infty$. If $W_0 < c$ no reduction in average risk could compensate for taking even a single observation, no matter what the value of h. Hence $\gamma(h) \leq 0$ for all h when $W_0 < c$, so that $A = B$. Since $B \leq \frac{W_0 g_0}{W_1 g_1}$, $B \to 0$ as $W_0 \to 0$. Hence $A \to 0$ as $W_0 \to 0$. It is evident from (4.9) that $\gamma(h \mid D(h))$ is non-decreasing with increasing W_0 (everything else fixed). Hence also

$$\gamma(h) = \sup_{D(h)} \gamma(h \mid D(h)),$$

is non-decreasing with increasing W_0, for fixed $h > \frac{W_0 g_0}{W_1 g_1}$ and fixed W_1. For a positive Δ sufficiently small and for any h such that $A \leq h < A + \Delta$, we have that

$$E_0 (\varphi \mid S^*(h)) > 0.$$

Hence, for such h, $\gamma(h, W_0, W_1)$ is strictly monotonically increasing with increasing W_0. Therefore A is (strictly) monotonically increasing with increasing W_0.

We now define the function $W_0(W_1, \delta)$ of the two positive arguments W_1, δ so that

$$A(W_0(W_1, \delta), W_1) = \delta.$$

By Lemma 5 such a function exists and is single-valued.

6. Properties of the function $W_0(W_1, \delta)$. LEMMA 6. *$W_0(W_1, \delta)$ is continuous in W_1.*

PROOF: Let

$$\lim_{N \to \infty} W_{1N} = W_1,$$

and suppose that the sequence $\{W_0(W_{1N}, \delta)\}$ did not converge. Suppose W_0' and W_0'' were two distinct limit points of this sequence. From the continuity of A (Lemma 4) it would follow that

$$A(W_0', W_1) = A(W_0'', W_1)$$

This, however, violates Lemma 5. The only remaining possibility to be considered is that

$$\lim_{N \to \infty} W_0(W_{1N}, \delta) = \infty.$$

If that were the case, then, since $A \geq \dfrac{W_0 g_0}{W_1 g_1}$, it would follow that $A \to \infty$, in violation of the fact that $A \equiv \delta$.

LEMMA 7. *We have, for fixed δ,*

$$\lim_{W_1 \to 0} W_0(W_1) = 0;$$

$$\lim_{W_1 \to \infty} W_0(W_1) = \infty.$$

PROOF: If, for small W_1, $W_0(W_1)$ were bounded below by a positive number, then, since $A \geq \dfrac{g_0 W_0(W_1, \delta)}{W_1 g_1}$, we could make A arbitrarily large by taking W_1 sufficiently small, in violation of the fact that $A \equiv \delta$. To prove the second half of the lemma, assume that $W_0(W_1)$ is bounded above as $W_1 \to \infty$. Then $B \left(\leq \dfrac{W_0 g_0}{W_1 g_1} \right)$ will approach zero as $W_1 \to \infty$. Let h be fixed so that $B < h < \delta$. Consider the totality of points ω for which there exists an integer $n^*(\omega)$ such that:

$$hr_{n^*} \leq B;$$

$$B < hr_j < \delta, \qquad j < n^*.$$

The conditional expected value of n^* in this totality, when H_0 is true, may be made arbitrarily large by making B sufficiently small. Hence, when W_1 is sufficiently large, for fixed but arbitrary $h < \delta$, the optimum procedure from the point of minimizing the average risk is to reject H_0 at once without taking any more observations. This, however, contradicts the fact that $h < \delta$, and proves the lemma.

LEMMA 8. *We have, for fixed $\delta > 1$,*

$$\lim_{W_1 \to 0} B(W_0(W_1, \delta), W_1) = \delta;$$

$$\lim_{W_1 \to \infty} B(W_0(W_1, \delta), W_1) = 0.$$

PROOF: By Lemma 7,

$$\lim_{W_1 \to 0} W_0(W_1) = 0.$$

When, for fixed c, both W_0 and W_1 are small enough, then, no matter what the value of h, $\gamma(h) < 0$. Hence $A = B$, which proves the first half of the lemma.

Let now $\{W_{1N}\}$ be a sequence such that $\lim W_{1N} = \infty$. Let $\delta > 1$. For the sake of brevity we write $B(W_{1N})$ instead of

$$B(W_0(W_{1N}\delta), W_{1N}).$$

Suppose that, for sufficiently large N, $B(W_{1N})$ is bounded below by a positive number. Hence, for sufficiently large N, the probability of rejecting H_1 when it it is true is bounded below by a positive number. Moreover, since $B \leq \frac{W_0 g_0}{W_1 g_1} \leq A$, it follows that, for N sufficiently large, $\frac{W_{0N} g_0}{W_{1N} g_1}$ is bounded above and below by positive constants. Thus, for large N the average risk of the test defined by $B(W_{1N})$, δ, is greater than $u g_1 W_{1N}$, where u is a positive constant which does not depend on N. Moreover, from the definition of $B(W_{1N})$, this risk is a minimum.

Let ϵ be a positive number such that $\epsilon\left(\frac{W_{0N} g_0}{W_{1N} g_1} + 1\right) < \frac{u}{2}$ for all N sufficiently large. Let V_1, V_2, with $0 < V_1 < 1 < V_2$, be two constants such that, for the sequential probability ratio test determined by them, both α_0 and α_1 are $< \epsilon$. Of course $E_0 n$ and $E_1 n$ are finite and determined by the test. For this test the average risk is less than

$$\begin{aligned} &\epsilon(g_0 W_{0N} + g_1 W_{1N}) + c g_0 E_0 n + c g_1 E_1 n \\ &< \frac{u}{2} g_1 W_{1N} + c g_0 E_0 n + c g_1 E_1 n \\ &< \frac{3u}{4} g_1 W_{1N}, \end{aligned}$$

for W_{1N} large enough. This however contradicts the fact that the minimum risk is $> u g_1 W_{1N}$, and proves the lemma.

7. Proof of the theorem. Let a given sequential probability ratio test S_0 be defined by B^*, A^*; $B^* < 1 < A^*$. Let $\alpha_i(S_0)$ be the probability, according to S_0, of rejecting H_i when it is true. Let c be fixed.

By Lemma 4, B is a continuous function of W_0 and W_1. Let $\delta = A^*$ in Lemma 8. Then there exists a pair $\overline{W}_0$, $\overline{W}_1$, with $\overline{W}_0 = W_0(\overline{W}_1, A^*)$, such that

$$A(\overline{W}_0, \overline{W}_1) = A^*;$$
$$B(\overline{W}_0, \overline{W}_1) = B^*.$$

Hence the average risk

$$\sum_i g_i [\overline{W}_i \alpha_i(S_0) + c E_i^0(n)],$$

corresponding to the sequential test S_0 is a mininium.

Now let S_1 be any other test for deciding between H_0 and H_1 and such that

$$\alpha_i(S_1) \leq \alpha_i(S_0), \quad \text{and} \quad E_i^1(n) \text{ exists } (i = 1, 2).$$

Then

$$\sum_i g_i [\overline{W}_i \alpha_i(S_0) + cE_i^0(n)] \leq \sum_i g_i [\overline{W}_i \alpha_i(S_1) + cE_i^1(n)].$$

Since $\alpha_i(S_1) \leq \alpha_i(S_0)$, we have

$$\sum_i g_i E_i^0(n) \leq \sum_i g_i E_i^1(n).$$

Now g_0, g_1 were arbitrarily chosen (subject, of course, to the obvious restrictions). Hence it must be that

$$E_i^0(n) \leq E_i^1(n).$$

This, however, is the desired result.

REFERENCES

[1] A. WALD, "Sequential tests of statistical hypotheses", *Annals of Math. Stat.*, Vol. 16 (1945), pp. 117–186.
[2] A. WALD, *Sequential Analysis*, John Wiley and Sons, Inc., New York, 1947.
[3] CHARLES STEIN, "A note on cumulative sums", *Annals of Math. Stat.*, Vol. 17 (1946), pp. 498–499.

REMARKS ON THE NOTION OF RECURRENCE

J. WOLFOWITZ

We give in several lines a simple proof of Poincaré's recurrence theorem.

THEOREM. *Let Ω be a point set of finite Lebesgue measure, and T a one-to-one measure-preserving transformation of Ω into itself.*[1] *Let $B \subset A \subset \Omega$ be measurable sets such that, if $b \in B$, $T^n b \notin A$ for all positive integral n. Then the measure $m(B)$ of B is 0.*

PROOF. First we show that, if $i < j$, $(T^i B)(T^j B) = 0$. Suppose $c \in T^j B$; then from the hypothesis on B it follows that j is the smallest integer such that $T^{-j} c \in A$. Hence $c \notin T^i B$. Now if $m(B) = \delta > 0$, Ω would contain infinitely many disjunct sets $T^n B$, each of measure δ. This contradiction proves the theorem.

The following generalization of the above theorem is trivially obvious: The result holds if we replace the hypothesis that T is measure-preserving by the following: If $m(D) > 0$, $\limsup_i m\{T^i(D)\} > 0$.

Received by the editors April 3, 1948.

[1] For a discussion in probability language see M. Kac, *On the notion of recurrence in discrete stochastic processes*, Bull. Amer. Math. Soc. vol. 53 (1947) pp. 1002–1010.

Another obvious generalization is this: Let C be the set of all points c of A such that $T^n c \in A$ for only finitely many n. Then $m(C)=0$ (for $C \subset \sum_{l=0}^{\infty} T^{-l}B$).

The following is a simple derivation of Kac's theorem on the mean recurrence time.[2]

THEOREM. *Let T above be metrically transitive. Let $a \in A-B$, and $n(a)$ be the smallest positive integer such that $T^n a \in A$. Let $m(A)>0$. Then*

$$\int_{A-B} n(a)dm = m(\Omega).$$

PROOF. Define $A_k = \{n(a)=k\}$. Let $i<j$, $i'<j'$, $j \neq j'$. We notice:

(a) $(T^i A_j)(T^{i'} A_{j'})=0$. For T has a single-valued inverse and $A_j A_{j'}=0$. If $T^i A_j$ and $T^{i'} A_{j'}$ had a point s in common, then $T^{-i}s \in A_j$, $T^{-i'}s \in A_{j'}$, in violation of the definition of j and j'.

(b) $$\int_{A-B} n(a)dm = m\left(\sum_{h=1}^{\infty}\sum_{l=0}^{h-1} T^l A_h\right).$$

(c) Metric transitivity implies that almost every point in Ω lies in some $T^l A_h$, that is, $m(\sum\sum T^l A_h)=m(\Omega)$.

This proves the desired result.

COLUMBIA UNIVERSITY

[2] Kac, loc. cit. Theorem 2.

Made in United States of America

Reprinted from THE ANNALS OF MATHEMATICAL STATISTICS
Vol. XX, No. 4, December, 1949

THE POWER OF THE CLASSICAL TESTS ASSOCIATED WITH THE NORMAL DISTRIBUTION

BY J. WOLFOWITZ

Columbia University

Summary. The present paper is concerned with the power function of the classical tests associated with the normal distribution. Proofs of Hsu, Simaika, and Wald are simplified in a general manner applicable to other tests involving the normal distribution. The set theoretic structure of several tests is characterized. A simple proof of the stringency of the classical test of a linear hypothesis is given.

1. Introduction. The present paper is concerned with the optimum properties, from the power function viewpoint, of the classical tests associated with the normal distribution. In 1941 Hsu [2] proved the result stated in Section 2 below, which is concerned with the general linear hypothesis (in this connection his paper [1] of 1938 will be of interest). Also in 1941 Simaika [3] proved similar results for the tests based on the multiple correlation coefficient and Hotelling's generalization of Student's t. In 1942, Wald [4] gave a generalization of Hsu's result.

In the present paper we give short and simple proofs of almost all these results, and a simple proof of the stringency property of the analysis of variance (Section 5). These proofs rest on theorems which characterize the set theoretic structure of the tests. Thus, while the proofs of Hsu, Simaika and Wald are rather elaborate and each problem is essentially attacked *de novo*, the methods of the present paper are in effect applicable to the classical tests based on the normal distribution. For these tests it will not be difficult to demonstrate the analogues of Theorems 1 and 3, and of the results of Hsu, Simaika, and Wald. In the present paper we first treat the general linear hypothesis, because it is the simplest problem, its solution is easiest to describe, and it admits Wald's integration theorem. Multivariate analogues of the latter are rather artificial and not as simple. We then discuss the problem of the multiple correlation coefficient, because it seems to be more difficult than that of Hotelling's T and indeed, to include all the essential multivariate difficulties. Theorems 6 and 7 are the analogues of 1 and 3, respectively, while Theorem 9 describes the essential property of the power function which is of interest to us. In other multivariate problems one will prove the analogues of Theorems 6, 7 and 9. A generally inclusive formulation is no doubt possible. Theorems 5 and 9 are slightly more general than the theorems of Hsu and Simaika.

Many of the statements below may be not valid on exceptional sets of measure zero. Usually this is so stated, but sometimes, for reasons of brevity or to avoid repetition, this qualification may be omitted. The reader will have no difficulty supplying it wherever necessary.

The author is indebted to Erich L. Lehmann of the University of California, who carefully read a first version of this paper. Theorem 4 below was arrived at independently by Professor Lehmann, with a somewhat different proof.

2. The general linear hypothesis. In canonical form the general linear hypothesis may be stated as follows: The chance variables

$$X_1, X_2, \cdots, X_{k+l}$$

have at $x_1, \cdots, x_{k+l}$, the density function

$$(2.1) \qquad (\sqrt{2\pi}\,\sigma)^{-(k+l)} \exp\left[-\frac{1}{2\sigma^2}\left\{\sum_1^k (x_i - \eta_i)^2 + \sum_{k+1}^{k+l} x_i^2\right\}\right] = f(\eta, \sigma)$$

with $\sigma, \eta_1, \cdots, \eta_k$ all unknown.

Let η be the vector $(\eta_1, \cdots, \eta_k)$. The null hypothesis H_0 states that

$$\eta_1 = \cdots = \eta_k = 0$$

and is to be tested with constant size $\alpha < 1$ (identically in σ).

Let D be any admissible critical region for testing H_0. If A is any event let $P\{A \mid \eta, \sigma\}$ denote the probability of A when η and σ are the parameters of (2.1). We have then

$$P\{D \mid 0, \sigma\} = \alpha$$

identically in σ, where 0 is the vector with k components all of which are zero. We now prove a property which characterizes all D. This theorem is due to Neyman and Pearson [12], and is given here only for completeness.

THEOREM 1. *The fraction of the surface area of the sphere*

$$\sum_1^{k+l} x_i^2 = c^2$$

which lies in D is α for almost all c.

PROOF. Let a be any positive integer, h a positive parameter, and $\psi(y)$ a measurable function of y defined for $y > 0$ and such that $0 \leq \psi(y) \leq 1$. In view of the distribution of ΣX_i^2, it will be enough to prove that, if

$$\frac{h^{a+1}}{\Gamma(a+1)} \int_0^\infty \psi(y) y^a e^{-hy}\, dy = \alpha$$

identically for all positive h, that then

$$\psi(y) = \alpha \text{ for almost all } y.$$

Write

$$(2.2) \qquad \frac{1}{\alpha\Gamma(a+1)} \int_0^\infty \psi(y) y^a e^{-hy}\, dy = h^{-(a+1)}.$$

Differentiating both members k times with respect to h and then setting $h = 1$

we obtain the following result. The function

$$\frac{1}{\alpha\Gamma(a+1)}\,\psi(y)y^{a}e^{-y}$$

is a density function with kth moment

$$\mu_k = (a+1)(a+2)\cdots(a+l).$$

The moments μ_k are the moments of the density function

$$\frac{1}{\Gamma(a+1)}\,y^{a}e^{-y}$$

They satisfy the Carleman criterion [5, p. 19, Th 1.10], and hence no essentially different distribution can have these moments. This proves the desired result.

THEOREM 2 (Wald). *Among all tests of the general linear hypothesis the analysis of variance test has the property that, for all positive d, the integral of its power on the surface* $\eta^2 = d^2$ *is a maximum.*

PROOF. Let c be any positive number. We have only to show that if we allocate to the critical region D of the test the fraction α of the surface area of the sphere

$$\sum_1^{k+l} x_i^2 = c^2 \tag{2.3}$$

for which

$$C = \frac{\sum_1^{k} x_i^2}{\sum_{k+1}^{k+l} x_i^2}$$

is as large as possible and that if we do this for all c, the desired maximum of the integral of the power will be achieved. If C is as large as possible so is

$$\frac{\sum_1^{k} x_i^2}{\sum_1^{k+l} x_i^2} = \frac{\sum_1^{k} x_i^2}{c^2}.$$

Let $a_1, \cdots, a_{k+l}$ be any point on the sphere (2.3). Let db be the differential of area on the surface $\eta^2 = d^2$. Then

$$\int_{\eta^2=d^2}\!\!\cdots\int f(\eta,\sigma)\,db = (\sqrt{2\pi}\,\sigma)^{-(k+l)}\exp\left\{-\frac{(c^2+d^2)}{2\sigma^2}\right\} \int_{\eta^2=d^2}\!\!\cdots\int \exp\left\{\frac{(\eta)'z}{\sigma^2}\right\} db, \tag{2.4}$$

where z is the vector $(a_1, \cdots, a_k)$ and $(\eta)'z$ is the scalar product of the two vectors. This last integral is easily seen to depend only upon $|z|$ and to be monotonically increasing in $|z|$. This proves the theorem.

COROLLARY (Hsu). *Among all tests of the general linear hypothesis whose power is a function of η^2 only, the analysis of variance is the most powerful.*

3. The set theoretic structure of tests whose power is a function only of η^2/σ^2. Wald's result (Theorem 2) cannot always be extended, in its simple form, to tests involving the multivariate normal distribution, but this can be done with Hsu's theorem (corollary to Theorem 2). In order to see what is involved we shall investigate the set theoretic structure of tests of the general linear hypothesis whose power is a function only of η^2/σ^2.

Let $q(x_1, \cdots, x_k)$ be the set of points in the region D whose first k coordinates are $x_1, \cdots, x_k$. Let $A(x_1, \cdots, x_k, \sigma)$ be the integral of

$$(2\pi\sigma^2)^{-(l/2)} \exp\left[-\frac{1}{2\sigma^2}\left\{\sum_{j=1}^{l} x_{k+j}^2\right\}\right]$$

with respect to $x_{k+1}, \cdots, x_{k+l}$, taken over $q(x_1, \cdots x_k)$. We first prove the following:

LEMMA. *Suppose the power of D is a function only of η^2/σ^2. Then for two points*

$$x_1, \cdots, x_k$$

and

$$x_1', \cdots, x_k'$$

such that

$$\sum_1^k x_i^2 = \sum_1^k x_i'^2 \tag{3.1}$$

we have

$$A(x_1, \cdots, x_k, \sigma) = A(x_1', \cdots, x_k', \sigma) \tag{3.2}$$

identically in σ, with the exception of a set of measure zero.

PROOF. Suppose the statement is false. Then under some orthogonal transformation T of $x_1, \cdots, x_k$ the region D would go over into a region D^* with the following property: Let $A^*(x_1, \cdots, x_k, \sigma)$ have the same definition for the region D^* as $A(x_1, \cdots, x_k, \sigma)$ has for D. Then on a set of positive measure[1] we would have

$$A(x_1, \cdots, x_k, \sigma) \gtrless A^*(x_1, \cdots, x_k, \sigma). \tag{3.3}$$

We shall now show that (3.3) results in a contradiction. We have

$$P\{D \mid \eta, \sigma\} = P\{D^* \mid T\eta, \sigma\} \tag{3.4}$$

identically in η. By the property of the region D, therefore, we have

$$P\{D \mid \eta, \sigma\} = P\{D \mid T^{-1}\eta, \sigma\}$$

[1] The situation here is similar to that described in footnote 3.

and hence

$$P\{D \mid \eta, \sigma\} = P\{D^* \mid \eta, \sigma\} \tag{3.5}$$

identically in η. Thus we obtain

$$\int (2\pi\sigma^2)^{-(k/2)} A(x_1, \cdots, x_k, \sigma) \exp\left[-\frac{1}{2\sigma^2}\left\{\sum_1^k (x_i - \eta_i)^2\right\}\right] dx_1 \cdots dx_k \tag{3.6}$$
$$\equiv \int (2\pi\sigma^2)^{-(k/2)} A^*(x_1, \cdots, x_k, \sigma) \exp\left[-\frac{1}{2\sigma^2}\left\{\sum_1^k (x_i - \eta_i)^2\right\}\right] dx_1 \cdots dx_k$$

with the integrations taking place over the entire space. Differentiating both members with respect to the components of η and setting $\eta = 0$, we obtain that the two density functions (for fixed σ)

$$(2\pi\sigma^2)^{-(k/2)}\alpha^{-1}A(x_1, \cdots, x_k, \sigma) \exp\left[-\frac{1}{2\sigma^2}\left\{\sum_1^k x_i^2\right\}\right]$$

and

$$(2\pi\sigma^2)^{-(k/2)}\alpha^{-1}A^*(x_1, \cdots, x_k, \sigma) \exp\left[-\frac{1}{2\sigma^2}\left\{\sum_1^k x_i^2\right\}\right]$$

have identical moments. We shall now argue that these moments satisfy the conditions of Cramér and Wold [7, Th. 2], so that the two density functions are essentially the same, in contradiction to (3.3). The Cramér-Wold theorem states the following: *Let $Y_1, \cdots, Y_k$ be k chance variables with a joint distribution function, and write*

$$\lambda_{2n} = \sum_{i=1}^k EY_i^{2n}.$$

Then the divergence of the series

$$\sum_{n=1}^\infty \lambda_{2n}^{-(1/2n)}$$

is sufficient to ensure that there exists essentially only one distribution which has these moments. We notice that the factor $1/\alpha$ of course makes no difference. If we set $A(x_1, \cdots, x_k, \sigma)$ and $A^*(x_1, \cdots, x_k, \sigma)$ both identically unity and consider the resulting moments which enter into the λ_{2n}, we see that these moments satisfy the Cramér-Wold condition. Now A and A^* are ≤ 1. Thus, using the true value of A can serve only to increase the value of $\lambda_{2n}^{-(1/2n)}$, so that the series will diverge a fortiori. This proves the lemma.

The following theorem helps to describe the set theoretic structure of tests whose power is a function only of $\lambda = \eta^2/\sigma^2$:

THEOREM 3. *Let D be a test whose power is a function only of λ. Let u be any positive number, and $D(x_1, \cdots, x_k, u)$ be the fraction of the "area" of the sphere $\Sigma_{j=1}^l x_{k+j}^2 = u^2$ occupied by points which are in D and whose first k coordinates are $x_1, \cdots, x_k$. If*

(3.7) $$\sum_1^k x_i^2 = \sum_1^k x_i'^2$$

then, except on a set of measure zero,

(3.8) $$D(x_1, \cdots, x_k, u) = D(x_1', \cdots, x_k', u).$$

PROOF. We shall show that, if the power of D is a function only of λ, the failure of (3.7) to imply (3.8) would contradict the preceding lemma. Suppose then that (3.8) is not true on a set of positive measure. Under some orthogonal transformation on $x_1, \cdots, x_k$ we obtain[2] a function $D^*(x_1, \cdots, x_k, u)$ which liffers from $D(x_1, \cdots, x_k, u)$ on a set of positive measure and such that, for almost every $x_1, \cdots, x_k$,

$$A(x_1, \cdots, x_k, \sigma) = K \int_0^\infty D(x_1, \cdots, x_k, u)\sigma^{-l} u^{l-1} e^{(-u^2)/2\sigma^2}\, du$$
$$= K \int_0^\infty D^*(x_1, \cdots, x_k, u)\sigma^{-l} u^{l-1} e^{(-u^2)/2\sigma^2}\, du$$

identically in σ, where K is a suitable constant of no interest to us. Multiplying by σ^l, differentiating repeatedly under the integral sign with respect to σ, and setting $\sigma = 1$, we obtain the result that the two density functions in u,

$$\frac{KD(x_1, \cdots, x_k, u)}{A(x_1, \cdots, x_k, 1)} u^{l-1} e^{(-u^2)/2}$$

and

$$\frac{KD^*(x_1, \cdots, x_k, u)}{A(x_1, \cdots, x_k, 1)} u^{l-1} e^{(-u^2)/2}$$

are identical except perhaps on a set of measure zero. This contradiction proves the theorem.

THEOREM 4. *A necessary and sufficient condition that the power of D be a function of λ only, is that, with the usual exception of a set of measure zero, $D(x_1, \cdots, x_k, u)$ be a function only of*

$$\frac{\sum_1^k x_i^2}{u^2}.$$

The proof of this theorem is not essentially different from that of the preceding theorem, and we shall therefore sketch it only briefly. Let Z be a transformation on $(x_1, \cdots, x_k, u) = (x, u)$ which consists of a rotation of the vector x, followed by a multiplication of u and the components of x by a positive constant c. If $D(x, u)$ is not a function of $\Sigma_1^k x_i^2/u^2$ alone, then, just as before[3], we can use some

[2] See footnote 1.

[3] This statement implies that a function of $x_1, \cdots, x_k, u$, which is invariant to within sets of measure zero under all transformations Z (the exceptional set may depend on the

transformation Z to give us a function $D^*(x, u)$ such that

$$D(x, u) \neq D^*(x, u)$$

on a set of positive measure, while

$$ED(x, u) = ED^*(x, u)$$

identically in η, σ. This yields a contradiction in the usual manner and proves the necessity of the condition.

To prove sufficiency, write $D(x, u) = \nu(\Sigma x_i^2/u^2) = \nu(v)$. Let $\gamma(v, \eta, \sigma)$ be the density function of v. Then

$$P\{D \mid \eta, \sigma\} = \int_0^\infty \nu(v)\gamma(v, \eta, \sigma)\, dv.$$

By hypothesis, $\nu(v)$ is a function only of v. We know [9, p. 140, eq. 101] that $\gamma(v, \eta, \sigma)$ is a function only of v and λ. Hence $P\{D \mid \eta, \sigma\}$ is a function only of λ. This completes the proof of the theorem.

THEOREM 5. *Among all tests of the general linear hypothesis which have the properties described in the conclusions of Theorems* 1 *and* 3, *the classical analysis of variance test is the most powerful.*

We shall omit the proof of this theorem, which is very similar to that of the more difficult Theorem 9 below.

Theorem 4 above shows that there exist regions D which satisfy the conclusions of Theorems 1 and 3 and such that $P\{D \mid \eta, \sigma\}$ is not a function of λ alone. It follows that the content of Theorem 5 is greater than that of Hsu's theorem (Corollary to Theorem 2).

It is instructive to note that Hsu's theorem follows almost immediately from Theorem 4 and the form of $\gamma(v, \lambda)$. For let λ be fixed but arbitrary. One verifies immediately from the form of $\gamma(v, \lambda)$ that

$$\frac{\gamma(v, \lambda)}{\gamma(v, 0)}$$

is, for fixed λ, a monotonically increasing function of v. This, by Neyman's lemma, immediately proves Hsu's result.

4. The multiple correlation coefficient. We shall now apply our methods to a multivariate test. For typographic ease we shall conduct the discussion for the

transformation), is a function of $\frac{\Sigma x_i^2}{u^2}$, except on a set of measure zero. This statement would be completely trivial were it not for the exceptional sets; in any case it must be well known to set theorists. The author constructed an unnecessarily long proof of it, and believes that a more expeditious proof can be constructed using the ideas of [11, page 91, Theorem 11.1, and page 318, p. 7]. Professor C. M. Stein of the University of California has informed the author that this result is a special case of one established by himself and G. H. Hunt in a forthcoming paper. For these reasons the proof is omitted. (See also [13, page 27, Lemma 9.1].)

case of three variates, but the reader will observe that the procedure is really perfectly general.

The chance variables $\{Y_{ij}\}$, $i = 1, 2, 3$, $j = 1, \cdots, n$, have the density function

$$g(B) = (2\pi)^{(-3n)/2} (| B |)^{n/2} \exp \left\{ -\tfrac{1}{2} \sum_{j=1}^{n} \sum_{i,l=1}^{3} b_{il} y_{ij} y_{lj} \right\} \tag{4.1}$$

where 1) $B = \{b_{il}\}$ is a positive definite (symmetric) 3×3 matrix, 2) y_{ij} is the value assumed by Y_{ij}. The null hypothesis H_0 asserts that a given multiple correlation coefficient is zero, say that of Y_1 with Y_2 and Y_3, i.e.,

$$b_{12} = b_{21} = b_{13} = b_{31} = 0. \tag{4.2}$$

The test is to be made on the level of significance α, i.e., if B_0 is any matrix which satisfies (4.2), and if G is a critical region for testing H_0, then

$$P\{G \mid B_0\} = \alpha \tag{4.3}$$

where the symbol in the left member means the probability of G according to $g(B_0)$.

Write

$$ns_{ij} = \sum_{k=1}^{n} y_{ik} y_{jk}$$

$$S = \begin{Bmatrix} s_{22} & s_{23} \\ s_{32} & s_{33} \end{Bmatrix}.$$

Let $M(c_{11}, C)$ be the manifold in the $3n$-space of

$$y_{11}, \cdots, y_{ik}, \cdots, y_{3n}$$

where $s_{11} = c_{11}$, $S = C$. First we prove the following:

THEOREM 6. *Any region G which satisfies (4.3) must have the property that the fraction of the area of $M(c_{11}, C)$ which lies in G is α, for any positive c_{11} and any positive definite 2×2 matrix $C = \{c_{ij}\}$.* (We remind the reader that exceptional sets of measure zero are not precluded).

PROOF. Let $\psi(c_{11}, C)$ be the fraction of the area of $M(c_{11}, C)$ in G. Recall equation (4.3) and the fact that s_{11}, s_{22}, s_{23}, s_{33} are sufficient statistics for the elements of B_0. On the manifold $M(c_{11}, C)$ the conditional density is uniform. Employing Wishart's distribution [6] we conclude that

$$K' \int \psi(s_{11}, S) \mid B_0 \mid N \mid S \mid^{(n-3)/2} s_{11}{}^{(n-2)/2} \cdot \exp \left[-\frac{n}{2} \{b_{11} s_{11} + b_{22} s_{22} + 2b_{23} s_{23} + b_{33} s_{33}\} \right] ds_{11}\, ds_{22}\, ds_{23}\, ds_{33} \equiv \alpha \tag{4.4}$$

where K' is a suitable constant which need not concern us. Here the symbol

"≡" means identically in b_{11}, b_{22}, b_{23}, b_{33}, provided only that $b_{11} > 0$, $b_{22} > 0$ $b_{22}b_{33} - b_{23}^2 > 0$. Of course s_{11} is distributed independently of s_{22}, s_{23}, s_{33}. Proceeding as in section 2, we can, by differentiation with respect to the b's, obtain all the moments of the s_{ij}'s. Now let the b's take any admissible constant values. The moments of the s_{ij}'s are then seen to satisfy the criterion of Cramér and Wold [7, Th. 2], and consequently essentially uniquely determine the distribution of the s_{ij}. The desired conclusion follows as before.

The six parameters which uniquely determine the trivariate normal distribution (of Y_1, Y_2, Y_3) with zero means may be taken to be the following:

1) The covariance matrix $\{\sigma_{ij}\}$, $i, j = 2, 3$, of Y_2 and Y_3.

2) The partial regression coefficients β_2, β_3, of Y_1 on Y_2 and Y_3. These are defined as follows: Let $E(Y_1 \mid Y_2 = y_2, Y_3 = y_3)$ denote the conditional expected value of Y_1, given $Y_2 = y_2$, $Y_3 = y_3$. Then

$$E(Y_1 \mid Y_2 = y_2, Y_3 = y_3) = \beta_2 y_2 + \beta_3 y_3 .$$

3) The conditional variance ω^2 of Y_1, given $Y_2 = y_2$, $Y_3 = y_3$.

The population multiple correlation coefficient $\bar{R}$ of Y_1 with Y_2 and Y_3 is then defined by

$$\frac{\bar{R}^2\omega^2}{(1 - \bar{R}^2)} = \beta_2^2\sigma_{22} + 2\beta_2\beta_3\sigma_{23} + \beta_3^2\sigma_{33}.$$

The six parameters above may be chosen arbitrarily, provided only that $\{\sigma_{ij}\}$ is positive definite. $\bar{R}$ and ω are, by definition, non-negative.

Let y_i be the column vector $y_{i1}, \cdots, y_{in}$; let y_i' be its transpose, and let y denote the point $y_{11}, y_{12}, \cdots, y_{1n}, y_{21}, \cdots, y_{3n}$ in $3n$-space. Let $z(y) = z(y_1, y_2, y_3)$ be the component of y_1 in the plane of y_2 and y_3; let $r = | z(y) |$ and θ the angle between z and y_2, measured positively say in the direction of y_3. Finally let h be the absolute value of the vector $y_1 - z(y_1, y_2, y_3)$.

We intend now to investigate the set theoretic structure of tests whose power is a function only of $\bar{R}$, and for this purpose prove the following:

Theorem 7. *Let H be a region whose power is a function only of $\bar{R}$. Let $V(h, r, \theta, s_{22}, s_{23}, s_{33})$ be the fraction of the "volume" of the manifold on which $h, r, \theta, s_{22}, s_{23}, s_{33}$ are fixed which is contained in H. With the usual exception of a set of measure zero, for fixed $h, r, s_{22}, s_{23}, s_{33}$, the quantity V above is constant for all θ.*

Later, after this theorem is proved, we shall write V without exhibiting θ. This procedure is justified by Theorem 7.

Proof. Suppose the theorem false, and proceed as in Theorem 3. A suitable[4] rotation of the radius vector $z(y)$ implies an orthogonal transformation T on the generic point y which leaves h, r, s_{22}, s_{23}, and s_{33} unaltered, and takes the region H into a region H^* such that H and H^* differ on a set of positive measure. T leaves R invariant, hence leaves invariant $\bar{R}$ which uniquely determines the distribution

[4] See footnote 1.

of R. Hence an argument almost the same as that which led us to (3.5) yields the conclusion that the power of H and the power of H^* are equal, identically in B. Proceeding as in Theorem 3, we obtain two essentially different density functions in h, r, θ, s_{22}, s_{23}, s_{33}, whose integrals over the entire space are identical in the elements of B. From these functions we obtain two different density functions in $s_{ij}(i, j = 1, 2, 3)$, with identical moments (obtained by differentiation with respect to the elements of B). The rest of the proof is essentially no different from that of Theorem 3.

THEOREM 8. *In order that the power of H be a function of $\bar{R}$ alone, it is necessary and sufficient that, with the usual exception of a set of measure zero,* $V(h, r, s_{22}, s_{23}, s_{33})$ *be a function only of* h/r (i.e., *of* R).

The proof of this theorem is essentially the same as the proof of Theorem 4. The place of the transformation Z is taken by one which consists of any linear transformation on the vectors y_2 and y_3, the addition of a constant angle to θ (rotation of $z(y)$), and multiplication of the vector y_1 by a positive scalar c. This transformation leaves $\bar{R}$ invariant. In the proof of sufficiency we use the distribution of R (see, for example, [10, p. 384, equation (15.55)]). The remainder of the proof is essentially the same as that of Theorem 4.

THEOREM 9. *Among all tests H which have the properties described in the conclusions of Theorems* 6 *and* 7, *the classical test based on R is the most powerful.*

As a corollary to this theorem we have the following result due to Simaika [3]: Of all tests H whose power is a function of $\bar{R}$ only, the classical test based on R is the most powerful.

Simaika's result also follows easily from Theorem 8 and the density function of R in the same manner that Hsu's result followed from Theorem 4 and the density function of v.

In the course of the proof of Theorem 9, the various symbols W, with or without subscripts, will denote suitable functions of the variables exhibited, and the various symbols k, with or without subscripts, will denote suitable constants.

We have that

$$P\{H \mid B\} = \int_H (2\pi)^{(-3n)/2} \mid B \mid^{n/2} \exp\left\{-\tfrac{1}{2}\sum_{j=1}^{n} y_j' B y_j\right\} dy_{11} \cdots dy_{3n}$$

$$= \int_H (2\pi\omega^2)^{(-n)/2} \exp\left[-\frac{1}{2\omega^2}\{y_1 - (\beta_2 y_2 + \beta_3 y_3)\}^2\right] \cdot$$

$$\cdot W_0(s_{22}, s_{23}, s_{33}, \{\sigma_{ij}\})\, dy_{11} \cdots dy_{3n} = (2\pi\omega^2)^{(-n)/2} \int_H \exp\left\{\frac{1}{\omega^2}(\beta_2 y_2 + \beta_3 y_3)'z\right\} \cdot \tag{4.5}$$

$$\exp\left[-\frac{1}{2\omega^2}\{y_1^2 + \beta_2^2 s_{22} + 2\beta_2\beta_3 s_{23} + \beta_3^2 s_{33}\}\right] \cdot$$

$$\cdot W_0(s_{22}, s_{23}, s_{33}, \{\sigma_{ij}\})\, dy_{11} \cdots dy_{3n}.$$

Now $(\beta_2 y_2 + \beta_3 y_3)'z$ is a function only of β_2, β_3, s_{22}, s_{23}, s_{33}, r, and θ. Also

$h^2 + r^2 = s_{11} = y_1^2$. Thus

$$P\{H \mid B\} = \int V(h, r, s_{22}, s_{23}, s_{33}) W_1(h, r, s_{22}, s_{23}, s_{33}, \{B\})$$

$$\cdot \exp\left\{\frac{1}{\omega^2}(\beta_2 y_2 + \beta_3 y_3)'z\right\} d\theta\, dh\, dr\, ds_{22}\, ds_{23}\, ds_{33} = \int V(h, r\ s_{22}, s_{23}, s_{33})$$

$$\cdot W_1(h, r, s_{22}, s_{23}, s_{33}, \{B\})(4hr)^{-1} \exp\left\{\frac{1}{\omega^2}(\beta_2 y_2 + \beta_3 y_3)'z\right\} \tag{4.6}$$

$$\cdot\, d\theta\, dh^2\, dr^2\, ds_{22}\, ds_{23}\, ds_{33} = \int V\left(\sqrt{y_1^2 - r^2}, r, s_{22}, s_{23}, s_{33}\right)$$

$$\cdot W_2\left(\sqrt{y_1^2 - r^2}, r, s_{22}, s_{23}, s_{33}, \{B\}\right) \exp\left\{\frac{1}{\omega^2}(\beta_2 y_2 + \beta_3 y_3)'z\right\}$$

$$d\theta\, dr^2\, dy_1^2\, ds_{22}\, ds_{23}\, ds_{33}.$$

Integrating with respect to θ and designating

$$W_2 \int \exp\left\{\frac{1}{\omega^2}(\beta_2 y_2 + \beta_3 y_3)'z\right\} d\theta$$

by $W(\sqrt{y_1^2 - r^2}, r, s_{22}, s_{23}, s_{33}, \{B\})$ we observe that just as in (2.4), W is monotonically increasing in r (all other variables fixed). Thus we have

$$P\{H \mid B\} = \int VW\, dr^2\, dy_1^2\, ds_{22}\, ds_{23}\, ds_{33}. \tag{4.7}$$

In constructing H only the function V is at our disposal, and this subject to the limitations imposed by the conclusions of Theorems 6 and 7 and the fact that $h^2 + r^2 = y_1^2 = s_{11}$. The function W is not within our control at all. With y_1^2, s_{22}, s_{23}, s_{33} fixed, W is monotonically increasing with r. To maximize the power it is therefore best to distribute the "mass" so that V is as large as possible for large values of r and hence of R. This implies the classical test and proves the theorem.

5. Stringency of the classical tests. Wald [8] calls a test T_1 "most stringent" if the following is true: Let $\{T\}$ be the totality of tests. Let θ be the generic point in the parameter space, and $P\{T \mid \theta\}$ be the power of T at the point θ. Let T_2 be any test other than T_1. Then

$$\sup_\theta\, [\sup_{\{T\}} P\{T \mid \theta\} - P\{T_1 \mid \theta\}] \le \sup_\theta\, [\sup_{\{T\}} P\{T \mid \theta\} - P\{T_2 \mid \theta\}].$$

Of course, we have omitted to specify the totality $\{T\}$. One can admit all tests whose size $\le \alpha$, a given constant between 0 and 1, or restrict one's self to tests whose size is exactly α. We shall do the latter.

Under these circumstances we shall prove that the classical test of a linear hypothesis is most stringent. Our proof will occupy but a few lines, and is an easy consequence of the structure of the classical tests as described in the lemma of section 2. The result itself is a special case of an unpublished theorem due to G. H. Hunt and C. M. Stein, and all priority on this result is theirs.

Return then to the notation of section 2. Let σ be fixed at any arbitrary positive value, and the surface

$$\eta^2 = c_0^2$$

be that one on which

$$\omega_1(\eta) = \sup_{\{T\}} P\{T \mid \eta\} - P\{L_1 \mid \eta\}$$

is a maximum, where L_1 is the classical test of the linear hypothesis. It is clear that this maximum is actually achieved, and that $\omega_1(\eta)$ is a constant on the surface $\eta^2 = c_0^2$. Let L_2 be any other test (of size α), and $\omega_2(\eta)$ be the corresponding function for L_2. We have only to show that on the surface $\eta^2 = c_0^2$ we cannot have everywhere $\omega_2(\eta) < \omega_1(\eta)$, and our proof is complete. If everywhere on the surface $\eta^2 = c_0^2$ we had $\omega_2(\eta) < \omega_1(\eta)$, we would have, also on the same surface, $P\{L_2 \mid \eta\} > P\{L_1 \mid \eta\}$. This would, however, violate Wald's Theorem 2 (section 2) and proves the desired result.

REFERENCES

[1] P. L. Hsu, "Notes on Hotelling's generalized T," *Annals of Math. Stat.*, Vol. 9 (1938) p. 231.

[2] P. L. Hsu, "Analysis of variance from the power function standpoint," *Biometrika*, Vol. 32 (1941), p. 62.

[3] J. B. Simaika, "On an optimum property of two important statistical tests," *Biometrika*, Vol. 32 (1941), p. 70.

[4] A. Wald, "On the power function of the analysis of variance test," *Annals of Math. Stat.*, Vol. 13 (1942), p. 434.

[5] J. A. Shohat and J. D. Tamarkin, *The Problem of Moments*, The American Mathematical Society, New York, 1943.

[6] John Wishart, "The generalized product moment distribution, etc.," *Biometrika*, Vol. 20A (1928), p. 32.

[7] H. Cramér and H. Wold, "Some theorems on distribution functions," *Lond. Math. Soc. Jour.*, Vol. 11 (1936).

[8] A. Wald, "Tests of statistical hypotheses concerning several parameters when the number of observations is large," *Am. Math. Soc. Trans.*, Vol. 54 (1943), p. 426.

[9] P. C. Tang, "The power function of the analysis of variance etc.," *Stat. Res. Memoirs*, Vol. 2 (1938) (University of London), p. 126.

[10] M. G. Kendall, *The Advanced Theory of Statistics*, Vol. 1, Charles Griffin and Company, London, 1945.

[11] S. Saks, *Theory of the Integral*, (Second Edition), G. E. Stechert and Company, New York, 1937.

[12] J. Neyman and E. S. Pearson, "On the problem of the most efficient tests of statistical hypotheses," *Roy. Soc. London Phil. Trans.*, Ser. A, Vol. 231 (1933), pp. 289–337.

[13] Eberhard Hopf, *Ergodentheorie*, Chelsea, New York, 1948.

Made in United States of America

Reprinted from The Annals of Mathematical Statistics
Vol. 21, No. 1, March, 1950

BAYES SOLUTIONS OF SEQUENTIAL DECISION PROBLEMS

By A. Wald and J. Wolfowitz

Columbia University

Summary. The study of sequential decision functions was initiated by one of the authors in [1]. Making use of the ideas of this theory the authors succeeded in [4] in proving the optimum character of the sequential probability ratio test. In the present paper the authors continue the study of sequential decision functions, as follows:

a) The proof of the optimum character of the sequential probability ratio test was based on a certain property of Bayes solutions for sequential decisions between two alternatives, the cost function being linear. This fundamental property, the convexity of certain important sets of a priori distributions, is proved in Theorem 3.9 in considerable generality. The number of possible decisions may be infinite.

b) Theorem 3.10 and section 4 discuss tangents and boundary points of these sets of a priori distributions.

(These results for finitely many alternatives were announced by one of us in an invited address at the Berkeley meeting of the Institute of Mathematical Statistics in June, 1948)[1]

c) Theorem 3.6 is an existence theorem for Bayes solutions. Theorem 3.7 gives a necessary and sufficient condition for a Bayes solution. These theorems generalize and follow the ideas of Lemma 1 of [4]

d) Theorems 3.8 and 3.8.1 are continuity theorems for the average risk function. They generalize Lemma 3 in [4]

e) Other theorems give recursion formulas and inequalities which govern Bayes solutions.

1. Introduction. In a previous publication of one of the authors [1] the decision problem was formulated as follows: Let $X = \{x_i\}$ $(i = 1, 2, \cdots$, ad inf.) be a sequence of chance variables. An observation on X is given by a sequence $x = \{x_i\}$ $(i = 1, 2, \cdots$, ad inf.) of real values, where x_i denotes the observed value of X_i. A sequence x is also called a sample or sample point, and the totality M of all possible sample points x is called the sample space. Let $G(x)$ denote the probability that $X_i < x_i$ for $i = 1, 2, \cdots$, ad inf.; i.e., G is the cumulative distribution function of X. In a statistical decision problem G is assumed to be unknown. It is merely known that G is an element of a given class Ω of distribution functions. There is given, furthermore, a space D^* whose elements d represent the possible decisions that can be made in the problem under consideration.

---

[1] A brief statement of some of the results of the present paper is to be found in the authors' paper of the same name in the *Proc. Nat. Acad. Sci. U. S. A.*, Vol. 35 (1949), pp. 99–102.

The problem is to construct a function $d = D(x)$, called the decision function, which associates with each sample point x an element d of D^* so that the decision $d = D(x)$ is made when x is observed.

Occasionally we shall use the symbol D to denote a decision function $D(x)$. This will be done especially when we want to emphasize that we mean the whole decision function and not merely a particular value of it corresponding to some particular x.

If $d = D(x)$ is the decision function adopted and if $x^0 = \{x_i^0\}$ $(i = 1, 2, \cdots)$ is the particular sample point observed, the number of components of x^0 we have to observe in order to reach a decision is equal to the smallest positive integer $n = n(x^0)$ with the property that $D(x) = D(x^0)$ for any x for which $x_1 = x_1^0, \cdots, x_n = x_n^0$. If no finite n exists with the above property, we put $n(x) = \infty$. If $d(x)$ is equal to a constant d, we put $n(x) = 0$. We shall call $n(x)$ the number of observations required by D when x is the observed sample. Of course, $n(x)$ depends also on the decision rule D adopted. To put this in evidence, we shall occasionally write $n(x, D)$ instead of $n(x)$. If D_0 is a decision function such that $n(x, D_0)$ has a constant value over the whole sample space M, we have the classical non-sequential case. If $n(x, D_0)$ is not constant, we shall say that D_0 is a sequential decision function.

In the remainder of this section we shall sketch briefly some of the fundamental notions of the theory without regard to regularity conditions. The latter will be discussed in the next section.

In [1] a weight function $W(G, d)$ was introduced which expresses the loss suffered by the statistician when G is the true distribution of X and the decision d is made. Let $c(n)$ denote the cost of making n observations; i.e., $c(n)$ is the cost of observing the values of $X_1, \cdots, X_n$. Then, if the decision function $d = D(x)$ is adopted and G is the true distribution of X, the expected value of the loss due to possible erroneous decisions plus the expected cost of experimentation is given by

$$r(G, D) = \int_M W[G, D(x)]\, dG(x) + \int_M c[n(x, D)]\, dG(x). \tag{1.1}$$

The above expression is called the risk when D is the decision function adopted and G is the true distribution.

Let ξ be an a priori probability distribution on Ω; i.e., ξ is a probability measure defined over a suitably chosen Borel field[2] of subsets of Ω. Then the expected value of $r(G, D)$ is given by

$$r(\xi, D) = \int_\Omega r(G, D)\, d\xi. \tag{1.2}$$

---

[2] A Borel field is an aggregate of sets such that a) the null set is a member of the field, b) the complement with respect to the entire space (here M) is a member of the field, c) the sum of denumerably many members of the field is itself in the field.

The above expression is called the risk when ξ is the a priori distribution on Ω and D is the decision function adopted.

We shall say that the decision function D_0 is a Bayes solution relative to the a priori distribution ξ if

$$r(\xi, D_0) \leqq r(\xi, D) \text{ for all } D. \tag{1.3}$$

If there existed an a priori distribution on Ω and if this distribution were known, we could put ξ equal to this a priori distribution and a Bayes solution relative to ξ would provide a very satisfactory solution of the decision problem. In most applications, however, not even the existence of an a priori distribution can be postulated. Nevertheless, the study of Bayes solutions corresponding to various a priori distributions is of great interest in view of some results given in [1]. It was shown in [1] that under rather general conditions the class C of the Bayes solutions corresponding to all possible a priori distributions ξ has the following property: If D_1 is a decision function that is not an element of C, there exists a decision function D_2 in C such that

$$r(G, D_2) \leqq r(G, D_1) \text{ for all } G \tag{1.4}$$

and

$$r(G, D_2) < r(G, D_1) \text{ for at least one } G. \tag{1.5}$$

It was furthermore shown in [1] that under general conditions a minimax solution D_0 of the decision problem is also a Bayes solution corresponding to some a priori distribution ξ. By a minimax solution we mean a decision function D_0 such that, for all D

$$\operatorname*{Sup}_G r(G, D_0) \leqq \operatorname*{Sup}_G r(G, D). \tag{1.6}$$

2. Regularity conditions and other assumptions. We shall make the following assumptions:

ASSUMPTION 1. *The chance variables are identically and independently distributed. The common distribution is either discrete or absolutely continuous.*

Let $p(a \mid F)$ denote the elementary probability law of X_i when F is the distribution of X_i; i.e., when F is discrete, $p(a \mid F)$ is the probability that $X_i = a$, and when F is absolutely continuous, $p(a \mid F)$ is the probability density of X_i at a.

In the space M of sequences x let B be the smallest Borel field which contains all sets of points x which are defined by the relations

$$x_i < a_i \qquad i = 1, 2, \cdots \text{ ad inf.},$$

where the a_i are real numbers or $+\infty$. Each admissible[3] F induces a probability measure $F^*(B)$ on M; the totality of these probability measures is Ω. Let H^*

[3] An F or F* is admissible if F* is in Ω.

be a given Borel field of subsets of Ω. The only subsets of Ω which we shall discuss in this paper will be members of H^*, and all probability measures on Ω which we shall discuss will be measurable (H^*). This will henceforth be assumed without further repetition.

Let A^* be any set in H^*, and A the set of F which corresponds to the F^* in A^*. The sets A form a Borel field, say H. By definition, the probability measure of a set A according to a probability measure $\xi(H^*)$ on Ω is to be the same as the probability measure of A^* according to ξ.

Let $M \times \Omega$ be the Cartesian product of M and Ω ([5], page 82), and K be the smallest Borel field of subsets of $M \times \Omega$ which contains the Cartesian product of any member of B by any member of H^*.

For a given decision function $d = D(x)$, $W(F, D(x))$ is a function of F and x. Hereafter in this paper we shall limit ourselves to functions $D(x)$ such that $W(F, D(x))$ is measurable (K), and $n(x, D)$ is measurable (B).

It is true that in Section 1, W was given as a function of G, the distribution of X. Because of Assumption 1, $G = F^*$, and there is a one-to-one correspondence between F and F^*. Thus we may, in appropriate places, interchange them freely.

Assumption 2. *For every real a, except possibly on a Borel set*[4] *whose probability is zero according to every admissible F,* $p(a \mid F)$ *exists and is a function of a and F which is measurable (K). If the admissible distributions F are discrete, there exists a fixed sequence* $\{b_i\}$ $(i = 1, 2, \cdots, ad\ inf.)$ *of real values such that* $\sum_{i=1}^{\infty} p(b_i \mid F) = 1$ *for all admissible F.*

Assumption 3. *$W(F, d)$ is bounded. For every d in D^*, $W(F, d)$ is a function of F which is measurable (H).*

In what follows ξ will always denote a probability measure (H^*) on Ω. Thus

$$W(\xi, d) = \int_{\Omega} W(F, d)\, d\xi$$

exists.

Assumption 4. *The function* $c(n) = cn$. *Without loss of generality we may take* $c = 1$, *so that* $c(n) = n$.

We shall introduce the following convergence definition in the space D^*: the sequence $\{d_i\}$ converges to d_0 if

$$\lim_{i \to \infty} W(F, d_i) = W(F, d_0)$$

uniformly in the admissible F's.

Assumption 5. *The space D^* is compact in the sense of the above convergence definition.*

One can easily verify that, if $\lim_{i \to \infty} d_i = d_0$, then

$$\lim_{i \to \infty} W(\xi, d_i) = W(\xi, d_0);$$

[4] A Borel set is a member of the smallest Borel field which contains all the open sets of the real line.

i.e., $W(\xi, d)$ is a continuous function of d. Thus, because of Assumption 5, the minimum of $W(\xi, d)$ with respect to d exists.

We shall now show that, under the above conditions

$$\int_M W[F^*, D(x)]\, dF^*(x) \tag{2.1}$$

exists and is a function of F^* measurable (H^*). For any j let R_j be the set in B such that $n(x, D) = j$. Then it is enough to show that, for any j,

$$\int_{R_j} W[F^*, D(x)]\, dF^*(x) \tag{2.2}$$

exists and is a function of F^* measurable (H^*).

In the discrete case, the integral (2.2) is equal to the sum[5]

$$\sum_{(x_1,\ldots,x_j)\in R_j} W[F^*, D(x)]p(x_1 \mid F) \cdots p(x_j \mid F). \tag{2.3}$$

For fixed values of $x_1, \cdots, x_j$, the expression under the summation sign is obviously a function of F^* measurable (H^*). Since, because of Assumption 2, there are only countably many points $(x_1, \cdots, x_j)$ in R_j, the sum (2.3) must be a function of F^* measurable (H^*).

In the absolutely continuous case, the integral (2.2) is equal to (2.4)

$$\int_{R_j} W[F^*, D(x)] \prod_{i=1}^{j} p(x_i \mid F)\, d\nu(j) \tag{2.4}$$

where $\nu(j)$ is Borel measure in the j-dimensional Euclidean space. The integrand is measurable (K). Hence, the integral (2.4) exists and is a function of F^* measurable (H^*) (see [5], Chapter III, Theorems 9.3 and 9.8).

3. Some results concerning Bayes solutions. If ξ is the a priori probability measure on Ω, the a posteriori probability of a subset ω of Ω for given values $x_1, \cdots, x_m$ of the first m chance variables is given by

$$\xi(\omega \mid \xi, x_1, \cdots, x_m) = \frac{\int_\omega p(x_1 \mid F) \cdots p(x_m \mid F)\, d\xi}{\int_\Omega p(x_1 \mid F) \cdots p(x_m \mid F)\, d\xi} \tag{3.1}$$

Let

$$\rho_0(\xi) = \operatorname*{Min}_d W(\xi, d). \tag{3.2}$$

For any positive integral value m, let $\rho_m(\xi)$ denote the infimum of $r(\xi, D)$ with respect to D where D is restricted to decision functions for which $n(x, D) \leq m$ for all x. For any positive integer m, let $d = D^m(x)$ denote a decision function

[5] Because of the definition of R_j we may, in the expressions (2.3) and (2.4), proceed as if R_j were a Borel set in j-dimensional Euclidean space.

D for which $n(x, D) \leqq m$ for all x. Thus, we can write

$$\rho_m(\xi) = \operatorname*{Inf}_{D^m} r(\xi, D^m) \quad (m = 1, 2, \cdots, \text{ad inf.}). \tag{3.3}$$

Let

$$\rho(\xi) = \operatorname*{Inf}_{D} r(\xi, D). \tag{3.4}$$

We shall first prove several theorems concerning the functions $\rho_0(\xi)$, $\rho_m(\xi)$, and $\rho(\xi)$.

THEOREM 3.1. *The following recursion formula holds:*[6]

$$\rho_{m+1}(\xi) = \text{Min}\left[\rho_0(\xi), 1 + \int_{-\infty}^{\infty} \rho_m(\xi_a)\, p(a \mid \xi)\, da\right] \tag{3.5}$$

$$(m = 0, 1, 2, \cdots, \text{ad inf.})$$

where

$$\xi_a(\omega) = \xi(\omega \mid \xi, a) \text{ and } p(a \mid \xi) = \int_\Omega p(a \mid F)\, d\xi. \tag{3.6}$$

PROOF: Let $\rho_m^*(\xi)$ ($m = 1, 2, \cdots$, ad inf.) denote the infimum of $r(\xi, D)$ with respect to D where D is subject to the restriction that $n(x, D) \geqq 1$ and $\leqq m$ for all x. Clearly,

$$\rho_{m+1}(\xi) = \text{Min}[\rho_0(\xi), \rho_{m+1}^*(\xi)]. \tag{3.7}$$

Let $\rho_m^*(\xi \mid a)$ denote the infimum with respect to D of the conditional risk (conditional expected value of $W[F, D(x)] + n(x, D)$) when the first observation x_1 on X_1 is a and D is restricted to decision functions for which $n(x, D) \geqq 1$ and $\leqq m$ for all x. Let $\bar{D}(m)$ be the temporary generic designation of such a decision function. Let $\bar{D}(m \mid a)$ be the decision function which is obtained from $\bar{D}(m)$ when the first observation is a. Finally let $r(\xi, D \mid a)$ be the conditional risk when the a priori distribution function is ξ, D is the decision function and requires at least one observation, and the first observation is a. We then have that

$$r(\xi, \bar{D}(m + 1) \mid a) = r(\xi_a, \bar{D}(m + 1 \mid a)) + 1.$$

Hence

$$\rho_{m+1}^*(\xi \mid a) = \rho_m(\xi_a) + 1. \tag{3.8}$$

The unconditional quantity $\rho_{m+1}^*(\xi)$ must clearly be equal to the average value of the infimum of the conditional risk. Thus we have

$$\rho_{m+1}^*(\xi) = \int_{-\infty}^{\infty} \rho_{m+1}^*(\xi \mid a) p(a \mid \xi) da. \tag{3.9}$$

---

[6] If the distribution of X is discrete, the integration with respect to a is to be replaced by summation with respect to a. This remark refers also to subsequent formulas.

Equation (3.5) follows from (3.7), (3.8) and (3.9).

THEOREM 3.2. *The function $\rho(\xi)$ satisfies the following equation:*

$$\rho(\xi) = \text{Min}\left[\rho_0(\xi), \int_{-\infty}^{\infty} \rho(\xi_a)p(a \mid \xi)\,da + 1\right]. \tag{3.10}$$

The proof of this theorem is omitted, since it is essentially the same as that of Theorem 3.1.

THEOREM 3.3.[7] *The following inequalities hold:*

$$0 \leqq \rho_m(\xi) - \rho(\xi) \leqq \frac{W_0^2}{m} \qquad (m = 1, 2, \cdots, \text{ad inf.}) \tag{3.11}$$

where W_0 is the least upper bound of $W(F, d)$.

PROOF: Let $\{D_i\}$ ($i = 1, 2, \cdots$, ad inf.) be a sequence of decision functions such that

$$\lim_{i \to \infty} r(\xi, D_i) = \rho(\xi). \tag{3.12}$$

Let, furthermore, $P_i(\xi)$ denote the probability that at least m observations will be made when D_i is the decision function adopted and ξ is the a priori probability measure on Ω. Since $\rho(\xi) \leqq W_0$ and since

$$r(\xi, D_i) \geqq mP_i(\xi), \tag{3.13}$$

it follows from (3.12) that

$$\limsup_{i \to \infty} P_i(\xi) \leqq \frac{W_0}{m}. \tag{3.14}$$

Let D_i^m be the decision function obtained from D_i as follows: $D_i^m(x) = D_i(x)$ for all x for which $n(x, D_i) \leqq m$. $D_i^m(x)$ is equal to a fixed element d_0 for all x for which $n(x, D_i) > m$.[8]
Clearly,

$$r(\xi, D_i^m) \leqq r(\xi, D_i) + P_i(\xi)W_0. \tag{3.15}$$

From (3.12), (3.14) and (3.15) it follows that

$$\limsup_{i \to \infty} r(\xi, D_i^m) \leqq \rho(\xi) + \frac{W_0^2}{m}. \tag{3.16}$$

Since $\rho_m(\xi)$ cannot exceed the left hand member of (3.16), the second half of (3.11) follows from (3.16). The first half of (3.11) is obvious.

---

[7] This theorem is essentially the same as Lemma 2.1 in [6].

[8] We verify that $W(F, D_i^m)$ is measurable (K), as follows: Consider the set V of couples (F, x) such that $W(F, D_i^m(x)) < c$, where c is some real constant. We want to show that $V \epsilon K$. For this purpose let V_0 be the set of couples (F, x) such that $W(F, D_i(x)) < c$. Then $V_0 \epsilon K$. Let V_1 be the set of x's such that $n(x, D_i) \leqq m$. Then $V_1 \epsilon B$, $(\Omega \times V_1) = V_2 \epsilon K$, $V_0V_2 \epsilon K$. Let $V_3 = M - V_1$. For every $x \epsilon V_3$ we have $W(F, D_i^m(x)) = W(F, d_0)$. Let V_4 be the set of F's such that $W(F, d_0) < c$. Then $V_4 \epsilon H$ by Assumption 3. Finally we have $V = V_0V_2 + V_4 \times V_3$, so that $V \epsilon K$.

The immediate consequence of Theorem 3.3 is the relation[9]

$$\lim_{m\to\infty} \rho_m(\xi) = \rho(\xi). \tag{3.17}$$

THEOREM 3.4. *If ξ_1 and ξ_2 are two probability measures on Ω such that*[10]

$$\frac{\xi_1(\omega)}{\xi_2(\omega)} \leqq 1 + \epsilon \text{ for all } \omega, \tag{3.18}$$

then

$$\rho(\xi_1) \leqq (1 + \epsilon)\rho(\xi_2). \tag{3.19}$$

PROOF: It follows from (3.18) that

$$r(\xi_1, D) \leqq (1 + \epsilon)r(\xi_2, D) \text{ for all } D. \tag{3.20}$$

Hence, (3.19) must hold.

The above theorem permits the computation of a simple and in many cases useful lower bound of $\int_{-\infty}^{\infty} \rho(\xi_a)p(a \mid \xi)\, da$ as follows:

For any real value a, let ϵ_a be a non-negative value (not necessarily finite) determined such that

$$\frac{\xi(\omega)}{\xi_a(\omega)} \leqq 1 + \epsilon_a \text{ for all } \omega. \tag{3.21}$$

Then

$$\int_{-\infty}^{\infty} \rho(\xi_a)\, p(a \mid \xi)\, da \geqq \int_{-\infty}^{\infty} \frac{\rho(\xi)}{1 + \epsilon_a}\, p(a \mid \xi)\, da = \rho(\xi) \int_{-\infty}^{\infty} \frac{p(a \mid \xi)}{1 + \epsilon_a}\, da. \tag{3.22}$$

Since $\epsilon_a \geqq 0$ and since $\rho_0(\xi) \geqq \rho(\xi)$, we obviously have

$$\rho(\xi) \int_{-\infty}^{\infty} \frac{p(a \mid \xi)}{1 + \epsilon_a}\, da \geqq \rho(\xi) - \left[1 - \int_{-\infty}^{\infty} \frac{p(a \mid \xi)}{1 + \epsilon_a}\, da\right]\rho_0(\xi). \tag{3.23}$$

Hence, we obtain the inequality

$$\int_{-\infty}^{\infty} \rho(\xi_a)\, p(a \mid \xi)\, da \geqq \rho(\xi) - \rho_0(\xi)\left[1 - \int_{-\infty}^{\infty} \frac{p(a \mid \xi)}{1 + \epsilon_a}\, da\right]. \tag{3.24}$$

An upper bound of the left hand member in (3.24) is obtained by replacing ρ by ρ_0; i.e.,

$$\int_{-\infty}^{\infty} \rho(\xi_a)p(a \mid \xi)\, da \leqq \int_{-\infty}^{\infty} \rho_0(\xi_a)p(a \mid \xi)\, da. \tag{3.25}$$

[9] A proof of (3.17) is contained implicitly in the work of Arrow, Blackwell and Girshick ([2], Section 1.3).

[10] The left member of (3.18) is defined to be equal to 1 when $\xi_1(\omega) = \xi_2(\omega) = 0$.

The bounds given in (3.24) and (3.25) may be useful in constructing Bayes solutions, since the following theorem holds:

THEOREM 3.5. *If*

$$\rho_0(\xi) > \int_{-\infty}^{\infty} \rho_0(\xi_a)p(a \mid \xi)\, da + 1, \tag{3.26}$$

then $\rho(\xi) < \rho_0(\xi)$. *If*

$$\rho_0(\xi)\left[1 - \int_{-\infty}^{\infty} \frac{p(a \mid \xi)}{1 + \epsilon_a}\, da\right] < 1, \tag{3.27}$$

then $\rho(\xi) = \rho_0(\xi)$.

The above theorem is an immediate consequence of (3.10), (3.24) and (3.25).

A decision procedure relative to a given a priori probability measure ξ_0 will be given with the help of the function $\rho(\xi)$ as follows: If $\rho(\xi_0) = \rho_0(\xi_0)$, take a final decision d for which $W(\xi_0, d)$ is minimized. If $\rho(\xi_0) < \rho_0(\xi_0)$, take an observation on X_1 and compute the a posteriori probability measure ξ_1. If $\rho(\xi_1) = \rho_0(\xi_1)$, stop experimentation with a final decision d for which $W(\xi_1, d)$ is minimized. If $\rho(\xi_1) < \rho_0(\xi_1)$, take an observation on X_2 and compute the a posteriori probability measure ξ_2 corresponding to the observed values of X_1 and X_2, and so on. The above decision procedure will be shown later to be a Bayes solution. Theorem 3.5 permits one to decide whether $\rho(\xi) < \rho_0(\xi)$ or $= \rho_0(\xi)$ whenever ξ satisfies (3.26) or (3.27). Theorem 3.5 will be useful when the class of all ξ's for which neither (3.26) nor (3.27) holds is small.

For the purposes of the next theorem let $\hat{D}$ designate the decision procedure described in the preceding paragraph. (We shall shortly show that $\hat{D}$ is a decision function in the sense of our definition.)

Let $\hat{D}^0$ be the decision procedure where the first observation is taken and then one proceeds according to $\hat{D}$.

We shall now prove that $\hat{D}$ and $\hat{D}^0$ are Bayes solutions. More precisely, we shall prove the following theorem:[11]

THEOREM 3.6. *For any* ξ, $\hat{D}$ *and* $\hat{D}^0$ *as defined above are decision functions. Let* D *be any decision function for which* $n(x, D) \geqq 1$ *and let*

$$\rho^*(\xi) = \operatorname*{Inf}_D r(\xi, D).$$

Then

$$r(\xi, \hat{D}) = \rho(\xi)$$

and

$$r(\xi, \hat{D}^0) = \rho^*(\xi).$$

[11] This theorem follows also from some earlier more general existence theorems ([6], Theorems 2.4 and 3.3). (See also [4], Lemma 1.) The validity of Theorem 3.6 was proved also by Arrow, Blackwell and Girshick [2].

In view of this theorem, the operation "infimum with respect to D" in the definitions of $\rho(\xi)$, and $\rho^*(\xi)$ can be replaced by "minimum with respect to D."

First we shall establish the measurability properties of $\hat{D}$ and $\hat{D}^0$. Since the proofs are similar, we restrict ourselves to consideration of $\hat{D}$. Let $\xi_{x_1,\ldots,x_m}$ be the a posteriori distribution (3.1). From the (B) measurability of $\rho_0(\xi_{x_1,\ldots,x_m})$ and $\rho(\xi_{x_1,\ldots,x_m})$ it follows easily that $n(x, \hat{D})$ is measurable (B). It remains to prove that $W(F, \hat{D}(x))$ is measurable (K). For this purpose, let $L^i = (d_1^i, \cdots, d_{k_i}^i)$ be a sequence $\frac{1}{i}$ dense in D^*, i.e., for any $d \in D^*$ there exists a $g \in D^*$ such that $g \in L^i$ and $| W(F, d) - W(F, g) | < \frac{1}{i}$ uniformly in F. (The existence of such a sequence follows from Assumption 5.) Let now $D_i(x)$ be a decision function defined as follows:

$$n(x, D_i) = n(x, \hat{D}).$$

Suppose $n(x, \hat{D}) = m$ when the observations are $x_1, \cdots, x_m$. We define $D_i(x)$ to be such that $D_i(x)$ is an element of L^i and

$$W(\xi_{x_1,\ldots,x_m}, D_i(x)) = \min_{d \in L^i} W(\xi_{x_1,\ldots,x_m}, d), \tag{3.28}$$

i.e., $D_i(x)$ takes the minimizing value of d. For any fixed d, the set of x's satisfying the equation $D_i(x) = d$ is without difficulty shown to be (B) measurable. Since $D_i(x)$ assumes only a finite number of values in D^*, it follows from Assumption 3 that $W(F, D_i(x))$ is measurable (K). Now

$$\lim_{i \to \infty} W(F, D_i(x)) = W(F, \hat{D}(x)),$$

so that $W(F, \hat{D}(x))$ is measurable (K).

We shall now prove that $\hat{D}$ is a Bayes solution, i.e., that

$$\rho(\xi) = r(\xi, \hat{D}). \tag{3.29}$$

In a similar way it can be proved that

$$\rho^*(\xi) = r(\xi, \hat{D}^0). \tag{3.30}$$

If $\rho_0(\xi) = \rho(\xi)$, there can be no better decision function (from the point of view of reducing the risk) than $\hat{D}$, i.e., $\hat{D}$ is a Bayes solution. Suppose then that

$$\rho_0(\xi) > \rho(\xi). \tag{3.31}$$

If (3.31) holds and $\hat{D}$ is not a Bayes solution, there exists a decision function $\bar{D}_1$ such that

$$r(\xi, \bar{D}_1) < r(\xi, \hat{D}) \tag{3.32}$$

and

$$r(\xi, \bar{D}_1) < \frac{\rho_0(\xi) + \rho(\xi)}{2}. \tag{3.33}$$

Now $\bar{D}_1$ must require that at least one observation be taken, else (3.33) could not hold. Thus $\hat{D}$ and $\bar{D}_1$ both require at least one observation.

Suppose one observation is taken. Let $r(\xi, D \mid a)$ denote the conditional risk of proceeding according to D when ξ is the a priori distribution and a is the first observation. For a given D we have that $r(\xi, D \mid a)$ is a function only of ξ_a. In particular $r(\xi, \hat{D} \mid a)$ and $r(\xi, \bar{D}_1 \mid a)$ are functions only of ξ_a.

We can now apply to $r(\xi, \hat{D} \mid a)$ and $r(\xi, \bar{D}_1 \mid a)$ the same argument that was applied above to $r(\xi, \hat{D})$ and $r(\xi, \bar{D}_1)$, and conclude again as follows: whenever $\rho_0(\xi_a) = \rho(\xi_a)$ (when one takes no more observations according to $\hat{D}$), taking additional observations cannot diminish the conditional risk below $r(\xi, \hat{D} \mid a)$ ($\bar{D}_1$ may require an additional observation without having

$$r(\xi, \bar{D}_1 \mid a) > r(\xi, \hat{D} \mid a).$$

This can happen when $\rho_0(\xi_a) = \rho^*(\xi_a)$). Whenever $\rho_0(\xi_a) > \rho(\xi_a)$ (when $\hat{D}$ requires us to take another observation) two cases may occur: either a) $\bar{D}_1$ requires us to take another observation, in which case its decision is the same as that of $\hat{D}$, or b) $\bar{D}_1$ requires us to stop taking observations. There exists then another decision function whose conditional risk is less than

$$\frac{\rho_0(\xi_a) + \rho(\xi_a)}{2} + 1.$$

Both this decision function and $\hat{D}$ require that another observation be taken. We conclude that up to and including the first observation, $\hat{D}$ coincides either with $\bar{D}_1$ or with another decision function $\bar{D}_2$ whose risk is not greater than that of $\bar{D}_1$.

We continue in this manner for 2, 3, $\cdots$ observations. The above argument is always valid because of Assumption 4 and because the past history of the process (the sequence of observations) enters only through the a posteriori probability. Thus we conclude that for any positive integer k there exists a decision function $\bar{D}_k$ such that up to and including the k-th observation $\hat{D}$ gives the same decision as $\bar{D}_k$ and the risk corresponding to $\bar{D}_k$ does not exceed the risk corresponding to $\bar{D}_1$. Since $\lim_{k\to\infty} r(\xi, \bar{D}_k) \geqq r(\xi, \hat{D})$, (3.32) cannot hold. Hence (3.29) holds and $\hat{D}$ is a Bayes solution.

For any probability measure ξ on Ω one of the following three conditions must hold:

(1) $\text{Min}_d\, W(\xi, d) < r(\xi, D)$ for any D for which $n(x, D) \geqq 1$.

(2) $\text{Min}_d\, W(\xi, d) \leqq r(\xi, D)$ for all D for which $n(x, D) \geqq 1$, and the equality sign holds for at least one D with $n(x, D) \geqq 1$.

(3) There exists a D with $n(x, D) \geqq 1$ such that $\text{Min}_d\, W(\xi, d) > r(\xi, D)$.

In view of Theorem 3.6, the conditions (1), (2) and (3) can be expressed by: (1) $\rho_0(\xi) < \rho^*(\xi)$, (2) $\rho_0(\xi) = \rho^*(\xi)$ and (3) $\rho_0(\xi) > \rho^*(\xi)$, respectively.

We shall say that a probability measure ξ on Ω is of the first type if it satisfies (1), of the second type if it satisfies (2), and of the third type if it satisfies (3). Since the a posteriori probability defined in (3.1) is also a probability measure on Ω, any a posteriori probability measure will be one of the three types mentioned above.

We shall now prove the following characterization theorem:

THEOREM 3.7.[12] *A necessary and sufficient condition for a decision function $d = D_0(x)$ to be a Bayes solution relative to a given a priori distribution ξ_0 is that the following three relations be fulfilled for any sample point x, except perhaps on a set whose probability measure is zero when ξ_0 is the a priori distribution in Ω:*

(a) *For any $m < n(x, D_0)$, the a posteriori distribution $\xi(\omega \mid \xi_0, x_1, \cdots, x_m)$ is either of the second or of the third type,*

(b) *For $m = n(x, D_0)$, the a posteriori distribution $\xi(\omega \mid \xi_0, x_1, \cdots, x_m)$ is either of the first or the second type,*

(c) *For $m = n(x, D_0)$, we have*

$$\operatorname*{Min}_d W(\xi_{x_1,\dots,x_m}, d) = W(\xi_{x_1,\dots,x_m}, D_0(x))$$

where $\xi_{x_1,\dots,x_m}$ stands for an a priori distribution that is equal to the a posteriori distribution corresponding to $\xi_0, x_1, \cdots, x_m$.

PROOF: We shall omit the proof of the sufficiency of the conditions (a), (b) and (c), since it is essentially the same as that of Theorem 3.6. To prove the necessity of these conditions, let $d = D_0(x)$ be a decision function and let M^* denote the set of all sample points x for which at least one of the relations (a), (b) and (c) is violated. First, we shall show tht M^* is a set measurable (B). Let M_1^* be the set of all x's for which (a) is violated, M_2^* the set of all x's for which (b) is violated, and M_3^* the set of all x's for which (c) is violated. Clearly, M^* is shown to be measurable (B) if we can show that $M_i^* (i = 1, 2, 3)$ is measurable (B). Let $M_{ir}^* (r = 1, 2, \cdots, \text{ad inf})$ denote the subset of M_i^* for which the first violation of the corresponding condition occurs for the sample $x_1, \cdots, x_r$. We merely have to show that M_{ir}^* is measurable (B) for all i and r. The measurability of M_{3r}^* follows from the fact that $\operatorname{Min}_d W(\xi_{x_1,\dots,x_r}, d)$ and

$$W[\xi_{x_1,\dots,x_r}, D_0(x)]$$

are functions of x measurable (B). To show the measurability of M_{1r}^* and M_{2r}^*, it is sufficient to show that the set of all samples $x_1, \cdots, x_r$ for which $\xi_{x_1,\dots,x_r}$ is of type $i (i = 1, 2, 3)$ is measurable (B). But this follows from the fact that $\rho_0(\xi_{x_1,\dots,x_r})$ and $\rho^*(\xi_{x_1,\dots,x_r})$ are functions of $(x_1, \cdots, x_r)$ measurable (B). Hence, M^* is proved to be measurable (B).

For any x in M^* let $m(x)$ be the smallest positive integer such that at least one of the relations (a), (b) and (c) is violated for the finite sample

$$x_1, x_2, \cdots, x_{m(x)}.$$

Clearly, if x is a point in M^*, then also any sample point y is in M^* for which $y_1 = x_1, \cdots, y_{m(x)} = x_{m(x)}$. Let x^0 be any particular sample point in M^* and let $r(\xi_0, D_0, x_1^0, \cdots, x_{m(x^0)}^0)$ denote the conditional risk when ξ_0 is the a priori

[12] See also the proof of Lemma 1 in [4].

distribution in Ω, D_0 is the decision function adopted and the first $m(x^0)$ observations are equal to $\overset{0}{x}_1, \cdots, \overset{0}{x}_{m(x^0)}$, respectively; i.e., $r(\xi_0, D_0, \overset{0}{x}_1, \cdots, \overset{0}{x}_{m(x^0)})$ is the conditional expected value of $W(F, D_0(x)) + n(x, D_0)$, when ξ_0 is the a priori distribution in Ω, D_0 is the decision function adopted and $\overset{0}{x}_1, \cdots, \overset{0}{x}_{m(x^0)}$ are the first $m(x^0)$ observations.

Let $D_1(x)$ be the decision function determined as follows: for any x not in M^* we put $D_1(x) = D_0(x)$. For any x in M^*, let $n(x_1, D_1)$ be equal to the smallest integer $n(x) \geqq m(x)$ for which

$$\rho_0(\xi_{x_1,\ldots,x_{n(x)}}) = \rho(\xi_{x_1,\ldots,x_{n(x)}})$$

and the value of $D_1(x)$ is determined so that condition (c) of our theorem is fulfilled. Since, for any positive integer m, the subset of M^* where $m(x) = m$ is (B) measurable, $D_1(x)$ has the proper measurability properties. Applying Theorem 3.6, we see that

$$r(\xi_0, D_1, x_1, \cdots, x_{m(x)}) = \rho(\xi_{x_1,\ldots,x_{m(x)}}) \tag{3.34}$$

for any x in M^*. On the other hand, since D_0 violates at least one of the conditions (a), (b), and (c) at every point x in M^*, we have

$$r(\xi_0, D_0, x_1, \cdots, x_{m(x)}) > \rho(\xi_{x_1,\ldots,x_{m(x)}}) \tag{3.35}$$

for every x in M^*. If the probability measure of M^* is positive when ξ_0 is the a priori probability measure, the above two relations imply that

$$r(\xi_0, D_0) > r(\xi_0, D_1).$$

Thus, D_0 is not a Bayes solution and the proof of Theorem 3.7 is complete.

We shall now prove the following continuity theorem.[13]

THEOREM 3.8. *Let $\{\xi_i\}$ $(i = 0, 1, 2, \cdots, ad\ inf.)$ be a sequence of probability measures on Ω such that*

$$\lim_{i=\infty} \frac{\xi_i(\omega)}{\xi_0(\omega)} = 1 \text{ uniformly in } \omega. \tag{3.36}$$

Then

$$\lim_{i=\infty} \rho(\xi_i) = \rho(\xi_0). \tag{3.37}$$

PROOF: It follows from (3.36) that for any $\epsilon > 0$, we have for almost all values i

$$\frac{\xi_i(\omega)}{\xi_0(\omega)} < 1 + \epsilon \text{ and } \frac{\xi_0(\omega)}{\xi_i(\omega)} < 1 + \epsilon \text{ for all } \omega. \tag{3.38}$$

Our theorem is an immediate consequence of (3.38) and Theorem 3.4.

[13] A proof of this theorem for finite Ω was given by G. W. Brown and is included in [2]. See also Lemma 3 in [4].

A stronger continuity theorem is the following:

THEOREM 3.8.1. *Let* $\{\xi_i\}$, $(i = 0, 1, 2, \cdots, ad\ inf.)$ *be a sequence of probabitity measures on* Ω *such that*

$$\lim_{i \to \infty} \xi_i(\omega) = \xi_0(\omega)$$

uniformly in ω. *Then* (3.37) *holds.*

PROOF: It follows from (3.11) that

$$\lim_{m \to \infty} \rho_m(\xi) = \rho(\xi)$$

uniformly in ξ. Hence it is sufficient to prove that, under the conditions of the theorem,

$$\lim_{i \to \infty} \rho_m(\xi_i) = \rho_m(\xi_0)$$

for any m. Let $D^m(x)$ denote a decision function for which $n(x, D^m) \leqq m$ for all x. It follows that, for a fixed m, $r(F, D^m)$ is bounded, uniformly in F and D^m (Assumptions 3 and 4). From the hypothesis on $\{\xi_i\}$ it then follows that

$$\lim_{i \to \infty} r(\xi_i, D^m) = r(\xi_0, D^m)$$

uniformly in D^m. From this the desired result follows readily.

A class C of probability measures ξ on Ω will be said to be convex if for any two elements ξ_1 and ξ_2 of C and for any positive value $\lambda < 1$, the probability measure $\xi = \lambda\xi_1 + (1 - \lambda)\xi_2$ is an element of C.

For any element d_0 of D, let C_{i,d_0} denote the class of all probability measures ξ of type i $(i = 1, 2, 3)$ for which $W(\xi, d_0) = \operatorname{Min}_d W(\xi, d)$. Let C_d denote the set-theoretical sum of $C_{1,d}$ and $C_{2,d}$. We shall now prove the following theorem.

THEOREM 3.9. *For any element* d, *the classes* $C_{1,d}$ *and* C_d *are convex.*[14]

Let ξ_1 and ξ_2 be two elements of $C_{1,d}$. Then for any decision function $D(x)$ which requires at least one observation we have

$$W(\xi_1, d) < r(\xi_1, D) \text{ and } W(\xi_2, d) < r(\xi_2, d). \tag{3.39}$$

Let $\xi = \lambda\xi_1 + (1 - \lambda)\xi_2$ where λ is a positive number <1. Clearly,

$$W(\xi, d) = \lambda W(\xi_1, d) + (1 - \lambda) W(\xi_2, d) \tag{3.40}$$

and

$$r(\xi, D) = \lambda r(\xi_1, D) + (1 - \lambda) r(\xi_2, D). \tag{3.41}$$

From (3.39), (3.40) and (3.41) we obtain

$$W(\xi, d) < r(\xi, D) \quad \text{and} \quad W(\xi, d) = \operatorname{Min}_{d^*} W(\xi, d^*). \tag{3.42}$$

Hence ξ is an element of $C_{1,d}$ and the convexity of $C_{1,d}$ is proved. The convexity of C_d can be proved in the same way by replacing $<$ by $\leqq$ in (3.39) and (3.42).

---

[14] See also Lemma 2 in [4].

We shall say that a set L of probability measures ξ is a linear manifold if for any two elements ξ_1 and ξ_2 of L, $\xi = \alpha\xi_1 + (1 - \alpha)\,\xi_2$ is also an element of L for any real value α for which $\alpha\xi_1 + (1 - \alpha)\,\xi_2$ is a probability measure. A linear manifold L will be said to be tangent to C_d if the intersection of L and $C_{2,d}$ is not empty, but the intersection of L and $C_{1,d}$ is empty.

For any decision function $D(x)$ and for any element d of D^*, let $L(D, d)$ denote the linear manifold consisting of all ξ which satisfy the equation

$$W(\xi, d) = r(\xi, D). \tag{3.43}$$

THEOREM 3.10. *Let ξ_0 be an element of $C_{2,d}$ and let $D_0(x)$ be a decision function that requires at least one observation and is such that $W(\xi_0, d) = r(\xi_0, D_0)$. Then the linear manifold $L(D_0, d)$ is tangent to C_d.*

PROOF: ξ_0 is obviously an element of $L(D_0, d)$. Thus the intersection of $L(D_0, d)$ and $C_{2,d}$ is not empty. For any element ξ_1 of $C_{1,d}$ we have $W(\xi_1, d) < r(\xi_1, D)$ for any D that requires at least one observation. Hence, $W(\xi_1, d) < r(\xi_1, D_0)$ and, therefore, ξ_1 cannot be an element of $L(D_0, d)$. This proves our theorem.

4. Applications to the case where Ω and D^* are finite. In this section we shall apply the general results of the preceding section to the following special case: the space Ω consists of a finite number of elements, $F_1, \cdots, F_k$ (say), and the space D^* consists of the elements $d_1, \cdots, d_k$ where d_i denotes the decision to accept the hypothesis H_i that F_i is the true distribution. Let

$$W(F_i, d_j) = W_{ij} = 0 \text{ for } i = j \text{ and } >0 \text{ for } i \neq j. \tag{4.1}$$

It will be sufficient to discuss the cases $k = 2$ and $k = 3$, since the extension to $k > 3$ will be obvious. We shall first consider the case $k = 2$. In this case any a priori distribution ξ is represented by two numbers g_1 and g_2 where g_i is the a priori probability that F_i is true ($i = 1, 2$). Thus, $g_i \geqq 0$ and $g_1 + g_2 = 1$. Let ξ_i denote the a priori distribution corresponding to $g_i = 1$ ($i = 1, 2$). Clearly C_{d_1} contains ξ_1 but not ξ_2, and C_{d_2} contains ξ_2 but not ξ_1. Because of Theorems 3.9 and 3.7, C_{d_1} and C_{d_2} are closed and convex. Furthermore, we obviously have

$$g_2 W_{21} \leqq g_1 W_{12} \text{ for all } \xi \text{ in } C_{d_1} \tag{4.2}$$

and

$$g_2 W_{21} \geqq g_1 W_{12} \text{ for all } \xi \text{ in } C_{d_2}. \tag{4.3}$$

Let $\xi_0 = (g_1^0, g_2^0)$ be the a priori distribution for which

$$g_2^0 W_{21} = g_1^0 W_{12}. \tag{4.4}$$

It follows from (4.2) and (4.3) that there exist two positive numbers c' and c'' such that

$$0 < c' \leqq g_2^0 \leqq c'' < 1 \tag{4.5}$$

and such that the class C_{d_1} consists of all ξ for which $g_2 \leqq c'$, and the class C_{d_2} consists of all ξ for which $g_2 \geqq c''$.

Thus, the following decision procedure will be a Bayes solution relative to the a priori distribution $\xi = (g_1, g_2)$: *If $g_2 \leqq c'$ or $\geqq c''$, do not take any observations and make the corresponding final decision. If $c' < g_2 < c''$, continue taking observations until the a posteriori probability of H_2 is either $\geqq c''$ or $\leqq c'$. If this a posteriori probability is $\geqq c''$, accept H_2, and if it is $\leqq c'$, accept H_1.*

The a posteriori probability of H_2 after the first m observations have been made is given by

$$(4.6) \qquad g_{2m} = \frac{g_2 p(x_1 \mid F_2) \cdots p(x_m \mid F_2)}{g_1 p(x_1 \mid F_1) \cdots p(x_m \mid F_1) + g_2 p(x_1 \mid F_2) \cdots p(x_m \mid F_2)}.$$

If $c' < g_2 < c''$ and if the probability (under F_1 as well as under F_2) is zero that $g_{2m} = c'$ or $= c''$ for some m, then it follows from Theorem 3.8 that the above described Bayes solution is essentially unique; i.e., any other Bayes solution can differ from the one given above only on a set whose probability measure is zero under both F_1 and F_2.

Provided that at least one observation is made, one can easily verify that the above described Bayes solution is identical with a sequential probability ratio test for testing H_2 against H_1. The sequential probability ratio test is defined as follows (see [3]): Two positive constants A and B ($B < A$) are chosen. Experimentation is continued as long as the probability ratio

$$(4.7) \qquad \frac{p_{2m}}{p_{1m}} = \frac{p(x_1 \mid F_2) \cdots p(x_m \mid F_2)}{p(x_1 \mid F_1) \cdots p(x_m \mid F_1)},$$

satisfies the inequality $B < \frac{p_{2m}}{p_{1m}} < A$. If $\frac{p_{2m}}{p_{1m}} \geqq A$, accept H_2. If $\frac{p_{2m}}{p_{1m}} \leqq B$, accept H_1. The Bayes solution described above coincides with this probability ratio test for properly chosen values of the constants A and B.

The results described above for $k = 2$ are essentially the same as those contained in Lemmas 1 and 2 of an earlier publication [4] of the authors.

We shall now discuss the case $k = 3$. Any a priori distribution ξ can be represented by a point with the barycentric coordinates g_1, g_2 and g_3, where g_i is the a priori probability of $H_i (i = 1, 2, 3)$. The totality of all possible a priori distributions ξ will fill out the triangle T with the vertices 0_1, 0_2 and 0_3 where 0_i represents the a priori distribution corresponding to $g_i = 1$ (see Figure 1).

Clearly, the vertex 0_i is contained in C_{d_i}. Thus, because of Theorem 3.9, $C_{d_i} (i = 1, 2, 3)$ is a convex subset of T containing the vertex 0_i, as indicated in Figure 1.

If one of the components of $\xi = (g_1, g_2, g_3)$ is zero, say $g_i = 0$, then H_i can be disregarded and the problem of constructing Bayes solutions reduces to the previously considered case where $k = 2$. Thus, in particular, the determination of the boundary points $P_1, P_2, \cdots, P_6$ of C_{d_1}, C_{d_2} and C_{d_3} which are on the boundary of the triangle T, reduces to the previously considered case, $k = 2$.

It follows from Theorems 3.8 and 3.9 that the intersection of C_{d_i} with any straight line T_i through 0_i is a closed segment. One endpoint of this segment is, of course, 0_i. Let B_i denote the other endpoint. It follows from Theorem 3.7 that B_i must be a point of C_{2,d_i}. Any interior point of 0_iB_i can be shown to be an element of C_{1,d_i}. The proof of this is very similar to that of Theorem 3.9.

We shall now show how tangents to the sets C_{d_1}, C_{d_2} and C_{d_3} can be constructed at the boundary points $P_1, P_2, \cdots, P_6$. Consider, for example, the boundary point P_1 of C_{d_1} that lies on the line $0_1 0_2$. Let ξ_1 be the a priori distribution represented by the point P_1. Since the a priori probability of H_3 is zero according to ξ_1, we can disregard H_3 in constructing Bayes solutions relative to ξ_1. Let $D_1(x)$ be a sequential probability ratio test for testing H_1 against H_2

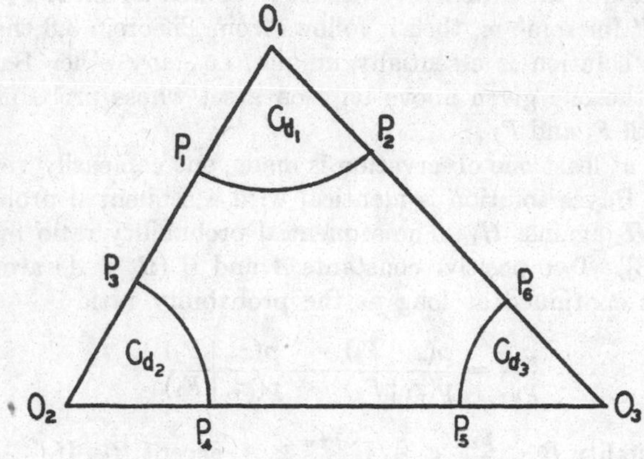

FIG. 1

which requires at least one observation and which is a Bayes solution relative to ξ_1. Since ξ_1 is a boundary point, such a decision function D_1 exists. Thus, we have

$$W(\xi_1, d_1) = r(\xi_1, D_1) = \operatorname*{Inf}_D r(\xi_1, D). \tag{4.8}$$

Let α_{ij} denote the probability of accepting H_j when H_i is true and D_1 is the decision function adopted. Let, furthermore, n_i denote the expected number of observations required by the decision procedure when F_i is true and D_1 is adopted. Then, for any a priori distribution $\xi = (g_1, g_2, g_3)$ we have

$$r(\xi, D_1) = \sum_{i,j} g_i W_{ij}\alpha_{ij} + \sum_i g_i n_i \tag{4.9}$$

and

$$W(\xi, d_1) = \sum_i g_i W_{i1}. \tag{4.10}$$

Thus, the linear manifold $L(D_1, d_1)$ is simply the straight line given by the equation

$$\sum_i g_i W_{i1} = \sum_{i,j} g_i W_{ij} \alpha_{ij} + \sum_i g_i n_i . \tag{4.11}$$

This straight line goes through P_1 and, because of Theorem 3.10, it is tangent to C_{d_1}. Tangents at the same points $P_2, \cdots, P_6$ can be constructed in a similar way.

The convexity properties of the sets $C_{d_i} (i = 1, 2, \cdots, k)$ were established by the authors prior to the more general results described in Sections 2 and 3 and were stated by one of the authors in an address given at the Berkeley meeting of the Institute of Mathematical Statistics, June, 1948. More general results when Ω and D^* are finite, admitting also non-linear cost functions, were obtained later by Arrow, Blackwell and Girshick [2].

REFERENCES

[1] A. Wald, "Foundations of a general theory of sequential decision functions," *Econometrica*, Vol. 15 (1947), pp. 279–313.

[2] K. J. Arrow, D. Blackwell, M. A. Girshick, "Bayes and minimax solutions of sequential decision problems," *Econometrica*, Vol. 17 (1949), pp. 213–244.

[3] A. Wald, "*Sequential analysis*," John Wiley & Sons, New York, 1947.

[4] A. Wald and J. Wolfowitz, "Optimum character of the sequential probability ratio test," *Annals of Math. Stat.*, Vol. 19 (1948), pp. 326–339.

[5] S. Saks, "*Theory of the integral*," Hafner Publishing Company, New York.

[6] A. Wald, "Statistical decision functions," *Annals of Math. Stat.*, Vol. 20 (1949), pp. 165–205.

Made in United States of America

Reprinted from THE ANNALS OF MATHEMATICAL STATISTICS
Vol. XXI, No. 2, June, 1950

MINIMAX ESTIMATES OF THE MEAN OF A NORMAL DISTRIBUTION WITH KNOWN VARIANCE

BY J. WOLFOWITZ[1]

Columbia University

Summary. It is proved that the classical estimation procedures for the mean of a normal distribution with known variance are minimax solutions of properly formulated problems. A result of Stein and Wald [1] is an immediate consequence. Other such optimum properties follow. Sequential and non-sequential problems can be treated in this manner. Interval and point estimation are discussed.

1. Sequential estimation by an interval of given length l. In this section we shall consider the problem of sequentially estimating the mean of a normal distribution with known variance by an interval of fixed length l. Without loss of generality we shall take the known variance to be unity. Such a sequential estimation procedure, which we shall designate generically by G, is a rule which says a) when to terminate taking random, independent observations on the normal chance variable with unknown mean $\xi(-\infty < \xi < \infty)$ and variance 1, and when this termination is to occur after the observations $x_1, \cdots, x_n$ have been obtained, gives b) the center of the estimating interval of length l as a function of $x_1, \cdots, x_n$. Let $\alpha(\xi, G)$ be the probability under G that the estimating interval will contain ξ, and let $n(\xi, G)$ be the expected number of observations when ξ is the mean and G is the estimation procedure (It is assumed that G is such that $\alpha(\xi, G)$ and $n(\xi, G)$ exist for all ξ).

Define

$$q(\xi, G) = 1 - \alpha(\xi, G),$$

and for fixed $c > 0$

$$W(\xi, G) = q(\xi, G) + cn(\xi, G). \tag{1.1}$$

Let $C(N, l)$ ($l > 0$, N a positive integer) be the classical non-sequential estimation procedure where one takes the fixed number N of observations, and estimates the mean by the interval $\left(\bar{x} - \frac{l}{2}, \bar{x} + \frac{l}{2}\right)$, where $\bar{x}$ is the sample mean. For p such that $0 < p \leq 1$, let $C(p, N, l)$ be the following estimation procedure: A chance experiment with two outcomes, N and $N + 1$, of respective probabilities p and $1 - p$, is performed. One then proceeds according to $C(i, l)$, where $i(= N, N + 1)$ is the outcome of the experiment. Finally define

$$M(y) = \frac{1}{\sqrt{2\pi}} \int_y^\infty e^{-\frac{1}{2}z^2}\, dz.$$

---

[1] Research under a contract with the Office of Naval Research.

218

Let us assume for a moment that the unknown ξ is itself a chance variable, normally distributed with mean zero and variance σ^2, and let us obtain a procedure G which minimizes

$$E\{q(\xi, G) + c\,n(\xi, G)\} = \frac{1}{\sqrt{2\pi}\sigma} \int_{-\infty}^{\infty} \{q(y, G) + cn(y, G)\} \exp\left[\frac{-y^2}{2\sigma^2}\right] dy. \tag{1.2}$$

Let $x_1, \cdots, x_m$ be m independent observations on a normal chance variable with mean ξ and variance 1. Let

$$\bar{x} = \frac{\sum_1^m x_i}{m}.$$

The a posteriori distribution of ξ, given $x_1, \cdots, x_m$, is easily verified (or see [1], eqs. (19) and (20)) to be normal with mean

$$\bar{x}\left[1 + \frac{1}{m\sigma^2}\right]^{-1} \tag{1.3}$$

and variance

$$\left[m + \frac{1}{\sigma^2}\right]^{-1}. \tag{1.4}$$

Thus if we stop after m observations the best procedure from the point of view of minimizing (1.2) is to put the center of the estimating interval of length l at the point (1.3). The conditional expected value of $q(\xi)$ is then

$$Q(x_1, \cdots, x_m \mid \sigma^2) = 2M\left(\frac{l}{2}\sqrt{m + \frac{1}{\sigma^2}}\right). \tag{1.5}$$

Thus $Q(x_1, \cdots, x_m)$ is a function only of m and σ^2. Define

$$R(m, \sigma^2) = 2M\left(\frac{l}{2}\sqrt{m + \frac{1}{\sigma^2}}\right) - 2M\left(\frac{l}{2}\sqrt{m + 1 + \frac{1}{\sigma^2}}\right). \tag{1.6}$$

We note that $R(m, \sigma^2)$ is, for fixed σ, a decreasing function of m. We conclude that a best decision as to whether or not to take another observation must be based on the value of $R(m, \sigma^2)$. If $R(m, \sigma^2) > c$ take another observation; if $R(m, \sigma^2) < c$ do not take another observation; if $R(m, \sigma^2) = c$ take either action at pleasure. Hence, if c is such that $R(N, \sigma^2) \leq c \leq R(N - 1, \sigma^2)$, a best procedure from the point of view of minimizing (1.2) is to take exactly N observations. This integer N is a function of c and σ^2, thus: $N(c, \sigma^2)$. In the next paragraph we shall show that $N(c, \sigma^2)$ can be defined for every positive c and σ^2. It is clearly a function which takes at most two values. We shall denote by $G(\sigma^2)$ the estimation procedure described above which minimizes (1.2). It consists of taking the fixed number $N(c, \sigma^2)$ of observations and putting the center of the estimating interval of length l at the point (1.3). Where $N(c, \sigma^2)$ is double-valued we may take either value at pleasure. We verify that the value of (1.2) is the same for either choice.

We now verify that $N(c, \sigma^2)$ can be defined for all positive c and σ^2. We have remarked earlier that $R(m, \sigma^2)$ is, for fixed σ^2, a monotonically decreasing function of m. We note that

$$\lim_{m\to\infty} R(m, \sigma^2) = 0.$$

When $c > R(0, \sigma^2)$ we take no observations whatever and take $\bar{x} \equiv 0$. When $c = R(0, \sigma^2)$ we take zero or one observation at pleasure.

Without difficulty we compute

$$W(\xi, G(\sigma^2)) = W(\xi, \sigma^2) = cN + M\left(\sqrt{N}\,\frac{l}{2}\left[1 + \frac{1}{N\sigma^2}\right] - \frac{\xi}{\sqrt{N}\,\sigma^2}\right)$$
$$+M\left(\sqrt{N}\,\frac{l}{2}\left[1 + \frac{1}{N\sigma^2}\right] + \frac{\xi}{\sqrt{N}\,\sigma^2}\right)$$

where for typographical simplicity we have written N for $N(c, \sigma^2)$. For fixed c and σ^2 the minimum of $W(\xi, \sigma^2)$ occurs at $\xi = 0$. Also $W(0, \sigma^2)$ is a monotonically increasing function of σ^2. If $N(c, \infty) > 0$ then, as $\sigma^2 \to \infty$ it approaches the limit

$$cN(c, \infty) + 2M\left(\frac{l}{2}\sqrt{N(c, \infty)}\right),$$

which is the constant value of

$$W(\xi, C(N(c, \infty), l)).$$

We therefore conclude that $C(N(c, \infty), l)$ is a minimax estimating procedure of type G, i.e.,

$$W(\xi, C(N(c, \infty), l)) = \inf_{G} \sup_{\xi} W(\xi, G)$$

for any $c > 0$. (The case $N(c, \infty) = 0$ may be verified separately. We define $\bar{x} \equiv 0$ for $C(0, l)$).

Conversely, let N_0 be a given non-negative integer. Then $C(N_0, l)$ is a minimax estimating procedure G for all $W(\xi, G)$ for which c satisfies

$$R(N_0, \infty) \le c \le R(N_0 - 1, \infty).$$

(We define $R(-1, \infty) = \infty$.) Thus we can say: For every $c > 0$ there exists a classical estimation procedure $C(N, l)$ with integral N such that

$$W(\xi, C(N, l)) = \inf_{G} \sup_{\xi} W(\xi, G).$$

For every integral N we can find at least one $c > 0$ such that the above equation holds. A method of finding N, given c, and of finding c, given N, has been described above. (We have taken the liberty of calling $C(0, l)$ a classical procedure.

Let α_0 be a given number such that

$$1 - 2M\left(\frac{l}{2}\right) \le \alpha_0 < 1.$$

Define p_0, $0 < p_0 \le 1$, and a positive integral N_0 uniquely by

$$\alpha_0 = p_0 \left(1 - 2M\left(\sqrt{N_0}\,\frac{l}{2}\right)\right) + (1 - p_0)\left(1 - 2M\left(\sqrt{N_0 + 1}\,\frac{l}{2}\right)\right).$$

Let

$$c_0 = R(N_0, \infty).$$

For $c = c_0$ we verify readily that both $C(N_0, l)$ and $C(N_0 + 1, l)$ are minimax estimating procedures G, so that

$$\begin{aligned} W(\xi, C(N_0, l)) &= W(\xi, C(N_0 + 1, l)) \\ &= p_0 W(\xi, C(N_0, l)) + (1 - p_0) W(\xi, C(N_0 + 1, l)) \\ &= (1 - \alpha_0) + c_0[p_0 N_0 + (1 - p_0)(N_0 + 1)] \\ &= (1 - \alpha_0) + c_0[N_0 + (1 - p_0)]. \end{aligned}$$

Therefore, for any G whatever,

$$\begin{aligned} (1 - \alpha_0) + c_0[N_0 + (1 - p_0)] &\le \sup_\xi \{q(\xi, G) + c_0 n(\xi, G)\} \\ &\le \sup_\xi q(\xi, G) + c_0 \sup_\xi n(\xi, G). \end{aligned}$$

Hence

$$\sup_\xi q(\xi, G) \le 1 - \alpha_0$$

implies

$$\sup_\xi n(\xi, G) \ge N_0 + (1 - p_0),$$

a result first proved by Stein and Wald [1].

Also

$$\sup_\xi n(\xi, G) \le N_0 + (1 - p_0)$$

implies

$$\sup_\xi q(\xi, G) \ge 1 - \alpha_0,$$

a result also proved in [1].

2. A sequential upper bound for the mean. The fact that in the last section l was a constant made matters simpler, as we see when we begin to consider the problem of a sequential upper bound for $\xi(-\infty < \xi < \infty)$. This of course means that we wish to use as estimating interval the interval $(-\infty, L(x_1, \cdots, x_n))$ where L is a function of the observations $x_1, \cdots, x_n$, and n (a chance variable) is the number of observations before the process of taking observations is terminated. What is wanted now is a suitable definition of the "length" of this interval. Also we shall admit the possibility that it might be in some sense advantageous to have intervals of varying length; this poses the problem of optimum choice of the function $L(x_1, \cdots, x_n)$.

As before, let ξ be the mean of a normal distribution with unit variance. Let T be the generic estimation procedure which consists of a rule for terminating the taking of observations, and of a function $L_T(x_1, \cdots, x_n)$ which is used to estimate ξ by the interval $(-\infty, L_T)$. Define

$$q(\xi, T) = P\{L_T \leq \xi\},$$
$$\lambda(\xi, T) = E(L_T - \xi)^2,$$

and

$$W(\xi, T) = q(\xi, T) + k\lambda(\xi, T) + cn(\xi, T), \tag{2.1}$$

where c and k are positive constants. (We admit only such T for which the quantities q, λ, and n are defined for all real ξ.) As before, let us temporarily assume that ξ is normally distributed with mean zero and variance σ^2, and set ourselves the task of minimizing

$$\frac{1}{\sqrt{2\pi}\,\sigma} \int_{-\infty}^{\infty} W(y, T)\, e^{-(y^2/2\sigma^2)}\, dy = W^*(T, \sigma^2) \tag{2.2}$$

with respect to T. In the next paragraph we digress for a moment to derive a needed elementary inequality.

Let us prove that, if h, h_1, and h_2 are non-negative, and

$$h^2 = p\, h_1^2 + (1 - p)\, h_2^2, \tag{2.3}$$

where $0 < p < 1$, then

$$M(h) \leq p\, M(h_1) + (1 - p)\, M(h_2). \tag{2.4}$$

Hold h and p fixed. The desired result is obviously true when $h_1 = h_2 = h$. Let h_1 and h_2 vary, subject to (2.3). Then

$$\frac{dh_2}{dh_1} = \frac{-ph_1}{(1 - p)h_2}.$$

Also

$$\frac{pdM(h_1)}{dh_1} = \frac{-p}{\sqrt{2\pi}} e^{-\frac{1}{2}h_1^2}$$

and

$$(1 - p)\frac{dM(h_2)}{dh_1} = (1 - p)\frac{dM(h_2)}{dh_2}\frac{dh_2}{dh_1} = \frac{ph_1}{\sqrt{2\pi}h_2} e^{-\frac{1}{2}h_2^2}.$$

Thus the derivative of the right member of (2.4) with respect to h_1 is 0 when $h_1 = h$, positive when $h_1 > h$, and negative when $h_1 < h$. From this we get (2.4).

Let T be any estimation procedure and $L_T(x_1, \cdots, x_n)$ its associated function. Write

$$l_T(x_1, \cdots, x_n) = L_T(x_1, \cdots, x_n) - \bar{x}\left[1 + \frac{1}{n\sigma^2}\right]^{-1}.$$

If $n = m$ and $x_1, \cdots, x_m$ is the sample obtained, we have that the conditional expected value of $W^*(T, \sigma^2)$ is

$$M\left(l_T(x_1, \cdots, x_m)\sqrt{m + \frac{1}{\sigma^2}}\right) + cm + kE(U_m^* + l_T(x_1, \cdots, x_m))^2, \tag{2.5}$$

where U_m^* is a normally distributed chance variable with mean zero and variance $\left(m + \frac{1}{\sigma^2}\right)^{-1}$. The last term in (2.5) is therefore

$$k\left[\left(m + \frac{1}{\sigma^2}\right)^{-1} + l_T^2(x_1, \cdots, x_m)\right].$$

This is an even function of l_T, while the first term of (2.5) is a monotonically decreasing function of l_T. Thus (2.5) and hence $W^*(T, \sigma^2)$ will be minimized by taking l_T non-negative. Now take the expected value of (2.5) over the set of samples where $n = m$. Application of the result of the preceding paragraph to the finite sums which approximate the integral gives the result that $W^*(T, \sigma^2)$ is minimized when $l_T(x_1, \cdots, x_m)$ is a function only of m. Hence we may restrict ourselves to consideration of procedures T for which (2.5) takes the value

$$M\left(\sqrt{m + \frac{1}{\sigma^2}}\, l_T(m)\right) + cm + k\left[\left(m + \frac{1}{\sigma^2}\right)^{-1} + \{l_T(m)\}^2\right]. \tag{2.6}$$

For any such procedure T, since k and c are fixed positive numbers (and σ^2 is held fixed for the present), the expression (2.6) takes its minimum for some value of m. Thus, in our quest for a procedure T which will minimize $W^*(T, \sigma^2)$ we may restrict ourselves to procedures of *fixed* sample size. This fixed sample size and the (constant) value of l_T are functions of k, c, and σ^2. For fixed m,

$$M\left(\sqrt{m + \frac{1}{\sigma^2}}\, l^0\right) + k(l^0)^2$$

has an absolute minimum at l_m, say, since it is a continuous function of $l^0 (l^0 \geq 0)$ which approaches ∞ with l^0. The case $m = 0$ must be considered. (In this event $\bar{x} \equiv 0$.) Now consider the sequence

$$\left\{M\left(\sqrt{m + \frac{1}{\sigma^2}}\, l_m\right) + cm + k\left[\left(m + \frac{1}{\sigma^2}\right)^{-1} + l_m^2\right]\right\}$$

for $m = 0, 1, 2, \cdots$ ad inf. This sequence condenses only at ∞. Hence there exists a value $N(k, c, \sigma^2)$ of m for which the elements of this sequence have a minimum value. We may choose $N(k, c, \sigma^2)$ so that $\lim_{\sigma^2 \to \infty} N(k, c, \sigma^2)$ exists. (We verify easily that this is always possible.) Designate this limit by $N(k, c, \infty)$, and the associated l by $l(k, c, \infty)$. The l associated with $N(k, c, \sigma^2)$ will be designated by $l(k, c, \sigma^2)$. Thus a best procedure for minimizing $W^*(T, \sigma^2)$ is to take the fixed number $N(k, c, \sigma^2)$ observations, and to use, as upper bound for ξ, the quantity

$$\bar{x}\left[1 + \frac{1}{\sigma^2 N(k, c, \sigma^2)}\right]^{-1} + l(k, c, \sigma^2).$$

We see readily that

$$l(k, c, \infty) = \lim_{\sigma^2=\infty} l(k, c, \sigma^2)$$

and that

$$M(\sqrt{N(k, c, \infty)}\ l(k, c, \infty)) = \lim_{\sigma^2=\infty} M\left(\sqrt{N(k, c, \sigma^2) + \frac{1}{\sigma^2}}\ l(k, c, \sigma^2)\right).$$

Let $T(\sigma^2)$ be the procedure described above which is a best procedure T in the sense of minimizing $W^*(T, \sigma^2)$ when σ^2 is the variance of ξ.

We now compute $W(\xi, T(\sigma^2))$ and obtain

$$\text{(2.7)}\quad \begin{aligned} W(\xi, T(\sigma^2)) &= cN + k\left[\frac{N\sigma^4}{(1 + N\sigma^2)^2} + \left(l - \frac{\xi}{1 + N\sigma^2}\right)^2\right] \\ &\quad + M\left(\frac{1 + N\sigma^2}{\sqrt{N}\sigma^2}\left[l - \frac{\xi}{1 + N\sigma^2}\right]\right), \end{aligned}$$

where for brevity we have written N and l for $N(k, c, \sigma^2)$ and $l(k, c, \sigma^2)$. Let

$$l - \frac{\xi}{1 + N\sigma^2} = x, \qquad \frac{1 + N\sigma^2}{\sqrt{N}\sigma^2} = \sqrt{N} + \epsilon.$$

Then

$$\text{(2.8)}\quad W = cN + k\left[\frac{1}{(\sqrt{N} + \epsilon)^2} + x^2\right] + M([\sqrt{N} + \epsilon]\, x),$$

$$\text{(2.9)}\quad \frac{\partial W}{\partial x} = 2kx - \frac{(\sqrt{N} + \epsilon)}{\sqrt{2\pi}} \exp\left[-\tfrac{1}{2}\{(\sqrt{N} + \epsilon)^2 x^2\}\right].$$

The second term above is always of the same sign and the exponential decreases as $|x|$ increases. Thus $\partial W/\partial x = 0$ has the unique positive root x^*. Put x^* for x in W (in 2.8) and call the result W^*. W is a continuous function of x and approaches ∞ as $|x| \to \infty$. Since the root x^* is unique it follows that W^* is the minimum value of W with respect to x. Now $N(k, c, \sigma^2)$ is constant for σ^2 sufficiently large. Hence, for such σ^2, we have

$$\text{(2.10)}\quad \begin{aligned} \frac{\partial W^*}{\partial \epsilon} &= \frac{-2k}{(\sqrt{N} + \epsilon)^3} + 2kx^* \frac{dx^*}{d\epsilon} - \frac{dx^*}{d\epsilon}\frac{(\sqrt{N} + \epsilon)}{\sqrt{2\pi}} \exp\left[-\tfrac{1}{2}\{(\sqrt{N} + \epsilon)^2 x^{*2}\}\right] \\ &\quad - \frac{x^*}{\sqrt{2\pi}} \exp\left[-\tfrac{1}{2}\{(\sqrt{N} + \epsilon)^2 x^{*2}\}\right] \\ &= \frac{-2k}{(\sqrt{N} + \epsilon)^3} - \frac{x^*}{\sqrt{2\pi}} \exp\left[-\tfrac{1}{2}\{(\sqrt{N} + \epsilon)^2 x^{*2}\}\right] \end{aligned}$$

since x^* is the root of $\partial W/\partial x = 0$. Also ϵ is positive and, for σ^2 sufficiently large, approaches zero monotonically as σ^2 approaches ∞. For $\epsilon > 0$ we have that $\partial W^*/\partial \epsilon < 0$, since $x^* > 0$. We conclude: For σ^2 sufficiently large,

$$\min_{\xi} W(\xi, T(\sigma^2))$$

increases monotonically with σ^2 and approaches

$$cN + k\left[\frac{1}{N} + \{x_N(k)\}^2\right] + M(\sqrt{N}\, x_N(k)),$$

where N is short for $N(k, c, \infty)$ and $x_N(k)$ is the unique positive root of the equation in x

$$2kx = \frac{\sqrt{N}}{\sqrt{2\pi}} \exp\left[-\tfrac{1}{2}Nx^2\right].$$

Going back to the definition of $l(k, c, \infty)$ we see that the latter satisfies the equation in l:

$$\frac{d}{dl}\{M(\sqrt{N}\, l) + kl^2\} = 0.$$

Hence

$$x_N(k) = l(k, c, \infty).$$

Thus the classical estimation procedure C_0 where one takes the fixed number $N(k, c, \infty)$ of observations and uses as upper bound for the mean $\bar{x} + l(k, c, \infty)$ is a minimax procedure T, i.e.,

$$W(\xi, C_0) = \inf_T \sup_{\xi} W(\xi, T).$$

For fixed N, $x_N(k)$ decreases monotonically from $+\infty$ to 0 as k increases from 0 to $+\infty$. Hence, for given positive integral N_0 and $l^* > 0$, there is a unique positive value k_0 such that $x_{N_0}(k_0) = l^*$. Consider the expression

$$B(m) = M(\sqrt{m}\, x_m(k_0)) + cm + k_0\left[\frac{1}{m} + \{x_m(k_0)\}^2\right], \tag{2.11}$$

where m is a positive, continuous variable. We have

$$\frac{dB(m)}{dm} = c - \frac{k_0}{m^2} + \frac{dx_m(k_0)}{dm}\,\frac{\partial}{\partial x_m(k_0)}\left\{M(\sqrt{m}\, x_m(k_0)) + k_0[x_m(k_0)]^2\right\} + \frac{\partial M(\sqrt{m}\, x_m(k_0))}{\partial m}. \tag{2.12}$$

The third term of the right member is identically zero because

$$2k_0 x_m(k_0) = \frac{\sqrt{m}}{\sqrt{2\pi}} \exp\left\{-\tfrac{1}{2}m[x_m(k_0)]^2\right\}. \tag{2.13}$$

Further we have

$$\frac{d^2B(m)}{dm^2} = \frac{2k_0}{m^3} - \frac{d}{dm}\left\{\frac{m^{-\frac{1}{2}}x_m(k_0)}{2\sqrt{2\pi}}\, e^{-\frac{1}{2}mx_m^2(k_0)}\right\} \tag{2.14}$$

$$= \frac{2k_0}{m^3} - \frac{k_0\, d\{m^{-1}(x_m(k_0))^2\}}{dm}.$$

For typographic simplicity we shall use y for $x_m(k_0)$ in the computations of the next few lines. From (2.13) we obtain

$$\log 2k_0 + \log y = -\log\sqrt{2\pi} + \tfrac{1}{2}\log m - \tfrac{1}{2}\, m\, y^2,$$

$$\frac{1}{y}\frac{dy}{dm} = \frac{1}{2m} - \frac{y^2}{2} - my\frac{dy}{dm},$$

$$\frac{dy}{dm} = \frac{y(1 - my^2)}{2m(1 + my^2)}.$$

Hence

$$\frac{d^2B(m)}{dm^2} = 2k_0 m^{-3} + k_0 m^{-2} y^2 - 2k_0 m^{-1} y \frac{dy}{dm}$$

$$= 2k_0 m^{-3} + k_0 m^{-2} y^2 - \frac{k_0 y^2(1 - my^2)}{m^2(1 + my^2)} \tag{2.15}$$

$$= 2k_0 m^{-3} + \frac{2y^4 k_0}{m(1 + my^2)} > 0.$$

Since $c > 0$, we have

$$\lim_{m=0} B(m) = \lim_{m=\infty} B(m) = +\infty.$$

Hence there exists a value of m for which $B(m)$ takes its minimum value. If in $d\,B\,(m)/dm$ we put $m = N_0$ and set the resulting expression equal to zero, we obtain an equation in c whose unique solution c_0, if it is positive, assures us that, when $c = c_0$ and $k = k_0$, $B\,(m)$ takes its minimum at $m = N_0$. A simple computation gives

$$c_0 = \frac{k_0}{N_0^2} + \frac{l^* \exp\{-\frac{1}{2}N_0 l^{*2}\}}{2\sqrt{2\pi N_0}} > 0. \tag{2.16}$$

Actually we are interested in considering $B\,(m)$ only for positive integral values of m. We see readily that the minimum of $B\,(m)$ occurs then at $m = N_0$ when c is such that

$$c_1\,(N_0\,,\,k_0) \le c \le c_2\,(N_0\,,\,k_0), \tag{2.17}$$

with c_1 and c_2 roots of the following equations in c:

$$B\,(N_0) = B\,(N_0 + 1),$$

$$B\,(N_0) = B\,(N_0 - 1).$$

(If $N_0 = 1$, then $c_2 = \infty$.)

Let C_0 (N_0, l^*) be the classical (non-sequential) procedure where one takes N_0 observations and uses $\bar{x} + l^*$ as upper bound for the mean. Choose $k = k_0$ and c such that (2.17) is satisfied. Then

$$W(\xi, C_0(N_0, l^*)) = cN_0 + k_0\left(\frac{1}{N_0} + l^{*2}\right) + M(\sqrt{N_0}\, l^*)$$

identically in ξ. $C_0(N_0, l^*)$ is a procedure T such that

$$W(\xi, C_0) = \inf_T \sup_\xi W(\xi, T). \tag{2.18}$$

Whenever c and k are given, the N and l of the minimax solution may be obtained as follows: First we obtain an integer N such that

$$c_1(N, k) \leq c \leq c_2(N, k).$$

Knowing N and k we can then solve for l.

The results of this section may be summarized as follows: For every positive c and k there exists a classical estimation procedure $C_0(N, l)$ with positive integral N and $l > 0$ such that (2.18) holds. Conversely, for every such pair (N, l) there exists a positive pair (c, k) so that (2.18) holds. A method of finding one member of the pair of couples (c, k) and (N, l) when the other is given, has been indicated above.

Let T_1 be any procedure for giving an upper bound for ξ. We shall say that T_1 is optimum if for any other procedure T_2 such that

$$\sup_\xi q(\xi, T_2) \leq \sup_\xi q(\xi, T_1),$$

$$\sup_\xi \lambda(\xi, T_2) \leq \sup_\xi \lambda(\xi, T_1),$$

we have

$$\sup_\xi n(\xi, T_2) \geq \sup_\xi n(\xi, T_1)$$

It is easy to prove that the classical procedure C_0 with any positive l and positive integral N is optimum by using the results of the last paragraph. For let $1 - \alpha = M\,(l\sqrt{N})$ and let k and c be the corresponding parameters. We have then

$$\sup_\xi q(\xi, T_2) + k \sup_\xi \lambda(\xi, T_2) + c \sup_\xi n(\xi, T_2) \geq \sup_\xi \{q(\xi, T_2)$$

$$+ k\,\lambda(\xi, T_2) + cn(\xi, T_2)\} \geq (1 - \alpha) + k\left(\frac{1}{N} + l^2\right) + cN.$$

Since $\sup q(\xi, T_2) \leq (1 - \alpha)$ and $\sup \lambda(\xi, T_2) \leq 1/N + l^2$, we must have

$$\sup_\xi n(\xi, T_2) \geq N,$$

which is the desired result.

In a general unprecise way we may say that an estimation procedure is the better the smaller the three quantities

$$\beta_1(T) = \sup_\xi q(\xi, T), \qquad \beta_2(T) = \sup_\xi \lambda(\xi, T), \qquad \beta_3(T) = \sup_\xi n(\xi, T).$$

We can now assert the following: No sequential procedure T can be superior to the classical fixed sample procedure C in the sense that

$$\beta_i(T) \leq \beta_i(C) \qquad \text{for } i = 1, 2, 3$$

and the inequality sign holds for at least one i.

In concluding this section we may remark that the case $\alpha \leq \frac{1}{2}$, i.e., $l \leq 0$, may be handled in the same manner as above except that we use $M(-l\sqrt{m})$ in place of $M(l\sqrt{m})$.

3. Miscellaneous results; point estimation. Without going into the necessarily involved details, we content ourselves with pointing out that the problem of estimating sequentially the mean of a normal distribution by a finite interval of length not specified in advance, can be solved in similar fashion. As before let ξ be the unknown mean of a normal distribution with unit variance, where ξ may be any real value. We want to estimate by an interval

$$(L_1(x_1, \cdots, x_n), \qquad L_2(x_1, \cdots, x_n)).$$

Let c, k_1, and k_2 be positive constants and consider the problem of minimizing the supremum with respect to ξ of

$$1 - P\{L_1 < \xi < L_2 \mid G^1\} + cn(\xi, G^1) \\ + k_1 E[(L_1 - \xi)^2 \mid G^1] + k_2 E[(L_2 - \xi)^2 \mid G^1],$$

where G^1 is the generic designation of the estimation procedure. As before, employ an a priori normal distribution of ξ with mean zero and variance σ^2, and let $\sigma^2 \to \infty$. A fixed sample size procedure will be a minimax solution. It will possess optimum properties similar to those described in the preceding sections. The problem of minimizing the supremum with respect to ξ of

$$1 - P\{L_1 < \xi < L_2 \mid G^1\} + cn(\xi, G^1) + kE\{(L_2 - L_1)^2 \mid \xi, G^1\}$$

can be treated similarly.

Suppose the sample size is fixed in advance. The problem of finding an estimate which will minimize

$$\sup_\xi[1 - P\{L_1 < \xi < L_2 \mid G^1\} + k_1E\{(L_1 - \xi)^2 \mid G^1\} + k_2E\{(L_2 - \xi)^2 \mid G^1\}]$$

or

$$\sup_\xi[1 - P\{L_1 < \xi < L_2 \mid G^1\} + kE\{(L_2 - L_1)^2 \mid \xi, G^1\}]$$

can be treated by the method of the preceding sections.

The problem of estimating (sequentially or with fixed sample size) the means of a multivariate normal distribution with known covariance matrix can be treated in similar fashion.

Suppose it is desired to estimate sequentially the mean ξ $(-\infty < \xi < \infty)$ of a normal distribution with unit variance by means of a chance point

$\hat{\xi}(x_1, \cdots, x_n)$. Let $R(\xi, \xi^1)$ be the Wald risk function (cf. [2]), a non-negative function which measures the loss incurred in using the particular value ξ^1 as an estimate when ξ is the actual value. The functions $\hat{\xi}(x_1, \cdots, x_n)$ and $R(\xi, \xi^1)$ must have suitable measurability properties for which we refer the reader to [2]. Let us seek a procedure ξ^* such that

$$\sup_{\xi}[E\{R(\xi, \xi^*)\} + cn(\xi, \xi^*)] = \inf_{\hat{\xi}} \sup_{\xi}[E\{R(\xi, \hat{\xi})\} + c\, n(\xi, \hat{\xi})].$$

Here $n(\xi, \hat{\xi})$ is the average number of observations under $\hat{\xi}$ when ξ is the "true" mean. The procedure ξ^* will be called a minimax solution. We shall assume that $R(a, b)$ is a monotonically non-decreasing function of $| a - b |$, and that there exists a positive number g such that

$$\int_0^\infty R(0, x) \exp\left\{\frac{-x^2}{2g}\right\} dx < \infty .$$

As examples of functions with these properties we may cite

$$R(a, b) = | a - b |,$$

$$R(a, b) = (a - b)^2.$$

As before, assume temporarily that ξ is normally distributed with mean zero and variance σ^2. We verify without difficulty that a solution $\hat{\xi} = \xi_0$ which minimizes

$$\frac{1}{\sqrt{2\pi}\sigma} \int_{-\infty}^\infty [E\{R(\xi, \hat{\xi})\} + cn(\xi, \hat{\xi})] \exp\left\{-\frac{1}{2}\frac{\xi^2}{\sigma^2}\right\} d\xi$$

is the following: n is identically a suitable constant, say N, and ξ_0 is $\bar{x}(1 + 1/N\sigma^2)^{-1} = \bar{x}h$ say, so that $h < 1$. For this solution we have

$$E\{R(\xi, \xi_0)\} + cn(\xi, \xi_0) = cN + \frac{\sqrt{N}}{\sqrt{2\pi}} \int_{-\infty}^\infty R(\xi, \bar{x}h) \exp\left\{-\frac{N}{2}(\bar{x} - \xi)^2\right\} d\bar{x} .$$

Write $u = \bar{x} - \xi$. Then

$$R(\xi, \bar{x}h) = R(\xi, h[\xi + u]) = R(0, hu - [1 - h]\xi),$$

$$\int_{-\infty}^\infty R(\xi, \bar{x}h) \exp\left\{-\frac{N}{2}(\bar{x} - \xi)^2\right\} d\bar{x}$$

$$= \int_{-\infty}^\infty R(0, hu - [1 - h]\xi) \exp\left\{-\frac{Nu^2}{2}\right\} du$$

$$= \int_{-\infty}^\infty R(0, v) \exp\left\{-\frac{N}{2h^2}(v + [1 - h]\xi)^2\right\} \frac{1}{h}\, dv.$$

Because of the assumptions on the function R the last expression is a minimum when $\xi = 0$. We may always choose N such that, for large enough σ^2, the integer N is a constant, say N_0. Also $h \to 1$ as $\sigma^2 \to \infty$. Thus we conclude that the following is a minimax solution: $n = N_0$ and $\hat{\xi} = \xi^* = \bar{x}$. If any estimation procedure $\hat{\xi}$ is such that $\sup_{\xi} n(\xi, \hat{\xi}) \leq N_0$ then

$$\sup_{\xi} E\{R(\xi, \hat{\xi})\} \geq E\{R(\xi, \xi^*)\}.$$

If $\hat{\xi}$ is such that

$$\sup_{\xi} E\{R(\xi, \hat{\xi})\} \leq E\{R(\xi, \xi^*)\},$$

then

$$\sup_{\xi} n(\xi, \hat{\xi}) \geq N_0 .$$

If the restrictions imposed above on R are satisfied and if the sample must always be of given size N, the above argument still holds when $1/N \leq g$, and shows that the estimate $\bar{x}$ minimizes

$$\sup_{\xi} E\{R(\xi, \hat{\xi})\}$$

with respect to $\hat{\xi}$.

REFERENCES

[1] C. Stein and A. Wald, "Sequential confidence intervals for the mean of a normal distribution with known variance," *Annals of Math. Stat.*, Vol. 18 (1947), pp. 427–433.

[2] A. Wald, "Statistical decision functions," *Annals of Math. Stat.*, Vol. 20 (1949), pp. 165–205.

Reprinted from The Annals of Mathematical Statistics
Vol. 22, No. 1, March, 1951

ELIMINATION OF RANDOMIZATION IN CERTAIN STATISTICAL DECISION PROCEDURES AND ZERO-SUM TWO-PERSON GAMES[1]

By A. Dvoretzky,[2] A. Wald,[3] and J. Wolfowitz[3]

Institute for Numerical Analysis and Columbia University

Summary. The general existence of minimax strategies and other important properties proved in the theory of statistical decision functions (e.g., [3]) and the theory of games (e.g., [5]) depends upon the convexity of the space of decision functions and the convexity of the space of strategies. This convexity can be obtained by the use of randomized decision functions and mixed (randomized) strategies. In Section 2 of the present paper the authors state the extension (first announced in [1]) of a measure theoretical result known as Lyapunov's theorem [2]. This result is applied in Section 3 to the statistical decision problem where the number of distributions and decisions is finite. It is proved that when the distributions are continuous (more generally, "atomless," see footnote 7 below) randomization is unnecessary in the sense that every randomized decision function can be replaced by an equivalent nonrandomized decision function. Section 4 extends this result to the case when the decision space is compact. Section 5 extends the results of Section 3 to the sequential case. Sections 6 and 7 show, by counterexamples, that the results of Section 3 cannot be extended to the case of infinitely many distributions without new restrictions.[4] Section 8 gives sufficient conditions for the elimination of randomization under maintenance of ϵ-equivalence. Section 9 concludes with a restatement of the results in the language of the theory of games.

1. Introduction. We shall consider the following statistical decision problem: Let x be the generic point in an n-dimensional Euclidean[5] space R, and let Ω be a given class of cumulative distribution functions $F(x)$ in R. The cumulative distribution function $F(x)$ of the vector chance variable $X = (X_1, \cdots, X_n)$ with range in R is not known. It is known, however, that F is an element of the given class Ω. There is also given a space D whose elements d represent the possible decisions that can be made by the statistician in the problem under consideration. Let $W(F, d, x)$ denote the "loss" when F is the true distribution of

[1] The main results of this paper were announced without proof in an earlier publication [1] of the authors.

[2] On leave of absence from the Hebrew University, Jerusalem, Israel.

[3] Research under a contract with the Office of Naval Research.

[4] The impossibility of such an extension is related to the failure of Lyapunov's theorem when infinitely many measures are considered. (cf. A. Lyapunov, "Sur les fonctions-vecteurs complètement additives," *Izvestiya Akad. Nauk SSSR. Ser. Mat.*, Vol. 10 (1946), pp. 277–279.)

[5] The restriction to a Euclidean space is not essential (see [1]).

X, the decision d is made and x is the observed value of X. We shall define the distance between two elements d_1 and d_2 of D by

$$(1.1) \qquad \rho(d_1, d_2) = \operatorname*{Sup}_{F,x} | W(F, d_1, x) - W(F, d_2, x) |.$$

Let B be the smallest Borel field of subsets of D which contains all open subsets of D as elements. Let B_0 be the totality of Borel sets of R. We shall assume that $W(F, d, x)$ is bounded[6] and, for every F, a function of d and x which is measurable $(B \times B_0)$. By a decision function $\delta(x)$ we mean a function which associates with each x a probability measure on D defined for all elements of B. We shall occasionally use the symbol δ_x instead of $\delta(x)$ when we want to emphasize that x is kept fixed. A decision function $\delta(x)$ is said to be nonrandomized if for every x the probability measure $\delta(x)$ assigns the probability one to a single point d of D. For any measurable subset D^* of D (D^* an element of B), the symbol $\delta(D^* \mid x)$ will denote the probability measure of D^* according to the set function $\delta(x)$. It will be assumed throughout this paper that for any given D^* the function $\delta(D^* \mid x)$ is a Borel measurable function of x. The adoption of a decision function $\delta(x)$ by the statistician means that he proceeds according to the following rule: Let x be the observed value of X. The element d of the space D is selected by an independent chance mechanism constructed in such a way that for any measurable subset D^* of D the probability that the selected element d will be included in D^* is equal to $\delta(D^* \mid x)$.

Given the sample point x and given that $\delta(x)$ is the decision function adopted, the expected value of the loss $W(F, d, x)$ is given by

$$(1.2) \qquad W^*(F, \delta, x) = \int_D W(F, d, x)\, d\delta_x .$$

The expected value of the loss $W(F, d, x)$ when F is the true distribution of X and $\delta(x)$ is the decision function adopted (but x is not known) is obviously equal to

$$(1.3) \qquad r(F, \delta) = \int_R W^*(F, \delta, x)\, dF(x).$$

The above expression is called the risk when F is true and δ is adopted.

We shall say that the decision functions $\delta(x)$ and $\delta^*(x)$ are equivalent if

$$(1.4) \qquad r(F, \delta^*) = r(F, \delta) \qquad \text{for all } F \text{ in } \Omega.$$

We shall say that $\delta(x)$ and $\delta^*(x)$ are strongly equivalent if for every measurable subset D^* of D we have

$$(1.5) \qquad \int_R \delta(D^* \mid x)\, dF(x) = \int_R \delta^*(D^* \mid x)\, dF(x) \qquad \text{for all } F \text{ in } \Omega.$$

[6] The restriction of boundedness is not essential (see [1]).

If δ and δ^* are strongly equivalent, they are equivalent for any loss function which is a function of F and d only.

For any positive ϵ, we shall say that $\delta(x)$ and $\delta^*(x)$ are ϵ-equivalent if

$$(1.6) \qquad | r(F, \delta) - r(F, \delta^*) | \leqq \epsilon \qquad \text{for all } F \text{ in } \Omega,$$

and strongly ϵ-equivalent if

$$(1.7) \qquad \left| \int_R \delta(D^* \mid x)\, dF(x) - \int_R \delta^*(D^* \mid x)\, dF(x) \right| \leqq \epsilon$$

for all measurable D^* and for all F in Ω.

In Section 2 we state a measure-theoretical result first announced in [1] and proved in [6]. This result is then used in Section 3 to prove that for every decision function there exists an equivalent, as well as a strongly equivalent, nonrandomized decision function δ^*, if Ω and D are finite and if each element $F(x)$ of Ω is atomless.[7] This result is extended in Section 4 to the case where D is compact. Section 5 deals with the sequential case for which similar results are proved. A precise definition of a sequential decision function is given in Section 5.

The finiteness of Ω is essential for the validity of the results given in Sections 2–5. The examples given in Section 6 show that even when Ω is such a simple class as the class of all univariate normal distributions with unit variance, there exist decision functions δ such that no equivalent nonrandomized decision functions exist. In Section 7, an example is given where a decision function δ and a positive ϵ exist such that no nonrandomized decision function δ^* is ϵ-equivalent to δ.

In Section 8, sufficient conditions are given which guarantee that for every δ and for every $\epsilon > 0$ there exists a nonrandomized decision function δ^* which is ϵ-equivalent to δ.

2. A measure-theoretical result. Let $\{y\} = Y$ be any space and let $\{S\} = \mathfrak{S}$ be a Borel field of subsets of Y. Let $\mu_k(S)(k = 1, \cdots, q)$ be a finite number of real-valued, σ-finite and countably additive set functions defined for all $S \in \mathfrak{S}$. The following theorem was stated by the authors [1]:

THEOREM 2.1. *Let $\delta_j(y)$ $(j = 1, 2, \cdots, m)$ be real non-negative $\mathfrak{S}$-measurable functions satisfying*

$$(2.1) \qquad \sum_{j=1}^{m} \delta_j(y) = 1$$

for all $y \in Y$. Then if the set functions $\mu_k(S)$ are atomless there exists a decomposition of Y into m disjoint subsets $S_1, \cdots, S_m$ belonging to $\mathfrak{S}$ having the property

[7] A set function μ defined on a Borel field $\mathfrak{S}$ is called atomless if it has the following property: If for some $S \in \mathfrak{S}$, $\mu(S) \neq 0$, then there exists an $S' \subset S$ such that $S' \in \mathfrak{S}$ and such that $\mu(S') \neq \mu(S)$ and $\mu(S') \neq 0$. A cumulative distribution function is called atomless if its associated set function is atomless.

that

$$\int_Y \delta_j(y)\, d\mu_k(y) = \mu_k(S_j) \qquad (j = 1, \cdots, m;\ k = 1, \cdots, q). \tag{2.2}$$

If $\delta_j^*(y) = 1$ *for all* $y \in S_j$ *and* $= 0$ *for any other* $y (j = 1, \cdots, m)$, *then the above equation can be written as*

$$\int_Y \delta_j(y)\, d\mu_k(y) = \int_Y \delta_j^*(y)\, d\mu_k(y) \qquad (j = 1, \cdots, m;\ k = 1, \cdots, q). \tag{2.3}$$

This theorem is an extension of a result of A. Lyapunov [2] and is basic for deriving most of the results of the present paper.

3. Elimination of randomization when Ω and D are finite and each element $F(x)$ of Ω is atomless. In this section we shall assume that Ω consists of the elements $F_1(x), \cdots, F_p(x)$ and D of the elements $d_1, \cdots, d_m$. Moreover, we assume that $F_i(x)$ is atomless for $i = 1, \cdots, p$. A decision function $\delta(x)$ is now given by a vector function $\delta(x) = [\delta_1(x), \cdots, \delta_m(x)]$ such that

$$\delta_j(x) \geqq 0, \qquad \sum_{j=1}^{m} \delta_j(x) = 1 \tag{3.1}$$

for all $x \in R$. Here $\delta_j(x)$ is the probability that the decision d_j will be made when x is the observed value of X. The risk when F_i is true and the decision function $\delta(x)$ is adopted is now given by

$$r(F_i, \delta) = \sum_{j=1}^{m} \int_R W(F_i, d_j, x)\delta_j(x)\, dF_i(x). \tag{3.2}$$

A nonrandomized decision function $\delta^*(x)$ is a vector function whose components $\delta_j^*(x)$ can take only the values 0 and 1 for all x.

For any measurable subset S of R let

$$\nu_{ij}(S) = \int_S W(F_i, d_j, x)\, dF_i(x) \qquad (i = 1, \cdots p;\ j = 1, \cdots, m). \tag{3.3}$$

Then the measures $\nu_{ij}(S)$ are finite, atomless and countably additive. Using these set functions, equation (3.2) can be written as

$$r(F_i, \delta) = \sum_{j=1}^{m} \int_R \delta_j(x)\, d\nu_{ij}(x). \tag{3.4}$$

Replacing in Theorem 2.1 the space Y by R, the set of measures $\{\mu_1, \cdots, \mu_q\}$ by the set $\{\nu_{ij}\} (i = 1, \cdots, p;\ j = 1, \cdots, m)$, it follows from Theorem 2.1 that there exists a nonrandomized decision function $\delta^*(x)$ such that

$$\int_R \delta_j(x)\, d\nu_{ij}(x) = \int_R \delta_j^*(x)\, d\nu_{ij}(x) \qquad (i = 1, \cdots, p;\ j = 1, \cdots, m). \tag{3.5}$$

This immediately yields the following theorems:

THEOREM 3.1. *If Ω and D are finite and if each element $F(x)$ of Ω is atomless, then for any decision function $\delta(x)$ there exists an equivalent nonrandomized decision function $\delta^*(x)$.*

Putting $W(F, d, x) = 1$ identically in F, d and x, equation (3.5) immediately yields the following theorem:

THEOREM 3.2. *If Ω and D are finite and if each element $F(x)$ of Ω is atomless, then for any decision function $\delta(x)$ there exists a strongly equivalent nonrandomized decision function $\delta^*(x)$.*

4. Elimination of randomization when Ω is finite, D is compact and each element $F(x)$ of Ω is atomless. Again, let $\Omega = \{F_1, \cdots, F_p\}$ where the distributions F are atomless. If the loss $W(F, d, x)$ does not depend on x, the finiteness of Ω implies that D is at least conditionally compact with respect to the metric (1.1) (see Theorem 3.1 in [3]). We postulate that D is compact (but permit the loss to depend on x), and shall prove that if $\delta(x)$ is any decision function, there exists a nonrandomized decision function $\delta^*(x)$ such that $\delta^*(x)$ is equivalent to $\delta(x)$, i.e.,

$$r_i(\delta) = r_i(\delta^*) \qquad (i = 1, \cdots, p), \tag{4.1}$$

where $r_i(\delta)$ stands for $r(F_i, \delta)$.

Since D is compact there exists an infinite sequence of decompositions of the space D into a finite number of disjoint nonempty measurable sets, the lth decomposition to be $C(1, 1, \cdots, 1), \cdots, C(k_1, \cdots, k_l)$ with the properties:

(a) Any two sets C which have the same number of indices not all identical, are disjoint.

(b) The sum of all sets with the same number l of indices is D ($l = 1, 2, \cdots$ ad inf.).

(c) If the sequence of indices of one set C constitutes a proper initial part of the sequence of indices of another set C, the first set includes the second.

(d) The diameters of all sets with l indices are bounded above by $h(l)$ and

$$\lim_{l\to\infty} h(l) = 0.$$

Let l be fixed and define

$$\Delta_{m_1,\cdots,m_l}(x) = \delta[C(m_1, \cdots, m_l) \mid x]. \tag{4.2}$$

Define, furthermore,

$$W_i[x, C(m_1, \cdots, m_l)] = \frac{1}{\Delta_{m_1\cdots m_l}(x)} \int_{C(m_1,\cdots,m_l)} W(F_i, d, x)\, d\delta_x \qquad \text{if } \Delta_{m_1\cdots m_l}(x) > 0, \tag{4.3}$$

$$= 0 \qquad \text{if } \Delta_{m_1\cdots m_l}(x) = 0.$$

Clearly,

$$r_i(\delta) = \sum_{m_l=1}^{k_l} \cdots \sum_{m_1=1}^{k_1} \int_R W_i[x, C(m_1, \cdots, m_l)]\Delta_{m_1\cdots m_l}(x)\, dF_i(x). \tag{4.4}$$

Considering a decision space D_l with elements $d_{m_1\cdots m_l}$ $(m_i = 1, \cdots, k_i;$ $i = 1, \cdots, l)$ and putting the loss $W(F_i, d_{m_1\cdots m_l}, x) = W_i[x, C(m_1, \cdots, m_l)]$, equations (3.3) and (3.5) imply that there exists a finite sequence of measurable functions $\bar{\Delta}_{m_1\cdots m_l}(x)$ $(m_1 = 1, \cdots, k_1; \cdots; m_l = 1, \cdots, k_l)$ such that

$$\bar{\Delta}_{m_1\cdots m_l}(x) = 0 \text{ or } 1 \qquad \text{for all } x, \tag{4.5}$$

$$\sum_{m_l} \cdots \sum_{m_1} \bar{\Delta}_{m_1\cdots m_l}(x) = 1 \qquad \text{for all } x, \tag{4.6}$$

$$\bar{\Delta}_{m_1\cdots m_l}(x) = 0 \qquad \text{whenever } \Delta_{m_1\cdots m_l}(x) = 0, \tag{4.7}$$

and

$$\begin{aligned} &\int_R W_i[x, C(m_1, \cdots, m_l)]\bar{\Delta}_{m_1\cdots m_l}(x)\, dF_i(x) \\ &\qquad = \int_R W_i[x, C(m_1, \cdots, m_l)\Delta_{m_1\cdots m_l}(x)\, dF_i(x). \end{aligned} \tag{4.8}$$

Let now $\bar{\delta}(x)$ be the decision function for which

$$\bar{\delta}[C(m_1, \cdots, m_l) \mid x] = \bar{\Delta}_{m_1\cdots m_l}(x) \tag{4.9}$$

and for any measurable subset $D_{m_1\cdots m_l}$ of $C(m_1, \cdots, m_l)$

$$\bar{\delta}[D_{m_1\cdots m_l} \mid x]\bar{\Delta}_{m_1\cdots m_l}(x) = \frac{\delta(D_{m_1\cdots m_l} \mid x)}{\delta[C(m_1, \cdots, m_l) \mid x]}, \tag{4.10}$$

where $\dfrac{\delta(D_{m_1\cdots m_l} \mid x)}{\delta[C(m_1, \cdots, m_l) \mid x]}$ is defined to be $= 0$ when $\delta[C(m_1, \cdots, m_l) \mid x] = 0$.

It then follows from (4.4) and (4.8) that

$$r_i(\delta) = r_i(\bar{\delta}). \tag{4.11}$$

Applying the above result for $l = 1$, we conclude that there exists a decision function $\delta^1(x)$ with the following properties: The choice among the C's with one index is nonrandom. The decision, once given the C (with one index) chosen, is made according to $\delta(x)$. We have $\delta^1[C(m_1) \mid x] = 0$ whenever $\delta[C(m_1) \mid x] = 0$ and

$$r_i(\delta) = r_i(\delta^1) \qquad (i = 1, \cdots, p).$$

Repeat the above procedure for every C with two indices, using $W_i\{x, C(m_1, m_2)\}$ as weight function and $\delta^1(x)$ as the decision function. We conclude that there exists a decision function $\delta^2(x)$ with the following properties: The choice among the C's with two indices is nonrandom. $\delta^2[C(m_1, m_2) \mid x] = 0$ whenever $\delta^1[C(m_1, m_2) \mid x] = 0$. The decision, once given the C (with two indices) chosen, is made according to $\delta^1(x)$ and, therefore, in accordance with $\delta(x)$. We have

$$\int_R \int_{C(m_1)} W(F_i, d, x)\, d\delta_x^1\, dF_i(x) = \int_R \int_{C(m_1)} W(F_i, d, x)\, d\delta_x^2\, dF_i(x) \quad \begin{matrix} (m_1 = 1, 2, \cdots, k_1) \\ (i = 1, \cdots, p). \end{matrix}$$

Repeat the above procedure for all C's with l indices, $l = 3, 4, \cdots$ ad inf. At the lth stage we obtain a decision function $\delta^l(x)$ with the following properties: The decision among the C's with l indices is nonrandom. $\delta^l[C(m_1, \cdots, m_l) \mid x] = 0$ whenever $\delta^{l-1}[C(m_1, \cdots, m_l) \mid x] = 0$. The decision, once given the chosen C with l indices, is made according to $\delta(x)$. We have

$$\int_R \int_{C(m_1, \cdots, m_{l-1})} W(F_i, d, x)\, d\delta_x^{l-1}\, dF_i(x) = \int_R \int_{C(m_1, \cdots, m_{l-1})} W(F_i, d, x)\, d\delta_x^l\, dF_i(x) \quad \begin{pmatrix} i = 1, \cdots, p \\ m_1 = 1, \cdots, k_1 \\ \vdots \\ m_{l-1} = 1, \cdots, k_{l-1} \end{pmatrix}.$$

Hold x fixed and let $C(x; l)$ be that C with l indices for which

$$\int_{C(x;l)} d\delta_x^l = 1.$$

Then $C(x; l + 1)$ is a proper subset of $C(x; l)$ for every positive l. The sequence $C(x; l)$, $l = 1, 2, \cdots$, determines, because D is compact, a unique limit point $c(x)$ such that any neighborhood of $c(x)$ contains almost all sets $C(x; l)$. Hence the sequence of probability measures $\delta_x^l(l = 1, 2, \cdots$, ad inf.) converges to a limit probability measure δ_x^* which assigns probability one to any measurable set which contains the point $c(x)$. Since $W(F_i, d, x)$ is continuous in d, we have

$$\text{(4.12)} \qquad \lim_{l=\infty} \int_D W(F_i, d, x)\, d\delta_x^l = \int_D W(F_i, d, x)\, d\delta_x^*$$

for any x.

Now let x vary over R. It follows from (4.12) and the boundedness of $W(F, d, x)$ that $\lim_{l=\infty} r_i(\delta^l) = r_i(\delta^*)$. Since $r_i(\delta^l) = r_i(\delta)$, also $r_i(\delta^*) = r_i(\delta)$ $(i = 1, \cdots, p)$. Thus the probability measures $\delta^*(x)$ constitute the desired nonrandomized decision function.

It remains to show that for any measurable subset D^* of D, the function $\delta^*(D^* \mid x)$ is a measurable function of x. The measurability of $\delta^*(D^* \mid x)$ can easily be shown for any D^*, if it is shown for all closed sets D^*, since every measurable set can be attained by a denumerable number of Borel operations (denumerably infinite sums and complements) starting with closed sets. Thus we shall assume that D^* is closed. For any positive ρ let D_ρ^* be the sum of all open spheres with center in D^* and radius ρ. It is easy to see that

$$\delta^*(D_{2\rho}^* \mid x) \geqq \liminf_{l=\infty} \delta^l(D_\rho^* \mid x) \geqq \delta^*(D^* \mid x).$$

Since $\lim_{\rho=0} \delta^*(D_{2\rho}^* \mid x) = \delta^*(D^* \mid x)$, it follows from the above relation that

$$\lim_{\rho=0} \liminf_{l} \delta^l(D_\rho^* \mid x) = \delta^*(D^* \mid x).$$

Since $\delta^l(D_\rho^* \mid x)$ is a measurable function of x, the measurability of $\delta^*(D^* \mid x)$ is proved.

5. Elimination of randomization in the sequential case. In this section we shall consider the following sequential decision problem: Let $X = \{X_n\}$ ($n = 1, 2, \cdots$, ad inf.) be a sequence of chance variables. Let x be the generic point in the space $\bar{R}$ of all infinite sequences of real numbers, i.e., $x = \{x_n\}$ ($n = 1, 2, \cdots$, ad inf.) where each x_n is a real number. It is known that the distribution function $F(x)$ of X is an element of Ω, where Ω consists of a finite number of distribution functions $F_1(x), \cdots, F_p(x)$, and that the distribution function of X_1 is continuous according to $F_i(x)$, $i = 1, \cdots, p$. The statistician is assumed to have a choice of a finite number of (terminal) decisions $d_1, \cdots, d_m$, i.e., the space D consists of the elements $d_1, d_2, \cdots, d_m$. A decision rule δ is now given by a sequence of nonnegative, measurable functions $\delta_{\nu t}(x_1, \cdots, x_t)$ ($\nu = 0, 1, \cdots, m$; $t = 1, 2, \cdots$, ad inf.) satisfying

$$\sum_{\nu=0}^{m} \delta_{\nu t}(x_1, \cdots, x_t) = 1 \tag{5.1}$$

for $-\infty < x_1, \cdots, x_t < \infty$. The decision rule δ is defined in terms of the functions $\delta_{\nu t}$ as follows: After the value x_1 of X_1 has been observed, the statistician decides either to continue experimentation and take another observation, or to stop further experimentation and adopt a terminal decision $d_j (j = 1, \cdots, m)$ with the respective probabilities $\delta_{01}(x_1)$ and $\delta_{j1}(x_1)$ ($j = 1, \cdots, m$). If it is decided to continue experimentation, a value x_2 of X_2 is observed and it is again decided either to take a further observation or adopt a terminal decision $d_j (j = 1, \cdots, m)$ with the respective probabilities $\delta_{02}(x_1, x_2)$ and $\delta_{j2}(x_1, x_2) (j = 1, \cdots, m)$, etc. The decision rule is called nonrandomized if each $\delta_{\nu t}$ can take only the values 0 and 1.

Let $v_{i\nu t}(x_1, \cdots, x_t)$ represent the sum of the loss and the cost of experimentation when F_i is true, the terminal decision d_ν is made and experimentation is terminated with the tth observation

$$(\nu = 1, 2, \cdots, m; i = 1, \cdots, p; t = 1, 2, \cdots, \text{ad inf.}).$$

The functions $v_{i\nu t}(x_1, \cdots, x_t)$ are assumed to be finite, nonnegative and measurable. We shall consider only decision rules δ for which the probability is one that experimentation will be terminated at some finite stage. The risk (expected loss plus expected cost of experimentation) when F_i is true and the rule δ is adopted is then given by

$$(5.2)\quad r_i(\delta) = \sum_{t=1}^{\infty}\sum_{\nu=1}^{m}\int_{R_t} v_{i\nu t}(x_1, \cdots, x_t)\delta_{01}(x_1)\delta_{02}(x_1, x_2) \cdots \delta_{0(t-1)}(x_1, \cdots, x_{t-1}) \cdot \delta_{\nu t}(x_1, \cdots, x_t)\, dF_{it}(x_1, \cdots, x_t),$$

where R_t is the t-dimensional space of $x_1, \cdots, x_t$ and $F_{it}(x_1, \cdots, x_t)$ is the cumulative distribution function of $X_1, \cdots, X_t$ when F_i is the distribution function of X.

We shall say that the decision rules δ^1 and δ^2 are equivalent if $r_i(\delta^1) = r_i(\delta^2)$ for $i = 1, \cdots, p$. We shall say that δ^1 and δ^2 are strongly equivalent if

$$(5.3)\quad \int_{R_t} v_{i\nu t}(x_1, \cdots, x_t)\delta_{01}^1(x_1) \cdots \delta_{0(t-1)}^1(x_1, \cdots, x_{t-1})\delta_{\nu t}^1(x_1, \cdots, x_t)\, dF_{it} = \int_{R_t} v_{i\nu t}(x_1, \cdots, x_t)\delta_{01}^2(x_1) \cdots \delta_{0(t-1)}^2(x_1, \cdots, x_{t-1})\delta_{\nu t}^2(x_1, \cdots, x_t)\, dF_{it}$$

for $i = 1, 2, \cdots, p$; $\nu = 1, \cdots, m$ and $t = 1, 2, \cdots$, ad inf.

Clearly, if δ^1 and δ^2 are strongly equivalent and if the functions $v_{i\nu t}(x_1, \cdots, x_t)$ reduce to constants $v_{i\nu t}$, then δ^1 and δ^2 are equivalent for all possible choices of the constants $v_{i\nu t}$.

Let

$$(5.4)\quad \varphi_i(x, \delta) = \sum_{t=1}^{\infty}\sum_{\nu=1}^{m} v_{i\nu t}(x_1, \cdots, x_t)\delta_{01}(x_1) \cdots \delta_{0(t-1)}(x_1, \cdots, x_{t-1})\delta_{\nu t}(x_1, \cdots, x_t).$$

We shall prove the following lemma:

Lemma 5.1. *Let δ be a decision rule for which $\varphi_i(x, \delta) < \infty$ for all x, except perhaps on a set of x's whose probability is zero according to every distribution function $F_i(x)(i = 1, \cdots, p)$. Let τ and T be given positive integers. Then there exists a decision function $\bar{\delta}$ with the following properties:*

$$(5.5)\quad \bar{\delta}_{\nu\tau}(x_1, \cdots, x_\tau) = 0 \text{ or } 1, \quad \sum_{\nu=0}^{m} \bar{\delta}_{\nu\tau}(x_1, \cdots, x_\tau) = 1,$$

for every point in $R_\tau(\nu = 0, 1, \cdots, m)$,

$$(5.6)\quad \bar{\delta}_{\nu t}(x_1, \cdots, x_t) = \delta_{\nu t}(x_1, \cdots, x_t) \qquad (\nu = 0, 1, \cdots, m;\ t \neq \tau),$$

$$(5.7)\quad r_i(\delta) = r_i(\bar{\delta}) \qquad (i = 1, \cdots, p),$$

$$(5.8)\quad \int_{R_t} v_{i\nu t}\delta_{01} \cdots \delta_{0(t-1)}\delta_{\nu t}\, dF_{it} = \int_{R_t} v_{i\nu t}\bar{\delta}_{01} \cdots \bar{\delta}_{0(t-1)}\bar{\delta}_{\nu t}\, dF_{it} \qquad (\nu = 1, \cdots, m;\ t = 1, \cdots, T),$$

$$(5.9)\quad \varphi_i(x, \bar{\delta}) < \infty,$$

for all x except perhaps on a set whose probability is zero according to every distribution $F_i(x)(i = 1, \cdots, p)$.

PROOF. We can write $\varphi_i(x, \delta)$ as follows:

$$\begin{aligned}\varphi_i(x, \delta) &= \sum_{t=1}^{\tau-1} \sum_{\nu=1}^{m} v_{i\nu t}(x_1, \cdots, x_t)\delta_{01} \cdots \delta_{0(t-1)}\delta_{\nu t} \\ &\qquad + \sum_{t=\tau}^{\infty} \sum_{\nu=0}^{m} g_{i\nu\tau t}(x_1, \cdots, x_t)\delta_{\nu\tau}, \end{aligned} \tag{5.10}$$

where $g_{i\nu\tau t}(x_1, \cdots, x_t)$ does not depend on $\delta_{0\tau}, \delta_{1\tau}, \cdots, \delta_{m\tau}$. The first double sum reduces to zero when $\tau = 1$. Clearly, if a $\bar{\delta}$ with the desired properties exists, then

$$\begin{aligned}\varphi_i(x, \bar{\delta}) &= \sum_{t=1}^{\tau-1} \sum_{\nu=1}^{m} v_{i\nu t}(x_1, \cdots, x_t)\delta_{01} \cdots \delta_{0(t-1)}\delta_{\nu t} \\ &\qquad + \sum_{t=\tau}^{\infty} \sum_{\nu=0}^{m} g_{i\nu\tau t}(x_1, \cdots, x_t)\bar{\delta}_{\nu\tau}. \end{aligned} \tag{5.11}$$

For any subset S of R, let

$$\mu_{i\nu\tau t}(S) = \int_S g_{i\nu\tau t}(x_1, \cdots, x_t)\, dF_i \qquad (t = \tau, \tau + 1, \cdots, T), \tag{5.12}$$

and

$$\mu_{i\nu\tau}(S) = \int_S \left[\sum_{t=T+1}^{\infty} g_{i\nu\tau t}(x_1, \cdots x_t) \right] dF_i. \tag{5.13}$$

The measures $\mu_{i\nu\tau t}$ are not defined if $\tau > T$. Clearly, the measures

$$\mu_{i\nu\tau t}(\nu = 0, 1, \cdots, m; t = \tau, \tau + 1, \cdots, T)$$

and the measures $\mu_{i\nu\tau}(\nu = 1, \cdots, m)$ are nonnegative, countably additive and σ-finite. Since for any x for which $\varphi_i(x, \delta) < \infty$ and $\delta_{0\tau} > 0$, the sum

$$\sum_{t=T+1}^{\infty} g_{i0\tau t}(x_1, \cdots, x_t) < \infty,$$

it follows from the assumptions of Lemma 5.1 that $\mu_{i0\tau}$ is σ-finite over the space R' consisting of all x for which $\delta_{0\tau} > 0$. Of course, $\mu_{i0\tau}$ is nonnegative and countably additive. Let R'' be the set of all points x for which $\delta_{0\tau} = 0$. We put

$$\bar{\delta}_{0\tau}(x_1, \cdots, x_\tau) = 0 \quad \text{for all} \quad x \text{ in } R''. \tag{5.14}$$

Application of Theorem 2.1 to each of the spaces R' and R'' shows that there exist measurable functions $\bar{\delta}_{\nu\tau}(x_1, \cdots, x_\tau)(\nu = 0, 1, \cdots, m)$ such that in addiion to (5.14) the following conditions hold:

$$\bar{\delta}_{\nu\tau} = 0 \quad \text{or} \quad 1(\nu = 0, 1, \cdots m) \quad \text{and} \quad \sum_{\nu=0}^{m} \bar{\delta}_{\nu\tau} = 1 \quad \text{for all } x, \tag{5.15}$$

$$\int_R \delta_{\nu\tau}\, d\mu_{i\nu\tau t} = \int_R \bar{\delta}_{\nu\tau}\, d\mu_{i\nu\tau t} \tag{5.16}$$
$$(i = 1, \cdots, p; \nu = 0, 1, \cdots m; t = \tau, \tau + 1, \cdots, T),$$

$$\int_R \delta_{\nu\tau}\, d\mu_{i\nu\tau} = \int_R \bar{\delta}_{\nu\tau}\, d\mu_{i\nu\tau} \qquad (i = 1, \cdots, p; \nu = 0, 1, \cdots m). \tag{5.17}$$

Lemma 5.1 is a simple consequence of the equations (5.14)–(5.17).

For any positive integer u, we shall say that a decision rule δ is truncated at the uth stage if $\delta_{0u'} = 0$ for $u' \geqq u$ identically in x.

THEOREM 5.1. *If δ is truncated at the uth stage there exists a nonrandomized decision rule δ^* that is strongly equivalent to δ.*

PROOF. It is sufficient to prove Theorem 5.1 in the case where $\delta_{\nu t} = 0$ for $t > u$ and $\nu \neq 1$ and $\delta_{1t} = 1$ for $t > u$. Clearly, $\varphi_i(x, \delta) < \infty$ for all x. Putting $\tau = 1$ and $T = u$ in Lemma 5.1, this lemma implies the existence of a decision rule δ^1 with the following properties: (a) δ^1 is strongly equivalent to δ; (b) $\delta^1_{\nu 1} = 0$ or 1 ($\nu = 0, 1, \cdots, m$); (c) $\delta^1_{\nu t} = \delta_{\nu t}$ for $\nu = 0, 1, \cdots, m$ and $t > 1$. Applying Lemma 5.1 to δ^1 and putting $\tau = 2$ and $T = u$, we see that there exists a decision rule δ^2 with the following properties: (a) δ^2 is strongly equivalent to δ^1; (b) $\delta^2_{\nu 2} = 0$ or 1 ($\nu = 0, 1, \cdots, m$); (c) $\delta^2_{\nu t} = \delta^1_{\nu t}$ for $\nu = 0, 1, \cdots, m$ and $t \neq 2$. Continuing this procedure, at the uth step we obtain a decision rule δ^u that is nonrandomized and is strongly equivalent to all the preceding ones. This proves our theorem.

We shall say that two decision rules δ^1 and δ^2 are strongly equivalent up to the Tth stage if

$$\begin{aligned}&\int_{R_t} v_{i\nu t}(x_1, \cdots, x_t)\delta^1_{01} \cdots \delta^1_{0(t-1)}\delta^1_{\nu t}\, dF_{it}\\ &\qquad = \int_{R_t} v_{i\nu t}(x_1, \cdots, x_t)\delta^2_{01} \cdots \delta^2_{0(t-1)}\delta^2_{\nu t}\, dF_{it}\end{aligned} \tag{5.18}$$
$$\text{for} \quad i = 1, \cdots, p; \nu = 1, \cdots, m \quad \text{and} \quad t = 1, \cdots, T.$$

Furthermore, we shall say that a decision rule δ is nonrandomized up to the stage T if $\delta_{\nu t} = 0$ or 1 for $\nu = 0, 1, \cdots, m$ and $t = 1, \cdots, T$.

We now prove the following theorem.

THEOREM 5.2. *If δ is a decision rule for which $\varphi_i(x, \delta) < \infty$, except perhaps on a set of x's of probability zero according to every $F_i(x)(i = 1, \cdots, p)$, then there exists a nonrandomized decision rule δ^* that is equivalent to δ.*

PROOF. Let $\{\epsilon_i\}$ and $\{\eta_i\}(i = 1, 2, \cdots$, ad inf.) be two sequences of positive numbers such that $\lim_{i=\infty} \epsilon_i = 0$ and $\lim_{i=\infty} \eta_i = \infty$. Let T_1 be a positive integer such that

$$r_i(\delta) - \sum_{t=1}^{T_1}\sum_{\nu=1}^{m}\int_{R_t} v_{i\nu t}(x_1, \cdots, x_t)\delta_{01} \cdots \delta_{0(t-1)}\delta_{\nu t}\, dF_{it} < \epsilon_1 \quad \text{if} \quad r_i(\delta) < \infty, \tag{5.19}$$

and

$$\sum_{t=1}^{T_1} \sum_{\nu=1}^{m} \int_{R_t} v_{i\nu t}(x_1, \cdots, x_t)\delta_{01} \cdots \delta_{0(t-1)} \delta_{\nu t} \, dF_{it} > \eta_1 \quad \text{if} \quad r_i(\delta) = \infty. \tag{5.20}$$

Let δ^1 be a decision rule such that $\varphi_i(x, \delta^1) < \infty$ (except perhaps on a set of probability measure zero); δ^1 is equivalent to δ; δ^1 is strongly equivalent to δ up to the T_1th stage; δ^1 is nonrandomized up to the T_1th stage and $\delta^1_{\nu t} = \delta_{\nu t}$ for $t > T_1$. The existence of such a decision rule follows from a repeated application of Lemma 5.1. In general, after $\delta^1, \cdots, \delta^j$ and $T_1, \cdots, T_j$ are given, let δ^{j+1} be a decision rule such that $\varphi_i(x, \delta^{j+1}) < \infty$ (except perhaps on a set of probability measure zero); δ^{j+1} is equivalent to δ^j; δ^{j+1} is strongly equivalent to δ^j up to the T_{j+1}th stage, where T_{j+1} is a positive integer chosen so that $T_{j+1} > T_j$ and (5.19) and (5.20) hold with δ replaced by δ^j, ϵ_1 replaced by ϵ_{j+1} and η_1 replaced by η_{j+1}; δ^{j+1} is nonrandomized up to the stage T_{j+1}; $\delta^{j+1}_{\nu t} = \delta^j_{\nu t}$ for $t \leqq T_j$ and $\delta^{j+1}_{\nu t} = \delta_{\nu t}$ for $t > T_{j+1}$. The existence of such a decision rule δ^{j+1} follows again from a repeated application of Lemma 5.1.

Let δ^* be the decision rule given by the equations

$$\delta^*_{\nu t} = \delta^t_{\nu t} \qquad (\nu = 0, 1, \cdots, m; t = 1, 2, \cdots, \text{ ad inf.}). \tag{5.21}$$

It follows easily from the above stated properties of the decision rules δ^j ($j = 1, 2, \cdots$, ad inf.) that δ^* is nonrandomized and $r_i(\delta^*) = r_i(\delta) (i = 1, \cdots, p)$. This completes the proof of Theorem 5.2.

6. Examples where admissible[8] decision functions do not admit equivalent nonrandomized decision functions. In this section we shall construct examples which show that there exist admissible decision functions $\delta(x)$ which do not admit equivalent nonrandomized decision functions $\delta^*(x)$.

EXAMPLE 1. Let X be a normally distributed chance variable with unknown mean θ and variance unity. This means that Ω is the totality of all univariate normal distributions with unit variance. Suppose we wish to test the hypothesis H_0 that the true mean θ is rational on the basis of a single observation x on X. Thus, D consists of two elements d_1 and d_2 where d_1 is the decision to accept H_0 and d_2 is the decision to reject H_0. For any decision function $\delta(x)$, let $\delta_1(x)$ denote the value of $\delta(d_1 \mid x)$. Let the loss be zero when a correct decision is made, and the loss be one when a wrong decision is made. Then the risk when θ is the true mean and the decision function $\delta(x)$ is adopted is given by

$$r(\theta, \delta) = \frac{1}{\sqrt{2\pi}} \int_{-\infty}^{\infty} e^{-\frac{1}{2}(x-\theta)^2} \delta_1(x) \, dx \qquad \text{when } \theta \text{ is irrational,} \tag{6.1}$$

$$r(\theta, \delta) = \frac{1}{\sqrt{2\pi}} \int_{-\infty}^{\infty} e^{-\frac{1}{2}(x-\theta)^2} (1 - \delta_1(x)) \, dx \qquad \text{when } \theta \text{ is rational.} \tag{6.2}$$

[8] A decision function with risk function $r(F)$ is called admissible if there exists no other decision function with risk function $r'(F)$ such that $r'(F) \leqq r(F)$ for every $F \in \Omega$, and the inequality sign holds for at least one $F \in \Omega$.

Let $\delta_1^0(x) = \frac{1}{2}$ for all x. Clearly,

$$r(\theta, \delta^0) = \tfrac{1}{2} \tag{6.3}$$

for all θ. We shall now show that $\delta^0(x)$ is an admissible decision function. For suppose that there exists a decision function $\delta'(x)$ such that

$$r(\theta, \delta') \leqq r(\theta, \delta^0) = \tfrac{1}{2} \tag{6.4}$$

for all θ, and

$$r(\theta_1, \delta') < r(\theta_1, \delta^0) = \tfrac{1}{2} \tag{6.5}$$

for some value θ_1. Suppose first that θ_1 is rational. Since the integrals in (6.1) and (6.2) are continuous functions of θ, for an irrational value θ_2 sufficiently near to θ_1 we shall have $r(\theta_2, \delta') > \frac{1}{2}$ which contradicts (6.4). Thus, θ_1 cannot be rational. In a similar way, one can show that θ_1 cannot be irrational. Hence, the assumption that a decision function $\delta'(x)$ satisfying (6.4) and (6.5) exists leads to a contradiction and the admissibility of $\delta^0(x)$ is proved.

Let now $\delta^*(x)$ be any decision function for which

$$r(\theta, \delta^*) = r(\theta, \delta^0) \tag{6.6}$$

for all θ. Now (6.6) implies that

$$\frac{1}{\sqrt{2\pi}} \int_{-\infty}^{\infty} e^{-\frac{1}{2}(x-\theta)^2} (\delta_1(x) - \delta_1^*(x))\, dx = 0 \tag{6.7}$$

identically in θ. Since $\delta_1(x) - \delta_1^*(x)$ is a bounded function of x, it follows from the uniqueness properties of the Laplace transform that (6.7) can hold only if $\delta_1(x) - \delta_1^*(x) = 0$ except perhaps on a set of measure zero. Hence, no nonrandomized decision function $\delta^*(x)$ can satisfy (6.6).

In the above example, the distributions consistent with the hypothesis H_0 which is to be tested (normal distributions with rational means) are not well separated from the alternative distributions (normal distributions with irrational means). One might think that this is perhaps the reason for the existence of an admissible decision function δ^0 such that no nonrandomized decision function δ^* can have as good a risk function as δ^0 has. That this need not be so, is shown by the following:

EXAMPLE 2. Suppose that X is a normally distributed chance variable with mean θ and variance unity. The value of θ is unknown. It is known, however, that the true value of θ is contained in the union of the two intervals $[-2, -1]$ and $[1, 2]$. Suppose that we want to test the hypothesis that θ is contained in the interval $[-2, -1]$ on the basis of a single observation x on X. Suppose, furthermore, that the chance variable X itself is not observable and only the chance variable $Y = f(X)$ can be observed where $f(x) = x$ when $|x| < 1$, and $= |x|$ when $|x| \geqq 1$. Let the loss be zero when a correct decision is made, and one when a wrong decision is made. For any decision function $\delta(y)$, let

$\delta_1(y)$ denote the value of $\delta(d_1 \mid y)$ where d_1 denotes the decision to accept H_0. Let $\delta^0(y)$ be the following decision function:

$$\begin{aligned} \delta_1^0(y) &= 1 \quad \text{when } -1 < y < 0 \\ &= 0 \quad \text{when } \ 0 \leqq y < 1 \qquad (6.8) \\ &= \tfrac{1}{2} \quad \text{when } \ y \geqq 1. \end{aligned}$$

First we shall show that $\delta^0(y)$ is an admissible decision function. For this purpose, consider the following probability density function $g(\theta)$ in the parameter space: $g(\theta) = \frac{1}{2}$ when $-2 \leqq \theta \leqq -1$ or $1 \leqq \theta \leqq 2$, $= 0$ for all other θ. If we interpret $g(\theta)$ as the a priori probability distribution of θ, the a posteriori probability of the θ-interval $[-2, -1]$ is greater (less) than the a posteriori probability of the θ-interval $[1, 2]$ when $-1 < y < 0$ $(0 < y < 1)$, and the a posteriori probabilities of the two intervals are equal to each other when $y = 0$ or $y \geqq 1$. Hence, $\delta^0(y)$ is a Bayes solution relative to the a priori distribution $g(\theta)$, i.e.,

$$\int_{-2}^{-1} r(\theta, \delta^0)\, d\theta + \int_1^2 r(\theta, \delta^0)\, d\theta \leqq \int_{-2}^{-1} r(\theta, \delta)\, d\theta + \int_1^2 r(\theta, \delta)\, d\theta \qquad (6.9)$$

for any decision function δ. Suppose now that δ is a decision function for which $r(\theta, \delta) \leqq r(\theta, \delta^0)$ for all θ. It then follows from (6.9) that $r(\theta, \delta) < r(\theta, \delta^0)$ can hold at most on a set of θ's of measure zero. Since, as can easily be verified, $r(\theta, \delta)$ and $r(\theta, \delta^0)$ are continuous functions of θ, it follows that $r(\theta, \delta) = r(\theta, \delta^0)$ everywhere and the admissibility of δ^0 is proved.

Let now $\delta'(y)$ be any decision function for which $r(\theta, \delta') = r(\theta, \delta^0)$ for all θ, i.e.,

$$\frac{1}{\sqrt{2\pi}} \int_{-\infty}^{\infty} e^{-\frac{1}{2}(x-\theta)^2} [\delta^0(y) - \delta'(y)]\, dx = 0 \qquad \text{for all } \theta. \qquad (6.10)$$

Since $\delta_1^0(y) - \delta_1'(y)$ is a bounded function of x, it follows from the uniqueness properties of the Laplace transform that (6.10) can hold only if $\delta_1^0(y) = \delta_1'(y)$ except perhaps on a set of measure zero. Thus, no nonrandomized decision function δ^* exists such that $r(\theta, \delta^*) = r(\theta, \delta^0)$ for all θ.

7. Compactness of Ω in the ordinary sense is not sufficient for the existence of ϵ-equivalent nonrandomized decision functions. Let $\Omega = \{F\}$ be the totality of density functions[9] on the interval $0 \leqq x \leqq 1$ for which $F(x) \leqq c$ for every x, where c is some positive constant greater than 2. The sample space will be the interval $0 \leqq x \leqq 1$. We shall say that the sequence $F_1, F_2, \cdots$ converges to F if

$$\lim_{\to\infty} \int_{-\infty}^{x} F_n(y)\, dy = \int_{-\infty}^{x} F(y)\, dy$$

[9] Here $F(x)$ denotes a density function. This represents a change in notation from preceding sections.

for every real x. The set Ω is compact in the sense of the above convergence definition.[10] Let A be a fixed interval $a_1 \leqq x \leqq a_2$ where $0 < a_1 < a_2 < 1$. Let $D = \{d_1, d_2\}$ and define W as follows:

$$W(F, d_1) + W(F, d_2) = 1,$$

$$W(F, d_1) = 0 \text{ or } 1$$

according as the probability of A under F is rational or not. For any decision function $\delta(x)$, let $\delta_1(x)$ denote the probability assigned to d_1 by $\delta(x)$, i.e., $\delta_1(x) = \delta(d_1 \mid x)$.

Let $\delta'(x)$ be the decision function for which $\delta_1'(x) \equiv \frac{1}{2}$. We shall prove that $\delta'(x)$ is an admissible decision function. For suppose there exists a decision function $\delta^0(x)$ such that

$$r(F, \delta^0) \leqq r(F, \delta') = \tfrac{1}{2} \tag{7.1}$$

for every F, and for F_0 we have

$$r(F_0, \delta^0) < r(F_0, \delta'). \tag{7.2}$$

Now, if $F_i \rightarrow F_0$ and $W(F_i, d_1) = W(F_0, d_1)$ for every i, then $r(F_i, \delta) \rightarrow r(F_0, \delta)$ for every decision function $\delta(x)$, and, in particular, for $\delta^0(x)$. If $F_i \rightarrow F_0$ and $W(F_i, d_1) + W(F_0, d_1) = 1$ for every i, then $r(F_i, \delta) \rightarrow 1 - r(F_0, \delta)$ for every decision function $\delta(x)$ and, in particular, for $\delta^0(x)$. Clearly, we can construct two sequences of functions F such that each sequence converges to F_0, the probability of A according to every member of the first sequence is rational, and the probability of A according to every member of the second sequence is irrational. Because of (7.2) it follows that inequality (7.1) will be violated for almost every member of one of these two sequences. Hence δ' is admissible.

Let us now prove that there cannot exist a nonrandomized decision function $\delta^*(x)$ such that

$$r(F, \delta^*) \leqq r(F, \delta') + \tfrac{1}{4} = \tfrac{3}{4} \tag{7.3}$$

for every $F \in \Omega$. Suppose there were such a decision function $\delta^*(x)$. Let H be the set of x's where $\delta_1^*(x) = 1$, and let $\bar{H}$ be the complement of H with respect to the interval [0, 1]. If H is a set of measure zero or one then obviously (7.3) is violated for some F. Thus, it is sufficient to consider the case when H is a set of positive measure $\alpha < 1$. Suppose for a moment that $\alpha > \frac{1}{2}$. Let G be the density which is zero on $\bar{H}$ and constant on H. From (7.3) it follows that $P\{A \mid G\}$ is rational. There exists a density $G' \in \Omega$ such that $P\{H \mid G'\} > \frac{3}{4}$ and $P\{A \mid G'\}$ is irrational. But then (7.3) is violated for G'. If $\alpha \leqq \frac{1}{2}$, let $\bar{G}$ be the density which is zero on H and constant on $\bar{H}$. From (7.3) it follows that $P\{A \mid \bar{G}\}$ is irrational. There exists a density $\bar{G}' \in \Omega$ such that $P\{\bar{H} \mid \bar{G}'\} >$

[10] The cumulative distribution functions are well-known to be compact in the usual convergence sense. Since the densities are bounded above the limit cumulative distribution function must be absolutely continuous.

$\frac{3}{4}$ and $P\{A \mid \bar{G}'\}$ is rational. But then (7.3) is violated for $\bar{G}'$. Thus (7.3) can never hold for every $F \epsilon \Omega$ and the desired result is proved.

8. Sufficient conditions for the existence of ϵ-equivalent nonrandomized decision functions. In this section we shall consider the nonsequential decision problem (as described in the introduction), and we shall give sufficient conditions for the existence of ϵ-equivalent nonrandomized decision functions. We shall consider the following four metrics in the space Ω:

$$\rho_1(F_1, F_2) = \operatorname{Sup}_S \left| \int_S dF_1 - \int_S dF_2 \right| \tag{8.1}$$

when S is any measurable subset of R,

$$\rho_2(F_1, F_2) = \operatorname{Sup}_{d,x} | W(F_1, d, x) - W(F_2, d, x) |, \tag{8.2}$$

$$\rho_3(F_1, F_2) = \rho_1(F_1, F_2) + \rho_2(F_1, F_2), \tag{8.3}$$

$$\rho_4(F_1, F_2) = \operatorname{Sup}_\delta | r(F_1, \delta) - r(F_2, \delta) | . \tag{8.4}$$

First we prove the following lemma:

LEMMA 8.1. *If Ω is conditionally compact in the sense of the metric ρ_3, then it is conditionally compact in the sense of the metric ρ_4.*

PROOF. Let $\{F_i\}(i = 1, 2, \cdots$, ad inf.) be a Cauchy sequence in the sense of the metric ρ_3, i.e.,

$$\lim_{i,j\to\infty} \rho_3(F_i, F_j) = 0. \tag{8.5}$$

It follows from (8.5) and (8.3) that $W(F_i, d, x)$ converges, as $i \to \infty$, to a limit function $W(d, x)$ uniformly in d and x, i.e.,

$$\lim_{i\to\infty} W(F_i, d, x) = W(d, x) \tag{8.6}$$

uniformly in d and x. Hence

$$\lim_{i\to\infty} \int_D W(F_i, d, x)\, d\delta_x = \int_D W(d, x)\, d\delta_x \tag{8.7}$$

uniformly in x and δ. Because of (8.5), we have

$$\lim_{i,j\to\infty} \rho_1(F_i, F_j) = 0. \tag{8.8}$$

Hence there exists a distribution function $F_0(x)$ (not necessarily an element of Ω) such that

$$\lim_{i\to\infty} \rho_1(F_i, F_0) = 0. \tag{8.9}$$

It follows from (8.7) and (8.9) that

$$\lim_{i\to\infty} \int_R \left[\int_D W(F_i, d, x)\, d\delta_x\right] dF_i(x) = \int_R \left[\int_D W(d, x)\, d\delta_x\right] dF_0(x) \tag{8.10}$$

uniformly in δ. Hence $\{F_i\}$ is a Cauchy sequence in the sense of the metric ρ_4 and Lemma 8.1 is proved.

Next we prove

LEMMA 8.2. *If D is conditionally compact in the sense of the metric* (1.1) *and if δ is any decision function, then for any $\epsilon > 0$ there exists a finite subset D^1 of D and a decision function δ^1 such that $\delta^1(D^1 \mid x) = 1$ identically in x and δ^1 is ϵ-equivalent to δ.*

PROOF. Since D is conditionally compact, it is possible to decompose D into a finite number of disjoint subsets $D_1, \cdots, D_u$ such that the diameter of D_j is less then $\epsilon(j = 1, \cdots, u)$. Let d_j be an arbitrary but fixed point of $D_j(j = 1, \cdots, u)$ and let $\delta^1(x)$ be the decision function determined by the condition

$$\delta^1(d_j \mid x) = \delta(D_j \mid x) \qquad (j = 1, \cdots, u). \tag{8.11}$$

Clearly

$$\left| \int_D W(F, d, x)\, d\delta_x - \int_D W(F, d, x)\, d\delta_x^1 \right| \leqq \epsilon \tag{8.12}$$

for all F and x. Hence,

$$| r(F, \delta^1) - r(F, \delta) | \leqq \epsilon \tag{8.13}$$

for all F and our lemma is proved.

We are now in a position to prove the main theorem.

THEOREM 8.1. *If the elements $F(x)$ of Ω are atomless, if Ω is conditionally compact in the sense of the metrics ρ_1 and ρ_2, and if D is conditionally compact in the the sense of the metric* (1.1), *then for any $\epsilon > 0$ and for any decision function $\delta(x)$ there exists an ϵ-equivalent nonrandomized decision function $\delta^*(x)$.*

PROOF. Because of Lemma 8.2, it is sufficient to prove our theorem for finite D. Thus, we shall assume that D consists of the elements $d_1, \cdots, d_m$. It is easy to verify that conditional compactness of Ω in the sense of both metrics ρ_1 and ρ_2 implies conditional compactness in the sense of the metric ρ_3, and because of Lemma 8.1, also in the sense of the metric ρ_4. Thus, conditional compactness of Ω in the sense of the metrics ρ_1 and ρ_2 implies the existence of a finite subset $\Omega^* = \{F_1, \cdots, F_k\}$ of Ω such that Ω^* is $\epsilon/2$-dense in Ω in the sense of the metric ρ_4. Let δ^* be a nonrandomized decision function that is equivalent to δ if Ω is replaced by Ω^*. The existence of such a δ^* follows from Theorem 3.1. Since Ω^* is $\epsilon/2$-dense in Ω (in the sense of the metric ρ_4), we have

$$| r(F, \delta^*) - r(F, \delta) | \leqq \epsilon \quad \text{for all} \quad F \text{ in } \Omega \tag{8.14}$$

and our theorem is proved.

We shall now introduce some notions with the help of which we shall be able to strengthen Theorem 3.1. For any measurable subset S of R, let

$$r(F, \delta \mid S) = \int_S \left[\int_D W(F, d, x)\, d\delta_x \right] dF(x). \tag{8.15}$$

We shall refer to the above expression as the contribution of the set S to the risk. For any S we shall consider the following four metrics in Ω:

$$\rho_{1S}(F_1, F_2) = \operatorname{Sup}_S \left| \int_{S^*} dF_1 - \int_{S^*} dF_2 \right| \tag{8.16}$$

where S^* is any measurable subset of S,

$$\rho_{2S}(F_1, F_2) = \operatorname{Sup}_{d, x \in S} \mid W(F_1, d, x) - W(F_2, d, x) \mid, \tag{8.17}$$

$$\rho_{3S}(F_1, F_2) = \rho_{1S}(F_1, F_2) + \rho_{2S}(F_1, F_2), \tag{8.18}$$

$$\rho_{4S}(F_1, F_2) = \operatorname{Sup}_\delta \mid r(F_1, \delta \mid S) - r(F_2, \delta \mid S) \mid. \tag{8.19}$$

Finally let the metric $\rho_S(d_1, d_2)$ in D be defined by

$$\rho_S(d_1, d_2) = \operatorname{Sup}_{F, x \in S} \mid W(F, d_1, x) - W(F, d_2, x) \mid. \tag{8.20}$$

We shall now prove the following stronger theorem:

THEOREM 8.2. *Let all elements F of Ω be atomless. If there exists a decomposition of R into a sequence $\{R_i\}$ $(i = 1, 2, \cdots, ad\ inf.)$ of disjoint subsets such that Ω is conditionally compact in the sense of the metrics ρ_{1R_i} and ρ_{2R_i} for each i, and such that D is conditionally compact in the sense of the metric ρ_{R_i} for each i, then for any $\epsilon > 0$ and for any decision function δ there exists an ϵ-equivalent nonrandomized decision function δ^*.*

PROOF. Let $\{R_i\}$ be a decomposition of R for which the conditions of our theorem are fulfilled. Let $\{\epsilon_i\}$ be a sequence of positive numbers such that $\sum_{i=1}^{\infty} \epsilon_i = \epsilon$. Let $\delta^1(x)$ be a decision function such that $\delta_1(x) = \delta(x)$ for any x not in R_1, $\delta^1(x)$ is nonrandomized over R_1 (for any x in R_1, $\delta^1(x)$ assigns the probability one to a single point d in D) and such that

$$\mid r(F, \delta \mid R_1) - r(F, \delta^1 \mid R_1) \mid \leqq \epsilon_1 \quad \text{for all } F. \tag{8.21}$$

The existence of such a decision function δ^1 follows from Theorem 8.1 (replacing R by R_1). After $\delta^1, \cdots, \delta^{i-1}$ have been defined $(i \geqq 1)$, let δ^i be a decision function such that δ^i is nonrandomized over R^i, $\delta^i(x) = \delta^{i-1}(x)$ for all x in $\bigcup_{j=1}^{i-1} R_j$, $\delta^i(x) = \delta(x)$ for all x in $R - \bigcup_{j=1}^{i} R_j$ and such that

$$\mid r(F, \delta^i \mid R_i) - r(F, \delta \mid R_i) \mid \leqq \epsilon_i \quad \text{for all } F \text{ in } \Omega. \tag{8.22}$$

The existence of such a decision function δ^i follows again from Theorem 8.1. Clearly $\delta^i(x)$ converges to a limit $\delta^*(x)$, as $i \to \infty$. This limit decision function $\delta^*(x)$ is obviously nonrandomized and satisfies the conditon

$$|r(F, \delta \mid R_i) - r(F, \delta^* \mid R_i)| \leqq \epsilon_i \tag{8.23}$$

for all i and F. Theorem 8.2 is an immediate consequence of this.

The conditions of Theorem 8.2 will be fulfilled for a wide class of statistical decision problems. For example, this is true for the decision problems which satisfy the following six conditions:

CONDITION 1. *The sample space R is a finite dimensional Euclidean space. All elements $F(x)$ of Ω are absolutely continuous.*

CONDITION 2. *Ω admits a parametric representation, i.e., each element F of Ω is associated with a parametric point θ in a finite dimensional Euclidean space E.*

We shall denote the density function $p(x)$ corresponding to the parameter point θ by $p(x, \theta)$.

CONDITION 3. *The set of parameter points θ which correspond to all elements F of Ω is a closed subset of E.*

We shall call this set of all parameter points θ the parameter space. Since there is a one-to-one correspondence between the elements F of Ω and the points θ of the parameter space, there is no danger of confusion if we denote the parameter space also by Ω.

CONDITION 4. *The density function $p(x, \theta)$ is continuous in $\theta \in \Omega$ for every x.*

CONDITION 5. *The loss $W(\theta, d)$ when θ is true and the decision d is made does not depend on x. D is conditionally compact in the sense of the metric $\rho(d_1, d_2) = \sup_\theta |W(\theta, d_1) - W(\theta, d_2)|$.*

CONDITION 6. *For any bounded subset M of R, we have* $\lim_{\{|\theta| \to \infty,\ \theta \in \Omega\}} \int_M p(x, \theta)\, dx = 0.$

We shall now show that the conditions of Theorem 8.2 are fulfilled for any decision problem that satisfies Conditions 1–6. Let S_i be the sphere in R with center at the origin and radius i. Let $R_1 = S_1$ and $R_i = S_i - \bigcup_{j=1}^{i-1} R_j (i = 1, 2, \cdots$, ad inf.). Condition 5 implies that D is conditionally compact in the sense of the metric ρ_{R_i} for all i. It follows from Condition 5 and Theorem 2.1 in [3] that Ω is conditionally compact in the sense of the metric $\rho(\theta_1, \theta_2) = \sup_d |W(\theta_1, d) - W(\theta_2, d)|$. Hence Ω is conditionally compact in the sense of the metric ρ_{2R_i} for each i. It remains to be shown that Ω is conditionally compact in the sense of the metric ρ_{1R_i} for each i. For this purpose, consider any sequence $\{\theta_j\} (j = 1, 2, \cdots$, ad inf.) of parameter points. There are 2 cases possible: (a) $\{\theta_j\}$ admits a subsequence that converges in the Euclidean sense to a finite point θ_0; (b) $\lim_{j=\infty} |\theta_j| = \infty$. Let us consider first the case (a) and let $\{\theta'_j\}$ be a subsequence of $\{\theta_j\}$ which converges to a finite point θ_0. It then follows from Condition 4 and a theorem of Robbins [4] that $\{\theta'_j\}$ is a Cauchy subsequence in the sense of the metric ρ_{1R_i} for each i. In case (b), Condition 6 implies that the sequence $\{\theta_j\}$ is a Cauchy sequence in the sense of the metric ρ_{1R_i} for each i. Thus, Ω is conditionally compact in the sense of the metric ρ_{1R_i}. This completes the proof of our assertion that a decision problem that satisfies Conditions 1–6, satisfies also the conditions of Theorem 8.2.

9. Application to the theory of games. Translation of the results of Section 2 into the language of the theory of games is immediate and we shall do this only very briefly. The function $W(F_i, d_j, x)$ $(i = 1, \cdots, p; j = 1, \cdots, m; x \epsilon R)$, of Section 1 is now called the pay-off function of a zero-sum two-person game. The game is played as follows: Player I selects one of the integers $1, \cdots, p$, say i, without communicating his choice to player II. A random observation $x \epsilon R$ on a chance variable whose distribution function is F_i is obtained and communicated to player II. The latter chooses one of the integers $1, \cdots, m$, say j. The game now ends with the receipt by player I and player II of the respective sums $W(F_i, d_j, x)$ and $-W(F_i, d_j, x)$. Randomized (mixed) and nonrandomized (pure) strategies are defined in the same manner as the corresponding decision functions in Section 1. When the distribution functions $F_i(x)(i = 1, \cdots, p)$ are all atomless the obvious analogues of Theorems 3.1 and 3.2 hold.

It should be remarked that the usual definition of randomized (mixed) strategy is not as general as the one given above. In the usual definition player II chooses, by a random mechanism independent of the random mechanism which yields the point x, some one of a (usually finite) number of nonrandomized (pure) strategies, and then plays the game according to the nonrandomized strategy selected. In our definition (used in [3]) the random choice is allowed to depend on x. Clearly our method of randomization includes the usual one as a special case. The relation between the two methods of randomization will be discussed by two of the authors in a forthcoming paper [7].

Suppose that the number of possible decisions is at most denumerable, and that the decision procedure consists in choosing at random and in advance of the observations, one of a finite number of nonrandomized decision functions. The sample space can be divided into an at most denumerable number of sets in each of which only a finite number of decisions is possible (the possible decisions vary from set to set). In each set our results are applicable. Since the number of sets is denumerable the resultant decision function is measurable. We conclude: It follows from our results that if a decision procedure consists of selecting with preassigned probabilities one of a finite number of nonrandomized decision functions with the number of possible decisions at most denumerably infinite, and if the possible distributions are finite in number and atomless, then there exists an equivalent nonrandomized decision function. More general results can be obtained for this case (where one chooses at random and in advance of the observations, one of a finite number of nonrandomized decision functions). By application of the methods of Sections 4 and 8 the requirement that the number of possible decisions be denumerable can be easily removed. The procedures are straightforward and we omit them.

REFERENCES

[1] A. Dvoretzky, A. Wald, and J. Wolfowitz, "Elimination of randomization in certain problems of statistics and of the theory of games," *Proc. Nat. Acad. Sci.*, Vol. 36 (1950), pp. 256–260.

[2] A. Lyapunov, "Sur les fonctions-vecteurs complètement additives," *Izvestiya Akad. Nauk SSSR. Ser. Mat.*, Vol. 4 (1940), pp. 465–478.

[3] A. Wald, *Statistical Decision Functions*, John Wiley & Sons, 1950.

[4] Herbert Robbins, "Convergence of distributions," *Annals of Math. Stat.*, Vol. 19 (1948), pp. 72–75.

[5] J. von Neumann and O. Morgenstern, *Theory of Games and Economic Behavior*, Princeton University Press, 1944.

[6] A. Dvoretzky, A. Wald, and J. Wolfowitz, "Relations among certain ranges of vector measures," *Pacific Journal of Mathematics* (1951).

[7] A. Wald and J. Wolfowitz, "Two methods of randomization in statistics and the theory of games," *Annals of Mathematics*, to be published.

ANNALS OF MATHEMATICS
Vol. 53, No. 3, May, 1951

TWO METHODS OF RANDOMIZATION IN STATISTICS AND THE THEORY OF GAMES[1]

BY A. WALD AND J. WOLFOWITZ

(Received August 11, 1950)

1. Introduction

The problem of statistical decisions has been formulated by one of the authors and described, for example, in [1]. We proceed to describe such a formulation in semi-intuitive terms before we state our problem precisely in Section 2. The purpose of this rough description is to describe the motivation of the problem.

Ω is a given collection (finite or infinite) of distribution functions (or probability measures) F, defined on a Borel field B of a space X. A chance variable K with range in X is distributed according to one of the distributions in Ω, but this distribution is unknown to the statistician (or player II of a two-person game). The distribution of K in statistics is determined by the actual problem; in a game it is determined by player I. The statistician (henceforth "player II" is always to be understood to follow "statistician" in parentheses) is required to make a decision, i.e., choose a point d in a given space D^*. His loss is a function of d and the actual distribution F_0 of K, and knowledge of the latter would enable him to minimize the loss. In order to obtain information on F_0 the statistician proceeds to take, seriatim, independent observations on K. These observations have a cost which is to be added to the loss of the statistician. Usually this cost increases with the number of observations and in meaningful problems the loss and cost functions are formulated so as to make it unprofitable for the statistician to take infinitely many observations. (The loss and cost functions may also be functions of the observations; this will in no way change the problem of the present paper.) Thus the statistician has to strike a balance between the cost of ignorance of F_0 and the cost of the observations which presumably furnish information about F_0. A non-randomized sequential decision function is a rule (a function of the observations) which tells the statistician a) when to stop taking further observations, b) what decision to make when he has stopped taking observations.

What we shall call "special randomization" was introduced by v. Neumann into the theory of games [2]. (It applies equally well to statistical decisions; the purpose in both cases is to make a certain space convex). A sequential decision function randomized in the special sense may, for the purposes of this section, be described as follows: There is given a collection Γ, finite or not, of nonrandomized sequential decision functions, and a probability distribution η on Γ. (If the collection is not denumerable questions of measure will arise; such questions

[1] Presented to the International Congress of Mathematicians at Boston, Massachusetts, on September 1, 1950. Research under a contract with the Office of Naval Research and Development.

are relegated to the later sections of this paper.) The statistician performs a random experiment with distribution function η which yields, as an observation, a non-randomized sequential decision function, say f. He then proceeds to act according to f, i.e., he makes observations on K, stops and makes a decision according to f.

What we shall call "general randomization" was introduced by one of the authors (see, for example, [1]). Let s be an order to the statistician to take another observation. Denote by D the set consisting of D^* and s. Let C be a Borel field on D such that the set consisting of the single point s is an element of C. A sequential decision function randomized in the general sense (hereafter it may be referred to simply as a randomized sequential decision function or r.s.d.f.) may be described roughly as follows: There is given a set of probability measures on C, each measure being a function of a possible sequence of observations. The statistician obtains the first observation.[2] This gives him the probability measure, say μ_1, associated with this particular observation. He performs a chance experiment with probability distribution μ_1 which gives him a point in D. If this point is $d \neq s$, the statistician terminates taking observations and makes the decision d. If the point obtained is s, the statistician takes another observation. The r.s.d.f. supplies him with another probability measure, say μ_2, which is a function of the first two observations. The statistician then performs a chance experiment with distribution μ_2, etc., etc.

Consider now the space D^*. In many problems it is convenient to define a metric on D^*. (For example, an "intrinsic" metric is defined in [1]. This particular metric can always be defined; it depends on the loss function.) In this paper we shall assume that D^* is metric, separable, and complete (i.e., every Cauchy sequence possesses a limit point). The set of Borel sets of D^* will be called C^*. The smallest Borel field which contains all the elements of C^* and the set which consists of the single element s will be the set C.

We shall say that two sequential decision functions are equivalent if the probability of obtaining the observations $(x_1, \cdots, x_n)$ and then choosing an element in the set $c \epsilon C$ is the same for both decision functions, identically in c, $(x_1, \cdots, x_n)$, and the distribution of K. It follows at once from the definitions that to every sequential decision function randomized in the special sense there exists an equivalent sequential decision function randomized in the general sense. The purpose of the present paper is to prove that, when D^* is metric, separable, and complete, and C is as defined in the preceding paragraph, the converse is true, namely, every sequential decision function randomized in the general sense is equivalent to one randomized in the special sense. A precise statement and the proof are given in the next section. A special case of this result was proved in [1].

The distribution function F_0 of K will not appear in the proof to be given below; the result is valid identically in F_0. Hence F_0 need not be a member of

---

[2] Even the decision whether to take the first observation at all can be made random. If this is so, only a slight and obvious change is necessary in the subsequent arguments.

any particular class. In particular, it is unnecessary that the observations on K be independent; indeed, F_0 can be the distribution function of an infinite sequence of chance variables, not necessarily independent, and with ranges in different spaces.

2. Proof of equivalence

We begin with the necessary definitions. The conditions on D^* and the definition of C have been given above.

Let $X_1, X_2, \cdots$ be an infinite sequence of abstract spaces, with $x_1, x_2, \cdots$, respectively, their generic points. Let Y denote the Cartesian product $X_1 \times X_2 \times \cdots$ (cf. [3], page 82), and let y be the generic designation of the sequence $(x_1, x_2, \cdots)$. Let $B_1, B_2, \cdots$ be Borel fields on $X_1, X_2, \cdots$, respectively, and let B be the smallest Borel field on Y which contains as elements every infinite Cartesian product $b_1 \times b_2 \times \cdots$, where $b_i \in B_i$, $i = 1, 2, \cdots$.

By a randomized sequential decision function (r.s.d.f.) δ is meant a set of probability measures $\delta(y, m; c)$ with the following properties: 1) For every $y \in Y$ and every positive integral m, $\delta(y, m; c)$ is a non-negative, completely additive set function defined for every set $c \in C$ and such that $\delta(y, m; D) = 1$, 2) If the first m coordinates of y and y' are the same then

$$\delta(y, m; c) \equiv \delta(y', m; c),$$

3) The measures $\delta(y, m; c)$ are such that, for any $c \in C$ and any m, $\delta(y, m; c)$, regarded as a function of y, is measurable B.

By a non-randomized sequential decision function (n.r.s.d.f.) γ is meant a set of probability measures $\gamma(y, m; c)$ which fulfill the requirements imposed upon a r.s.d.f. and in addition are such that, for every y and m, there exists a single element $d[y, m]$ of D such that $\gamma(y, m; d[y, m]) = 1$. We will find it typographically simpler therefore on many occasions to write γ as a point function with values in D, thus:

$$\gamma(y, m) = d[y, m].$$

Both forms will be employed without danger of confusion.

Let α be any sequential decision function, randomized or not. We define

$$\pi_\alpha(y, m; c) = \alpha(y, m; c) \prod_{i=1}^{m-1} \alpha(y, i; s) \tag{1}$$

for $m > 1$, and

$$\pi_\alpha(y, 1; c) = \alpha(y, 1; c). \tag{2}$$

Let $\delta(y, m; c)$ be any given r.s.d.f. We shall construct 1) a set of n.r.s.d.f.'s $\{\gamma(y, m \mid a)\}$, where a, which serves as an index to distinguish different functions γ from each other, is a point in a space H^* on which is defined a Borel field H, 2) a probability measure μ defined on H, such that

$$\pi_\delta(y, m; c) = \int_{H^*} \pi_{\gamma(y,m|a)}(y, m; c)\, d\mu \tag{3}$$

identically in y, m, and c.

In our construction the space H^* will be defined as follows: Let I_m be the interval $0 \leqq a_m < 1$, $m = 1, 2, \cdots$. H^* will be the infinite Cartesian product $I_1 \times I_2 \times \cdots$. The generic point of H^* will be

$$a = (a_1, a_2, \cdots).$$

Let L_m be the totality of Lebesgue measurable subsets of I_m. Then H will be the smallest Borel field on H^* which contains as an element every infinite Cartesian product $l_1 \times l_2 \times \cdots$ where $l_m \in L_m$, $m = 1, 2, \cdots$. The measure μ of such a Cartesian product will be defined as the product of the Lebesgue measures of its factors. It is well known that the definition of μ on such Cartesian products is sufficient to determine μ for every element of H.

Because of the assumptions on D we can construct a denumerable number of sets

$$\{A_{j_1 j_2 \cdots j_k}\},$$

$j_i = 1, 2, \cdots$ ad inf. for $i = 1, 2, \cdots k$, $k = 1, 2, \cdots$ ad inf. with the following properties:

a) every A is a member of C,
b) $\{s\} = A_1 = A_{11} = A_{111} = \cdots$,
c) A's with the same number of indices are disjoint,
d) if the indices of one set A form an initial sequence of the indices of another set A, the first set contains the second set,
e) the diameter of a set A with k indices ($k = 1, 2, \cdots$ ad inf.) is $\leqq 2^{-k}$
f) the union of all A's with the same number of indices is D.

We now define, for fixed y and m, a function $\psi_k(y, m, a_m)$ of a_m, $0 \leqq a_m < 1$, with range in D. Let $d(A_{j_1 \cdots j_k})$ be a fixed point in the closure $\bar{A}_{j_1 \cdots j_k}$ of $A_{j_1 \cdots j_k}$. Let

$$(4)\quad \begin{aligned} J(y, m; A_{j_1 \cdots j_k}) = {} & \sum_{i=1}^{j_1-1} \delta(y, m; A_i) + \sum_{i=1}^{j_2-1} \delta(y, m; A_{j_1 i}) \\ & + \sum_{i=1}^{j_3-1} \delta(y, m; A_{j_1 j_2 i}) + \cdots + \sum_{i=1}^{j_k-1} \delta(y, m; A_{j_1 j_2 \cdots j_{(k-1)} i}). \end{aligned}$$

For a_m such that

$$(5)\quad J(y, m; A_{j_1 \cdots j_k}) \leqq a_m < J(y, m; A_{j_1 \cdots j_k}) + \delta(y, m; A_{j_1 \cdots j_k})$$

we define

$$(6)\quad \psi_k(y, m, a_m) = d(A_{j_1 \cdots j_k})$$

(For $a_m < \delta(y, m; s)$ we define $\psi_k(y, m, a_m) = s$.) We define the function $\psi(y, m, a_m)$ by

$$\psi(y, m, a_m) = \lim_{k \to \infty} \psi_k(y, m, a_m).$$

Finally, we define

(7) $$\gamma(y, m \mid a) = \psi(y, m, a_m)$$

for every y, m, and a.

Waiving for the moment all questions of measurability, let us prove that (3) holds. For this purpose fix y, m, and c at y_0, m_0, and c_0, respectively. We have that

$$\Pi_{\gamma(y, m|a)}(y_0, m_0; c_0)$$

is one or zero according as the following conditions are or are not satisfied:

(8) $$a_m < \delta(y_0, m; s) \qquad m < m_0$$

(9) $$\psi(y_0, m_0, a_{m_0}) \in c_0 .$$

These are conditions on

(10) $$a_1, \cdots, a_{m_0}$$

only. Thus the right member of (3) is simply the Lebesgue measure of the points (10) which satisfy conditions (8) and (9). Obviously it remains only to prove that the Lebesgue measure of the points a_{m_0} for which (9) holds is

$$\delta(y_0, m_0; c_0).$$

Let G be an open set such that

(11) $$\delta(y_0, m_0; G) = \delta(y_0, m_0; \bar{G})$$

where $\bar{G}$ is the closure of G. Let β_k be the union of all those A's with k indices whose closures have at least one point in common with $\bar{G}$. Then

(12) $$\lim_{k\to\infty} \delta(y_0, m_0, \beta_k) = \delta(y_0, m_0, \bar{G}).$$

Since for any a_{m_0} for which

(13) $$\psi(y_0, m_0, a_{m_0}) \in \bar{G}$$

holds, inequality (5) must hold for some term $A_{j_1 \cdots j_k}$ which appears in the sum β_k, the Lebesgue outer measure of the set of points a_{m_0} satisfying (13) cannot exceed $\delta(y_0, m_0, \beta_k)$ for every k. Hence, because of (12), this outer measure is not greater than $\delta(y_0, m_0; \bar{G}) = \delta(y_0, m_0; G)$. Now $G' = D - \bar{G}$ is also an open set such that

$$\delta(y_0, m_0; \bar{G}') = \delta(y_0, m_0; G').$$

Hence the Lebesgue outer measure of the points a_{m_0}, $0 \leqq a_{m_0} < 1$, such that

$$\psi(y_0, m_0, a_{m_0}) \in G',$$

is not greater than $\delta(y_0, m_0, G') = 1 - \delta(y_0, m_0, \bar{G})$. But the function $\psi(y_0, m_0, a_{m_0})$ is defined for every a_{m_0}, $0 \leqq a_{m_0} < 1$. Hence the sum of the two outer measures must be one. Consequently the set of points a_{m_0}, $0 \leqq a_{m_0} < 1$, such that (13) holds, is measurable and has Lebesgue measure $\delta(y_0, m_0; \bar{G})$. This proves (3) for the case that c_0 is a set G which satisfies (11).

Now every closed set can be represented as the limit of a descending sequence of sets of type G which satisfy (11). Hence (3) holds whenever c_0 is a closed set, from which it follows that (3) holds for any c_0 in C.

The function $\psi_k(y, m, a_m)$ is, for fixed m and a_m, a measurable function of y, i.e., those points y for which $\psi_k(y, m, a_m)$ is in some fixed member of C constitute a set which is in B. Hence $\psi(y, m, a_m)$ is a measurable function of y. This shows that the $\gamma(y, m \mid a)$ fulfill the required measurability conditions. Our proof is now complete.

COLUMBIA UNIVERSITY

REFERENCES

[1] WALD, A., *Statistical Decision Functions*, Ann. Math. Statist., Vol. 20 (1949), pp. 165–205.
[2] v. NEUMANN, J., *Zur Theorie der Gesellschaftsspiele*, Math. Ann., Vol. 100 (1928), pp. 295–320.
[3] SAKS, STANISLAW, Theory of the Integral, Stechert & Company, New York, 1937.

SUMS OF RANDOM INTEGERS REDUCED MODULO m

By A. Dvoretzky and J. Wolfowitz

1. Introduction. Let

$$X_1, X_2, \cdots, X_n, \cdots \tag{1}$$

be an infinite sequence of independent random variables which assume *only integral values*. Put

$$S_n = X_1 + \cdots + X_n \qquad (n = 1, 2, \cdots), \tag{2}$$

and let m be any fixed integer greater than 1. Denote S_n reduced mod m by Y_n; i.e., Y_n is a random variable which assumes only the values $0, 1, \cdots, m - 1$ with the respective probabilities

$$P_n(j) = \text{Prob}\{S_n \equiv j \ (\text{mod } m)\} \qquad (j = 0, 1, \cdots, m - 1).$$

It is easily seen that under quite general assumptions Y_n is equidistributed in the limit; i.e.,

$$\lim_{n=\infty} P_n(j) = \frac{1}{m} \qquad (j = 0, 1, \cdots, m - 1). \tag{3}$$

In this paper we obtain necessary and sufficient conditions for the validity of (3) in terms of the distribution functions of (1). We also derive various sufficient conditions for (3), distinguish between essential and accidental equidistribution in the limit, and remark upon the rapidity of approach to equidistribution and some related problems.

The special case $m = 2$ has been considered, in a somewhat different setting, by H. B. Horton [1] who applied his results to obtain a method of generating "random numbers". Our results may also be used for the same purpose, that is, to construct a device for effectuating physically, so to speak, a random variable which assumes all integral values from 0 to $m - 1$ with the same probability $1/m$. Taking large m and employing a suitable transformation, it is possible to obtain a physical device for "generating" a random variable whose cumulative distribution function approximates a given one to any prescribed degree of accuracy.

Since this paper was written there has appeared a paper by Horton and Smith [2] dealing with the same subject. In it the authors, generalizing Horton's previous work [1], obtain sufficient conditions for the validity of (3). These are special cases of sufficient conditions which we deduce from the necessary and sufficient ones.

Received October 5, 1949; in revised form, November 19, 1949.

2. Necessary and sufficient conditions for equidistribution in the limit. There is no loss of generality in assuming the random variables (1) reduced mod m. Therefore, we consider henceforth the X_n as *taking the values* $0, 1, \cdots, m-1$ *only* and put

$$p_n(j) = \text{Prob}\,\{X_n = j\} \qquad (j = 0, 1, \cdots, m-1).$$

The necessary and sufficient conditions for equidistribution in the limit are given by

THEOREM 1. *S_n is in the limit equidistributed* mod m; *i.e.*, (3) *holds if and only if*

$$\prod_{n=1}^{\infty} \sum_{j=0}^{m-1} p_n(j) \exp\left(\frac{2\pi irj}{m}\right) = 0 \qquad (r = 1, 2, \cdots, m-1). \tag{4}$$

Proof. Writing E for the m-th root of unity $\exp(2\pi i/m)$, we put

$$t_n(r) = \sum_{j=0}^{m-1} p_n(j)E^{rj}, \qquad T_n(r) = \sum_{j=0}^{m-1} P_n(j)E^{rj} \qquad (r = 0, 1, \cdots, m-1). \tag{5}$$

It is easily seen that $T_{n+1}(r) = T_n(r)t_{n+1}(r)$ from which

$$T_n(r) = \prod_{\nu=1}^{n} t_\nu(r) \tag{6}$$

follows. From this we obtain that (3) implies (4).

On the other hand, we have

$$P_n(j) = \frac{1}{m}\left(1 + \sum_{r=1}^{m-1} T_n(r)E^{-rj}\right) \qquad (j = 0, 1, \cdots, m-1). \tag{7}$$

From this it follows that (4) implies (3). Hence, (3) and (4) are equivalent.

Theorem 1 is, of course, a result of the Fourier analysis appropriate to our problem. Other results of a kindred nature may be found in Lévy's paper [3].

3. Some sufficient conditions for equidistribution in the limit. From Theorem 1 we immediately deduce the following corollary.

THEOREM 2. *Let*

$$\mu_n = \min_{0 \le j \le m-1} p_n(j) \qquad (n = 1, 2, \cdots), \tag{8}$$

then

$$\sum_{n=1}^{\infty} \mu_n = \infty \tag{9}$$

implies (3).

Indeed, for $1 \le r \le m-1$ we have

$$| t_n(r) | = \left| \sum_{j=0}^{m-1} (p_n(j) - \mu_n)E^{rj} + \mu_n \sum_{j=0}^{m-1} E^{rj} \right| = \left| \sum_{j=0}^{m-1} (p_n(j) - \mu_n)E^{rj} \right|$$

$$\le \sum_{j=0}^{m-1} (p_n(j) - \mu_n) = 1 - m\mu_n .$$

Thus, (9) implies (4) and hence, by Theorem 1, (3).

The special case of Theorem 2 obtained by replacing (9) by the stronger requirement $\liminf_{n\to\infty} \mu_n > 0$ is due to Horton and Smith [2]. (As a matter of fact their method of proof—similar to the one we use in §4—yields also Theorem 2).

A somewhat deeper consequence of Theorem 1 is the following.

Theorem 3. *Let*

$$Q_n(r) = \max_{0 \le v \le m-1} \sum_{rj \equiv rv \pmod m} p_n(j) \qquad (r = 1, 2, \cdots, m-1; n = 1, 2, \cdots).$$

Then (3) *holds if*

$$\sum_{n=1}^{\infty} (1 - Q_n(r)) = \infty \qquad (r = 1, 2, \cdots, m-1). \tag{10}$$

Remark. For $m > 2$ the $m-1$ conditions (10) are not independent. It is easily seen that the number of independent conditions equals the number of different prime factors of m. More precisely, if $s_1, \cdots, s_\sigma$ are the different prime factors of m then the conditions (10) for $r = m/s_v$, $v = 1, \cdots, \sigma$, are independent but already entail (10) for all $1 \le r \le m-1$.

Proof of Theorem 3. Let $q_0, q_1, \cdots, q_{m-1}$ be non-negative numbers whose sum equals 1, and denote by Q the largest among them. Let t be any real number and let $E^{\varphi_0}, E^{\varphi_1}, \cdots, E^{\varphi_{m-1}}$ denote $E^t, E^{1+t}, \cdots, E^{m-1+t}$ rearranged according to increasing distances from the point 1. Then the real part of $\sum_{k=0}^{m-1} q_k E^{k+t}$ cannot exceed the larger of the two numbers

$$Q \cos \frac{2\pi\varphi_0}{m} + (1-Q) \cos \frac{2\pi\varphi_1}{m}, \qquad (1-Q) \cos \frac{2\pi\varphi_0}{m} + Q \cos \frac{2\pi\varphi_1}{m}.$$

But neither of these numbers exceeds the positive square root of

$$(1-Q)^2 + Q^2 + 2Q(1-Q) \cos \frac{2\pi}{m} = 1 - 2Q(1-Q)\left(1 - \cos \frac{2\pi}{m}\right).$$

Therefore, since $Q \ge 1/m$, the real part in question is bounded by $1 - C_m(1-Q)$, where C_m denotes a positive number depending only on m. This being true for all real t, the same must hold for the absolute value; i.e., we have

$$\left| \sum_{k=0}^{m-1} q_k E^k \right| \le 1 - C_m(1-Q) \qquad (C_m > 0). \tag{11}$$

Applying (11) to the definition of $t_n(r)$ we obtain

$$| t_n(r) | \leq 1 - C_{m_r}(1 - Q_n(r)),$$

where $m_r = m/(m, r)$. Thus (10) implies (4) and, hence, (3) and the proof is complete.

Theorem 3 contains the preceding one as a special case. It is worthwhile to reformulate it as follows.

The divergence of $\sum_{n=1}^{\infty} (1 - Q_n(r))$ is equivalent to the existence of pairs j_n and j'_n with $0 \leq j_n, j'_n \leq m - 1$ satisfying the conditions that $j'_n - j_n$ is not a multiple of m_r and $\sum_{n=1}^{\infty} \min (p_n(j_n), p_n(j'_n)) = \infty$. Putting

$$\delta_n(\lambda) = \max_{0 \leq j \leq m-1} \min (p_n(j), p_n(j + \lambda)) \qquad (\lambda = 1, 2, \cdots, m),$$

where $p_n(j + \lambda)$ denotes $p_n(j + \lambda - m)$ whenever $j + \lambda \geq m$, the above conditions are equivalent to the existence of a λ with m_r not a divisor of λ for which $\sum_{n=1}^{\infty} \delta_n(\lambda)$ is divergent.

Let $\{\lambda_1, \lambda_2, \cdots, \lambda_l\}$ be the set of those λ $(1 \leq \lambda \leq m)$ which satisfy

$$\sum_{n=1}^{\infty} \delta_n(\lambda) = \infty. \tag{12}$$

This set is certainly non-void since $\lambda = m$ satisfies (12). Denoting by Λ the greatest common divisor of this set the preceding considerations show that Theorem 3 may be restated as

THEOREM 4. *If* $\Lambda = 1$ *then* (3) *holds.*

As an immediate corollary we obtain

THEOREM 5. *Let*

$$M_n = \max_{j=0,\cdots,m-1} p_n(j) \qquad (n = 1, 2, \cdots)$$

then the divergence of $\sum_{n=1}^{\infty} (1 - M_n)$ *implies* (3) *whenever* m *is prime.*

4. **Rate of approach to equidistribution.** Let $P'_n(j) = P_n(j) - 1/m$, and $\alpha_n = \max_{0 \leq j \leq m-1} | P'_n(j) |$. Since

$$P'_{n+1}(j) = \sum_{k=0}^{m-1} P'_n(k) p_{n+1}(j - k) \tag{13}$$

(where $p_{n+1}(j - k)$ is defined as $p_{n+1}(j - k + m)$ if $j < k$), we have $\alpha_{n+1} \leq \alpha_n$. Thus the numbers $\max_j | P_n(j) - 1/m |$ always form a monotone non-increasing sequence. In particular if Y_n is equidistributed for some value of n, it remains so from that value on.

We shall use (13) to establish the following result.

THEOREM 6. *Let U_n denote the sum of the $[m/2]$ largest numbers among $p_n(0)$, $p_n(1), \cdots, p_n(m-1)$ and L_n the sum of the $[m/2]$ smallest among them. Then*

$$\alpha_{n+1} \leq (U_{n+1} - L_{n+1})\, \alpha_n, \tag{14}$$

where $\alpha_n = \max_{0 \leq j \leq m-1} | P_n(j) - 1/m |$.

A similar result, with α_n replaced by $\max_{0 \leq j \leq j' \leq m-1} | P_n(j') - P_n(j) |$, has been established by Horton and Smith [2].

Proof. Let $a_1, a_2, \cdots, a_m$ be real numbers satisfying

$$| a_j | \leq a, \qquad \sum a_j = 0 \qquad (j = 1, 2, \cdots, m), \tag{15}$$

and $b_1, b_2, \cdots, b_m$ a set of non-negative numbers satisfying

$$b_1 \geq b_2 \geq \cdots \geq b_m . \tag{16}$$

Putting

$$A = a \sum_{j=1}^{[m/2]} b_j - a \sum_{j=1}^{[m/2]} b_{m+1-j}, \quad B = \sum_{j=1}^{m} a_j b_j,$$

we have $A - B = \sum_{j=1}^m c_j b_j$, where, by (15), the first $[m/2]$ of the c_j are non-negative, the last $[m/2]$ are non-positive, and $\sum_{j=1}^m c_j = 0$. Let C denote the sum of the positive c_j and $k = [m/2]$ or $[m/2] + 1$, according to whether $c_{[m/2]+1}$ is or is not negative. Then by (16)

$$\sum_{j=1}^{m} c_j b_j \geq C b_k - C b_{k+1} \geq 0,$$

whence $A \geq B$. In the same way we obtain $B \geq -A$ and thus $| B | \leq A$.

We now apply the above with b_j as the j-th largest among $p_{n+1}(0), \cdots, p_{n+1}(m-1)$ and a_j as the corresponding one among $P'_n(0), \cdots, P'_n(m-1)$ which multiplies it in (13). The role of a may be taken over by α_n and we see that the right member of (13) cannot exceed $A = \alpha_n(U_{n+1} - L_{n+1})$ in absolute value. This is precisely the statement (14).

As an immediate consequence of Theorem 6 we have

THEOREM 7. *If the series $\sum_{n=1}^{\infty} (1 - U_n)$ is divergent then* (3) *holds.*

5. Essential and accidental equidistribution in the limit. As already noted in §4, if Y_N is equidistributed then, no matter what $X_{N+1}, X_{N+2}, \cdots$ are, $Y_{N+1}, Y_{N+2}, \cdots$ are also equidistributed and, *a fortiori*, (3) holds. However, if $X_{N+1}, X_{N+2}, \cdots$ have distributions which are too heavily concentrated (an extreme example of this is given by $p_n(0) = 1$ for $n > N$) then the convergence to (3) is due entirely to Y_N being equidistributed, i.e., to the N first random variables $X_1, X_2, \cdots, X_N$, and may thus be affected by the omission (or replacement) of these.

We say that the convergence to equidistribution is *essential* if it cannot be altered by replacing any finite number of the X_n by any other random variables (which of course also assume only integral values). Otherwise the convergence to equidistribution is called *accidental.*

Let N be a positive integer and put

$$\bar{S}_n = X_{N+1} + \cdots + X_{N+n} \qquad (n = 1, 2, \cdots).$$

It is quite clear that the convergence to equidistribution is essential if and only if we have

$$\lim_{n\to\infty} \text{Prob}\,\{\bar{S}_n \equiv j \pmod m\} = \frac{1}{m} \qquad (j = 0, 1, \cdots, m-1) \tag{17}$$

for all N.

The possibility of (3) holding "accidentally" is responsible for conditions of the type given in §3 never being necessary for equidistribution in the limit. In case of Theorem 1 this is automatically taken care of since the relations (4) may be satisfied in two ways—either the product is divergent or it converges to the value zero; these correspond respectively to essential and accidental equidistribution in the limit.

If we, however, limit ourselves to the essential case we have the following complement to §3.

THEOREM 8. *The conditions of Theorems* 3, 4 *and* 5 *are necessary as well as sufficient for essential equidistribution in the limit.*

The sufficiency is evident since the divergence of a series is unaffected by changes in a finite number of terms.

To prove the necessity we start with Theorem 5. A cyclic permutation of the values assumed by X_n induces a cyclic permutation of the values assumed by Y_n and thus leaves unaffected the relations (3) and (17). We may therefore assume $p_n(0) = M_n$ for all n.

But then

$$\text{Prob}\,\{\bar{S}_n = 0\} = \prod_{\nu=1}^{n} M_{N+\nu} = \prod_{\nu=1}^{n} (1 - (1 - M_{N+\nu})) \geq \prod_{\nu=N+1}^{\infty} (1 - (1 - M_\nu))$$

and if $\sum (1 - M_n)$ is convergent, the last member can be made arbitrarily near to 1 by taking N large enough, thus contradicting (17).

Theorems 3 and 4 being equivalent it is sufficient to deal with the second. But Λ is necessarily a divisor of m and thus, if $\Lambda > 1$, it is sufficient to consider the values of X_n and S_n reduced mod Λ in order to reduce the case of Theorem 4 to that of Theorem 5.

6. **Examples.** The $m - 1$ conditions (4) are not independent and, besides, they are not formulated in the most suitable form for application. We therefore illustrate the passage to handier formulations by considering separately the cases $m = 2$, 3, and 4.

(a) $m = 2$. In this case (4) reduces to

$$\prod |\, p_n(0) - p_n(1)\,| = \prod (1 - 2\mu_n) = 0$$

which at once yields

If $m = 2$ then (3) holds if, and only if, one, at least, of the X_n is equidistributed, or $\sum \mu_n$ is divergent (or both).

(b) $m = 3$. By Theorem 5, the divergence of $\sum (1 - M_n)$ implies (3). Writing $\omega = \exp(2\pi i/3)$, we have on the other hand

$$| p_n(0)\omega + p_n(1)\omega^2 + p_n(2) | \geq M_n - \tfrac{1}{2}(1 - M_n) = 1 - \tfrac{3}{2}(1 - M_n).$$

Hence, if $M_n > \frac{1}{3}$ for all n and $\sum (1 - M_n)$ is convergent, (4) and hence, by Theorem 1, (3), cannot hold. Thus we have

If $m = 3$, (3) holds if, and only if, one at least of the X_n is equidistributed, or $\sum (1 - M_n)$ is divergent (or both).

We notice that this formulation is also valid for $m = 2$.

(c) $m = 4$. In this case (4) reduces to the two conditions

$$\prod_{n=1}^{\infty} \{(p_n(0) - p_n(2))^2 + (p_n(1) - p_n(3))^2\} = 0 \tag{18}$$

and

$$\prod_{n=1}^{\infty} | p_n(0) - p_n(1) + p_n(2) - p_n(3) | = 0. \tag{19}$$

The factors of (18) satisfy, for all n for which $M_n \geq \frac{1}{2}$,

$$[1 - 2(1 - M_n)]^2 = [M_n - (1 - M_n)]^2 \leq \{ \ \}$$
$$\leq M_n^2 + (1 - M_n)^2 = 1 - 2M_n(1 - M_n) \leq 1 - \tfrac{1}{2}(1 - M_n).$$

Thus $\sum (1 - M_n) = \infty$ implies (18) while if $\sum (1 - M_n)$ is convergent (18) cannot hold unless at least one of its factors vanishes.

Putting $\Delta_n = \min (p_n(0) + p_n(2), p_n(1) + p_n(3))$, (19) may be written as $\prod (1 - 2\Delta_n) = 0$. Thus, (19) holds if $\sum \Delta_n = \infty$ while otherwise it cannot hold unless $\Delta_n = \frac{1}{2}$ for some n. Since $\Delta_n \leq 1 - M_n$ the above can be summed up as follows:

If $m = 4$ then (3) holds if and only if either 1) $\sum \Delta_n = \infty$, or 2) $\Delta_n = \frac{1}{2}$ for at least one n and $\sum (1 - M_n) = \infty$, or 3) $\Delta_n = \frac{1}{2}$ for at least one n and $p_n(0) = p_n(2)$, $p_n(1) = p_n(3)$ for at least one value of n.

References

1. H. Burke Horton, *A method for obtaining random numbers*, Annals of Mathematical Statistics, vol. 19(1948), pp. 81–85.
2. H. Burke Horton and R. Tynes Smith III, *A direct method for producing random digits in any number system*, Annals of Mathematical Statistics, vol. 20(1949), pp. 82–90.
3. Paul Lévy, *L'addition des variables aléatoires définies sur une circonférence*, Bulletin de la Société Mathématique de France, vol. 67(1939), pp. 1–41.

Institute for Advanced Study
and
Columbia University.

Reprinted from THE ANNALS OF MATHEMATICAL STATISTICS
Vol. 23, No. 3, September, 1952

STOCHASTIC ESTIMATION OF THE MAXIMUM OF A REGRESSION FUNCTION[1]

By J. KIEFER AND J. WOLFOWITZ

Cornell University

1. Summary. Let $M(x)$ be a regression function which has a maximum at the unknown point θ. $M(x)$ is itself unknown to the statistician who, however, can take observations at any level x. This paper gives a scheme whereby, starting from an arbitrary point x_1, one obtains successively $x_2, x_3, \cdots$ such that x_n converges to θ in probability as $n \to \infty$.

2. Introduction. Let $H(y \mid x)$ be a family of distribution functions which depend on a parameter x, and let

$$M(x) = \int_{-\infty}^{\infty} y \, dH(y \mid x). \tag{2.1}$$

We suppose that

$$\int_{-\infty}^{\infty} (y - M(x))^2 \, dH(y \mid x) \leqq S < \infty, \tag{2.2}$$

and that $M(x)$ is strictly increasing for $x < \theta$, and $M(x)$ is strictly decreasing for $x > \theta$. Let $\{a_n\}$ and $\{c_n\}$ be infinite sequences of positive numbers such that

$$c_n \to 0, \tag{2.3}$$

$$\sum a_n = \infty, \tag{2.4}$$

$$\sum a_n c_n < \infty, \tag{2.5}$$

$$\sum a_n^2 c_n^{-2} < \infty. \tag{2.6}$$

(For example, $a_n = n^{-1}$, $c_n = n^{-1/3}$.)

We can now describe a recursive scheme as follows. Let z_1 be an arbitrary number. For all positive integral n we have

$$z_{n+1} = z_n + a_n \frac{(y_{2n} - y_{2n-1})}{c_n}, \tag{2.7}$$

where y_{2n-1} and y_{2n} are independent chance variables with respective distributions $H(y \mid z_n - c_n)$ and $H(y \mid z_n + c_n)$. Under regularity conditions on $M(x)$ which we shall state below we will prove that z_n converges stochastically to θ (as $n \to \infty$).

The statistical importance of this problem is obvious and need not be discussed. The stimulus for this paper came from the interesting paper by Robbins and Monro [1] (see also Wolfowitz [2]).

---

[1] Research under contract with the Office of Naval Research. Presented to the American Mathematical Society at New York on April 25, 1952.

While we have no need to postulate the existence of the derivative of $M(x)$ (indeed, $M(x)$ can be discontinuous), the spirit of our regularity assumptions postulated below is as follows. (a) If $M(x)$ did have a derivative it would be zero at $x = \theta$. Hence we would have expected the derivative not to be too large in a neighborhood of $x = \theta$. (b) If, at a distance from θ, $M(x)$ were very flat, then movement towards θ would be too slow. Hence outside of a neighborhood of $x = \theta$ we would have liked the absolute value of the derivative to be bounded below by a positive number. (c) If $M(x)$ rose too steeply in places we might through mischance get a movement of z_n which would throw us far out from θ. If there were many such steep places z_n could be made to approach $+\infty$ or $-\infty$ with positive probability. We would therefore have postulated a Lipschitz condition.

From the mathematical point of view it would be aesthetic to weaken the conditions. From the practical point of view it might be objected that these conditions prevent $M(x)$ from being a function which flattens out toward the x-axis, for example, $M(x) = e^{-x^2}$, or from being a function which drops off steadily faster to $-\infty$, for example, $M(x) = -x^2$. Now in any practical situation one can always give a priori an interval $[C_1, C_2]$ such that $C_1 \leqq \theta \leqq C_2$. It will be sufficient if our conditions are fulfilled in this interval.

Suppose, however, that some $z_n \pm c_n$ falls outside the interval $[C \quad C_2]$ and one cannot take an observation at that level. If one then moves z_n so that the offending $z_n \pm c_n$ is at C_1 or C_2, as the case may be, and proceeds as directed by (2.7), then our conclusion remains valid.

We postulate the following regularity conditions on $M(x)$.

CONDITION 1. There exist positive β and B such that

$$|x' - \theta| + |x'' - \theta| < \beta \text{ implies } |M(x') - M(x'')| < B|x' - x''|. \tag{2.8}$$

CONDITION 2. There exist positive ρ and R such that

$$|x' - x''| < \rho \text{ implies } |M(x') - M(x'')| < R. \tag{2.9}$$

CONDITION 3. For every $\delta > 0$ there exists a positive $\pi(\delta)$ such that

$$|z - \theta| > \delta \text{ implies } \inf_{\frac{1}{2}\delta > \epsilon > 0} \frac{|M(z + \epsilon) - M(z - \epsilon)|}{\epsilon} > \pi(\delta). \tag{2.10}$$

3. Proof that z_n converges stochastically to 0. Let

$$b_n = E(z_n - \theta)^2, \tag{3.1}$$

$$U_n(z) = (z - \theta)\, E\{y_{2n} - y_{2n-1} \mid z_n = z\}, \tag{3.2}$$

$$U_n^+(z) = \tfrac{1}{2}(U_n(z) + |U_n(z)|),\ U_n^-(z) = \tfrac{1}{2}(U_n(z) - |U_n(z)|), \tag{3.3}$$

$$P_n = E(U_n^+(z_n)),\ N_n = E(U_n^-(z_n)), \tag{3.4}$$

$$e_n = E(y_{2n} - y_{2n-1})^2. \tag{3.5}$$

From (2.7) we have

$$b_{n+1} = b_n + 2\frac{a_n}{c_n}(P_n + N_n) + \frac{a_n^2}{c_n^2}e_n. \tag{3.6}$$

Adding the expressions obtained from (3.6) for $b_{j+1} - b_j$ for $1 \leqq j \leqq n$, we obtain

$$b_{n+1} = b_1 + 2\sum_{j=1}^{n}\frac{a_j}{c_j}P_j + 2\sum_{j=1}^{n}\frac{a_j}{c_j}N_j + \sum_{j=1}^{n}\frac{a_j^2}{c_j^2}e_j. \tag{3.7}$$

Noting that $U_n^+(z) \geqq 0$ and that $U_n^+(z) > 0$ implies that $|z - \theta| < c_n$ because $M(x)$ is monotonic for $x < \theta$ and for $x > \theta$, it follows from (2.8) that, for all n for which $c_n < \frac{1}{2}\beta$, we have

$$0 \leqq U_n^+(z) < 2Bc_n^2 \tag{3.8}$$

It follows from (2.5) and (3.8) that the positive-term series

$$\sum_{n=1}^{\infty}\frac{a_n}{c_n}P_n \tag{3.9}$$

converges, say to α. From (2.9) we have

$$[M(z_n + c_n) - M(z_n - c_n)]^2 < R^2 \tag{3.10}$$

for n sufficiently large. Also for large enough n,

$$\begin{aligned} &E\{(y_{2n} - y_{2n-1})^2 \mid z_n\} \\ &= E\{(y_{2n} - M(z_n + c_n))^2 + (y_{2n-1} - M(z_n - c_n))^2 \mid z_n\} \\ &\qquad + [M(z_n + c_n) - M(z_n - c_n)]^2 \leqq 2S + R^2 \end{aligned} \tag{3.11}$$

by (2.2) and (3.10). Hence for large enough n

$$E[y_{2n} - y_{2n-1}]^2 \leqq 2S + R^2. \tag{3.12}$$

Consequently from (2.6) we obtain that the positive-term series

$$\sum_{n=1}^{\infty}\frac{a_n^2}{c_n^2}e_n \tag{3.13}$$

converges, say to γ. Hence, since $b_{n+1} \geqq 0$, it follows from (3.7) that

$$2\sum_{j=1}^{n}\frac{a_j}{c_j}N_j \geqq -b_1 - 2\alpha - \gamma > -\infty, \tag{3.14}$$

so that the negative-term series

$$\sum_{n=1}^{\infty}\frac{a_n}{c_n}N_n \tag{3.15}$$

converges.

Let

$$K_n = \left| \frac{M(z_n + c_n) - M(z_n - c_n)}{c_n} \right|. \tag{3.16}$$

Then

$$E\{K_n \mid z_n - \theta \mid \} = \frac{P_n - N_n}{c_n}. \tag{3.17}$$

From the convergence of (3.9) and (3.15) and the divergence of $\sum a_n$, it follows that

$$\liminf_{n\to\infty} E\{K_n \mid z_n - \theta \mid \} = 0. \tag{3.18}$$

Let $n_1 < n_2 < \cdots$ be an infinite sequence of positive integers such that

$$\lim_{j\to\infty} E\{K_{n_j} \mid z_{n_j} - \theta \mid \} = 0. \tag{3.19}$$

We assert that $(z_{n_j} - \theta)$ converges stochastically to zero as $j \to \infty$. For if not, there would exist two positive numbers δ and ϵ and a subsequence $\{t_j\}$ of $\{n_j\}$ such that, for all j,

$$P\{ \mid z_{t_j} - \theta \mid > \delta\} > \epsilon, \tag{3.20}$$

which implies that

$$E\{K_{t_j} \mid z_{t_j} - \theta \mid \} \geqq \delta\epsilon\pi\left(\frac{\delta}{2}\right) > 0 \tag{3.21}$$

for all j for which $c_{t_j} < \frac{1}{2}\delta$. But (3.21) contradicts (3.19) and the stochastic convergence to zero of $(z_{n_j} - \theta)$ is proved.

Let η and ϵ be arbitrary positive numbers. The proof of the theorem will be complete if we can show the existence of an integer $N(\eta, \epsilon)$ such that

$$P\{ \mid z_n - \theta \mid > \eta\} \leqq \epsilon \text{ for } n > N(\eta, \epsilon). \tag{3.22}$$

Let s be a positive number such that

$$\frac{s^2 + s}{\eta^2} < \frac{\epsilon}{2}. \tag{3.23}$$

Because z_{n_j} converges stochastically to θ there exists an integer N_0 such that

$$P\{\mid z_{N_0} - \theta \mid \geqq s\} < \frac{\epsilon}{2}. \tag{3.24}$$

We may also choose N_0 so large that

$$c_n < \min\left(\frac{\rho}{2}, \frac{\beta}{2}\right) \text{ for all } n \geqq N_0, \tag{3.25}$$

and

$$\sum_{n=N_0}^{\infty} \frac{a_n^2}{c_n^2} < \frac{s}{2R^2 + 4S}, \tag{3.26}$$

and

$$\sum_{n=N_0}^{\infty} a_n c_n < \frac{s}{8B}. \tag{3.27}$$

Proceeding in a manner similar to that used to obtain (3.7), we have, for each $n > N_0$,

$$\begin{aligned} E\{(z_n - \theta)^2 \mid z_{N_0} = z\} &= (z - \theta)^2 + 2 \sum_{j=N_0}^{n-1} \frac{a_j}{c_j} E\{U_j \mid z_{N_0} = z\} \\ &\quad + \sum_{j=N_0}^{n-1} \frac{a_j^2}{c_j^2} E\{(y_{2j} - y_{2j-1})^2 \mid z_{N_0} = z\} \end{aligned} \tag{3.28}$$

$$\leqq (z - \theta)^2 + 2 \sum_{j=N_0}^{\infty} \frac{a_j}{c_j} E\{U_j^+ \mid z_{N_0} = z\} + (R^2 + 2S) \sum_{j=N_0}^{\infty} \frac{a_j^2}{c_j^2} < (z - \theta)^2 + s.$$

Using (3.23), (3.28), and Tchebycheff's inequality, we have

$$P\{| z_n - \theta | > \eta \,\big|\, | z_{N_0} - \theta | < s\} < \frac{\epsilon}{2}. \tag{3.29}$$

The inequalities (3.24) and (3.29) show that (3.22) holds for $N(\eta, \epsilon) = N_0$, and the proof is complete.

4. Further problems. The following remarks about further problems apply also to [1].

A. An obvious problem is to determine sequences $\{c_n\}$ and $\{a_n\}$ which would be optimal in some reasonable sense.

B. An important problem is to determine a stopping rule, that is, a rule by which the statistician decides when he is sufficiently close to θ.

C. This problem is a combination of *B* and a generalization of *A*, that is, to determine an optimal procedure with its stopping rule.

REFERENCES

[1] H. Robbins and S. Monro, "A stochastic approximation method," *Annals of Math. Stat.*, Vol. 22 (1951), pp. 400–407.

[2] J. Wolfowitz, "On the stochastic approximation method of Robbins and Monro," *Annals of Math. Stat.*, Vol. 23 (1952), pp. 457–461.

KONINKL. NEDERL. AKADEMIE VAN WETENSCHAPPEN - AMSTERDAM
Reprinted from Proceedings, Series A, **56**, No. 2 and Indag. Math., **15**, No. 2, 1953

MATHEMATICS

THE METHOD OF MAXIMUM LIKELIHOOD AND THE WALD THEORY OF DECISION FUNCTIONS

BY

J. WOLFOWITZ

(Communicated by Prof. D. VAN DANTZIG at the meeting of December 20, 1952)

1. The purpose of the present note is to help unite two currents of thought in mathematical statistics. The new theory of statistical decision functions founded by WALD embraces the entire theory of statistical inference, but hitherto there has not been any special contact between it and the theory of maximum likelihood estimators. In this note we show that the theory of decision functions explains why the maximum likelihood estimator is asymptotically efficient. The argument is essentially this:

a) the theory of decision functions shows that a complete class of estimators for the problem we shall shortly describe consists entirely of Bayes solutions (a complete class is, roughly speaking, the class of the only solutions which the statistician need consider seriously);

b) asymptotically every Bayes solution for suitable frequency functions becomes a maximum likelihood estimator. Thus asymptotically the maximum likelihood estimator is efficient because it has no competitors which the statistician need seriously consider.

This argument is developed in Sections 2 and 3, is non-rigorous throughout, and even the theorem of Section 2 is not precisely formulated. The reason for proceeding in this manner is that the several parts of this kind of argument have already been repeated in the literature many times. We content ourselves therefore with giving, in Section 2, an infallible prescription for the precise formulation and proof of such a theorem.

Section 4 is written to provide some heuristic considerations on the very important problem [1]) which was posed by NEYMAN and SCOTT [5].

2. Let $f(x, \theta)$ be the frequency function of a chance variable on which we have n independent observations $x_1, \ldots, x_n$ (i.e., $x_1, \ldots, x_n$ are n independent chance variables with the same frequency function $f(x, \theta)$, where θ is an unknown parameter to be estimated [2]). Suppose for the moment that θ is itself a chance variable with the (a priori) density function $g(\theta)$. Then the a posteriori (conditional) density of θ, given

[1]) Results on this problem obtained by the present author will appear shortly in the Skandinavisk Aktuarietidskrift and elsewhere.

[2]) Hereafter we shall write as if $f(x, \theta)$ is a density function. Obvious changes will be needed if $f(x, \theta)$ is a (discrete) probability function.

$x_1, \ldots, x_n$, is of course given by

$$(1) \qquad h_g(\theta \mid x_1, \ldots, x_n) = \frac{g(\theta)\, f(x_1, \theta) \ldots f(x_n, \theta)}{\int g(\theta)\, f(x_1, \theta) \ldots f(x_n, \theta)\, d\theta}.$$

The phenomenon we want to describe now was first noted by v. MISES [1], but it will be simpler to describe the result of KOLMOGOROFF [3]. Suppose

$$f(x, \theta) = \frac{1}{\sqrt{2\pi}} e^{-\frac{1}{2}(x-\theta)^2}.$$

Then

$$\prod_{i=1}^{n} f(x_i, \theta) = t(x_1, \ldots, x_n) \exp\left\{-\frac{n}{2}(\bar{x} - \theta)^2\right\}$$

where $t(x_1, \ldots, x_n)$ is a function of $x_1, \ldots, x_n$ only, and $\bar{x} = n^{-1} \Sigma^i x_i$. When n is large and $g(\theta)$ has desirable regularity properties (according to KOLMOGOROFF, if it has a bounded first derivative and $g(\bar{x}) \neq 0$), the numerator of the right member of (1) and the integrand in the denominator of the right member of (1) are negligible outside of a small neighborhood of $\bar{x}$, within which $g(\theta)$ is practically constant. One can then obtain that, asymptotically, the a posteriori density function of θ is normal, with mean $\bar{x}$ and variance n^{-1}, *no matter what the a priori density function $g(\theta)$ is*, i.e., *the influence of $g(\theta)$ vanishes in the limit*. (This fact was proved earlier by v. MISES [1] for the binomial distribution.)

It would seem plausible that these results should hold in great generality, and this is indeed the case. Let $\hat{\theta}(x_1, \ldots, x_n)$ be the maximum likelihood estimator of θ, i.e., the function which maximizes $\Pi^i f(x_i, \theta)$ with respect to θ. If regularity conditions obtain which make $\hat{\theta}$ a consistent estimator and permit one essentially to ignore higher order terms in the Taylor series for $\Sigma^i \log f(x_i, \theta)$ (a standard statistical argument to which we give references below) one has then, with probability arbitrarily close to one for n sufficiently large,

$$(2) \quad \left\{ \begin{aligned} \sum_{i=1}^{n} \log f(x_i, \theta) &\sim \sum_{i=1}^{n} \log f(x_i, \hat{\theta}) + \tfrac{1}{2}\left[\sum_{i=1}^{n} \frac{\partial^2 \log f(x_i, \theta)}{\partial \theta^2}\right]_{\theta = \hat{\theta}} (\theta - \hat{\theta})^2 \\ &\sim \sum_{i=1}^{n} \log f(x_i, \hat{\theta}) - \tfrac{1}{2}\, n c(\hat{\theta})\, (\theta - \hat{\theta})^2 \end{aligned} \right.$$

where

$$c(\theta) = \int \left(\frac{\partial \log f(x, \theta)}{\partial \theta}\right)^2 f(x, \theta)\, dx.$$

From (2) one obtains

$$(3) \qquad \prod_{i=1}^{n} f(x_i, \theta) \sim \prod_{i=1}^{n} f(x_i, \hat{\theta}) \cdot e^{-\frac{1}{2} n c(\hat{\theta})(\theta - \hat{\theta})^2}.$$

Insert this in (1). The effect of the fact that n is large is, exactly as before, to cause the numerator and the integrand in the denominator of (1) to become negligible outside of a small neighborhood of $\hat{\theta}$, within which $g(\theta)$ is "practically" constant. Just as before, therefore, under suitable regularity conditions and with the usual meaning of the word "asymptotic", one could rigorously obtain a simple result like the following: *Let θ_0 be any fixed value of θ. As $n \to \infty$ the probability (according to the probability measure whose density function is $\Pi_i\, f(x_i, \theta_0)$)* approaches one that the a posteriori density function $h_g(\theta | x_1, \ldots, x_n)$ *(which, as a function of $x_1, \ldots, x_n$, is a chance function) is asymptotically the normal density* function with mean $\hat{\theta}$ and variance $\{nc(\hat{\theta})\}^{-1}$.

3. Consider now the problem of estimating θ as a decision problem in the sense of WALD (which it of course is; see for example [6]) with the weight function

$$W(\theta, \delta(x^*)) = K(\theta - \delta(x^*))^2,$$

where K is a constant and $\delta(x^*)$ is the estimate of θ from the sample $x^* = x_1, \ldots, x_n$. (Actually it would be sufficient to take for W any function which is a monotonically non-decreasing function of $|\theta - \delta(x^*)|$ (cf. WOLFOWITZ [7]).) It has been proved by WALD under various conditions ([6]) that the totality of Bayes solutions consitute a complete class. The intuitive (and geometric) basis of this result of WALD's is that the Bayes solutions correspond to certain obviously "efficient" points on the boundary of the convex body which is the totality of all risk points (see WALD and WOLFOWITZ [8]). A Bayes solution with respect to an a priori density function $g(\theta)$ is one which minimizes

$$\int W(\theta, \delta(x^*)) \left(\prod_{i=1}^{n} f(x_i, \theta)\right) g(\theta) \mathrm{d}x_1 \ldots \mathrm{d}x_n \mathrm{d}\theta \tag{4}$$

with respect to $\delta(x^*)$. A Bayes solution (minimizing $\delta_g(x^*)$) is, for our weight function, obviously given by

$$\delta_g(x^*) = \int \theta h_g(\theta | x^*) \mathrm{d}\theta, \tag{5}$$

i.e., the mean of the a posteriori distribution. In view of our earlier remarks one could obtain that asymptotically the only Bayes solution with respect to a priori density functions $g(\theta)$ which satisfy appropriate requirements is the maximum likelihood estimator $\hat{\theta}$ (the usual exceptional sets must be admitted).

Suppose the regularity conditions on $f(x, \theta)$ are such that the distribution of $\sqrt{nc(\theta)}\,(\hat{\theta} - \theta)$ is asymptotically normal, with mean zero and variance one. Then there cannot be an estimator θ^* of θ which is also asymptotically normally distributed about θ with a variance never greater than that of $\hat{\theta}$, and less than the variance of $\hat{\theta}$ on a θ-set of positive measure. For then $\hat{\theta}$ would not be a Bayes solution with respect to one of our $g(\theta)$, in violation of the above. Indeed there cannot exist an estimator θ^{**} with the following properties: 1) except on a θ-set of measure zero the variance $n^{-1}\{\sigma_1^2(\theta) + o(1)\}$ of θ^{**} is finite, 2) on a θ-set of positive measure $\sigma_1^2(\theta) < \{c(\theta)\}^{-1}$, the variance of $\hat{\theta}$ being $\{1 + o(1)\}\,\{nc(\theta)\}^{-1}$. For as before we could construct a density function $g(\theta)$ such that

$$\int \sigma_1^2(\theta) g(\theta) \mathrm{d}\theta < \int \frac{1}{c(\theta)}\, g(\theta) \mathrm{d}\theta,$$

in violation of the fact that $\hat{\theta}$ is a Bayes solution (asymptotically). Thus $\hat{\theta}$ is asymptotically uniformly efficient in the classical sense.

It remains only to give a prescription for making rigorous the discussion in Section 2. One postulates regularity conditions on $f(x, \theta)$ to ensure that $\hat{\theta}$ is a consistent estimator and that $\sqrt{n}\,(\hat{\theta} - \theta)$ is asymptotically normally distributed. A good set would be those of CRAMÉR ([4]), page 500 et seq.). Then an additional condition is added so that the stochastic convergence to zero of $(\hat{\theta} - \theta)$ and the approach to the normal distribution of the distribution of $\sqrt{n}\,(\hat{\theta} - \theta)$ are both uniform in θ. One is then approximately at the point where KOLMOGOROFF ([3]) begins and one postulates his conditions, again with due regard to uniformity (in θ). Of course, reasonable care will need to be exercised with some minor questions. For example, the argument requires not only that the a posteriori density should approach the normal density, but also that the mean of the a posteriori distribution approach θ, the mean of the normal density. There are many ways of achieving this; a standard one is to ensure that the $(1 + \varepsilon)^{\text{th}}$ moment, $\varepsilon > 0$, is bounded uniformly in n.

4. Consider now the following estimation problem first posed by NEYMAN and SCOTT [5]: Let $n_1, n_2, \ldots$ be an infinite sequence of integers. Let $F(x|\theta, a)$ be a distribution function which depends upon the parameters θ and a. Let $\{x_{ij}\}$, $i = 1, 2, \ldots$, ad. inf., $j = 1, \ldots, n_i$, be a sequence of independent chance variables such that $x_{il}, \ldots, x_{in_i}$ have the same distribution function $F(x|\theta_0, a_i)$. The constants $\theta_0, a_1, a_2, \ldots$, ad. inf. are unknown. The problem now is to construct a sequence of consistent estimators $M_j(x_{11}, \ldots, x_{jn_j})$ of $\theta_0 (j = 1, 2, \ldots$, ad. inf.). NEYMAN and SCOTT have shown by an example that the maximum likelihood estimator, even if consistent, need not be efficient.

Before attempting to explain this fact let us describe another phenomenon also first noticed by v. MISES [9]. Let $G(x)$ be the distribution function of a chance variable Z, and let $G(0) = 1 - G(1) = 0$. Let $z_1, z_2, \ldots$ be a sequence of independent observations on Z. Let $y_1, y_2, \ldots$ be a sequence of independent chance variables such that

$$\mathrm{P}\{y_i = 1\} = z_i, \ \mathrm{P}\{y_i = 0\} = 1 - z_i.$$

What is the a posteriori distribution of $n^{-1}\Sigma^i z_i$, given $n^{-1}\Sigma^i y_i$? Or rather, what is the asymptotic a posteriori distribution as $n \to \infty$? VON MISES [9] proves that the asymptotic a posteriori distribution is again normal, but *with mean and variance which in general are functions of the a priori distribution function $G(x)$, i.e., the influence of the a priori distribution does not disappear with increasing n.*

The dependence, described above, of the asymptotic a posteriori distribution on the a priori distribution, may help to explain the phenomenon pointed out by NEYMAN and SCOTT. For let us try to generate the complete class of estimators for our previous weight function by using all a priori distribution functions. Let α_i be the incidental parameter of the ith group of observations. There would be no point in using an a priori distribution of $\theta, \alpha_1, \ldots, \alpha_m$. Since the individual n_i will in general not approach infinity, we could not expect asymptotic results independent of the a priori distribution. The simplest procedure would be to assume an a priori density function

$$g(\theta, \alpha) = g_1(\theta) g_2(\alpha | \theta),$$

with $g_1(\theta)$ the marginal density of θ, and $g_2(\alpha|\theta)$ the conditional density of α given θ, and to envision the operation of $g(\theta, \alpha)$ as follows: Nature [3]) uses a chance mechanism with density function $g_1(\theta)$ to obtain the value θ_0 which she will present to the statistician. Before each group of n_i observations, $i = 1, 2, \ldots$, Nature selects the parameter α_i of the chance variables to be presented to the statistician by means of an independent chance mechanism with density function $g_2(\alpha|\theta_0)$. If now the maximum likelihood estimator $\hat{\theta}$ were to be asymptotically uniformly efficient in the classical sense, it would be asymptotically the only Bayes solution with respect to all suitable a priori densities $g(\theta)$. This will in general not be so, because the Bayes estimators for our problem are the means of the a posteriori distributions, and the result of v. Mises [9] shows that the mean of the a posteriori distribution is in general a function of the a priori distribution.

5. Let us now return to the equation (3). The result of Neyman [10] says essentially that a necessary and sufficient condition for J to be a sufficient statistic for θ is that the frequency function (likelihood) of $x_1, \ldots, x_n$ be expressible as a product of two factors, one a function of the observations only, and the other a function of J and θ only. The equation (3) could be made to show that this is asymptotically so for $J = \hat{\theta}$ (with probability close to 1). Thus under suitable regularity conditions, $\hat{\theta}$ would be asymptotically sufficient, i.e., the ratio of the values of the frequency function at any two points of a contour $\hat{\theta} =$ constant of the maximum likelihood estimator $\hat{\theta}$, asymptotically would not depend on θ, or, to put it differently, the conditional distribution on a contour $\hat{\theta} =$ constant of the maximum likelihood estimator $\hat{\theta}$ would be asymptotically independent of θ.

The fact that $\hat{\theta}$ is asymptotically sufficient explains the result of Wald [11]. When his c (in [11]) is very small the probability is large that the number of observations required will exceed a large number. Since $\hat{\theta}$ is asymptotically sufficient it follows that, when the number of observations is large, the stopping rule will be a function of $\hat{\theta}$ only. Also, when the

[3]) It is our opinion that, in most practical problems, it is inadmissible to consider θ as a chance variable. Nor do we subscribe to the naively teleological view of Nature which a literal reading of this passage would imply. The use of the a priori distribution is a purely mathematical device, and we put the matter as we do only in the interest of simplicity of explanation.

number of observations is large $\hat{\theta}$ will be practically normally distributed with mean θ and variance $\{nc(\theta)\}^{-1}$. A variation of the argument in [7] (which is conducted for the case when the chance variables are exactly normally distributed) then will give WALD's result.

REFERENCES

1. MISES, R. v., "Fundamentalsätze der Wahrscheinlichkeitsrechnung" Mathematische Zeitschrift, 4, 1–97 (1919).
2. ———, "Wahrscheinlichkeitsrechnung und ihre Anwendungen", (MARY S. ROSENBERG, Publisher, New York, 1945).
3. KOLMOGOROFF, A. N., "Determination of the center of dispersion and of the accuracy, on the basis of a finite number of observations" (Russian) Izvyestia Akademiya Nauk SSSR, Mathematical Series, 6, 3–32 (1942).
4. CRAMER, H., "Mathematical methods of statistics", (Princeton University Press, Princeton, 1946).
5. NEYMAN, J, and E. L. SCOTT, "Consistent estimates based on partially consistent observations" Econometrica, 16, 1–32 (1948)
6 WALD, A., "Statistical decision functions", (JOHN WILEY and Sons, New York, 1950)
7 WOLFOWITZ, J, "Minimax estimates of the mean of a normal distribution with known variance", Ann Math. Stat., 21, 218–230 (1950).
8. WALD, A., and J. WOLFOWITZ, "Characterization of the minimal complete class of decision functions when the number of distributions and decisions is finite", Proceedings of the Second Berkeley Symposium on Mathematical Statistics and Probability, pp. 149–158, (University of California Press, Berkeley and Los Angeles, 1951).
9. MISES, R. v., "A modification of Bayes' problem", Ann. Math. Stat., 9, 256–259 (1938).
10. NEYMAN, J., "Su una teorema concernente le cosidette statistische sufficienti", Giornale del Istituto Italiano degli Attuari, 6, 320 (1935).
11. WALD, A., "Asymptotic minimax solutions of sequential point estimation problems", Proceedings of the Second Berkeley Symposium on Mathematical Statistics and Probability, pp. 1–11 (University of California Press, Berkeley and Los Angeles, 1951).

Estimation by the Minimum Distance Method[1]

By J. WOLFOWITZ

(Received May 10, 1953)

1. Introduction

The purpose of the present paper is to develop further the method of estimation first employed by the author in [1]. This method is also applicable to testing statistical hypotheses, but we will defer consideration of such applications to a later publication.

In statistics a sequence of chance variables is called a sequence of consistent estimators of θ if it converges stochastically to θ. In this paper we shall construct sequences of chance variables which converge to θ with probability one. It seems not unreasonable, therefore, to call such estimators super-consistent. A great utility of the minimum distance method is that, in a wide variety of problems, it will furnish super-consistent estimators even when classical methods, like the maximum likelihood method, fail to give consistent estimators. The problem treated in [1] and to be described briefly below is a case in point; in [1] are given for the first time super-consistent estimators of both parameters α and β without any assumptions about the distribution of ξ other than the indispensable one of non-normality.[2]

The method which we shall describe and refer to as the minimum distance method also yields consistent (super-consistent) estimators when the likelihood method or other methods give consistent estimators. The problem arises which estimator is to be preferred. The question of efficiency will not be discussed in the present paper. Where an efficient estimator is known the problem is, of course, settled. It should be borne in mind, however, that a criterion of efficiency may have considerable arbitrariness about it, and that an estimator efficient under one criterion need not be so under another criterion.

In Section 5 we discuss the important problem first formulated by Neyman and Scott in their interesting and important paper [2]. The problems of Sections 6 and 7 fall into a class described in [2]. In Section 6 we give a brief discussion of the essential difference between the two classes of problems and the consequently

1) Work done under partial support of the U. S. Office of Naval Research.

2) Another example is furnished by the author's recent work on estimation in non-parametric stochastic difference equations, to be submitted to the Annals of Mathematical Statistics.

different application of the minimum distance method.

The definition of distance used throughout the present paper is that given in Section 2. It is invariant under (integral) linear transformations of the chance variables and for this reason is easy to manipulate in certain proofs. However, we wish to stress here what was already stated in [1], that our method can be applied with very many definitions of distance, and is in no way tied to any particular definition of distance. One of the obvious problems connected with the question of efficiency is what definition of distance, if any, will yield efficient estimators.

As we have stated earlier, the estimators to be given in this paper not only converge stochastically to their respective parameters, but converge with probability one. We have decided to prove the stronger results here because the proofs are scarcely lengthened thereby. We wish to remark here that the estimators of [1] also converge with probability one, although this claim is not made in [1]. The proofs of this stronger result are almost the same as they are given in [1]. One has only to use the strong law of large numbers and the strong form of the Glivenko-Cantelli theorem where in [1] we used the corresponding weak forms. The appropriate lemmas of [1] also hold in this stronger form.

In accordance with this procedure, the result of [3], which is important for the present method of estimation, is there given in the stronger version.

2. Preliminary considerations

Let $T_1(x)$ and $T_2(x)$ be any two distribution functions. The distance $\delta(T_1, T_2)$ between them is to be defined by

$$\delta(T_1, T_2) = \sup_x |T_1(x) - T_2(x)| . \tag{2.1}$$

Let n be any positive integer and $s_1, \ldots\ldots, s_n$ any real numbers. By the *empiric* distribution function of $s_1, \ldots\ldots, s_n$ we shall mean a monotonically non-decreasing function of variation one with jumps of size $\frac{1}{n}$ at each of the points $s_1, \ldots\ldots, s_n$ and whose value at infinity is one. An empiric distribution function is, of course, a distribution function. To settle the ambiguity of definition at a point of discontinuity we shall always assume that, for any distribution function $T(x)$ and for all x, $T(x) = T(x^-)$.

Let T^* be a class distribution function (d. f.' s). The distance of the d. f. T from the class T^* is defined by

$$\delta(T, T^*) = \inf_{T' \varepsilon T^*} \delta(T, T') . \tag{2.2}$$

Let T^{**} be another class of d.f.'s. The distance between T^* and T^{**} is given by

$$\delta(T^*, T^{**}) = \inf_{T' \epsilon T^*} \delta(T', T^{**}). \tag{2. 3}$$

Let (u, v) be jointly normally (not necessarily independently) distributed chance variables with means zero and unknown covariance matrix, the pair (u, v) distributed independently of the chance variable ξ, whose distribution is completely unknown except for the fact that it is known not to be normal. Let

$$X = \xi + u \tag{2. 4}$$

$$Y = \alpha + \beta\xi + v \tag{2. 5}$$

where α and β are unknown constants which may be any real numbers. We may also have $|\beta| = \infty$. The treatment of this case, described in [1], will apply throughout the present paper; in the interests of simplicity we shall assume throughout the present paper that $|\beta| < \infty$.

Let (x_i, y_i), $i = 1, \ldots\ldots, n$, be n independent observations on (X, Y). The problem considered in [1] is to construct two functions, or rather two sequences of functions

$$a_n(x_1, \ldots\ldots, x_n, y_1, \ldots\ldots, y_n)$$

$$b_n(x_1, \ldots\ldots, x_n, y_1, \ldots\ldots, y_n) \qquad n = 1, 2, \ldots\ldots, \text{ad inf.}$$

of the arguments exhibited, such that a_n and b_n will converge stochastically, as $n \to \infty$, to α and β, respectively. (The stochastic convergence must therefore occur for any α and β, any covariance matrix of u and v, and any non-normal distribution of ξ).

Let c_1 and c_2 be any real numbers and $A_n(x|c_1, c_2)$ be the empiric distribution function of

$$\{y_i - c_1 - c_2 x_i\}, \qquad i = 1, \ldots\ldots, n.$$

Let N^* be the class of all normal distribution functions with mean zero. Define a_n and b_n as any Borel measurable functions of the arguments $x_1, \ldots\ldots, y_1, \ldots\ldots, y_n$ such that

$$\delta(A_n(x|a_n, b_n), N^*) < \frac{1}{n} + \inf_{c_1, c_2} \delta(A_n(x|c_1, c_2), N^*). \tag{2. 6}$$

It is shown in [1] that with this definition of a_n and b_n, the latter converge stochastically to α and β, respectively. We remind the reader of our remarks in Section 1, to the effect that the proof of [1], with very minor changes, yields the result that $(a_n - \alpha)$ and $(b_n - \beta)$ converge to zero with probability one.

The choice of functions of the observations so as to minimize the distance of an appropriate empiric d. f. from an appropriate class of d. f.'s characterizes the method to which this paper is devoted.

3. On the estimation of variances when fitting a straight line with both variables subject to error

Before proceeding to the main problem of this paper we wish to make a few remarks about the estimation of variances in the problem discussed in the preceding section which we neglected to make in [1].

In the notation of Section 2 let $D_n(x)$ be any member of N^* such that

$$(3.1)\qquad \delta(A_n(x|a_n, b_n), D_n(x)) < \frac{1}{n} + \delta(A_n(x|a_n, b_n), N^*)$$

and such that the left member of (3. 1) is a Borel measurable function of $x_1, \ldots\ldots, x_n, y_1, \ldots\ldots, y_n$. We want to show now that the variance d_n^* of $D_n(x)$ converges to the variance of $(v-\beta u)$ with probability one.

Let $E_n(x)$ be the d. f. of $Y-a_n-b_nX$. Write for short $B_n(x)$ for $A_n(x|a_n, b_n)$.

In [1] we proved that $\delta(B_n, E_n)$ and $\delta(B_n, D_n)$ converge to zero stochastically. If one uses the corresponding strong laws the same proof gives that the convergence is with probability one. Hence $\delta(E_n, D_n)$ converges to zero with probability one. Since (a_n-a) and $(b_n-\beta)$ converge to zero with probability one it follows that $\delta(E_n, Q)$ converges to zero w. p. 1 (with probability one), where Q is the d. f. of $v-\beta u$. Hence $\delta(D_n, Q)$ approaches zero w.p. 1. From this the desired result follows immediately.

We also take this opportunity to clear up a misunderstanding which may have arisen about a remark made in [1]. There we stated (page 136): "To avoid the trivial we assume that no linear combination of u and v with at least one coefficient not zero has zero variance." The fact is that the estimators a_n and b_n converge to a and β, respectively, with probability one, whether or not the assumption in the sentence quoted holds or not. The point we tried to make was that the desired result follows trivially if the assumption does not hold. Let us clarify the matter here by proving the result explicitly.

Consider the chance variable $(v-Bu)$. This can have variance zero only for a) no value of B or b) all value of B or c) exactly one value of B. Case a) was regarded in [1] as the only non-trivial one and disposed of there. Case b) is truly trivial, because then the "errors" u and v are both identically zero. Let us now show that case c) is also very easy. There are several ways of doing this, and we shall adopt the briefest. Let t be such that $(v-tu)$ has variance zero, so that $(v-tu)$ is zero with probability one. Now

$$\begin{aligned} Y-a_n-b_nX &= (v-b_nu)+(a-a_n)+(\beta-b_n)\xi \\ &= (v-tu)+(t-b_n)u+(a-a_n)+(\beta-b_n)\xi \\ &= (t-b_n)u+(a-a_n)+(\beta-b_n)\xi \end{aligned}$$

with probability one. Since u and ξ are independently distributed we obtain from the results of [3] the stronger form of equation (4. 6) of [1], i.e., that

$$P\{\lim_{n\to\infty} \delta(B_n(x), E_n(x)) = 0\} = 1 .$$

This is the only place in [1] where the assumption that the variance of $(v-Bu)$ has a positive lower bound is used, and we have just seen that the result can be obtained easily without this assumption. If $|\beta| = \infty$ and one uses $(u-Bv)$ instead the procedure is the same.

4. A Property of the Distance Function

Let $\{Y_{ij}\}$, $i = 1, \ldots, n$; $j = 1, \ldots, m_i$, be independently distributed chance variables. Let $G_i(x)$ be the distribution function of Y_{ij}, $j = 1, \ldots, m_i$, and let $G_i^*(x)$ be the empiric distribution function of $Y_{i1}, \ldots, Y_{im_i}$. Define

$$G^n(x) = \frac{\sum_{i=1}^n m_i G_i(x)}{\sum_{i=1}^n m_i}$$

and

$$G^{n*}(x) = \frac{\sum_{i=1}^n m_i G_i^*(x)}{\sum_{i=1}^n m_i}.$$

We shall now prove the following extension of the theorem of Glivenko-Cantelli.

THEOREM 4. 1. *We have*

$$P\{\lim_{n\to\infty} \delta(G^n(x), G^{n*}(x)) = 0\} = 1 . \tag{4. 1}$$

Proof of the theorem. Let $\varepsilon > 0$ be arbitrary. For every n we can find a fixed number $M(\varepsilon)$ (independent of n) of points $a_1 < a_2 < \ldots < a_M$ (the points themselves may depend on n) with the following properties: Let $a_0 = -\infty$ and $a_{M+1} = +\infty$. Then

$$G^n(a_j) - G^n(a_{j-1}^+) \leq \frac{\varepsilon}{2}, \qquad j = 1, \ldots, M(\varepsilon)+1 . \tag{4. 2}$$

Suppose first that $G^n(x)$ is continuous for every n. Obviously

$$E\, G^{n*}(x) = G^n(x) \tag{4. 3}$$

for all x, and

$$E(G^n(x)-G^{n*}(x))^4 = O\Big(\big(\sum_{i=1}^{n} m_i\big)^{-2}\Big) = O(n^{-2}) \tag{4. 4}$$

uniformly in x. Hence, for any $\eta > 0$ and $j = 1, 2, \ldots\ldots, M(\varepsilon)$,

$$P\{|G^n(a_j)-G^{n*}(a_j)| > \eta\} \leq \frac{1}{\eta^4}\cdot O(n^{-2}) \tag{4. 5}$$

by Chebyshev's inequality. Since the series $\sum^{\infty} n^{-2}$ converges and η was arbitrary, we have, from the Borel-Cantelli lemma, that

$$|G^n(a_j)-G^{n*}(a_j)|, \qquad j = 1, \ldots\ldots, M(\varepsilon), \tag{4. 6}$$

converge to zero with probability one. From this and (4. 2) it follows that, w. p. 1,

$$(\limsup_{n\to\infty})\delta(G^n(x), G^{n*}(x)) \leq \varepsilon\,.$$

Since ε was arbitrary the desired result follows.

When $G^n(x)$ need not be continuous the proof is the same, except that one replaces (4. 6) by

$$|G^n(a_j)-G^{n*}(a_j)| \text{ and } |G^n(a_j^+)-G^{n*}(a_j^+)|, \quad j=1,\ldots M(\varepsilon). \tag{4. 7}$$

The right members of (4. 4) and (4. 5) are bounded uniformly in the G_i's, and $M(\varepsilon)$ does not depend upon the G_i's. This yields an important fact which we state as

THEOREM 4. 2. *Let ε and η be arbitrary positive numbers. There exists an integer $N(\eta,\varepsilon)$ which does not depend upon the G_i's and such that*

$$P\{\delta(G^n(x), G^{n*}(x)) < \eta \text{ for all } n > N(\eta,\varepsilon)\} > 1-\varepsilon\,. \tag{4. 8}$$

Suppose now that, when the values of

$$Y_{11}, \ldots\ldots, Y_{(i-1)m_{(i-1)}} \tag{4. 9}$$

are given, $Y_{i1}, \ldots\ldots, Y_{im_i}$ are independently distributed with the same distribution function

$$G_i(x|Y_{11}, \ldots\ldots, Y_{(i-1)m_{(i-1)}})$$

which depends upon the values of the Y's in (4. 9). (The definition when $i = 1$ is obvious). We now define

$$G^n(x) = \frac{\sum_{i=1}^{n} m_i G_i(x|Y_{11}, \ldots\ldots, Y_{(i-1)m_{(i-1)}})}{\sum_{i=1}^{n} m_i} \tag{4. 10}$$

with the definition of $G^{n*}(x)$ as before. Let us now try to extend Theorem 4. 1

to this case. The difficulty then is that $a_1, \ldots\ldots, a_M$ may be functions of the Y's. Suppose however that, for every $\varepsilon > 0$, there exists an integer $M(\varepsilon)$ such that, for almost all n, there exist points $a_1, \ldots\ldots, a_{M(\varepsilon)}$ (which may depend on n but are not functions of the Y's) with the property that

$$(4.\ 11) \qquad G^n(a_j) - G^n(a_{j-1}) \leq \frac{\varepsilon}{2} \qquad (j = 1, \ldots\ldots, M(\varepsilon)+1)$$

with $a_0 = -\infty$, $a_{M+1} = +\infty$. When this is so we shall say that $\{G^n(x)\}$ has the finite property.

It is not difficult to see that, if $\{G^n(x)\}$ has the finite property, the proof of Theorem 4. 1 goes through with little change. We may therefore state

THEOREM 4. 3. *If $\{G^n(x)\}$ has the finite property then Theorems 4. 1. and 4. 2 hold even when the Y's are not independent.*

We now cite a simple example where $\{G^n(x)\}$ has the finite property. Let $H(x)$ be a continuous d. f., and let

$$G_i(x \mid Y_{11}, \ldots\ldots, Y_{(i-1)m_{(i-1)}}) = H(x + g_i(Y_{11}, \ldots\ldots, Y_{(i-1)m_{i-1}}))$$

where the g_i are uniformly bounded. It is easy to see that $\{G^n(x)\}$ has the finite property.

5. The problem of estimation of structual parameters

Let $\{X_{ij}\}$, $(i = 1, \ldots\ldots, n;\ j = 1, \ldots\ldots, m_i)$ be independently distributed chance variables. Let $F_i(x \mid \theta, \alpha_i)$ be the distribution function of $X_{i1}, \ldots\ldots, X_{im_i}$. The parameters θ, α_i upon which this d. f. depends are unknown; for simplicity we take them to be scalars although our results are equally valid for vectors. We see that the parameter θ occurs in every group of X's ($X_{i1}, \ldots\ldots, X_{im_i}$ constitute the ith group) and was called by Neyman and Scott ([2]) "structural." The parameter α_i occurs only in the ith group and was called "incidental." The fact that there are as many incidental parameters as groups makes the problem of estimating the structural parameter much more difficult. This interesting and important problem was formulated in the interesting paper of Neyman and Scott ([2]). These authors proved, inter alia, that the maximum likelihood estimator of the structural prarameter need not be consistent, and when consistent, need not be efficient. We shall show that, under very weak regularity conditions, the minimum distance method will furnish super-consistent estimators of the structural parameter (or parameters).

Let T be a (given) set within which θ is known to lie (of course T may be the whole line). Write

$$\bar{\alpha}_n = (\alpha_1, \alpha_2, \ldots\ldots, \alpha_n)$$

and

$$\bar{\alpha} = (\alpha_1, \alpha_2, \ldots\ldots).$$

Let A_n be the set within which $\bar{\alpha}_n$ is known to lie, and A the set within which $\bar{\alpha}$ is known to lie. Let $F_i^*(x)$ be the empiric distribution function of X_{i1}, X_{i2} $\ldots\ldots, X_{im_i}$, and define

$$B^n(x) = \frac{\sum_{i=1}^{n} m_i F_i^*(x)}{\sum_{i=1}^{n} m_i}.$$

Let $\bar{\alpha}_n' = (\alpha_1', \cdots, \alpha_n')$, and define

$$C^n(x|\theta', \bar{\alpha}_n') = \frac{\sum_{i=1}^{n} m_i F_i(x|\theta', \alpha_i')}{\sum_{i=1}^{n} m_i}.$$

Let $\theta_n^*, \alpha_{1n}^*, \cdots, \alpha_{nn}^*$ be Borel measurable functions of $X_{11}, \cdots, X_{nm_n}$ such that (writing $\alpha_n^* = (\alpha_{1n}^*, \cdots, \alpha_{nn}^*)$) $\theta_n^* \varepsilon T$, $\alpha_n^* \varepsilon A_n$, and

$$(5.\ 1) \qquad \delta(C^n(x|\theta_n^*, \alpha_n^*), B^n(x)) < \frac{1}{n} + \inf_{\theta' \varepsilon T, \alpha_n' \varepsilon A_n} \delta(C^n(x|\theta', \bar{\alpha}_n'), B^n(x)).$$

The estimator θ_n^* is a minimum dis'ance estimator. We shall show under certain regularity conditions that it is a super-consistent estimator of θ.

We shall say that the sequence $\{F_i(x|\theta, \alpha_i), m_i\}$ has the asymptotic positive distance property with respect to (T, A) if there exists a function $l(d)$, defined and positive in the domain $d > 0$, and a positive integer $N_0(d)$ which is a function of d, such that $\theta' \varepsilon T$ and $|\theta - \theta'| > d$ imply that

$$(5.\ 2) \qquad \inf_{\bar{\alpha}_n' \varepsilon A_n} \delta\left(\frac{\sum_{i=1}^{n} m_i F_i(x|\theta', \alpha_i')}{\sum_{i=1}^{n} m_i}, \frac{\sum_{i=1}^{n} m_i F_i(x|\theta, \alpha_i)}{\sum_{i=1}^{n} m_i}\right) > l(d)$$

for all $n \geq N_0(d)$.

THEOREM 5. 1. *If $\{F_i(x|\theta, \alpha_1), m_i\}$ has the asymptotic positive distance property with respect to (T, A) then θ_n^* is a super-consistent estimator of θ.*

Proof. Suppose θ_n^* is not a supper-consistent estimator of θ. Then there exist positive η and ε such that

$$(5.\ 3) \qquad P\{\limsup_n |\theta_n^* - \theta| > \eta\} > \varepsilon.$$

Here the probability is of course that determined by $(\theta, \bar{\alpha}$.

Let

$$D^n(x) = \frac{\sum_{i=1}^{n} m_i F_i(x|\theta_n^*, \alpha_{in}^*)}{\sum_{i=1}^{n} m_i},$$

and

$$F^n(x) = \frac{\sum_{i=1}^{n} m_i F_i(x|\theta, \alpha_i)}{\sum_{i=1}^{n} m_i}.$$

From Theorem 4. 1 we obtain that

(5. 4) $$P\{\lim \delta(B^n(x), F^n(x)) = 0\} = 1.$$

Since

(5. 5) $$\delta(B^n(x), F^n(x)) + \frac{1}{n} \geq \delta(B^n(x), D^n(x))$$

we obtain from (5. 4) that

(5. 6) $$P\{\lim \delta(B^n(x), D^n(x)) = 0\} = 1.$$

From (5. 3) we obtain that

(5. 7) $$P\{\limsup_n \delta(D^n(x), F^n(x)) \geq l(\eta)\} > \varepsilon$$

From (5. 4) and (5. 6) we obtain

(5. 8) $$P\{\lim \delta(D^n(x), F^n(x)) = 0\} = 1.$$

The contradiction between (5. 7) and (5. 8) proves the theorem.

As an example, let T be the real line, A_n that orthant of Euclidean n-space where all coordinates are positive, and

$$F_i(x|\theta, \alpha_i) = \frac{1}{\sqrt{2\pi}\alpha_i} \int_{-\infty}^{x} \exp\left\{-\frac{1}{2}\left(\frac{y-\theta}{\alpha_i}\right)^2\right\} dy.$$

(This is Example 1 of [2].) Suppose, for example, that

(5. 9) $$\sup_n \frac{\sum_{i=1}^{n} m_i \alpha_i^2}{\sum_{i=1}^{n} m_i} = K^2 < \infty.$$

(This is, of course, a further restriction on A). Then it is easy to verify that $\{F_i(x|\theta, \alpha_i), m_i\}$ has, for any $(\theta, \bar{\alpha})$, the asymptotic positive distance property with respect to (T, A). A simple computation gives that, for any $\theta, \bar{\alpha}$, and n,

$$l(d) > \frac{4}{9\sqrt{2\pi}} \int_0^{d/5K} \exp\left\{\frac{-y^2}{2}\right\} dy.$$

6. Fitting a straight line when ξ is not a chance variable

Let us return to the problem treated in [1] and described in Section 2, and see how it fits into the formulation by Neyman and Scott of the problem described in Section 5. The parameters α and β are, of course, structural parameters. The successive independent observations on ξ, say $\xi_1, \xi_2, \ldots\ldots$, are the incidental parameters. The reason why the treatment of the problem in [1] is so much more difficult than that of the problem of Section 5 is this: In the latter the distribution functions are determined by the parameters θ, α. In the process of estimation one can study the distribution implied by the estimator. In the problem of [1] the distribution of ξ is completely unknown, so that its "contaminating" effect is not so obvious. In this problem our estimator is obtained by studying the change in the empiric d. f., whereas in the problem of Section 5 the estimator is obtained by studying the changes in the (population) d. f.

In the formulation of the problem of Section 5 the α_i were constants, unknown to the statistician, but given in advance. In the problem of Section 2 they are independent observations on a chance variable ξ. Such random observations can be expected to show more "regularity" than an arbitrarily selected sequence. In this section we will study the problem of obtaining super-consistent estimators of α and β when $\xi_1, \xi_2, \ldots$ is a sequence of constants fixed in advance but unknown to the statistician. (Hence a_n and b_n may not depend on them). It will be seen that our treatment of this problem includes as a special case the problem where ξ is a chance variable.

Let $N(x|f_1, f_2)$ be the normal d. f. with mean f_1 and variance f_2. Let the (unknown) covariance matrix of u and v be

$$\begin{Bmatrix} \sigma_1^2 & \rho\sigma_1\sigma_2 \\ \rho\sigma_1\sigma_2 & \sigma_2^2 \end{Bmatrix}.$$

(6. 1)
$$\sigma^2(c_2) = \sigma_2^2 - 2c_2\rho\sigma_1\sigma_2 + c_2^2\sigma_1^2$$

Since

(6. 2)
$$y_i - c_1 - c_2 x_i = (v_i - c_2 u_i) + (\alpha - c_1) + (\beta - c_2)\xi_i$$

where (u_i, v_i) is the ith independent observation on (u, v), the d. f. of $y_i - c_1 - c_2 x_i$ is

$$N(x|(\alpha - c_1) + (\beta - c_1)\xi_i, \sigma^2(c_2)).$$

We define a_n and b_n exactly as in [1] or Section 2. We shall now show under suitable regularity conditions on the sequence

$$\xi = \xi_1, \xi_2, \ldots\ldots$$

that a_n and b_n converge w. p. 1 to α and β, respectively.

We require of ξ that it be such that the following two conditions are satisfied:

Condition I). For any $\eta > 0$ we have

(6. 3)

$$P\left\{\lim_{n\to\infty} \sup_{c_1, c_2} \delta\left(A_n(x|c_1, c_2), n^{-1}\sum_{i=1}^{n} N(x|(\alpha-c_1)+(\beta-c_2)\xi_i, \sigma^2(c_2))\right) = 0\right\} = 1 .$$

Condition II). For any $d > 0$ there exists an $l(d) > 0$ such that

$$\text{(6. 4)} \qquad \liminf_{n\to\infty}\left[\inf \delta\left(N^*, n^{-1}\sum_{i=1}^{n} N(x|(\alpha-c_1)+(\beta-c_2)\xi_i, \sigma^2(c_2))\right)\right] \geq l(d)$$

where the infimum inside the bracket is taken over all c_1 and c_2 which satisfy

$$\text{(6. 5)} \qquad |\alpha-c_1|+|\beta-c_2| \geq d .$$

These conditions must appear strange and formidable, bnt we shall show that this impression is misleading and that the conditions are natural and not difficult to meet.

Under these conditions a_n and b_n converge w. p. 1 to α and β, respectively. The proof follows the same essential lines as the proof in [1], and will be omitted. One employs Theorem 4. 1 of the present paper where [1] uses the the Glivenko-Cantelli theorem.

Instead of Conditions I and II it may be easier to verify Conditions III and IV below. Their relation will be discussed below.

Let $Z_n(x)$ be the empiric d. f. of $\xi_1, \ldots, \xi_n$.

Condition III). For any $\eta>0$ there exists $M(\eta)>0$ such that

$$\text{(6. 6)} \qquad \liminf_{n\to\infty}[Z_n(M(\eta))-Z_n(-M(\eta))] > 1-\eta .$$

Condition III requires that the ξ_i do not go off to infinity too fast. If one examines carefully the proofs of Lemmas 3 and 4 of [1] it is seen that it is essentially this property which is furnished there by the chance variable X_1. By using the method of proof of Lemma 1 of [1], but replacing the Glivenko-Cantelli theorem by Theorem 4. 1 of the present paper, and making use of Condition III, one can prove (6. 3). The details, while many, are not essentially different from those of the proof of Lemma 1 of [1], and hence will be omitted.

Let $\delta^*(T_1(x), T_2(x))$ be the Lévy distance between the d. f.'s $T_1(x)$ and $T_2(x)$. This is defined as the infimum of all positive h such that

$$T_1(x-h)-h \leq T_2(x) \leq T_1(x+h)+h$$

for $-\infty < x < \infty$. We define

$$\delta^*(T, N^*) = \inf_{T' \varepsilon N^*} \delta^*(T, T') .$$

Condition IV). We have

$$\liminf_{n\to\infty} \delta^*(Z_n(x), N^*) > 0\,. \tag{6. 7}$$

By a compactness argumens similar to that by which Lemma 2 of [1] is proved it may be shown that conditions III and IV imply (6. 5). We omit the details which we leave to the reader.

Thus we have replaced Conditions I and II by Conditions III and IV which imply them. The latter conditions are not at all formidable and in very many instances are easy to verify. Finally, if $\xi_1, \xi_2, \ldots$ are independent chance variables with the same non-normal distribution as ξ, Conditions III and IV (and hence Conditions I and II) are fulfilled with probability one. For, by the Glivenko-Cantelli theorem,

$$\delta(Z_n(x), Z(x)) \to 0$$

with probability one, where $Z(x)$ is the (non-normal) d. f. of ξ. Conditions III and IV follow immediately from this. Thus Conditions I and II are eminently reasonable. The main result of [1], quoted in Section 2, can thus be made to follow from the result proved in this section. Of course, we have leaned heavily on the method and proofs of [1].

The extension to higher dimensions where one fits a hyperplane instead of of a line is immediate, and proceeds as sketched in [1].

Let us now try to apply the argument of Section 3 to the present problem. It is seen that the role of $E_n(x)$ in Section 3 is now played by

$$n^{-1}\sum_{i=1}^{n} N(x|(\alpha-c_1)+(\beta-c_2)\xi_i, \sigma^2(c_2))\,.$$

It can be verified that, if Condition III holds and $(a_n-\alpha)$ and $(b_n-\beta)$ converge to zero w. p. 1, the argument of Section 3 holds. Hence we may assert that, under these conditions, the variance d_n^* of $D_n(x)$ converges to the variance of $(v-\beta u)$ with probability one.

We may state this result in the following form: Suppose Conditions III and IV are fulfilled. Let c_3 be a positive number, and $A_n(x|c_1, c_2, c_3)$ be the empiric d. f. of

$$\left\{\frac{y_i-c_1-c_2x_i}{c_3}\right\}, \qquad i = 1, \ldots\ldots, n.$$

Let a_n, b_n, c_n be any Borel measurable functions of $(x_1, \ldots, x_n, y_1, \ldots, y_n)$ such that $c_n > 0$ and

$$\delta(A_n(x|a_n, b_n, c_n), N(x|0, 1)) < \frac{1}{n} + \inf_{c_1, c_2, c_3} \delta(A_n(x|c_1, c_2, c_3), N(x|0, 1)). \tag{6. 8}$$

Then the chance variables

$$(a_n-\alpha),\ (b_n-\beta),\ (c_n-\sigma^2(\beta))$$

all converge w. p. 1 to zero.

7. Fitting a straight line when the errors have different variances

The formulation of of the present problem is essentially that of Example 4 of [2] where the ξ_i are chance variables. The more general case posed there, that of several parameters, can be treated in the same way by the methods of this section, but in order not to make the exposition too cumbersome we limit ourselves to the case of one parameter. We shall discuss the case where the ξ_i's are constants given in advance, and give sufficient conditions for our method to yield super-consistent estimators.

Let (u_i, v_i), $i=1, 2, \ldots$, ad inf., be independently distributed chance variables with the same bivariate normal distribution. Let

$$Eu_i = Ev_i = 0$$
$$Eu_i^2 = \sigma_1^2,\ Ev_i^2 = \sigma_2^2,\ Eu_iv_i = \rho\sigma_1\sigma_2$$

The constants σ_1^2, σ_2^2, $\rho\sigma_1\sigma_2$ are unknown to the statistician. Let

$$\bar{\xi} = \xi_1, \xi_2, \ldots\ldots$$

and

$$\bar{d} = d_1, d_2, \ldots\ldots$$

be two infinite sequences of real numbers, with $\bar{d}$ known to the statistician and $\bar{\xi}$ unknown to the statistician. Finally we define, for all i,

$$y_i = \beta d_i\xi_i + v_i$$
$$x_i = \xi_i + u_i$$

where β is a constant, unknown to the statistician, which may be any real number. The problem now is to obtain, for every n, a function b_n of $x_1, \ldots, x_n$, $y_1, \ldots, y_n$ such that b_n converges to β w. p. 1.

We shall assume that the following two conditions are satisfied: Condition I) We have

$$(7.1)\qquad P\left\{\limsup_{n\to\infty}\ \sup_{c_1}\ \delta\left(A_n(x|c_1),\ n^{-1}\sum_{i=1}^{n} N(x|(\beta-c_1)d_i\xi_i,\ w_i')\right) = 0\right\} = 1 .$$

where $A_n(x|c_1)$ is the empiric d. f. of

$$\{y_i - c_1d_ix_i\}, \qquad i = 1, \ldots, n; -\infty < c_1 < \infty$$

and

$$w_i' = w_i(c_1, \sigma_1, \sigma_2, \rho)$$

where

(7. 2) $$w_i(c_1, c_2, c_3, c_4) = c_3^2 + c_1^2 d_i^2 c_2^2 - 2c_1 d_i c_2 c_3 c_4 .$$

Condition II) There exists, for any $d>0$, an $l(d)>0$ such that

(7. 3) $$\liminf_{n\to\infty}\left[\inf \delta\left(n^{-1}\sum_{i=1}^{n} N(x|(\beta-c_1)d_i\xi_i,\ w_i(c_1, \sigma_1, \sigma_2, \rho)),\right.\right.$$
$$\left.\left. n^{-1}\sum_{i=1}^{n} N(x|0,\ w_i(c_1, c_2, c_3, c_4))\right)\right] \geq l(d)$$

where the infimum inside the bracket is taken over all c_1, c_2, c_3, c_4 such that $|c_1-\beta|\geq d$, $c_2\geq 0$, $c_3\geq 0$, and $|c_4|\leq 1$.

We now define b_n, s_{1n}, s_{2n}, r_n, four Borel measurable functions of $(x_1,\dots, x_n, y_1,\dots, y_n)$ such that $s_{1n}\geq 0$, $s_{2n}\geq 0$, $|r_n|\leq 1$, and such that

(7. 4) $$\delta\left(A_n(x|b_n),\ n^{-1}\sum_{i=1}^{n} N(x|0,\ w_i(b_n, s_{1n}, s_{2n}, r_n))\right)$$
$$< \frac{1}{n} + \inf \delta\left(A_n(x|c_1),\ n^{-1}\sum_{i=1}^{n} N(x|0,\ w_i, (c_1, c_2, c_3, c_4))\right)$$

where the infimum is taken over all (c_1, c_2, c_3, c_4) such that $c_2\geq 0$, $c_3\geq 0$, and $|c_4|\leq 1$.

We shall now prove that, if Conditions I and II are fulfilled, b_n converges to β w. p. 1.

Suppose that, for $q_1 > 0$ and $q_2 > 0$,

(7. 5) $$P\{\limsup_{n\to\infty} |b_n-\beta| > q_1\} > q_2 .$$

Condition II then implies that

(7. 6) $$P\left\{\liminf_{n\to\infty} \delta\left(n^{-1}\sum_{i=1}^{n} N(x|(\beta-b_n)d_i\xi_i,\ w_i(b_n, \sigma_1, \sigma_2, \rho)),\right.\right.$$
$$\left.\left. n^{-1}\sum_{i=1}^{n} N(x|0,\ w_i(b_n, s_{1n}, s_{2n}, r_n))\right) \geq l(q_1)\right\} > q_2 .$$

From Theorem 4. 1 it follows that

(7. 7) $$P\left\{\lim_{n\to\infty} \delta\left(A_n(x|\beta),\ n^{-1}\sum_{i=1}^{n} N(x|0,\ w_i(\beta, \sigma_1, \sigma_2, \rho))\right) = 0\right\} = 1 .$$

From the definition of b_n, s_{1n}, s_{2n}, and r_n it follows that

(7. 8) $$\delta\left(A_n(x|b_n),\ n^{-1}\sum_{i=1}^{n} N(x|0,\ w_i(b_n, s_{1n}, s_{2n}, r_n))\right)$$
$$< \frac{1}{n} + \delta\left(A_n(x|\beta),\ n^{-1}\sum_{i=1}^{n} N(x|0,\ w_i(\beta, \sigma_1, \sigma_2, \rho))\right) .$$

From (7. 7) and (7. 8) it follows that

(7. 9) $$P\left\{\lim_{n\to\infty} \delta\left(A_n(x|b_n),\ n^{-1}\sum_{i=1}^{n} N(x|0,\ w_i(b_n, s_{1n}, s_{2n}, r_n))\right) = 0\right\} = 1 .$$

From Condition I we obtain that

$$(7.\ 10)\quad P\Big\{\lim_{n\to\infty}\delta\Big(A_n(x|b_n),\ n^{-1}\sum_{i=1}^{n}N(x|(\beta-b_n)d_i\xi_i,\ w_i(b_n,\sigma_1,\sigma_2,\rho))\Big)=0\Big\}=1.$$

Hence from (7. 9) and (7. 10) we obtain that

$$(7.\ 11)\quad P\Big\{\lim_{n\to\infty}\delta\Big(n^{-1}\sum_{i=1}^{n}N(x|0,\ w_i(b_n,s_{1n},s_{2n},r_n)),$$
$$n^{-1}\sum_{i=1}^{n}N(x|(\beta-b_n)d_i\xi_i,\ w_i(b_n,\sigma_1,\sigma_2.\rho))\Big)=0\Big\}=1.$$

This contradicts (7. 6) and proves that b_n converges to β w.p. 1.

Let $V_n(x)$ be the empiric d.f. of $\{d_i\xi_i\}$, $i=1,\dots,n$. Let $V_n'(x)$ be the empiric d.f. of $\{d_i\}$, $i=1,\dots,n$. Consider the following two conditions: Condition III) For any $\eta>0$ there exists an $M(\eta)>0$ such that

$$(7.\ 12)\qquad \liminf_{n\to\infty}[V_n(M(\eta))-V_n(-M(\eta))]>1-\eta.$$

Condition IV) For any $\eta>0$ there exists an $M'(\eta)>0$ such that

$$(7.\ 13)\qquad \liminf_{n\to\infty}[V_n'(M'(\eta))-V_n'(-M'(\eta))]>1-\eta.$$

Following essentially the same line of argument as in Lemmas 3 and 4 of [1] one can prove that the satisfaction of Condition III implies the satisfaction of Condition I. (Theorem 4. 1 of the present paper should be employed instead of the theorem of Glivenko-Cantelli in this argument.) It is in general much easier to investigate whether Condition III is satisfied or not.

It is easy to prove, if Condition IV is satisfied and ξ_1, ξ_2,... are independently and identically distributed chance variables, that Condition III is satisfied with probability one.

CORNELL UNIVERSITY
AND
UNIVERSITY OF ILLINOIS

REFERENCE

[1] J. WOLFOWITZ: "Consistent estimators of the parameters of a linear structural relation," *Skand. Aktuarietidskrift*, 1952, pp. 132-151.

[2] J. NEYMAN and E. SCOTT: "Consistent estimators based on partially consistent observations," *Econometrica*, Vol. 16, 1948, pp. 1-32.

[3] J. WOLFOWITZ: "Generalization of the theorem of Glivenko-Cantelli," submitted to the *Rendiconti Cir. Mat. di Palermo*.

ON THE THEORY OF QUEUES WITH MANY SERVERS(1)

BY

J. KIEFER AND J. WOLFOWITZ

1. Introduction. The physical original of the mathematical problem to which this paper is devoted is a system of s "servers," who can be machines in a factory, ticket windows at a railroad station, salespeople in a store, or the like. Individuals (clients) who are to be served by these servers arrive at random and the duration of anyone's service (e.g., stay at the ticket window) is a chance variable whose distribution function may be arbitrary. The phrase "at random" used above is not to be interpreted to mean that the interval between successive arrivals is to have an exponential distribution. The assumption of an exponential or other special distribution for either the interval between arrivals or the service time of an individual or both usually makes the problem much easier. We also allow the distribution of the interval between arrivals to be arbitrary. The queue discipline is "first come, first served." The system is described precisely in §2.

In this system the waiting time of the individual who is ith in order of arrival, i.e., the time which elapses between his arrival and the beginning of his service, is a chance variable whose distribution function depends upon i. In §3 we prove that this distribution function approaches a limit as $i \to \infty$. This limit may not be a distribution function because its variation may be less than one. We assume that the expected value of the time interval between the arrivals of successive clients and the expected value of the service time of an individual both exist. In terms of these one defines a quantity ρ in §6. The situation may then be classified according as $\rho < 1$ or $\rho \geqq 1$. In the former and interesting case the limiting function is a distribution function (§6), in the latter case it is not a distribution function (§7). The limiting function is (a marginal function) obtained from a function which satisfies an integral equation derived in §3. This integral equation is satisfied by a unique distribution function on s-space when $\rho < 1$, and by no distribution function when $\rho \geqq 1$ (§8). These results for the case of one server were obtained by Lindley [1]. The problem when there are many servers offers many difficulties not present when there is only one server. The methods of the present paper are different from those of [1]. The proof of the result of §7, that the limit is not a distribution function when $\rho \geqq 1$, is obtained by reducing the problem to the case $s=1$ by using our lemma of §4, and then employing the corresponding result of [1]; except for this argument our paper is self-contained. For special distributions of the time between successive arrivals

Presented to the Society, October 24, 1953; received by the editors August 13, 1953.

(1) Research under contract with the Office of Naval Research.

and of the service time the results of the present paper have been obtained by various authors (we refer the reader to [5] and [6] which contain extensive bibliographies). The methods of these authors make use of their special assumptions in an essential way. The novelty of the results of the present paper lies in the fact that no restrictions are imposed on the distributions, with the exception of the assumption of finite first moment(2). Thus the results of the present paper include the corresponding ones of previous papers as special cases(3).

Mathematically speaking, our study is one of the ergodic character of the waiting time in our system, and the conditions under which the distribution of the latter approaches stability. Our problem can be reduced, and actually is so reduced by us, to studying a random walk in s-space with certain impassable but not absorbing barriers. We actually show that, when $\rho<1$, the distribution function of the particle engaged in the random walk approaches a limiting distribution which is the same no matter what the original starting point of the particle (§8).

Perhaps our principal device is to dominate the stochastic process to be studied by a lattice process to which we then apply available theorems from the theory of Markoff processes with discrete time parameter and denumerably many states. This device makes possible the argument of §6 and is also employed in §8. We are of the opinion that this device could be applied to other ergodic problems connected with random walks.

When the original process is a lattice process, i.e., when the chance variables R_1 and g_1 (defined in §2) take, with probability one, only values which are integral multiples of some positive number c, and when $\rho<1$, the limiting probabilities (which are shown to exist in §6) are reciprocals of certain mean recurrence times (this follows from the application of Theorem 2 of Chapter 15 of [3] to the argument of §6). Monte Carlo methods (see, e.g., [4]) may perhaps then be profitably employed to solve the integral equation (3.8).

It would be very desirable and interesting to solve the integral equation (3.8), at least for interesting or important functions G and H (see §2). This, however, is likely to be very difficult. Even in the simplest case, when $s=1$, the equation becomes the Wiener-Hopf equation, which has been of considerable interest to physicists but has been solved only in special cases. Some special cases of the equation (3.8) are discussed in [5], [6], and [1]. It may also interest the pure analyst that one can, by probabilistic methods

(2) Under stronger assumptions (e.g., existence of all moments), F. Pollaczek in recent notes (C. R. Acad. Sci. Paris vol. 236 (1953) pp. 578–580, 1469–1470) gives formally an integral equation for the Laplace transform of F (to be defined below), but does not consider the questions of the present paper.

(3) In a paper to be published elsewhere which makes extensive use of the present paper, the authors obtain, under minimal conditions, theorems on convergence of the mean of various variables connected with the queueing process.

like ours, prove the existence or non-existence of distribution function solutions of (3.8).

Finally, in §9 we discuss the limiting distribution (as $i\rightarrow\infty$) of the queue size, i.e., of the number waiting to be served when the service of the ith customer begins.

We are obliged to Professor J. L. Doob for helpful discussions.

2. Description of the system. The system consists of s ($\geqq 1$) machines, $M_1, \cdots, M_s$. The ith individual arrives at time t_i ($\geqq 0$), with, of course, $t_i \leqq t_{i+1}$. If all machines are in service at his arrival he takes his place in the queue. His service begins as soon as at least one machine is unoccupied, and all individuals with smaller indices have been or are being served. If more than one machine becomes unoccupied at the time when it is the ith individual's turn to be served, we shall assume, for definiteness, that he takes his place at the unoccupied machine with smallest index.

Let $t_0=0$, $g_i=t_i-t_{i-1}$ for all $i\geqq 1$. We assume that the g_i are independently and identically distributed chance variables; let $G(z)=P\{g_1\leqq z\}$, where $P\{\ \}$ is the probability of the relation in braces. Throughout the paper we assume that $G(0)<1$; the case $G(0)=1$ is too trivial to discuss. We assume $Eg_1<\infty$.

Let R_i be the length of time the ith person spends being serviced by a machine. We assume that the R_i are independently and identically distributed chance variables, distributed independently of the g_i; let $H(z)=P\{R_1\leqq z\}$. We assume $ER_1<\infty$. We also assume $H(0)<1$, the case $H(0)=1$ being trivial.

Let $w_{i1}+t_i$ be the time at which service of the ith individual begins; w_{i1} is his waiting time. Then the ith individual leaves his machine at the time $w_{i1}+t_i+R_i$.

Let $u_{ij}+t_i$ be the time at which the jth machine finishes serving the last of those among the first $(i-1)$ individuals which it serves. Let $u'_{ij}=\max(0, u_{ij})$. Let $w_{i1}, \cdots, w_{is}$ be the quantities $u'_{i1}, \cdots, u'_{is}$ arranged in order of increasing size. It is easy to see that this definition of w_{i1} coincides with the former.

Let

$$F_i(x_1, \cdots, x_s) = P\{w_{i1}\leqq x_1, \cdots, w_{is}\leqq x_s\}. \tag{2.1}$$

If ever $x_j>x_{j+1}$ we may, since $w_{ij}\leqq w_{i(j+1)}$, replace x_j by x_{j+1} in both members of (2.1) without changing the value of either.

Write $w_i=(w_{i1}, \cdots, w_{is})$. The earliest times at which the various machines could attend to the $(i+1)$st individual are $t_i+w_{i1}+R_i$, $t_i+w_{i2}, \cdots, t_i+w_{is}$. If t_{i+1} is greater than or equal to any of these quantities the $(i+1)$st individual finds at least one machine unoccupied at his arrival and does not have to wait at all. If t_{i+1} is less than all these quantities the $(i+1)$st individual has to wait for the first machine to be unoccupied. Since $t_{i+1}=t_i+g_{i+1}$, w_{i+1} is obtained from w_i as follows: Subtract g_{i+1} from *every* component of $(w_{i1}+R_i, w_{i2}, w_{i3}, \cdots, w_{is})$. Rearrange the resulting quantities in ascending order and replace all negative quantities by zero. The ensuing result is w_{i+1}.

3. Recursion formula for F_i. Existence of the limit of F_i as $i\to\infty$. Let $\phi_j(a, b, c)$, $j=1, \cdots, s$, be the value of $w_{(i+1),j}$ when $w_i=a$, $R_i=b$, $g_{i+1}=c$. If d is a point in s-space we shall say that $a\leqq d$ if every coordinate of a is not greater than the corresponding coordinate of d. If now $a\leqq d$ then obviously

$$\phi_j(a, b, c) \leqq \phi_j(d, b, c)$$

for $1\leqq j\leqq s$. Applying this argument k times we obtain the following result: Let $R_{i+j-1}=b_{i+j-1}$, $g_{i+j}=c_{i+j}$, $j=1, \cdots, k$. Let $w_{i+k}=e_1$ when $w_i=a_1$, and let $w_{i+k}=e_2$ when $w_i=a_2$. Then $a_1\leqq a_2$ implies $e_1\leqq e_2$.

Let S be the totality of points $(x_1, x_2, \cdots, x_s)$ of Euclidean s-space such that $0\leqq x_1\leqq x_2\leqq \cdots \leqq x_s$. Let x and y be generic points of S. For $i\geqq 1$, let

$$F_i(x\mid y) = P\{w_i \leqq x\mid w_1 = y\}.$$

Let 0 be the origin in s-space. Then

$$F_i(x\mid 0) = F_i(x)$$

and

$$F_{i+1}(x) = \int F_i(x\mid y)dF_2(y). \tag{3.1}$$

The conclusion of the preceding paragraph enables us to conclude that $y_1\in S$, $y_2\in S$, $y_1\leqq y_2$, imply

$$F_i(x\mid y_1) \geqq F_i(x\mid y_2)$$

for every x and every i. Now

$$F_{i+1}(x) - F_i(x) = \int [F_i(x\mid y) - F_i(x\mid 0)]dF_2(y). \tag{3.2}$$

Since the integrand is never positive we have that

$$F_{i+1}(x) \leqq F_i(x) \tag{3.3}$$

for all x and i. From (3.3) it follows that $F_i(x)$ approaches a limit, say $F(x)$, which is nondecreasing in every component of x, continuous to the right, and assigns non-negative measure to all rectangles. It need not, however, be a distribution function, i.e., its variation over S (hence over all of s-space) may be less than one.

Write

$$\phi(a, b, c) = (\phi_1(a, b, c), \cdots, \phi_s(a, b, c)). \tag{3.4}$$

For given $x\in S$, b, c, let $\psi(x, b, c)$ be the totality of points $y\in S$ such that

$\phi(y, b, c) \leqq x$ and $\in S$. Obviously

$$F_{(i+1)}(x) = \int P_i\{\psi(x, b, c)\} dH(b) dG(c) \tag{3.5}$$

where P_i is the measure according to F_i. This equation determines each F_i uniquely by recursion, since of course $F_1(0) = 1$. For $s = 1$ and $x \geqq 0$, $\psi(x, b, c)$ is $\{y | 0 \leqq y \leqq x - b + c\}$. Hence (3.5) becomes, for $x \geqq 0$,

$$F_{(i+1)}(x) = \int F_i(x - b + c) dH(b) dG(c), \tag{3.6}$$

an equation due to Lindley [1]. (In (3.6) it is understood that $F_i(x - b + c) = 0$ whenever $x - b + c < 0$.) For $s = 2$ and $x = (x_1, x_2) \in S$ we have that $\psi(x, b, c)$ is the set of points $y \in S$ such that $y \leqq (x_1 - b + c, x_2 + c)$, together with the set of points $y \in S$ such that $y \leqq ([\min(x_2 - b + c, x_1 + c)], x_1 + c)$. We extend the definition of $F_i(y_1, y_2)$ to all of s-space in the natural way as follows: $F_i(y_1, y_2) = 0$ if either y_1 or $y_2 < 0$, $F_i(y_1, y_2) = F_i(y_2, y_2)$ if $y_1 > y_2$. Then for all (x_1, x_2) in S and $s = 2$, (3.5) becomes

$$\begin{aligned} F_{(i+1)}(x_1, x_2) = \int [&F_i(x_1 - b + c, x_2 + c) + F_i(x_2 - b + c, x_1 + c) \\ &- F_i(x_1 - b + c, x_1 + c)] dH(b) dG(c). \end{aligned} \tag{3.7}$$

In general, when (3.5) is written in the form of (3.6) and (3.7) the integrand contains $(2^s - 1)$ terms. With the integrand in this form let $i \to \infty$ in (3.5). By Lebesgue's bounded convergence theorem we obtain for $x \in S$,

$$F(x) = \int P_\infty\{\psi(x, b, c)\} dH(b) dG(c) \tag{3.8}$$

where P_∞ is the measure according to $F(x)$. (When $s = 1$ or 2 equation (3.8) becomes (3.6) and (3.7) with the subscripts of F deleted.) This is an integral equation satisfied by $F(x)$. We shall later prove that, when $\rho < 1$, $F(x)$ is a distribution function (d.f.), and the only d.f. over S which satisfies (3.8). Moreover, we shall prove that, when $\rho \geqq 1$, $F(x)$ is not a d.f., and (3.8) has no solution which is a d.f. over S.

We remark that (3.8) implies that if $F(x)$ is a d.f., the latter defines a stationary absolute probability distribution for our (Markoff) stochastic process, i.e., if w_1 is distributed according to $F(x)$ then w_i has this distribution for every value of i.

Write $\bar{x}_1 = (x_1, \infty, \cdots, \infty)$, $F_i^*(x_1) = F_i(\bar{x}_1)$. Then, from (3.5), the Lebesgue bounded convergence theorem, and the structure of ψ, it follows that

$$F^*_{(i+1)}(x_1) = \int P_i\{\psi(\bar{x}_1, b, c)\} dH(b) dG(c). \tag{3.9}$$

We proved earlier that

$$F_i(x) \geqq F_{(i+1)}(x) \tag{3.10}$$

for every i and x. In (3.10) let the last $(s-1)$ coordinates of x approach infinity. We obtain

$$F_i^*(x_1) \geqq F_{(i+1)}^*(x_1). \tag{3.11}$$

We conclude that $\lim_{i\to\infty} F_i^*(x_1)$ exists; call it $F^*(x_1)$, say. Clearly we have

$$F^*(x_1) \geqq F(\bar{x}_1). \tag{3.12}$$

We shall prove in §5 that equality holds in (3.12). It will then follow from (3.8) that

$$F^*(x_1) = \int P_\infty\{\psi(\bar{x}_1, b, c)\} dH(b)dG(c). \tag{3.13}$$

4. An essential lemma. In this section we shall prove the following

LEMMA.

$$\lim_{y'\to\infty} \liminf_{i\to\infty} P\{w_{i,s} - w_{i,1} \leqq y'\} = 1$$

for $s>1$.

Proof. Let

$$B_i = (s-1)w_{i,s} - \sum_{j=1}^{s-1} w_{i,j} \tag{4.1}$$

for $i \geqq 1$. It follows easily from the way in which $w_{(i+1)}$ is obtained from w_i that

$$B_{(i+1)} \leqq \begin{cases} B_i - R_i & \text{when } R_i \leqq w_{is} - w_{i1}, \\ B_i - s(w_{is} - w_{i1}) + (s-1)R_i \leqq (s-1)R_i & \text{when } R_i \geqq w_{is} - w_{i1}. \end{cases} \tag{4.2}$$

In either case we have

$$B_{(i+1)} \leqq \max(B_i - R_i, (s-1)R_i). \tag{4.3}$$

Applying (4.3) to B_i we obtain

$$B_{(i+1)} \leqq \max(B_{i-1} - R_{i-1} - R_i, (s-1)R_{i-1} - R_i, (s-1)R_i). \tag{4.4}$$

Continuing in this manner and noting that $B_1 = 0$ we obtain

$$\begin{aligned} B_{i+1} \leqq \max\, [&(s-1)R_i, (s-1)R_{i-1} - R_i, (s-1)R_{i-2} - R_{i-1} - R_i, \\ &\cdots, (s-1)R_1 - R_2 - \cdots - R_i] = Y_i \text{ (say).} \end{aligned} \tag{4.5}$$

Since the R_j are independently and identically distributed, we may interchange indices j and $i-j+1$, $j=1, \cdots, i$, in the middle member of (4.5) without altering its distribution. Hence, setting $h=(s-1)^{-1}$, we have

$$(4.6)\quad P\{Y_n \leqq y'\} = P\{R_1 \leqq hy', R_2 \leqq h(R_1 + y'), \cdots, R_n \leqq h(R_1 + \cdots + R_{n-1} + y')\}$$

(where, for $n=1$, we replace $R_1 + \cdots + R_{n-1}$ by 0). Since $B_i \geqq w_{is} - w_{i1}$, our proof will be complete if we show that

$$(4.7)\quad \lim_{y'\to\infty} \liminf_{n\to\infty} P\{Y_n \leqq y'\} = 1.$$

From the strong law of large numbers we have

$$(4.8)\quad P\left\{\lim_{n\to\infty} \frac{1}{n}\sum_{i=1}^{n} R_i = ER_1,\ \lim_{n\to\infty} \frac{R_n}{n} = 0\right\} = 1.$$

Now, for $y' \geqq 0$,

$$(4.9)\quad \begin{aligned} &P\{Y_n \leqq y'\} \\ &\geqq P\{R_n \leqq h(R_1 + \cdots + R_{n-1} + y') \text{ for } n = 1, 2, \cdots, \text{ad inf.}\} \\ &= P\left\{\frac{R_n}{hn} \leqq \frac{R_1 + \cdots + R_{n-1}}{n} + \frac{y'}{n} \text{ for } n = 1, 2, \cdots, \text{ad inf.}\right\}. \end{aligned}$$

Because of (4.8), for any $\epsilon > 0$ there exists an integer N such that

$$P\left\{\frac{R_n}{hn} \leqq \frac{R_1 + \cdots + R_{n-1}}{n} \text{ for } n > N\right\} > 1 - \epsilon.$$

Clearly, there is a value y_0' such that for $y' > y_0'$

$$P\left\{\frac{R_n}{hn} \leqq \frac{R_1 + \cdots + R_{n-1}}{n} + \frac{y'}{n} \text{ for } n \leqq N\right\} > 1 - \epsilon.$$

Equation (4.7) is an immediate consequence.

5. **Certain immediate consequences of the lemma.**

(A) $F^*(x_1) = F(\bar{x}_1)$.

Proof. Let $x(x_1, y')$ be the point $x_1, y', \cdots, y'$. From the lemma it follows at once that for any $\epsilon > 0$ and i and y' sufficiently large we have

$$(5.1)\quad |P\{w_{i1} \leqq x_1\} - P\{w_{i1} \leqq x_1, w_{i2} \leqq y', \cdots, w_{is} \leqq y'\}| < \epsilon.$$

Let $i \to \infty$. We obtain

$$(5.2)\quad |F^*(x_1) - F(x(x_1, y'))| \leqq \epsilon.$$

Let $y' \to \infty$. We obtain

$$(5.3)\quad |F^*(x_1) - F(\bar{x}_1)| \leqq \epsilon.$$

Since ϵ was arbitrary the desired result follows.

(B) Either F and F^* are both distribution functions or neither is a distribution function.

This follows from the fact that (A) above implies that $\lim F(x)$ as all coordinates of x approach infinity is the same as $\lim F^*(x_1)$ as x_1 approaches infinity.

6. **Proof that F is a distribution function when $\rho<1$.** We define

$$\rho = (ER_1)(sEg_1)^{-1}. \tag{6.1}$$

We shall now prove that, if $\rho<1$, $F(x)\to 1$ as all coordinates of x approach infinity. Then, by (3.12), we have $\lim F^*(x_1)=1$ as $x_1\to\infty$.

I. We show that it is sufficient to prove this result in a "dominating" case where, for some $c>0$,

$$\sum_{i=0}^{\infty} P\{R_1 = ci\} = \sum_{i=0}^{\infty} P\{g_1 = ci\} = 1. \tag{6.2}$$

Let $[a]$ be the largest integer $\leqq a$ and for some one $c>0$ define, for all i,

$$g_i' = c\left[\frac{g_i}{c}\right], \qquad R_i' = c\left[\frac{R_i}{c}\right] + c.$$

Then $g_i' \leqq g_i$ and $R_i' \geqq R_i$. Let w_i' be the same function of $\{g_j'\}$ and $\{R_j'\}$ that w_i is of $\{g_j\}$ and $\{R_j\}$. It follows from an argument like that of §3 that $w_i' \geqq w_i$ for all i. Hence if we can show that

$$\lim_{y'\to\infty} \liminf_{i\to\infty} P\{w_{i1}' \leqq y', \cdots, w_{is}' \leqq y'\} = 1 \tag{6.3}$$

it will follow that

$$\lim_{y'\to\infty} F(y', y', \cdots, y') = 1 \tag{6.4}$$

which is the desired result.

We have

$$Eg_j' \geqq Eg_j - c, \qquad ER_j' \leqq ER_j + c.$$

Hence, if c is sufficiently small, $(ER_1')(sEg_1')^{-1}<1$.

In the remainder of this section we assume that (6.2) is satisfied witn $\rho<1$, and we shall prove (6.3) for this process.

II. We show that (6.3) is valid if $P\{R_1=0\}>0$. We recall that $P\{g_1=0\}<1$. Hence, for any i and integral $a_1, \cdots, a_s$, with $0\leqq a_1 \leqq a_2 \leqq \cdots \leqq a_s$,

$$P\{w_{(i+j)} = 0 \text{ for some } j \mid w_{i,1} = a_1c,\ w_{i,2} = a_2c, \cdots, w_{i,s} = a_sc\}$$

is positive. Let Z be the totality of all points $(a_1c, a_2c, \cdots, a_sc)$ with integral

a's such that $0 \leqq a_1 \leqq a_2 \leqq \cdots \leqq a_s$. Let z be a generic point of Z. The points z are the states of our Markoff process $\{w_i\}$. The preceding argument shows that the origin 0 and all the points z which can be reached by the process $\{w_i\}$ with positive probability form a chain C which is irreducible. C is aperiodic, since

$$P\{w_{i+1} = 0 \mid w_i = 0\} \geqq P\{R_1 = 0\} > 0. \tag{6.5}$$

The desired result then follows by III below.

III. We shall show that, if C is aperiodic and irreducible, (6.3) holds. From §3 or [3, Chap. 15, Theorems 1 and 2], it follows that

$$\lim_{i\to\infty} P\{w_i = z\}$$

exists for all z in C; call it $f(z)$. From the theorems of [3] cited above it follows that

$$\sum_{z\in C} f(z) = 0 \quad \text{or} \quad 1. \tag{6.6}$$

Our result is proved if we show that the sum in (6.6) is 1. Suppose it were 0; every $f(z)$ is then zero. We would then have

$$F(y', y', \cdots, y') = 0 \tag{6.7}$$

for every y'. Hence from (A), §5, we obtain, using (6.7),

$$F^*(x_1) = 0 \tag{6.8}$$

for every x_1. From the definition of w_i and the fact that $\rho < 1$ we obtain that there exists an $M > 0$ such that, whenever

$$M \leqq a_1 \leqq a_2 \leqq \cdots \leqq a_s, \tag{6.9}$$

we have

$$E\left\{\sum_{j=1}^{s} w_{(i+1),j} \,\middle|\, w_{ij} = a_j,\ j = 1, \cdots, s\right\} < \sum_{j=1}^{s} a_j - \delta \tag{6.10}$$

for some $\delta > 0$. It is to be noted that, whether (6.9) holds or not, the left member of (6.10) is never greater than

$$\sum_{j=1}^{s} a_j + ER_1. \tag{6.11}$$

Since $F^*(x_1) \equiv 0$ we can find an $N > 0$ such that for $i \geqq N$ we have

$$P\{w_{i,1} < M\} < \frac{\delta}{\delta + ER_1}. \tag{6.12}$$

Then, for $i \geqq N$, we have

$$E\left\{\sum_{j=1}^{s} w_{i+1,j} - \sum_{j=1}^{s} w_{i,j}\right\}$$

$$(6.13) \qquad = E\left\{E\left\{\sum_{j=1}^{s} w_{i+1,j} - \sum_{j=1}^{s} w_{i,j} \,\middle|\, w_{ij}, j = 1, \cdots, s\right\}\right\}$$

$$< \frac{\delta}{\delta + ER_1}(ER_1) + \frac{ER_1}{\delta + ER_1}(-\delta) = 0.$$

Hence, for $i>0$,

$$(6.14) \qquad E\sum_{j=1}^{s} w_{i+N,j} < E\sum_{j=1}^{s} w_{N,j}$$

so that $E\sum_{j=1}^{s} w_{i,j}$ is bounded uniformly in i. This contradicts (6.7) and proves III.

IV. We now suppose $P\{R_1=0\}=0$, and we construct a suitable "dominating" process for which we can prove results analogous to II and III.

Let k be a positive integer such that $(sk-1)>0$,

$$(6.15) \qquad P\{R_1 \geqq (sk-1)c\} = 1,$$

and

$$(6.16) \qquad P\{R_1 = (sk-1)c\} > 0.$$

If necessary, we can decrease the c of (6.2) so that such a k can always be found. If now

$$(6.17) \qquad P\{g_1 \geqq skc\} > 0,$$

then it is clear that

$$(6.18) \qquad P\{w_{i+j} = 0 \text{ for some } j > 0 \mid w_i = z\} > 0$$

for every z in C. Hence C is irreducible. It is also aperiodic, because

$$(6.19) \qquad P\{w_{i+1} = 0 \mid w_i = 0\} > 0.$$

Hence the desired result follows by III. We therefore assume that (6.17) does not hold, i.e., that

$$(6.20) \qquad P\{g_1 \leqq (sk-1)c\} = 1.$$

Let m be the largest integer for which

$$P\{g_1 = mc\} > 0.$$

Let A_1 be the set of α (say) non-negative integers $j<m$ such that

$$P\{g_1 = jc\} = 0, \qquad j \in A_1.$$

Let $\{g_i'\}$ be independently and identically distributed chance variables with the following distribution:

$$P\{g_1' = jc\} = \lambda, \qquad j \in A_1,$$
$$P\{g_1' = mc\} = P\{g_1 = mc\} - \alpha\lambda,$$
$$P\{g_1' = jc\} = P\{g_1 = jc\}, \qquad j \neq m, j \notin A_1.$$

Here λ is a small positive number, whose choice will be more fully described shortly, but which should in any case be such that $P\{g_1 = mc\} - \alpha\lambda > 0$ and $\lambda < P\{R_1 = (sk-1)c\}$.

Let A_2 be the totality of integers $j > (sk-1)$ such that

$$P\{R_1 = jc\} = 0, \qquad j \in A_2.$$

Let $\{R_i'\}$ be independently and identically distributed chance variables, independent of $\{g_i'\}$ and with the following distribution:

$$P\{R_1' = jc\} = \frac{\lambda}{2^i}, \qquad j \in A_2,$$

$$P\{R_1' = jc\} = P\{R_1 = jc\}, \qquad j \neq (sk-1), j \notin A_2,$$
$$P\{R_1' \geqq (sk-1)c\} = 1.$$

We choose $\lambda > 0$ so small that

$$(ER_1')(sEg_1')^{-1} < 1.$$

Any such λ will suffice.

Let $\{w_i'\}$ be the same functions of $\{R_i', g_i'\}$ as w_i are of $\{R_i, g_i\}$. Let F_i' and F' be the corresponding functions for the primed w_i. Comparing corresponding sequences in the manner of §3 we obtain that

$$F_i'(x) \leqq F_i(x) \qquad \text{for every } x.$$

Hence

$$F'(x) = \lim F_i'(x) \leqq F(x) \qquad \text{for every } x.$$

If, therefore, we prove the desired result for the system $\{w_i'\}$ we have a fortiori proved the desired result for the system $\{w_i\}$. We may therefore drop the accents and henceforth assume that

$$(6.21) \qquad P\{g_1 = jc\} > 0, \qquad j = 0, \cdots, (sk-1),$$

$$(6.22) \qquad P\{g_1 \leqq (sk-1)c\} = 1,$$

$$(6.23) \qquad P\{R_1 = jc\} > 0, j = (sk-1), (sk), (sk+1), \cdots,$$

$$(6.24) \qquad P\{R_1 \geqq (sk-1)c\} = 1.$$

(We note that these imply that we are in the case $s > 1$, for $s = 1$ would violate the requirement that $\rho<1$.)

Let

$$z^* = (skc, skc, \cdots, skc).$$

Let

$$z = (a_1c, a_2c, \cdots, a_sc)$$

with $0 \leqq a_1 \leqq a_2 \leqq \cdots \leqq a_s$ be any point in Z. Let

$$L = w_{is} + skc.$$

If $R_{i+j-1} = L - w_{ij}$ and $g_{i+j}=0$, $j=1, \cdots, s$, an event of positive probability, we have

$$w_{i+s} = (L, L, \cdots, L).$$

If $R_{i+s+j-1} = (sk-1)c$ and $g_{i+s+j} = kc$, $j=1, \cdots, s$, again an event of positive probability, we have, since $L \geqq skc$, that

$$w_{i+2s} = (L-c, L-c, \cdots, L-c).$$

Applying the above argument a_s times we conclude that, for any z and i, $P\{w_{i+j}=z^* \text{ for some } j\geqq 0 \mid w_i = z\}>0$. Let D be the set of all points in Z which can be reached from z^* with positive probability. The above argument shows that the states of D form an irreducible Markoff chain. This chain is aperiodic, because a modification of the above argument shows (using (6.23)) that there exists a number N such that, whatever be $n \geqq N$, there is a positive probability of moving from z^* back to z^* in exactly n steps.

If now, with probability one, $w_i \in D$ for some i, an argument similar to that of III applies and the desired result is proved.

V. We now prove that, with probability one, $w_i \in D$ for some i.

From (6.23) and the fact that $P\{g_1=0\}>0$ it follows that any point $z=(a_1c, \cdots, a_sc) \in Z$ such that $(2sk-1) \leqq a_1$ is a member of D. We now note that the probability of entering D in at most s steps from any point z not in D is bounded below by a number (say) $\mu>0$, independently of z (not in D). To see this, we note that this can be accomplished in at most s steps where each $R=2skc$ and each $g=0$. From this it follows that the probability of entering D for some i is one.

The proof of the result of this section is now complete.

7. Proof that F is not a distribution function when $\rho \geqq 1$. To prove this result we must in addition assume that, when $\rho=1$,

$$P\{R_i - sg_i = 0\} < 1. \tag{7.1}$$

For if (7.1) does not hold we have, for some $e>0$,

$$P\{g_1 = e\} = P\{R_1 = se\} = 1. \tag{7.2}$$

(Hence $\rho=1$ here.) Therefore, with probability one,

$$\begin{aligned} w_1 &= (0, 0, \cdots, 0),\\ w_2 &= (0, 0, \cdots, 0, (s-1)e),\\ w_3 &= (0, 0, \cdots, 0, (s-2)e, (s-1)e),\\ &\vdots\\ w_s &= (0, e, 2e, \cdots, (s-1)e),\\ w_i &= w_s, \qquad i > s. \end{aligned}$$

Hence a limiting distribution function F does exist.

We therefore assume that $\rho \geqq 1$ and (7.1) holds. We shall show that $F(x) \equiv 0$, and hence (see §5, A) that $F^*(x_1) \equiv 0$.

Let $\{L_i\}$ be a sequence of chance variables defined as follows: $L_1=0$ with probability one. For $i \geqq 1$

$$L_{i+1} = \max(0, L_i + R_i - sg_{i+1}).$$

Thus L_i is the waiting time of the ith individual in a system such as described in §2 where $s=1$, the service time of the ith individual is R_i, and the interval between the ith arrival and $(i+1)$st arrival is sg_{i+1}. In this system $\rho \geqq 1$, so that the theorem of Lindley (which treats the case $s=1$) is applicable, i.e.,

$$\lim_{x'\to\infty} \lim_{i\to\infty} P\{L_i \leqq x'\} = 0. \tag{7.3}$$

Now, if $0 \leqq a_1 \leqq a_2 \leqq \cdots \leqq a_s$, and b, c, and d are non-negative numbers with $b \leqq \sum_{j=1}^{s} a_j$, we clearly have $\max(0, a_1+c-d) + \sum_{j=2}^{s} \max(0, a_j-d) \geqq \max(0, b+c-sd)$. We conclude, using induction, that for all $i \geqq 1$,

$$L_i \leqq \sum_{j=1}^{s} w_{i,j} \tag{7.4}$$

with probability one. It follows from (7.3) and (7.4) that

$$\begin{aligned} \lim_{x'\to\infty} F(x', x', \cdots, x') &= \lim_{x'\to\infty} \lim_{i\to\infty} P\{w_{i,j} \leqq x', j = 1, \cdots, s\}\\ &\leqq \lim_{x'\to\infty} \lim_{i\to\infty} P\left\{\sum_{j=1}^{s} w_{i,j} \leqq sx'\right\} \qquad (7.5)\\ &\leqq \lim_{x'\to\infty} \lim_{i\to\infty} P\{L_i \leqq sx'\} = 0, \end{aligned}$$

which proves the desired result.

8. Proof that $\lim_{i\to\infty} F_i(x|y)$ exists and is independent of y. Uniqueness of the solution of the integral equation (3.8). Suppose first that $\rho \geqq 1$. Since $F_i(x|y) \leqq F_i(x)$ for every i and every x and $y \in S$, it follows from the results of §7 that

$$\lim_{i\to\infty} F_i(x \mid y) = 0$$

when $\rho \geqq 1$. We shall shortly show that, when $\rho \geqq 1$, (3.8) has no solution which is a distribution function over S.

Assume, therefore, that $\rho < 1$, which is the interesting case. We shall show that, for all x and y in S, the ergodic property

$$\lim_{i\to\infty} F_i(x \mid y) = F(x) \tag{8.1}$$

holds. From this it follows easily that (3.8) has at most one solution which is a distribution function over S (thus, by §6, it has exactly one such solution). For, suppose, to the contrary, that there were another such distribution function, say $V(x)$. It is clear then that, if w_1 is distributed according to $V(x)$, so are $w_2, w_3, \cdots$, so that $V(x)$ is the limiting distribution. On the other hand, it follows from (8.1) and the Lebesgue bounded convergence theorem applied to

$$V(x) = \int F_i(x \mid y) dV(y)$$

that

$$V(x) = \int F(x) dV(y) = F(x)$$

which is the desired result. (Thus we have proved that, when $\rho < 1$, $F(x)$ is the unique stationary absolute probability distribution; see the paragraph following equation (3.8).)

Conversely, if (3.8) has a solution V which is a distribution function over S, then

$$F(x) = \lim_{i\to\infty} F_i(x) \geqq \lim_{i\to\infty} \int F_i(x \mid y) dV(y) = V(x),$$

so that (from the result of §7) $\rho < 1$ and hence $V(x) = \lim_{i\to\infty} F_i(x)$ is the unique solution of (3.8).

Denote by $[a, b]$ and $[c, d]$ the smallest closed intervals for which

$$P\{a \leqq R_1 \leqq b\} = P\{c \leqq g_1 \leqq d\} = 1.$$

We shall conduct the proof separately for several cases.

Case 1: $b > sc$. Let $b - sc = 2\nu > 0$. Then, for any positive integer n,

$$P\{w_{sn,1} > \nu n\} = q_n > 0.$$

Fix n. For any x and $\delta > 0$ there exists an integer M such that, for all $j \geqq M$

and $k>0$, we have

$$|F_j(x) - F_{j+k}(x)| < q_n\delta.$$

We recall that, if $y_1 \leqq y_2$, $y_1 \in S$, $y_2 \in S$, we have, for all i,

$$F_i(x \mid y_1) \geqq F_i(x \mid y_2).$$

Hence, for $j \geqq M$, we have

$$\begin{aligned} 0 &\leqq q_n[F_j(x) - F_j(x \mid (nv, nv, \cdots, nv))] \\ &\leqq \int [F_j(x) - F_j(x \mid y)]dF_{sn}(y) \qquad (8.2) \\ &< q_n\delta + \int [F_{j+sn-1}(x) - F_j(x \mid y)]dF_{sn}(y) = q_n\delta. \end{aligned}$$

Therefore

$$0 \leqq [F_j(x) - F_j(x \mid y)] < \delta \qquad (8.3)$$

for all $y \leqq (nv, nv, \cdots, nv)$, $y \in S$, and all $j \geqq M$. Since x, n, and δ were arbitrary this proves (8.1) for Case 1.

Case 2: $a<d$. Let y be any point in S, and

$$p_n(y) = P\{w_n = 0;\ w_i \neq 0,\ i < n \mid w_1 = y\}.$$

We shall show that, for all y in S,

$$\sum_{n=1}^{\infty} p_n(y) = 1. \qquad (8.4)$$

This is sufficient to prove the desired result, because

$$F_i(x \mid y) - \sum_{n \leqq i} p_n(y)F_{i-n}(x) \qquad (8.5)$$

then approaches zero as $i \to \infty$.

To prove (8.4) we proceed as in §6 to construct a "dominating" random walk on a lattice. The walk begins at a point on the lattice all of whose coordinates are no less than the corresponding ones of y. As in §6 one proves that with probability one the walk enters an irreducible aperiodic chain. Since $a<d$ this chain contains the origin. Since $\rho<1$ and $F(x)$ is a distribution function this chain constitutes a positive recurrent class. For an irreducible, recurrent class (8.4) must hold for all y in the class. Since the walk enters the class with probability one, (8.4) holds for all y in S.

Case 3: $c=d \leqq a=b \leqq sc$. In this case we have $P\{R_1=b\}=P\{g_1=c\}=1$. Since $\rho<1$, we also have $b<sc$.

Let $y^*=(y_1, \cdots, y_s)$ be any point in S, and let $y''=(y_s, \cdots, y_s)$. Given the process $\{R_i, g_i\}$, let w_i^* be the position of w_i if $w_1=y^*$, let w_i' be the position of w_i if $w_1=0$, and let w_i'' be the position of w_i if $w_1=y''$. Clearly, $w_i \leqq w_i^* \leqq w_i''$ for all i with probability one, and for $i \geqq s$ we have $w_i' = \bar{w}$, where $\bar{w}$ is defined by

$$(8.6) \qquad \bar{w} = (0, u_{s-1}, \cdots, u_2, u_1),$$

where

$$u_j = \max(b - jc, 0).$$

We shall show that $w_i'' = \bar{w}$ with probability one for i sufficiently large, which implies that for sufficiently large i with probability one, $w_i^* = \bar{w}$, and proves the desired result.

It is clear that, for all i,

$$(8.7) \qquad \begin{aligned} w''_{i+1,j} &= \max(0, w''_{i,j+1} - c) \qquad \text{for } 1 \leqq j < s, \\ w''_{i+1,s} &= w''_{i,1} + b - c. \end{aligned}$$

For $n \geqq 0$ and $0 \leqq i \leqq s-1$, we evidently have

$$(8.8) \qquad w''_{1+ns+i,1} \leqq w''_{ns+1,1}.$$

Let N be a positive integer such that $y_s - N(sc-b) \leqq 0$. Then $w''_{ns+1,1}=0$ for $n \geqq N$, and hence from (8.8) we have $w''_{i,1}=0$ for $i \geqq Ns+1$. It follows from (8.7) that $w_i'' = \bar{w}$ for $i \geqq (N+1)s$.

Case 4: $d \leqq a$, $b \leqq sc$, *and either* $a<b$ *or* $c<d$. Let u_j be as in (8.6) and for $\epsilon>0$ define $\bar{w}^\epsilon = (0, u^\epsilon_{s-1}, \cdots, u^\epsilon_2, u^\epsilon_1)$, where $u^\epsilon_j = \max(0, u_j - \epsilon)$. From the definition of b and c we have that, for every $\epsilon>0$,

$$P\{w_{s,j} \geqq u^\epsilon_{s-j+1}, j = 1, \cdots, s\} = \gamma > 0.$$

An argument like that of Case 1 (with γ for q_n and F_s for F_{sn}) then shows that

$$\lim_{i\to\infty} F_i(x \mid y) = F(x)$$

for all $y \in \Gamma^\epsilon = \{y \mid y \in S, y \leqq \bar{w}^\epsilon\}$ and all x.

Let y be any point in S, and

$$p^\epsilon_n(y) = P\{w_n \in \Gamma^\epsilon, w_i \notin \Gamma^\epsilon, i < n \mid w_1 = y\}.$$

We shall show that, for some $\epsilon>0$ and all $y \in S$, we have

$$(8.9) \qquad \sum_{n=1}^{\infty} p^\epsilon_n(y) = 1;$$

from this and the result of the previous paragraph, the desired result is proved in the manner of the first paragraph of Case 2.

Let

(8.10) $$E = \{y \mid y = (y_1, \cdots, y_s) \in S;\ y_1 = 0;\ y_s \leqq (s-1)b\}.$$

In order to prove (8.9) for some $\epsilon>0$ and for all $y \in S$, it clearly suffices to show that

(8.11) $$P\{w_i \in E \text{ for infinitely many } i \mid w_1 = y\} = 1$$

for all $y \in S$, and that there exists a positive integer M and positive numbers α and ϵ such that

(8.12) $$P\{w_M \in \Gamma^{\epsilon} \mid w_1 = y\} > \alpha$$

for all $y \in E$.

We first prove (8.11). To this end, let $y = (y_1, \cdots, y_s)$ be any fixed point in S. Since we have always assumed $d>0$, we have in Case 4 that $a>0$. It follows from equation (4.3) that for $n>(s-1)y_s/a$ we have

(8.13) $$P\{B_n \leqq (s-1)b \mid w_1 = y\} = 1.$$

Let $\{e_i\}$, $\{f_i\}$ be any sequences of non-negative numbers, and let $\{v_i\}$ be the corresponding sequence of values of $\{w_i\}$ when $w_1 = y^*$, $R_i = e_i$, and $g_i = f_i$. Then, if $v_{i1}=0$ for only finitely many i, it would follow that $\liminf_{n\to\infty} (1/n) \sum_{i=1}^{n} (e_i - sf_{i+1}) \geqq 0$. However, since $\rho<1$, the strong law of large numbers implies that

$$P\left\{\lim_{n\to\infty} \frac{1}{n} \sum_{i=1}^{n} (R_i - sg_{i+1}) = ER_1 - sEg_1 < 0\right\} = 1.$$

Hence

(8.14) $$P\{w_{i,1} = 0 \text{ for infinitely many } i \mid w_1 = y^*\} = 1$$

for all $y^* \in S$. Equation (8.14) is a fortiori true for the original process, and (8.11) is an immediate consequence of (4.1), (8.13), and (8.14).

It remains to prove (8.12). We recall that in Case 4 we have $c<b\leqq sc$ and that there are numbers b', c' such that $P\{R_1 \leqq b'\} = p>0$, $P\{g_1 \geqq c'\} = q>0$, and $b'-c' = b-c-\epsilon$ for some $\epsilon>0$. An obvious modification of the argument of Case 3 (put b', c' for b, c) shows that if $w_1 = y \in E$ and if $R_j = b'$ and $g_{j+1} = c'$ for $1 \leqq j < M$ where M/s is the greatest integer contained in $2 + (s-1)b/(sc'-b')$, then

(8.15) $$w_{M,i} \leqq \max(0, b' - (s+1-i)c') \leqq u^{\epsilon}_{s-i+1}.$$

Equation (8.15) is a fortiori true if $R_j \leqq b'$ and $g_{j+1} \geqq c'$ for $1 \leqq j < M$ (the argument being similar to that of §3). We conclude that (8.12) is satisfied for ϵ and M as defined here and for $\alpha = (pq)^{M-1}$.

9. **Distribution of the number of individuals waiting in the queue.** In order to avoid trivial cases and the circumlocutions required to dispose of them, we shall assume in this section that $G(0)=0$. This means that the prob-

ability is zero that two or more individuals arrive simultaneously.

Let Q_i be the number of arrivals in the open time interval (t_i, t_i+w_{i1}); i.e., Q_i is the number of individuals in the queue waiting to be served, just before the service of the ith individual begins.

Since $g_{i+1}, g_{i+2}, \cdots$ are independent of t_i, we have

$$(9.1)\qquad \begin{aligned} P\{Q_i \geqq n\} &= \int P\{g_1 + g_2 + \cdots + g_n < a\} dF_i^*(a) \\ &= \int G^{n*}(a-) dF_i^*(a), \end{aligned}$$

where $G^{n*}(a)$ denotes the n-fold convolution of $G(a)$ with itself. Since $F_i^*(a)$ tends nonincreasingly to $F^*(a)$ as $i \to \infty$ for all a, and since $G^{n*}(a-)$ is continuous from the left, we obtain, in the case $\rho < 1$,

$$(9.2)\qquad \lim_{i\to\infty} P\{Q_i \geqq n\} = \int G^{n*}(a-) dF^*(a).$$

If $\rho \geqq 1$, equation (9.1) shows that $\lim_{i\to\infty} P\{Q_i \geqq n\} = 1$ for all n, except in the trivial case where $P\{R_1 - sg_1 = 0\} = 1$.

References

1. D. V. Lindley, *The theory of queues with a single server*, Proc. Cambridge Philos. Soc. vol. 48 (1952), part 2, pp. 277–289.

2. J. V. Uspensky, *Introduction to mathematical probability*, New York, McGraw-Hill, 1937.

3. W. Feller, *An introduction to probability theory and its applications*, New York, Wiley, 1950.

4. N. Metropolis and S. Ulam, *The Monte Carlo method*, Journal of the American Statistical Association vol. 44 (1949) pp. 335–341.

5. D. G. Kendall, *Stochastic processes occurring in the theory of queues and their analysis by the method of the imbedded Markov chain*, Ann. Math. Statist. vol. 24 (1953) pp. 338–354.

6. ———, *Some problems in the theory of queues*, Journal Royal Statistical Society (B) vol. 13 (1951) pp. 151–173 and 184–185.

Cornell University,
Ithaca, N. Y.

Reprinted from THE ANNALS OF MATHEMATICAL STATISTICS
Vol. 27, No. 3, September, 1956

ASYMPTOTIC MINIMAX CHARACTER OF THE SAMPLE DISTRIBUTION FUNCTION AND OF THE CLASSICAL MULTINOMIAL ESTIMATOR

BY A. DVORETZKY,[1] J. KIEFER,[1] AND J. WOLFOWITZ[2]

Cornell University

0. Summary. This paper is devoted, in the main, to proving the asymptotic minimax character of the sample distribution function (d.f.) for estimating an unknown d.f. in $\mathfrak{F}$ or $\mathfrak{F}_c$ (defined in Section 1) for a wide variety of weight functions. Section 1 contains definitions and a discussion of measurability considerations. Lemma 2 of Section 2 is an essential tool in our proofs and seems to be of interest per se; for example, it implies the convergence of the moment generating function of G_n to that of G (definitions in (2.1)). In Section 3 the asymptotic minimax character is proved for a fundamental class of weight functions which are functions of the maximum deviation between estimating and true d.f. In Section 4 a device (of more general applicability in decision theory) is employed which yields the asymptotic minimax result for a wide class of weight functions of this character as a consequence of the results of Section 3 for weight functions of the fundamental class. In Section 5 the asymptotic minimax character is proved for a class of integrated weight functions. A more general class of weight functions for which the asymptotic minimax character holds is discussed in Section 6. This includes weight functions for which the risk function of the sample d.f. is not a constant over $\mathfrak{F}_c$. Most weight functions of practical interest are included in the considerations of Sections 3 to 6. Section 6 also includes a discussion of multinomial estimation problems for which the asymptotic minimax character of the classical estimator is contained in our results. Finally, Section 7 includes a general discussion of minimization of symmetric convex or monotone functionals of symmetric random elements, with special consideration of the "tied-down" Wiener process, and with a heuristic proof of the results of Sections 3, 4, 5, and much of Section 6.

1. Introduction and Preliminaries.

Throughout this paper we shall denote by $\mathfrak{F}$ the class of all univariate d.f.'s and by $\mathfrak{F}_c$ the subclass of continuous members of $\mathfrak{F}$ (for the sake of definiteness, members of $\mathfrak{F}$ will be considered continuous on the right). Let R^n denote n-dimensional Euclidean space, and let G be any subspace of the space of all real-valued functions on R^1. For simplicity we assume $\mathfrak{F} \subset G$, although it is really only necessary that G contain the function S_n, defined below, for every $x^{(n)}$. Let B be the smallest Borel field on G such that every element of $\mathfrak{F}$ is an element of B and such that, for every positive integer k and all sets of real numbers $\{t_1, \cdots, t_k\}$ and $\{a_1, \cdots, a_k\}$ with $t_1 < t_2 < \cdots <$

Received May 31, 1955. Revised October 5, 1955.

[1] Research sponsored by the Office of Naval Research.

[2] The research of this author was supported in part by the United States Air Force under Contract No. AF18(600)-685 monitored by the Office of Scientific Research.

t_k, the set $\{g \mid g \varepsilon G; g(t_1) < a_1, \cdots, g(t_k) < a_k\}$ is in B. (Thus, we might have $G = \mathfrak{F}$ and B the Borel field generated by open sets in the common metric topology.) Let D_n be the class of all real-valued functions ϕ_n on $B \times R^n$ with the following properties: for each $x^{(n)} \varepsilon R^n$, $\phi_n(\cdot; x^{(n)})$ is a probability measure (B) on G; and for each $\Delta \varepsilon B$, $\phi_n(\Delta; \cdot)$ is Borel-measurable on R^n.

The problem which confronts the statistician may now be described. Let $X_1, \cdots, X_n$ be independently and identically distributed according to some d.f. F about which it is known only that $F \varepsilon \mathfrak{F}_c$ (or even $F \varepsilon \mathfrak{F}$). The statistician is to estimate F. Write $X^{(n)} = (X_1, \cdots, X_n)$. Having observed $X^{(n)} = x^{(n)} = (x_1, \cdots, x_n)$, the statistician uses the decision function ϕ_n as follows: a function $g \varepsilon G$ is selected by means of a randomization according to the probability measure $\phi_n(\cdot; x^{(n)})$ on G; the function g so selected (which need not be a member of $\mathfrak{F}$) is then the statistician's estimate of the unknown F. It is desirable to select a procedure ϕ_n which may be expected to yield a g which will lie close to the true F, whatever the latter may be; the term "close" will be made precise in succeeding sections. We note that the decision procedure ϕ_n^* which for each $x^{(n)}$ assigns probability one to the "sample d.f." S_n defined by

$$S_n(x) = (\text{number of } x_i \leqq x)/n$$

is a member of D_n.

We now turn (in this and the four succeeding paragraphs) to measure-theoretic considerations which are relevant to this paper. Our point of view is to waste as little space as possible on these considerations, since our results hold under any measurability assumptions which imply the meaningfulness of certain probabilities and integrals involving elements ϕ of D_n, and, in fact, our results hold even if these are interpreted as inner measures and integrals (which will be proper ones when $\phi = \phi_n^*$), as we shall now see.

In Sections 3, 4, and 6 we shall be concerned, for a given n, $\phi \varepsilon D_n$, $r > 0$, and $F \varepsilon \mathfrak{F}$, with the probability that, when the procedure ϕ_n is used and the X_i have d.f. F, the selected estimate g of F will satisfy the inequality

$$\sup_x | g(x) - F(x) | > r.$$

We shall denote this probability by

$$P_{F,\phi}\{\sup_x | g(x) - F(x) | > r\}. \tag{1.1}$$

It is clear when $\phi = \phi_n^*$ that this probability is well defined. This probability will also be meaningful if G is sufficiently regular; for example, if G consists of functions continuous on the right, the supremum in the displayed expression is unchanged if it is taken over rational x, and the probability in question is well defined. For our considerations it is not even necessary to restrict G in this way; we need not concern ourselves with questions of measurability of

$$\sup_x | g(x) - F(x) |,$$

since the optimal properties proved for ϕ_n^* hold if the supremum is taken only over the rationals (this last supremum is never greater than the supremum over all x and is equal to the latter when $g = S_n$). Thus, for arbitrary G and ϕ, the "probability" expression displayed above may be interpreted with the supremum taken over the rationals (or, alternately, as an inner measure, or as the infimum over all positive integers k and sets of real numbers $t_1, \cdots, t_k$ of

$$P_{F,\phi}\{\max_{1 \leqq i \leqq k} | g(t_i) - F(t_i) | > r\}).$$

In Sections 4 and 6 expressions such as

$$\int W(r)\, d_r P_{F,\phi} \left\{\sup_x | g(x) - F(x) | \leqq r\right\} \tag{1.2}$$

appear, the integral being taken over the nonnegative reals with $W \geqq 0$ and nondecreasing. The probability appearing here is to be interpreted as unity minus the probability previously displayed in (1.1), but the integral is to be interpreted as including a term $\gamma \lim_{r\to\infty} W(r)$ if $\gamma > 0$, where

$$\gamma = \lim_{r\to\infty} P_{F,\phi}\left\{\sup_x | g(x) - F(x) | > r\right\}.$$

In Sections 5 and 6 we will encounter such expressions as

$$r(F, \phi) = E_{F,\phi} \int W(g(t) - F(t), F(t))\, dF(t), \tag{1.3}$$

or such an expression with the first two symbols (operations) interchanged. Here $W(x, t)$ is defined for x real and $0 \leqq t \leqq 1$, is measurable (in the Borel sense on R^2), is nonnegative, and for each t is even in x and nondecreasing in x for $x \geqq 0$. $E_{F,\phi}$ is the operation of expectation when the procedure ϕ is used and the X_i have d.f. F. If $\phi = \phi_n^*$, $r(F, \phi)$ is clearly well defined. For other ϕ, any of a number of general assumptions on W and G will suffice to make the integral meaningful; for example, if W is continuous, $F \varepsilon \mathfrak{F}_c$, and G consists of functions continuous on the right, then the integral is determined by the values of g on the rationals, and $r(F, \phi)$ is meaningful. Weaker assumptions may be made, and, in fact, one could treat $r(F, \phi)$ as an inner integral (which is a proper integral when $\phi = \phi_n^*$) and still obtain the optimum properties for ϕ_n^* which are derived in this paper.

Finally, in Sections 3, 4, 5, and 6, the method of proof used involves integration of expressions such as (1.1), (1.2), and (1.3) with respect to probability measures ξ_{kn} on $\mathfrak{F}_c$. These ξ_{kn} will always be measures (B) and, in fact, will be of a very simple form. Sometimes the order of integration will be interchanged in these sections. If $\phi = \phi_n^*$, the above operations are all easily justified. For other ϕ these operations may be justified, as in the previous three paragraphs, by suitable regularity assumptions on G and W; or, again, the integrals in question may be considered as inner integrals.

2. Two Lemmas. In this section we shall state two lemmas (and a corollary to the second) which will be used to prove the results of Sections 3 and 4, respectively. Lemma 1 is due to Anderson [8], while Lemma 2 is derived from results of Smirnoff [9].

For any set S in R^n and any n-vector ρ, we write $S + \rho = \{x \mid x - \rho \varepsilon S\}$. Denote m-dimensional Lebesgue measure by μ_m. The case of Anderson's result which will be of use to us is the following:

LEMMA 1. *Let P be a (possibly degenerate)*[3] *normal probability measure on R^n with means zero, and let T be any convex body in R^n which is symmetric about the origin. Then $P(T) \geqq P(T + \rho)$ for all ρ.*

We shall also use (in Section 5) the trivial fact that the result of Lemma 1 holds for $n = 1$ when P is a normal probability measure truncated at $(-\beta, \beta)$ for $\beta > 0$. In Section 7 we shall mention briefly an application of the more general form of Lemma 1 given in [8].

Before stating Lemma 2, we shall introduce some notation. Let U denote the uniform d.f. (i.e., the d.f. whose density with respect to μ_1 is unity) on [0, 1], and write, for $r \geqq 0$,

$$G_n(r) = P_U\left\{\sup_{0 \leqq x \leqq 1} | S_n(x) - x | \leqq r/\sqrt{n}\right\},$$

$$G_{k,n}(r) = P_U\left\{\max_{1 \leqq i \leqq k} | S_n(i/(k+1)) - i/(k+1) | \leqq r/\sqrt{n}\right\},$$

$$G(r) = 1 - 2\sum_{m=1}^{\infty} (-1)^{m+1} e^{-2m^2r^2}, \tag{2.1}$$

$$H_n(r) = P_U\left\{\sup_{0 \leqq x \leqq 1} [S_n(x) - x] \leqq r/\sqrt{n}\right\},$$

$$H_{k,n}(r) = P_U\left\{\max_{1 \leqq i \leqq k} [S_n(i/(k+1)) - i/(k+1)] \leqq r/\sqrt{n}\right\},$$

$$H(r) = 1 - e^{-2r^2}$$

Then

$$G_{k,n}(r) \geqq G_n(r) \quad \text{and} \quad H_{k,n}(r) \geqq H_n(r) \tag{2.2}$$

for all k, n, r. Moreover,

$$\lim_{k\to\infty} G_{k,n}(r) = G_n(r), \qquad \lim_{k\to\infty} H_{k,n}(r) = H_n(r), \tag{2.3}$$

and ([1], [2], [3])

$$\lim_{k\to\infty}\lim_{n\to\infty} G_{k,n}(r) = \lim_{n\to\infty} G_n(r) = G(r), \qquad \lim_{k\to\infty}\lim_{n\to\infty} H_{k,n}(r) = \lim_{n\to\infty} H_n(r) = H(r). \tag{2.4}$$

---

[3] The fact that the measure need not be n-dimensional necessitates only trivial modifications of the argument in [8].

We shall now prove the following:

LEMMA 2. *There exists a finite positive constant c such that*

$$1 - H_n(r) < ce^{-2r^2} \tag{2.5}$$

and

$$1 - G_n(r) < ce^{-2r^2} \tag{2.6}$$

hold for all $r \geq 0$ and all positive integers n.

An immediate consequence is

COROLLARY 2. *If $W(r)$ is any nondecreasing nonnegative function defined for $r > 0$, then*

$$\lim_{n\to\infty} \int_0^\infty W(r)\, dH_n(r) = \int_0^\infty W(r)\, dH(r) \tag{2.7}$$

and

$$\lim_{n\to\infty} \int_0^\infty W(r)\, dG_n(r) = \int_0^\infty W(r)\, dG(r). \tag{2.8}$$

Indeed, the lim inf of the integral on the left side of (2.7) or (2.8) is always $\geq$ the respective integral on the right side. Now, if $\int_0^\infty W(r)re^{-2r^2}\, dr = \infty$, then by (2.1), the integrals on the right side of (2.7) and (2.8) are both infinite and thus (2.7) and (2.8) hold in this case. If, on the other hand,

$$\int_0^\infty W(r)re^{-2r^2}\, dr < \infty,$$

then Corollary 2 follows from (2.4), (2.5), and (2.6), and in this case both sides of (2.7) and (2.8) are finite.

PROOF OF LEMMA 2. Since $1 - G_n(r) \leq 2\,(1 - H_n(r))$, it suffices to prove (2.5). We shall deduce (2.5) from the explicit expression for $1 - H_n(r)$ given by Smirnoff [9]. Obviously, $1 - H_n(r) = 0$ for $r \geq \sqrt{n}$, while for $0 < r < \sqrt{n}$, equation (50) of [9] asserts

$$1 - H_n(r) = (1 - r/\sqrt{n})^n + r\sqrt{n} \sum_{j=[r\sqrt{n}]+1}^{n-1} Q_n(j, r), \tag{2.9}$$

where $[x]$ denotes the greatest integer $\leq x$ and

$$Q_n(j, r) = \binom{n}{j} (j - r\sqrt{n})^j (n - j + r\sqrt{n})^{n-j-1} n^{-n}. \tag{2.10}$$

In what follows we may, and do, restrict ourselves to $0 < r < \sqrt{n}$.

Taking logarithms and differentiating, it is seen that the maximum of $(1 - r/\sqrt{n})^n e^{2r^2}$ occurs at $r = 0$; hence,

$$\left(1 - \frac{r}{\sqrt{n}}\right)^n e^{2r^2} < 1. \tag{2.11}$$

A simple computation yields for all j with $r\sqrt{n} < j < n$,

$$\frac{d}{dr}\log Q_n(j,r) = \frac{-rn^2}{(j-r\sqrt{n})(n-j+r\sqrt{n})} - \frac{\sqrt{n}}{n-j+r\sqrt{n}}$$

$$< \frac{-4r}{1-\frac{4}{n^2}\left(\frac{n}{2}-j+r\sqrt{n}\right)^2}$$

$$< -4r - \frac{16r}{n^2}\left(\frac{n}{2}-j+r\sqrt{n}\right)^2,$$

which on integrating gives

$$Q_n(j,r) < Q_n(j,0)\exp\left[-2r^2 - \frac{8r^2}{n^2}\left(\frac{n}{2}-j+\frac{2r\sqrt{n}}{3}\right)^2 - \frac{4r^4}{9n}\right], \tag{2.12}$$

as well as

$$Q_n(j,r) < c_1 Q_n(j,1)\exp\left[-2r^2 - \frac{8r^2}{n^2}\left(\frac{n}{2}-j+\frac{2r\sqrt{n}}{3}\right)^2 - \frac{4r^4}{9n}\right] \tag{2.13}$$

for $r \geqq 1$; here c_1 denotes a universal finite constant (and similarly, c_2, c_3, c_4, c_5 in the sequel).

We divide the sum of (2.9) into two parts: $\sum'$ will denote summation over those j for which

$$\left|j-\frac{n}{2}\right| \leqq \frac{n}{4} \tag{2.14}$$

and $\sum''$ will denote summation over the remaining values. It follows immediately from Stirling's formula that

$$Q_n(j,0) < c_2 n^{-3/2}$$

for j satisfying (2.14). Hence we have from (2.12),

$$\sum{}' Q_n(j,r) < \frac{c_2}{n^{3/2}}e^{-2r^2}\sum{}'\exp\left[-8r^2\left(\frac{1}{2}+\frac{2r}{3\sqrt{n}}-\frac{j}{n}\right)^2\right]$$

$$< \frac{2c_2}{n^{3/2}}e^{-2r^2}\sum_{j=0}^{\infty}e^{-8r^2j^2/n^2}$$

$$< \frac{2c_2}{\sqrt{n}}e^{-2r^2}\left(\frac{1}{n}+\int_0^\infty e^{-8r^2t^2}\,dt\right)$$

$$< \frac{c_3}{r\sqrt{n}}e^{-2r^2}.$$

Hence,

$$r\sqrt{n}\sum{}' Q_n(j,r) < c_3 e^{-2r^2}. \tag{2.15}$$

Let us now deal with the j occurring in $\sum''$, i.e., those for which (2.14) does not hold. If $2r\sqrt{n}/3 \leqq n/8$, then the second term in the exponent in (2.13) is $\leqq -(r^2/8)$ while otherwise $r > 3\sqrt{n}/16$ and the last term in the exponent in (2.13) is $< -(4/9)(3/16)^2r^2$. Thus, in both cases we have for $r > 1$,

$$Q_n(j, r) < c_1 Q_n(j, 1)e^{-2r^2}e^{-c_4r^2} < \frac{c_5}{r} Q_n(j, 1)e^{-2r^2}.$$

Hence we have from (2.9).

$$r\sqrt{n}\sum'' Q_n(j, r) < c_5e^{-2r^2}\sqrt{n}\sum'' Q_n(j, 1) < c_5e^{-2r^2}. \tag{2.16}$$

(2.11), (2.15), and (2.16) imply (2.5) for $1 < r < \sqrt{n}$ and thus obviously for all r.

3. Asymptotic minimax character of ϕ_n^* for a fundamental class of weight functions. In this section we shall prove the asymptotic minimax character of ϕ_n^* (as $n \to \infty$) in a sense which is fundamental in that the minimax character relative to all reasonable weight functions of a certain type will follow (in Section 4) from the results of the present section. We shall now prove the following strong property of ϕ_n^*:

THEOREM 3. *For every value* $r > 0$,

$$\lim_{n\to\infty} \frac{\sup_{F\varepsilon\mathfrak{F}_c} P_F\{\sup_x | S_n(x) - F(x) | > r/\sqrt{n}\}}{\inf_{\phi\varepsilon D_n} \sup_{F\varepsilon\mathfrak{F}_c} P_{F,\phi}\{\sup_x | g(x) - F(x) | > r\sqrt{n}\}} = 1. \tag{3.1}$$

In fact, the probability in the numerator of (3.1) is independent of F for $F \varepsilon \mathfrak{F}_c$ and is no greater for any $F \varepsilon \mathfrak{F} - \mathfrak{F}_c$ than for $F \varepsilon \mathfrak{F}_c$ (see [1]); as an immediate consequence of Theorem 3, we thus have

COROLLARY 3. *The result of Theorem* 3 *holds if* $\mathfrak{F}_c$ *is replaced by* $\mathfrak{F}$ *in its statement.*

We also remark that (3.9) and (3.20) below may be used to give an explicit bound on the departure of ϕ_n^* from minimax character; the integer N of (3.9) may be computed explicitly by merely keeping track of the constants which go into various error orders in the proof which follows; an explicit estimate of departure for $n \leqq N$ could be given similarly. With slightly more difficulty such a bound could also be computed in the cases treated in Sections 4, 5, and 6.

In order to prove (3.1), we shall exhibit a sequence $\{\xi_{kn}\}$ of a priori probability measures on $\mathfrak{F}_c$ such that, letting A_k (k a positive integer) denote the set consisting of the k points $i/(k + 1)$ (for $1 \leqq i \leqq k$), we have

$$\begin{aligned}\lim_{k\to\infty}\lim_{n\to\infty}\inf_{\phi\varepsilon D_n} &\int P_{F,\phi}\{\sup_{a\varepsilon A_k} | g(a) - F(a) | > r/\sqrt{n}\}\, d\xi_{kn} \\ &= \lim_{k\to\infty}\lim_{n\to\infty}\int P_F\{\sup_{a\varepsilon A_k} | S_n(a) - F(a) | > r/\sqrt{n}\}\, d\xi_{kn} \\ &= \lim_{k\to\infty}\lim_{n\to\infty} P_U\{\sup_{a\varepsilon A_k} | S_n(a) - a | > r/\sqrt{n}\},\end{aligned} \tag{3.2}$$

where U is the uniform distribution on [0, 1]. Now, the expression under the limit operations on the left side of (3.2) is, for each n and k, obviously no greater than the denominator of (3.1) for the same n. On the other hand, the right side of (3.2) is equal to the (positive) limit as $n \to \infty$ of the numerator of (3.1), by (2.4). Hence, (3.2) implies (3.1).

In order to prove (3.2), we shall for each k limit ourselves to measures ξ_{kn} which assign probability one to distribution functions in $\mathfrak{F}_c$ of the form

$$F_k(x) = \sum_{i=1}^{k+1} p_i\, U_{ik}(x), \qquad p_i > 0, \sum p_i = 1, \tag{3.3}$$

where $U_{ik}(x)$ is the uniform probability distribution on the interval $[(i-1)/(k+1), i/(k+1)]$. For fixed k and n, it is easily seen that a sufficient statistic for the vector $\{p_i\}$ (and thus, for the family of F_k's of the form (3.3)) is given by the vector $T_k^{(n)} = \{T_{k1}^{(n)}, T_{k2}^{(n)}, \cdots, T_{k,k+1}^{(n)}\}$, where $T_{ki}^{(n)}$ is equal to the number of components of $X^{(n)}$ which lie in the interval $[(i-1)/(k+1), i/(k+1)]$. Hence, the validity of (3.2) will be implied by the following stronger result: Let B_k be the family of vectors $\pi = \{p_i, 1 \leqq i \leqq k+1\}$ satisfying $p_i \geqq 0, \sum p_i = 1$; $T_k^{(n)}$ has the multinomial distribution arising from n observations on $k+1$ types of objects, according to some $\pi \in B_k$, i.e., for integers $x_i \geqq 0$ with $\sum_1^{k+1} x_i = n$,

$$P_\pi\{T_{ki}^{(n)} = x_i, 1 \leqq i \leqq k+1\} = \frac{n!}{x_1! \cdots x_{k+1}!} p_1^{x_1} \cdots p_{k+1}^{x_{k+1}}; \tag{3.4}$$

$\mathcal{E}_n$ is the class of all (possibly randomized) vector estimators

$$\psi_n = \{\psi_{n1}, \cdots, \psi_{n,k+1}\}$$

of $\pi = \{p_i\}$ based on $T_k^{(n)}$ (ψ_n need not take on values in B_k); the ξ_{kn} are probability measures on B_k, which will be chosen so that

$$\begin{aligned} &\lim_{n\to\infty} \inf_{\psi_n \in \mathcal{E}_n} \int P_\pi\left\{\sup_i \left| \sum_{j=1}^{i} (\psi_{nj} - p_j)\right| > r/\sqrt{n}\right\} d\xi_{kn} \\ &\quad = \lim_{n\to\infty} \int P_\pi\left\{\sup_i \left| \sum_{j=1}^{i} (T_{kj}^{(n)}/n - p_j)\right| > r/\sqrt{n}\right\} d\xi_{kn} \\ &\quad = \lim_{n\to\infty} P_{V_k}\left\{\sup_i \left| \sum_{j=1}^{i} (T_{kj}^{(n)}/n - 1/(k+1))\right| > r/\sqrt{n}\right\}, \end{aligned} \tag{3.5}$$

where $V_k = \{1/(k+1), \cdots, 1/(k+1)\} \in B_k$. Taking limits as $k \to \infty$ (we have seen that this limit exists for the last expression of (3.5)), we see that the demonstration of (3.5) will imply that of (3.2). If we prove (3.5) with $\mathcal{E}_n$ replaced by the class of *non*randomized ψ_n, then (3.5) will a fortiori be true in the form stated above. Hence, in what follows, all ψ_n will be nonrandomized.

Some intuitive remarks are in order regarding the choice of ξ_{kn} (and the m_{kn} defining it) in the next paragraph. For simplicity, let us consider the case $k = 1$. We are then faced with a binomial estimation problem. The classical estimator of the parameter p_1 is asymptotically normal with maximum variance at $p_1 = \frac{1}{2}$ (this is V_1; in general, the corresponding phenomenon which concerns us occurs at $\pi = V_k$). In order to obtain our asymptotic Bayes result (3.5), we want ξ_{1n} to approximate a uniform measure on an interval of p_1 which has the following properties: on the one hand, the width e_n of this interval, when multiplied by $\sqrt{n}$, must tend to infinity with n; on the other hand, the width itself must tend to zero. In terms of the parameter $\sqrt{n}(p_1 - \frac{1}{2})$ and random variable $(T_{11}^{(n)} - n/2)/\sqrt{n}$, we will then be faced, asymptotically, with the problem of estimating the mean of a normal distribution (where, asymptotically, all real values are possible for the mean, with a uniform a priori distribution over a region whose width $\sqrt{n}e_n$ tends to ∞) with *almost constant variance*. The classical estimator will then be asymptotically Bayes for our weight function. Since a uniform a priori distribution would be slightly less simple to use (in keeping track of limits), we use instead one of the form (3.6) below; but the choice of the parameter m_{kn} therein is motivated by the remarks above.

Let $m = m_{k,n} =$ (greatest integer $\leqq n^{1/4}/k^2$), let $\epsilon = \epsilon_{k,n} = m/n$, and let $\xi_{k,n}$ be the probability measure on B_k which is given rise to by the probability density function

$$h_{k,n}(p_1, \cdots, p_k) = C_{k,n}\left[\left(1 - \sum_1^k p_i\right)\prod_{i=1}^k p_i\right]^m \tag{3.6}$$

with respect to Lebesgue measure on the k-simplex $\{0 \leqq \sum_1^k p_i \leqq 1,\ p_i \geqq 0\ (1 \leqq i \leqq k)\}$ and is zero elsewhere. Here

$$C_{k,n} = \Gamma([m + 1][k + 1])/[\Gamma(m + 1)]^{k+1}.$$

Let $Y_{ki}^{(n)} = T_{ki}^{(n)}/n$. Let $\bar{\delta}_i = p_i - 1/(k + 1)$. The a posteriori density of $\bar{\delta}_1, \cdots, \bar{\delta}_k$, given that $Y_{ki}^{(n)} = y_i$ $(1 \leqq i \leqq k)$ (for possible values of the set $\{y_i\}$) when $\xi_{k,n}$ is the a priori probability measure on B_k is (the domain being obvious)

$$f_{k,n}(\delta_1, \cdots, \delta_k \mid y_1, \cdots, y_k) = \left[C_1 \prod_{i=1}^{k+1}\left(\delta_i + \frac{1}{k+1}\right)^{y_i+\epsilon}\right]^n, \tag{3.7}$$

where we have written $\delta_{k+1} = 1 - \sum_1^k \delta_i$ and $y_{k+1} = 1 - \sum_1^k y_i$ for typographical simplicity; here $(C_1)^n = \Gamma([m + 1][k + 1] + n)/\prod_1^{k+1}\Gamma(m + 1 + ny_i)$. Let $\bar{\eta}_i = \bar{\delta}_i - Y_{ki}^{(n)} + 1/(k + 1)$. Then the a posteriori density of $\bar{\eta}_1, \cdots, \bar{\eta}_k$ under the same conditions is (the domain again being obvious)

$$\begin{aligned} f_{k,n}^*(\eta_1, \cdots, \eta_k \mid y_1, \cdots, y_k) &= [g_{k,n}(\eta_1, \cdots, \eta_k \mid y_1, \cdots, y_k)]^n \\ &= \left[C_1 \prod_{i=1}^{k+1} (y_i + \eta_i)^{y_i+\epsilon}\right]^n, \end{aligned} \tag{3.8}$$

where $\eta_{k+1} = -\sum_1^k \eta_i$.

We shall now prove that, for each k and each r^* with $0 < r^* < \infty$, we have for $n > N(k, r^*)$ (the latter will be defined below)

$$(3.9)\qquad \begin{aligned} &E_t P_a^* \left\{ \sup_i \left| \sum_{j=1}^{i} (p_j - Y_{kj}^{(n)}) \right| \leq \frac{r}{\sqrt{n}} \right\} \\ &\qquad \geq E_t P_a^* \left\{ \sup_i \left| \sum_{j=1}^{i} (p_j - \psi_{nj}) \right| \leq \frac{r}{\sqrt{n}} \right\} - n^{-1/9} \end{aligned}$$

for all r with $0 \leq r \leq r^*$ and all ψ_{ni} (not necessarily positive or summing to unity); here P_a^* denotes a posteriori probability of π (i.e., of $\{p_j\}$) when (3.6) is the a priori distribution, while E_t denotes expectation with respect to the measure on $B_k \times R^{k+1}$ given by (3.6) and (3.4). Noting that the second integral in (3.5) is unity minus the left side of (3.9) and that for each k the left side of (3.9) tends to a limit as $n \to \infty$ (this will follow from (3.20) below), we see that (3.9) actually implies that the first and second expressions of (3.5) are equal for each k. On the other hand, the limiting joint distribution function of the set of random variables $\{\sqrt{n}[Y_{ki}^{(n)} - 1/(k+1)], 1 \leq i \leq k\}$ under V_k is well known to be that whose density is given in (3.20), below, if we set all $y_i = 1/(k+1)$ and let $n \to \infty$ in the latter; since (3.20), which is the asymptotic a posteriori joint density of the $(p_i - T_{ki}^{(n)}/n)$, is continuous in the y_i, and since the $Y_{ki}^{(n)}$ tend in probability (according to (3.6) and (3.4)) to $1/(k+1)$ as $n \to \infty$, it follows that the second and third expressions of (3.5) are equal. (This last follows also from the continuity in π of $\lim_{n\to\infty} P_\pi\{\ \}$ in the second expression of (3.5) and the fact that $\lim_{n\to\infty} \xi_{kn}(J) = 1$ for any neighborhood J of V_k.) Thus, our theorem will be proved if we prove (3.9), and we now turn to this proof.

In this demonstration our calculations will be performed under the conditions

$$(3.10)\qquad \begin{aligned} &|y_i - 1/(k+1)| < 1/2(k+1) && (1 \leq i \leq k+1), \\ &|\eta_i| < n^{-3/8}/4k(k+1) && (1 \leq i \leq k+1), \\ &n > k^{40}. \end{aligned}$$

All orders $O(\cdot)$ will be uniform in the variables not appearing in the arguments. By (3.8),

$$(3.11)\qquad \log g_{k,n} = \log C_1 + \sum_1^{k+1} (y_i + \epsilon) \log y_i + \sum_1^{k+1} (y_i + \epsilon) \log\left(1 + \frac{\eta_i}{y_i}\right).$$

From (3.10), we have

$$(3.12)\qquad \left|\frac{\eta_i}{y_i}\right| < \frac{1}{2kn^{3/8}} \leq \frac{1}{2},$$

and hence

$$\log\left(1 + \frac{\eta_i}{y_i}\right) = \frac{\eta_i}{y_i} - \frac{\eta_i^2}{2y_i^2} + \theta_i \frac{\eta_i^3}{y_i^3},$$

with

$$(3.13)\qquad |\theta_i| < 1, \qquad (1 \leq i \leq k+1).$$

Now, writing

$$(y_i + \epsilon) \log \left(1 + \frac{\eta_i}{y_i}\right) = \eta_i - \frac{\eta_i^2}{2y_i} + \theta_i \frac{\eta_i^3}{y_i^2} + \epsilon \frac{\eta_i}{y_i} \left(1 - \frac{\eta_i}{2y_i} + \theta_i \frac{\eta_i^2}{y_i^2}\right)$$

and remarking that $\sum_1^{k+1} \eta_i = 0$, that by (3.10) and (3.13)

$$\left| \sum_1^{k+1} \theta_i \frac{\eta_i^3}{y_i^2} \right| < (k+1) \frac{4(k+1)^2}{64k^3(k+1)^3 n^{9/8}} < \frac{1}{n^{9/8}},$$

and that by (3.10), (3.12), and the definition of ϵ

$$\epsilon \left| \sum_1^{k+1} \frac{\eta_i}{y_i} \left(1 - \frac{\eta_i}{2y_i} + \theta_i \frac{\eta_i^2}{y_i^2}\right) \right| < 2\epsilon \sum_1^{k+1} \left| \frac{\eta_i}{y_i} \right| < \frac{2}{k^2 n^{3/4}} \cdot \frac{k+1}{2kn^{3/8}} \leqq \frac{2}{n^{9/8}},$$

we obtain

$$\sum_1^{k+1} (y_i + \epsilon) \log \left(1 + \frac{\eta_i}{y_i}\right) = -\frac{1}{2} \sum_1^{k+1} \frac{\eta_i^2}{y_i} + \frac{3\theta}{n^{9/8}}, \tag{3.14}$$

with $|\theta| < 1$. Combining (3.14) and (3.11), we have

$$\log g_{k,n} = \log C_1 + \sum_1^{k+1} (y_i + \epsilon) \log y_i - \frac{1}{2} \sum_1^{k+1} \frac{\eta_i^2}{y_i} + O(n^{-9/8}) \tag{3.15}$$

Next, we note that

$$(C_1)^n \prod_1^{k+1} y_i^{ny_i+m} = p_{n+(k+1)m}^{(k)}(ny_1 + m, \cdots, ny_{k+1} + m; y_1, \cdots, y_{k+1}), \tag{3.16}$$

where $p_N^{(k)}(w_1, \cdots, w_{k+1}; q_1, \cdots, q_{k+1})$ is the (multinomial) probability that among N independent, identically distributed random variables taking on the value i with probability $q_i(\sum_1^{k+1} q_i = 1, q_i \geqq 0)$, there will be w_i taking on the value $i(\sum w_i = N)$. Using the familiar representation of this probability in terms of binomial probabilities, the definition of m, the inequalities (3.10), and the estimate for binomial probabilities

$$\begin{aligned} &p_N^{(1)}(Np + t\sqrt{Np(1-p)}, N(1-p) - t\sqrt{Np(1-p)}; p, 1-p) \\ &\qquad = [2\pi Np(1-p)]^{-1/2} e^{-t^2/2}[1 + O(N^{-1/2})] \end{aligned} \tag{3.17}$$

for $|t| < C_6$ and $|p - \frac{1}{2}| < C_7 < \frac{1}{2}$ (given in [5], p. 135), we obtain (with a conservative estimate of error)

$$(C_1)^n \prod_1^{k+1} y_i^{ny_i+m} = (1 + O(n^{-1/8}))(2\pi n)^{-k/2} \prod_1^{k+1} y_i^{-1/2}. \tag{3.18}$$

Hence, in the region (3.10) we obtain from (3.15) and (3.18), writing again $\eta_{k+1} = -\sum_1^k \eta_i$ and $y_{k+1} = 1 - \sum_1^k y_i$,

$$\begin{aligned} &f_{k,n}^*(\eta_1, \cdots, \eta_k \mid y_1, \cdots, y_k) \\ &\qquad = (1 + O(n^{-1/8}))(2\pi n)^{-k/2} \left(\prod_1^{k+1} y_i\right)^{-1/2} \exp\left(-\frac{n}{2} \sum_1^{k+1} \frac{\eta_i^2}{y_i}\right). \end{aligned} \tag{3.19}$$

For the corresponding a posteriori joint density of $\bar{\gamma}_i = \sqrt{n}\bar{\eta}_i$, $i = 1, \cdots, k$, in the region (3.10), we thus obtain (writing $\gamma_{k+1} = -\sum_1^k \gamma_i$)

$$(1 + O(n^{-1/8}))(2\pi)^{-k/2}\left(\prod_1^{k+1} y_i\right)^{-1/2} \exp\left(-\frac{1}{2}\sum_1^{k+1}\frac{\gamma_i^2}{y_i}\right). \tag{3.20}$$

Except for the first factor, this is a k-dimensional normal distribution centered at the origin. Note also that the probability assigned by this density to the complement of the region $|\bar{\eta}_i| < n^{-3/8}/4k(k+1)$ of (3.10) (for a single i) is (by Chebychev's inequality) $\leqq [1 + O(n^{-1/8})]O(k^4 n^{-1/4})$, so that the probability of the above inequality on the $\bar{\eta}_i$ for all i according to (3.20) (using $k < n^{1/40}$) is at least $1 - O(n^{-1/8})$. Also, the p_i or (3.6) have means $1/(k+1)$ and variances $O(m^{-1}k^{-2}) = O(n^{-1/4})$, while $Y_{ki}^{(n)}$ (given the p_i) has mean p_i and variance $O(n^{-1})$, whatever the p_i may be. Hence, for a single i, the probability (according to (3.6) and (3.4)) that $|Y_{ki}^{(n)} - 1/(k+1)| < \frac{1}{2}(k+1)$ is

$$\geqq P\{|p_i - 1/(k+1)| < \tfrac{1}{4}(k+1)\} \times P\{|Y_{ki}^{(n)} - p_i| < \tfrac{1}{4}(k+1)\,|p_i\}$$
$$\geqq 1 - k^2\{O(n^{-1/4}) + O(n^{-1})\}.$$

The probability that $|Y_{ki}^{(n)} - 1/(k+1)| < 1/2(k+1)$ for all i is thus

$$\geqq 1 - k^3 O(n^{-1/4}) \geqq 1 - O(n^{-1/8}).$$

We conclude, then, that the region of $Y_{ki}^{(n)}$, $\bar{\eta}_i$ $(1 \leqq i \leqq k+1)$ specified in (3.10) (putting $Y_{ki}^{(n)}$ for y_i and $\bar{\eta}_i$ for η_i), and hence where (3.20) holds, has probability $1 - O(n^{-1/8})$ according to (3.6) and (3.4).

Now, for fixed $r^* > 0$, let $N_1(k, r^*)$ be such that if $n > N_1(k, r^*)$, then

$$8r^* n^{-1/2} < n^{-3/8}/4k(k+1)$$

and the probability under (3.20) that all $|\bar{\eta}_i|$ are $< n^{-3/8}/16k(k+1)$ is $\geqq 1 - n^{-1/9}/2$; clearly, such a number $N_1(k, r^*)$ exists. For $0 < r \leqq r^*$, let T_r be the region where $|\sum_{j=1}^i \gamma_j| \leqq r$, $i = 1, \cdots, k+1$. Note that T_r is contained in the region where $|\gamma_j| \leqq 2r^*$ for all j. If ρ is any vector all of whose $(k+1)$ components are $\leqq n^{1/8}/8k(k+1)$ and if $n > N_1(k, r^*)$, then T_r and $T_r + \rho$ both lie entirely in the region of (3.10) (where (3.20) holds), whose probability according to (3.6) and (3.4) is $1 - O(n^{-1/8})$. Write C_r and D_r for the events in brackets on the left and right sides of (3.9), and define $L = L(X_1, \cdots, X_n, \psi_n)$ to be 1 or 0 according to whether or not

$$\max_j \sqrt{n}\,|Y_{kj}^{(n)} - \psi_{nj}| \leqq n^{1/8}/8k(k+1).$$

From the previous remarks of this paragraph and Lemma 1 we conclude that

$$E_t[L \cdot P_a^*\{C_r\}] \geqq E_t[L \cdot P_a^*\{D_r\}] - n^{-1/9}/3 \tag{3.21}$$

for $0 < r \leqq r^*$, $n > N(k, r^*)$, and all ψ_n, where $N(k, r^*)$ is chosen (as it clearly may be because $n^{-1/8} = o(n^{-1/9})$) to be enough larger than $N_1(k, r^*)$ to give the term $n^{-1/9}/3$ in (3.21). On the other hand, if any component of ρ has magnitude $> n^{1/8}/8k(k+1)$, then with probability $1 - O(n^{-1/8})$ according to

(3.4) and (3.6), $T_r + \rho$ has a posteriori probability $<n^{-1/9}/2$. Hence, the $N(k, r^*)$ above may clearly also be chosen so large that

$$E_t[(1 - L)P^*_a\{D_r\}] - 2n^{-1/9}/3 \leqq 0 \tag{3.22}$$

for $0 < r \leqq r^*$, $n > N(k, r^*)$, and all ψ_n. Equation (3.9) follows from (3.21) and (3.22), completing the proof of Theorem 3.

We remark that ϕ_n^* will not be minimax in the sense of Theorem 3 for all r and fixed finite n. The first nontrivial case is that of $n = 3$. A tiresome but straightforward computation in this case shows that, among the procedures ϕ_c which for a given number c $(0 \leqq c \leqq \frac{1}{2})$ assign probability one to

$$g_c(x) = \begin{cases} 0 & \text{if } x > Z_1, \\ c & \text{if } Z_1 \leqq x < Z_2, \\ 1 - c & \text{if } Z_2 \leqq x < Z_3, \\ 1 & \text{if } Z_3 \leqq x, \end{cases}$$

where the Z_i are the ordered $X_i^{(3)}$, the expression $P_U\{\sup_x | g_c(x) - x | \leqq z\}$ is maximized for $\frac{1}{6} \leqq z \leqq \frac{1}{3}$ at $c = \frac{1}{3}$ (i.e., by ϕ_3^*), for $\frac{1}{3} \leqq z \leqq \frac{1}{2}$ by $c = z$, and for $\frac{1}{2} < z \leqq 1$ by any $c \geqq 1 - z$ (for $z \leqq \frac{1}{6}$, all values of c give probability zero). Similar remarks apply to the problems considered in the next three sections. For example, $E_U\{\sup_x | g_c(x) - x |\}$ in the above example is minimized by $c = [33 - 3(17)^{1/2}]/52 = 0.397$. Similar calculations are more easily made in the case studied in Section 4 (where the distribution of the maximum deviation need not be calculated), and such calculations may be found in the reference cited at the end of that section.

4. Other loss functions which are functions of distance. In this section we show that the asymptotic minimax character of ϕ_n^* proved in Section 3 may be extended to a broad class of weight functions. It turns out that it is unnecessary to start anew in order to prove this; the class of weight functions considered in Section 3 (see below) is the basic class in the sense that the minimax character relative to many other weight functions may be concluded from the results of Section 3 and the integrability result given in Corollary 2. It is clear that the method of attack used here, i.e., of carrying out the detailed proof of the minimax character for the basic class of weight functions and then extending to other weight functions, can be stated as a general theorem to apply to other statistical problems; we shall not bother to state this obvious extension in a general setting.

Throughout this section W will represent any nonnegative function defined on the nonnegative reals which is nondecreasing in its argument, not identically zero (the case $W \equiv 0$ is trivial), and which satisfies

$$\int_0^\infty W(r)re^{-2r^2}\,dr < \tag{4.1}$$

The main result of this section is the following:

THEOREM 4. *Under the above assumptions on* W,

$$\lim_{n\to\infty} \frac{\sup_{F\varepsilon\mathfrak{F}_c} \int W(r)\, d_r P_F\{\sup_x |\, S_n(x) - F(x) \,| < r/\sqrt{n}\}}{\inf_{\phi\varepsilon D_n} \sup_{F\varepsilon\mathfrak{F}_c} \int W(r)\, d_r P_{F,\phi}\, \{\sup_x |\, g(x) - F(x) \,| < r/\sqrt{n}\}} = 1. \tag{4.2}$$

As in Section 3 (and for the same reason), an immediate corollary is

COROLLARY 4. *The result of Theorem 4 holds if* $\mathfrak{F}_c$ *is replaced by* $\mathfrak{F}$ *in its statement.*

PROOF OF THEOREM 4. By a reduction like that of Section 3, it is seen that (4.2) will be proved if, for the sequence $\{\xi_{kn}\}$ of Section 3, we can prove the following three statements, (4.3), (4.4), and (4.5):

$$\begin{aligned}\lim_{n\to\infty} \inf_{\phi\varepsilon D_n} &\int W(r)\, d_r P_{F,\phi}\, \{\max_{a\varepsilon A_k} |\, g(a) - F(a) \,| < r/\sqrt{n}\}\, d\xi_{kn} \\ &= \lim_{n\to\infty} \int W(r)\, d_r P_F\, \{\max_{a\varepsilon A_k} |\, S_n(a) - F(a) \,| < r/\sqrt{n}\}\, d\xi_{kn}\end{aligned} \tag{4.3}$$

for each positive integer k;

$$\begin{aligned}\lim_{n\to\infty} &\int W(r)\, d_r P_F\, \{\max_{a\varepsilon A_k} |\, S_n(a) - F(a) \,| < r/\sqrt{n}\}\, d\xi_{kn} \\ &= \lim_{n\to\infty} \int W(r)\, d_r P_U\, \{\max_{a\varepsilon A_k} |\, S_n(a) - a \,| < r/\sqrt{n}\}\end{aligned} \tag{4.4}$$

for each positive integer k;

$$\begin{aligned}0 < \lim_{k\to\infty} \lim_{n\to\infty} &\int W(r)\, d_r P_U\, \{\sup_{a\varepsilon A_k} |\, S_n(a) - a \,| < r/\sqrt{n}\} \\ &= \lim_{n\to\infty} \int W(r)\, d_r P_U\, \{\sup_{0\leq x\leq 1} |\, S_n(x) - x \,| < r/\sqrt{n}\} < \infty.\end{aligned} \tag{4.5}$$

(This includes, of course, proving the existence of the indicated limits.)

Firstly, (4.5) is an immediate consequence of (4.1), (2.4), (2.2), the continuity of G and of the d.f. $\lim_{n\to\infty} G_{k,n}$, and of Corollary 2.

In order to prove (4.4), we note first that, for fixed k and any $F \varepsilon \mathfrak{F}_c$, we have (similarly to (2.2)) the inequality $P_F\{\max_{a\varepsilon A_k} |\, S_n(a) - F(a) \,| \leq r/\sqrt{n}\} \geq G_n(r)$. Hence, by Corollary 2, the integral with respect to r on the left side of (4.4) is bounded uniformly in n and F. On the other hand, given any $\epsilon > 0$, there exists an integer N_0 such that, for $n > N_0$, ξ_{kn} assigns probability at least $1 - \epsilon$ to a set of F for which the expressions $P_F\{\ \}$ and $P_U\{\ \}$ of (4.4) differ by less than ϵ for all r (this rests on the continuity in π, for π in a neighborhood of V_k, of the normal approximation (for large n) to the joint distribution of the random variables $\sqrt{n}(Y_{ki}^{(n)} - p_i)$, $1 \leq i \leq k$). Since $P_U\{\ \}$ is continuous in r, (4.4) follows.

Finally, we must prove (4.3). Consider any fixed k. Write $P_n^*(r; x^{(n)}, \phi)$ for the probability, calculated according to the a posteriori probability distribution of π (given that $X^{(n)} = x^{(n)}$ and when ξ_{kn} is the a priori probability measure on B_k) and the probability measure $\phi(\cdot; x^{(n)})$ on G (where $\phi \varepsilon D_n$ and perhaps $\phi = \phi_n^*$) of the set of (g, π) in $G \times B_k$ for which $\max_{a \varepsilon A_k} |g(a) - F(a)| < r/\sqrt{n}$. If (4.3) is false, there exists a value $\epsilon > 0$ such that, for every positive N, there is an $n > N$ and a $\phi_n \varepsilon D_n$ for which (the operation E_t being as defined in Section 3)

$$(4.6)\quad E_t \int W(r)\, d_r P_n^*(r; X^{(n)}, \phi_n) < E_t \int W(r)\, d_r P_n^*(r; X^{(n)}, \phi_n^*) - 2\epsilon.$$

It is clear from the preceding paragraphs that there is a real number $q > 0$ such that $W(q) > 0$ and

$$(4.7)\qquad E_t \int_q^\infty W(r)\, d_r P_n^*(r; X^{(n)}, \phi_n^*) < \epsilon$$

for all n. Write $W_q(r) = \min(W(r), W(q))$ Then (4.6) and (4.7) imply

$$(4.8)\qquad E_t \int_0^\infty W_q(r)\, d_r \{P_n^*(r; X^{(n)}, \phi_n) - P_n^*(r; X^{(n)}, \phi_n^*)\} < -\epsilon.$$

Since $W_q(r) \leqq W(q)$, the integral on the left side of (4.8) is $\geqq -W(q)$. Hence, (4.8) implies that, with probability at least $\epsilon/W(q)$ (under (3.6) and (3.4)), $X^{(n)}$ will be such that

$$(4.9)\qquad \int_0^\infty W_q(r)\, d_r \{P_n^*(r; X^{(n)}, \phi_n) - P_n^*(r; X^{(n)}, \phi_n^*)\} < -\epsilon.$$

Let $\epsilon' = \epsilon/2W(q)$. The discussion of the previous paragraph shows that we can find an R^* and M such that, for $n > M$, the probability (under (3.6) and (3.4)) will be $>1 - \epsilon'$ that $X^{(n)}$ will be such that

$$(4.10)\qquad P_n^*(R^*; X^{(n)}; \phi_n^*) > 1 - \epsilon'.$$

Let $\gamma_n = \sup_r\{P_n^*(r; X^{(n)}, \phi_n) - P_n^*(r; X^{(n)}, \phi_n^*)\}$. We shall show below that (4.9) implies

$$(4.11)\qquad \gamma_n > \epsilon'.$$

Then (4.10) and (4.11) (the latter of which is an event of probability at least $2\epsilon'$ according to (3.6) and (3.4)) will imply that for each $N > M$ there is an $n > N$ and a $\phi_n \varepsilon D_n$ for which, with probability $>\epsilon'$ according to (3.6) and (3.4), $X^{(n)}$ will be such that

$$(4.12)\qquad \{P_n^*(r; X^{(n)}, \phi_n) - P_n^*(r; X^{(n)}, \phi_n^*)\} > \epsilon'$$

for some r with $0 \leqq r < R^*$ (here r depends on n, $\phi^{(n)}$, $X^{(n)}$). This contradicts the fact that, with probability $1 - O(n^{-1/8})$ according to (3.6) and (3.4), the region T_r of the last paragraph of Section 3 was seen to maximize with respect to ρ (uniformly in $0 \leqq r \leqq R^*$), to within an (added) error of $O(n^{-1/9})$, the a posteriori probability of $T_r + \rho$. Thus, it remains only to prove that (4.9) implies (4.11). For fixed n, ϕ_n, $X^{(n)}$, abbreviate the bracketed expression in (4.9) as $B(r) - C(r)$. Let

$$B^*(r) = \begin{cases} 0 & \text{if } r \leqq 0. \\ \min (C(r) + \gamma_n, 1) & \text{if } r > 0. \end{cases} \tag{4.13}$$

Clearly, $B(r) \leqq B^*(r)$. Hence, since $W_q(r)$ is nondecreasing in r, we have

$$\int W_q(r)\, dB(r) \geqq \int W_q(r)\, dB^*(r). \tag{4.14}$$

Let α be the infimum of values r for which $B^*(r) = 1$. From (4.9), (4.14), and the fact that $B^*(r) - C(r)$ is constant for $0 < r < \alpha$, we obtain

$$\begin{aligned} \epsilon &< \int_{0-}^{\infty} W_q(r)\, d(C(r) - B(r)) \\ &\leqq \int_{0-}^{\infty} W_q(r)\, d(C(r) - B^*(r)) \\ &\leqq \int_{0+}^{\alpha-} W_q(r)\, d(C(r) - B^*(r)) + W(0)\gamma_n + W(q)\gamma_n \\ &\leqq 2W(q)\gamma_n, \end{aligned} \tag{4.15}$$

which proves (4.11) and thus completes the proof of Theorem 4.

5. Integral weight functions. In this section we consider weight functions W_n^* arising from integration of a function W in the following manner:

$$W_n^*(F, g) = \int_{-\infty}^{\infty} W(\sqrt{n}[g(x) - F(x)], F(x))\, dF(x). \tag{5.1}$$

Here $W(y, z)$, which is defined for y real and $0 \leqq z \leqq 1$, is nonnegative and is symmetric in y and nondecreasing in y for $y \geqq 0$; it may be thought of as a measure of the contribution to W_n^* arising from a deviation of $y\sqrt{n}$ of the estimator g from the true F at an argument x for which $F(x) = z$. Typical W's which might be of interest are $W(y, z) = |y|^p$ or 0 according to whether or not $a \leqq z \leqq b$ (here $p > 0$ and $0 \leqq a < b \leqq 1$), $W(y, z) = 0$ or 1 according to whether $|y| \leqq a$ or $|y| > a$ where a is a suitably chosen constant, $W(y, z) = y^2/z(1 - z)$, etc.

We now turn to considerations of the asymptotic minimax character of ϕ_n^* with respect to a sequence of risk functions $r_n(F, \phi) = E_{F,\phi} W_n^*(F, g)$, where $\phi \varepsilon D_n$. (The remainder of the present paragraph will be somewhat heuristic in order to compare the present problem with those of Sections 3 and 4; the statement and proof of Theorem 5 begin in the next paragraph.) These considerations are much easier than those of the previous two sections, since in obtaining a

Bayes solution with respect to the a priori probability measure ξ_{kn} of Section 3 it will suffice (as will be seen below) to minimize with respect to ϕ, *for each fixed* x (more precisely, for each irrational x),

$$(5.2) \quad r_{kn}(x, \phi, t_k^{(n)}) = \int_{B_k} E_\phi W(\sqrt{n}[g(x) - F(x, \pi)], F(x, \pi))\, d_\pi \xi_{kn}^*(\pi; x, t_k^{(n)});$$

here B_k is as in Section 3, $F(x, \pi)$ denotes the distribution function of (3.3) for a given value of $\pi = (p_1, \cdots, p_{k+1})$, and for any measurable subset B of B_k we set

$$(5.3) \qquad \xi_{kn}^*(B, x, t_k^{(n)}) = \frac{\int_B f(x, \pi) P_\pi\{t_k^{(n)}\}\, d\xi_{kn}(\pi)}{\int_{B_k} f(x, \pi) P_\pi\{t_k^{(n)}\}\, d\xi_{kn}(\pi)},$$

where ξ_{kn} is given by (3.6) of Section 3, $f(x, \pi) = dF(x, \pi) / dx$ (this derivative exists for x irrational), and $P_\pi\{t_k^{(n)}\} = P_\pi\{T_{ki}^{(n)} = t_{ki}^{(n)}, 1 \leq i \leq k + 1\}$ is the probability function defined in (3.4). (Of course, ϕ in (5.2) may randomize over many g, which accounts for the presence of the E_ϕ operation.) Thus, present considerations will involve only the obtaining of a (univariate) normal approximation to the a posteriori distribution (more precisely, to a slight modification (5.3) of it) of $F(x, \pi)$ *for fixed irrational* x, which is much easier than the multivariate approximation (3.20) which it was necessary to obtain in Section 3. (We shall actually use (3.20), which implies easily the needed univariate approximation; however, the latter could have been obtained more easily directly.) The above remarks will be made precise in what follows. We hereafter denote the infimum of $r_{kn}(x, \phi, t_k^{(n)})$ over all ϕ in D_n by $r_{kn}^*(x, t_k^{(n)})$. The set of reals

$$\{z \mid 0 < z < \epsilon \text{ or } 1 - \epsilon < z < 1\}$$

will be denoted by I_ϵ for $0 < \epsilon < \frac{1}{2}$.

We now state Theorem 5. Our statement of this theorem is not the most general possible. (The set I_ϵ may be replaced by other sets where $W(y, z)$ is large, the continuity conditions on W may be weakened by considering continuous approximations to (a more general) measurable W, the integrability condition may be weakened, and W may be replaced by a distribution (rather than a density) in z so as to obtain results, e.g., on the estimation of F at a finite number of quantiles.) Rather, it is stated in a form which allows W to be any of the functions which would usually be of interest in applications, e.g., any of those functions given at the end of the first paragraph of this section, etc. (It should be noted that if the assumptions of Theorem 5 below were altered by deleting (5.5) and putting $\epsilon = 0$ in (5.4), then such weight functions as $y^2/z(1 - z)$ would be excluded. The circumlocution of including the condition (5.5) could be avoided in such cases if one could obtain a sufficiently strong bound on

$$P_U\{\sqrt{n}[S_n(x) - x] > r\sqrt{x(1 - x)}\}$$

which is independent of x. The difficulty of obtaining such an approximation is discussed in [4], p. 285.)

THEOREM 5. *Let* $W(y, z) \geqq 0$ *be defined for* $0 \leqq y < \infty, 0 < z < 1$ *and assume that* $W(y, z)$ *is monotone nondecreasing in* y *and (to avoid trivialities) that* $W(y, z)$ *is not almost everywhere zero (in the two-dimensional Lebesgue sense). Suppose further that* (a) *to every* $z', 0 < z' < 1$, *not belonging to an exceptional set of linear measure zero, and every* $\delta > 0$ *there corresponds* $\epsilon(\delta, z') > 0$ *with the property that the set of* y *for which* $W(y, z)$ *is discontinuous for at least one* z *satisfying* $| z - z' | < \epsilon(\delta, z')$ *has exterior (linear Lebesgue) measure smaller than* δ. *Suppose also that* (b) *for each* ϵ *with* $0 < \epsilon < \frac{1}{2}$ *there is a function* $V(y, \epsilon)$ *such that* $W(y, z) \leqq V(y, \epsilon)$ *for* $\epsilon < z < 1 - \epsilon$ *and* $0 \leqq y < \infty$ *and such that*

$$\int_0^\infty V(y, \varepsilon) y e^{-2y^2}\, dy < \infty. \tag{5.4}$$

Suppose, finally, that (c)

$$\lim_{\epsilon \to 0} \sup_n \int_{I_\epsilon} E_U W(\sqrt{n}[S_n(x) - x], x)\, dx = 0. \tag{5.5}$$

Then

$$\lim_{n \to \infty} \frac{\sup_{F \epsilon \mathfrak{F}_c} r_n(F, \phi_n^*)}{\inf_{\phi \epsilon D_n} \sup_{F \epsilon \mathfrak{F}_c} r_n(F, \phi)} = 1. \tag{5.6}$$

PROOF. $r_n(F, \phi_n^*)$ is, of course, independent of F for F in $\mathfrak{F}_c$. Because of Corollary 2 and the assumptions of Theorem 5, the numerator of (5.6) approaches a finite positive limit, say L, as $n \to \infty$. For any δ with $0 < \delta < L$ we may choose ϵ so small that $r_{n,\epsilon}(F, \phi_n^*)$ tends to a limit $> L - \delta$ when $n \to \infty$, where $r_{n,\epsilon}$ is the risk function corresponding to loss function $W_\epsilon(y, z)$ defined by

$$W_\epsilon(y, z) = \begin{cases} W(y, z) & \text{if } z \not\in I_\epsilon, \\ 0 & \text{if } z \in I_\epsilon. \end{cases} \tag{5.7}$$

It clearly suffices to prove (5.6) with r_n replaced by $r_{n,\epsilon}$. We hereafter drop the subscript ϵ on W_ϵ and $r_{n,\epsilon}$ and (because of (5.7)) may restate what is to be proved as (5.6) under the continuity assumption (a) on W and (replacing (5.4) and (5.5)) the assumption that $W(y, z) \leqq V(y)$ for $0 \leqq y < \infty$ and $0 < z < 1$, where

$$\int_0^\infty V(y) y e^{-2y^2}\, dy < \infty. \tag{5.8}$$

In what follows we denote (for fixed k, n, irrational x) by $P_x^*\{A\}$ the probability of any event A which is expressed in terms of $T_k^{(n)}$ when the probability function of $T_k^{(n)}$ is given by

$$P\{T_{ki}^{(n)} = t_i, 1 \leqq i \leqq k+1\}$$
$$(5.9) \qquad = \frac{1}{d(k, n, x)} \int_{B_k} f(x, \pi) P_\pi\{T_{ki}^{(n)} = t_i, 1 \leqq i \leqq k+1\}\, d\xi_{kn}(\pi),$$

where P_π is given by (3.4) and $d(k, n, x)$ is the sum over all $(t_1, \cdots, t_{k+1})$ of the integral on the right side of (5.9). Expectation with respect to the probability function (5.9) will be denoted by E_x^*. We now have

$$\int r_n(F, \phi)\, d\xi_{kn} = \int_0^1 \int_{B_k} E_{\pi,\phi} W(\sqrt{n}[g(x) - F(x, \pi)], F(x, \pi))\, d\xi_{kn}(\pi)\, dx$$
$$(5.10) \qquad = \int_0^1 E_x^* r_{kn}(x, \phi, T_k^{(n)})\, d(k, n, x)\, dx,$$

where the last integration (and each integration which follows) is over irrational x. Hence, in order to prove (5.6), it suffices to show that (5.8) and our continuity assumption on W imply that

$$\lim_{k\to\infty} \lim_{n\to\infty} \int_0^1 E_x^* r_{kn}^*(x, T_k^{(n)})\, d(k, n, x)\, dx$$
$$(5.11) \qquad = \lim_{n\to\infty} \int_0^1 E_U W(\sqrt{n}[S_n(x) - x], x)\, dx,$$

since the right side of (5.11) is the limit of the finite positive numerator of (5.6).

Let x be an irrational number, $0 < x < 1$, which is a *non*exceptional z' of our continuity assumption (a). For fixed k with $1/(k+1) < \min(x, 1-x)$, we may write $x = (i_0 + t)/(k+1)$ with $1 \leqq i_0 \leqq k-1$ and $0 \leqq t < 1$. Write $q(r, \sigma^2) = (2\pi\sigma)^{-1/2} \exp(-r^2/2\sigma^2)$. We shall show that, given any x and k as above and any $\epsilon' > 0$, there is an integer $N = N(\epsilon', x, k)$ such that for $n > N$ we have $|d(k, n, x) - 1| < \epsilon'$ and such that, for $n > N$, P_x^* assigns probability at least $1 - \epsilon'$ to a set of $T_k^{(n)}$ values for which

$$(5.12) \qquad r_{kn}^*(x, T_k^{(n)}) + \epsilon' > \int_{-\infty}^{\infty} W(y, x) q(y, x(1-x) + h)\, dy,$$

where $h = (t^2 - t)/(k+1)$. But, for fixed irrational and nonexceptional x, the right side of (5.12) tends, as $k \to \infty$ (and thus, $h \to 0$), to the limit as $n \to \infty$ of the integrand in the right-hand member of (5.11). The integral of this limit is, by (5.8), the same as the right-hand member of (5.11). Thus, using (5.12) and applying Fatou's lemma to the left side of (5.11), we conclude that (5.11) will be proved if we demonstrate the statement of the sentence containing (5.12).

For fixed x and k as above and for any $\epsilon > 0$, ξ_{kn} assigns to the set of π for which $|f(x, \pi) - 1| < \epsilon$ a probability which tends to unity as $n \to \infty$. It follows that $d(k, n, x) \to 1$ as $n \to \infty$ and that (noting the relationship between ξ_{kn}^* and the f_k^* of Section 3), for any $\epsilon > 0$ and for n sufficiently large, P_x^* assigns probability at least $1 - \epsilon$ to a set of values $t_k^{(n)}$ of $T_k^{(n)}$ for which, writing $y_i = t_{ki}^{(n)}/n$, the joint density function of the $\bar{\gamma}_i = \sqrt{n}(p_i - y_i)$ $(1 \leqq i \leqq k)$ according to $\xi_{kn}^*(\cdot, x, t_k^{(n)})$ in a spherical region centered at 0 in the space of the $\bar{\gamma}_i$ and of probability at least $1 - \epsilon$ according to ξ_{kn}^* is at least

$$(1 - \epsilon)(2\pi)^{-k/2}\left(\prod_1^{k+1} y_i\right)^{-1/2} \exp\left(-\frac{1}{2}\sum_1^{k+1} \gamma_i^2/y_i\right). \tag{5.13}$$

Now, in the notation of Section 3, for $1 \leqq i \leqq k$,

$$F(i/(k+1), \pi) = p_1 + \cdots + p_i\,.$$

Hence, if $T_{ki}^{(n)} = ny_i$ $(1 \leqq i \leqq k)$, we have (because of the form of (3.3))

$$\begin{aligned} F(x, \pi) = p_1 + \cdots + p_{i_0} + tp_{i_0+1} &= (y_1 + \cdots + y_{i_0} + ty_{i_0+1}) \\ &\quad + (\bar{\gamma}_1 + \cdots + \bar{\gamma}_{i_0} + t\bar{\gamma}_{i_0+1})/\sqrt{n}. \end{aligned}$$

Now, $T_{ki}^{(n)}/n$ tends in probability (according to P_x^*) to $1/(k+1)$, and expression (5.13) with $\epsilon = 0$ is continuous in the y_i (in the region where all $y_i > 0$). Moreover, if we had $\epsilon = 0$ in (5.13) and assumed the validity of this expression for all values of the γ_i and put all $y_i = 1/(k+1)$, then $(\bar{\gamma}_1 + \cdots + \bar{\gamma}_{i_0})$ and $\bar{\gamma}_{i_0+1}$ would, according to (5.13), have a bivariate normal density function with means zero and covariance matrix

$$\frac{1}{(k+1)^2}\begin{pmatrix} i_0(k+1-i_0) & -i_0 \\ -i_0 & k \end{pmatrix}. \tag{5.14}$$

The corresponding density function of $\bar{\gamma}_i + \cdots + \bar{\gamma}_{i_0} + t\bar{\gamma}_{i_0+1}$ would then be normal with mean zero and variance

$$[i_0(k+1-i_0) - 2ti_0 + t^2k]/(k+1)^2 = x(1-x) + h. \tag{5.15}$$

Hence, if we carry through this last argument with the actual form of (5.13) and its region of validity, we conclude that, for any $\epsilon'' > 0$ and for n sufficiently large, P_x^* assigns probability at least $1 - \epsilon''$ to a set of values $t_k^{(n)}$ of $T_k^{(n)}$ for which, on a real interval centered at 0 and of probability at least $1 - \epsilon''$ according to $\xi_{kn}^*(\cdot, x, t_k^{(n)})$, this last measure induces a distribution function J for

$$\sqrt{n}[F(x, \pi) - (y_1 + \cdots + y_{i_0} + ty_{i_0+1})] = \Lambda_x \text{ (say)}$$

whose absolutely continuous component has a corresponding density (the derivative of J) whose magnitude is at least

$$(1 - \epsilon'')q(\lambda, x(1-x) + h) \tag{5.16}$$

almost everywhere on this interval of λ-values.

Next, we note that

$$(5.17)\qquad W(\sqrt{n}[g(x) - F(x, \pi)], F(x, \pi)) = W(\rho - \Lambda_x, x + \mu + \Lambda_x/\sqrt{n}),$$

where $\rho = \sqrt{n}[g(x) - (y_1 + \cdots + y_{i_0} + ty_{i_0+1})]$ and

$$\mu = -x + (y_1 + \cdots + y_{i_0} + ty_{i_0+1}).$$

For fixed x and k as above, denote by α the right side of (5.12). Let β be such that the right side of (5.12) is at least $\alpha - \epsilon'/4$ if the limits of integration are changed to $(-\beta, \beta)$. Let $c = W(\beta, x)$. Let the δ of our assumption (a) be

$$\epsilon'/8cq(0, x(1 - x) + h),$$

and let $z' = x$ where x is nonexceptional. The set $0 \leqq y \leqq \beta$, $|z - x| \leqq \epsilon(\delta, x)$ minus a suitable countable set of open intervals of total length $<\delta$ covering the points of discontinuity is closed and bounded. Hence, W is uniformly continuous on this set. Hence, there is a value $\epsilon_1 > 0$ such that $W(y, z) \geqq W(y, x) - \epsilon'/4$ for $|x - z| \leqq \epsilon_1$ and $0 \leqq y \leqq \beta$ but y not in the excluded set. If $0 \leqq y \leqq \beta$ and y is in the exceptional set, y is in a maximal subinterval of the exceptional set of either the form $a < y < b$ with $a > 0$ or else of the form $0 \leqq y < b$. Define $\bar{W}(y, x) = W(a, x)$ in the former case and $\bar{W}(y, x) = 0$ in the latter. If $0 \leqq y \leqq \beta$ but y is not exceptional, define $\bar{W}(y, x) = W(y, x)$. If $y > \beta$, define $\bar{W}(y, x) = \bar{W}(\beta, x)$. Finally, set $\bar{W}(-y, x) = \bar{W}(y, x)$. The function $\bar{W}$ so defined is symmetric in y, nondecreasing in y for $y \geqq 0$, and has the property that

$$(5.18)\qquad W(y, z) \geqq \bar{W}(y, x) - \epsilon'/4 \qquad \text{for } |x - z| \leqq \epsilon_1 \text{ and all } y,$$

and also that

$$(5.19)\qquad \int_{-\beta}^{\beta} \bar{W}(y, x)q(y, x(1 - x) + h)\, dy \geqq \alpha - \epsilon'/2.$$

Now, let $N = N(\epsilon', x, k)$ be such that, for $n > N$ and with $\epsilon'' = \epsilon'/4(\alpha + 1)$, the conclusion (5.16) holds with the λ-interval including the interval $(-\beta, \beta)$, and such that $|d(k, n, x) - 1| < \epsilon'$ for $n > N$. Write $\bar{\mu}$ for the random variable defined by putting $T_{ki}^{(n)}/n$ for y_i in the definition of μ. Since $\bar{\mu}$ tends to zero in probability (according to P_x^*) as $n \to \infty$, we may also suppose N to be such that, for $n > N$, $P_x^*\{|\bar{\mu}| + \beta/\sqrt{n} < \epsilon_1\} \geqq 1 - \epsilon''$. Next, we recall the statement made immediately following the statement of Lemma 1, that for $n = 1$ the conclusion of Lemma 1 holds if the normal probability density is replaced by one truncated at $(-\beta, \beta)$. We also note that the integral (with respect to λ) of this truncated density multiplied by $\bar{W}(\rho - \lambda, x)$ is easily seen (by an argument like that used to deduce (4.11) from (4.9)) to be minimized at $\rho = 0$. We note, as in previous sections, that if (5.12) is true under the restriction to nonrandomized ϕ (in the definition of r^*), then (5.12) is a fortiori true without this restriction. Thus, from (5.2), (5.16), (5.17), (5.18), and (5.19), we have for $n > N$ that, with P_x^*-probability at least

(5.20) $$1 - 2\epsilon'' > 1 - \epsilon',$$

$T_k^{(n)}$ will be such that

$$\begin{aligned} &r_{kn}^*(x, T_k^{(n)}) \\ &\geqq \inf_\rho \int_{-\beta}^{\beta} W(\rho - \lambda, x + \mu + \lambda/\sqrt{n})(1 - \epsilon'')q(\lambda, x(1 - x) + h)\, d\lambda \end{aligned}$$

(5.21) $$\begin{aligned} &\geqq (1 - \epsilon'') \inf_\rho \int_{-\beta}^{\beta} [\bar{W}(\rho - \lambda, x) - \epsilon'/4]q(\lambda, x(1 - x) + h)\, d\lambda \\ &= (1 - \epsilon'') \int_{-\beta}^{\beta} [\bar{W}(-\lambda, x) - \epsilon'/4]q(\lambda, x(1 - x) + h)\, d\lambda \\ &\geqq (1 - \epsilon'')(\alpha - 3\epsilon'/4) > \alpha - \epsilon'. \end{aligned}$$

This completes the proof of (5.12) and thus of Theorem 5.

We have not stated a corollary to Theorem 5 of the type given after Theorems 3 and 4. For $F \varepsilon \mathfrak{F} - \mathfrak{F}_c$, a weight function of the form (5.1) seems less meaningful because the loss contributed at a saltus x of F is measured by $W(y, z)$, where $z = F(x + 0)$. There are also certain technical difficulties in that the numerator of (5.6) need no longer be the same if $\mathfrak{F}_c$ is replaced by $\mathfrak{F}$. We shall not bother with the circumlocutions (e.g., additional restrictions on W) necessary to obtain a corollary from Theorem 5 in the same trivial manner as such corollaries were obtained from Theorems 3 and 4.

Theorem 5 implies certain much weaker results which, for special forms of W, may also be obtained from results obtained by Aggarwal [6]. He considers only the class C_n of procedures which with probability one set $g(x) = c_i^{(n)}$ for $Z_i^{(n)} \leqq x < Z_{i+1}^{(n)}$, where the $\{Z_i^{(n)}\}$ are the ordered values of the $\{X_i^{(n)}\}$. (Such procedures have constant risk for $F \varepsilon \mathfrak{F}_c$ and W_n^* of the form (5.1).) For the special functions $W(y, z) = |y|^r$ and $W(y, z) = |y|^r/z(1 - z)$ (r a positive integer), he obtains the best $c_i^{(n)}$ explicitly in a few cases and in the other cases characterizes them as the solutions of certain equations. In the former cases ϕ_n^* may be seen to be asymptotically best in C_n. This result is an immediate consequence of Theorem 5, where the result is proved for the class D_n of all procedures, of which the class C_n is a small subclass.

6. Other loss functions; multinomial estimation problems. The results obtained in the previous three sections may be extended to a more general class of loss functions to which the same methods of proof may be seen to apply. Thus, for example, in Sections 3 and 4 we could consider the maximum deviation over a set of x values for which $F(x)$ is in a specified *sub*set of the unit interval (this will involve techniques like those used in Section 5); the formulation of Theorem 5 already includes weight functions which may (e.g.) vanish for certain values of $F(x)$, and other modifications (e.g., to consider a finite set of points) are mentioned in the paragraph preceding Theorem 5. We may also consider (in Section 4) loss functions such as $W_1(r_1) + W_2(r_2)$ where r_1 and r_2 are the maximum deviations over two (not necessarily disjoint) sets of the type mentioned above and W_1 and W_2 are functions of the type considered in Section 4. Linear combinations of loss functions of this last type and the type considered in Section 5 may similarly be treated. In all of the above we may replace $\sup_x |g(x) - F(x)|$ by $\sup_x [|g(x) - F(x)|h(F(x))]$, where h is any nonnegative function (suitably regular), without any difficulty; this includes as a special case maximization over a subset as described above.

Thus, it appears that our results hold for a very general class of weight functions. It would of course be of interest to subsume all cases under one unified criterion and one method of proof. In the portion of Section 7 which is devoted to heuristic remarks, such a criterion (symmetry and convexity of a certain functional) is indicated; unfortunately, it does not include all cases treated above (e.g., the result of Section 3, which is apparently somewhat deeper), some of which will be seen in Section 7 to be slightly more difficult to handle than the symmetric convex functionals. A more general class Ω of monotone functionals for which (perhaps under slight regularity conditions) our results would seem likely to hold, and which includes the weight functions of Sections 3, 4, and 5 as well as those of the previous paragraph, is also indicated in Section 7. In the present context, this class consists of nonnegative functionals W of the function $|\delta|$ defined by $\delta(y) = g(F^{-1}(y)) - y$, $0 \leqq y \leqq 1$ (where we suppose for simplicity that the possible c.d.f.'s F under consideration are members of $\mathfrak{F}_c$ which are for each F strictly increasing for $\sup F^{-1}(0) \leqq y \leqq \inf F^{-1}(1)$) for which $W(|\delta_1(y)|) \leqq W(|\delta_2(y)|)$ whenever $|\delta_1(y)| \leqq |\delta_2(y)|$ for $0 \leqq y \leqq 1$. However, at this writing it is not evident how to give a rigorous *unified* proof (as distinguished from the heuristic one of Section 7) even for the class of weight functions which are convex symmetric functionals (of δ, in the present context), let alone to give one for the class Ω.

Another modification is to consider $\sup_x |g(x) - F(x)|h(x)$ above instead of $\sup_x |g(x) - F(x)|h(F(x))$. In this case ϕ_n^* will not have constant risk over $\mathfrak{F}_c$. However, this case is easily treated as follows: suppose for simplicity that h is continuous and bounded (the unbounded case is trivial and may be treated by a similar argument). Let J be an interval in which h is entirely within a prescribed $\epsilon > 0$ of $\sup_x h(x)$. We may for simplicity suppose J to be the unit interval. Then the risk function of ϕ_n^* will attain a value close to its maximum for $F = U$. The argument of Sections 3 and 4 may now be applied. In a similar manner we may consider in Section 5 loss functions for which the risk function of ϕ_n^* is not a constant; for example, (5.1) could be replaced by

$$W_n^*(F, g) = \int_{-\infty}^{\infty} W(\sqrt{n}[g(x) - F(x)], x)\, d\mu(x) \tag{6.1}$$

for a specified function W and measure μ satisfying certain regularity conditions.

An interesting question is whether or not our results can be extended to yield a *sequential* asymptotic minimax character, e.g., in the sense of Wald [7]. This is too large a topic to be discussed thoroughly in this paragraph, but a few indicative comments are in order. An essential idea present in the form of the ξ_{kn} of Sections 3, 4, and 5 is that, when k is large, a certain multinomial estimation problem is *almost as difficult* as the problem of estimating F. This suggests that, when the weight function considered here is such that the corresponding multinomial problem has (perhaps only asymptotically) a fixed sample-size minimax estimator (among all sequential estimators), then we may conclude that the fixed sample-size procedure ϕ_n^* is asymptotically minimax among all *sequential* procedures. An examination of [7] shows that such an asymptotic sequential minimax property for the multinomial problem will often be easy to prove using methods like Wald's.

Finally, the methods of this paper (without the limit considerations as $k \to \infty$) may be used to prove certain asymptotic minimax results for the estimation of the parameter π of the multinomial distribution (3.4) as $n \to \infty$, for any fixed k. To see this, we note that, under fairly general conditions of monotonicity and symmetry of the weight function (similar to those of Sections 3 to 5), the limiting risk function of $T_k^{(n)}/n$ as $n \to \infty$ will be continuous in a neighborhood of the point of B_k at which its maximum is achieved. Hence, for any $\epsilon > 0$, there will exist an interior point V_k^* of B_k in a neighborhood of which the limiting risk function of $T_k^{(n)}/n$ is continuous and at which point the limiting risk of $T_k^{(n)}/n$ is within ϵ of its maximum. One can then find a sequence $\{\xi_{kn}\}$ of a priori distributions on B_k (similar to the sequence used in Sections 3, 4, and 5) which assigns to any neighborhood of V_k^* a probability approaching one as $n \to \infty$ and which "shrinks down" on V_k^* at a slow enough rate (see the remarks of the paragraph preceding that containing (3.6)) to make the a posteriori probability distribution of the $\sqrt{n}(p_j - T_{kj}^{(n)}/n)$ normal with mean 0 so that $T_k^{(n)}/n$ is asymptotically Bayes with respect to $\{\xi_{kn}\}$, with integrated risk approaching the limiting risk of $T_k^{(n)}/n$ at V_k^*. The asymptotic minimax character of $T_k^{(n)}/n$ follows. We need not detail the wide variety of weight functions for which this optimum asymptotic property of the classical multinomial estimator $T_k^{(n)}/n$ follows from the methods and the results of the three previous sections as well as of the present section. It is perhaps worth while to remark that, although the results in Sections 3, 4, and 5 are stated in terms of deviations of sums $\psi_{n1} + \psi_{n2} + \cdots + \psi_{nj}$ of components ψ_{ni} of the estimator ψ_n from $p_1 + p_2 + \cdots + p_j$ $(1 \leq j \leq k + 1)$, the given proofs apply with only trivial modifications to weight functions depending on differences $\psi_{nj} - p_j$. Thus, for example, for any set of numbers $c_j > 0$, the asymptotic minimax character of $T_k^{(n)}/n$ for estimating $\pi \varepsilon B_k$ for the risk function

$$(6.2) \qquad r_n(\pi, \psi_n) = 1 - P_\pi\{|\psi_{nj} - p_j| \leq c_j/\sqrt{n}, (1 \leq j \leq k + 1)\}$$

follows from the asymptotic normality of the a posteriori distribution, noted above, and from the convexity and symmetry about ψ_n of the set of π (in R^{k+1}, not B_k) satisfying the inequalities in brackets in (6.2). (It is clear from this example that V_k^* need not be the V_k of Section 3.) The result for other risk functions follows similarly, using the methods and results of Sections 4, 5, and 6.

These asymptotic results for the multinomial estimation problem do not seem to have appeared previously in the literature. As indicated two paragraphs above, some of these multinomial results may also be extended to sequential problems.

7. Convex functionals and monotone functionals of stochastic processes; heuristic considerations. The first part of this section will be devoted to some simple remarks concerning convex symmetric functionals of random elements; these remarks will then be applied to give a short heuristic argument for many of the results obtained in previous sections.

Let $B = \{b\}$ be a linear space (or system) and ζ a random element with range in B and having a symmetric distribution, i.e., such that whenever A is a measurable subset of B so is $-A$ and $P\{\zeta \varepsilon A\} = P\{\zeta \varepsilon -A\}$. Let ω be a measurable real-valued convex functional on B which is symmetric ($\omega(b) = \omega(-b)$ for $b \varepsilon B$) and convex ($\omega(\lambda b_1 + (1 - \lambda)b_2) \leqq \lambda\omega(b_1) + (1 - \lambda)\omega(b_2)$ for $0 < \lambda < 1$ and $b_1, b_2 \varepsilon B$). We now note that, since $\min_b \omega(b) = \omega(0) > -\infty$ so that the expected value $E\omega(\zeta)$ is always defined, we may conclude that

$$E\omega(\zeta) = \min_{b \varepsilon B} E\omega(\zeta + b) \tag{7.1}$$

from the equation (implied by symmetry of P)

$$E\omega(\zeta + b) = E\omega(-\zeta + b) = \tfrac{1}{2}E\{\omega(\zeta + b) + \omega(-\zeta + b)\} \tag{7.2}$$

and the equation (implied by symmetry and convexity of ω)

$$\omega(\zeta + b) + \omega(-\zeta + b) = \omega(\zeta + b) + \omega(\zeta - b) \geqq 2\omega(\zeta). \tag{7.3}$$

We shall now apply (7.1) to the "tied-down" Wiener process (see [2]). B is now the space of continuous functions $b(t)$ on the unit interval $0 \leqq t \leqq 1$. The probability measure P assigns probability one to the subset B_0 of elements b of B satisfying $b(0) = b(1) = 0$. The measurable sets are generated by all sets of the form $\{b \mid b(t_0) < a_0\}$ for $0 \leqq t_0 \leqq 1$ and a_0 real. The joint distribution of $\zeta(t_1), \cdots, \zeta(t_n)$ for any $0 \leqq t_1 \leqq \cdots \leqq t_n \leqq 1$ is normal with $E\zeta(s) = 0$ and $E\{\zeta(s)\zeta(t)\} = \min(s, t) - st$ for $0 \leqq s, t \leqq 1$. Note that the distribution of ζ is symmetric.

Let W be any symmetric real-valued convex function on R^1. Then

$$W(\max_t |b(t)|)$$

is a convex functional of b and (7.1) implies that

$$EW(\max_t |\zeta(t)|) \leqq EW(\max_t |\zeta(t) + \rho(t)|) \tag{7.4}$$

for all continuous functions ρ. Generalizations of this result to the case where $\max_t$ is replaced by $\max_t h(t)$ in the manner of Section 6, or where ρ is allowed to be of a more general class than the continuous functions, are easily achieved by adjoining additional functions to B. One may also note that, for every $r > 0$,

$$P\{\max_t |\zeta(t)| > r\} \leqq P\{\max_t |\zeta(t) + \rho(t)| > r\} \tag{7.5}$$

for all continuous (or more general, as noted above) functions ρ. However, this cannot be proved in the same manner as (7.4), since the characteristic function of the subset of B for which $\max_t |b(t)| > r$ is not a convex functional on B. The validity of (7.5) follows, however, from (2.4) and Lemma 1. This strong result of a domination of an entire distribution function in the sense of (7.5) is deeper than the result (7.4); for (7.5) requires (in the proof of Lemma 1 in [8]) not merely the symmetry of the probability distribution, but also the convexity for every $u > 0$ of the set where the joint density of $\zeta(t_1), \cdots, \zeta(t_k)$ (for any k and $t_1, \cdots, t_k$) is $\geqq u$. (Note, for example, that it is not necessarily true for a symmetrically distributed real-valued random variable X that $P\{|X| > r\} \leqq P\{|X + \rho| > r\}$ for all real ρ.) Similarly, the result (7.4) for real functions $W(z)$ on the nonnegative reals which are nondecreasing in z for $z \geqq 0$ (but not necessarily convex) is a consequence of (7.5) but cannot be proved directly in the manner of (7.4) for convex W. Thus, to summarize, (7.4) for convex W follows from the symmetry of the probability measure, while in proving (7.4) for nondecreasing W (and, in particular, (7.5)) we use the additional assumption on the probability measure which is used to prove Lemma 1. We note that it has not been necessary to assume any integrability condition here.

It is interesting to note that, for the special case of a linear function $\rho(t) = c + dt$, the right side of (7.5) is given by formula (4.3) of [2] with $a = r - c - d$, $b = r - c$, $\alpha = r + c + d$, $\beta = r + c$ (unless $a \geqq 0$, $b > 0$, $\alpha \geqq 0$, $\beta > 0$, the probability in question is unity; our $\zeta(t)$ is Doob's $X(t)$). It does not seem completely apparent from the form of (4.3) of [2] that this expression, with the above substitutions, is a minimum for $c = d = 0$. The same is true of expectations with respect to the d.f. (4.3) of [2] of functions W of the type considered above.

Next, let the real-valued function $W(y, z)$ be symmetric and convex in y for each z $(-\infty < y < \infty,\ 0 \leqq z \leqq 1)$ and satisfy obvious measurability conditions. Let μ be any measure on the unit interval. Then

$$\omega(b) = \int_0^1 W(b(t), t)\, d\mu(t) \tag{7.6}$$

is a convex functional on B and hence (7.1) holds. In this case the result for the case where $W(y, z)$ is symmetric in y and nondecreasing in y for $y \geqq 0$ (but not necessarily convex for each z) is not much more difficult, although it cannot be handled by using (7.1): we need only apply Lemma 1 for $n = 1$ for each fixed z in this case in order to obtain the desired result.

More general convex functionals (such as combinations of the two varieties as treated in Section 6) or nonconvex functionals with certain monotonicity properties may be handled, similarly, by using (7.1) or consequences of Lemma 1 similar to (7.5), respectively. It is possible that the conclusion (7.1) holds for the class Ω of (not necessarily convex) functionals ω which are nonnegative, for which $\omega(\zeta) = \omega(|\zeta|)$, and for which $\omega(|\zeta_1|) \leqq \omega(|\zeta_2|)$ whenever $|\zeta_1(t)| \leqq |\zeta_2(t)|$ for all t. Similarly, results on processes other than the tied-down Wiener process, whose distributions are symmetrical or also satisfy the property which (as mentioned above) is used in [8] in proving the more general form of Lemma 1, may be obtained by using (7.1) or the generalization of Lemma 1 in [8], respectively;

We now turn to a heuristic argument for the results obtained in previous sections (except for certain results of Section 6, as noted below). This discussion may also be thought of as an outline of one intuitive explanation of why these results hold, the epsilontics and use of Bayes solutions in the previous sections supplying the needed rigor. However, the discussion which follows does not use Bayes solutions, and it would certainly be worth while to obtain an independent argument which would show that in the limit one need only consider "limiting" decision procedures of the type considered below, and thus to conclude that the argument which follows can be made rigorous by means of only brief additions.

In the previous sections we were concerned with estimating (for various weight functions) an unknown element F of $\mathfrak{F}_c$. Denote by $g_n(x; X^{(n)})$ such an estimator of $F(x)$ based on $X^{(n)}$ (for notational simplicity, we have considered a nonrandomized $\phi_n \varepsilon D_n$). Suppose we could show that for our considerations, at least asymptotically, it is only necessary to consider functions g_n of the form

$$g_n(x; X^{(n)}) = \psi_n(S_n(x)),$$

i.e., procedures in the class C_n mentioned in the last paragraph of Section 5 (this is one of two crucial gaps in our heuristic argument, for it is not obvious how to give a short proof, which in no way depends on the results of Sections 3 to 5, of this supposition). Procedures in C_n will have constant risk for $F \varepsilon \mathfrak{F}_c$ and for any of the weight functions of Sections 3, 4, and 5 (and those of Section 6 for which ϕ_n^* was not remarked to have constant risk). Thus, we may consider the distribution of the random function $\psi_n(S_n(t)) - t$ for $0 \leqq t \leqq 1$ where $F = U$. Write $\psi_n(z) = z + \rho_n(z)/\sqrt{n}$. Then

$$\sqrt{n}[\psi_n(S_n(t)) - t] = \sqrt{n}[S_n(t) - t] + \rho_n(S_n(t)).$$

If we could now suppose (and this is the other crucial nonrigorous development) that there is a sequence $\{\psi_n\}$ of minimax procedures (for $n = 1, 2, \cdots$) such that the corresponding sequence $\{\rho_n(z)\}$ has a continuous limit $\rho(z)$ uniformly in z as $n \to \infty$, and note that $\rho_n(S_n(t))$ would then be bounded (for n sufficiently large) and would tend to $\rho(t)$ with probability one as $n \to \infty$, then by [2] and [3] our consideration of $\sqrt{n}[\psi_n(S_n(t)) - t]$ would be reduced, asymptotically, to that of $\zeta(t) + \rho(t)$. The earlier comments of this section would then yield the desired asymptotic minimax properties of ϕ_n^*.

REFERENCES

[1] A. N. Kolmogorov, "Sulla determinazione empirica di una legge di distribuzione," *Inst. Ital. Atti. Giorn.*, Vol. 4 (1933), pp. 83–91.

[2] J. L. Doob, "Heuristic approach to the Kolmogorov-Smirnov theorems," *Ann. Math. Stat.*, Vol. 20 (1949), pp. 393–403.

[3] M. Donsker, "Justification and extension of Doob's heuristic approach to the Kolmogorov-Smirnov theorems," *Ann. Math. Stat.*, Vol. 23 (1952), pp. 277–281.

[4] PAUL LÉVY, *Théorie de L'Addition des Variables Aléatoires*, Gauthier-Villars, Paris (1937).
[5] J. V. USPENSKY, *Introduction to Mathematical Probability*, McGraw-Hill Book Co., New York (1937).
[6] O. P. AGGARWAL, "Some minimax invariant procedures for estimating a c.d.f.," *Ann. Math. Stat.*, Vol. 26 (1955), pp. 450–463.
[7] A. WALD, "Asymptotic minimax solutions of sequential point estimation problems," *Proceedings of the Second Berkeley Symposium on Mathematical Statistics and Probability*, University of California Press (1951), pp. 1–11.
[8] T. W. ANDERSON, "The integral of a symmetric unimodal function," *Proc. Amer. Math. Soc.*, Vol. 6 (1955), pp. 170–176.
[9] N. V. SMIRNOFF, "Approach of empiric distribution functions," *Uspyekhi Matem. Nauk.*, Vol. 10 (1944), pp. 179–206.

Reprinted from THE ANNALS OF MATHEMATICAL STATISTICS
Vol. 27, No. 4, December, 1956

CONSISTENCY OF THE MAXIMUM LIKELIHOOD ESTIMATOR IN THE PRESENCE OF INFINITELY MANY INCIDENTAL PARAMETERS

BY J. KIEFER[1] AND J. WOLFOWITZ[2]

Cornell University

Summary. It is shown that, under usual regularity conditions, the maximum likelihood estimator of a structural parameter is strongly consistent, when the (infinitely many) incidental parameters are independently distributed chance variables with a common unknown distribution function. The latter is also consistently estimated although it is not assumed to belong to a parametric class. Application is made to several problems, in particular to the problem of estimating a straight line with both variables subject to error, which thus after all has a maximum likelihood solution.

1. Introduction. Let $\{X_{ij}\}$, $i = 1, \cdots, n$, $j = 1, \cdots, k$, be chance variables such that the frequency function of $X_{i1}, \cdots, X_{ik}$ is $f(x \mid \theta, \alpha_i)$ when θ and α_i are given, and thus depends upon the unknown (to the statistician) parameters θ and α_i. The parameter θ, upon which all the distributions depend, is called "structural"; the parameters $\{\alpha_i\}$ are called "incidental". Throughout this paper we shall assume that the X_{ij} are independently distributed when $\theta, \alpha_1, \cdots, \alpha_n$, are given, and shall consider the problem of consistently estimating θ (as $n \to \infty$). The chance variables $\{X_{ij}\}$ and the parameters θ and $\{\alpha_i\}$ may be vectors. However, for simplicity of exposition we shall throughout this paper, except in Example 2, assume that they are scalars. Obvious changes will suffice to treat the vector case.

Very many interesting problems are subsumed under the above formulation. Among these is the following:

$$f(x \mid \theta, \alpha_i) = (2\pi\theta)^{-k/2} \exp\left\{\frac{-\sum_j (x_{ij} - \alpha_i)^2}{2\theta}\right\}. \tag{1.1}$$

Suppose now that the $\{\alpha_i\}$ are considered as unknown constants and we form in the usual manner the likelihood function

$$(2\pi\theta)^{-kn/2} \exp\left\{-\frac{1}{2\theta}\sum_{i,j} (X_{ij} - \alpha_i)^2\right\} \tag{1.2}$$

corresponding to (1.1). Then the maximum likelihood (m.l.) estimator of θ is

$$\frac{\sum_{i,j} (X_{ij} - \bar{X}_i)^2}{kn} \tag{1.3}$$

Received September 28, 1955.

[1] Research sponsored by the Office of Naval Research.

[2] The research of this author was supported in part by the United States Air Force under Contract AF 18(600)-685, monitored by the Office of Scientific Research.

with $\bar{X}_i = k^{-1}\sum_j X_{ij}$, and is obviously not consistent. This example is due to Neyman and Scott [1], who used it to prove that the m.l. estimator[3] need not be consistent when there are infinitely many incidental parameters (constants). The latter authors, to whom the names "structural" and "incidental" are due, seem to have been the first to formulate the general problem. Special forms of the problem, like Example 2 below, had been studied for a long time (e.g., Wald [2] and the literature cited there).

The general fact that, when the $\{\alpha_i\}$ are unknown constants, the m.l. estimator of θ need not be consistent, is certainly basically connected with the fact that, since there are only a constant number of observations which involve a particular α_i, it is in general impossible to estimate the $\{\alpha_i\}$ consistently. Now there are many meaningful and practical statistical problems where the $\{\alpha_i\}$ are not arbitrary constants but independently and identically distributed chance variables with distribution function (df) G_0 (unknown to the statistician). The question then arises whether the m.l. method, which does not always yield a consistent estimator when there are infinitely many incidental constants, and does yield consistent estimators in the classical parametric case where there are no incidental parameters, will give a consistent estimator in this case, where the $\{\alpha_i\}$ are independent chance variables with the common df G_0. This note is devoted to this question.

The answer is affirmative. Not only is the m.l. estimator of θ strongly consistent (i.e., converges to θ with probability one) under reasonable regularity conditions, but also the m.l. estimator of G_0 converges to G_0 at every point of continuity of the latter, with probability one (w.p.1). This is the more striking when one recalls that G_0 does not belong to a parametric class, i.e., a set of df's indexed by a finite number of parameters. (If G_0 were a member of such a given class, the problem would fall completely in the domain of classical maximum likelihood.) The interest of the present authors was originally in estimating θ. That G can also be estimated by the m.l. method is a felicitous by-product of our investigation. A heuristic explanation of the present result may be this: A sequence of chance variables is more "regular" than an arbitrary sequence of numbers. In the present procedure one does not attempt to determine the particular values of the chance variables $\{\alpha_i\}$, but only their distribution function; thus, we seek the m.l. estimator of the "parameter" $\gamma = (\theta, G)$ based on a sequence of independent random variables whose common distribution function is indexed by γ.

In sections 3, 4, and 5, the results are applied to three problems which seem to be of interest per se. Among these is the problem of fitting a straight line with both variables subject to normal error. This problem has a very long history and has been the subject of many investigations (see, for example [2], [7], [4], and the literature cited there); it seems interesting that it can, after all, be treated by the m.l. method. The verification of the regularity assumptions or the formulation of not too onerous conditions for them to be verified is sometimes not entirely ob-

[3] Throughout this paper, for the sake of brevity, we use the term "estimator" to mean "sequence of estimators for $n = 1, 2, \cdots$."

vious, and the verification of these assumptions (in the form used in Section 2) constitutes the main difficulty of the paper. As is explained in detail below, the fact that these assumptions imply the general consistency result of Section 2 follows from a modification of the proof of [5]. Professor Herbert Robbins has kindly called our attention to his abstract in *Ann. Math. Stat.*, vol. 21 (1950), p. 314, Abstract 35, which states that the m.l. estimator of G is consistent. Since nothing further has appeared on this subject, the intended restrictions under which the statement is true, and the intended method of proof, are unknown to the present authors. This seems to be the second instance in the literature where the m.l. estimator has been used to estimate an entire df which is not assumed to belong to a class depending only on a finite number of real parameters. The first instance of the employment of such an estimator is the classical estimation of a df by its empiric df (shown to be asymptotically optimal in [3]), which is its m.l. estimator (see the paragraph preceding the lemma in Section 2). The only other instance of the estimation of a df in the nonparametric case seems to be that of the estimation of identifiable df's in stochastic structures such as those of the present paper by means of the minimum distance method [4].[4] (The latter requires regularity conditions weaker than those of the present paper. Compare, for example, [4] with Example 2 below; see also Example 3a.)

In connection with these examples, and also in Section 6, we give some examples which illustrate the fact that the classical m.l. estimator may not be consistent, even in parametric examples which lack the pathological discontinuity sometimes present in hitherto published examples.

Section 6 also contains the statement of a simple device which can be used in the classical parametric case as well as in the case studied in this paper, to prove consistency of the m.l. estimator in some cases where the assumptions used in published proofs of consistency are not satisfied.

The proof in Section 2 is a modification of Wald's [5], and its fundamental ideas are to be found in [5]; for this reason some of its details will be omitted. Wald states in his paper that his method applies more generally when his Assumption 9 is fulfilled. However, this assumption is not fulfilled in our problem *ab initio* and some technical modifications have to be made. One obstacle to extending Wald's proof to our problem is in establishing an analogue of (16) in [5]; one "neighborhood of infinity" does not always seem to suffice. Also some changes in the assumptions are made necessary by the nature of our problem. The results of the present paper can be extended in the usual manner to abstract spaces, but we forego this. It should also be remarked that in [6] Wald studied the present problem of estimating a structural parameter.

The attitude towards the $\{\alpha_i\}$, i.e., whether they are to be regarded as unknown constants or identically and independently distributed chance variables or something else, seems to vary with the author and sometimes even within the

---

[4] A paper entitled "The minimum distance method," which gives the details and proofs of the results announced in [4], is scheduled for publication in a forthcoming issue of these Annals.

publications of the same author. For example, Wald [2], in his treatment of the problem of fitting a straight line mentioned above, considers the $\{\alpha_i\}$ as unknown constants; and Neyman and Scott, in their general formulation of the problem given in [1] and described at the beginning of the present section, also consider the $\{\alpha_i\}$ as unknown constants. On the other hand, Neyman in his treatment [7] of the straight line problem treats the α_i as independently and identically distributed chance variables. Also Neyman and Scott [8] criticize Wald's solution [2] on the ground that the conditions he postulates on the sequence of constants $\{\alpha_i\}$ are such that they are unlikely to be satisfied when the $\{\alpha_i\}$ are independently and identically distributed chance variables. Our own point of view and perhaps also that of the other writers cited, is that one need not insist on any one formulation to the exclusion of all others. There are certainly reasonable statistical problems where the $\{\alpha_i\}$ may be looked upon as independently and identically distributed chance variables, and consequently the problem of the present paper is statistically meaningful and interesting. This is also the attitude implicit in [4] and [9].

2. Proof of consistency. As we have stated earlier, the essential idea of the proof comes from [5]. A compactification device has to be employed because the space Γ defined below may not be compact.

We postulate that the following assumptions are fulfilled (see also the paragraph preceding the lemma at the end of this section):

ASSUMPTION 1: $f(x \mid \theta, \alpha)$ is a density with respect to a σ-finite measure μ on a Euclidean space of which x is the generic point. (This is also Wald's Assumption 1.)

Let Ω be the space of possible values of θ, and let A be the space of values which α_i can take. (Both Ω and A are measurable subsets of Euclidean spaces, f is jointly measurable in x and α for each θ, and we hereafter denote by $\theta_t^{(s)}$ $(1 \leqq s \leqq r)$ the components of a point θ_t in Ω and by $|\alpha|$ the Euclidean distance from the origin of a point $\alpha \in A$; τ will denote Lebesgue measure on A.) Let $\Gamma = \{G\}$ be a given space of (cumulative) distributions of α_i. Let θ_0, G_0 be, respectively, the "true" value of the parameter θ and the "true" distribution of α_i. It is assumed that $\theta_0 \in \Omega$ and $G_0 \in \Gamma$. Let $\gamma = (\theta, G)$ be the generic point in $\Omega \times \Gamma$. We define

$$f(x \mid \gamma) = \int_A f(x \mid \theta, z)\, dG(z) \tag{2.1}$$

and $\gamma_0 = (\theta_0, G_0)$. In the space $\Omega \times \Gamma$ we define the metric

$$\begin{aligned} \delta(\gamma_1, \gamma_2) &= \delta([\theta_1, G_1], [\theta_2, G_2]) \\ &= \sum_{s=1}^{r} |\operatorname{arc\,tan} \theta_1^{(s)} - \operatorname{arc\,tan} \theta_2^{(s)}| \qquad (2.2) \\ &\qquad + \int_A |G_1(z) - G_2(z)|\, e^{-|z|}\, d\tau(z). \end{aligned}$$

Let $\bar{\Omega} \times \bar{\Gamma}$ be the completed space of $\Omega \times \Gamma$ (the space together with the limits of its Cauchy sequences in the sense of the metric (2.2)). Then $\bar{\Omega} \times \bar{\Gamma}$ is compact.

ASSUMPTION 2 (Continuity Assumption): It is possible to extend the definition of $f(x \mid \gamma)$ so that the range of γ will be $\bar{\Omega} \times \bar{\Gamma}$ and so that, for any $\{\gamma_i\}$ and γ^* in $\bar{\Omega} \times \bar{\Gamma}$, $\gamma_i \to \gamma^*$ implies

$$f(x \mid \gamma_i) \to f(x \mid \gamma^*), \tag{2.3}$$

except perhaps on a set of x whose probability is 0 according to the probability density $f(x \mid \gamma_0)$. (The exceptional x-set may depend on γ^*; $f(x \mid \gamma^*)$ need not be a probability density function.) (This assumption corresponds to Wald's continuity Assumptions 3 and 5.)

ASSUMPTION 3: For any γ in $\bar{\Omega} \times \bar{\Gamma}$ and any $\rho > 0$, $w(x \mid \gamma, \rho)$ is a measurable function of x, where

$$w(x \mid \gamma, \rho) = \sup f(x \mid \gamma'),$$

the supremum being taken over all γ' in $\bar{\Omega} \times \bar{\Gamma}$ for which $\delta(\gamma, \gamma') < \rho$. (This assumption is made for the reasons given by Wald. See his remarks following Assumption 8 in [5].)

ASSUMPTION 4 (Identifiability Assumption): If γ_1 in $\bar{\Omega} \times \bar{\Gamma}$ is different from γ_0, then, for at least one y,

$$\int_{-\infty}^{y} f(x \mid \gamma_1)\, d\mu \neq \int_{-\infty}^{y} f(x \mid \gamma_0)\, d\mu, \tag{2.4}$$

the integral being over those x all of whose components are $\leqq$ the corresponding components of y. (This is the same as Wald's Assumption 4.)

Let X be a chance variable with density $f(x \mid \gamma_0)$. *The operator E will always denote expectation under γ_0*; γ_0 will always, of course, be a member of $\Omega \times \Gamma$.

ASSUMPTION 5 (Integrability Assumption): For any γ in $\bar{\Omega} \times \bar{\Gamma}$ we have, as $\rho \downarrow 0$,

$$\lim E\left[\log \frac{w(X \mid \gamma, \rho)}{f(X \mid \gamma_0)}\right]^+ < \infty. \tag{2.5}$$

(This assumption is implied by assumptions corresponding to Wald's Assumptions 2 and 6.)

For any γ in $\bar{\Omega} \times \bar{\Gamma}$ other than γ_0, define $v = \log f(X, \gamma) - \log f(X, \gamma_0)$. We begin the proof of consistency by showing that

$$Ev < 0. \tag{2.6}$$

First, if γ is in $\Omega \times \Gamma$, $Ee^v \leqq 1$. Hence from (2.3) and Fatou's lemma it follows that, for any γ in $\bar{\Omega} \times \bar{\Gamma}$,

$$Ev \leqq Ee^v \leqq 1. \tag{2.7}$$

If v is $-\infty$ with probability one according to $f(x \mid \gamma_0)$, then (2.6) is obvious. Suppose therefore that $v > -\infty$ with positive probability according to

$f(x \mid \gamma_0)$. Then, by Jensen's inequality and (2.7),

$$Ev \leqq \log Ee^v \leqq 0, \tag{2.8}$$

and the first equality sign can hold only if v is a constant c with probability one according to $f(x \mid \gamma_0)$. If the first equality sign does not hold (2.6) follows at once. Consider, therefore, the constant c. If $c < 0$ then (2.6) holds. If $c > 0$ then (2.8) is violated. We cannot have $c = 0$ because of Assumption 4. This proves (2.6).

Now, as $\rho \downarrow 0$, for $\gamma' \neq \gamma_0$,

$$\lim E\left[\log \frac{w(X \mid \gamma, \rho)}{f(X \mid \gamma_0)}\right]^+ = E\left[\log \frac{f(X \mid \gamma)}{f(X \mid \gamma_0)}\right]^+ \tag{2.9}$$

by (2.3), (2.5), and Lebesgue's dominated convergence theorem. Also,

$$\lim E\left[\log \frac{w(X \mid \gamma, \rho)}{f(X \mid \gamma_0)}\right]^- = E\left[\log \frac{f(X \mid \gamma)}{f(X \mid \gamma_0)}\right]^-, \tag{2.10}$$

since the integrand of the left member decreases monotonically to the integrand of the right member. Hence, as $\rho \to 0$,

$$\lim E\left[\log \frac{w(X \mid \gamma, \rho)}{f(X \mid \gamma_0)}\right] = E \log \frac{f(X \mid \gamma)}{f(X \mid \gamma_0)} < 0 \tag{2.11}$$

by (2.6). Just as in [5] (see also [12]) it may then be shown that, for any positive ρ, there exists an $h(\rho)$, $0 < h(\rho) < 1$, such that the probability is one that, for all n sufficiently large,

$$\sup \left\{ \frac{\prod_{i=1}^{n} f(X_i \mid \gamma)}{\prod_{i=1}^{n} f(X_i \mid \gamma_0)} \right\} < h^n, \tag{2.12}$$

the supremum being taken over all γ in $\bar{\Omega} \times \bar{\Gamma}$ for which $\delta(\gamma, \gamma_0) > \rho$, and where $X_1, X_2, \cdots$ are independent chance variables with the common density $f(x \mid \gamma_0)$.

Let $L(x_1, \cdots, x_n \mid \gamma) = \prod_1^n f(x_i \mid \gamma)$. A *modified m.l. estimator* is defined to be a sequence of μ-measurable functions $\{\hat{\gamma}_n\}$ such that

$$L(x_1, \cdots, x_n \mid \hat{\gamma}_n(x_1, \cdots, x_n)) \geqq c \sup_\gamma L(x_1, \cdots, x_n \mid \gamma)$$

for almost all (μ) $x_1, \cdots, x_n$ for each n, where c is a positive number (the supremem is over $\Omega \times \Gamma$); for $c = 1$, this of course defines an m.l. estimator. (We shall not be concerned in this paper with conditions which ensure the existence of such measurable functions, although reasonable conditions are not difficult to formulate.) We also define a *neighborhood m.l. estimator* to be a sequence of μ-measurable functions $\{\gamma_n^*\}$ such that there exists a sequence of positive numbers ϵ_n with $\lim_{n\to\infty} \epsilon_n = 0$ for which $\sup_{\gamma \in \Pi_n} L(x_1, \cdots, x_n \mid \gamma) = \sup_\gamma L(x_1, \cdots, x_n \mid \gamma)$ for almost all $(\mu) x_1, \cdots, x_n$, where Π_n is the set of all γ in $\Omega \times \Gamma$ for which

$\delta(\gamma, \gamma_n^*(x_1, \cdots, x_n)) < \epsilon_n$. (Thus, neighborhood m.l. estimators exist in some cases where m.l. and modified m.l. estimators do not; this will be useful in making clear certain examples below where the lack of consistency is not merely due, as it might seem, to the fact that no strict m.l. or modified m.l. estimator exists.)

The above result (2.12) implies the strong convergence of m.l., modified m.l., and neighborhood m.l. estimators (in the respective cases where they exist). The component of the estimator which estimates G_0 converges to it at all its points of continuity w.p.1.

We remark that the above proof actually demonstrates consistency if, in the definition of m.l. estimator (or its variants), the supremum is taken over $\bar{\Omega} \times \bar{\Gamma}$ instead of over $\Omega \times \Gamma$ or, in fact, over any subset of $\bar{\Omega} \times \bar{\Gamma}$ containing γ_0. This last fact implies that if consistency is verified in an example where $\Omega = \Omega_1$, $\Gamma = \Gamma_1$, then it automatically holds in the example where $\Omega = \Omega_2$, $\Gamma = \Gamma_2$, whenever $\Omega_2 \subset \Omega_1$ and $\Gamma_2 \subset \Gamma_1$. In particular, this remark applies to the examples of Sections 3, 4, and 5.

We remark that Assumption 1 is not really essential in the above proof. Let P_γ denote the probability measure of X when γ is the true parameter value, and let $d(x, \gamma, \gamma_0) = r(x, \gamma, \gamma_0)/[1 - r(x, \gamma, \gamma_0)]$, where $r(x, \gamma, \gamma_0)$ denotes a Radon-Nikodym derivative of P_γ with respect to $P_\gamma + P_{\gamma_0}$ at the point x. If, for each $\gamma_0 \in \Omega \times \Gamma$, Assumptions 2 and 3 are satisfied when $f(x \mid \gamma)$ is replaced by $d(x, \gamma, \gamma_0)$, if (2.4) is replaced by the condition that $d(x, \gamma, \gamma_0) = 1$ does not hold on a set of probability one under γ_0 for any γ, and if $f(x \mid \gamma)/f(x \mid \gamma_0)$ is replaced by $d(x, \gamma, \gamma_0)$ (with a similar replacement for $w(x \mid \gamma, \gamma_0)$) in Assumption 5 and in the argument of the section, then (2.12) (with the replacement noted above) will still hold. An m.l. estimator $\hat{\gamma}$ is now defined to be one for which $\sup_\gamma \prod_1^n d(X_i, \gamma, \hat{\gamma}) = 1$ (with an analogous definition of modified and neighborhood m.l. estimator). We have not stated our assumptions and result (2.12) in this more general setting above because the stated form of the assumptions will suffice in most applications and will be easier to verify than assumptions stated in terms of $d(x, \gamma, \gamma_0)$ (which must be verified for each γ_0). As an example of the use of the more general result just cited, consider the problem of estimating the df F of a sequence of independent identically distributed discrete random variables, it being assumed that the true probability measure P_F (corresponding to the df F) satisfies

$$\sum_x P_F(x) \log P_F(x) > -\infty,$$

the sum being over all points x for which $P_F(x) > 0$. Then the assumptions are easily seen to be satisfied, and we may conclude that the sample df, which is the m.l. estimator, is a consistent estimator of F, a well-known result which does not usually seem to be considered as an example of the consistency of the m.l. estimator. (Of course, even if no restrictions of discreteness or logarithmic summability are placed on P_F, the sample df is still consistent and, as pointed out in the introduction, this is the m.l. estimator. However, Assumption 5 is not satisfied in this case.)

Before proceeding to our examples in subsequent sections, we prove a simple lemma which will be useful later in verifying Assumption 5.

LEMMA. *If $f(z_1, \cdots, z_k)$ is a bounded probability density function with respect to Lebesgue measure μ on Euclidean k-space R^k, and if*

$$\int_{|z_i|>1} (\log |z_i|) f \, d\mu < \infty \qquad (1 \leq i \leq k), \tag{2.13}$$

then

$$-\int_{R^k} f \log f \, d\mu < \infty. \tag{2.14}$$

PROOF: If we prove that (2,13) implies (2.14) when f is replaced by cf in these equations, where $c > 0$, then the lemma is clearly proved. Thus, since f was assumed bounded, we may hereafter assume $f \leq (2e)^{-1}$. (The new f need not have integral unity.) Let

$$g(z_1, \cdots, z_k) = f(z_1, \cdots, z_k) + \prod_{i=1}^{k} (z_i^2 + 1)^{-1}. \tag{2.15}$$

Clearly, (2.13) is true with f replaced by g. Moreover, since $g(z_1, \cdots, z_k) < e^{-1}$ outside of a sufficiently large sphere about the origin, and since $-f \log f < -g \cdot \log g$ if $0 < f < g < e^{-1}$, it suffices to prove (2.14) with f replaced by g, assuming g bounded and (2.13) with f replaced by g. By (2.13), we have

$$\int_{R^k} g \log \prod_{i=1}^{k} (1 + z_i^2)^{\frac{1}{2}} \, d\mu < \infty. \tag{2.16}$$

Thus, it suffices to prove the finiteness of

$$\begin{aligned} -\int_{R^k} g \log g \, d\mu &- \int_{R^k} g \log \prod_{i=1}^{k} (1 + z_i^2)^{\frac{1}{2}} \, d\mu \\ &= \int_{R^k} g \log \prod_{i=1}^{k} (1 + z_i^2)^{\frac{1}{2}} \left\{ \frac{-\log [g \prod (1 + z_i^2)^{\frac{1}{2}}]}{\log \prod (1 + z_i^2)^{\frac{1}{2}}} \right\} d\mu. \end{aligned} \tag{2.17}$$

The fact that $g(z_1, \cdots, z_k) \geq \prod (z_i^2 + 1)^{-1}$ (see (2.15)) implies easily that the bracketed expression in (2.17) is ≤ 1; by (2.16), this completes the proof of the lemma.

3. Example 1. Structural location parameter, incidental scale parameter. Let k be a positive integer, let μ be Lebesgue measure on Euclidean k-space, let g be a univariate probability density function with respect to Lebesgue measure, and let

$$f(x_i \mid \theta, \alpha_i) = \frac{1}{\alpha_i^k} \prod_{j=1}^{k} g\left(\frac{x_{ij} - \theta}{\alpha_i}\right), \tag{3.1}$$

where $x_i = (x_{i1}, \cdots, x_{ik})$. (Thus, observations are taken in groups of $k \geqq 1$, the value of the incidental parameter being the same within each group. The (unconditional) density of $X_i = (X_{i1}, \cdots, X_{ik})$ is given by Equation (2.1). Thus, the X_i are independent, but, for fixed i, the $X_{ij}(j = 1, \cdots, k)$ need not be independent.) Here Ω is the real line. Some further assumptions on g will be made below; we remark here that the important case

$$g(x) = (2\pi)^{-\frac{1}{2}}e^{-(x^2/2)} \tag{3.2}$$

will satisfy our assumptions. (See also (3.4) below.)

The cases $k = 1$ and $k > 1$ are essentially different. In Example 1a the consistency of the m.l. estimator will be proved for $k = 1$ assuming that A is the set of values $\alpha \geqq c$ where c is a known positive constant, and it is pointed out that the property of consistency of the m.l. estimator *does not hold* without this assumption. The proof of consistency in Example 1a is actually carried out for $k \geqq 1$ since this requires little additional space and will save space in Example 1b where we may refer back to 1a for proofs. In Example 1b we prove consistency of the m.l. estimator in the case $k > 1$ without assuming $\alpha \geqq c > 0$.

Example 1a. We assume that $k \geqq 1$ and that A is the set of all real values $\alpha \geqq c$ where c is a known *positive* constant. In the case $k = 1$, this assumption on A can be weakened slightly to an assumption on the behavior of $G(\alpha)$ as $\alpha \to 0$; however, some such assumption is necessary for consistency, since the last example of Section 6 shows that, even in cases where Γ is restricted to a simple parametric class of df's on a set of positive reals which is *not* bounded away from zero, it can happen that no m.l. or modified m.l. estimator exists and that there are neighborhood m.l. estimators which are not consistent.

We now state our assumptions on g and G_0. They seem reasonable and are in a form which makes brief proofs possible; they undoubtedly can be weakened. (These last remarks apply also to Examples 2 and 3. See also the first part of Section 6 for one method by which we can prove the results of our examples under weaker conditions.) We hereafter assume

(a) $\sup_x g(x) < \infty$;

(b) g is lower semicontinuous and for every $\epsilon > 0$ there is a continuous function $h_\epsilon \geqq g$ for which $\int [h_\epsilon(x) - g(x)]\, dx < \epsilon$;

(c) $\lim_{|x| \to \infty} g(x) = 0$;

(3.3) (d) $-\int_{-\infty}^{\infty} g(x)[\log |x|]^+ \, dx > -\infty$;

(e) $\int_{-\infty}^{\infty} |x|^{it} g(x)\, dx \neq 0$ for almost all real t;

(f) $g(x) > 0$ for almost all x in some open interval whose closure contains the point $x = 0$.

We note that, in addition to being satisfied in the case (3.2), Assumption (3.3) is also satisfied in such important cases as

(3.4)
(a) $g(x) = 1/\pi(1 + x^2)$;
(b) $g(x) = 1$ if $|x| < \frac{1}{2}$ and $g(x) = 0$ otherwise;
(c) $g(x) = e^{-x}$ if $x > 0$ and $g(x) = 0$ otherwise.

Of course, if g does not satisfy (3.3) but if there is a function g^* satisfying (3.3) and for which $g(x) = g^*(x)$ almost everywhere, then we may carry out our considerations replacing g by g^*.

We assume that Γ consists of all G such that

$$\int_c^\infty (\log \alpha) \, dG(\alpha) < \infty, \tag{3.5}$$

where c is the constant used before in the definition of A. For example, G belongs to Γ if, for some positive constants b and ϵ,

$$1 - G(\alpha) < \frac{b}{\log \alpha (\log\log \alpha)^{1+\epsilon}} \tag{3.6}$$

for $\alpha > e^e$; integration by parts will verify that (3.6) implies (3.5). Condition (3.5) is weaker than the requirement that any positive (not necessarily integral) movement of G be finite.

We now verify the assumptions of Section 2. We complete the definition of f for (θ, α) in $\bar{\Omega} \times \bar{A}$ by setting $f(x \mid \theta, \alpha) = 0$ whenever $\theta = \pm\infty$ or $\alpha = \infty$. For $(\theta, G) \in \bar{\Omega} \times \bar{\Gamma}$, we then define $f(x \mid \theta, G)$ by (2.1). (We remark that $\bar{\Gamma}$ obviously contains all df's on $\bar{A}$.) Assumption 1 is obviously satisfied. Assumption 3 follows from the fact that (3.3) implies that $f(x \mid \theta, G)$ is for each x lower semicontinuous in (θ, G) (in the sense of the metric δ) on $\bar{\Omega} \times \bar{\Gamma}$, and the fact that $\bar{\Omega} \times \bar{\Gamma}$ is separable. Write $h_\epsilon(x_i \mid \theta, \alpha) = \alpha^{-k} \prod_{j=1}^k h_\epsilon[(x_{ij} - \theta)/\alpha]$. In order to verify Assumption 2, we note that, by the lower semicontinuity in (θ, G) of $f(x \mid \theta, G)$ and by the Helly-Bray theorem, we have (assuming, as we may, that the h_ϵ of (3.3) (b) satisfies $\lim_{|x| \to \infty} h_\epsilon(x) = 0$) that $(\theta_i, G_i) \to (\theta^*, G^*)$ as $i \to \infty$ implies

$$\begin{aligned} f(x \mid \theta^*, G^*) &\leq \liminf_{i \to \infty} \int f(x \mid \theta_i, \alpha) \, dG_i \leq \limsup_{i \to \infty} \int f(x \mid \theta_i, \alpha) \, dG_i \\ &\leq \lim_{i \to \infty} \int h_\epsilon(x \mid \theta_i, \alpha) \, dG_i = \int h_\epsilon(x \mid \theta^*, \alpha) \, dG^*. \end{aligned} \tag{3.7}$$

Since the last member of (3.7) is greater than or equal to the first for all x and since their difference has integral $< \epsilon^k$ (with respect to μ), Assumption 2 follows at once.

In verifying Assumption 4, it clearly suffices to prove that, if $f(x \mid \theta_0 . G_0) =$

$f(x \mid \theta_1, G_1)$ for almost all x, where $(\theta_i, G_i) \varepsilon \Omega \times \Gamma$ for $i = 0, 1$, then $(\theta_0, G_0) = (\theta_1, G_1)$. If an interval $0 < x < \epsilon$ satisfies (3.3) (f), there is a value β such that

$$P\{X_{1j} \leqq t \text{ for } 1 \leqq j \leqq k \mid \theta_0, G_0\} \leqq \beta$$

is satisfied (whatever be G_0) if and only if $t \leqq \theta_0$, a similar assertion holding if the interval $-\epsilon < x < 0$ satisfies (3.3) (f). Hence, it suffices to prove the above assertion when $\theta_0 = \theta_1$, since it cannot hold when $\theta_0 \neq \theta_1$. Let H_i be the df of the random variable $\log \alpha_1$ when G_i is the df of the random variable α_1; i.e., $H_i(t) = G_i(e^t)$. Then, putting $g^*(z) = e^z[g(e^z) + g(-e^z)]$ (g^* is the density of $\log |U|$ when g is the density of U), it suffices to prove that, if H_0 and H_1 are not identical, then $p_1(z_1, \cdots, z_k)$ and $p_2(z_1, \cdots, z_k)$ are not identical for almost all $(z_1, \cdots, z_k)$, where

$$p_i(z_1, \cdots, z_k) = \int_{-\infty}^{\infty} \prod_{j=1}^{k} g^*(z_j - \beta)\, dH_i(\beta). \tag{3.8}$$

Let g^{**} be the density function of $\sum_1^k Z_j/k$ when the Z_j are independent random variables with common density g^*. The above assertion is then implied by the assertion that the function

$$q(r) = \int_{-\infty}^{\infty} g^{**}(r - \beta)\, dH(\beta) \tag{3.9}$$

uniquely determines the df H. But if A, B, C are the characteristic functions of q, g^{**}, H, respectively, then $B(t) \neq 0$ for almost all t by (3.3) (e) and hence $C(t)$ is determined for those t for which $B(t) \neq 0$ by $C(t) = A(t)/B(t)$ and elsewhere by continuity. Thus, Assumption 4 is verified.

It remains to verify Assumption 5. Since $f(x \mid \theta, G)$ is uniformly bounded in x, θ, G, Assumption 5 will clearly be satisfied if

$$E \log f(X_1 \mid \theta_0, G_0) > -\infty. \tag{3.10}$$

Since the left side of (3.10) does not depend on θ_0, we may assume $\theta_0 = 0$. By (3.3) (d) and (3.5), we have

$$\begin{aligned} E[\log \mid X_{11} \mid]^+ &= E\left[\log \frac{|X_{11}|}{\alpha_1} + \log \alpha_1\right]^+ \\ &\leqq E\left[\log \frac{|X_{11}|}{\alpha_1}\right]^+ + E[\log \alpha_1]^+ < \infty; \end{aligned} \tag{3.11}$$

equation (3.10) is a consequence of (3.11) and the lemma at the end of Section 2.

This completes our verification of the fact that the assumptions of Section 2 are implied by (3.3) and (3.5).

Example 1b. We now assume $k > 1$. A is the set of all positive α, while Γ is the set of all df's G on A satisfying

$$\int_0^{\infty} |\log \alpha \mid dG(\alpha) < \infty. \tag{3.12}$$

We assume that g satisfies (3.3) (some alterations could be made here but, for the sake of brevity, we forego making them) and also that

$$\text{(3.13)}\qquad \begin{aligned} &\text{(a)}\quad \lim_{|x|\to\infty} xg(x) = 0;\\ &\text{(b)}\quad \sup_{x_1}\,[\min_{r<j} |x_{1r} - x_{1j}|]^k \prod_{j=1}^{k} g(x_{1j}) < \infty. \end{aligned}$$

Assumption (3.13) is easily verified, for example, in cases (3.2) and (3.4).

We now verify the assumptions of Section 2. We define $f(x \mid \theta, \alpha) = 0$ whenever $\theta = \pm\infty$ or $\alpha = 0$ or ∞; $f(x \mid \theta, G)$ is then defined by (2.1) for $(\theta, G) \,\varepsilon\, \bar{\Omega} \times \bar{\Gamma}$. Assumptions 1, 3, and 4 are verified exactly as in Example 1a. In verifying Assumption 2, we may follow the demonstration of Example 1a, noting only that the h_ϵ of (3.3) (b) may (because of (3.13) (a)) clearly be assumed to satisfy $\lim_{|x|\to\infty} xh_\epsilon(x) = 0$, so that for every x none of whose components is θ^*,

$$\text{(3.14)}\qquad \lim_{\substack{i\to\infty\\ \alpha\to 0}} h_\epsilon(x \mid \theta_i, \alpha) = 0;$$

thus, for almost all (μ) x, the Helly-Bray theorem may still be used at the last step of (3.7), no difficulty being caused by the possibility that $\lim\inf_{i\to\infty} G_i(0) < G^*(0)$.

It remains to verify Assumption 5. Now, $f(x \mid \theta, G)$ is no longer uniformly bounded as it was in Example 1a. However, by (3.13) (b), there is a constant B such that, for all $x_1 = (x_{11}, \cdots, x_{1k})$ none of whose components are equal, every $\theta \,\varepsilon\, \Omega$, and every $\alpha \,\varepsilon\, A$,

$$\text{(3.15)}\qquad \begin{aligned} f(x_1 \mid \theta, \alpha) &= [\min_{r<s} |x_{1r} - x_{1s}|]^{-k} \left\{ [\min_{r<s} |y_{1r} - y_{1s}|]^k \prod_{j=1}^{k} g(y_j) \right\}\\ &\leqq B[\min_{r<s} |x_{1r} - x_{1s}|]^{-k}, \end{aligned}$$

where $y_{1r} = (x_{1r} - \theta)/\alpha$. Hence, for almost all x_1,

$$\text{(3.16)}\qquad \begin{aligned} \sup_{\substack{\theta\varepsilon\Omega\\ \alpha\varepsilon A}} \log f(x_1 \mid \theta, \alpha) &\leqq \log B + k \max_{r<s} \log [1/|x_{1r} - x_{1s}|]\\ &\leqq \log B + k \sum_{\substack{r,s\\ r<s}} [\log (1/|x_{1r} - x_{1s}|)]^+. \end{aligned}$$

Now, by (3.3) (a), there is a value B' such that $g(z) \leqq B'$ for all z. Hence, by (3.12), B_1 denoting a finite constant, we have

$$\text{(3.17)}\qquad \begin{aligned} E[\log(1/|X_{11} - X_{12}|)]^+ &\leqq E[\log 1/\alpha_1]^+ + E[\log(\alpha_1/|X_{11} - X_{12}|)]^+\\ &\leqq B_1 - 2\int_{-\infty}^{\infty} g(z_2) \int_{z_2}^{z_2+1} B' \log(z_1 - z_2)\, dz_1\, dz_2\\ &= B_1 + 2B' < \infty. \end{aligned}$$

From (3.16) and (3.17), we obtain

$$E \sup_{\gamma \in \bar{\Omega} \times \bar{\Gamma}} \log f(X_1 \mid \gamma) < \infty. \tag{3.18}$$

Assumption 5 is a consequence of (3.18) and of (3.10), the latter of which is proved exactly as in Example 1a. This completes the verification of the assumptions of Section 2 in Example 1b.

The discrete analogue of Example 1 can be carried out similarly by letting x, θ, α take on only rational values; this is, however, of less practical importance. The multivariate extension of Example 1 (X_{ij} a vector) may also be carried out similarly.

4. Example 2. The straight line with both variables subject to error.

In this section we shall treat the case $k = 1$ of fitting a straight line with both variables subject to normal error, a famous problem with a long history.

We consider a system $\{(X_{i1}, X_{i2})\}, i = 1, 2, \cdots$, of independent chance 2-vectors (the two components X_{i1}, X_{i2} need not be independent for fixed i). We have $\theta = (\theta_1, \theta_2)$, Ω the entire plane, $\theta_0 = (\theta_{10}, \theta_{20})$, A the entire line. Γ is the totality of all non-normal (univariate) distributions G (a chance variable which is constant with probability one is to be considered normally distributed with variance zero) which satisfy

$$\int (\log \mid \alpha \mid)^+ dG(\alpha) < \infty.$$

It is known to the statistician that

$$X_{i1} = \alpha_i + u_i,$$

$$X_{i2} = \theta_{10} + \theta_{20}\alpha_i + v_i,$$

where (u_i, v_i) are jointly normally distributed chance variables with means zero, each pair (u_i, v_i) distributed independently of every other pair and of the independent chance variables $\{\alpha_i\}$, with a common covariance matrix which is unknown to the statistician.

It is known (see [10]) that the distribution of (X_{i1}, X_{i2}) then determines θ_0 uniquely, but in general not G_0, the "true" df of α_i, or the "true" covariance matrix

$$\begin{Vmatrix} d_{11}^0 & d_{12}^0 \\ d_{12}^0 & d_{22}^0 \end{Vmatrix}$$

of (u_i, v_i). However, a "canonical" complex is determined. (See [4].)

Complete the spaces Ω, A, and Γ to obtain $\bar{\Omega}$, $\bar{A}$ and $\bar{\Gamma}$. The space $\bar{\Gamma}$ contains all normal distributions on A, but this will cause us no trouble in estimating θ_0, as we shall soon see.

Let D be the space of all triples (d_{11}, d_{12}, d_{22}) such that

$$d_{11} \geqq \lambda_{11} > 0, \qquad d_{22} \geqq \lambda_{22} > 0,$$
$$d_{11} d_{22} - d_{12}^2 \geqq \lambda_{12} > 0,$$

where $\lambda_{11}, \lambda_{12}, \lambda_{22}$, are given positive numbers. (This will be discussed further below.) We define a metric in D in the same way that one is defined on Ω. Let $\bar{D}$ be the completed space. We shall assume that the "true" triple $d_{11}^0, d_{12}^0, d_{22}^0$ is in D.

The place of $\bar{\Omega} \times \bar{\Gamma}$ in Section 2 and in Example 1 will now be taken by $\bar{\Omega} \times \bar{\Gamma} \times \bar{D}$. We therefore define

$$\gamma = (\theta_1, \theta_2, G, d_{11}, d_{12}, d_{22})$$

as the generic point in $\bar{\Omega} \times \bar{\Gamma} \times \bar{D}$.

Let $f(x_1, x_2 \mid \theta_1, \theta_2, \alpha, d_{11}, d_{12}, d_{22})$ be the joint density function of (X_{i1}, X_{i2}) when $\theta = (\theta_1, \theta_2)$, $\alpha_i = \alpha$, and the covariance matrix of (u_i, v_i) is

$$\begin{Bmatrix} d_{11} & d_{12} \\ d_{12} & d_{22} \end{Bmatrix}$$

(μ is Lebesgue measure in the plane). If, in the above, θ is in $\bar{\Omega} - \Omega$ or α is in $\bar{A} - A$ or (d_{11}, d_{12}, d_{22}) is in $\bar{D} - D$, we define f to be zero. Finally we define

$$f(x_1, x_2 \mid \gamma) = \int_{\bar{A}} f(x_1, x_2 \mid \theta_1, \theta_2, \alpha, d_{11}, d_{12}, d_{12})\, dG(\alpha).$$

It is known ([10] and [4]) that all γ in the same canonical class, and only such, define the same $f(x_1, x_2 \mid \gamma)$ (of course, to within a set of μ-measure zero). Two members of the same canonical class have the same $\theta = (\theta_1, \theta_2)$ but different G's and d_{ij}'s. We shall estimate only θ_0. For an estimator of the entire canonical complex by the minimum distance method under necessary assumptions only, see [4].[5] In Section 5 below will be found an explanation of why the entire canonical complex cannot be estimated by the m.l. method.

From the definition of $f(x_1, x_2 \mid \gamma)$ it follows immediately that Assumptions 1, 2, and 3 of Section 2 are satisfied. Since we are estimating only θ_0, it is sufficient to verify Assumption 4 only for θ_0 and $\theta^* \neq \theta_0$, i.e., if we write the γ_0 and γ_1 of (2.4) as

$$\gamma_0 = (\theta_{10}, \theta_{20}, G_0, d_{11}^0, d_{12}^0, d_{22}^0),$$
$$\gamma_1 = (\theta_1^*, \theta_2^*, G_1, d_{11}, d_{12}, d_{22}),$$

only $\theta_0 = (\theta_{10}, \theta_{20})$ has to be different from the corresponding $\theta^* = (\theta_1^*, \theta_2^*)$. Now we know that G_0 is in Γ, hence is not normal and assigns probability one to A. If G_1 is also in Γ then Assumption 4 follows at once from the results of

[5] See footnote 4.

Reiersøl [10] or from [11]. If G_1 assigns probability less than one to A, $f(x \mid \gamma_1)$ assigns probability less than one to the Euclidean plane of (x_1, x_2). If G_1 is normal and assigns probability one to A, then (X_{i1}, X_{i2}) are jointly normal under γ_1, but not under γ_0. Thus Assumption 4 is always satisfied.

To verify Assumption 5 we proceed essentially as in Example 1, and use the lemma at the end of Section 2. Assumption 5 is satisfied if

$$E \log f(X_{i1}, X_{i2} \mid \gamma_0) > -\infty.$$

By the lemma this will follow if we prove

$$E\{\log |X_{ij}|\}^+ < \infty$$

for $j = 1, 2$. Now

$$\begin{aligned} E\{\log |X_{i1}|\}^+ &\leqq E\{\log [\,|X_{i1} - \alpha_i| + |\alpha_i|\,]\}^+ \\ &\leqq E\{\log [\,|X_{i1} - \alpha_i| + 1]\} + E\{\log |\alpha_i|\}^+ \\ &= E\{\log [\,|u_i| + 1]\} + E\{\log |\alpha_i|\}^+ \\ &< \infty. \end{aligned}$$

Similarly,

$$\begin{aligned} E\{\log |X_{i2}|\}^+ &\leqq E\{\log [\,|X_{i2} - \theta_{10} - \theta_{20}\alpha_i| + |\theta_{10} + \theta_{20}\alpha_i|\,]\}^+ \\ &\leqq E \log [\,|X_{i2} - \theta_{10} - \theta_{20}\alpha_i| + 1] + E\{\log |\theta_{10} + \theta_{20}\alpha_i|\}^+ \\ &\leqq E \log [\,|v_i| + 1] + \{\log |\theta_{10}|\}^+ + \mathrm{E} \log [1 + |\theta_{20}\alpha_i|\,] \\ &< \infty. \end{aligned}$$

Thus we have shown, under our assumptions on Γ and D, that Assumptions 1 through 5 of Section 2 are satisfied, so that the m.l. estimator of θ_0 converges strongly to θ_0 as $n \to \infty$.

The assumption on D (that d_{11}, d_{22}, and $d_{11} d_{22} - d_{12}^2$ are bounded away from zero) cannot be entirely dispensed with. For if D consists of all triples for which d_{11}, d_{22}, and $d_{11} d_{22} - d_{12}^2$ are positive, if S_n is the sample df of $x_{11}, \cdots, x_{n1}$, and if $\hat{\gamma}_\epsilon$ is the complex $(0, 0, S_n, \epsilon, 0, \sum_1^n x_{i2}^2)$, then it is easily verified that $\lim_{\epsilon \to 0} L((x_{11}, x_{12}), \cdots, (x_{n1}, x_{n2}) \mid \hat{\gamma}_\epsilon) = \infty$; thus, no m.l. or modified m.l. estimator exists, and there are neighborhood m.l. estimators which are not consistent (for θ).

The case $k > 1$ is much simpler to treat than the above case. It is easy to see that then the covariance matrix of (u_i, v_i) is uniquely determined, and from this it follows easily that the whole complex γ is uniquely determined. The problem can be treated in a manner similar to that of Examples 1b and 3b.

The problem of this section with the distribution of (u_i, v_i) other than normal may also be treated by the m.l. method, as in Examples 1 and 3. The last paragraph of Section 3 applies also to the present example.

5. Example 3. Structural scale parameter, incidental location parameter. We consider here the case of a structural scale parameter and an incidental location parameter; this reverses the roles of the two parameters of Example 1. Thus, we suppose μ to be Lebesgue measure on R^k and

$$f(x_i \mid \theta, \alpha) = \frac{1}{\theta^k} \prod_{j=1}^{k} g\left(\frac{x_{ij} - \alpha}{\theta}\right). \tag{5.1}$$

The cases $k = 1$ and $k > 1$ are essentially different, and we consider them separately.

Example 3a. *The case* $k = 1$. This example is another simple one where no m.l. estimator is consistent, and also shows, in a simpler setting, why in Example 2 the m.l. method was incapable of estimating the components of the canonical complex other than θ. Since Example 3a is intended to illustrate the *failure* of the m.l. method in certain situations, we shall for simplicity assume that g is given by (3.2); examples with other g (e.g., (3.4)) may be treated similarly. Ω may be taken to be any specified set of positive numbers containing more than one point; for the sake of brevity, we assume that Ω contains its greatest lower bound c (say) (and thus, that $c > 0$), but it is easy to carry through a similar demonstration (with modified or neighborhood m.l. estimators in place of m.l. estimators) when $c \not\in \Omega$. Γ is taken to be the class of all df's G on the real line for which $\int [\log |\alpha|]^+ dG(\alpha) < \infty$ and such that G has no normal component; i.e., no G in Γ can be represented as the convolution of two df's, one of which is normal with positive variance. (Γ may be further restricted, e.g., by the condition that for each G there is a bounded set outside of which G has no variation.)

All assumptions of Section 2 are easily verified except Assumption 4; there is no difficulty of identifiability in $\Omega \times \Gamma$, but there clearly *is* in $\bar{\Omega} \times \bar{\Gamma}$. Consider now the expression

$$\prod_{i=1}^{n} \int_{-\infty}^{\infty} \frac{1}{(2\pi)^{\frac{1}{2}} c} e^{-(1/2c^2)(x_i - s)^2} \, dM(s). \tag{5.2}$$

It is clear that the maximum of (5.2) with respect to M can be achieved only by an M which assigns probability one to the interval (min $(x_1, \cdots, x_n)$, max $(x_1, \cdots, x_n)$) and hence which has no normal component. This discussion of the expression (5.2) shows that, for every n, any m.l. estimator (the fact that the maximum is attained is easily verified) of (θ, G) subject to our assumption $\theta \geq c$ *always* estimates θ to be c. Thus, no m.l. estimator of (θ, G) is consistent (unless $\theta = c$).

To summarize the result of this example, then, the m.l. method is incapable of estimating consistently the normal component of the df of the sequence $\{X_i\}$ of independent identically distributed random variables because, in every neighborhood of a point (θ, G) with $\theta > c$, there are points with $\theta = c$ (and for which the likelihood is larger).

Let N_σ denote the normal df with mean 0 and variance σ^2, and let $H_1 * H_2$ denote the convolution of the two df's H_1 and H_2.

It is interesting to note that, without any assumption on Γ (except the necessary identifiability assumption that G_0 has no normal component), the minimum distance method is capable of estimating (θ_0, G_0) consistently [4]. The difficulty noted above for the m.l. estimator is avoided by noting the *rate* at which the sample df S_n converges to the df $N_{\theta_0} * G_0$ of X_1 and estimating θ_0 *not* by the value t for which $N_t * H$ is closest to S_n for some normal-free H (this would encounter the same difficulty as the m.l. estimator, since, the smaller t is taken, the closer can $N_t * H$ be made to approximate S_n), but as the *largest* value for which there is an $N_t * H$ suitably close to S_n ("suitably" is connected with the rate mentioned above.)

One could modify the example as considered above so as not to require G_0 to have no normal component, and try then to escape the difficulty of nonidentifiability by asking for an estimator of the canonical representation of (θ, G), this representation consisting of two df's, the normal and nonnormal components of $N_\theta * G$. The previous demonstration then shows that no m.l. estimator of the canonical representation estimates it consistently, and thus illustrates, in a simpler setting than that of Example 2 with $k = 1$, why the m.l. estimator could not be used in Example 2 to estimate the components of the canonical complex other than θ.

We remark that it is easy in many cases such as that of the present example to prove a result such as the one that, (t_n, H_n) denoting an m.l. estimator of (θ_0, G_0) after n observations, the df $N_{t_n} * H_n$ converges w.p.1 to $N_{\theta_0} * G_0$ as $n \to \infty$. Such a property is much weaker than that of the consistency of the m.l. estimator, and does not lie much deeper than the Glivenko-Cantelli theorem.

Example 3b. *The case* $k > 1$. We assume f to be given by (5.1) with $k > 1$. The function g is assumed to satisfy the conditions (a), (b), (c), and (d) of (3.3); conditions (a) and (b) of (3.13), and

$$\int_{-\infty}^{\infty} e^{itx} g(x)\, dx \neq 0 \text{ for almost all real } t. \tag{5.3}$$

(As in Example 1a, weaker conditions could be assumed here if we assumed also $\theta \geqq c > 0$; the above conditions are analogous to those of Example 1b.) Thus, for example, (3.2) and (3.4) satisfy these assumptions. Ω is the set of all values $\theta > 0$, while A is the real line and Γ is the set of all df's G on A for which

$$\int_{-\infty}^{\infty} [\log |\alpha|]^+ \, dG(\alpha) < \infty. \tag{5.4}$$

We now verify the assumptions of Section 2. We define $f(x \mid \theta, \alpha) = 0$ when $\theta = 0$ or ∞ or $\alpha = \pm\infty$. The definition of $f(x \mid \theta, G)$ for $(\theta, G) \in \bar{\Omega} \times \bar{\Gamma}$ is then given by (2.1). Assumptions 1, 2, and 3 are now verified as in Example 1b, interchanging the roles of θ and α in the latter (including the definition of $h_s(x \mid \theta, \alpha)$) and noting that (3.14)) still holds for almost all (μ) x, with this interchange. In order to verify Assumption 4, we note, for $(\theta, G) \varepsilon \Omega \times \Gamma$, that θ is determined by the density function of $X_{11} - X_{12}$ and that, for almost all real t, the characteristic function of G is then given by $B(t/k, \cdots, t/k)/[C(\theta t/k)]^k$ where $B(t_1, \cdots, t_k)$ is the characteristic function of $X_{11}, \cdots, X_{1k}$ and $C(t)$ is the characteristic function of g.

Finally, Assumption 5 is a consequence of equation (3.18), which is proved in the present case exactly as in Example 1b (using (3.15), (3.16), and (3.17), with α_1 replaced by θ in the latter), and of equation (3.10) (with f defined by (5.1)). Equation (3.10) in the present example is a consequence of the lemma at the end of Section 2 and of

$$(5.5) \quad \begin{aligned} E\{\log |X_{11}|\}^+ &\leqq E\{\log [\,|X_{11} - \alpha_1| + |\alpha_1|\,]\}^+ \\ &\leqq E \log [\,|X_{11} - \alpha_1| + 1] + E\{\log |\alpha_1|\}^+ < \infty. \end{aligned}$$

This completes the verification of the assumptions of Section 2 in Example 3b. The last paragraph of Section 3 applies also to the present example.

6. The Classical case. Miscellaneous remarks. It does not seem to have been noticed in the literature that a simple device exists for proving consistency of the m.l. estimator in certain cases where the regularity conditions of published proofs fail. This device may be used in the case studied in the present paper (to prove consistency in the examples under weaker conditions than those stated) as well as in the classical parametric case. We now illustrate this device in an example of the latter case.

When Γ consists only of distributions which give probability one to a single point, the problem of the present paper becomes the classical problem of estimating the parameter θ and the parameter σ (say) to which G_0 gives probability one. If θ may be any real value and σ any positive value, then the function $(1/\sigma)g((x - \theta)/\sigma)$ of Section 3 does not satisfy Wald's integrability condition or the corresponding condition of any other published proof; one verifies easily that (2.5) is not satisfied for any point in the (θ, σ) half-plane which lies on the line $\sigma = 0$. (The line $\sigma = 0$ has to be added to Ω in the process of forming $\bar{\Omega}$. As in earlier sections, we assume the true σ_0 to be >0.) Often, however, when the observations are considered as if they were taken in groups of two or more, the integrability condition will be satisfied. Such is the case, for example, with the density function

$$\frac{1}{\pi} \frac{\sigma}{\sigma^2 + (x_1 - \theta)^2} \cdot \frac{1}{\pi} \frac{\sigma}{\sigma^2 + (x_2 - \theta)^2}$$

and the normal density function

$$\frac{1}{(2\pi)^{\frac{1}{2}}\sigma} \exp\left\{-\frac{1}{2}\frac{(x_1 - \theta)^2}{\sigma^2}\right\} \cdot \frac{1}{(2\pi)^{\frac{1}{2}}\sigma} \exp\left\{-\frac{1}{2}\frac{(x_2 - \theta)^2}{\sigma^2}\right\}.$$

(Of course the estimator from the normal distribution is known to be consistent, but this does not alter the validity of the example.) In such cases it follows from Wald's proof [5] (using the compactification device used above) or from the result of Example 1b that the m.l. sequence of estimators considered only after an even number of observations in consistent, and from this it is an easy matter to show that the entire m.l. sequence of estimators is consistent.

We shall now discuss the integrability conditions of [5] and of the present paper. The integrability condition (2.5) involves the difference of two logarithms; the integrability condition as given by Wald in [5] requires the finiteness of the expected value of each logarithm. The form (2.5) is satisfied whenever the condition of [5] is, and has one other advantage which we shall now illustrate by an example. Let the observed chance variable X have density function $\theta e^{-\theta x}$ for $x > 0$ and zero elsewhere. The parameter θ is unknown and Ω is the positive half-line, so that $\bar{\Omega}$ contains the point $\theta = 0$. One verifies easily that the condition of [5], and hence (2.5), are satisfied. Suppose now that, instead of observing X, one observes $Y = e^{(e^X)}$, which therefore has the density function

$$\frac{\theta}{x} (\log x)^{-\theta-1}$$

for $x > e$, and zero elsewhere. One readily verifies that, when $\theta < 1$,

$$E \log \left\{ \frac{\theta}{Y} (\log Y)^{-\theta-1} \right\} = -\infty,$$

so that the condition of [5] is not satisfied when $0 < \theta_0 < 1$. Thus, whether the condition of [5] is satisfied depends in this instance on whether one observes X or Y; this is an unfortunate circumstance, since the estimation problems are in simple correspondence. On the other hand, condition (2.5) is invariant under one-to-one transformation of the observed chance variable because the numerator and denominator of the ratio in (2.5) are multiplied by the same Jacobian. (In particular, therefore, the chance variable Y satisfies (2.5).)

Without resorting to artificial or pathologic examples as is sometimes done in the literature, it is still easy to give instances where the m.l. method does not give consistent estimators in the classical parametric case. For example, consider the density function

$$\frac{1}{2(2\pi)^{\frac{1}{2}}} \exp\{-\tfrac{1}{2}(x-\theta)^2\} + \frac{1}{2(2\pi)^{\frac{1}{2}}\sigma} \exp\left\{-\frac{1}{2}\frac{(x-\theta)^2}{\sigma^2}\right\}$$

of the sequence of independent and identically distributed chance variables $X_1, X_2, \cdots$. Here θ and σ are the unknown parameters, θ may be any real number and σ any positive number. It is easy to see that the supremum of the likelihood function is almost always infinite, no m.l. or modified m.l. estimator exists, and there are neighborhood m.l. estimators (where θ_0 is estimated by X_1, say) which are obviously not consistent.

REFERENCES

[1] J. NEYMAN AND E. L. SCOTT, "Consistent estimates based on partially consistent observations" *Econometrica*, Vol. 16 (1948), pp. 1–32.

[2] A. Wald, "The fitting of straight lines if both variables are subject to error," *Ann. Math. Stat.*, Vol. 11 (1940), pp. 284–300.
[3] A. Dvoretzky, J. Kiefer and J. Wolfowitz, "Asymptotic minimax character of the sample distribution function and of the classical multinomial estimator," *Ann. Math. Stat.*, Vol. 27 (1956), pp. 642–669.
[4] J. Wolfowitz, "Estimation of the components of stochastic structures," *Proc. Nat. Acad. Sci., U.S.A.*, Vol. 40, No. 7 (1954), pp. 602–606.
[5] A. Wald, "Note on the consistency of the maximum likelihood estimate," *Ann. Math. Stat.*, Vol. 20 (1949), pp. 595–601.
[6] A. Wald, "Estimation of a parameter when the number of unknown parameters increases indefinitely with the number of observations," *Ann. Math. Stat.*, Vol. 19 (1948), pp. 220–227.
[7] J. Neyman, "Existence of consistent estimates of the directional parameter in a linear structural relation between two variables," *Ann. Math. Stat.* Vol. 22 (1951), pp. 497–512.
[8] J. Neyman and E. L. Scott, "On certain methods of estimating the linear structural relation between two variables," *Ann. Math. Stat.* Vol. 22 (1951), pp. 352–361.
[9] J. Wolfowitz, "Estimation by the minimum distance method," *Ann. Inst. Stat. Math.* Tokyo, Vol. 5 (1953), pp. 9–23.
[10] O. Reiersøl, "Identifiability of a linear relation between variables which are subject to error," *Econometrica*, Vol. 18 (1950), pp. 375–389.
[11] J. Wolfowitz, "Consistent estimators of the parameters of a linear structural relation," *Skand. Aktuarietids.* (1952), pp. 132–151.
[12] J. Wolfowitz, "On Wald's proof of the consistency of the maximum likelihood estimate," *Ann. Math. Stat.*, Vol. 20 (1949), pp. 602–603.

Reprinted from THE ANNALS OF MATHEMATICAL STATISTICS
Vol. 28, No. 1, March, 1957

THE MINIMUM DISTANCE METHOD[1]

BY J. WOLFOWITZ

Cornell University

1. Summary and Introduction. The present paper gives the formal statements and proofs of the results illustrated in [1]. In a series of papers ([2], [3], [4]) the present author has been developing the minimum distance method for obtaining strongly consistent estimators (i.e., estimators which converge with probability one). The method of the present paper is much superior, in simplicity and generality of application, to the methods used in the papers [2] and [4] cited above. Roughly speaking, the present paper can be summarized by saying that, in many stochastic structures where the distribution function (d.f.) depends continuously upon the parameters and d.f.'s of the chance variables in the structure, those parameters and d.f.'s which are identified (uniquely determined by the d.f. of the structure) can be strongly consistently estimated by the minimum distance method of the present paper. Since identification is obviously a necessary condition for estimation by *any* method, it follows that, in many actual statistical problems, identification implies estimatability by the method of the present paper.

Thus problems of long standing like that of Section 5 below are easily solved. For this problem the whole canonical complex (Section 6 below; see [1]) has never, to the author's knowledge, been estimated by any other method. The directional parameter of the structure of Section 4 seems to be here estimated for the first time.

As the identification problem is solved for additional structures it will be possible to apply the minimum distance method. The proofs in the present paper are of the simplest and most elementary sort.

In Section 8 we treat a problem in estimation for nonparametric stochastic difference equations. Here the observed chance variables are not independent, but the minimum distance method is still applicable. The treatment is incomparably simpler than that of [4], where this and several other such problems are treated. The present method can be applied to the other problems as well.

Application of the present method is routine in each problem as soon as the identification question is disposed of. In this respect it compares favorably with the method of [4], whose application was far from routine.

As we have emphasized in [1], the present method can be applied with very many definitions of distance (this is also true of the earlier versions of the minimum distance method). The definition used in the present paper has the convenience of making a certain space conditionally compact and thus eliminating the need for certain circumlocutions. Since no reason is known at present for

Received January 31, 1956.

[1] Research under contract with the Office of Naval Research.

preferring one definition of distance to another we have adopted a convenient definition. It is a problem of great interest to decide which, if any, definition of distance yields estimators preferable in some sense. The definition of distance used in this paper was employed in [9].

As the problem is formulated in Section 2 below (see especially equation (2.1), the "observed" chance variables $\{X_i\}$ are known functions (right members of (2.1)) of the "unobservable" chance variables $\{Y_i\}$ and of the unknown constants $\{\theta_i\}$. In the problems treated in [3], [9], and [11], it is the distribution of the observed chance variables which is a known function of unobservable chance variables and of unknown constants, and not the observed chance variables themselves. However, the latter problems can easily be put in the same form as the former problems. Moreover, in the method described below the values of the observed chance variables are used only to estimate the distribution function of the observed chance variables (by means of the empiric distribution function). Consequently there is no difference whatever in the treatment of the problems by the minimum distance method, no matter how the problems are formulated.

The unobservable chance variables $\{Y_i\}$ correspond to what in [11] are called "incidental parameters"; the unknown constants $\{\theta_i\}$ are called in [11] "structural parameters". In [9] there is a discussion of the fact that in some problems treated in the literature the incidental parameters are considered as constants and in other problems as chance variables. In contradistinction to the present paper [3] (in particular its Section 5) treats the incidental parameters as unknown constants. The fundamental idea of both papers is the same: The estimator is chosen to be such a function of the observed chance variables that the d.f. of the observed chance variables (when the estimator is put in place of the parameters and distributions being estimated) is "closest" to the empiric d.f. of the observed chance variables. The details of application are perhaps easier in the present paper; the problems are different and of interest per se.

2. The minimum distance method. Let m, m', k, k', be integers such that $0 \leqq m \leqq m'$, $0 \leqq k \leqq k'$. For $j = 1, 2, \cdots$ ad inf. let $(Y_{j1}, \cdots, Y_{jk'})$ be independent, identically distributed vector chance variables with the common d.f. G_0 which is unknown to the statistician. The constants $\theta_1, \cdots, \theta_{m'}$, are also unknown to the statistician. It is known that, for $j = 1, 2, \cdots$ ad inf.,

$$X_{ji} = t_i(Y_{j1}, \cdots, Y_{jk'}, \theta_1, \cdots, \theta_{m'}) \qquad i = 1, \cdots, h \tag{2.1}$$

where the t_i, for $i = 1, \cdots, h$, are known Borel-measurable functions of the arguments exhibited. Define the common d.f. of $(Y_{j1}, \cdots, Y_{jk})$, $j = 1, 2, \cdots$ ad. inf., by

$$G(y_1, \cdots, y_k) = G_0(y_1, \cdots, y_k, +\infty, \cdots, +\infty)$$

Let $\theta = (\theta_1, \cdots, \theta_{m'})$. Let $A = \{(\bar{\alpha}, g)\}$ be a space of couples $(\bar{\alpha}, g)$, the first member of which is a real m'-dimensional vector $(\alpha_1, \cdots, \alpha_{m'})$, and the second member of which is a k'-dimensional d.f. It is known that (θ, G_0) is in A. On A we define a metric δ as follows:

$$\delta([\bar{\alpha}_1, g_1], [\bar{\alpha}_2, g_2]) = \sum_{j=1}^{m'} |\arc\tan \alpha_{1j} - \arc\tan \alpha_{2j}| + \int |g_1(z) - g_2(z)| e^{-|z|} dz_1 \cdots dz_{k'} \tag{2.2}$$

where

$$\bar{\alpha}_i = \alpha_{i1}, \cdots, \alpha_{im'} \qquad i = 1, 2$$
$$z = z_1, \cdots, z_{k'}$$

We shall also use δ to denote a metric on any Euclidean space or on any space of d.f.'s of the same dimensionality. In that case δ is to be understood as the expression corresponding, respectively, to the first or second term of the right member of (2.2).

Our problem is to give (strongly) consistent estimators of G and $(\theta_1, \cdots, \theta_m)$, i.e., for $n = 1, 2, \cdots$ ad inf., to construct measurable functions (Q_{n1}, Q_{n2}) from hn-dimensional Euclidean space (of $X_{11}, \cdots, X_{1h}, \cdots, X_{n1}, \cdots, X_{nh}$) to A such that, *whatever be* (θ, G_0) (in A), we have, with probability one (w.p. 1), both

$$Q_{n1}^{(j)} \rightarrow \theta_j, \qquad j = 1, \cdots, m$$

where $Q_{n1}^{(j)}$ is the jth component of Q_{n1}, and

$$Q_{n2}(y_1, \cdots, y_k, +\infty, \cdots, +\infty) \rightarrow G(y_1, \cdots, y_k)$$

at every point of continuity of the latter.

Let $J(\bar{\alpha}, g)$ be the (h-dimensional) d.f. of $(X_{j1}, \cdots, X_{jh})$ when $\theta = \bar{\alpha}$ and $G_0 = g$. In this notation the generic point in h-space is suppressed since it will rarely come explicitly into play, and the emphasis is on the fact that this is a transformation from A into the space of h-dimensional d.f.'s. We shall make the following *Identification and Continuity (I.C.) Assumption*:

Let $\{\bar{\alpha}_i, g_i\}$ be any Cauchy sequence (i.e., as $i \rightarrow \infty$, $\delta[\bar{\alpha}_i, g_i], [\bar{\alpha}_{i+n}, g_{i+n}]) \rightarrow 0$ uniformly in n) such that

$$\delta(J(\bar{\alpha}_i, g_i), J(\theta, G_0)) \rightarrow 0 \tag{2.3}$$

as $i \rightarrow \infty$. Then, as $i \rightarrow \infty$,

$$\alpha_{ij} \rightarrow \theta_j, \qquad j = 1, \cdots, m \tag{2.4}$$

(α_{ij} is the jth component of $\bar{\alpha}_i$) and

$$g_i(y_1, \cdots, y_k, +\infty, \cdots, +\infty) \rightarrow G(y_1, \cdots, y_k) \tag{2.5}$$

at every point of continuity of the latter.

Let

$$C_n = \{(X_{j1}, \cdots, X_{jh}), j = 1, \cdots, n\}$$

and $F_n(C_n)$ be the empiric d.f. of C_n, i.e., an h-dimensional d.f. such that its value at $(y_1, \cdots, y_h)$ is $1/n$ times the number of elements in C_n for which $X_{ji} < y_i$, $i = 1, \cdots, h$. Let $\gamma(n)$ be any positive function of n which approaches zero as $n \to \infty$. Let $S_n(C_n) = (\theta_n^*, G_{0n}^*)$ be any function from the hn-dimensional Euclidean space of C_n to the space A which is measurable and such that

$$\delta(J(\theta_n^*, G_{0n}^*), F_n(C_n)) < \inf_{(\bar{\alpha}, g)\varepsilon A} \delta(J(\bar{\alpha}, g), F_n(C_n)) + \gamma(n) \tag{2.6}$$

$S_n(C_n)$ is a minimum distance estimator, for which the following holds:

THEOREM. *If the I.C. Assumption holds, then, with probability one (w.p. 1),*

$$\theta_{nj}^* \to \theta_j, \qquad j = 1, \cdots, m \tag{2.7}$$

(θ_{nj}^* is the jth *component of θ_n^*) and*

$$G_{0n}^*(y_1, \cdots, y_k, +\infty, \cdots, +\infty) \to G(y_1, \cdots, y_k) \tag{2.8}$$

at every continuity point of the latter.

(In view of (2.7) and (2.8) it is actually the appropriate components of $S_n(C_n)$ which could be called minimum distance estimators of the θ_j and G).

PROOF: By the Glivenko-Cantelli theorem we have that, w.p.1,

$$\delta(F_n(C_n), J(\theta, G_0)) \to 0 \tag{2.9}$$

as $n \to \infty$. Hence, w.p.1,

$$\inf_{(\bar{\alpha}, g)\varepsilon A} \delta(F_n(C_n), J(\bar{\alpha}, g)) \to 0 \tag{2.10}$$

Hence, w.p.1,

$$\delta(J(\theta_n^*, G_{0n}^*), J(\theta, G_0)) \to 0. \tag{2.11}$$

Since the space A is (sequentially) conditionally compact (with respect to the metric δ) it follows that, at every sample point and from every subsequence of $S_n(C_n)$, $n = 1, 2, \cdots$, we can select a Cauchy subsequence. For every such sequence the relation corresponding to (2.11) holds, except on a set of sample points of probability zero. When the relation corresponding to (2.11) holds, then, by the I.C. Assumption, relations corresponding to (2.7) and (2.8) hold. Thus, we have proved that, except on a set of sample points of probability zero, *every* subsequence of $S_n(C_n)$ contains a subsequence for which the equations corresponding to (2.7) and (2.8) hold. But this easily implies the theorem.

3. Discussion of the Identification and Continuity Assumption. We have seen that the proof of the strong consistency of the minimum distance estimator follows almost trivially from the I.C. Assumption. Let us now examine this assumption more carefully.

The constants $(\theta_1, \cdots, \theta_m)$ and the d.f. G, which belong to the "structure" (system) (2.1), are said to be "identified in A" if, when $(\bar{\alpha}, g)$ is in A, and

$$J(\theta, G_0) = J(\bar{\alpha}, g) \tag{3.1}$$

identically in the h arguments, then

$$\theta_i = \alpha_i \qquad i = 1, \cdots, m \tag{3.2}$$

$$G(y_1, \cdots, y_k) = g(y_1, \cdots, y_k, +\infty, \cdots, +\infty) \tag{3.3}$$

identically in $y_1, \cdots, y_k$ (of course (θ, G_0) is in A). It is obvious that identification in A is an indispensable condition for our problem of estimating consistently the constants $\theta_1, \cdots, \theta_m$ and the d.f. G, for no particular value of the sequence

$$\{(X_{j1}, \cdots, X_{jh}), j = 1, 2, \cdots\}$$

can furnish more information than the function $J(\theta, G_0)$ itself.

In most, if not all, actual statistical problems, J will be a continuous function on A, i.e., whenever

$$(\bar{\alpha}_i, g_i) \to (\bar{\alpha}, g) \text{ in } A,$$

then

$$\delta(J(\bar{\alpha}_i, g_i), J(\bar{\alpha}, g)) \to 0. \tag{3.4}$$

We shall assume that this is so in the remainder of this section. Then the following considerations will help to understand the I.C. Assumption and to furnish a convenient way of proving that it is satisfied.

Let $\bar{C}_1$ be the map of A under J, i.e.,

$$\bar{C}_1 = \{J(\bar{\alpha}, g), (\bar{\alpha}, g)\varepsilon A\}.$$

Let $\{\bar{\alpha}_i', g_i'\}$ be *any* Cauchy sequence in A which does *not* have a limit in A, and for which

$$J(\{\bar{\alpha}_i', g_i'\}) = \lim_{i\to\infty} J(\bar{\alpha}_i', g_i')$$

exists. Let $\bar{C}_2$ be the totality of all such $J(\{\bar{\alpha}_i', g_i'\})$. ($\bar{C}_1$ and $\bar{C}_2$ need not be disjoint).

The *indispensable* condition of identification[2] in A may be stated as follows: If

$$(\bar{\alpha}_i, g_i) \to (\bar{\alpha}, g) \text{ in } A \tag{3.5}$$

and

$$J(\bar{\alpha}_i, g_i) \to J(\theta, G_0) \tag{3.6}$$

then

$$\alpha_j = \theta_j, \qquad j = 1, \cdots, m \tag{3.7}$$

$$g(y_1, \cdots, y_k, +\infty, \cdots, +\infty) = G(y_1, \cdots, y_k) \tag{3.8}$$

identically in $y_1, \cdots, y_k$. If *either* of the two following conditions is *also* met the I.C. Assumption is fulfilled:

---

[2] We remind the reader that J is assumed to be continuous on A.

A) $J(\theta, G_0)$ is not a member of $\bar{C}_2$

B) If $J(\theta, G_0) = J(\{\bar{\alpha}_i', g_i'\})$ is[3] in $\bar{C}_2$, then $\alpha_{ij}' \to \theta_j$, $j = 1, \cdots, m$, and $g_i'(y_1, \cdots, y_k, +\infty, \cdots, +\infty) \to G(y_1, \cdots, y_k)$ at every point of continuity of the latter.

Thus the I.C. Assumption is, for most A to be encountered in actual problems where J is continuous, not much more onerous than the indispensable identification condition. In the important examples to be discussed below condition A, and hence the I.C. Assumption, will hold.

4. A linear relationship between two chance variables subject to independent errors. We illustrate the last two sections by application to the following very important structure: Suppose it is known to the statistician that, for $j = 1, 2, \cdots$, ad inf.,

$$X_{j1} = \xi_j + v_{j1} \tag{4.1}$$

$$X_{j2} = \alpha + \beta\xi_j + v_{j2} \tag{4.2}$$

where α and β are constants[4] unknown to the statistician, and $\{v_{j1}\}$, $\{v_{j2}\}$, and $\{\xi_j\}$ are sequences of independent, identically distributed chance variables, with respective d.f.'s L_1, L_2, L_3, say, which are unknown to the statistician. The different sequences are known to be independent of each other. We shall consider first the problem of estimating β.

Let d be the generic designation of a complex

$$\{a, b, p_1, p_2, p_3\}$$

whose first two elements are real numbers, and whose last three elements are one-dimensional d.f.'s. Let

$$d_0 = \{\alpha, \beta, L_1, L_2, L_3\}$$

The symbol $J(d)$ will have the same meaning as in Section 2.

We shall assume that A is the totality of all d's such that g_3 is not a normal d.f.; for the purposes of this definition and elsewhere in this paper a d.f. which assigns probability one to a single point is to be considered normal (with variance zero). It was proved by Reiersol [5] that β is identified in A; actually an examination of his proof (especially equation (4.5)) shows that Reiersol proved somewhat more, namely that, if

$$J(d_0) = J([a^0, b^0, p_1^0, p_2^0, p_3^0]) \tag{4.3}$$

---

[3] The preceding symbol has been defined in the first displayed equation which precedes (3.5).

[4] This formulation does not include the case when the regression line is parallel to the axis of X_2, i.e., when $\beta = \infty$; in that case $X_1 = \text{constant} + v_1$, $X_2 = \xi + v_2$. This omission is made only in the interest of simplicity. We invite the reader to verify that the formulation where this case is also a possibility can be treated by the methods of the present paper in exactly the same way as this is done in Sections 4, 5, and 6.

where a^0 and b^0 are finite, p_1^0 and p_2^0 are d.f.'s, but the d.f. p_3^0 is not *required* to be not normal, then $b^0 = \beta$ and p_3^0 *must* be not normal. Thus $(a^0, b^0, p_1^0, p_2^0, p_3^0)$ *is* in A.

It is obvious that $J(d)$ is a continuous function of d (on A).

Let $\{d_i = (a_i, b_i, p_{i1}, p_{i2}, p_{i3}), i = 1, 2, \cdots\}$ be *any* Cauchy sequence in A which does *not* have a limit in A and which is such that

$$J(\{d_i\}) = \lim_{i\to\infty} J(d_i) \tag{4.4}$$

exists. Let $d^* = (a^*, b^*, p_1^*, p_2^*, p_3^*)$ be such that $\delta(d_i, d^*) \to 0$. Then at least one of the following four properties must hold:

1) p_3^* is a normal d.f., a^* and b^* are both finite, and p_1^* and p_2^* have variation one.

2) p_3^* is a normal d.f., either $a^* = \pm\infty$ or $b^* = \pm\infty$ or both, and p_1^* and p_2^* have variation one.

3) p_3^* is a non-normal d.f. (therefore has variation one), either $a^* = \pm\infty$ or $b^* = \pm\infty$ or both, and both p_1^* and p_2^* have variation one.

4) the variation of at least one of p_1^*, p_2^*, p_3^* is less than one.

We shall now show that the I.C. Assumption is satisfied in the present problem. We shall try first to show that $J(d_0)$ is not in $\bar{C}_2$; we will be able to to achieve this except for one obstacle which we will treat somewhat differently. Suppose then that $J(d_0) = J(\{d_i\})$ were[5] in $\bar{C}_2$. Then d^* could not have the first of the above properties, because of Reiersol's result cited above. If d^* had property 2 above then either the variation of $J(\{d_i\})$ would be less than one (which cannot be true of $J(d_0)$) or else $J(\{d_i\})$ is the same as $J(a'', b'', p_1'', p_2'', p_3''])$, where a'' and b'' are finite, p_3'' assigns probability one to a single point and is therefore normal, and p_1'' and p_2'' have variation one (which cannot be true of $J(d_0)$ because of Reiersol's result cited above). If d^* had property 3 above then $J(\{d_i\})$ would be of variation less than one, which cannot be true of $J(d_0)$. If d^* had property 4 above then either $J(\{d_i\})$ has variation less than one (in which case $J(d_0)$ is not in $\bar{C}_2$) or else $J(\{d_i\}) = J(d_0)$ is in $\bar{C}_2$! To see how this can happen we note that, if z and z' are any real numbers,

$$\xi_j + v_{j1} = (\xi_j + z) + (v_{j1} - z)$$

$$\alpha + \beta\xi_j + v_{j2} = (\alpha + z' - \beta z) + \beta(\xi_j + z) + (v_{j2} - z').$$

If either z or z' or both approach $\pm\infty$ the variations of some or all of p_1^*, p_2^*, p_3^* will be zero. This difficulty is easily overcome. One can show that in this case condition B of Section 3 holds. A method which is essentially the same but formally simpler is the following: Without changing the problem or any loss of generality we can reduce the set A so as to prevent the occurrence of this case. We simply define A as the totality of all d's which, in addition to the conditions previously imposed, satisfy the requirement that the smallest median of both p_1

[5] The preceding symbol was defined in (4.4).

and p_2 is zero. It is clear that the estimation of β is not affected by this additional restriction, and that, under this restriction, $J(d_0)$ cannot be in $\bar{C}_2$. (The definition of d_0, but not the value of β, may be affected by this restriction).

It is obvious that, unless the space A is suitably reduced, the parameter α cannot be identified. In [5] Reiersol states the result that, if the space A is that subset of the originally defined $A = \{d\}$ where the d's are subject to the further restriction

$$\text{median of } p_1 = \text{median of } p_2 = 0, \tag{4.5}$$

then α is also identified on (the new) A. It seems to the writer that one must make precise which median is meant in order to make the proof of [5] go through. Either of the following conditions, for example, will permit the proof of [5] to go through:

$$p_1 \text{ and } p_2 \text{ each have zero as the unique median} \tag{4.6}$$

$$p_1 \text{ and } p_2 \text{ have zero as the smallest (largest) median.} \tag{4.7}$$

What will suffice is a condition such that, if $p_1(x)$, $p_2(x)$ are the third and fourth components of a point in A, $p_1(x + c_1)$, $p_2(x + c_2)$ cannot be the third and fourth elements of any point in A unless $c_1 = c_2 = 0$.

Suppose, for example, one adopts the restriction (4.6) above. Then α is identified on (the new) A, by the result of [5]. Let $\bar{A}$ be the totality of limit points of A which are not in A. $\bar{A}$ will include points whose third and fourth elements will not have zero as a *unique* median. In order to show, *just as before*, that $J(d_0)$ is not in $\bar{C}_2$, we need the additional result analogous to the one implicit in [5] about β and cited above, namely that, if $J(d_0) = J(\{d_i\})$, then the first element of d^* is α. However, this result does not seem to be implicit in [5] under the condition (4.6), and a stronger condition may be needed.

5. A linear relationship between two chance variables whose errors are jointly normally distributed. The following structure is a very famous one with a long history of study (see [2] and [5], for example). Let it be known to the statistician that X_{j1} and X_{j2} satisfy (4.1) and (4.2) respectively, that α and β are unknown constants,[6] that the two sequences $\{\xi_j\}$ and $\{v_{j1}, v_{j2})\}$, $j = 1, 2, \cdots$, ad inf., of independent and identically distributed chance variables are distributed independently of each other, and that the common distribution of $\{(v_{j1}, v_{j2})\}$ is normal, with zero means and covariances $\sigma_{11}(= E(v_{j1})^2)$, $\sigma_{12}(= E(v_{j1}v_{j2}))$, and $\sigma_{22}(= E(v_{j2}^2))$, unknown to the statistician. Designate the common unknown d.f. of $\{\xi_j\}$ by L.

Let m be the generic designation of a complex

$$(a, b, c_{11}, c_{12}, c_{22}, l)$$

[6] See footnote 4.

such that a, b are real numbers, c_{11}, c_{12}, and c_{22} are real numbers such that the matrix

$$\begin{Bmatrix} c_{11} & c_{12} \\ c_{12} & c_{22} \end{Bmatrix}$$

is non-negative definite, and l is a one-dimensional non-normal d.f. Let A be the totality of all m. It follows from the results of Reiersol ([5]) and the Cramér-Lévy theorem ([7], p. 52, Th. 19) that, if

$$J(\mu) = J(m^0)$$

where

$$\mu = (\alpha, \beta, \sigma_{11}, \sigma_{12}, \sigma_{22}, L),$$

$$m^0 = (a^0, b^0, c_{11}^0, c_{12}^0, c_{22}^0, l^0),$$

μ is of course in A, and m^0 satisfies all the requirements imposed on the elements of A except that l^0 is not *required* to be not normal, then l^0 *must* be not normal and $\alpha = a^0$, $\beta = b^0$.

Obviously $J(m)$ is a continuous function of m on A. We will show that condition A of Section 3 is satisfied, so that the I.C. Assumption is fulfilled, and the minimum distance estimator of α and β is strongly consistent.

Let $\bar{A}$ for the present problem be defined as in Section 4. If a point $m^{00} = (a^{00}, b^{00}, c_{11}^{00}, c_{12}^{00}, c_{22}^{00}, l^{00})$ is in $\bar{A}$, and, for a sequence $\{m_i\}$ in A, $m_i \to m^{00}$ and $J(m_i) \to J(\{m_i\})$ in $\bar{C}_2$, at least one of the following must be true:

1) l^{00} is a normal d.f., and c_{11}^{00} and c_{22}^{00} are finite
2) l^{00} is a d.f. which is not normal, and either $a^{00} = \pm\infty$ or $b^{00} = \pm\infty$ or both
3) either $c_{11}^{00} = \infty$ or $c_{22}^{00} = \infty$ or both
4) the variation of l^{00} is less than one

Suppose $J(\mu)$ were in $\bar{C}_2$ and $= J(\{m_i\})$. If the first of the conditions above held then either $J(\{m_i\})$ would be of variation less than one, or $J(\{m_i\})$ would be normal, neither of which can be true of $J(\mu)$. If one of conditions 2, 3, and 4 held, then the variation of $J(\{m_i\})$ would be less than one, which of course cannot be true of $J(\mu)$. This completes the proof that $J(\mu)$ is not in $\bar{C}_2$.

6. Estimation of the remainder of the structure of Section 5. Let $H(y)$ be any one-dimensional d.f. The Gaussian component of H is the largest value of λ for which H can be expressed as the convolution of a normal d.f. with variance λ, and another d.f. H is said to have no Gaussian component if its Gaussian component is zero.

The elements σ_{11}, σ_{12}, σ_{22}, L of μ are not, in general, uniquely determined by $J(\mu)$. Among the, in general, infinitely many m such that $J(m) = J(\mu)$, there is exactly one, say

$$\mu_0 = (\alpha, \beta, \sigma_{11}^0, \sigma_{12}^0, \sigma_{22}^0, L_0),$$

which is such that L_0 has no Gaussian component. We have $\sigma_{11}^0 + \sigma_{22}^0 > c_{11} + c_{22}$ for any other complex m such that $J(m) = J(\mu_0) = J(\mu)$; all such complexes are readily determinable from μ_0. These remarks follow from (4.1), (4.2), and the results of Reiersol [5]. We shall call the complex μ_0 "canonical" and estimate *all* its components in a strongly consistent manner. Of course α and β have already been estimated in Section 5; the present method will also estimate them, inter alia.

Let $Z_1, \cdots, Z_n$ be any independent chance variables with the common df. $H(z)$ and the empiric d.f. $H_n(z)$. Let $d(n)$ be any positive function defined on the positive integers such that $d(n) \to 0$ as $n \to \infty$ and

$$P\{\delta(H(z), H_n(z)) > d(n) \text{ for infinitely many } n\} = 0.$$

There are many such functions; it is easy to verify that $n^{-1/10}$ is such a function, but this is a crude result. For $H(z)$ continuous and one-dimensional and δ the Fréchet distance between two d.f.'s there is available the sharp result of Chung [6], according to which the function

$$\left(\frac{\log \log n}{cn}\right)^{1/2}$$

is a function $d(n)$ if $0 < c < 2$.

Let

$$U(C_n) = (a^*(n), b^*(n), c_{11}^*(n), c_{12}^*(n), c_{22}^*(n), L_n^*) \tag{6.1}$$

be any function from the $2n$-dimensional space of C_n to the space $A = \{m\}$ which is measurable and such that

$$\delta(J(U(C_n)), F_n(C_n)) < d(n) \tag{6.2}$$

and

$$c_{11}^*(n) + c_{22}^*(n) + \gamma(n) > \sup (c_{11} + c_{22}) \tag{6.3}$$

where the supremum operation in the right member is performed over all m such that

$$\delta(J(m), F_n(C_n)) < d(n). \tag{6.4}$$

When there is no m which satisfies (6.4) let $U(C_n)$ be defined in any manner provided it is measurable. It will follow from the general considerations of the next section that

$$\delta(U(C_n), \mu_0) \to 0 \tag{6.5}$$

w.p.1. Thus the elements of $U(C_n)$ are strongly consistent estimators of the elements of the canonical complex.

7. The method of the maximum index. We shall now generalize the considerations of the preceding section.

Consider the structure (2.1), and the totality of $(\bar{\alpha}, g)$ in A such that $J(\bar{\alpha}, g) =$

$J(\bar{\alpha}^*, G^*)$; call this totality $T(\bar{\alpha}^*, G^*)$. In every $T(\bar{\alpha}, g)$ let there be defined a unique member called the canonical complex of $T(\bar{\alpha}, g)$; we may denote this element by $D(\bar{\alpha}, g)$. If $(\bar{\alpha}_1, g_1)$ and $(\bar{\alpha}_2, g_2)$ are such that $J(\bar{\alpha}_1, g_1) = J(\bar{\alpha}_2, g_2)$, then we must have $D(\bar{\alpha}_1, g_1) = D(\bar{\alpha}_2, g_2)$. Suppose that there is defined on A a real-valued function $\psi(\bar{\alpha}, g)$ such that, whenever

$$(\bar{\alpha}_i, g_i) \to (\bar{\alpha}, g) \text{ in } A, \tag{7.1}$$

then

$$\liminf_{i\to\infty} \psi(\bar{\alpha}_i, g_i) \leqq \psi(\bar{\alpha}, g) \tag{7.2}$$

and such that, whenever $(\bar{\alpha}^*, g^*)$ is a canonical complex,

$$\psi(\bar{\alpha}^*, g^*) > \psi(\bar{\alpha}, g) \tag{7.3}$$

for every other $(\bar{\alpha}, g)$ in $T(\bar{\alpha}^*, g^*)$.

Let $d(n)$ be any function defined on the positive integers such that $d(n) \to 0$ as $n \to \infty$ and

$$P\{\delta(J(\theta, G_0), F_n(C_n)) > d(n) \text{ for infinitely many } n\} = 0 \tag{7.4}$$

Let $U(C_n) = (\theta_n^{**}, G_{0n}^{**})$ be any function from the hn-dimensional Euclidean space of C_n to the space A, which is measurable and such that

$$\delta(J(\theta_n^{**}, G_{0n}^{**}), F_n(C_n)) < d(n) \tag{7.5}$$

and

$$\psi(\theta_n^{**}, G_{0n}^{**}) + \gamma(n) > \sup \psi(\bar{\alpha}, g) \tag{7.6}$$

where the supremum in the right member is over all $(\bar{\alpha}, g)$ such that

$$\delta(J(\bar{\alpha}, g), F_n(C_n)) < d(n). \tag{7.7}$$

When there is no $(\bar{\alpha}, g)$ which satisfies (7.7) let $(\theta_n^{**}, G_{0n}^{**})$ be defined in any manner provided it is measurable. We will call $U(C_n)$ a maximum index estimator (of $D(\theta, G_0)$) and prove the following

THEOREM. *If J is a continuous function on A, i.e., whenever $(\bar{\alpha}_i, g_i) \to (\bar{\alpha}, g)$ in A, $J(\bar{\alpha}_i, g_i) \to J(\bar{\alpha}, g)$, and if $J(\theta, G_0)$ is not in $\bar{C}_2$, then, w.p.1,*

$$\delta(U(C_n), D(\theta, G_0)) \to 0, \tag{7.8}$$

so that $U(C_n)$ is a strongly consistent estimator of $D(\theta, G_0)$.

PROOF: Obviously $\delta(J(\theta_n^{**}, G_{0n}^{**}), F_n(C_n)) \to 0$, w.p. 1. Hence

$$\delta(J(\theta_n^{**}, G_{0n}^{**}), J(D[\theta, G_0])) \to 0, \qquad \text{w.p.1.}$$

If (7.8) were not true w.p.1, then, with positive probability, we may choose a Cauchy subsequence (A is conditionally compact; the particular sequence may depend upon the sample point in the probability space) which converges to a point $(\bar{\alpha}', g')$ in A (since $J(\theta, G_0)$ is not in $\bar{C}_2$) and $(\bar{\alpha}', g')$ is not $D[\theta, G_0]$.

It is impossible that, with positive probability,

$$\psi(D[\theta, G_0]) > \psi(\bar{\alpha}', g'), \tag{7.9}$$

because of (7.2) and the fact that, w.p.1, $\delta(J(D[\theta, G_0]), F_n(C_n))$ is eventually less than $d(n)$.

Suppose that, with positive probability,

$$\psi(D[\theta, G_0]) < \psi(\bar{\alpha}', g'). \tag{7.10}$$

Then $(\bar{\alpha}', g')$ would not be in $T(\theta, G_0)$ and $J(\bar{\alpha}', g')$ and $J(\theta, G_0)$ would not be identical. Since J is continuous we must have that, for the Cauchy subsequence, $\lim_{i\to\infty} J(\theta^{**}_{n_i}, G^{**}_{0n_i}) = J(\bar{\alpha}', g')$. Since $\delta(J(\theta^{**}_n, G^{**}_{0n}), J(\theta, G_0)) \to 0$ w.p.1, it follows that $J(\bar{\alpha}', g')$ and $J(\theta, G_0)$ are identical, contradicting the above. Hence (7.10) cannot occur.

Suppose then that, with positive probability,

$$\psi(D[\theta, G_0]) = \psi(\bar{\alpha}', g') \tag{7.11}$$

but $(\bar{\alpha}', g')$ were not $D(\theta, G_0)$. Because of the maximizing property of ψ (on each T) it would follow that $(\bar{\alpha}', g')$ is not in $T(\theta, G_0)$. But then $J(\bar{\alpha}', g')$ and $J(\theta, G_0)$ could not be identical. We have already seen that this cannot be. This leaves, as the only remaining possibility, that $(\bar{\alpha}', g')$ is $D(\theta, G_0)$, a contradiction which proves the theorem.

It is easy to verify that the postulated conditions are verified in the problem of Section 6. We have already seen that there $J(\theta, G_0)$ is not in $\bar{C}_2$. Let

$$\psi(a, b, c_{11}, c_{12}, c_{22}, l) = c_{11} + c_{22}.$$

Then, in any $T(m)$, ψ attains its unique maximum on the canonical complex. The function ψ is obviously continuous on A. Thus it satisfies the requirements of the theorem of the present section.

8. Application to stochastic difference equations. Let it be known to the statistician that $u_0, u_1, u_2, \cdots$ are independent chance variables with the common one-dimensional d.f. G, which is unknown to the statistician. Also it is known that, for $j = 1, 2, \cdots$

$$X_j = u_j + \alpha u_{j-1} \tag{8.1}$$

where α is a constant less than one in absolute value but otherwise unknown to the statistician. The problem is to estimate α consistently, under minimal assumptions on G.

Let q be the generic designation of a couple (a, L), with a real and less than one in absolute value, and L a one-dimensional d.f. which does not assign probability one to a single point. Let $A = \{q\}$. Let $J(q)$ be the d.f. of (X_1, X_2) when $\alpha = a$ and $G = L$. Let F_n be the two-dimensional empiric d.f. of

$$\{(X_{2i-1}, X_{2i}), i = 1, 2, \cdots n\}. \tag{8.2}$$

Finally let $q_0 = (\alpha, G)$; of course, q_0 is in A.

If G were to assign probability one to a single point then it is obvious that α would not be identified. The necessary condition, that G not assign probability one to a single point, is also sufficient, and the d.f. of (X_1, X_2) then determines $\alpha(|\alpha| < 1)$ uniquely. Even more: Let q' be a couple (a', L'), where $|a'| \leqq 1$, and L' is a d.f. which does not assign probability one to a single point. Suppose that $J(q_0) = J(q')$. Then $a' = \alpha$, hence is less than one in absolute value, and q' must be in A. For it follows from Theorem 1 of [10] that, if α were not uniquely determined, G would have to be normal. The possibility that G is normal and $\alpha \neq a'$ is then easily eliminated. (In [1] through an oversight it is erroneously stated that the d.f. of X_1 already determines α uniquely. Attention has been called to this error in, e.g., [8], page 211, footnote 6.)

Although the members of the sequence (8.2) are not independent, the two sequences made up of alternate members of this sequence are sequences of independent chance variables, and it is easy to show, as was done in [4], that

$$\delta(J(q_0), F_n) \to 0. \tag{8.3}$$

The minimum distance estimator of α is obtained in the usual manner. Let $S_n = (\alpha_n^*, G_n^*)$ be any function from the $2n$-Euclidean space of $(X_1, \cdots, X_{2n})$ to the space A which is measurable and such that

$$\delta(J(\alpha_n^*, G_n^*), F_n) < \inf_{q \epsilon A} \delta(J(q), F_n) + \gamma(n). \tag{8.4}$$

Then α_n^* is a minimum distance estimator of α, to which it converges w.p.1.

To prove the latter we have only to show that $J(\alpha, G) = J(q_0)$ is not in $\bar{C}_2$. Let $\bar{A}$ be as defined in Section 4. Any member $\bar{q} = (\bar{a}, \bar{L})$ of $\bar{A}$ has one of the following properties:

1) $\bar{L}$ assigns probability one to a single point.

2) $\bar{L}$ is a d.f. which does not assign probability one to a single point, and $\bar{a} = \pm 1$.

3) $\bar{L}$ has variation less than one.

Suppose $J(q_0) = J(\bar{q})$. Then $\bar{q}$ cannot have the first of these properties, because then X_i = constant w.p.1. Also $\bar{q}$ cannot have the second of these properties, by the result described in the third paragraph of this section. If $\bar{q}$ had the third of these properties then either $J(\bar{q})$ would have variation less than one or $J(\bar{q})$ would assign probability one to a single point, neither of which can be true of $J(q_0)$. Hence $J(q_0)$ is not in $\bar{C}_2$.

The author is grateful to Professors Henry Teicher and Lionel Weiss for reading the manuscript.

REFERENCES

[1] J. Wolfowitz, "Estimation of the components of stochastic structures," *Proc. Nat. Acad. Sci.*, Vol. 40 (1954), pp. 602–606.

[2] J. Wolfowitz, "Consistent estimators of the parameters of a linear structural relation," *Skand. Aktuarietids.* 1952, pp. 132–151.

[3] J. Wolfowitz, "Estimation by the minimum distance method," *Ann. Inst. Stat. Math.* (Japan), Vol. 5 (1953), pp. 9–23.

[4] J. Wolfowitz, "Estimation by the minimum distance method in non-parametric stochastic difference equations," *Ann. Math. Stat.*, Vol. 25 (1954), pp. 203–217.

[5] O. Reiersol, "Identifiability of a linear relation between variables which are subject to error," *Econometrica*, Vol. 18 (1950), pp. 375–89.

[6] K. L. Chung, "An estimate concerning the Kolmogoroff distribution," *Trans. Amer. Math. Soc.*, Vol. 67 (1949), pp. 36–50.

[7] H. Cramér, *Random Variables and Probability Distributions*, Cambridge University Press, 1937.

[8] M. Kac, J. Kiefer, and J. Wolfowitz, "On tests of normality and other tests of goodness of fit based on distance methods," *Ann. Math. Stat.*, Vol. 26 (1955), pp. 189–211.

[9] J. Kiefer and J. Wolfowitz, "Consistency of the maximum likelihood estimator in the presence of infinitely many incidental parameters" *Ann. Math. Stat.*, Vol. 27 (1956), pp. 887–906.

[10] H. Teicher, "Identification of a certain stochastic structure," *Econometrica*, Vol. 24 (1956), pp. 172–177.

[11] J. Neyman and E. L. Scott, "Consistent estimators based on partially consistent observations," *Econometrica*, Vol. 16 (1948), pp. 1–32.

Reprinted from Illinois Journal of Mathematics
Vol. 1, No. 4, December 1957

THE CODING OF MESSAGES SUBJECT TO CHANCE ERRORS[1]

BY J. WOLFOWITZ

1. The transmission of messages

Throughout this paper we assume that all "alphabets" involved contain exactly two symbols, say 0 and 1. What this means will be apparent in a moment. This assumption is made only in the interest of simplicity of exposition, and the changes needed when this assumption is not fulfilled will be obvious.

Suppose that a person has a vocabulary of S words (or messages), any or all of which he may want to transmit, in any frequency and in any order, over a "noisy channel". For example, S could be the number of words in the dictionary of a language, provided that it is forbidden to coin words not in the dictionary. What a "noisy channel" is will be described in a moment. Here we want to emphasize that we do not assume anything about the frequency with which particular words are transmitted, nor do we assume that the words to be transmitted are selected by any random process (let alone that the distribution function of the random process is known). Let the words be numbered in some fixed manner. Thus transmitting a word is equivalent to transmitting one of the integers $1, 2, \cdots, S$.

We shall now explain what is meant by a "noisy channel" of memory m. A sequence of $(m + 1)$ elements, each zero or one, will be called an α-sequence. A function p, defined on the set of all α-sequences, and such that always $0 \leqq p \leqq 1$, is associated with the channel and called the channel probability function. A sequence of n elements, each of which is zero or one, will be call an x-sequence. To describe the channel, it will be sufficient to describe how it transmits any given x-sequence, say x_1. Let α_1 be the α-sequence of the first $(m + 1)$ elements of x_1. The channel "performs" a chance experiment with possible outcomes 1 and 0 and respective probabilities $p(\alpha_1)$ and $(1 - p(\alpha_1))$, and transmits the outcome of this chance experiment. It then performs another chance experiment, independently of the first, with possible outcomes 1 and 0 and respective probabilities $p(\alpha_2)$ and $(1 - p(\alpha_2))$, where α_2 is the α-sequence of the $2^{\text{nd}}, 3^{\text{rd}}, \cdots, (m + 2)^{\text{nd}}$ elements of the sequence x_1. This is repeated until $(n - m)$ independent experiments have been performed. The probability of the outcome one in the ith experiment is $p(\alpha_i)$, where α_i is the α-sequence of the $i^{\text{th}}, (i + 1)^{\text{st}}, \cdots, (i + m)^{\text{th}}$ elements of x_1. The x-sequence x_1 is called the transmitted sequence. The chance sequence $Y(x_1)$ of outcomes of the experiments in consecutive order is called the received sequence. Any sequence of $(n - m)$ elements, each zero or one, will be called a y-sequence. Let y_1 be any y-sequence. If $P\{Y(x_1) = y_1\} > 0$ (the symbol

---

Received January 26, 1957; received in revised form May 15, 1957.

[1] Research under contract with the Office of Naval Research.

$P\{\quad\}$ denotes the probability of the relation in braces), we shall say that y_1 is a possible received sequence when x_1 is the transmitted sequence.

Let λ be a positive number which it will usually be desired to have small. A "code" of length t is a set $\{(x_i, A_i)\}$, $i = 1, \cdots, t$, where (a) each x_i is an x-sequence, (b) each A_i is a set of y-sequences, (c) for each i

$$P\{Y(x_i) \epsilon A_i\} \geqq 1 - \lambda,$$

(d) $A_1, \cdots, A_t$ are disjoint sets. The coding problem which is a central concern of the theory of transmission of messages may be described as follows: For given S, to find an n and then a code of length S. The practical applications of this will be as follows: When one wishes to transmit the ith word, one transmits the x-sequence x_i. Whenever the receiver receives a y-sequence which is in A_j, he always concludes that the jth word has been sent. When the receiver receives a y-sequence not in $A_1 \cup A_2 \cup \cdots \cup A_S$, he may draw any conclusion he wishes about the word that has been sent. The probability that any word transmitted will be correctly received is $\geqq 1 - \lambda$.

When such a code is used, s/n is called the "rate of transmission," where $s = \log S$. (All logarithms which occur in the present paper are to the base 2.) Except for certain special[2] functions p, one can find a code for any s, provided that one is willing to transmit at a sufficiently small rate; for the law of large numbers obviously applies, and by sufficient repetition of the word to be transmitted, one can insure that the probability of its correct reception exceeds $1 - \lambda$. The practical advantages of a high rate of transmission are obvious. If there were no "noise" (error in transmission) and signals were received exactly as sent, then s symbols zero or one would suffice to transmit any word in the vocabulary, and one could transmit at the rate one. The existence of an error of transmission means that the sequences to be sent must not be too similar in some reasonable sense, lest they be confused as a result of transmission errors. When n is sufficiently large, we can find $S = 2^s$ sufficiently dissimilar sequences. The highest possible rate of transmission obviously depends on the channel probability function.

2. The contents of this paper

The fundamental ideas of the present subject and paper are due to the fundamental and already classical paper [1] of Shannon. Theorem 1 below was stated and proved by Shannon. However, the latter permits the use of what are called "random codes," and indeed proves Theorem 1 by demonstrating the existence of a random code with the desired property. It seems to the present writer questionable whether random codes are properly codes at all. The definition of a code given in Section 1 of the present paper does not admit random codes as codes; what we have called a code is called in the literature of communication theory an "error correcting" code. In any case,

[2] For example, if $p(\alpha_1) = p(\alpha_2)$, then α_1 and α_2 are indistinguishable in transmission.

the desirability of proving the existence of an error correcting code which would satisfy the conclusion of Shannon's Theorem 1 has always been recognized and well understood (see, for example, [8], Section 3).

The achievement of such a proof is due to Feinstein [2] and Khintchine [4]. The latter utilized an idea from the earlier, not entirely rigorous and without gaps, work of Feinstein, to prove, in full rigor, the general Theorem 3 below. In the present paper, starting from first principles in Section 3, we give already in Section 5 a short and simple proof of Theorem 1. We then return to the subject in Section 8 to prove Theorem 3. Even after allowance is made for the fact that Lemmas 8.2 and 8.3 are not proved here, it seems that our proofs have something to offer in simplicity and brevity.

Theorem 2 for general memory m was stated by Shannon in [1]. Khintchine in [4] pointed out that neither the argument of [1] nor any of the arguments to be found in the literature constitute a proof or even the outline of a proof; he also pointed out the desirability of proving the result and mentioned some of the difficulties. In the present paper we give what seems to be the first proof of Theorem 2. We have reason to believe that it is possible to treat the case of general finite memory along the same lines.[3]

The notion of extending the result for stationary Markov chains (Theorem 1) to stationary, not necessarily Markovian processes (Theorem 3) is due to McMillan [5]. The difficult achievement of carrying out this program correctly and without gaps is due to Khintchine [4]. The theorem we cite below as Lemma 8.3 is due to McMillan. Lemma 8.2 is due to Khintchine.

In [4] Khintchine acknowledges his debt to the paper [2] of Feinstein, although he states that its argument is not exact and that it deals largely with the case of zero memory (and only with Theorem 1 of course). The main idea of [2] seems, to the present writer, to be the ingenious one of proving an inequality like (5.4) below. This pretty idea is employed in the present paper; we find it possible to dispense with many of the details which occur in this connection in [2] and [4].

Shannon and all other writers cited above employ the law of large numbers. The simple notion of δx-sequences and the sequences they generate, which so simplifies our proof below and makes the proof of Theorem 2 possible, also enables us to use Chebyshev's inequality instead of the law of large numbers in Theorems 1 and 2. This has the incidental effect of slightly improving Theorems 1 and 2 over Shannon's original formulation by replacing $o(n)$ terms by $O(n^{1/2})$ terms.

This entire paper is self-contained except for the following incidental remark which we make here in passing: The quantity called $e(n)$ in [2] (the maximum

[3] *Added in proof*. A sequel to the present paper, which has been accepted for publication by this Journal, gives an upper bound on the length of a code for any memory m. When $m = 0$ this bound is the same as that given by Theorem 2. The proof of this result is different from that of Theorem 2.

probability of incorrectly receiving any word) is shown there, for $m = 0$, to approach zero "faster than $1/n$". Using the arguments of the present paper and the inequality (96) of page 288 of [9], one can prove easily for any m that

$$e(n) < c_1 n^{-1/2} e^{-c_2 n},$$

where c_1 and c_2 are positive constants.

3. Combinatorial preliminaries

Let x be any x-sequence and α be any α-sequence. Let $N(\alpha \mid x)$ be the number of elements in x such that each, together with the m elements of x which follow it, constitute the sequence α. Let δ and δ_2 be fixed positive numbers. Let π be any nonnegative function defined on the set of all α-sequences such that

$$\sum_\alpha \pi(\alpha) = 1.$$

We shall say that an x-sequence x is a $\delta\pi x$-sequence if

$$| N(\alpha \mid x) - n\pi(\alpha) | \leqq \delta n^{1/2} \tag{3.1}$$

for every α-sequence.

A y-sequence y will be said to be generated by the x-sequence x if (1) y is a possible received sequence when x is the transmitted sequence, (2) for any α-sequence α_1 the following is satisfied: Let $j(1), \cdots, j(N(\alpha_1 \mid x))$ be the serial numbers of the elements of x which begin the sequence α_1 (e.g., the elements in the places with serial numbers $j(1), j(1) + 1, \cdots, j(1) + m$, constitute the sequence α_1). Then the number $N(\alpha_1, y \mid x)$ of elements one among the elements of y with serial numbers $j(1), \cdots, j(N(\alpha_1 \mid x))$ satisfies

$$| N(\alpha_1, y \mid x) - N(\alpha_1 \mid x)p(\alpha_1) | \leqq \delta_2[N(\alpha_1 \mid x)(p(\alpha_1))(1 - p(\alpha_1))]^{1/2}. \tag{3.2}$$

Let $M(x)$ denote the number of y-sequences generated by x.

Whenever in this paper the expression $0 \log 0$ occurs, it is always to be understood as equal to zero. We remind the reader that all logarithms occurring in this paper are to the base 2. For any x-sequence x we define $H_x(Y)$, the conditional entropy of $Y(x)$, by

$$\begin{aligned} H_x(Y) = & -(1/n) \sum_\alpha N(\alpha \mid x) p(\alpha) \log p(\alpha) \\ & -(1/n) \sum_\alpha N(\alpha \mid x)(1 - p(\alpha)) \log (1 - p(\alpha)). \end{aligned} \tag{3.3}$$

Lemma 3.1. *For any δ_2 there exists a $K_1 > 0$ such that, for any n and any x-sequence x,*

$$M(x) < 2^{nH_x(Y) + K_1 n^{1/2}}. \tag{3.4}$$

Proof. Let θ_2 be a generic real number $\leqq \delta_2$ in absolute value. Let y be any y-sequence generated by x. Then

$$\begin{aligned}
\log P\{Y(x) = y\} &= \textstyle\sum_\alpha N(\alpha \mid x)p(\alpha)\log p(\alpha) \\
&\quad + \textstyle\sum_\alpha N(\alpha \mid x)(1 - p(\alpha))\log(1 - p(\alpha)) \\
&\quad + \textstyle\sum_\alpha \theta_2(\alpha)[N(\alpha \mid x)p(\alpha)(1 - p(\alpha))]^{1/2}\log p(\alpha) \\
&\quad - \textstyle\sum_\alpha \theta_2(\alpha)[N(\alpha \mid x)p(\alpha)(1 - p(\alpha))]^{1/2}\log(1 - p(\alpha)) \\
&> -nH_x(Y) + \delta_2 n^{1/2}\textstyle\sum_\alpha \log p(\alpha) \\
&\quad + \delta_2 n^{1/2}\textstyle\sum_\alpha \log(1 - p(\alpha)) = -nH_x(Y) - K_1 n^{1/2},
\end{aligned} \tag{3.5}$$

with

$$K_1 = -\delta_2\textstyle\sum_\alpha \log p(\alpha) - \delta_2\textstyle\sum_\alpha \log(1 - p(\alpha)).$$

(Here the first summation is over all α such that $p(\alpha) > 0$, and the second summation is over all α such that $p(\alpha) < 1$.) The lemma follows at once from (3.5).

LEMMA 3.2. *Let $\lambda > 0$ be any number. Then, for δ_2 larger than a bound which depends only upon λ, we have, for any n and any x-sequence x,*

$$P\{Y(x) \text{ is a sequence generated by } x\} > 1 - \tfrac{1}{2}\lambda. \tag{3.6}$$

There then exists a $K_2 > 0$ which depends only on δ_2 such that, for any n and any x-sequence x,

$$M(x) > 2^{nH_x(Y) - K_2 n^{1/2}}. \tag{3.7}$$

Proof. (3.6) follows at once from Chebyshev's inequality. As in (3.5) we have, for any y-sequence y generated by x,

$$\log P\{Y(x) = y\} < -nH_x(Y) + K_1 n^{1/2}. \tag{3.8}$$

From (3.6) and (3.8) we have at once that

$$M(x) > (1 - \tfrac{1}{2}\lambda)2^{nH_x(Y) - K_1 n^{1/2}}. \tag{3.9}$$

Then (3.7) follows at once from (3.9).

4. Preliminaries on Markov chains[4]

Let $X_1, X_2, \cdots$ be a stationary, metrically transitive Markov chain with two possible states, 0 and 1; we shall call this the X process, for short. Suppose

[4] Since we do not assume that the words to be transmitted are chosen by any random process or sent with any particular frequency, the introduction of the X process is not a necessity. The lemmas which involve the X process are of purely combinatorial character (e.g., Lemmas 4.1 and 6.1). The X process serves merely as a device for stating or proving certain combinatorial facts. The reader is invited to verify that this entire paper could be written without the introduction of the X process. In that case the $Y(x)$ process would take the place of the Y process. Only the entropies $H(Y)$ and $H_X(Y)$ need be introduced, and this can be done by means of the $Y(x)$ process and δQx-sequences.

$$Q_i = P\{X_1 = i\}, \qquad i = 0, 1, \tag{4.1}$$

and

$$q_{ij} = P\{X_{k+1} = j \mid X_k = i\} \tag{4.2}$$

is the probability of a transition from state i to state j; $i, j = 0, 1$. For any α-sequence α define

$$Q(\alpha) = P\{(X_1, \cdots, X_{m+1}) = \alpha\}. \tag{4.3}$$

The function Q is a function which satisfies the requirements on the function π of Section 3. Let $\gamma < 1$ be any number. It follows at once from Chebyshev's inequality that, for any n and any δ greater than a lower bound which is a function only of γ and the q_{ij},

$$P\{(X_1, \cdots, X_n) \text{ is a } \delta Qx\text{-sequence}\} > \gamma. \tag{4.4}$$

By the Y process we shall mean the sequence $Y_1, Y_2, \cdots$, where Y_i is a chance variable which assumes only the values zero and one, and the conditional probability that $Y_i = 1$, given the values of $X_1, X_2, \cdots, Y_1, \cdots, Y_{i-1}$, is $p(X_i, \cdots, X_{i+m})$. Henceforth we write for short $X = (X_1, \cdots, X_n)$. Then the conditional distribution of $Y = (Y_1, \cdots, Y_{n-m})$, given X, is the same as that of the sequence received when X is the sequence transmitted. The Y process is obviously stationary, and, by Lemma 8.2 below (proof in [4], page 53), metrically transitive. The conditional entropy $H_X(Y)$ of the Y process relative to the X process is defined by

$$H_X(Y) = -\sum_\alpha Q(\alpha)p(\alpha) \log p(\alpha) - \sum_\alpha Q(\alpha)(1 - p(\alpha)) \log (1 - p(\alpha)).$$

One verifies easily that there exists a $K_3 > 0$ such that, for any δQx-sequence x,

$$| H_x(Y) - H_X(Y) | < K_3 \, \delta/n^{1/2}. \tag{4.5}$$

We at once obtain

LEMMA 4.1. *For any n and any δQx-sequence x, the inequalities* (3.4) *and* (3.7) *hold with $H_x(Y)$ replaced by $H_X(Y)$ and K_1 and K_2 replaced by K_1' and K_2', where K_1' and K_2' are positive numbers which depend only upon δ and δ_2.*

We define the chance variable (function of X) $P\{X\}$ as follows: when $X = x$, $P\{X\} = P\{X = x\}$. Similarly we define the chance variable (function of Y) $P\{Y\}$ as follows: when $Y = y$, $P\{Y\} = P\{Y = y\}$. We define the entropy $H(X)$ of the X process by

$$H(X) = \lim_{n\to\infty} -\frac{1}{n} E\,[\log P\{X\}].$$

This limit obviously exists.

Let the symbol $\sigma^2(\ )$ denote the variance of the chance variable in parentheses. We now prove

LEMMA 4.2. *We have*

$$E[\log P\{Y\}] = -Dn + D_0, \tag{4.6}$$

where D *is a nonnegative constant and* D_0 *is a bounded function of* n. *Also,*

$$\sigma^2(\log P\{Y\}) = O(n). \tag{4.7}$$

The quantity

$$D = \lim_{n\to\infty} -\frac{1}{n} E\,[\log P\{Y\}]$$

is called the entropy $H(Y)$ of the Y process.

Proof. We have

$$\log P\{Y\} = \sum_{i=1}^{n} \log P\{Y_i \mid Y_1, \cdots, Y_{i-1}\}. \tag{4.8}$$

Let α^* be some fixed α-sequence such that $Q(\alpha^*) > 0$. In the sequence $X_1, X_2, \cdots$, let $j(1), j(2), \cdots$ be the indices such that

$$(X_{j(i)}, X_{j(i)+1}, \cdots, X_{j(i)+m}) = \alpha^*, \qquad i = 1, 2, \cdots.$$

(These exist with probability one.) Let l^* be the smallest integer such that $j(l^*) \geqq n + 1$. (Again l^* is defined with probability one.) Define symbols such as

$$C_1 = \log P\{Y_1, \cdots, Y_{j(1)-1}\}$$

in the obvious manner analogous to that in which $\log P\{Y\}$ was defined. Since C_1 is a sum of quantities which enter into (4.8) and which are all zero or negative, it follows that $B_1 = EC_1$ could fail to exist only if it were $-\infty$. It will be seen that the latter cannot be. Define

$$C_2 = \sum_{i=n-m+1}^{j(l^*)-1} \log P\{Y_i \mid Y_1, \cdots, Y_{i-1}\}.$$

As before, $B_2 = EC_2$ either exists or is $-\infty$. It will be seen that $B_2 \neq -\infty$.

It is easy to see that

$$Ej(1) \leqq \text{a constant, independent of } n \tag{4.9}$$

$$El^* = 1 + nQ(\alpha^*) \tag{4.10}$$

$$E(j(i) - j(i-1)) = \text{a constant, independent of } n \text{ and } i. \tag{4.11}$$

$$E(j(l^*) - n) \leqq \text{a constant, independent of } n. \tag{4.12}$$

From the construction of $j(1), j(2), \cdots$ it follows that the chance variables

$$W_i = \log P\{Y_{j(i)}, \cdots, Y_{j(i+1)-1} \mid Y_1, \cdots, Y_{j(i)-1}\}, \qquad i = 1, 2, \cdots \tag{4.13}$$

are independently and identically distributed. Actually

$$W_i = \log P\{Y_{j(i)}, \cdots, Y_{j(i+1)-1}\}.$$

From Wald's equation ([10], Theorems 7.1 and 7.4), (4.9), (4.11), (4.12), and the fact that the chance variables W_i, C_1, and C_2 are sums of always non-positive and bounded chance variables which appear in the right member of (4.8), it follows that B_1, B_2, and $EW_i = w$ are all finite, and that B_1 and B_2 are bounded uniformly in n. Applying Wald's equation again we obtain, using (4.10), that

$$E \log P\{Y_1, \cdots, Y_{j(l^*)-1}\} = B_1 + nwQ(\alpha^*). \tag{4.14}$$

Hence

$$E \log P\{Y\} = B_1 + nwQ(\alpha^*) - B_2, \tag{4.15}$$

which proves (4.6).

Now

$$\begin{aligned} \log P\{Y\} - E \log P\{Y\} &= (C_1 - B_1) \\ &\quad + \left(\sum_{i=1}^{l^*-1} W_i - nwQ(\alpha^*)\right) - (C_2 - B_2) \\ &= (C_1 - B_1) + \left(\sum_{i=1}^{l^*-1} W_i - (l^* - 1)w\right) \\ &\quad + ((l^* - 1)w - nwQ(\alpha^*)) - (C_2 - B_2). \end{aligned} \tag{4.16}$$

Now we note that

$$\sigma^2(j(i) - j(i-1)) = \text{a constant, independent of } n. \tag{4.17}$$

Applying an argument like that which leads to Theorem 7.2 of [10], together with (4.17) and Schwarz's inequality, we obtain first that

$$\sigma^2(W_i) = \text{(a finite) constant}, \tag{4.18}$$

and then that

$$E\left(\sum_{i=1}^{l^*-1} W_i - (l^* - 1)w\right)^2 = \sigma^2(W_1)nQ(\alpha^*) \tag{4.19}$$

by (4.10) and

$$E([l^* - 1]w - E[l^* - 1]w)^2 = O(n) \tag{4.20}$$

(see [6], p. 263, equation (8.10)). Obviously

$$\sigma^2(C_1) \leqq \text{a constant independent of } n, \tag{4.21}$$

$$\sigma^2(C_2) \leqq \text{a constant independent of } n. \tag{4.22}$$

Now take the expected value of the squares of the first and third members of (4.16). Using (4.19), (4.20), (4.21), and (4.22) we obtain that the sum of the expected values of the squares which occur after squaring the third member of (4.16) is $O(n)$. The cross products have expected value $O(n)$ by the Schwarz inequality. This proves (4.7) and completes the proof of the lemma.

Another proof of the fact that the variance of $\log P\{Y\}$ is $O(n)$ can be based on the following: It is known from the theory of Markov chains ([7], page 173, equation (2.2)) that there exists a number h, $0 < h < 1$, such that the absolute value of the correlation coefficient between X_i and X_j is less than $h^{|i-j|}$. From the distribution of the Y_i it follows that a similar statement is true of the correlation coefficient between Y_i and Y_j and also of the correlation coefficient between

$$\log P\{Y_i \mid Y_1, \cdots, Y_{i-1}\}$$

and

$$\log P\{Y_j \mid Y_1, \cdots, Y_{j-1}\}.$$

Since $\log P\{Y\}$ can be written in the form (4.8), the desired conclusion can be deduced from the above.

An immediate consequence of Lemma 4.2 and Chebyshev's inequality is that, for any $\varepsilon' > 0$, there exists a $K_4 > 0$ such that, for any n,

$$P\{-nH(Y) - K_4 n^{1/2} < \log P\{Y\} < -nH(Y) + K_4 n^{1/2}\} > 1 - \varepsilon'. \tag{4.23}$$

The following lemma is now an immediate consequence of (4.23):

LEMMA 4.3. *Let $\varepsilon' > 0$ be any number, and $K_4 > 0$ be a number which, for any n, satisfies* (4.23). *For any n let B be any set of y-sequences such that*

$$P\{Y \in B\} > \gamma_1 > \varepsilon'.$$

Then the set B must contain at least

$$(\gamma_1 - \varepsilon')2^{nH(Y) - K_4 n^{1/2}}$$

y-sequences.

Proof. From (4.23) it follows that the y-sequences in B which satisfy the relationship in braces in (4.23) have probability greater than $\gamma_1 - \varepsilon'$. Since the probability of each such sequence is bounded above by $2^{-nH(Y) + K_4 n^{1/2}}$, the desired result follows.

5. The coding theorem

THEOREM 1. *Let $X_1, X_2, \cdots$ be a stationary, metrically transitive Markov chain with states 0 and 1 and notation as in Section 4. Let the Y process be as defined in Section 4. Let λ be an arbitrary positive number. There exists a $K > 0$ such that, for any n, there is a code of length at least*[5]

$$2^{n(H(Y) - H_X(Y)) - Kn^{1/2}} \tag{5.1}$$

The probability that any word transmitted according to this code will be incorrectly received is less than λ.

[5] An alternate and perhaps more graphic way to state Theorem 1 is to replace $(H(Y) - H_X(Y))$ in (5.1) by $C_1 = \max(H(Y) - H_X(Y))$, where the maximum is over all Markov processes X and their associated Y processes as defined in the statement of Theorem 1. It is obvious that this is an equivalent way of stating Theorem 1.

Proof. We may take $\lambda < \frac{1}{2}$. Let $\gamma < 1$ be any positive number. Let δ be sufficiently large so that (4.4) holds, and choose δ_2 sufficiently large so that (3.6) holds.

Let x_1 be any δQx-sequence, and A_1 any set of y-sequences generated by x_1 such that the following is satisfied for $i = 1$:

$$P\{Y(x_i) \text{ is a sequence generated by } x_i \text{ and not in } A_i\} < \tfrac{1}{2}\lambda. \tag{5.2}$$

Let x_2 be any other δQx-sequence for which we can find a set A_2 of y-sequences generated by x_2 such that A_1 and A_2 are disjoint and (5.2) is satisfied for $i = 2$. Continue in this manner as long as possible, i.e., as long as there exists another δQx-sequence, say x_i, and a set A_i of y-sequences generated by x_i such that $A_1, A_2, \cdots, A_i$ are all disjoint and A_i satisfies (5.2). Let

$$(x_1, A_1), \cdots\cdots, (x_N, A_N)$$

be the resulting code. We have to show that N is large enough.

Let x^* be any δQx-sequence (if one exists) not in the set $x_1, \cdots, x_N$. Then

$$P\{Y(x^*) \text{ is a sequence generated by } x^* \text{ and belongs to } (A_1 \cup A_2 \cup \cdots \cup A_N)\} \geqq \tfrac{1}{2}\lambda. \tag{5.3}$$

If this were not so, we could prolong the code by adding (x^*, A^*), where A^* is the totality of y-sequences generated by x^* and not in $A_1 \cup A_2 \cup \cdots \cup A_N$; this would violate the definition of N. From (4.4), (3.6), (5.2), and (5.3) it follows that

$$P\{Y \in (A_1 \cup A_2 \cup \cdots \cup A_N)\} > \tfrac{1}{2}\gamma\lambda. \tag{5.4}$$

Let the ε' of (4.23) and Lemma 4.3 be equal to $\frac{1}{4}\gamma\lambda$ and let $K_4 > 0$ be any number for which (4.23) is satisfied. It follows from Lemma 4.3 that the set $A_1 \cup A_2 \cup \cdots \cup A_N$ contains at least

$$\tfrac{1}{4}\gamma\lambda \cdot 2^{nH(Y) - K_4 n^{1/2}} \tag{5.5}$$

y-sequences. By Lemma 4.1 the number of y-sequences in $A_1 \cup A_2 \cup \cdots \cup A_N$ is at most

$$N \cdot 2^{nH_X(Y) + K_1' n^{1/2}}. \tag{5.6}$$

The desired result follows at once from (5.5) and (5.6), with

$$K = K_1' + K_4 - \log(\tfrac{1}{4}\gamma\lambda).$$

6. Further preliminaries[6]

The essential part of the present section is the second part of the inequality (6.13) below, which is basic in the proof of Theorem 2 of Section 7. Neither

[6] All the lemmas of this section are of purely combinatorial character. Lemma 6.3 could be easily proved by a purely combinatorial argument without any use of the Y process. This entire section is a concession to the conventional treatment of the subject. All that is needed for the statement and proof of Theorem 2 is the second part of (6.13) and a formal analytic definition of capacity. See also footnote 4.

Lemma 6.1 nor Lemma 6.2 is used in the sequel, and both are given only for completeness. The proof of Lemma 6.1 is omitted because it is very simple, and the proof of Lemma 6.2 is omitted because it involves some computation.

Let the X and Y processes be as defined in Section 4. Obviously

$$H(X) = -\sum_{i,j} Q_i \, q_{ij} \log q_{ij} . \tag{6.1}$$

We define the chance variables (functions of X and Y) $P\{Y \mid X\}$ and $P\{X \mid Y\}$ as follows: when $X = x$ and $Y = y$, $P\{Y \mid X\} = P\{Y = y \mid X = x\}$, and $P\{X \mid Y\} = P\{X = x \mid Y = y\}$. We verify easily that

$$H_X(Y) = \lim_{n\to\infty} -\frac{1}{n} E\,[\log P\{Y \mid X\}]. \tag{6.2}$$

We define $H_Y(X)$, the conditional entropy of the X process relative to the Y process, by

$$H_Y(X) = \lim_{n\to\infty} -\frac{1}{n} E\,[\log P\{X \mid Y\}]. \tag{6.3}$$

(We shall see in a moment ((6.5)) that this limit exists.) From the obvious relation

$$\log P\{X\} + \log P\{Y \mid X\} = \log P\{Y\} + \log P\{X \mid Y\}, \tag{6.4}$$

we obtain

$$H(X) + H_X(Y) = H(Y) + H_Y(X). \tag{6.5}$$

Throughout the rest of this section we assume that the memory $m = 0$, and that the X_i are independent, identically distributed chance variables. Hence

$$Q_i = q_{ji}, \qquad i, j = 0, 1. \tag{6.6}$$

Since $m = 0$, there are only two α-sequences, namely, (0) and (1), and $Q(i) = Q_i$, $i = 0, 1$. Write for short $Q(1) = q$. We assume that $0 < q < 1$. It seems reasonable in this case to denote what was called in Section 4 a "δQx-sequence" by the term "δqx-sequence", and we shall employ this usage (when $m = 0$ and the chance variables X_i are independent). We now give the values of the various entropies, inserting a zero in the symbol for entropy to indicate that $m = 0$ and the X_i are independently (and identically) distributed.

$$H(X_0) = -q \log q - (1 - q) \log (1 - q). \tag{6.7}$$

$$\begin{aligned} H(Y_0) = &- [qp(1) + (1 - q)p(0)] \log [qp(1) + (1 - q)p(0)] \\ &- [(1 - q)\,(1 - p(0)) + q\,(1 - p(1))] \cdot \\ &\cdot \log[(1 - q)\,(1 - p(0)) + q\,(1 - p(1))]. \end{aligned} \tag{6.8}$$

$$H_X(Y_0) = -qp(1)\log p(1) - (1-q)p(0)\log p(0)$$
$$(6.9) \qquad - q(1-p(1))\log(1-p(1))$$
$$- (1-q)(1-p(0))\log(1-p(0)).$$

From (6.5) we obtain

$$H_Y(X_0) = -(1-q)p(0)\log\frac{(1-q)p(0)}{[qp(1)+(1-q)p(0)]}$$
$$-qp(1)\log\frac{qp(1)}{[qp(1)+(1-q)p(0)]}$$
$$(6.10)$$
$$-q(1-p(1))\log\frac{q(1-p(1))}{[q(1-p(1))+(1-q)(1-p(0))]}$$
$$-(1-q)(1-p(0))\log\frac{(1-q)(1-p(0))}{[q(1-p(1))+(1-q)(1-p(0))]}.$$

The maximum, with respect to q, of

$$H(X_0) - H_Y(X_0) = H(Y_0) - H_X(Y_0)$$

is called the capacity (when $m = 0$) C_0 of the channel.

LEMMA 6.1. *There exists a $K_5 > 0$ such that, for any n, the number $M(\delta q)$ of δqx-sequences satisfies*

$$(6.11) \qquad 2^{nH(X_0)-K_5 n^{1/2}} < M(\delta q) < 2^{nH(X_0)+K_5 n^{1/2}}.$$

Let δ' be some fixed positive number such that any y-sequence which is generated by a δqx-sequence, cannot be generated by an x-sequence which is not a $\delta' qx$-sequence. Such a δ' exists; we have only to take δ' larger than a lower bound which is a function of q, δ, δ_2, $p(0)$, and $p(1)$. We have

LEMMA 6.2. *There exists a $K_6 > 0$ with the following property: Let y be any y-sequence which is generated by some δqx-sequence. Then the number $M'(y)$ of $\delta' qx$-sequences which generate y satisfies*

$$(6.12) \qquad 2^{nH_Y(X_0)-K_6 n^{1/2}} < M'(y) < 2^{nH_Y(X_0)+K_6 n^{1/2}}$$

We now prove

LEMMA 6.3. *There exists a $K_7 > 0$ such that, for any n, the number $M''(\delta q)$ of different y-sequences generated by all δqx-sequences satisfies*

$$(6.13) \qquad 2^{nH(Y_0)-K_7 n^{1/2}} < M''(\delta q) < 2^{nH(Y_0)+K_7 n^{1/2}}.$$

Proof. Let θ, with any subscript, denote a number not greater than one in absolute value. The chance variables $Y_1, Y_2, \cdots$ are independently and identically distributed. We have

$$P\{Y_1 = 1\} = qp(1) + (1-q)p(0) = u, \text{ say.}$$

If y is generated by a δqx-sequence, then the number V_1 of elements one in y is given by

$$V_1 = n(qp(1) + (1 - q)p(0)) + n^{1/2}(\theta_1\delta + 2\theta_2\delta_2) + 0(1).$$

Since

$$P\{Y = y\} = u^{V_1} (1 - u)^{n-V_1} > 2^{-nH(Y_0)-K_7 n^{1/2}}$$

for a suitable $K_7 > 0$, the second part of (6.13) follows at once.

The first part of (6.13) follows from Lemma 4.3. It may be necessary to increase the above K_7.

When $p(1) = p(0)$, $C_0 = 0$. One can verify that otherwise $C_0 > 0$. Incidentally, it follows from Lemmas 6.1 and 6.2 that $H(X_0) \geqq H_Y(X_0)$.

7. Impossibility of a rate of transmission greater than the capacity when $m = 0$

In this section we prove the following

THEOREM 2. *Let $m = 0$, and let λ, $1 > \lambda > 0$, be any given number. There exists a $K' > 0$ such that, for any n, any code with the property that the probability of transmitting any word incorrectly is $< \lambda$, cannot have a length greater than*

$$2^{nC_0 + K' n^{1/2}} \tag{7.1}$$

If $p(1) = p(0)$ and therefore $C_0 = 0$, the theorem is trivial. For then it makes no difference whether one transmits a zero or a one, and it is impossible to infer from the sequence received what sequence has been transmitted. We therefore assume henceforth that $C_0 > 0$.

It will be convenient to divide the proof into several steps. Let q_0 be the value of q which maximizes $H(Y_0) - H_X(Y_0)$. We shall have occasion to consider the various entropies as functions of q, which, in this section, we shall always exhibit explicitly, e.g., $H(Y; q)$.[7] Let δ and δ_2 be positive constants. Throughout this section it is to be understood that by the word "code" we always mean a code with the property that the probability of transmitting any word incorrectly is $< \lambda$.

LEMMA 7.1. *There exists a $K_8 > 0$ with the following property: Let n be any integer. Let $(x_1, A_1), \cdots, (x_N, A_N)$ be any code such that $x_1, \cdots, x_N$ are $\delta q_0 x$-sequences, and A_i, $i = 1, \cdots, N$, contains only y-sequences generated by x_i. Then*

$$N < 2^{nC_0 + K_8 n^{1/2}}. \tag{7.2}$$

Proof. It follows from (3.8) and (4.5) that there exists a $K_8' > 0$ such that the set $(A_1 \cup A_2 \cup \cdots \cup A_N)$ contains at least

[7] Naturally, this is the entropy of Y when the X's are independently distributed and $m = 0$.

(7.3) $$N \cdot 2^{nH_X(Y;q_0)-K_8' n^{1/2}}$$

sequences. By Lemma 6.3 it cannot contain more than

(7.4) $$2^{nH(Y;q_0)+K_7 n^{1/2}}$$

sequences. The lemma follows at once with $K_8 = K_7 + K_8'$.

LEMMA 7.2. *There exists a $K_9 > 0$ with the following property: Let n be any integer. Let $(x_1, A_1), \cdots, (x_N, A_N)$ be any code such that $x_1, \cdots, x_N$ are $\delta q_0 x$-sequences. Then*

(7.5) $$N < 2^{nC_0+K_9 n^{1/2}}.$$

(In other words, the conclusion of Lemma 7.1 holds even if the A_i, $i = 1, \cdots, N$, are not required to consist only of sequences generated by x_i.)

Proof. Let δ_2 be so large that (3.6) holds. From A_i, $i = 1, \cdots, N$, delete the y-sequences not generated by x_i; call the resulting set A_1'. The A_i', $i = 1, \cdots, N$, are of course disjoint. The set $(x_1, A_1'), \cdots, (x_N, A_N')$ fulfills all the requirements of a code except perhaps the one that the probability of correctly transmitting any word is $> 1 - \lambda$. However, from (3.6) it follows that the probability of correctly transmitting any word when this latter set is used is $> 1 - 3\lambda/2$. But now the result of Lemma 7.1 applies,[8] and the present lemma follows. (Of course the constant K_8 of Lemma 7.1 depends on λ, but this does not affect our conclusion.)

LEMMA 7.3. *There exists a constant $K_{10} > 0$ with the following property: Let q be any point in the closed interval* [0, 1], let n *be any integer, and let* $(x_1, A_1), \cdots, (x_N, A_N)$ *be any code such that $x_1, \cdots, x_N$ are $\delta q x$-sequences. Then*

(7.6) $$N < 2^{nC_0+K_{10} n^{1/2}}.$$

Proof. Let q', $0 < q' < \frac{1}{2}$, be such that $H(X; q) < \frac{1}{2} C_0$ if $q < q'$ or $q > 1 - q'$. If $q < q'$ or $q > 1 - q'$, the total number of all $\delta q x$-sequences is less than the right member of (7.6) for suitable $K_{10} > 0$. Then (7.6) holds a fortiori.

It remains to consider the case $q' \leqq q \leqq 1 - q'$. If now one applies the argument of Lemma 7.2 and considers how K_9 depends upon q, one obtains that there exists a positive continuous function $K_9(q)$ of q, $q' \leqq q \leqq 1 - q'$, such that, for any n,

$$N < 2^{n(H(Y;q)-H_X(Y;q))+K_9(q)n^{1/2}}$$

$$\leqq 2^{nC_0+K_9(q)n^{1/2}}.$$

We now increase, if necessary, the constant K_{10} of the previous paragraph so that it is not less than the maximum of $K_9(q)$ in the closed interval $[q', 1 - q']$, and obtain the desired result (7.6).

[8] Except when $3\lambda/2 \geqq 1$. In that case we choose δ_2 so large that the right member of (3.6) is $1 - \lambda/a$, where $a > 0$ is such that $\lambda + \lambda/a < 1$.

Proof of Theorem 2. Divide the interval [0, 1] into $J = n^{1/2}/2\delta$ intervals of length $2\delta/n^{1/2}$ and let $t_1, \cdots, t_J$ be the midpoints of these intervals. Let $(x_1, A_1), \cdots, (x_N, A_N)$ be any code. Then this code is the union of J codes $W_1, \cdots, W_J$ as follows: For $i = 1, \cdots, J$, W_i is that subset of the original code all of whose x-sequences are δt_i x-sequences. By Lemma 7.3 the length of W_i, $i = 1, \cdots, J$, is less than $2^{nC_0+K_{10}n^{1/2}}$ Hence the length N of the original code is less than

$$J \cdot 2^{nC_0+K_{10}n^{1/2}}$$

The theorem follows at once if K' is sufficiently large.

8. Extension to stationary processes

Throughout this section let $X_1, X_2, \cdots$ be a stationary, metrically transitive stochastic process such that X_i, $i = 1, 2, \cdots$, takes only the values one and zero. Define the Y process, $Q(\alpha)$, and $H_X(Y)$ exactly as in Section 4. Let $\varepsilon^* > 0$ be any number, no matter how small, and write $\delta^* = \varepsilon^* n^{1/2}$. Let $\gamma < 1$ be any positive number. From the ergodic theorem we obtain at once the following analogue of (4.4): For n sufficiently large,

$$P\{(X_1, \cdots, X_n) \text{ is a } \delta^*Qx\text{-sequence}\} > \gamma. \tag{8.1}$$

For δ_2 sufficiently large the inequalities (3.6), (3.4), and (3.7) hold exactly as before, and we obtain the following analogue of Lemma 4.1:

LEMMA 8.1. *For any $\varepsilon > 0$, ε^* sufficiently small, and δ_2 sufficiently large, we have, for n sufficiently large and any δ^*Qx-sequence x,*

$$2^{n(H_X(Y)-\varepsilon)} < M(x) < 2^{n(H_X(Y)+\varepsilon)}. \tag{8.2}$$

The following lemmas are proved in [4]:

LEMMA 8.2. *The process $Y_1, Y_2, \cdots$ is metrically transitive.*

LEMMA 8.3. *Let $Z_1, Z_2, \cdots$ be any stationary, metrically transitive stochastic process such that Z_i can take only finitely many values. Let $Z = (Z_1, \cdots, Z_n)$. Define the chance variable (function of Z) $P\{Z\}$ as follows: When Z is the sequence z, $P\{Z\} = P\{Z = z\}$. Then $-(1/n) \log P\{Z\}$ converges stochastically to a constant.*

For our Y process the constant limit of Lemma 8.3 is called the entropy $H(Y)$ of the Y process. This definition of $H(Y)$ is easily verified to be consistent with that of Section 4.

Lemma 8.3 implies the following analogue of (4.23) for our Y process: Let $\varepsilon' > 0$ be any number. Then, for n sufficiently large,

$$P\{-n(H(Y)+\varepsilon') < \log P\{Y\} < -n(H(Y)-\varepsilon')\} > 1 - \varepsilon'. \tag{8.3}$$

Exactly as (4.23) easily implies Lemma 4.3, so (8.3) implies

LEMMA 8.4. *Let $\varepsilon' > 0$ be any number and let n be sufficiently large for (8.3) to hold. Let B be any set of y-sequences such that*

$$P\{Y \in B\} > \gamma_1 > \varepsilon'.$$

Then the set B must contain at least

$$(\gamma_1 - \varepsilon')2^{n(H(Y)-\varepsilon')}$$

y-sequences.

Now the analogues of all the preliminaries needed to prove Theorem 1 have been established, and we have, by exactly the same proof,

THEOREM 3. *Let $X_1, X_2, \cdots$ be a stationary, metrically transitive stochastic process with states 0 and 1. Let the Y process be as defined in Section 4. Let λ and ε be arbitrary positive numbers. For any n sufficiently large there exists a code of length at least*

$$2^{n(H(Y)-H_X(Y)-\varepsilon)} \tag{8.4}$$

The probability that any word transmitted according to this code will be incorrectly received is less than λ.[9]

The author is grateful to Professor K. L. Chung and Professor J. Kiefer for their kindness in reading the manuscript and for interesting comments.

REFERENCES

1. C. E. SHANNON, *A mathematical theory of communication*, Bell System Tech. J., vol. 27 (1948), pp. 379–423, 623–656.
2. A. FEINSTEIN, *A new basic theorem of information theory*, Transactions of the Institute of Radio Engineers, Professional Group on Information Theory, 1954 Symposium on Information Theory, pp. 2–22.
3. A. KHINTCHINE, *The concept of entropy in the theory of probability*, Uspehi Matem. Nauk (N.S.), vol. 8 no. 3 (55), (1953), pp. 3–20.
4. ———, *On the fundamental theorems of the theory of information*, Uspehi Matem. Nauk (N.S.), vol. 11 no. 1 (67), (1956), pp. 17–75.
5. B. MCMILLAN, *The basic theorems of information theory*, Ann. Math. Statistics, vol. 24 (1953), pp. 196–219.
6. W. FELLER, *An introduction to probability theory and its applications*, New York, John Wiley and Sons, 1950.
7. J. L. DOOB, *Stochastic processes*, New York, John Wiley and Sons, 1953.
8. E. N. GILBERT, *A Comparison of signaling alphabets*, Bell System Tech. J., vol. 31 (1952), pp. 504–522.
9. PAUL LÉVY, *Théorie de l'addition des variables aléatoires*, Paris, Gauthier-Villars, 1937.
10. J. WOLFOWITZ, *The efficiency of sequential estimates and Wald's equation for sequential processes*, Ann. Math. Statistics, vol. 18 (1947), pp. 215–230.

CORNELL UNIVERSITY
ITHACA, NEW YORK

[9] An equivalent way of stating Theorem 3 is to replace $(H(Y) - H_X(Y))$ in (8.4) by $C_2 = \sup (H(Y) - H_X(Y))$, where the supremum operation is over all X processes as described in the theorem, each X process with its associated Y process. Obviously $C_0 \leqq C_1 \leqq C_2$ (C_1 is defined in footnote 5). When $m = 0$ it follows from Theorem 2 that $C_0 = C_1 = C_2$. Hence, when $m = 0$, Theorem 3 is actually weaker than Theorem 1.

ON THE DEVIATIONS OF THE EMPIRIC DISTRIBUTION FUNCTION OF VECTOR CHANCE VARIABLES

BY

J. KIEFER(1) AND J. WOLFOWITZ(2)

1. **Introduction.** Let F be a distribution function (d.f.) on Euclicean m-space, and let $X_1, X_2, \cdots, X_n$, be independent chance variables with the common d.f. F. The empiric d.f. S_n is a chance d.f. defined for any $x = (x_1, \cdots, x_m)$ as follows: $nS_n(x)$ is the number of X_i's, $i=1, \cdots, n$, such that, for $j=1, \cdots, m$, the jth component $X_i^{(j)}$ of X_i is less than x_j. Define

$$D_n = \sup_x |S_n(x) - F(x)|,$$

$$D_n^+ = \sup_x (S_n(x) - F(x)),$$

$$D_n^- = \sup_x (F(x) - S_n(x)),$$

and

$$G_n(r) = P\{D_n < r/n^{1/2}\},$$
$$G_n^+(r) = P\{D_n^+ < r/n^{1/2}\},$$
$$G_n^-(r) = P\{D_n^- < r/n^{1/2}\}.$$

In this paper we prove two theorems, of which the first is the following:

THEOREM 1. *For each m there exist positive constants c_0 and c such that, for all n, all F, and all positive r,*

$$1 - G_n(r) < c_0 e^{-cr^2}, \tag{1.1}$$

$$1 - G_n^+(r) < c_0 e^{-cr^2}, \tag{1.2}$$

$$1 - G_n^-(r) < c_0 e^{-cr^2}. \tag{1.3}$$

The nub of the theorem is, of course, that it sets a minimum rate at which $G_n(r)$, $G_n^+(r)$, and $G_n^-(r)$ go to one as $r \to \infty$, independent of n and F. It is rather curious that a bound independent of F can be given, since the limits of G_n, G_n^+, and G_n^- (as $n \to \infty$) depend on F for $m > 1$. The limits of G_n, G_n^+, and G_n^- for $m=1$ are of course known [1] and independent of F when F is continuous. The limits for $m > 1$ are at present writing unknown.

Received by the editors September 6, 1956.

(1) Research under contract with the Office of Naval Research.

(2) This research was supported by the United States Air Force under Contract No. AF(600)-685 monitored by the Office of Scientific Research.

Theorem 1 for $m=1$ was proved in [2] by a method which took as its point of departure an exact expression for $G_n^+(r)$ due to Smirnov [3]. No such formula is known for the case $m>1$. The method of the present paper is entirely different and does not use the result of [2](3). The extension of the result from $m=1$ to $m=2$ presents difficulties; the extension of the result for $m=2$ to larger values of m by our method of proof is obvious, and proceeds by induction on m. Theorem 1 is used in proving Theorem 2.

The constants c_0 and c in general depend upon m. We make no attempt in this paper to obtain the best possible constants or even to perform some tedious calculations which would improve them. At the end of the proof of Theorem 1 we calculate possible values of c and give some suggestions for improving the constants (it is shown in [2] that 2 is the best value of c for $m=1$; we also show at the end of the proof of Theorem 1 that $c<2$ for $m>1$). We also point out that the supremum operation can be performed over a larger class of sets without affecting the result.

Before stating Theorem 2 we introduce some additional notation. For fixed F and positive integral k, write A_k^m for the subset consisting of every point in Euclidean m-space for which, for $1\leq j\leq m$, the jth coordinate w_j satifies $F_j(w_j)\leq h_j/k\leq F_j(w_j+0)$ for any integers h_j, where F_j is the (marginal) d.f. of $X_i^{(j)}$. Write

$$D_{n,k} = \max_{x\in A_k^m} | S_n(x) - F(x) | .$$
$$D_{n,k}^+ = \max_{x\in A_k^m} [S_n(x) - F(x)],$$
$$D_{n,k}^- = \max_{x\in A_k^m} [F(x) - S_n(x)].$$

as well as

$$G_{n,k}(r) = P\{ D_{n,k} < r/n^{1/2}\},$$
$$G_{n,k}^+(r) = P\{ D_{n,k}^+ < r/n^{1/2}\},$$
$$G_{n,k}^-(r) = P\{ D_{n,k}^- < r/n^{1/2}\},$$

and

$$H_n(r, r') = P\{ D_n^+ < r/n^{1/2},\ D_n^- < r'/n^{1/2}\},$$
$$H_{n,k}(r, r') = P\{ D_{n,k}^+ < r/n^{1/2},\ D_{n,k}^- < r'/n^{1/2}\}.$$

We shall also denote by

$$G_{\infty,k},\qquad G_{\infty,k}^+,\qquad G_{\infty,k}^-,\qquad H_{\infty,k}$$

(3) In a first, unpublished, version of [2], a weaker result than that mentioned below as appearing in [2] for the case $m=1$, was proved by a method which has points in common with the present proof of Theorem 1; one idea used in this method is due to P. Erdös.

the respective limits as $n \to \infty$ of the d.f.'s $G_{n,k}$, $G^+_{n,k}$, $G^-_{n,k}$, and $H_{n,k}$; the existence of these limits is a consequence of the multivariate central limit theorem.

Our second result is

THEOREM 2. *For every m and F, there exists a d.f. G (resp., G^+, G^-, H) such that the sequence of d.f.'s G_n (resp., G^+_n, G^-_n, H_n) converges to G (resp., G^+, G^-, H) at every continuity point of the latter as $n \to \infty$ and such that the sequence of d.f.'s $G_{\infty,k}$ (resp., $G^+_{\infty,k}$, $G^-_{\infty,k}$, $H_{\infty,k}$) converges to this d.f. G (resp., G^+, G^-, H) at every continuity point of the latter as $k \to \infty$.*

It is obvious that G, G^+, G^-, and H cannot be degenerate unless F is. Of course, as noted above, these d.f.'s depend on F.

Theorem 2 generalizes the result of Donsker [4] for the case $m=1$; we remark that our proof starts ab initio and does not make use of Donsker's result or method. Donsker's result is needed to justify Doob's [5] computation of G, G^+, G^-, and H in the case $m=1$, and Theorem 2 could perhaps prove of similar use in the more difficult problem of computing these limiting d.f.'s when $m>1$ if this is to be done by consideration of a Gaussian process (depending on F) with m-dimensional time. Donsker's result was also used in [2] in the case $m=1$ for proving certain asymptotic optimum properties of S_n in estimating F.

(*Added in proof*: In another paper we shall prove that analogous optimality results hold for S_n when $m>1$, *even though D_n is no longer distribution free and the distribution theory of D_n is unknown.* These results follow from those of the present paper, arguments like those of [1], and the fact that the integral of a continuous bounded function with respect to G_n converges to that with respect to G uniformly in continuous F; the latter result will also appear in another paper.)

Some generalizations of Theorem 2 are mentioned at the end of §3.

2. **Proof of Theorem 1.** We shall give a detailed proof for $m=2$. As we have remarked earlier, the proof for general m is by induction on m and is obvious to carry out. We shall indicate below the point where induction would be used. The result for the case $m=1$ can be obtained by an argument similar to but simpler than that used below. Alternatively, it can be obtained from Lemma 2 of [2].

Throughout this section c_0 and c will be a generic notation for positive constants which do not depend on n, r, and F. Hence these symbols in different places will not, in general, stand for the same numbers. No confusion will be caused by this.

We have

$$(1 - G_n(r)) \leq (1 - G^+_n(r)) + (1 - G^-_n(r)).$$

We will content ourselves with proving (1.2). The proof of (1.3) follows in the same fashion, and (1.1) then follows from (1.2) and (1.3).

We shall assume that F is continuous. If this is not so (1.1), (1.2), and (1.3) hold a fortiori; the proof of this is the same as in the one-dimensional case and therefore obvious. Since F is continuous we may transform $X_i^{(1)}$ and $X_i^{(2)}$ separately so that the marginal d.f.'s F_j are uniform on $[0, 1]$ without changing $G_n(r)$, $G_n^+(r)$, or $G_n^-(r)$; we hereafter assume that the F_j are uniform on $[0, 1]$.

In the discussion which follows we shall always assume, to simplify the discussion, that, for any given number x_1, there is at most one i such that $X_i^{(1)}=x_1$. The probability that this be not so is zero.

In the course of the proof we shall always assume that $r<n^{1/2}$. The theorem is trivially true when this is not so.

If the theorem is true for all $r>R>0$, it is true for all $r\geqq 0$. One has only to enlarge c_0, if necessary, so that $c_0e^{-cR^2}>1$; the inequalities (1.1), (1.2), and (1.3) are then trivially true. It will therefore be sufficient to prove the theorem for all r sufficiently large, say $>R>3$. Then $n>R^2$.

Since n and r will be fixed in the present discussion we may allow ourselves the luxury of a notation simpler than that of the next section and not display all dependences on n and r. (We remind the reader that c and c_0 will not depend on n and r.) Define the events

$$L = \left\{S_n(x_1, x_2) - F(x_1, x_2) > \frac{r}{n^{1/2}} \text{ for some } x_1 \leqq \frac{1}{2}\right\},$$

$$L' = \left\{S_n(x_1, x_2) - F(x_1, x_2) < -\frac{r}{n^{1/2}} \text{ for some } x_1 \leqq \frac{1}{2}\right\},$$

$$B = \left\{S_n\left(\frac{1}{2}, x_2\right) - F\left(\frac{1}{2}, x_2\right) > \frac{r}{4n^{1/2}} \text{ for some } x_2\right\},$$

$$\bar{L} = \left\{S_n(x_1, x_2) - F(x_1, x_2) > \frac{r}{n^{1/2}} \text{ for some } x_1 > \frac{1}{2}\right\},$$

$$L^* = \left\{S_n(x_1, x_2) - F(x_1, x_2) > \frac{r}{n^{1/2}} \text{ for some } x_1 > \frac{1}{2},\ x_2 > \frac{1}{2}\right\}.$$

Define the chance variables (z_1, z_2) when the event L occurs (we shall not need them when L does not occur) as follows: First, z_1 is the infimum of those values x_1 $(\leqq 1/2)$ for which $\sup_{x_2}[S_n(x_1, x_2)-F(x_1, x_2)]>r/n^{1/2}$. There is then an i such that $X_i^{(1)}=z_1$. We define $z_2=X_i^{(2)}$. We now define the event $L(x_1, x_2)$ for any pair x_1, x_2, $0\leqq x_1\leqq 1/2$, $0\leqq x_2\leqq 1$, as follows: $L(x_1, x_2)$ is the subset of L where $z_1=x_1$, $z_2=x_2$.

Define $r(x_1, x_2)$ as

$$\frac{1}{n^{1/2}}\left[(\text{least integer} > nF(x_1, x_2) + n^{1/2}r) - nF(x_1, x_2)\right].$$

Then, for almost all values (x_1, x_2) of (z_1, z_2), the event $L(x_1, x_2)$ implies the event

$$\left\{S_n(x_1, x_2) - F(x_1, x_2) = \frac{r(x_1, x_2)}{n^{1/2}}\right\}. \tag{2.1}$$

Define, for any x_2, $0 \leqq x_2 \leqq 1$, the events

$$B(x_2) = \left\{S_n\left(\frac{1}{2}, x_2\right) - F\left(\frac{1}{2}, x_2\right) > \frac{r}{4n^{1/2}}\right\}$$

and

$$J = \bigcup_{0 \leq x_1 \leq 1/2} \bigcup_{0 \leq x_2 \leq 1} L(x_1, x_2) B(x_2).$$

Obviously

$$1 - G_n^+(r) \leqq P\{L\} + P\{\bar{L}\}.$$

Our immediate goal will now be to prove

$$P\{L\} < c_0 e^{-cr^2} \tag{2.2}$$

for all F, and for r sufficiently large, say $>R$.

We have

$$P\{L\} \cdot \text{ess. inf. } P\{J \mid z_1, z_2\} \leqq \int_{0 \leqq x_1 \leqq 1/2; 0 \leqq x_2 \leqq 1} P\{J \mid x_1, x_2\} d_{x_1, x_2} P\{z_1 < x_1, z_2 < x_2\}$$
$$= P\{J \cap L\} \leqq P\{J\} \leqq P\{B\}.$$

Hence

$$P\{L\} \leqq \frac{P\{B\}}{\text{ess. inf. } P\{J \mid z_1, z_2\}}. \tag{2.3}$$

Our plan to prove (2.2) is as follows: First, we shall prove that

$$\text{ess. inf. } P\{J \mid z_1, z_2\} \geqq 1/2. \tag{2.4}$$

Then we shall prove that

$$P\{B\} < c_0 e^{-cr^2} \tag{2.5}$$

for $r > R$ and $n > r^2$.

Suppose the event $L(x_1, x_2)$ has occurred. Since there is exactly one i such that $X_i^{(1)} = x_1$, and since

$$\sup_{x_2} [S_n(x_1', x_2) - F(x_1', x_2)] \leqq \frac{r}{n^{1/2}}$$

for $x_1' < x_1$, we have

$$nS_n(x_1, 1) < nF(x_1, 1) + rn^{1/2} + 1. \tag{2.6}$$

Hence the number N of $X_1, \cdots, X_n$ which have first coordinate greater than x_1 is at least

$$M = n(1 - x_1) - rn^{1/2} - 1. \tag{2.7}$$

First suppose that $M \leqq 0$. Then

$$\frac{r}{n^{1/2}} \geqq -\frac{1}{n} - (1 + x_1). \tag{2.8}$$

Obviously $0 \leqq r(x_1, x_2) - r \leqq 1/n^{1/2}$. The event $B(x_2)$ occurs when

$$S_n\left(\frac{1}{2}, x_2\right) > F\left(\frac{1}{2}, x_2\right) + \frac{r}{4n^{1/2}}. \tag{2.9}$$

From (2.1) and (2.8) we obtain that

$$\begin{aligned} S_n\left(\frac{1}{2}, x_2\right) &\geqq S_n(x_1, x_2) \geqq F(x_1, x_2) + \frac{r}{n^{1/2}} \\ &\geqq F(x_1, x_2) + (1 - x_1) - \frac{1}{n} \\ &= F\left(\frac{1}{2}, x_2\right) - \left[F\left(\frac{1}{2}, x_2\right) - F(x_1, x_2)\right] + (1 - x_1) - \frac{1}{n} \qquad (2.10) \\ &\geqq F\left(\frac{1}{2}, x_2\right) - \left(\frac{1}{2} - x_1\right) + (1 - x_1) - \frac{1}{n} \\ &= F\left(\frac{1}{2}, x_2\right) + \frac{1}{2} - \frac{1}{n}. \end{aligned}$$

Since $r/4n^{1/2} < 1/4$, it follows that, for $n > R^2 > 9$ (which is all we need consider), (2.9) holds.

Suppose now that $M > 0$. Let R_0 be the region in the x_1', x_2' plane defined by the inequalities

$$x_1 < x_1' \leqq 1/2, \qquad 0 \leqq x_2' \leqq x_2.$$

In order for $B(x_2)$ to occur it is sufficient, by (2.9) and (2.1), that, of the N chance variables among $X_1, \cdots, X_n$ whose first coordinates are greater than x_1, at least

$$n(F(1/2, x_2) - F(x_1, x_2)) - 3rn^{1/2}/4 \tag{2.11}$$

take on values in R_0. We shall compute a lower bound for the probability of this under the assumption that $N = M$ (>0). It will be easy to see that, if

$N > M$ or M is not an integer, the probability is a fortiori greater than this lower bound, which is $>1/2$ for $r > R$. This will prove (2.4).

If we define

$$p = \frac{F(1/2, x_2) - F(x_1, x_2)}{1 - x_1} \tag{2.12}$$

and

$$t = \frac{n(F(1/2, x_2) - F(x_1, x_2)) - 3rn^{1/2}/4 - Mp}{(Mp(1 - p))^{1/2}}, \tag{2.13}$$

we obtain that

$$t = \frac{r(p - 3/4) + p/n^{1/2}}{\left(\frac{Mp(1-p)}{n}\right)^{1/2}}.$$

Since $p < 1/2$ and $M/n \leqq 1$ it follows that $t < -r/4$ for $r > R$. The probability in question is the probability that, of M independent Bernoulli chance variables with common probability p of a "success," the number N^* of "successes" satisfy the inequality

$$\frac{N^* - EN^*}{(E(N^* - EN^*)^2)^{1/2}} > t. \tag{2.14}$$

This probability is greater than

$$P\left\{\frac{N^* - EN^*}{(E(N^* - EN^*)^2)^{1/2}} > -\frac{r}{4}\right\} \tag{2.15}$$

which, by Chebyshev's inequality, is greater than $1 - 16/r^2$, which, for $r > R$, is $>1/2$, as was to be proved. This proves (2.4).

From [6, p. 288, Equation (96)], it follows that, for $r > R$,

$$P\left\{\left|S_n\left(\frac{1}{2}, 1\right) - \frac{1}{2}\right| > \frac{r}{16n^{1/2}}\right\} < c_0e^{-cr^2}. \tag{2.16}$$

Suppose now that the event $\{|S_n(1/2, 1) - 1/2| \leqq r/16n^{1/2}\}$ occurs. Then the number n_1 of chance variables $X_1, \cdots, X_n$ with first coördinate not greater than 1/2 satisfies

$$\frac{n}{2} - \frac{rn^{1/2}}{16} = n_2 < n_1 < n_3 = \frac{n}{2} + \frac{rn^{1/2}}{16} < \frac{9n}{16}.$$

Let $T_{n_1}(x_2)$ denote $1/n_1$ multiplied by the number of chance variables $X_1, \cdots, X_n$ whose first coordinates are less than 1/2 and whose second coördinates are less than x_2. The relation

$$nS_n\left(\frac{1}{2}, x_2\right) - nF\left(\frac{1}{2}, x_2\right) > \frac{rn^{1/2}}{4} \tag{2.17}$$

implies the relation

$$n_1T_{n_1}(x_2) - n_1\left[2F\left(\frac{1}{2}, x_2\right)\right] > \frac{rn^{1/2}}{4} - \frac{rn^{1/2}}{8}F\left(\frac{1}{2}, x_2\right), \tag{2.18}$$

whose right member is not less than

$$\frac{3rn^{1/2}}{16} > \frac{r(n_3)^{1/2}}{4} > \frac{r(n_1)^{1/2}}{4}. \tag{2.19}$$

The theorem for the case $m=1$ implies that

$$P\left\{\sup_{x_2}\left[n_1T_{n_1}(x_2) - n_1\left(2F\left(\frac{1}{2}, x_2\right)\right)\right] > \frac{r(n_1)^{1/2}}{4}\right\} < c_0e^{-cr^2}. \tag{2.20}$$

Equations (2.16) to (2.20) prove (2.5) and hence (2.2).

In the proof of Theorem 1 for general m the induction on m would occur at this point. We have just used the theorem for $m=1$ to prove (2.5) for $m=2$. We can then use this to prove the result corresponding to (2.5) for $m=3$, and so on (x_2 represents all variables other than x_1 in this proof, when $m>2$).

Returning to the case $m=2$, the proof of

$$P\{L'\} < c_0e^{-cr^2} \tag{2.21}$$

is practically the same as that of (2.2), and will be omitted. We shall henceforth assume that (2.21) holds, and use this fact to prove that

$$P\{\bar{L}\} < c_0e^{-cr^2}. \tag{2.22}$$

First, applying the result (2.2) to the chance variables $X_1^*, X_2^*, \cdots$ defined by $X_i^* = (X_i^{(2)}, X_i^{(1)})$, we obtain for the original sequence $X_1, X_2, \cdots$ that

$$P\left\{S_n(x_1, x_2) - F(x_1, x_2) > \frac{r}{n^{1/2}} \text{ for some } x_2 \leq \frac{1}{2}\right\} < c_0e^{-cr^2}. \tag{2.23}$$

For any pair (x_1, x_2), $1/2<x_1\leq 1$, $1/2<x_2\leq 1$, we define the following regions in the x_1', x_2' plane:

$$U_1(x_1, x_2) = \{x_1', x_2' \mid x_1 < x_1' \leq 1, x_2 < x_2' \leq 1\},$$
$$U_2(x_1, x_2) = \{x_1', x_2' \mid x_1 < x_1' \leq 1, 0 \leq x_2' < x_2\},$$
$$U_3(x_1, x_2) = \{x_1', x_2' \mid 0 \leq x_1' < x_1, x_2 < x_2' \leq 1\},$$

and the following events for $i=1, 2, 3$:

$$Q_i = \left\{\text{for some } (x_1, x_2), \frac{1}{2} < x_1 \leqq 1, \frac{1}{2} < x_2 \leqq 1, \text{ the number of } (X_1, \cdots, X_n) \right.$$
$$\left. \text{in } U_i(x_1, x_2) \text{ minus expected number } < -\frac{rn^{1/2}}{3}\right\}.$$

Obviously

$$L^* \subset Q_1 \cup Q_2 \cup Q_3.$$

We shall prove

$$P\{Q_v\} < c_0 e^{-cr^2}, \qquad v = 1, 2, 3. \tag{2.24}$$

The result (2.24) follows for $v=1$ from the application of (2.21) to the sequence of chance variables $(1-X_i^{(2)}, 1-X_i^{(1)})$, for $v=2$ by the application of (2.21) to the sequence of chance variables $(1-X_i^{(1)}, X_i^{(2)})$, and for $v=3$ by the application of (2.21) (in the form (2.23)) to the sequence of chance variables $(X_i^{(1)}, 1-X_i^{(2)})$. Thus, (2.24) is proved, and this and (2.23) imply (2.22) and hence (1.2).

The proof of (1.3) is completed in a similar manner. Obviously (1.2) and (1.3) imply (1.1). This completes the proof of Theorem 1.

We shall now obtain explicit possible values for the constant c (c_0 could be obtained similarly, but this is of less interest). First consider the case $m=2$. In the definition of the set B, let us replace $r/4$ by $r/(2+\epsilon)$ with $\epsilon>0$; the proof of Theorem 1 then still holds, but will yield a larger value of c. Making appropriate changes in the argument, an analogue of (2.15) holds, as before. In (2.16) and what follows, put λ for $1/16$. The constant c on the right side of (2.16) then becomes $2\lambda^2$. The displayed inequality on n_1 becomes

$$n\left(\frac{1}{2} - \lambda\right) < \frac{n}{2} - r\lambda n^{1/2} < n_1 < \frac{n}{2} + r\lambda n^{1/2} < n\left(\lambda + \frac{1}{2}\right). \tag{2.25}$$

Equation (2.17) (with $1/(2+\epsilon)$ for $1/4$) and Equation (2.25) imply an analogue of (2.18) with (2.19) replaced by

$$rn^{1/2}\left(\frac{1}{2+\epsilon} - \lambda\right) > rn_1^{1/2}\left(\frac{1}{2+\epsilon} - \lambda\right) \Big/ \left(\lambda + \frac{1}{2}\right)^{1/2}.$$

The fact that we can take $c=2$ for $m=1$ implies, in place of (2.20),

$$P\left\{\sup_{x_2}\left| n_1 T_{n_1}(x_2) - 2n_1 F\left(\frac{1}{2}, x_2\right)\right| > rn^{1/2}\left(\frac{1}{2+\epsilon} - \lambda\right)\Big/\left(\lambda + \frac{1}{2}\right)^{1/2}\right\}$$
$$< c_0 \exp\left(-2r^2\left(\frac{1}{2+\epsilon} - \lambda\right)^2 \Big/ \left(\lambda + \frac{1}{2}\right)\right). \tag{2.26}$$

The minimum of the coefficients of r^2 in the exponents of (2.26) and the analogue of (2.16) is maximized when ϵ is small by taking $\lambda=.266$; this gives $c>.142+o(1)$ as $\epsilon\to 0$, in (2.5). The same value of c may be obtained similarly for L' in (2.21), and also in (2.23). If this value of c is multiplied by 1/9, we obtain a value applicable in (2.24), (2.22), and (1.2); a similar argument applies for (1.1). Thus, we also obtain $c>.0157-\epsilon'$ in (1.3) for $m=2$, where ϵ' is an arbitrary positive value. Thus, $c=.0157$ is a possible value in (1.3) for $m=2$.

For general m, we may similarly obtain a possible value for c. Let $2d_m^2$ be the value for c obtained by this argument for dimension m, with $d_m>0$. For dimension m we then obtain $2d_{m-1}^2(1/2-\lambda)^2$ for the coefficient of r^2 in (2.26) and thus the solution λ of the equation $2\lambda^2=2d_{m-1}^2(1/2-\lambda)^2/(\lambda+1/2)$ is the value of λ which maximizes the minimum of the coefficients of r^2 in (2.16) and (2.26). Rather than carry out the obvious analogue of the case $m=2$ in terms of this inexplicit λ, we shall obtain explicitly a slightly smaller value of c. This value is suggested by the fact that d_{m-1}, and hence the above λ, is small for $m>2$. Taking then for λ the value $d_{m-1}/2^{1/2}$, the two coefficients of r^2 are almost equal, the smaller (that of (2.26)) being

$$2d_{m-1}^2(1/2-d_{m-1}/2^{1/2})^2/(1/2+d_{m-1}/2^{1/2}).$$

The factor 1/9 above must be replaced by $(2^m-1)^{-2}$. Thus, we obtain for possible values of d_m and c (the ϵ' no longer being needed):

$$c = 2d_m^2 \quad \text{(dimension } m\text{)},$$

$$(2.27) \qquad d_m = \frac{d_{m-1}(1/2 - d_{m-1}/2^{1/2})}{(1/2 + d_{m-1}/2^{1/2})^{1/2}(2^m-1)} \qquad (d_2 = .088).$$

The above possible value for c is probably not a very good one ($c=.0157$ for $m=2$ and $c=.000107$ for $m=3$). It could be improved by considering $S_n(x)-F(x)$ at a large finite number of lines (in the case $m=2$, for example) instead of just on the line $x_1=1/2$; but this would be at the expense of more tedious computations. The value $c=2$ obtained in [2] for the case $m=1$ is the best possible in the sense that (1.3) is clearly false for any $c>2$ and any c_0. We next show that $c<2$ for $m>1$; i.e., (1.3) is false for $c=2$ and any c_0 when $m>1$ (in fact, this is so even if c_0 is permitted to depend on F).

In fact, consider the case $m=2$ and suppose F_1 is the d.f. which distributes all probability uniformly on the line L: $x_1+x_2=1$. Then $S_n(x, y)=0$ w.p. 1 if $x+y\leq 1$, and for $x+y>1$ we have $nS_n(x, y)=$ number of observations on L between $(1-y, y)$ and $(x, 1-x)$. Let $S_n^*(u)=S_n(u, 1)$. Of course, $S_n^*(u)$ is a univariate sample d.f. corresponding to the uniform d.f. on the unit interval. Denoting by D_n^+ and D_n^- the supremum positive and negative deviations of $S_n^*(u)$ from the function u, $0\leq u\leq 1$, we have, w.p.1.,

$$\sup_{x,y} | S_n(x, y) - F_1(x, y) | = \sup_{u,v} | (S_n^*(u) - u) - (S_n^*(v) - v) | = D_n^+ + D_n^-.$$

From [5] we obtain for the limiting d.f. of $n^{1/2}(D_n^+ + D_n^-)$ (note, e.g., Equation (4.6) of [7], which gives the limiting d.f. of $n^{1/2}(D_n^+ + D_n^-)/2$), as $r \to \infty$,

$$\lim_{n\to\infty} P\{n^{1/2}(D_n^+ + D_n^-) > r\} \sim 8r^2e^{-2r^2},$$

which demonstrates the impossibility of taking $c=2$. We note, in fact, that (1.3) cannot hold for $c=2$ and any c_0 and for all absolutely continuous F (or, instead, for all *discrete* F); this is obtained easily from the above result by taking a fixed r so large that $4r^2 > c_0$, a k so large that the above limiting probability for the case of the discrete approximation of F_1 is $> 6r^2e^{-2r^2}$, and an absolutely continuous d.f. F_2 whose probability is concentrated on such small spheres about the discrete points that the probability of a deviation $> r$ cannot be smaller for F_2 than for F_1.

The supremum operation involved in the definition of D_n, D_n^+, and D_n^- is over all sets of the form $x_j \leqq a_j$, $j=1, \cdots, m$, for all $a=(a_1, \cdots, a_m)$ in m-space. It is obvious that Theorem 1 applies also to the case when the supremum is taken over any of several larger classes of sets such as, for example, that which consists of all rectangular parallelepipeds with sides parallel to the coördinate planes. This will be of interest in statistical applications where it is often required or desired that the results be invariant under certain transformations of the chance variables, e.g., $X \to -X$.

3. **Proof of Theorem** 2. Let I^m denote the closed unit m-cell $\{x \mid 0 \leqq x_1, \cdots, x_m \leqq 1\}$. We shall first prove Theorem 2 for the case when F is continuous, and then, at the conclusion of the proof, we shall remark on how the proof proceeds for discontinuous F. As in §2, since F is now assumed continuous, it suffices to consider the case where all F_j's are uniform on $[0, 1]$, and we hereafter assume we are in this case. Write $Q_{k,0} = I^m$, and for $j > 0$ let $Q_{k,j}$ be the subset of I^m whose first j coördinates are integral multiples of $1/k$ (thus, $Q_{k,m} = A_k^m$). Write

$$D_{n,k,j}^+ = \sup_{x\in Q_{k,j}} [S_n(x) - F(x)],$$

$$D_{n,k,j}^- = \sup_{x\in Q_{k,j}} [F(x) - S_n(x)].$$

For fixed $d > 0$, r, and r', we shall show in the succeeding paragraphs that

$$\tag{3.1} \begin{aligned} &\limsup_{k\to\infty} \limsup_{n\to\infty} [P\{D_{n,k,j}^+ < r/n^{1/2};\ D_{n,k,j}^- < r'/n^{1/2}\} \\ &\quad - P\{D_{n,k,j-1}^+ < (r+d)/n^{1/2};\ D_{n,k,j-1}^- < (r'+d)/n^{1/2}\}] \leqq 0 \end{aligned}$$

for $1 \leqq j \leqq m$. Adding these inequalities over j yields

$$\text{(3.2)} \qquad \limsup_{k\to\infty} \limsup_{n\to\infty} [H_{n,k}(r, r') - H_n(r + md, r' + md)] \leqq 0.$$

We have remarked in §1 that $\lim_n H_{n,k} = H_{\infty,k}$ exists. Hence, writing r for $r+md$ and r' for $r'+md$, (3.2) becomes

$$\text{(3.3)} \qquad \limsup_{k\to\infty} H_{\infty,k}(r - md, r' - md) \leqq \liminf_{n\to\infty} H_n(r, r').$$

Write $H^*(r, r') = \limsup_k H_{\infty,k}(r, r')$. Since obviously $H_n(r, r') \leqq H_{n,k}(r, r')$, we obtain from (3.3),

$$\text{(3.4)} \quad H^*(r - md, r' - md) \leqq \liminf_{n\to\infty} H_n(r, r') \leqq \limsup_{n\to\infty} H_n(r, r') \leqq H^*(r, r').$$

Since H^* is clearly monotone and bounded, it is continuous except possibly on a denumerable set of lines parallel to the coördinate axes. Letting d tend to zero in (3.4) at continuity points (r, r') of H^*, we see that $\lim_n H_n(r, r')$ exists for all points (r, r') in the plane, except possibly on a denumerable set of lines parallel to the coördinate axes. This limit determines a left-continuous function H (say) which has variation one by Theorem 1 and which is clearly a d.f. Hence the sequence H_n converges to a d.f. H at every continuity point of the latter. Finally, since clearly we can also write, for all continuity points (r, r') of H,

$$\text{(3.5)} \qquad \begin{aligned} H(r, r') &\leqq \liminf_{k\to\infty} H_{\infty,k}(r, r') \\ &\leqq \limsup_{k\to\infty} H_{\infty,k}(r, r') \leqq H(r + md, r' + md), \end{aligned}$$

letting $d \to 0$ shows that $\lim_k H_{\infty,k}(r, r') = H(r, r')$ at all continuity points of the latter, and hence that $H_{\infty,k}$ converges to H at every continuity point of the latter as $k \to \infty$. Thus, the theorem for H will be proved if we can show (3.1), and the result for G, G^+, and G^- can be obtained easily from this result or else can be proved directly in the same manner as the result for H.

We now prove (3.1). Fix $d > 0$, r, r', and j. For h an integer, write $V_{k,j,h}$ for the subset of I^m where $h/k < x_j \leqq (h+1)/k$ and write $x^{(j)} = (x_1, \cdots, x_{j-1}, h/k, x_{j+1}, \cdots, x_m)$ if $x \in V_{k,j,h}$. Clearly, if the event

$$\Lambda_{n,k} = \left\{ \overset{+}{D}_{n,k,j} < r/n^{1/2},\ \overset{-}{D}_{n,k,j} < r'/n^{1/2} \right\}$$

occurs and the event

$$\overset{*}{\Lambda}_{n,k} = \left\{ \overset{+}{D}_{n,k,j-1} < (r + d)/n^{1/2},\ \overset{-}{D}_{n,k,j-1} < (r' + d)/n^{1/2} \right\}$$

does not occur, it is necessary that for some h with $0 \leqq h \leqq k-1$ the event

$$\Gamma_{n,k,h} = \left\{ \sup_{x \in V_{k,j,h}} \left| [S_n(x) - F(x)] - [S_n(x^{(j)}) - F(x^{(j)})] \right| > d/n^{1/2} \right\}$$

occurs. Write N_{nkjh}=number of X_i $(1 \leqq i \leqq n)$ whose values are in $V_{k,j,h}$. Now,

$$(3.6)\quad \begin{aligned} P\{\Gamma_{n,k,h}\} &\leqq P\left\{\sup_{V_{k,j,h}}\left|\frac{S_n(x)-S_n(x^{(j)})}{N_{nkjh}/n} - k\left[F(x)-F(x^{(j)})\right]\right| > dn^{1/2}/2N_{nkjh}\right\} \\ &\quad + P\left\{\left|\frac{kN_{nkjh}}{n}-1\right| \sup_{V_{k,j,h}}\left[F(x)-F(x^{(j)})\right] > d/2n^{1/2}\right\} \\ &= P\{Y_{nkjh}\} + P\{Z_{nkjh}\} \text{ (say).} \end{aligned}$$

Now, given that $N_{nkjh}=N$, the event Y_{nkjh} is a subset of the event that the maximum deviation of the empiric d.f. of N independent, identically distributed m-variate random variables from the corresponding theoretical d.f. is more than $(d/2)(n/N)^{1/2}/N^{1/2}$. Hence, by Theorem 1,

$$(3.7)\quad P\{Y_{nkjh} \mid N_{nkjh} = N\} < c_0 \exp\{-cd^2n/N\}$$

where c_0 and c are positive constants. Also, by Chebyshev's inequality,

$$(3.8)\quad P\{N_{nkjh} \leqq n/k + n/k^{1/2}\} > 1 - 1/n.$$

From (3.7) and (3.8) we obtain

$$(3.9)\quad P\{Y_{nkjh}\} < n^{-1} + c_0 \exp\{-cd^2k^{1/2}/2\}.$$

Also, an application of Chebyshev's inequality based on the fourth moment yields

$$(3.10)\quad \begin{aligned} P\{Z_{nkjh}\} &= P\left\{\left|\frac{N_{nkjh}}{n}-\frac{1}{k}\right| > d/2n^{1/2}\right\} \\ &< \frac{3n^2/k^2 + n/k}{n^4[d^4/16n^2]} = \frac{16}{d^4}\left[\frac{3}{k^2}+\frac{1}{nk}\right] \end{aligned}$$

From (3.6), (3.9), and (3.10) we obtain

$$(3.11)\quad P\{\Lambda_{n,k} - \Lambda^*_{n,k}\} < \frac{k}{n} + kc_0 \exp\{-cd^2k^{1/2}/2\} + \frac{16}{d^4}\left[\frac{3}{k}+\frac{1}{n}\right],$$

which proves (3.1) and, hence, Theorem 2 in the continuous case.

We now remark on the method of proving Theorem 2 when F has discontinuities. The conclusion follows by the same method as that used to prove Theorem 2 for continuous F, upon noting the manner in which any discontinuous F can be obtained from a continuous one by "lumping together" (in the same manner as that used to obtain Theorem 1 for discontinuous F) certain points in the domain of the latter.

Generalizations of the theorem may be obtained by noting that, as in the case of Theorem 1, the conclusion of Theorem 2 holds if the supremum of observed from theoretical frequency is taken over a larger class of sets than those of the form $x_j \leqq a_j$, $j=1, \cdots, m$ (for all a in m-space).

References

1. A. N. Kolmogorov, *Sulla determinazione empirica di una legge di distribuzione*, Inst. Ital. Atti. Giorn. vol. 4 (1933) pp. 83–91.

2. A. Dvoretzky, J. Kiefer, and J. Wolfowitz, *Asymptotic minimax character of the sample distribution function and of the classical multinomial estimator*, Ann. Math. Stat. vol. 27 (1956) pp. 642–669.

3. N. V. Smirnov, *Approximation of the distribution law of a random variable by empirical data*, Uspehi Matematičeskih Nauk vol. 10 (1944) pp. 179–206.

4. M. Donsker, *Justification and extension of Doob's heuristic approach to the Kolmogorov-Smirnov theorems*, Ann. Math. Stat. vol. 23 (1952) pp. 277–281.

5. J. L. Doob, *Heuristic approach to the Kolmogorov-Smirnov theorems*, Ann. Math. Stat. vol. 20 (1949) pp. 393–403.

6. P. Lévy, *Théorie de l'addition des variables aléatoires*, Paris, Gauthier-Villars, 1937.

7. M. Kac, J. Kiefer and J. Wolfowitz, *On tests of normality and other goodness of fit based on distance methods*, Ann. Math. Stat. vol. 26 (1955) pp. 189–211.

Cornell University,
Ithaca, N. Y.

Reprinted from The Annals of Mathematical Statistics
Vol. 30, No. 2, June, 1959

OPTIMUM DESIGNS IN REGRESSION PROBLEMS

By J. Kiefer[1] and J. Wolfowitz[2]

Cornell University

1. Introduction and Summary. Although regression problems have been considered by workers in all sciences for many years, until recently relatively little attention has been paid to the optimum design of experiments in such problems. At what values of the independent variable should one take observations, and in what proportions? The purpose of this paper is to develop useful computational procedures for finding optimum designs in regression problems of estimation, testing hypotheses, etc. In Section 2 we shall develop the theory for the case where the desired inference concerns just one of the regression coefficients, and illustrative examples will be given in Section 3. In Section 4 the theory for the case of inference on several coefficients is developed; here there is a choice of several possible optimality criteria, as discussed in [1]. In Section 5 we treat the problem of global estimation of the regression *function*, rather than of the individual coefficients.

We shall now indicate briefly some of the computational aspects of the search for optimum designs by considering the problem of Section 2 wherein the inference concerns one of k regression coefficients. For the sake of concreteness, we shall occasionally refer here to the example of polynomial regression on the real interval $[-1, 1]$, where all observations are independent and have the same variance. The quadratic case is rather trivial to treat by our methods, so we shall sometimes refer here to the case of cubic regression. In the latter case we suppose all four regression coefficients to be unknown, and we want to estimate or test a hypothesis about the coefficient a_3 of x^3. If a fixed number N of observations is to be taken, we can think of representing the proportion of observations taken at any point x by $\xi(x)$, where ξ is a probability measure on $[-1, 1]$. To a first approximation (which is discussed in Section 2), we can ignore the fact that in what follows $N\xi$ can take only integer values. We consider three methods of attacking the problem of finding an optimum ξ:

A. The direct approach is to compute the variance of the best linear estimator of a_3 as a function of the values of the independent variable at which observations are taken or, equivalently, as a function of the moments of ξ. Denoting by μ_i the ith moment of ξ, and assuming ξ to be concentrated entirely on more than three points (so that a_3 is estimable), we find easily that the *reciprocal* of

Received April 21, 1958; revised November 25, 1958.

[1] Research under contract with the Office of Naval Research.

[2] The research of this author was supported by the U. S. Air Force under Contract No. AF 18(600)-685, monitored by the Office of Scientific Research.

this variance is proportional to

$$\frac{\begin{array}{r}\mu_5^2(\mu_1^2 - \mu_2) + 2\mu_5(\mu_2^2\mu_3 + \mu_3\mu_4 - \mu_1\mu_3^2 - \mu_1\mu_2\mu_4) \\ - \mu_4^3 + \mu_4^2(\mu_2^2 + 2\mu_1\mu_3) - 3\mu_4\mu_2\mu_3^2 + \mu_3^4\end{array}}{\mu_4(\mu_2 - \mu_1^2) - \mu_3^2 - \mu_2^3 + 2\mu_1\mu_2\mu_3} + \mu_6$$

in the case of cubic regression.

The problem is to find a ξ on $[-1, 1]$ which maximizes this expression. Thus, this direct approach leads to a calculation which appears quite formidable. This is true even if one uses the remark on symmetry of the next paragraph and restricts attention to symmetrical ξ, so that $\mu_i = 0$ for i odd. For polynomials of higher degree or for regression functions which are not polynomials, the difficulties are greater.

B. The results of Section 2 yield the following approach to the problem: Let $c_0 + c_1x + c_2x^2$ be a best Chebyshev approximation to x^3 on $[-1, 1]$, i.e., such that the maximum over $[-1, 1]$ of $|x^3 - (c_0 + c_1x + c_2x^2)|$ is a minimum over all choices of the c_i, and suppose B is the subset of $[-1, 1]$ where the maximum of this absolute value is taken on. Then ξ must give measure one to B, and the weights assigned by ξ to the various points of B (there are four in this case) can be found either by solving the *linear* equations (2.10) or by computing these weights so as to make ξ a maximin strategy for the game discussed in Section 2. Two points should be mentioned:

(1) In the general polynomial case, where there are k parameters ($k = 4$ here), the results described in [10], p. 42, or in Section 2 below imply that there is an optimum ξ concentrated on at most k points. Thus, even if we use this result with the approach of the previous paragraph, we obtain the following comparison in a k-parameter problem in Section 2:

Method A: minimize a nonlinear function of $2k - 1$ real variables.

Method B: solve the Chebyshev problem and then solve $k - 1$ simultaneous *linear* equations.

The fact that the solution of the Chebyshev problem can often be found in the literature (e.g., [2]) makes the comparison of the second method with the first all the more favorable.

(2) Although the computational difficulty cannot in general be reduced further, in the case of polynomial regression on $[-1, 1]$ there is present a kind of symmetry (discussed in Section 2) which implies that there is an optimum ξ which is symmetrical about 0 and which is concentrated on four points; thus, in the case of cubic regression, this fact reduces the computation under Method A to a minimization in 3 variables, but Method B involves only the solution of a single linear equation.

C. A third method, which rests on the game-theoretic results of Section 2, and which is especially useful when one has a reasonable guess of what an optimum ξ is, involves the following steps: first guess a ξ, say ξ^*, and compute the minimum on the left side of (2.8); second, if this minimum is achieved for $c = c^*$, compute the square of the maximum on the right side of (2.9); then, if these two computations yield the same number, ξ^* is optimum. If one has a guess of a class of ξ's depending on one or several parameters, among which it is thought that there is an optimum ξ, then one can maximize over that class at the end of the first step and, the maximum being at ξ^*, go through the same analysis as above. This method is illustrated in Example 3.5 and Example 4. Of course, the remarks (1) and (2) of the previous paragraph can be used in applying Method C, as in these examples.

In the example of cubic regression just cited, the optimum procedure turns out to be $\xi(-1) = \xi(1) = \frac{1}{6}$, $\xi(\frac{1}{2}) = \xi(-\frac{1}{2}) = \frac{1}{3}$. It is striking that any of the commonly used procedures which take equal numbers of observations at equally spaced points on $[-1, 1]$ requires over 38% more observations than this optimum procedure in order to yield the same variance for the best linear estimator of a_3 (see Example 3.1); the comparison is even more striking for higher degree regression. The unique optimum procedure in the case of degree h is given by (3.3).

The comparison of a direct computational attack, analogous to that of A above, with the methods developed in Sections 4 and 5 for the problems considered there, indicates even more the inferiority of the direct attack. In particular cases, e.g., Example 5.1, special methods may prove useful.

Among recent work in the design of experiments we may mention the papers of Elfving [3], [4], Chernoff [5], Williams [11], Ehrenfeld [12], Guest [13], and Hoel [15]. Only Guest and Hoel explicitly consider computational problems of the kind discussed below. Our methods of employing Chebyshev and game theoretic results seem to be completely new. The results obtained in the examples below are also new, except for some slight overlap with results of [13] and [15], which is explicitly described below.

We shall consider elsewhere some further problems of the type considered in this paper.

2. The optimum design relative to 1 out of k regression coefficients. Let $f_1, \cdots, f_k$ be k real-valued functions on a given space $\mathfrak{X}$. Throughout this section we assume a topology is given on $\mathfrak{X}$ in which

(2.1) $\mathfrak{X}$ is compact; $f_1, \cdots, f_k$ are continuous.

We also assume

(2.2) $f_1, \cdots, f_k$ are linearly independent on $\mathfrak{X}$.

Since we will be considering a regression problem in which the f_i are known functions and $\sum_i a_i f_i$ is the regression function, (2.2) is really only an assumption of identifiability of the a_i which will avoid trivial circumlocutions. Without some assumption like the first part of (2.1), there may trivially exist procedures which estimate some of the regression coefficients with arbitrarily small variance, as can be seen in the example of estimation of the slope of a straight line on $\mathfrak{X}$ = real line. The assumption of continuity of the f_i can be somewhat weakened, as will be clear from our proofs.

We consider the following regression setup: For any point x (value of the independent variable) in $\mathfrak{X}$, one can observe a random variable Y_x for which

$$EY_x = \sum_1^k a_i f_i(x), \tag{2.3}$$
$$\mathrm{Var}(Y_x) = \sigma^2,$$

where $a = (a_1, \cdots, a_k)$ is the vector of regression coefficients, an unknown element of $\mathcal{A}$. The value of σ^2 will usually be unknown. (The case where σ^2 can depend on x in a way which is known except for a proportionality constant will be discussed in the last paragraph of this section.) An integer n is given (usually $n > k$), and the experimenter must select a collection $\mathrm{X} = (x_1, \cdots, x_n)$ of n points in $\mathfrak{X}$ at which the independent random variables $Y_{x_1}, \cdots, Y_{x_n}$ are to be observed. The x_i need not be distinct, but if $i \neq j$ and $x_i = x_j$ we shall still, without confusion, write Y_{x_i} and Y_{x_j} for two *independent* random variables.

Any X can be viewed as a measure η on $\mathfrak{X}$ which assigns to each point x a mass equal to the number of x_i in X which are equal to x. Dividing this measure by n, we obtain a discrete probability measure ξ on $\mathfrak{X}$ which assigns to each point of $\mathfrak{X}$ a measure equal to an integral multiple of $1/n$. In the present section (a similar discussion applying in Sections 4 and 5), we shall be concerned with choosing a ξ (hence, an X) to maximize a quantity of the form

$$\min_c \int_{\mathfrak{X}} H_c(x)\eta(dx) = \min_c \int_{\mathfrak{X}} nH_c(x)\xi(dx), \tag{2.4}$$

where the form of H_c is determined by the problem at hand. The fact that ξ can only take on multiples of $1/n$ as its values makes this problem of maximization quite unwieldy in general. We shall treat, instead, a problem whose solution will sometimes give a solution to the original problem and which will usually give a good approximation to the latter: *Find a probability measure ξ^* on $\mathfrak{X}$ for which the right side of* (2.4) *is a maximum*: i.e., we maximize (2.4) with no restriction on ξ. Of course, the maximum does not depend on n. Thus, if n is such that $n\xi^*$ takes on only integral values, this yields an exact solution η to the original problem. We shall see in Sections 3 and 5 that, in two typical examples, ξ^* takes on only values which are multiples of $1/(2k-2)$ (Example 3.1) or $1/k$ (Example 5.1), so that this situation is not vacuous. Moreover, there will typically be a ξ^* which is concentrated on approximately k points; thus, when $n\xi^*$ does not take on only integral values, obvious integral approximations η' to $n\xi^*$ will yield values of (2.4) whose ratio to the maximum tends to 1 as $n \to \infty$ (it is easy to give a bound on the difference of this ratio from unity). Thus, the characterization of a single ξ^* which yields an *almost* optimum design for all large n, in distinction to finding the best ξ which may depend in a complicated fashion on n, seems to be of practical value.

We therefore define Ξ to be the space of all discrete probability measures ξ on $\mathfrak{X}$. We could, more generally, specify a Borel field $\mathcal{B}$ on $\mathfrak{X}$ and let Ξ be the class of all measures ($\mathcal{B}$) on $\mathfrak{X}$; however, in all of our applications (see Theorem 2) it will suffice to let $\mathcal{B}$ consist of countable sets and their complements.

In the present section we are concerned with statistical inference about the single parameter a_k, where all a_i are assumed unknown. We shall give a precise definition of optimality in the next paragraph. What this definition means is that we restrict ourselves to designs for which a_k is estimable (i.e., for which there exist linear unbiased estimators of a_k; in practice, of course, n will have to be suitably large for there to exist such designs), and seek a design for which the linear unbiased estimator of a_k with minimum variance (best linear estimator, or b.l.e.) has a variance which is a minimum over all designs, within the approximation noted two paragraphs above. It is well known that such a design is optimum for problems of point estimation of a_k if the Y_x are assumed to be normal, in the sense that (for example) it yields a minimax procedure for any of a wide variety of weight functions; when the distributions of Y_x are assumed to belong to any larger class, the same result holds for the squared error loss function. For problems of interval estimation and hypothesis testing or m decisions, similar optimality results hold under normality if σ^2 is known. If σ^2 is unknown, such results hold provided every design for which a_k is estimable yields as many degrees of freedom to error as does the design we obtain; see Example 3.4 in Section 3 for further discussion.

We now define precisely the term "optimum" as used in this section. There are a few preliminaries. In the original description of a design, let X be a design for which a_k is estimable. Let $h_1, \cdots, h_{k-1}$ be numbers such that the function $f_k^* = f_k - \sum_1^{k-1} h_j f_j$ is orthogonal to f_i for $i < k$ in the sense that

$$\sum_{r=1}^{n} f_i(x_r) f_k^*(x_r) = 0, \qquad i < k. \tag{2.5}$$

Let $a^* = (a_1^*, \cdots, a_k^*)$ be such that $\sum_1^k a_i f_i = \sum_1^{k-1} a_i^* f_i + a_k^* f_k^*$; thus, $a_k^* = a_k$. For the least squares setup in terms of a^*, the orthogonality of f_k^* to the f_i for $i < k$ makes the last of the normal equations

$$\sum_{r=1}^{n} [f_k^*(x_r)]^2 a_k^* = \sum_{r=1}^{n} f_k^*(x_r) Y_{x_r}, \tag{2.6}$$

so that σ^2 times the *reciprocal* of the variance of the b.l.e. of $a_k^* = a_k$ is $\sum_r [f_k^*(x_r)]^2$. Since f_k^* is orthogonal to $f_1, \cdots, f_{k-1}$, this last sum is just the square of the distance of the n-vector $(f_k(x_1), \cdots, f_k(x_n))$ from the linear space spanned by the vectors $(f_i(x_1), \cdots, f_i(x_n))$ for $i < k$, namely,

$$\min_c \sum_r [f_k(x_r) - \sum_{j=1}^{k-1} c_j f_j(x_r)]^2, \tag{2.7}$$

where we have written c for $(c_1, \cdots, c_{k-1})$. Since (2.7) is σ^2 times the inverse of the variance of the b.l.e. of a_k, a design X will minimize that variance if it maximizes (2.7). Thus, finally, in terms of the probability measures ξ we have introduced above, we make the following

DEFINITION. *A measure ξ^* in Ξ is said to be an optimum design (for the pa-*

rameter a_k) if

$$\min_c \int [f_k(x) - \textstyle\sum_1^{k-1} c_j f_j(x)]^2 \xi^*(dx) = \max_{\xi \in \Xi} \min_c \int [f_k(x) - \textstyle\sum_1^{k-1} c_j f_j(x)]^2 \xi(dx). \tag{2.8}$$

For any ξ in Ξ, the ratio of the left side of (2.8) (with ξ for ξ^*) to the right will be called the efficiency $e(\xi)$ of ξ.

Of course, the practical meaning of efficiency is that, if one design has r times the efficiency of the second design, then the latter requires r times as many observations as the former in order to obtain the same value for the left side of (2.4). We note that it is a consequence of this definition that an optimum design is optimum for all values of σ^2.

The form of (2.8) is very suggestive of a game, and we shall exploit that fact presently. However, the main aspect of our technique for computing an optimum ξ^* has nothing to do with the game formulation, so we treat that aspect first. Our technique is to throw the main computational difficulties into a Chebyshev approximation problem, which can often be solved by standard methods and which, for many important $\{f_j\}$, even has a solution which can be found in the literature. We shall call $c^* = (c_1^*, \cdots, c_{k-1}^*)$ a *Chebyshev coefficient vector* if $\sum_1^{k-1} c_j^* f_j$ is a best approximation to f_k on $\mathfrak{X}$ in the sense of Chebyshev, i.e., in the uniform norm:

$$\min_c \max_{x \in \mathfrak{X}} |f_k(x) - \sum_1^{k-1} c_j f_j(x)| = \max_{x \in \mathfrak{X}} |f_k(x) - \sum_1^{k-1} c_j^* f_j(x)|. \tag{2.9}$$

Let $m(c^*)$ denote the right side of (2.9), and let $B(c^*)$ be the set of points x for which $|f_k(x) - \sum_1^{k-1} c_j^* f_j(x)| = m(c^*)$. Our first result gives a simple geometric sufficient condition for a ξ to be optimum; this is valid even without the conditions that yield the game-theoretic results of Theorem 2.

THEOREM 1. *If c^* is Chebyshev and $\xi(B(c^*)) = 1$ and*

$$\int [f_k(x) - \textstyle\sum_1^{k-1} c_j^* f_j(x)] f_i(x) \xi(dx) = 0 \tag{2.10}$$

for $i < k$, then ξ is optimum.

PROOF: According to (2.10), $\sum_1^{k-1} c_j^* f_j$ is the projection relative to ξ of f_k on the linear space spanned by $f_1, \cdots, f_{k-1}$. Hence, for any element ξ' of Ξ,

$$\begin{aligned} &\min_c \int [f_k(x) - \textstyle\sum_1^{k-1} c_j f_j(x)]^2 \xi(dx) \\ &= \int [f_k(x) - \textstyle\sum_1^{k-1} c_j^* f_j(x)]^2 \xi(dx) \\ &= [m(c^*)]^2 \geq \int [f_k(x) - \textstyle\sum_1^{k-1} c_j^* f_j(x)]^2 \xi'(dx) \\ &\geq \min_c \int [f_k(x) - \textstyle\sum_1^{k-1} c_j f_j(x)]^2 \xi'(dx), \end{aligned} \tag{2.11}$$

which proves the desired result.

The question arises as to whether there always exists a ξ which satisfies the hypotheses of Theorem 1 and whether, in fact, the conditions of the theorem are also *necessary* for a ξ to be optimum. There also arises the question of whether we can find a useful bound such that there is an optimum ξ which assigns positive probability to at most the number of points given by this bound. These questions can be answered directly algebraically, but since the results we require already appear in the literature in connection with the analysis of certain games, we shall therefore consider the following *zero-sum two-person game associated with the design problem*: player 1 (resp., 2) has $\mathfrak{X}$ (resp., $C =$ Euclidean $(k-1)$-space) as his space of pure strategies; the payoff function is $K(x, c) = [f_k(x) - \sum_1^{k-1} c_j f_j(x)]^2$; the space of mixed strategies of player 1 is Ξ, while that of player 2 is immaterial, since the convexity of K in C implies, according to Jensen's inequality, that for any randomized strategy of player 2 there is a nonrandomized strategy which is at least as good for all x. Of course the important thing is that an optimum (maximin) strategy for player 1 represents an optimum design. We now state the simple modifications of certain results of [6] which we require.

LEMMA. *The game of Ξ vs. C is determined, player 2 has a nonrandomized minimax strategy c^*, and player 1 has a maximin strategy ξ^* which is concentrated on at most $k - p$ points where p is the dimensionality of the convex set of nonrandomized minimax strategies of player 2.*

PROOF: Let C_N be the set of all c for which $c'c = \sum_1^{k-1} c_i^2 \leq N^2$, and let $\bar{C}_N$ be the complement of C_N. Since the f_i are linearly independent, there is a finite subset H of $\mathfrak{X}$ such that, for every c with $c'c = 1$, $\sum_1^{k-1} c_i f_i(x)$ is nonzero for at least one x in H. Hence, if ξ' assigns positive probability to each x in H, we clearly have $|\sum_1^{k-1} c_i \sum_{x \in H} f_i(x)\xi'(x)| > \epsilon > 0$ for all c such that $c'c = 1$, and thus this absolute value is $> N\epsilon$ for $c'c = N^2$. Since f_k is bounded, we conclude that $\inf_{c \in \bar{C}_N} K(\xi', c) \to \infty$ as $N \to \infty$. Hence, there is an N' such that for any c in $\bar{C}_{N'}$ there is a c' in $C_{N'}$ with $\sup_\xi K(\xi, c') < \sup_\xi K(\xi, c)$. Thus c^* is minimax if and only if $c^* \in C_{N'}$ and c^* is minimax when the space of player 2 is restricted to $C_{N'}$. Since C_N is compact and K is continuous, the game of Ξ vs. C_N is determined, and there exists, for all $N > N'$, a minimax strategy c^* which we can take to be a fixed member of $C_{N'}$. Let p be the dimension of the (convex) set of such minimax strategies in $C_{N'}$. There also exists a maximin strategy ξ_N^* for the game of Ξ vs. C_N, and by [6] we can for $N > N'$ take ξ_N^* to be concentrated on at most $k - p$ points. Let $\xi_j = [(j-1)\xi_j^* + \xi']/j$. Clearly, for each j there is an N_j such that $K(\xi_j, c^*) < K(\xi_j, c)$ for all c in $\bar{C}_{N_j}$. Thus, since ξ_j^* is maximal with respect to c^*, we have, for $N_j > N'$,

$$\sup_{\xi \in \Xi} \inf_{c \in C} K(\xi, c) \geq \inf_{c \in C} K(\xi_j, c) = \inf_{c \in C_{N_j}} K(\xi_j, c)$$

$$\geq \left(1 - \frac{1}{j}\right) \inf_{c \in C_{N_j}} K(\xi_j^*, c) = \left(1 - \frac{1}{j}\right) K(\xi_j^*, c^*) \tag{2.12}$$

$$= \left(1 - \frac{1}{j}\right) \sup_\xi K(\xi, c^*) \geq \left(1 - \frac{1}{j}\right) \inf_{c \in C} \sup_\xi K(\xi, c).$$

Letting $j \to \infty$, we see that the game of Ξ vs. C is determined, that c^* is minimax, and that if $\{\xi^*_{j_i}\}$ is a subsequence of the $\{\xi^*_j\}$ which converges to a limit ξ^* which is concentrated on no more than $k - p$ points (such a subsequence and limit exist, by the compactness of $\mathfrak{X}$) and c'' minimizes $K(\xi^*, c)$, we have

$$\sup_c \inf_{c\epsilon C} K(\xi, c) = \lim_{i\to\infty} \inf_{c\epsilon C} K(\xi_{j_i}, c) \leqq \lim_{i\to\infty} K(\xi_{j_i}, c'') \tag{2.13}$$
$$= K(\xi^*, c'') = \inf_{c\epsilon C} K(\xi^*, c),$$

so that ξ^* is maximin. Thus, the lemma is proved.

We mention in passing several other related points: The bound $k - p$ is indicated in [6] not to be the best possible and is reduced under conditions (c^* in the boundary of a compact C) for which it is difficult to find general counterparts here. Also, it is evident that c^* is unique ($p = 0$) if $K(x, c)$ is strictly convex in c, but strict convexity is clearly not a useful condition in our problem. If $\mathfrak{X}$ is not compact or the f_j are not continuous, suitable assumptions will still imply determinateness, but the other results will have to be stated in terms of ϵ-optimum strategies.

The above lemma indicates one method for trying to compute a ξ^*: For simplicity, assume $p = 0$ or that we have no knowledge of p. The ξ's on $\mathfrak{X}$ which are concentrated on at most k points form a $(2k - 1)$-parameter family. One can thus, in principle, maximize $\min_c K(\xi, c)$ with respect to these $2k - 1$ parameters and obtain an optimum ξ^*. As we have indicated in the introduction, this is usually an unrewarding task, and the method indicated in Theorem 1 seems far superior in practical examples. The consequences of the lemma for the method of Theorem 1 may be summarized as follows:

THEOREM 2. *If ξ is maximal with respect to c^* while c^* is minimal with respect to ξ, then ξ is optimum and c^* is Chebyshev. Every optimum ξ satisfies the conditions of Theorem 1 for every Chebyshev c^*. There exists an optimum ξ concentrated on at most $k - p$ points, where p is the dimensionality of the Chebyshev vectors.*

PROOF: The Chebyshev vectors clearly coincide with the minimax strategies. If ξ is maximin and c^* is minimax, then determinateness implies that c^* is minimal with respect to ξ, i.e., $\min_c K(\xi, c) = K(\xi, c^*)$. Thus, $\sum_1^{k-1} c_j^* f_j$ is the projection, relative to ξ, of f_k on the linear space spanned by $f_1, \cdots, f_{k-1}$, so that (2.10) clearly holds. Since, by (2.11), max min $K(\xi, c) = [m(c^*)]^2$ is the value of the game, $\xi(B_{c_*}) = 1$. The last assertion of the theorem is taken directly from the lemma, while the first is a general result in the theory of games. We note that any optimum ξ must give measure one to the intersection of all $B(c^*)$ for c^* Chebyshev.

We have mentioned, in Theorem 1 and in the second paragraph below the proof of the lemma, two computational approaches. The first sentence of Theorem 2 indicates a useful approach if one can make a good guess of ξ: guess a ξ' and compute $\min_c K(\xi', c) = K(\xi', c')$ (say); compute $\max_x K(x, c')$; if these two are equal, then ξ' is optimum. This is an approach which is standard in game theory and which has proved useful in many examples; it sometimes helps to let ξ' depend on a few parameters, with respect to which one maximizes

$\min_c K(\xi', c)$. A comparison of the various methods for obtaining an optimum ξ was given in an example in Section 1.

In the next section we shall give several examples of the computation of optimum ξ's. We shall not bother to list in detail all of the standard results in approximation theory which are useful in such computations. We mention here for future reference only the classical generalized Chebyshev theorem [2, p. 74], which states that if $\mathfrak{X}$ is a compact real interval and if no nontrivial linear combination of $f_1, \cdots, f_{k-1}$ has more than $k - 2$ zeros (in this case, these f_i are called a *Chebyshev system*), then the Chebyshev vector c^* is unique and is characterized by the fact that there are at least k points at which $f_k - \sum_1^{k-1} c_j^* f_j$ attains its maximum absolute deviation from zero, the maximum being taken on with successive alternations in sign. (The literature contains generalizations of this result to other spaces.)

Before proceeding further it is relevant here to point out the following connections with earlier results:

1) Elfving [3] considered the special case where $\mathfrak{X}$ contains a finite number of discrete points. It follows from his elegant geometrical argument that the optimum ξ is concentrated on at most k points and satisfies (2.10).

2) Consider the case of polynomial regression ($\mathfrak{X}$ a closed interval of the real line, $f_i(x) = x^{i-1}$). Then $p = 0$ by the Chebyshev theorem cited above. Theorem 2 then says, inter alia, that there exists an optimum ξ concentrated on at most k points. This result (for this important particular case) is already well known in the theory of moment problems ([10], p. 42). It holds *identically in* σ^2. If it did not hold for all σ^2 it would be useless in our problem when σ^2 is unknown. This result holds even when (2.18) below is true, with fixed v.)

We now give a simple result on the uniqueness of the optimum ξ^*.

THEOREM 3. *If $\mathfrak{X}$ is a compact real interval, $f_1, \cdots, f_{k-1}$ is a Chebyshev system, and $B(c^*)$ contains exactly k points, then the optimum ξ^* is unique.*

PROOF: Let $x_1, \cdots, x_k$ be the ordered members of $B(c^*)$, and let Q be the $(k - 1) \times k$ matrix whose (i, j)th element is $(-1)^j f_i(x_j)$. Let ξ denote a k-vector whose jth component is the number $\xi(x_j)$. According to (2.10), which, by Theorem 2, is necessary, and the Chebyshev theorem cited above, any optimum ξ must satisfy

$$Q\xi = 0. \tag{2.14}$$

(Of course, it must also satisfy $\xi(B(c^*)) = 1$.) Now, Q has rank $k - 1$, since, if it had smaller rank, a nontrivial weighted sum of rows of Q would be 0 and the f_i could not be a Chebyshev system. The linear equations (2.14) thus have a one-dimensional set of solutions ξ, and clearly at most one of these can be a probability measure. This completes the proof.

If $B(c^*)$ consists of more than k points, an analysis like that above will give information on how large the class of optimum ξ's can be.

Remark on symmetry (invariance): As we have indicated in Section 1, it will sometimes be easy, as in the case of polynomial regression, to infer that there is an optimum ξ with some symmetry property. Formally, suppose that there is a group G of transformations on $\mathfrak{X}$ such that for each g in G there is a transformation g' on $\mathfrak{A}$ such that, writing $(g'a)_i$ for the ith coordinate of $g'a$, we have $(g'a)_k = a_k$ for g in G and

$$\sum_i a_i f_i(x) = \sum_i (g'a)_i f_i(gx) \tag{2.15}$$

for all x and all $(a_1, \cdots, a_k)$. (One may let g' act on the vector of functions f_i instead of on $\mathfrak{a}$.) Then the problem in terms of the parameters $(g'a)_i$ and the independent variable gx coincides with the original problem. Hence, if ξ is optimum for the original problem, it is also optimum for the above problem in terms of gx and hence the measure ξ_g defined by

$$\xi_g(A) = \xi(g^{-1}A) \tag{2.16}$$

is optimum for the original problem in terms of x. Suppose for the moment that G contains a finite number, say L, of elements. Write

$$\bar{\xi} = \sum_{g \in G} \xi_g / L. \tag{2.17}$$

It is easy to prove that, if ξ is optimum, then so is $\bar{\xi}$; in fact, this is obvious statistically, since the variance of the average of L b.l.e.'s from the L independent experiments ξ_g with N/L observations each cannot be less than that of the b.l.e. from $\bar{\xi}$ with N observations (since $\bar{\xi}$ can be broken up into such experiments), but is clearly equal to the variance of the b.l.e. from ξ based on N observations. Thus, we have:

There exists an optimum design which is symmetric with respect to (invariant under) G.

The analogous result can be proved for G compact or satisfying conditions which yield the usual minimax invariance theorem in statistics; see, e.g., [7].

The fact that there exists an optimum symmetric design and an optimum design concentrated on (e.g.) k points does not imply the existence of an optimum design with both of these properties. For example, if $\mathfrak{X} = [-1, 1]$, $k = 2$, $f_1(x) = 1$, and $f_2(x) = x^2$, there is an optimum design concentrated on the two points 0 and 1, but the only symmetric design requires the three points 0, -1, and 1. However, in the event that g' does not act (as it does in the example just cited) as the identity for every g, we may be able to obtain some simplification. For example, without discussing the most general possibility, let us suppose that Q is a set of integers containing k and such that $(g'a)_j = a_j$ for all g if $j \varepsilon Q$, while $\sum_g (g'a)_j = 0$ for j not in Q. Consider the problem of finding an optimum design ξ on the space of equivalence classes of $\mathfrak{X}$ under the equivalence $x \sim x'$ if $x' = gx$ for some g, where the regression function is $\sum_{j \varepsilon Q} a_j f_j(x)$ (at the equivalence class of x). If there are q integers in Q, there is by Theorem 2 an optimum τ^* concentrated on at most q points. This τ^* corresponds to a unique symmetric (with respect to G) measure ξ^* on $\mathfrak{X}$, and it is easy to see that (2.10) is satisfied for *all* $i < k$. Thus, if there are L elements in G, this ξ^* is concentrated on at most qL points. For example, in the case of polynomial regression of even degree h $(= k - 1)$ on $[-1, 1]$, G contains two elements and the set Q corresponds to the $q = 1 + h/2$ even powers, and we obtain that there is a symmetric optimum ξ concentrated on at most $h + 2$ points. The actual case (see Ex. 3.1) is that there is a symmetric optimum ξ concentrated on $k = h + 1$ points; the previous argument did not give the best result because τ^* gave positive probability to the equivalence class of 0, which corresponds to only one point of $\mathfrak{X}$. The best result could, however, be obtained using another argument: since, according to Theorem 3, the optimum ξ is *unique*, our discussion of two paragraphs above implies that it must be symmetric, and it is thus concentrated on $h + 1$ points. Similarly, one could conclude that there is a symmetric optimum design concentrated on $h + 1$ points when h is odd, either by using Theorem 3, or else by invoking an obvious modification of the previous argument for the case when $(g'a)_k = \pm a_k$. A similar result holds in the setup of Ex. 3.5.

Remark on heteroscedasticity and variable cost: Suppose the second line of (2.3) is replaced by

$$\text{Var}(Y_x) = [v(x)\sigma]^2, \tag{2.18}$$

where v is a known positive continuous function on $\mathfrak{X}$. To avoid trivialities, assume $v(x)$ bounded away from 0. Then, replacing Y_x by $Y_x^* = Y_x/v(x)$ and $f_i(x)$ by $f_i^*(x) = f_i(x)/v(x)$, it is clear that the entire discussion of this section goes through exactly as before (i.e., assuming (2.3)) since the a_i for which $EY_x = \sum a_i f_i(x)$ are the same a_i as those for which $EY_x^* = \sum a_i f_i^*(x)$, and the latter setup satisfies the original condition (2.3) of this section.

If there is a cost $c(x)$ of taking an observation at the point x, and the total cost rather than the total number of observations is to be kept constant, it is easily seen that an optimum design is obtained by going through the analysis of this section with $v(x)$ above replaced by $v(x)[c(x)]^{1/2}$.

Similar remarks will apply to the problems considered in Sections 4 and 5.

3. Examples of optimum designs in the case of Section 2.

Example 3.1. *Polynomials on* $[\alpha, \beta]$. One of the most important practical examples is that where $\mathfrak{X}$ is the closed finite nondegenerate interval $[\alpha, \beta]$ of reals, $k = h + 1$ for some $h > 0$, and $f_j(x) = x^{j-1}$ for $1 \leq j \leq h + 1$; we hereafter write $b_{j-1} = a_j$, $b = (b_0, \cdots, b_h)$, $d_{j-1} = c_j$, and $d = (d_0, \cdots, d_{h-1})$. Thus, assuming that the regression function is a polynomial of degree $\leq h$, we may want to test the hypothesis that it is actually of degree $\leq h - 1$, i.e., that $b_h = 0$. (In Section 4 we consider the possibility of testing that the degree is $\leq h - m$ where m is specified). We first note that we can write

$$\sum_{j=0}^{h} b_j x^j = \sum_{j=0}^{h} b_j'[(2x - \alpha - \beta)/(\beta - \alpha)]^j,$$

where $b_h' = [(\beta - \alpha)/2]^h b_h$; since $(2x - \alpha - \beta)/(\beta - \alpha)$ takes on values in $[-1, 1]$, an optimum strategy for arbitrary $[\alpha, \beta]$ is immediately obtained by an obvious change in location and scale from an optimum strategy in the case $[-1, 1]$, and we may hereafter limit our attention to the latter. Next, we note that b_h is obviously not estimable unless ξ gives positive probability to at least $h + 1$ points (of course, in practice we need $n > h + 1$ if σ^2 is unknown and $n \geqq h + 1$ if σ^2 is known). Hence, by Theorem 2 (or by the result of [10] cited in Section 2) there exists an optimum ξ concentrated on exactly $(h + 1)$ points. We shall actually find a unique ξ^* which satisfies (2.8) and gives positive probability to exactly $h + 1$ points.[3] Thus, the phenomenon concerning degrees of freedom in the estimate of σ^2 which was discussed in the sixth paragraph of Section 2, and which is illustrated in Example 3.4 below, cannot occur in the present example.

The unique Chebyshev d^* (i.e., c^*) is well known in this example: $x^h - \sum_0^{h-1} d_j^* x^j$ is simply the hth Chebyshev polynomial (see, e.g., [2]),

$$\begin{aligned} x^h - \textstyle\sum_0^{h-1} d_j^* x^j &= 2^{1-h} \cos(h \cos^{-1} x) \\ &= 2^{-h}\{[x + (x^2 - 1)^{1/2}]^h + [x - (x^2 - 1)^{1/2}]^h\}. \end{aligned} \tag{3.1}$$

Moreover, $m(d^*) = 2^{1-h}$, and this extreme value is attained in magnitude (with successive alterations in sign) by $x^h - \sum_0^{h-1} d_j^* x_j$ at the $h + 1$ points

$$x_j = -\cos \frac{j\pi}{h}, \qquad 0 \leqq j \leqq h. \tag{3.2}$$

Thus, $B(d^*)$ consists of these $h + 1$ points. Moreover, the above d^* is the unique Chebyshev vector, since $x^0, x^1, \cdots, x^{h-1}$ form a Chebyshev system.

According to Theorem 3, the optimum ξ^* is unique. We now show that the unique optimum ξ^* is

$$\begin{aligned} \xi^*(-1) = \xi^*(1) &= \tfrac{1}{2}h, \\ \xi^*\left(\cos \frac{j\pi}{h}\right) &= 1/h, \qquad 1 \leqq j \leqq h - 1. \end{aligned} \tag{3.3}$$

To prove this, we shall verify (2.14) for $\xi = \xi^*$, since this is just (2.10), which by Theorems 1 and 2 is necessary and sufficient for an optimum ξ. Since the d_j^*'s of (3.1) are zero if $j + h$ is odd, the polynomial of (3.1) is clearly orthogonal (with respect to ξ^*) to x^t when $t + h$ is odd. When $t + h$ is even, we can combine the weights $\xi^*(-1)$ and $\xi^*(1)$ and rewrite (2.14) as

$$\sum_{j=0}^{h-1} (-1)^j \left(\cos \frac{\pi j}{h}\right)^t = 0. \tag{3.4}$$

Since $\cos^t \theta$ can be written as a linear combination of $\cos t\theta$, $\cos(t - 2)\theta, \cdots$ it suffices to prove (3.4) with $\cos^t (\pi j/h)$ replaced by $\cos (rj\pi/h)$, where $h + r$

[3] For $h = 1$ and 2, the solution is given in [14]. The general solution (3.3) of the problem of Example 3.1 for a design optimum in the sense of Section 2, is also given in the abstract [11] of the apparently contemporaneous work of E. J. Williams. The methods of this author are probably different from ours because he does not seem to use probability measures ξ. The authors are indebted to H. L. Lucas for calling their attention to [11] which appeared after submission of the present manuscript.

is even and $0 \leq r \leq h$. But for such r we have

$$\begin{aligned}\sum_{j=0}^{h-1} (-1)^j \cos (rj\pi/h) &= \operatorname{Re}\left\{\sum_{j=0}^{h-1} \exp [ji\pi(1 + r/h)]\right\} \\ &= \operatorname{Re}\left\{\frac{1 - \exp [i\pi(h + r)]}{1 - \exp [i\pi(1 + r/h)]}\right\} = 0.\end{aligned} \tag{3.5}$$

It is interesting to compare the design ξ^* of (3.3) with the often used design $\xi^{h,M}$ (say) which assigns measure $1/M$ to each of the values $(2i - M - 1)/(M - 1)$, $i = 1, 2, \cdots, M$; thus $\xi^{h,M}$ takes an equal number of observations at each of M equally spaced points ranging from -1 to 1. Of course, $M > h$. For such a design with M observations on the interval $[0, M - 1]$, Fisher [8, p. 153] has calculated the left side of (2.4) to be $(h!)^4M(M^2 - 1)(M^2 - 4) \cdots (M^2 - h^2)/(2h)!(2h + 1)!$ To obtain the corresponding quantity for the interval $[-1, 1]$, we must divide by $[(M - 1)/2]^{2h}$, and we must divide also by M in order to obtain the left side of (2.4) with η replaced by $\xi^{h,M}$. Since $[m(d^*)]^2 = 2^{2-2h}$, we obtain for the efficiency (see the definition following (2.8)) of $\xi^{h,M}$

$$e(\xi^{h,M}) = \frac{2^{4h-2}(h!)^4}{(2h)!(2h + 1)!} \prod_{i=1}^{h} \frac{M^2 - i^2}{(M - 1)^2}. \tag{3.6}$$

The best choice of M varies: it is $h + 1$ if $h = 1$ or 2, $h + 2$ if $h = 3$, etc. For the often used procedure $\xi^{h,h+1}$, we have

$$e(\xi^{h,h+1}) = \frac{2^{4h-2}(h!)^4}{(2h)!h^{2h}(h + 1)}. \tag{3.7}$$

Of course, (3.7) becomes 1 for $h = 1$, since $\xi^{1,2} = \xi^*$ for $h = 1$; for $h = 2$, (3.7) becomes 8/9, for $h = 3$ it is 256/405 (the best procedure, $\xi^{3,5}$, has efficiency .72), etc.; for large h, by Stirling's approximation, it is approximately $\pi^{3/2}h^{1/2}2^{2h-1}e^{-2h}$, which goes to zero very rapidly. For $\xi^{h,M}$ with $M \to \infty$, the efficiency (3.6) approaches $2^{4h-2} (h!)^4/(2h)!(2h + 1)!$, which as $h \to \infty$ is approximately $\pi/8$.

To the experimenter who protests at the above comparison that the design $\xi^{h,M}$ for some $M > h$ is more to his liking than is the ξ^* of (3.3) because the former will permit him to estimate regression coefficients a_j up to a_{M-1} (instead of up to a_h), we can only answer that his problem is not the one of the present example, that he is probably using a method of inference (to "choose the polynomial of correct degree") whose properties are questionable, and that a precise statement of his decision problem would probably lead to a procedure far superior to $\xi^{h,M}$. In Sections 4 and 5 we shall consider some other related problems which may be what the experimenter is faced with, rather than the problem of the present example. The problem of "fitting the polynomial of best degree" is more unwieldy, depending strongly on the somewhat arbitrary choice of losses which are to be assigned to errors in estimation as compared with the penalty for using a polynomial of large degree.

Example 3.2. *An example where* $p > 0$. It is easy to construct examples where the p of Theorem 2 is not 0 as it is in the case of a Chebyshev system. We illustrate the situation with a very simple example. Suppose $\mathfrak{X} = [-1, 1]$, $k = 3$, $f_1(x) = 1, f_2(x) = x^2, f_3(x) = x + 1$. The expression $x + 1 - c_1 - c_2x^2$ has, within $[-1, 1]$, derivative equal to 0 at $x = \frac{1}{2}c_2$ if $|c_2| \geqq \frac{1}{2}$ and is monotone on $\mathfrak{X}$ if $|c_2| < \frac{1}{2}$. Thus, a routine computation of $\max_x |x + 1 - c_1 - c_2x^2|$ leads to the conclusion that any c with $c_1 + c_2 = 1$ and $|c_2| \leqq \frac{1}{2}$ is Chebyshev; i.e., $p = 1$. Hence, $k - p = 2$, and indeed the design ξ^* for which $\xi^*(-1) = \xi^*(1) = \frac{1}{2}$ is optimum. The heart of the matter is that $(1, x^2)$ is not a Chebyshev system and that is is possible to estimate a_3 optimally without estimating a_2 at all.

Example 3.3. *An example where the optimum* ξ^* *is not unique*. There are many obvious examples of this kind, as we have indicated in the paragraph following the proof of Theorem 3. For example, one simple example is given by $\mathfrak{X} = [-1, 1]$, $k = 2$, $f_1(x) = 1$, $f_2(x) = 1 + \sin 10x$ (any ξ which assigns measure $\frac{1}{2}$ to each of the sets where $\sin 10x = 1$ or -1 satisfies (2.10)); an even more trivial one is $k = 1, f_1(x) = 1$, where every strategy is optimum.

Example 3.4. *An example where a nonoptimum* ξ *may be preferable*. This example illustrates the phenomenon alluded to in the text, wherein a design ξ which is not optimum in the sense defined in Section 2 may be preferable to an optimum design ξ^* for use (e.g., in testing a hypothesis about a_2) because the latter yields one less degree of freedom for the estimate of σ^2. Let ϵ be a fixed small positive number, and suppose that $\mathfrak{X}$ consists of the three integers 0, 1, and 2, that $k = 2$, and that $f_1(x) = x^2$ and $f_2(x) = 1 + (1 + \epsilon)x$. It is easily computed that the Chebyshev c^* is $1 + 3\epsilon/5$, that $B(c^*)$ consists of the points 1 and 2, that $m(c^*) = 1 + 2\epsilon/5$, and that the optimum ξ^* is given by $\xi^*(1) = 1 - \xi^*(2) = 4/5$. Thus, the efficiency of the design which takes all observations at $x = 0(\xi(0) = 1)$ and estimates a_2 in the obvious way, is $(1 + 2\epsilon/5)^{-2}$; when ϵ is small, this is more than offset by the extra degree of freedom for estimating σ^2 (e.g., 4 for the latter design against 3 for ξ^*, when 5 observations are taken), for the problem of testing a hypothesis about a_2 or giving a confidence interval on a_2.

Example 3.5. *A multidimensional example*. Let $\mathfrak{X}$ be the set of all points (x_1, x_2) in the Euclidean plane for which $|x_1| \leqq 1$ and $|x_2| \leqq 1$. Let $k = 6$ and suppose that the functions f_i are, in order. 1, x_1, x_2, x_1^2, x_2^2, and x_1x_2; thus, for example, we may be testing the hypothesis that a quadratic function of two variables has no interaction term $a_6x_1x_2$, i.e., that $a_6 = 0$. An easy approach to obtaining an optimum ξ is the third method mentioned in Section 2: An obvious guess of a ξ which might be optimum is that measure ξ' (say) which assigns probability $\frac{1}{4}$ to each corner of the square $\mathfrak{X}$. Thus, writing $c_1 + c_4 + c_5 = \bar{c}$, we see that $K(\xi', c)$ is symmetric in each of the variables c_2, c_3, and $\bar{c}$ (which are the only quantities on which it depends), so that $\min_c K(\xi', c) = K(\xi', c') = 1$ is attained for any c' for which $c_2 = c_3 = \bar{c} = 0$. Let c'' have all five of its components equal to zero. Then, clearly, $\max_x K(x, c'') = 1$. Thus, by the discussion following Theorem 2, we have proved that ξ' is optimum. Another way of verifying the optimality of

ξ' is to note that, in the terminology of the remark on symmetry of Section 2, G is the group of symmetries of the square, and an analogue of the last argument mentioned there for the case of polynomial regression with h odd obtains ξ' from the optimum design τ^* which assigns mass 1 to (1, 1) for the problem of estimating a_6 on $0 \leq x \leq y \leq 1$ when the regression function is a_6xy. We note that only a_2, a_3, a_6, and $a_1 + a_4 + a_5$ are estimable for this design. The fact that only four linearly independent estimable linear parametric functions exist here is reflected in the fact that, in the notation of Section 2, $p = 2$. This can be seen by noting that, if $c' = (\epsilon + \delta, 0, 0, -\epsilon, -\delta)$, where ϵ and δ are sufficiently small, then $\max_x K(x, c')$ is still equal to unity, so c' is Chebyshev.

Other examples. Many other examples of optimum designs can be obtained from the extensive literature on Chebyshev approximation problems. For example, Section 37 of [2] can be used to obtain such a design for the setup of Example 3.1 wherein f_k is altered to $f_k(x) = 1/(x - c)$ with $c > b$.

4. The case of several regression coefficients. We consider now the setup of (2.1)–(2.3) (see also (2.18)) in the case where we are interested in inference about more than one of the a_i. In some estimation problems, a treatment like that of Section 5, wherein the behavior of the function $\sum a_i f_i$ rather than that of the a_i themselves is considered, will seem appropriate. However, in most problems of testing hypotheses, as well as in many problems of estimation (especially where the inference is not about all of the a_i), the treatment of the present section may seem appropriate.

We must first choose a criterion of optimality of a design for a problem of estimation or testing hypotheses about s of the a_i, say $a_{k-s+1}, \cdots, a_k$. Of course, it is easy to specify a loss function and a criterion (minimax, etc.) for choosing a design and associated decision procedure; but, as shown in [1], such a simple criterion as that of maximizing the minimum power of a test on an appropriate contour (M-optimality) will usually lead to most unwieldy computations. Even the corresponding local criterion on the power near the null hypothesis (L-optimality) will lead to difficult computations. Two other criteria considered in [1] are D-optimality and E-optimality. In the present setting, n being fixed, a design d^* is said to be D-optimum if $a_{k-s+1}, \cdots, a_k$ are all estimable under d^* and if, among all designs for which these parameters are estimable, denoting by $\sigma^2 V_d$ the covariance matrix of the b.l.e.'s of these parameters when design d is used, det V_d is a minimum for $d = d^*$. A design is said to be E-optimum in the above setting if the maximum eigenvalue of V_d is a minimum for $d = d^*$. The relevance of these criteria for problems of testing hypotheses and of estimation was indicated in [1] and the reference cited there. It was shown that D-optimality is generally more meaningful. There is an additional reason why this is so in problems of the type considered here: Consider the polynomial setup of Example 3.1 for any value $k > 2(h > 1)$ and $s > 1$. It is clear that the change of scale $x' = hx$ does *not* leave invariant the criterion of E-optimality: a change in the scale of measurement can change the E-optimum design. This is unsatisfactory from both an intuitive point of view (the optimum design depends on the choice of a unit of scale) and from a practical one; one would have to table optimum designs in such problems, as a function of α, β. (A similar remark, of course, applies to L-optimum and M-optimum designs.) On the other hand, D-optimality is invariant under such transformations. The same result is true under a change of origin (or a change of both scale and origin) in this polynomial example: D-optimality is invariant, but E-optimality is not.

Thus, although D-optimality is not an appropriate criterion in all problems, for the reasons given in the previous paragraph it seems reasonable to investigate this criterion as a first attack on the problem of finding optimum designs. We shall thus develop a method for obtaining D-optimum designs in the remainder of this section, except that we shall indicate briefly at the end of this section how various other criteria can be treated similarly.

Proceeding as in Section 2, let h_{ij} be numbers such that, for $i \leq k - s < t$, the functions f_i are orthogonal to the functions $f_t^* = f_t - \sum_{j=1}^{k-s} h_{tj} f_j$ in the sense of (2.5), i.e.,

$$\sum_{r=1}^{n} f_i(x_n) f_t^*(x_r) = 0, \qquad i \leq k - s < t. \tag{4.1}$$

Then, as in the discussion of (2.6), we see that σ^2 times the inverse of the covariance matrix $\sigma^2 V_d$ of best linear estimators of $a_{k-s+1}, \cdots, a_k$ has elements $\sum_r f_i^*(x_r) f_j^*(x_r)$, $k - s < i, j \leq k$. For $t > k - s$, let $f_t^{**} = f_t^* - \sum_{j<t} g_{tj} f_j^*$ be orthogonal to f_j^* for $k - s < j < t$. Since the linear transformation which takes the f_t^* into the f_t^{**}, $k - s < t \leq k$, has determinant 1, and since $\sum_r f_i^{**}(x_r) f_j^{**}(x_r) = 0$ if $k - s < i < j$, we obtain

$$\det V_d^{-1} = \prod_{i>k-s} \sum_r [f_i^{**}(x_r)]^2. \tag{4.2}$$

Now, f_i^{**} is clearly f_i minus the projection of f_i on the linear space spanned by $f_1, f_2, \cdots, f_{i-1}$. Thus, the ith term in the product of (4.2) is just the expression of (2.7) with k replaced by i. Finally, then, making the same approximation as in Section 2 regarding the representation of the class of all designs by the class of all probability measures ξ on $\mathfrak{X}$, we have demonstrated, to within this approximation, the validity of the following definition, wherein $c^{(j)}$ denotes a vector $(c_1^{(j)}, \cdots, c_{j-1}^{(j)})$ of $j - 1$ components:

Definition. *A measure ξ^* in Ξ is said to be D-optimum (for the parameters $a_{k-s+1}, \cdots, a_k$) if*

$$\begin{aligned} &\prod_{j>k-s} \min_{c^{(j)}} \int [f_j(x) - \textstyle\sum_1^{j-1} c_i^{(j)} f_i(x)]^2 \xi^*(dx) \\ &\qquad = \max_{\xi \in \Xi} \prod_{j>k-s} \min_{c^{(j)}} \int [f_j(x) - \sum_1^{j-1} c_i^{(j)} f_i(x)]^2 \xi(dx) \end{aligned} \tag{4.3}$$

Of course, (4.3) reduces to (2.8) in the case $s = 1$. When f_1 is a constant, a ξ which is optimum for $s = k - 1$ is also optimum for $s = k$.

We note that it is a consequence of this definition that an optimum design is optimum for all values of σ^2.

For the special case where $s = k$ and $\mathfrak{X}$ consists of k points, it is easy to prove that the unique optimum ξ puts mass $1/k$ on each point. For if A is the matrix whose (i, j) element is $f_i(x_j)$ and B is the diagonal matrix with $\xi(x_j)$ the diagonal element in the jth row, an optimum design maximizes $\det(ABA') = (\det A)^2 \det B$. This argument has been employed by Hoel in the problem considered by him; see Example 4 below.

The methods of Section 2 do not directly yield anything here for the general problem. The analogue of Theorem 1 is essentially empty, since the various $B(c^{(j)})$'s for $c^{(j)}$ Chebyshev will not in general coincide. The game-theoretic approach is inapplicable because the product on the left side of (4.3) is not linear in ξ^*; moreover, the product of the integrals (before minimizing over the $c^{(j)}$) is not convex in the $c^{(j)}$'s, since u^2v^2 is not a convex function of u and v. The following analysis will, however, yield a method for obtaining an optimum ξ.

For $j > k - s$, let

$$F_j(\xi) = \min_{c^{(j)}} \int [f_j(x) - \sum_1^{j-1} c_i^{(j)} f_i(x)]^2 \xi(dx). \tag{4.4}$$

In s-dimensional Euclidean space R^s, let S be the set of all points $F(\xi) = (F_{k-s+1}(\xi), \cdots, F_k(\xi))$ for ξ in Ξ. Although S may not be convex, it possesses the following "upper convexity" property, which is all we require: For any ξ_1 and ξ_2 in Ξ and any λ with $0 < \lambda < 1$,

$$F_j(\lambda\xi_1 + (1 - \lambda)\xi_2) \geqq \lambda F_j(\xi_1) + (1 - \lambda)F_j(\xi_2) \tag{4.5}$$

for all $j > k - s$. In fact, (4.5) is an immediate consequence of the linearity in ξ of the integral of (4.4).

Let $u_{k-s+1}, \cdots, u_k$ be the coordinate functions of R^s. For $\delta > 0$, let G_δ be the set of all points in R^s with all coordinates positive and $\prod_j u_j \geqq \delta$. Let G_δ' be the subset of G_δ where $\prod_j u_j = \delta$. We note that G_δ is convex. Suppose that S is closed (this is easily proved from (4.4) if $\mathfrak{B}$ is large enough so that Ξ is compact; the modification which is needed if Ξ is not closed is trivial, anyway), and let δ_0 be the largest value of δ such that G_δ and S have a nonempty intersection. (Such a δ_0 exists since S has points with all coordinates positive.) If T is the convex hull of S, property (4.5) implies that δ_0 is also the largest value of δ such that G_δ and T have a nonempty intersection. Hence, applying the separation theorem for G_{δ_0} and T, we conclude that there is a hyperplane L with positive direction cosines such that L separates G_{δ_0} and S. Thus, any point $F(\xi^*)$ in $G_{\delta_0} \cap S$ clearly maximizes $\prod_j F_j(\xi)$ (i.e., that ξ^* satisfies (4.3)); and, for positive numbers λ_j' for which L is given by $\sum \lambda_j' U_j =$ constant, that point maximizes $\sum_j \lambda_j' F_j(\xi)$. Finally, since all points of G_δ' are extreme, L intersects G_{δ_0} in exactly one point, as does therefore S.

Before summarizing the above results, we note that, for $\lambda = (\lambda_{k-s+1}, \cdots, \lambda_k)$

with all $\lambda_i > 0$, the payoff function

$$K_\lambda(x, c) = \sum_{j>k-s} \lambda_j [f_j(x) - \sum_{i<j} c_i^{(j)} f_i(x)]^2, \tag{4.6}$$

where $c = (c^{(k-s+1)}, \cdots, c^{(k)})$, satisfies all of those conditions satisfied by the function K of Section 2 which were used in the proof of the game-theoretic results of the lemma there. Thus, that lemma is valid when K is replaced by K_λ.[4] The function K_λ is of course no longer in a form suitable to make use of Chebyshev approximation results. However, for any λ, if c_λ^* is minimax for the payoff function K_λ, we can still characterize maximin ξ_λ's in terms of the set $\bar{B}_\lambda(c_\lambda^*)$ (say), defined to be the set of x for which $K_\lambda(x, c_\lambda^*)$ achieves its maximum. With this interpretation of symbols, the analogue of (2.10) is proved here exactly as in (2.11).

We have thus proved that following,[5] where C now stands for the set of vectors $c = (c^{(k-s+1)}, \cdots, c^{(k)})$ and $c_\lambda^* = \{c_i^{(j)*}\}$ stands for a vector of this type:

THEOREM 4. *The game of Ξ vs. C with payoff function K_λ is determined. If ξ_λ is maximal with respect to c_λ^* while c_λ^* is minimal with respect to ξ_λ, then ξ_λ is maximin. Thus, if c_λ^* is minimax and*

$$\xi_\lambda(\bar{B}_\lambda(c_\lambda^*)) = 1 \tag{4.8}$$

and

$$\int [f_j(x) - \sum_{i<j} c_i^{(j)*} f_i(x)] f_i(x) \xi_\lambda(dx) = 0 \tag{4.9}$$

for $i < j$ and $k - s < j \leq k$, then ξ_λ is maximin; moreover, every maximin ξ_λ satisfies (4.8) *and* (4.9) *for every minimax c_λ^*. There is, to within a multiplicative constant, a unique value λ^* of λ such that $\prod_i F_i(\xi_\lambda)$ is a maximum for $\lambda = \lambda^*$ and some ξ_{λ^*}. Those ξ_{λ^*} which maximize $\prod_i F_i(\xi_{\lambda^*})$, and no other ξ's, are optimum. $F(\xi_{\lambda^*})$ is the same for any optimum ξ_{λ^*}.*

We now consider an example.

Example 4. Consider the setup of Example 3.1, where (see the end of the second paragraph of the present section) we may suppose $\alpha = -1$, $\beta = 1$. Suppose $k = 3$ $(h = 2)$, and $s = 2$; as we have remarked earlier, the optimum design obtained below will also obviously be D-optimum for the case $s = 3$. An

[4] That part ot the lemma which concerns the number $k - p$ is valid when k is replaced by $1 + s(2k - s - 1)/2$ $(= 1 +$ number of components of $c)$ in the statement of the lemma. However, this is of no use to us since it may be that no maximin strategy on the specified number of points is optimum. For example, in the set-up of Example 5.2 below with $s = k = 2$, one can verify that the λ^* of Theorem 4 is (15/4, 1), and that any ξ_λ^* with first and second components equal is maximin, but only (4/15, 4/15, 7/15) is optimum.

It is trivial that the optimum strategy need be concentrated on no more than $1 + k(k + 1)/2$ points. For the criterion of optimality (4.3) involves ξ only through the elements (5.2) below of the matrix $M(\xi)$. These matrices form a convex body of dimensionality at most $k(k + 1)/2$, spanned by matrices of ξ's concentrated on a single point Hence any $M(\xi)$ is a linear convex combination of at most $1 + k(k + 1)/2$ extreme elements.

[5] See also footnote 6.

elegant solution to this problem for general k and $s = k$, has been given by P. G. Hoel [15] (see also Example 5.1 below). The case $s < k - 1$ does not seem to yield to his attack. The present problem is discussed here as an illustration of our methods. We may take 1 and γ for the components of λ, and write $K'_\gamma(x, d) = (x - d'_0)^2 + \gamma(x^2 - d''_1 x - d''_0)^2$ in place of K_λ. For fixed γ, one may *guess* that there will be a maximin strategy ξ'_γ of the form $\xi'_\gamma(-1) = \xi'_\gamma(1) = \alpha_\gamma$, $\xi'_\gamma(0) = 1 - 2\alpha_\gamma$, for some α_γ. With respect to such a ξ_γ, the minimal strategy (which must merely satisfy the orthogonality relation (4.9)) is obviously $d'_0 = d''_1 = 0$, $d''_0 = 2\alpha_\gamma$. For this choice d_γ (say) of d, we obtain $K'_\gamma(\xi'_\gamma, d_\gamma) = \gamma[2\alpha_\gamma - 4\alpha_\gamma^2] + 2\alpha_\gamma$. This is maximized by $a_\gamma = \min(\frac{1}{2}, (\gamma + 1)/4\gamma)$, and for the strategy ξ_{γ^*} corresponding to this value of α_γ we obtain

$$\min_d K'_\gamma(\xi^*_\gamma, d) = \begin{cases} (\gamma + 1)^2/4\gamma & \text{if } \gamma > 1, \\ 1, & \text{if } \gamma \leqq 1. \end{cases} \tag{4.10}$$

On the other hand,

$$\min_d \max_x K'_\gamma(x, d) \leqq \max_x K'_\gamma(x, d_\gamma). \tag{4.11}$$

Since $K'_\gamma(x, d_\gamma)$ is convex in x^2, its maximum is attained at either $x^2 = 0$ or $x^2 = 1$, and an easy computation shows that the right side of (4.11) is in fact equal to the right side of (4.10). Thus, we have proved that ξ^*_γ is maximin. Finally, $F_2(\xi^*_\gamma)F_3(\xi^*_\gamma) = 4\alpha_\gamma^2(1 - 2\alpha_\gamma)$, which is maximized by $\alpha_\gamma = \frac{1}{3}$. Thus, an optimum design for this problem is $\xi(-1) = \xi(0) = \xi(1) = \frac{1}{3}$. Of course the optimum designs for a given set of f_i will depend on s, as exemplified by the different results obtained in Example 3.1 and Example 4.

We shall now mention briefly methods for obtaining designs which are optimum in two other senses. Although it is not difficult to characterize E-optimum procedures in *simple* examples, they often seem much harder to calculate than D-optimum ones. Somewhat easier is the characterization of that design which minimizes the maximum eigenvalue of the covariance matrix of best linear estimators of the regression coefficients of the f_i^{**} (the regression function being expressed in terms of the f_t for $t \leqq k - s$ and of the f_t^{**} for $t > k - s$); i.e., of $L_d V_d L'_d$ where L_d is a square matrix with ones on the main diagonal and zeros above it. (The f_t^{**} depend on the design, which indicates the intuitive weakness of this criterion; however, as pointed out in [1], the criterion of E-optimality, which has often been considered in the literature, suffers from a similar shortcoming.) Again making the approximation that we do not restrict $n\xi$ to be integer-valued, this criterion amounts to finding that ξ which maximizes $\min_{j>k-s} F_j(\xi)$, i.e., if δ' is the largest value of δ for which the orthant $H_\delta = \{\min_j u_j \geqq \delta\}$ intersects S nonvacuously, those ξ for which $F(\xi)$ is in $H_{\delta'} \cap S$ are the optimum procedures with respect to this criterion. Another criterion which has been considered in the literature, especially in estimation problems, is that of minimizing the "average variance", $\sigma^2 s^{-1}$ trace (V_d). Defining $F_j^*(\xi')$ to be the expression of (4.4) with the sum in the integrand taken only from 1 to $k - s$, this criterion amounts to minimizing $\sum_{j>k-s} F_j^*(\xi)$. Replacing S by the set of points $F^*(\xi) = (F_{k-s+1}^*(\xi), \cdots, F_k^*(\xi))$, and restricting the sum over i in (4.6) to values $\leq k - s$, this amounts to finding the maximin ξ's for a λ with all components equal. These maximin ξ's for the *original* S and K_λ (with all λ_j equal) would of course minimize the average variance of the b.l.e.'s of regression coefficients of the f_i^{**}; i.e., would minimize the trace of $L_d V_d L_d'$. Criteria like that of minimizing the average variance are subject to the same criticisms as E-optimality.

Remarks. As in the problem of Section 2, one can prove that the symmetry condition (2.15) and the obvious analogue of the condition of the line above (2.15) imply the existence of a symmetrical optimum ξ for any of the criteria considered in the present section. For example, from (4.5) it follows at once that, if ξ is D-optimum, then the symmetrical $\bar{\xi}$ defined by (2.17) is also D-optimum. Remarks analogous to those of Section 2 on the number of points at which a symmetrical optimum ξ will be concentrated, clearly hold in the problems of this section. We note that the choice of the form of ξ_γ' in Example 4 is motivated by symmetry considerations, although the optimum weights must be computed in any approach.

The remark concerning the modification of (2.18) applies also to the problems of this section.

5. Estimation of the whole regression function. In the setup described by (2.1–(2.3), suppose the problem is one of estimation concerning all the a_i. One approach has been indicated in Section 4. Another approach is to think of the problem not as one of estimating the parameters a_i, but rather as one of estimating the entire function $\sum a_i f_i$. Thus, if g is the estimate of $\sum a_i f_i$, it is desired to make some measure of the average deviation of g from $\sum a_i f_i$ small in some sense, by choosing an appropriate design. The most obvious possibilities of such measures are perhaps (1) $\sup_a EW(\sup_x | g(x) - \sum a_i f_i(x) |)$, where W is nondecreasing; (2) the integral with respect to some measure μ on $\mathfrak{X}$ of $\sup_a EW(| g(x) - \sum a_i f_i(x) |)$; (3) the supremum on $\mathfrak{X}$ and $\mathfrak{A}$ of $EW(| g(x) - \sum a_i f_i(x) |)$. Of these three possibilities, the first is perhaps the most meaningful for most applications (with perhaps the inclusion of a weight function $h(x)$ multiplying $| g(x) - \sum a_i f_i(x) |$) but is computationally much more difficult to treat than the others; the second possibility is by far the easiest computationally, but is least satisfactory from a practical point of view because of the necessity of choosing μ—for example, if $\mathfrak{X}$ is a line segment, the optimum design will not be invariant under homeomorphisms of $\mathfrak{X}$, if μ is always chosen to be Lebesgue measure; the third possibility is a compromise between the first two and, as a first attack on the problem, is what we consider in this section, with $W(t) = t^2$. We note that a remark of [9, p. 215] indicates that Box and Hunter are considering the second approach for certain polynomial multiple regression problems when $W(t) = t^2$ and μ is Lebesgue measure on a Euclidean set. We note that it is a consequence of all three definitions of optimality discussed in this paragraph that an optimum design is optimum for all values of σ^2.

We shall not have to concern ourselves here with the choice of the function g: for example, the remarks of Section 2 extend here to show that, for a given design, if $\hat{a}$ is the b.l.e. of a, then $\sup_{a,x} E_a[\sum a_i f_i(x) - g(x)]^2$ is a minimum for $g(x) = \sum \hat{a}_i f_i(x)$. We therefore assume this choice of g in what follows. Thus, we are led to consider the minimization with respect to the design d of the expression

$$\max_x E[\sum_i (\hat{a}_i - a_i) f_i(x)]^2 = \sigma^2 \max_x f(x)' V_d f(x), \tag{5.1}$$

where we have written $f(x)$ for the vector of $f_i(x)$'s. Using again the representation of a design as a measure ξ, the analogue of V_d is the inverse of the matrix $M(\xi)$ whose (i, j)th element is

$$m_{ij}(\xi) = \int f_i(x) f_j(x) \xi(dx). \tag{5.2}$$

Thus, making an approximation like that of Section 2 in not requiring $n\xi$ to be integral, we define a design ξ^* to be optimum for the problem of this section if $M(\xi^*)$ is nonsingular and

$$\max_x f(x)' M(\xi^*)^{-1} f(x) = \min_{\xi \epsilon \Xi} \max_x f(x)' M(\xi)^{-1} f(x). \tag{5.3}$$

It seems more difficult here than in Section 2 to give a useful general computing algorithm. We now describe one device which seems useful in many examples. Let D_ξ be a non-singular matrix such that the vector $g_\xi = D_\xi f$ consists of functions $g_{\xi,i}$ which are orthonormal with respect to ξ; it is clear that such a D_ξ exists for any ξ in Ξ for which $M(\xi)$ is non-singular. Since the (i, j)th element of $D_\xi M(\xi) D_\xi'$ is the integral with respect to ξ of $g_{\xi,i} g_{\xi,j}$, we obtain

$$f(x)' M(\xi)^{-1} f(x) = g_\xi'(x) (D_\xi M(\xi) D_\xi')^{-1} g_\xi(x) = \sum_i [g_{\xi,i}(x)]^2 \tag{5.4}$$

(Since the left side of (5.4) does not depend on D_ξ, neither does the right side; thus, in searching for a ξ which minimizes the maximum with respect to x of (5.4), it suffices to consider for each ξ only that D_ξ and g_ξ which are computationally most convenient.) Since the $g_{\xi,i}$'s are orthonormal with respect to ξ, the integral with respect to ξ of the last expression of (5.4) is k, and this cannot be greater than the maximum with respect to x of (5.4). Thus, a sufficient condition for a given ξ to be an optimum design is

$$\max_x \sum_{i=1}^{k} [g_{\xi,i}(x)]^2 = k. \tag{5.5}$$

Of course, a necessary condition for (5.5) to be satisfied is that ξ give measure one to the set of x where $\sum [g_{\xi,i}(x)]^2 = k$, and it is useful to keep this in mind in examples.

Suppose (5.5) is satisfied for a ξ' concentrated on k points, say $x_1, \cdots, x_k$. Then the $k \times k$ matrix whose (i, j)th element is $g_{\xi',i}(x_j)[\xi'(x_j)]^{\frac{1}{2}}$ has orthonormal rows and, hence, orthonormal columns: $\sum_i [g_{\xi',i}(x_j)]^2 \xi_j = 1$ for $1 \leq j \leq k$.

Hence, $\xi_j' > 0$, and by (5.5) $\xi_j' \geqq 1/k$. Hence, each ξ_j is $1/k$. We summarize our results.

THEOREM 5. *If* (5.5) *holds, then* ξ *is optimum.*[6] *If* (5.5) *holds for a* ξ *concentrated on* k *points, then* ξ *gives measure* $1/k$ *to each of these points.*

If the setup is that of Example 3.1 it follows from the results of [10] cited in Section 2 that there exists an optimum ξ concentrated on exactly k points. This will not be true in general (see Example 5.2 below).

In the special case where $\mathfrak{X}$ consists of k points, the argument of the paragraph preceding the present theorem, applied to the distribution $\xi_j = 1/k, j = 1, \cdots, k$, shows that (5.5) is satisfied for this distribution, and hence the latter is optimum. Combining this with a remark which follows (4.3) we conclude that, when $\mathfrak{X}$ consists of k points, the design which puts mass $1/k$ on each point is the unique optimum design according to both the definition (4.3) for $s = k$ (the problem of Section 4) and the definition (5.3) (the problem of the present section).

Example 5.1. *The setup of Example* 3.1. It is possible to solve this problem by our methods and such a solution was given in the original draft of this paper. In the meantime, however, a solution has been published by Guest [13], so that there is no point to repeating the details of our solution. An earlier discussion by Smith [14] gave details of designs up to $k = 7$. The optimum design assigns mass $1/k$ to the points $+1$, -1, and the roots of $L_k'(x) = 0$, where L_k' is the derivative of the Legendre polynomial. (5.5) is satisfied ([13], equation (10)). It therefore follows from Theorem 6 below that this design is also optimum in the sense of definition (4.3) for $s = k$ (problem of Section 4) for this setup; i.e., a special case of Theorem 6 asserts that Hoel's design [15] is the same as that of Guest [13].[7] This last fact was noted by Hoel through an examination of the explicit results in the polynomial case.

Example 5.2. This example illustrates the use of Theorem 5 where the optimum ξ is concentrated on more than k points and does not give equal measure to all of them. Let $k = 2$ and let $\mathfrak{X}$ consist of three points. Thus, we hereafter write the f_i and ξ and S as triples, where $S(x) = \sum_i [g_{\xi,i}(x)]^2$. Suppose $f_1 = (1, 1, 0)$ and $f_2 = (0, 1, 2)$. For $\xi = (\xi_1, \xi_2, \xi_3)$, we obtain easily

$$S = (\xi_1\xi_2 + 4\xi_2\xi_3 + 4\xi_1\xi_3)^{-1}(\xi_2 + 4\xi_3, \xi_1 + 4\xi_3, 4\xi_1 + 4\xi_2)$$

We have $\sum \xi_i S_i = 2$, identically in ξ. Suppose $\xi_1 = 0, \xi_2 > 0, \xi_3 > 0$. Then either 1) $S_2 = S_3$, in which case $\xi_2 = \xi_3 = \frac{1}{2}$ and $S_1 > S_2 = 2$, or 2) $\max(S_2, S_3) > 2$. Thus, in either case $\max_i S_i > 2$. A similar argument applies if either of the other ξ_i's is 0, and two ξ_i's can obviously not be 0. Thus, $\max_i S_i$ can be 2 only if all ξ_i are positive and all S_i are equal to 2. The unique optimum ξ is thus easily seen to be (4/15, 4/15, 7/15).

[6] The converse of this statement is true. In fact, it will be proved in a subsequent paper (the results were obtained too late for inclusion in the present paper) that the following three statements are equivalent: (a) the design ξ is optimum in the sense of Section 4 with $s = k$; (b) the design ξ is optimum in the sense of Section 5; (c) the design ξ satisfies (5.5).

[7] This is a special case (for polynomial regression) of the result described in footnote 6.

It is obvious how to give examples like those of Section 3 where the optimum ξ is not unique, etc.

The argument just after (2.17) is easily modified to apply to the expression on the left side of (5.1), so that we can again conclude that there exists an optimum symmetrical ξ if (2.15) and the obvious analogue of the condition of the line above (2.15) hold. The question of the number of points at which an optimum symmetrical ξ will be concentrated is difficult, as is the corresponding question for general optimum ξ.

The modification of (2.18) can be made in the problem of this section, exactly as in Section 2.

We shall conclude this section with a result which sheds some light on the connection between the problem of Section 4 for $s = k$ and the problem of the present section. This result has already been cited in Example 5.1.

THEOREM 6.[8] *If the design which puts mass* $1/k$ *on each of* k *points satisfies* (5.5) *and is optimum in the sense of* (5.3) (*problem of Section* 5), *then this design is also optimum in the sense of* (4.3) (*problem of Section* 4) *with* $s = k$.

PROOF: Let ξ_0 be a design, optimum in the sense of (5.3), such that ξ_0 assigns mass $1/k$ to each of the points $x_1, \cdots, x_k$ in $\mathfrak{X}$, and such that (5.5) is satisfied for $\xi = \xi_0$. Since a design optimum for the problem of Section 4 with $s = k$ is invariant under a linear transformation on the f_i, it will suffice to prove that ξ_0 is optimum for this problem assuming $f_i = g_{\xi_0,i}$; henceforth we make this assumption. Thus

$$\max_x \sum_i f_i^2(x) = k \tag{5.6}$$

and

$$m_{ij}(\xi_0) = \delta_{ij},$$

and we have to prove that ξ_0 maximizes $\det M(\xi)$. Now from (5.6) for any ξ we have

$$\sum_i m_{ii}(\xi) \leq k$$

and hence

$$\det M(\xi) \leq \prod_i m_{ii}(\xi) \leq 1 = \det M(\xi_0)$$

This proves the theorem.

The authors are obliged to Professor G. Elfving for helpful comments.

REFERENCES

[1] J. KIEFER, "On the nonrandomized optimality and randomized nonoptimality of symmetrical designs", *Ann. Math. Stat.*, Vol. 29 (1958), pp. 675–699.

[2] N. I. ACHIESER, *Theory of Approximation*, Ungar Pub. Co., New York, 1956.

[8] Theorem 6 is a very special case of the results announced in footnote 6.

[3] G. Elfving, "Optimum allocation in linear regression theory", *Ann. Math. Stat.*, Vol. 23 (1952), pp. 255–262.

[4] G. Elfving, "Geometric allocation theory", *Skand. Akt.* 1955, pp. 170–190.

[5] H. Chernoff, "Locally optimum designs for estimating parameter" *Ann. Math. Stat.*, Vol. 24 (1953), pp. 586–602.

[6] H. F. Bohnenblust, S. Karlin, and L. Shapley, "Games with continuous, convex payoff", *Ann. Math. Studies*, No. 24, pp. 181–192.

[7] J. Kiefer, "Invariance, minimax sequential estimation, and continuous time processes", *Ann. Math. Stat.*, Vol. 28 (1957), pp. 573–601.

[8] R. A. Fisher, *Statistical Methods for Research Workers*, tenth edition, Oliver and Boyd, Edinburg, 1946.

[9] G. E. P. Box and J. S. Hunter, "Multi-factor experimental designs for exploring response surfaces", *Ann. Math. Stat.* Vol. 28 (1957), pp. 195–241.

[10] J. A. Shohart and J. D. Tamarkin, *The Problems of Moments, Math. Surveys*, No. 1, Amer. Math. Soc., New York, 1943.

[11] E. J. Williams, "Optimum allocation for estimation of polynomial regression," (abstract) *Biometrics*, Vol. 14 (1958), p. 573.

[12] S. Ehrenfeld, "Complete class theorems in experimental design", *Proceedings of the Third Berkeley Symposium on Mathematical Statistics and Probability*, University of California Press, 1955.

[13] P. G. Guest, "The spacing of observations in polynomial regression", *Ann. Math. Stat.*, Vol. 29 (1958), pp. 294–299.

[14] K. Smith, "On the standard deviations of adjusted and interpolated values of an observed polynomial function and its constants and the guidance they give towards a proper choice of the distribution of observations", *Biometrika*, Vol. 12 (1918), pp. 1–85.

[15] P. G. Hoel, "Efficiency problems in polynomial estimation", *Ann. Math. Stat.*, Vol. 29 (1958), pp. 1134–46.

Reprinted from THE ANNALS OF MATHEMATICAL STATISTICS
Vol. 30, No. 2, June, 1959

ASYMPTOTIC MINIMAX CHARACTER OF THE SAMPLE DISTRIBUTION FUNCTION FOR VECTOR CHANCE VARIABLES

BY J. KIEFER[1] AND J. WOLFOWITZ[2]

Cornell University

Summary. The purpose of this paper is to prove Theorem 1 stated in Section 1 below and Theorem 2 of Section 6 and the results of Section 7. These theorems are the generalizations to vector chance variables of Theorems 4 and 5 and Section 6 of [1], and state that the sample distribution function (d.f.) is asymptotically minimax for the large class of weight functions of the type described below. The main difficulties are embodied in the proof of Theorem 1 (Sections 2 to 5), where the loss function is a function of the maximum difference between estimated and true d.f. The proof utilizes the results of [2] and is not a straightforward extension of the result of [1], because the sample d.f. is no longer "distribution free" (even in the limit), and hence it is necessary to prove the uniformity of approach, to its limit, of the d.f. of the normalized maximum deviation between sample and population d.f.'s (for a certain class of d.f.'s). The latter fact enables us essentially to infer the existence of a uniformly (with the sample number) approximately least favorable (to the statistician) d.f., by means of which the proof of the theorem is achieved. Theorem 2 (Section 6) considers loss functions of integral type, and more general loss functions are treated in Section 7.

1. Introduction and preliminaries. The problem of finding a reasonable estimator of an unknown distribution function (d.f.) F in one or more dimensions is an old one. In the one-dimensional case the first extensive optimality results were obtained in [1]. It was shown there that, although a minimax procedure for sample size n may depend on the weight function as well as on n, the sample d.f. ϕ_n^* is asymptotically minimax as $n \to \infty$ for a very large class of weight functions which includes almost any weight function of practical interest. Also, an exact minimax procedure is extremely tedious to calculate in most practical cases, and is less convenient to use in practice than is ϕ_n^*. Moreover, one can obtain from [1] a bound on the relative difference between the maximum losses which can be encountered from using ϕ_n^* or the actual minimax procedure, and for many common weight functions this bound indicates that ϕ_n^* is very close to being minimax for fairly small values of n.

For dimension $m > 1$ the minimax problem presents difficulties which are not present when $m = 1$. (An outline of the main ideas and difficulties encountered in the proofs when $m = 1$ or when $m > 1$ will be given in Section 4; the

Received April 21, 1958.

[1] The research of this author was under contract with the Office of Naval Research.

[2] The research of this author was supported by the U. S. Air Force under Contract No. AF 18(600)-685, monitored by the Office of Scientific Research.

proof there is completed in Section 5; additional considerations for various weight functions are outlined in Sections 6 and 7.) These difficulties stem from the fact that neither ϕ_n^* nor any other known procedure which seems a reasonable candidate for optimality, has the distribution-free property possessed by ϕ_n^* when $m = 1$. This fact has led investigators of the problem when $m > 1$ to try (unsuccessfully) to find reasonable distribution-free procedures. Such investigations now seem to have been aimed in the wrong direction; for the main result of the present paper is that ϕ_n^* *is still asymptotically minimax for a large class of weight functions, even though it is no longer distribution free.*

The proof of the result just stated presents new difficulties far greater than those encountered when $m = 1$. In order to describe these difficulties briefly, let us suppose for the moment that the risk function is the expected value (under the true F) of $n^{1/2}$ times the maximum absolute deviation between estimated and true d.f. The computation of this risk function or its limit as $n \to \infty$ for the sequence of procedures ϕ_n^* (or any other reasonable sequence procedures) is known to present formidable difficulties, even for very simple continuous F (e.g., the uniform distribution on the unit square when $m = 2$). Our method of proof circumvents such a computation by showing that, when n is suitably large, the risk function of ϕ_n^* is changed arbitrarily little from what it would be if the maximum deviation were taken over a large but *finite* set of points instead of over all of m-space (this uses a result of [2]). Thus, the problem is reduced to a multinomial problem, similar to the reduction of [1] when $m = 1$, and we can circumvent the explicit computation of the risk there in a manner like that used in the multinomial case in [1], and which will be described in Section 3 below. But there remains another difficulty: in order to use a Bayes technique like that of [1] to prove the asymptotic minimax character of ϕ_n^*, we must show that there is a d.f. F_δ at which the risk function of ϕ_n^* is almost a maximum for all sufficiently large n; i.e., that the location of some approximate maximum does not "wander around" too much with n. Because of the distribution-free nature of the chance loss (for many common loss functions) under ϕ_n^* when $m = 1$, the existence of such an F_δ was automatic there (any continuous d.f. could be used); for $m > 1$, our proof requires the result of Lemma 1 of Section 2 below to obtain the existence of such an F_δ, at least when F is restricted to belong to a class of d.f.'s which in Section 5 is seen to be dense enough in an appropriate sense to yield the desired result. Once such an F_δ is known to exist, a sequence of approximately least favorable a priori distributions can be constructed for the approximating multinomial problem in the manner of [1]; this will be described in Section 4.

Aside from the difficulties described in the previous paragraph, the proofs of minimax results when $m > 1$ are very similar to those when $m = 1$. Therefore, rather than to repeat all of the details of [1], in each of Sections 4, 6, and 7 we will first describe the idea of the proof and then will indicate the modifications needed in the proof of the corresponding section of [1] to make it apply when $m > 1$.

We now give the notation used in this paper. m will denote any positive integer, fixed throughout the sequel. $\mathfrak{F}$ denotes the class of all d.f.'s on Euclidean m-space R^m, and $\mathfrak{F}^c$ denotes the subclass of continuous members of $\mathfrak{F}$. Let D be any subclass of the space of real functions on R^m. For simplicity we assume $\mathfrak{F} \subset D$, although it is really only necessary that D contains every possible function of the form S_n (defined below), for all n and $z^{(n)}$. Let B be the smallest Borel field on D such that every element of $\mathfrak{F}$ is an element of B and such that, for every positive integer k, real numbers $a_1, \cdots, a_k$, and m-vectors $t_1, \cdots, t_k$, the set $\{g \mid g \in D; g(t_1) < a_1, \cdots, g(t_k) < a_k\}$ is in B. (For example, we might have $D = \mathfrak{F}$ and B the Borel sets of the usual metric topology.) Let $\mathfrak{D}_n$ be the class of all real functions ϕ_n on $B \times R^{mn}$ such that $\phi_n(\cdot\,; z)$ is a probability measure (B) on D for each z in R^{mn} and such that $\phi_n(\Delta\,; \cdot)$ is a Borel-measurable function on R^{mn} for each Δ in B.

We now describe the statistical problem. Let $Z_1, \cdots, Z_n$ be independently and identically distributed m-vectors, each distributed according to some d.f. F about which it is known only that $F \in \mathfrak{F}$ (or $\mathfrak{F}^c$ or some other suitably dense subclass of $\mathfrak{F}$). The statistician wants to estimate F. Write $Z^{(n)} = (Z_1, \cdots, Z_n)$ and $z^{(n)} = (z_1, \cdots, z_n)$, where $z_i \in R^m$. Having observed $Z^{(n)} = z^{(n)}$, the statistician uses some decision function ϕ_n (a member of $\mathfrak{D}_n$) as follows: a function $g \in D$ is selected by means of a randomization according to the probability measure $\phi_n(\cdot\,; z^{(n)})$ on D; the function g so selected (which need not even be a member of $\mathfrak{F}$) is then the statistician's estimate of the unknown F. It is desirable to select a procedure ϕ_n which may be expected to yield a g which will lie close to the true F, whatever it may be; the precise meaning of "close" will be reflected by a weight function $W_n(F, g)$ which measures the loss when F is the true distribution function and g is the estimate of it. The probability of making a decision in Δ when ϕ_n is used and F is the true d.f. is

$$\mu_{F,\phi_n}(\Delta) = \int \phi_n(\Delta, z^{(n)}) F\,(dz^{(n)}), \tag{1.1}$$

which, as a function of Δ, will be a probability measure on D (see the next paragraph). Denoting expectation of a function on D with respect to this measure by E_{F,ϕ_n} (the symbol P_{F,ϕ_n} is used analogously, and the subscript ϕ_n will be omitted when it is not relevant), the risk function of the procedure ϕ_n is defined by

$$r_n(F, \phi_n) = E_{F,\phi_n} W_n(F, g); \tag{1.2}$$

i.e., it is the expected loss when F is true and ϕ_n is used. A sequence $\{\phi_n'\}$ of procedures is said to be *asymptotically minimax* relative to a sequence W_n of weight functions and a subclass $\mathfrak{F}'$ of $\mathfrak{F}$ if

$$\lim_{n\to\infty} \frac{\sup_{F \in \mathfrak{F}'} r_n(F, \phi_n')}{\inf_{\phi_n \in \mathfrak{D}_n} \sup_{F \in \mathfrak{F}'} r_n(F, \phi_n)} = 1. \tag{1.3}$$

(We note that this is a stronger property than that obtained by suppressing the supreme operation in the numerator and asking that the upper limit as

$n \to \infty$ be $\leqq 1$ for each F; this latter asymptotic property is much easier to verify than (1.3).) A nonrandomized decision function is one which for each $z^{(n)}$ assigns probability one to a single element (depending on $z^{(n)}$) of D. By ϕ_n^* we denote the nonrandomized procedure which chooses as decision the "sample d.f." S_n defined by

$$S_n(z) = n^{-1} \text{ (number of } Z_i \leqq z, 1 \leqq i \leqq n),$$

where as usual $Z \leqq z$ means that each component of Z is $\leqq$ the corresponding component of z. We shall not explicitly display the dependence of the chance function S_n on $Z^{(n)}$.

Obvious measurability considerations arise in connection with (1.1), (1.2), etc. These are handled exactly as in Section 1 of [1].

We can now state the main result of this paper, whose proof will occupy the next four sections (modifications and extensions are considered in Sections 6 and 7).

THEOREM 1. *Suppose* $W_n(F, g) = W(n^{1/2} \sup_z |g(z) - F(z)|)$, *where, for* $r \geqq 0$, $W(r)$ *is continuous, nonnegative, monotonically nondecreasing, not identically zero, and satisfies*

$$\int_0^\infty rW(r)e^{-c_m' r^2}\, dr < \infty \tag{1.4}$$

where c_m' *is given by* (1.8). *Then* $\{\phi_n^*\}$ *is asymptotically minimax relative to* $\{W_n\}$ *and* $\mathfrak{F}$.

Before listing the results of [2] which will be used in the present paper, we introduce some additional notation. When Z_1 has d.f. F, define

$$D_n = \sup_{x \varepsilon R^m} |S_n(x) - F(x)|$$

and

$$G_n(r; F) = P_F\{D_n < r/n^{1/2}\}.$$

For k a positive integer, write A_k^m for the subset of $(k+1)^m$ points in the m-dimensional unit cube $I^m = \{x \mid 0 \leqq x \leqq 1, x \varepsilon R^m\}$ for which each coordinate is an integral multiple of $1/k$. Write

$$D_{n,k} = \sup_{x \varepsilon A_k^m} |S_n(x) - F(x)|$$

and

$$G_{n,k}(r; F) = P_F\left\{D_{n,k} < \frac{r}{n^{1/2}}\right\}.$$

We also write

$$G_{\infty,k}(r; F) = \lim_{n\to\infty} G_{n,k}(r; F);$$

the existence of this limit follows from the multivariate central limit theorem. Finally, let $\mathfrak{F}^*$ be the class of d.f.'s F which are in $\mathfrak{F}^c$ and for which each one-dimensional marginal d.f. of F is uniform on I^1. Clearly, if Z_1 has d.f. F in $\mathfrak{F}^c$, we can perform continuous transformations on the components of Z_1, so as to make the result have a d.f. F^* in $\mathfrak{F}^*$, without changing G_n. This fact will be used in the sequel.

The results of [2] which will be used in the present paper are the following (some of these results hold with little or no modification for F in $\mathfrak{F}$, but we need them here only for F in $\mathfrak{F}^*$):

A. (Theorem 2 of [2].) For F in $\mathfrak{F}^*$, there is a d.f. $G(\cdot\,;F)$ such that

$$\lim_{n\to\infty} G_n(r;F) = G(r;F) \tag{1.5}$$

at every continuity point of the latter. Moreover, for F in $\mathfrak{F}^*$,

$$\lim_{k\to\infty} G_{\infty,k}(r;F) = G(r;F) \tag{1.6}$$

and (obviously)

$$\lim_{k\to\infty} G_{n,k}(r;F) = G_n(r;F). \tag{1.7}$$

B. (Theorem 1 of [2].) There are positive constants c_m^* and c_m' (independent of n, F, and r) such that, for F in $\mathfrak{F}^*$, all n, and all $r \geqq 0$,

$$1 - G_n(r;F) < c_m^* e^{-c_m' r^2} \tag{1.8}$$

Further remarks on possible values of c_m' are contained in [1] and [2].

C. For each F in $\mathfrak{F}$ there is an F_1 in $\mathfrak{F}^*$ such that, for all n and r,

$$G_n(r;F_1) \leqq G_n(r;F). \tag{1.9}$$

(This is fairly obvious; see [2] for further discussion.)

Of course, (1.8) and (1.9) also hold in the limit; i.e., with the subscript n deleted.

D. (A consequence of (3.11) of [2].) For all F in $\mathfrak{F}^*$, and for each $d > 0$,

$$G_{n,k}(r;F) - G_n(r+d;F) < \frac{c_1 k}{n} + c_2 k \exp\{-c_3 d^2 k^{1/2}\} + \frac{c_4}{d^4}\left(\frac{1}{k} + \frac{1}{n}\right), \tag{1.10}$$

where the c_i are positive constants depending only on m.

A further result of [2] will be given in Lemma 2 of Section 2, after some additional notation has been introduced.

In most of the arguments of this paper we will be dealing with F's which are in $\mathfrak{F}^c$. To simplify the discussion in such cases, we shall always assume that, for every real number t and integer j, at most one Z_i has its jth coordinate equal to t. The probability that this be not so is zero.

2. Uniformity of approach of G_n to G in the subclass $\mathfrak{F}_\epsilon$. The purpose of this section is to prove Lemma 1 (stated below), which will be used in Section 4 to prove the existence of an F_δ with the properties described in Section 1, when F is restricted to belong to a suitable subclass $\mathfrak{F}_\epsilon'$ of $\mathfrak{F}$. This and the multinomial result of Section 3 will then be used in Section 4 to demonstrate Theorem 1 with $\mathfrak{F}$ replaced by $\mathfrak{F}'_\epsilon$. The proof of Theorem 1 is then completed in Section 5 by showing that $\mathfrak{F}'_\epsilon$ is suitably dense in $\mathfrak{F}$ as $\epsilon \to 0$. Thus, although by far the greatest amount of new effort needed to prove Theorem 1 when $m > 1$ over what is needed when $m = 1$, is contained in the arguments of the present section, the reader who is interested mainly in the ideas of the statistical proof may read the statement of Lemma 1 and then go on to Section 3.

We first introduce some notation which will be used in this and subsequent sections. Let ϵ be a small positive number and let r be a positive number, both of which will be fixed in the present section. Other ϵ's with subscripts will be used in this paper to denote positive variables which will approach zero. The symbol $o(1 \mid \epsilon_i)$ is to denote a quantity which, as ϵ_i approaches zero, approaches zero uniformly in all other relevant quantities. Sometimes the latter will be explicitly indicated. Thus $o(1 \mid \epsilon_i \mid n, F)$ denotes a quantity which approaches zero, uniformly for all n (sometimes for all large n) and for all F (either in $\mathfrak{F}$ or in some indicated subclass), as $\epsilon_i \to 0$. The symbol $o(1 \mid \epsilon_i, n \mid F)$ denotes a quantity which approaches zero as $\epsilon_i \to 0$, $n \to \infty$, uniformly in F (either in $\mathfrak{F}$ or some indicated subclass). The symbol $o(1 \mid n \mid F)$ denotes a quantity which approaches zero as $n \to \infty$, uniformly in F (either in $\mathfrak{F}$ or some indicated subclass). The symbol $o(1 \mid d, N(d) \mid \cdot, \cdot)$ is to mean a quantity which approaches zero as $d \to 0$ while n stays larger than a suitable function $N(d)$ of d (which may change in various appearances of the symbol, although we shall sometimes use N, N', etc., to denote several such symbols which arise in the proof of the same lemma), and the approach of this quantity to zero is uniform in all other relevant quantities, which may be indicated where the dots are. The symbols $o(1 \mid \epsilon_i, N(\epsilon_i) \mid \cdot, \cdot)$ and $o(1 \mid k, N(k) \mid \cdot, \cdot)$ (with $k \to \infty$) will be used similarly. Finally the symbol θ will always denote a generic quantity <1 in absolute value; two θ's in different places need not be the same. The quantity d will always be >0.

Let $\mathfrak{F}_\epsilon$ be the subclass of those d.f.'s F in $\mathfrak{F}^*$ which have a Lebesgue density f_F in the subset of all points in I^m where at least one coordinate is $\geqq 1 - \epsilon$, and such that $\frac{1}{2} \leqq f_F \leqq 2$ almost everywhere in this region. The proofs of this section actually hold when $\mathfrak{F}_\epsilon$ is replaced by a somewhat larger class; but this is of little importance, the main use of Lemma 1 being to prove Theorem 1. (The relationship of $\mathfrak{F}'_\epsilon$ to $\mathfrak{F}_\epsilon$ will be stated in Section 4.)

LEMMA 1. *We have, for each fixed m,*

$$|G_n(r; F) - G(r; F)| = o(1 \mid n \mid F \, \varepsilon \, \mathfrak{F}_\epsilon). \tag{2.1}$$

The proof of Lemma 1 will require several supplementary lemmas. The proofs for all $m > 1$ are essentially the same, but the proof is most easily written out and followed in the case $m = 2$. *Hence, throughout the remainder of this section we shall carry out all proofs in the case* $m = 2$. The modifications in the statements and proofs which are necessary when $m > 2$ will usually be completely obvious;

and we shall explicitly mention, at appropriate points in the argument, those modifications which are not completely obvious.

Thus, we can write in coordinates $Z_n = (X_n, Y_n)$ and $z = (x, y)$, throughout the remainder of the section. (In most of the corresponding arguments for the case of m components, x will stand for the first $m - 1$ components of z, and y will stand for the last component of z.)

The idea of the proof of Lemma 1 is that (1.10) should somehow be used to prove Lemma 5, which, by a suitable uniformity result (Lemma 7) on the approach of the multinomial distribution to its limit, will yield (2.1). What is needed to obtain Lemma 5 from (1.10) is Lemma 4, the idea of which is that if $n^{1/2}|S_n(z) - F(z)|$ attains the value r somewhere, then it is very likely to attain the value $r + d$ somewhere, if d is small; it is the structure of $\mathfrak{F}_\epsilon$ which is used, in (2.18), to prove this.

For $0 < \epsilon_1 < 1$ we define the events

$$L_1(\epsilon_1) = \{\sup_{\substack{0 \le x \le \epsilon_1 \\ 0 \le y \le 1}} |S_n(z) - F(z)| \ge r/n^{1/2}\}, \tag{2.2}$$

and

$$L_2(\epsilon_1) = \{\sup_{\substack{1-\epsilon_1 \le x \le 1 \\ 1-\epsilon_1 \le y \le 1}} |S_n(z) - F(z)| \ge r/n^{1/2}\}. \tag{2.3}$$

(For the case of vectors with m components, the supremum in (2.2) is taken over the set where at least one of the $m - 1$ components of x is $\le \epsilon_1$; in (2.3), it is taken over the set where all m components are $\ge 1 - \epsilon_1$.)

The next two lemmas lead up to Lemma 4.

LEMMA 2. *We have*

$$P_F\{L_1(\epsilon_1)\} = o(1 \mid \epsilon_1, n \mid F \varepsilon \mathfrak{F}^*), \tag{2.4}$$

and

$$P_F\{L_2(\epsilon_1)\} = o(1 \mid \epsilon_1, n \mid F \varepsilon \mathfrak{F}^*). \tag{2.5}$$

PROOF. An upper bound on the probability of $L_1(\epsilon_1)$ can be obtained from equations (3.6), (3.9), and (3.10) of [2], if, in the latter, we set $h = 0, j = 1, k = 1/\epsilon_1$ (the relevant argument of [2] is valid even if k is not an integer), $d = r$. We obtain

$$\begin{aligned} P_F\{L_1(\epsilon_1)\} &< \frac{1}{n} + c_o \exp\{-cr^2/2\epsilon_1^{1/2}\} \\ &\quad + \frac{16}{r^4}\left(3\epsilon_1^2 + \frac{\epsilon_1}{n}\right) = o(1 \mid \epsilon_1, n \mid F \varepsilon \mathfrak{F}^*). \end{aligned} \tag{2.6}$$

We shall now use an argument like that by which (2.22) of [2] was proved, in order to prove (2.5). The event $L_2(\epsilon_1)$ implies the occurrence of at least one of the following events:

$$L_2^1 = \left\{ \sup_{\substack{1-\epsilon_1 \leqq x \leqq 1 \\ 1-\epsilon_1 \leqq y \leqq 1}} \mid (\text{number of } Z_1, \cdots, Z_n \text{ which satisfy } 0 \leqq X_i \leqq x, y \leqq Y_i \leqq 1) - \text{expected number} \mid \geqq \frac{rn^{1/2}}{3} \right\},$$

$$(2.7) \quad L_2^2 = \left\{ \sup_{\substack{1-\epsilon_1 \leqq x \leqq 1 \\ 1-\epsilon_1 \leqq y \leqq 1}} \mid (\text{number of } Z_1, \cdots, Z_n \text{ which satisfy } x \leqq X_i \leqq 1, y \leqq Y_i \leqq 1) - \text{expected number} \mid \geqq \frac{rn^{1/2}}{3} \right\},$$

$$L_2^3 = \left\{ \sup_{\substack{1-\epsilon_1 \leqq x \leqq 1 \\ 1-\epsilon_1 \leqq y \leqq 1}} \mid (\text{number of } Z_1, \cdots, Z_n \text{ which satisfy } x \leqq X_i \leqq 1, 0 \leqq Y_i \leqq y) - \text{expected number} \mid \geqq \frac{rn^{1/2}}{3} \right\}.$$

The random variables in the original sequence $\{Z_j\}$ all have the same distribution as $Z_1 = (X_1, Y_1)$. Apply the argument by which (2.4) was obtained for sequences all of whose members have the same distribution as each of the following, in order: $(1 - Y_1, X_1)$, $(1 - X_1, 1 - Y_1)$, and $(1 - X_1, Y_1)$. We obtain that

$$(2.8) \quad P_F\{L_2^i\} = o(1 \mid \epsilon_1, n \mid F \in \mathfrak{F}^*), \qquad i = 1, 2, 3.$$

Hence (2.5) is verified.

Define the events

$$(2.9) \quad L_3(\epsilon_1) = \left\{ \sup_{0 \leqq x \leqq 1} |S_n(x, 1) - x| < \frac{1}{\epsilon_1 n^{1/2}} \right\}$$

and

$$(2.10) \quad L_4(\epsilon_1) = \left\{ \sup_{\substack{\epsilon_1 \leqq x \leqq 1 \\ 0 \leqq y \leqq 1-\epsilon_1}} |S_n(z) - F(z)| \geqq \frac{r}{n^{1/2}} \right\}.$$

From (1.8) we obtain that

$$(2.11) \quad P_F\{L_3(\epsilon_1)\} = 1 - o(1 |\epsilon_1| n, F \in \mathfrak{F}^*).$$

Write $L(\epsilon_1) = L_3(\epsilon_1) \cap L_4(\epsilon_1)$. Whenever $L(\epsilon_1)$ occurs we can define chance variables H and T as follows: $H = h$, $\epsilon_1 \leqq h \leqq 1$, and $T = t, 0 < t \leqq 1 - \epsilon_1$, if

$$(2.12) \quad |S_n(h, t) - F(h, t)| \geqq r/n^{1/2}$$

and

$$(2.13) \quad |S_n(h', t') - F(h', t')| < r/n^{1/2}$$

for $\epsilon_1 \leqq h' \leqq 1$, $0 \leqq t' < t$, as well as for $\epsilon_1 \leqq h' < h$, $t' = t$. (In the m-component case, h' has all $m - 1$ components $\geqq \epsilon_1$ and h can be specified by any rule which does not depend on y for $y > t$, and such that (2.12) holds.) Thus, if a horizontal line $y = t'$ is swept upward starting at $t' = 0$, the line $y = t$ is the first for which (2.12) can hold, and h is a well-defined value such that it does.

Lemma 3. *We have, for some* $N(d)$ and $\epsilon_1 = d^{1/4}$,

$$P_F\left\{\sup_{T<y\leqq 1} |S_n(H,y) - F(H,y)| \geqq \frac{r+d}{n^{1/2}} \,\middle|\, L(\epsilon_1)\right\} \tag{2.14}$$
$$= 1 - o(1 \mid d, N(d) \mid F \varepsilon \mathfrak{F}_\epsilon) .$$

Proof. We suppose that

$$S_n(H, T) - F(H, T) = r/n^{1/2} \tag{2.15}$$

and we will prove that, conditional on $L(\epsilon_1)$ occurring, the probability that

$$S_n(H, y) - F(H, y) \geqq (r + d)/n^{1/2} \tag{2.16}$$

for some y, $T < y \leqq 1$, is $1 - o(1 \mid d, N(d) \mid F \varepsilon \mathfrak{F}_\epsilon)$. This will be enough to prove (2.14), for (a) if the left member of (2.15) is greater than $r/n^{1/2}$ the result we want to prove is a fortiori true, and (b) if the left member of (2.15) is $\leqq -r/n^{1/2}$, it is proved, in the same way as below, that the probability (conditional on $L(\epsilon_1)$ occurring) that the left member of (2.16) be $\leqq -(r + d)/n^{1/2}$ for some y, $T < y \leqq 1$, is $1 - o(1 \mid d, N(d) \mid F \varepsilon \mathfrak{F}_\epsilon)$.

Define

$$n_1 = n(S_n(H, 1) - S_n(H, T)),$$
$$\bar{y} = \frac{F(H, y) - F(H, T)}{H - F(H, T)}, \tag{2.17}$$
$$n_2(\bar{y}) = n(S_n(H, y) - S_n(H, T)).$$

From (2.9), (2.10), (2.15), and the definition of $\mathfrak{F}_\epsilon$, we have, in $L(\epsilon_1)$, if $\epsilon_1 < \epsilon$,

$$n_1 = n(H - F(H, T)) + \frac{\theta n^{1/2}}{\epsilon_1} - rn^{1/2} > n\epsilon_1^2/2 - n^{1/2}\left(\frac{1}{\epsilon_1} + r\right), \tag{2.18}$$

which goes to ∞ as $n \to \infty$ (uniformly in H and $T \leqq 1 - \epsilon_1$), and is thus arbitrarily large for $n >$ some $N'(\epsilon_1)$. Using (2.15) we find that (2.16) is equivalent to

$$\frac{n_2(\bar{y})}{n_1} - \frac{n\bar{y}(H - F(H, T))}{n_1} \geqq \frac{dn^{1/2}}{n_1}. \tag{2.19}$$

From (2.18) we obtain that the probability that (2.19) occur for some $\bar{y}$, $0 \leqq \bar{y} \leqq 1$, is $\geqq$ the probability that, for some $\bar{y}$,

$$\frac{n_2(\bar{y})}{n_1} - \bar{y} \geqq \frac{2d\,\epsilon_1^{-1}}{n_1^{1/2}} + \frac{4\epsilon_1^{-2}\bar{y}}{n_1^{1/2}}, \tag{2.20}$$

provided that ϵ_1 is small enough and $n >$ a suitable $N_1(\epsilon_1)$.

Now set

$$\epsilon_1 = d^{1/4} \tag{2.21}$$

and suppose that d is small enough that (2.20) holds when ϵ_1 is given by (2.21).

Let $\{W(t), 0 \leqq t < \infty\}$ be the separable (Wiener) process with independent, normally distributed increments, $W(o) = 0$, $E(W(t)) = 0$, $\text{Var}(W(t)) = t$. Given H, T, and n_1, the left member of (2.20) clearly is distributed as the difference between a sample d.f. and the uniform d.f. on the one-dimensional interval $0 \leqq \bar{y} \leqq 1$, when the sample d.f. is that of n_1 independent, uniformly distributed random variables. It follows from [3] and [4] and the fact that $n_1 \to \infty$ as $n \to \infty$ that, under (2.21), the conditional probability (given that $L(\epsilon_1)$ occurs) that (2.20) hold for some $\bar{y}$ approaches, uniformly in $\mathfrak{F}_\epsilon$, as $n \to \infty$,

$$(2.22) \qquad P\{W(t) \geqq 2d^{3/4} + (2d^{3/4} + 4d^{-1/2})t \text{ for some } t > 0\}.$$

The latter is, by [4], equation (4.2),

$$(2.23) \qquad \exp\{-2(2d^{3/4})(2d^{3/4} + 4d^{-1/2})\}$$

which approaches one as $d \to 0$. Hence, for d sufficiently small and $n >$ some $N(d)$, the conditional probability (given that $L(\epsilon_1)$ occurs) that (2.16) holds for some $y > T$ is arbitrarily close to 1, uniformly in $\mathfrak{F}_\epsilon$. This proves (2.14).

LEMMA 4. *We have, for some* $N(d)$,

$$(2.24) \qquad G_n(r + d; F) - G_n(r; F) = o(1 \mid d, N(d) \mid F \,\varepsilon\, \mathfrak{F}_\epsilon).$$

PROOF. Substituting (2.21) into (2.2), (2.4), (2.9), (2.10) and (2.11) (none of which previously depended on d in any way), and using Lemma 3, we have

$$(2.25) \qquad P_F\{r \leqq \sup_{\substack{0 \leqq x \leqq 1 \\ 0 \leqq y \leqq 1 - d^{1/4}}} \sqrt{n}|S_n(z) - F(z)| \leqq r + d\} = o(1 \mid d, N(d) \mid F \,\varepsilon\, \mathfrak{F}_\epsilon),$$

where of course the $N(d)$ may differ from that of Lemma 3. Now, the definition of $\mathfrak{F}_\epsilon$ is such that, by interchanging the roles of x and y, we obtain, in the same way that (2.25) was obtained,

$$(2.26) \qquad P_F\{r \leqq \sup_{\substack{0 \leqq x \leqq 1 - d^{1/4} \\ 0 \leqq y \leqq 1}} \sqrt{n}|S_n(z) - F(z)| \leqq r + d\} = o(1 \mid d, N(d) \mid F \,\varepsilon\, \mathfrak{F}_\epsilon).$$

(In the case of vectors with m components, there are $m - 2$ additional analogues of (2.26).) Finally, substituting (2.21) into (2.5) and combining the result with (2.25) and (2.26), we obtain (2.24).

LEMMA 5. *We have*

$$(2.27) \qquad 0 \leqq G_{n,k}(r; F) - G_n(r; F) = o(1 \mid k, N'(k) \mid F \,\varepsilon\, \mathfrak{F}_\epsilon).$$

PROOF. The left side of (2.27) is trivial. Adding (1.10) and (2.24), we have

$$(2.28) \qquad \begin{aligned} G_{n,k}(r; F) - G_n(r; F) &\leqq o(1 \mid d, N(d) \mid F \,\varepsilon\, \mathfrak{F}_\epsilon) \\ &\quad + c_2 k \exp\{-c_3 d^2 k^{1/2}\} + c_4/d^4 k + c_1 k/n + c_4 / d^4 n. \end{aligned}$$

Let $\epsilon' > 0$ be given arbitrarily. Let $d_1 > 0$ be such that the first term on the right side of (2.28) is $< \epsilon'/3$ if $d = d_1$ and $n > N(d_1)$. Let k_1 be such that the sum of the next two terms on the right side of (2.28) is $<\epsilon'/3$ when $d = d_1$ and $k > k_1$. For $k > k_1$, let $N'(k)$ be $>N(d_1)$ and be such that the sum of the last two terms of (2.28) is $<\epsilon'/3$ when $d = d_1$ and $n > N'(k)$. Then, putting $d = d_1$, we have that the right side of (2.28) is $<\epsilon'$ when $k > k_1$ and $n > N'(k)$. Thus, Lemma 5 is proved.

The discussion which immediately follows, as well as Lemma 6, leads up to the proof of Lemma 7.

There are k^2 cells into which I^2 is divided by the lines $x = i/k$, $y = j/k$, $i, j = 0, 1, \cdots, k$. (There are, of course, k^m cells in the case of vectors with m components.) Number the cells as follows: The cell bounded by $x = (i - 1)/k$, $x = i/k$, $y = (j - 1)/k$, $y = j/k$, is to be called the (i, j) cell. Write

$$\pi_{Fij} = P_F\{Z_1 \,\varepsilon\, \text{cell } (i, j)\}.$$

Write $(i', j') \leqq (i, j)$ if $i' \leqq i, j' \leqq j$, and write $(i', j') < (i, j)$ if $(i', j') \leqq (i, j)$ and either $i' < j$ or $j' < j$. Let $\bar{H}$ be any collection of cells. For any fixed (i_0, j_0) not in $\bar{H}$, there clearly exist integers c_{ij} (depending only on $\bar{H}$ and (i_0, j_0)) such that we can write

$$F(i_0/k, j_0/k) = \sum_{\substack{(i,j) \leqq (i_0,j_0) \\ (i,j)\,\varepsilon H}} c_{ij}F(i/k, j/k) + \sum_{\substack{(i,j) \leqq (i_0,j_0) \\ (i,j)\,\not\varepsilon H}} c_{ij}\,\pi_{Fij}, \tag{2.29}$$

identically in F (i.e., in the π_{Fij}).

Let $\epsilon_2 > 0$ be given. Call the cell (i, j) regular if $\pi_{Fij} \geqq \epsilon_2$ and $(i, j) \neq (k, k)$. Call the cell (i, j) singular if $\pi_{Fij} < \epsilon_2$ and $(i, j) \neq (k, k)$. Let $\bar{H}_F$ be the collection of regular cells, let (i_0, j_0) be singular under F, and let the c_{ij} be as in (2.29). Denote a summation over the region $(i, j) \leqq (i_0, j_0)$, $(i, j) \,\varepsilon\, \bar{H}_F$ by $\sum^{(F, i_0, j_0)}$. Then, clearly,

$$|F(i_0/k, j_0/k) - \textstyle\sum^{(F,i_0,j_0)} c_{ij}F(i/k, j/k)| < h(k)\epsilon_2, \tag{2.30}$$

where h is a suitable positive function of k alone, which can be chosen so that (2.30) is valid for every ϵ_2, every F, and every (i_0, j_0) singular for such an F; here the $\bar{H}_F$ depends on the F and ϵ_2 being considered, but the c_{ij} depend on these quantities only through $\bar{H}_F$.

Define $Q_{F,n}(\epsilon_2)$ to be the probability that

$$|S_n(i/k, j/k) - F(i/k, j/k)| < r/n^{1/2} \text{ for all } (i, j) \text{ in } \bar{H}_F,$$

$$|\textstyle\sum^{(F,i_0,j_0)} c_{ij}[S_n(i/k, j/k) - F(i/k, j/k)]| < r/n^{1/2} \tag{2.31}$$

$$\text{for all } (i, j) \neq (k, k) \text{ and not in } \bar{H}_F.$$

The proof of the next lemma is actually valid when $\mathfrak{F}^*$ is replaced by the class of all d.f.'s on I^2.

LEMMA 6. *We have*

$$|Q_{F,n}(\epsilon_2) - G_{n,k}(r; F)| = o(1 \mid \epsilon_2, N(\epsilon_2) \mid F \,\varepsilon\, \mathfrak{F}^*) \tag{2.32}$$

PROOF. Define, for (i_0, j_0) singular,

$$U = S_n(i_0/k, j_0/k) \tag{2.33}$$

and

$$V = \Sigma^{(F, i_0, j_0)} c_{ij} S_n(i/k, j/k). \tag{2.34}$$

Let B be the event defined by

$$B = \{|(U - V) - E(U - V)| < \epsilon_2^{1/4}/n^{1/2}\}. \tag{2.35}$$

Now, $U - V$ is just the last sum of (2.29) with π_{Fij} replaced by n^{-1} (number of $Z_1, \cdots, Z_n$ falling in cell (i, j)). Hence, $U - V$ has variance $<h'(k)\epsilon_2/n$, where h' is a suitable positive function of k. Thus, by Chebyshev's inequality,

$$P_F\{B\} > 1 - h'(k)\epsilon_2^{1/2}. \tag{2.36}$$

Of course, the definition of B depends on F, was well as on (i_0, j_0); but, again, h' can be chosen so that (2.36) holds for all F. Consider the events

$$A_1 = \{|U - EU| \geqq r/n^{1/2}, |V - EV| < r/n^{1/2}\} \tag{2.37}$$

and

$$A_2 < \{|U - EU| < r/n^{1/2}, |V - EV| \geqq r/n^{1/2}\}. \tag{2.38}$$

The definition of these events also depends on F and (i_0, j_0). Define $P_{Ft} = P_F\{A_t\}$, $t = 1, 2$. We are first going to show that, for all F for which (i_0, j_0) is singular,

$$P_{Ft} = o(1 \mid \epsilon_2, N(\epsilon_2) \mid F \,\varepsilon\, \mathfrak{F}^*), \qquad \text{for } t = 1, 2. \tag{2.39}$$

In proving this, let W stand for U in the case $t = 1$ and for V in the case $t = 2$. Then W is n^{-1} times the sum of n independent, identically distributed random variables, each bounded in absolute value by some constant L (independent of F). Let σ^2 be the variance and β_3 the absolute third moment about its expected value of each summand (i.e., of $W - EW$ when $n = 1$). Now, if $\sigma^2 < \epsilon_2^{1/8}$, Chebyshev's inequality yields $P_{Ft} < r^{-2}\epsilon_2^{1/8}$, so that (2.39) is verified in that case. On the other hand, if $\sigma^2 \geqq \epsilon_2^{1/8}$, by (2.36) we have

$$\begin{aligned} P_{Ft} &\leqq P_F\{B \cap A_t\} + P_F\{\bar{B}\} < P_F\{B \cap A_t\} + h'(k)\epsilon_2^{1/2} \\ &\leqq P_F\{r \leqq n^{1/2}|W - EW| \leqq r + \epsilon_2^{1/4}\} + h'(k)\epsilon_2^{1/2} \\ &\leqq P_F\{r/\sigma \leqq n^{1/2}|W - EW| / \sigma \leqq r/\sigma + \epsilon_2^{3/16}\} + h'(k)\epsilon_2^{1/2}. \end{aligned} \tag{2.40}$$

By the Berry-Esseen estimate (see, e.g., [5]) and the fact that $\beta_3/\sigma^3 \leqq L/\sigma$, we have from (2.40) for all F for which $\sigma^2 \geqq \epsilon_2^{1/8}$,

$$P_{Ft} \leqq h'(k)\epsilon_2^{1/2} + \epsilon_2^{3/16} + c_5 L n^{-1/2}\epsilon_2^{-1/16}, \tag{2.41}$$

where c_5 is a positive constant. Thus, (2.39) is proved.

Lemma 6 follows at once from (2.39).

The proof of the next lemma is also valid when $\mathfrak{F}^*$ is replaced by the class of all d.f.'s on I^2.

LEMMA 7. *For any fixed positive integer k, we have*

$$|G_{\infty,k}(r; F) - G_{n,k}(r; F)| = o(1|n|F \varepsilon \mathfrak{F}^*). \tag{2.42}$$

PROOF. Let $\epsilon_3 > 0$ be given arbitrarily. Choose ϵ_2 so small and $N(\epsilon_2)$ so large that, for this value of ϵ_2, the left side of (2.32) is $\leqq \epsilon_3/4$ for all F when $n > N(\epsilon_2)$. We shall show below that, writing $Q_F(\epsilon_2) = \lim_{n\to\infty} Q_{F,n}(\epsilon_2)$, we have

$$|Q_F(\epsilon_2) - Q_{F,n}(\epsilon_2)| < \epsilon_3/2 \tag{2.43}$$

for n sufficiently large, uniformly in F. Hence, we shall have, for n sufficiently large, uniformly in F, that the left side of (2.42) is no greater than

$$\begin{aligned} &|G_{\infty,k}(r; F) - Q_F(\epsilon_2)| + |Q_F(\epsilon_2) - Q_{F,n}(\epsilon_2)| \\ &\qquad + |Q_{F,n}(\epsilon_2) - G_{n,k}(r; F)| < \epsilon_3/4 + \epsilon_3/2 + \epsilon_3/4 = \epsilon_3, \end{aligned} \tag{2.44}$$

and (2.42) will be proved.

We shall now fix $\bar{H}$ and prove that (2.43) holds, uniformly in all F for which $\bar{H}_F = \bar{H}$, for n sufficiently large. Since k is fixed, the number of possible choices of $\bar{H}$ is finite, so that Lemma 7 will be proved.

Consider the joint distribution of the $n^{1/2}(S_n(i/k, j/k) - F(i/k, j/k))$ for all regular (i, j), which, as $n \to \infty$, approaches a multivariate normal distribution. Since $\pi_{Fij} \geqq \epsilon_2$ for any regular point it follows that the determinant of the covariance matrix of the $n^{1/2}(S_n(i/k, j/k) - F(i/k, j/k))$ (for regular (i, j)) is bounded away from 0 (and, of course, from ∞ as well) by a function of ϵ_2, uniformly in all F for which $\bar{H}_F = \bar{H}$. It follows from [6], page 121, that the maximum of the absolute value of the difference between the joint d.f. of these $n^{1/2}(S_n(i/k, j/k) - F(i/k, j/k))$ and their limiting multivariate normal d.f. is less than $n^{-1/2}M(\epsilon_2)$, where M is a real function of ϵ_2 only. The maximum of the density of this limiting normal d.f. is a real function only of ϵ_2, say $M'(\epsilon_2)$. Thus, the statements in the last two sentences are uniform in all F for which $\bar{H}_F = \bar{H}$.

It follows from (1.8) that the probability of a sufficiently large cube C in the space of the $n^{1/2}(S_n(i/k, j/k) - F(i/k, j/k))$ (for all regular (i, j)) which is centered at the origin, is greater than $1 - \epsilon_3/12$ uniformly in F and n. Hence this is also true of the limiting multivariate normal d.f. of the $n^{1/2}(S_n(i/k, j/k) - F(i/k, j/k))$.

Consider the region R in the space of these $n^{1/2}(S_n(i/k, j/k) - F(i/k, j/k))$, which is defined by (2.31) and whose probability is $Q_{F,n}(\epsilon_2)$. The region $R \cap C$ is a bounded polyhedron and can be approximated from within by a finite union R_1 of "rectangles" with sides parallel to the coordinate planes, such that the volume of the region $R_2 = [(R \cap C) - R_1]$ is $< \epsilon_4$, where $\epsilon_4 > 0$ is such that $\epsilon_4 M'(\epsilon_2) < \epsilon_3/12$. The set R_2 can be covered by a finite union R_3 of rectangles with sides parallel to the coordinate planes whose total volume is $< 2\epsilon_4$. Let m_3 be the number of rectangles in R_3, and m_1 be the number of rectangles in R_1.

The probability of R_3 according to the limiting normal d.f. is less than

$$2\epsilon_4 M'(\epsilon_2) < \epsilon_3/6. \tag{2.45}$$

The probability $P_F\{R_2\}$ of the region R_2 according to F is $<P_F\{R_3\}$, which, by the aforementioned result of Bergstrom [6], differs from the probability of R_3 according to the limiting normal d.f. by less than $4m_3M(\epsilon_2)n^{-1/2}$. Hence,

$$P_F\{R_2\} < \epsilon_3/6 + 4m_3M(\epsilon_2)n^{-1/2}. \tag{2.46}$$

Also, by Bergstrom's result just cited, the probability of R_1 according to the limiting normal d.f. differs from $P_F\{R_1\}$ by less than $4m_1M(\epsilon_2)n^{-1/2}$. Since the sum of this and the second term in the right member of (2.46) can be made less than $\epsilon_3/6$ by making n sufficiently large, it follows from the present paragraph and the previous two paragraphs that (2.43) holds for n sufficiently large, uniformly in all F for which $\bar{H}_F = \bar{H}$. This completes the proof of Lemma 7.

Proof of Lemma 1. Let $\epsilon_5 > 0$ be chosen arbitrarily. Choose k' such that the right side of (2.27) is $<\epsilon_5/3$ for $k = k'$ and $n > N'(k')$. In particular, $0 \leqq G_{\infty,k'}(r; F) - G(r; F) \leqq \epsilon_5/3$. Choose N to be $>N'(k')$ and such that, for $k = k'$, the left member of (2.42) is $<\epsilon_5/3$ for $n > N'$. Then, for $n > N'$ and all F in $\mathfrak{F}_\epsilon$, we have

$$\begin{aligned} |G_n(r; F) - G(r; F)| &\leqq |G_n(r; F) - G_{n,k'}(r; F)| \\ &+ |G_{n,k'}(r; F) - G_{\infty,k'}(r; F)| + |G_{\infty,k'}(r; F) - G(r; F)| < \epsilon_5 . \end{aligned} \tag{2.47}$$

Since ϵ_5 was arbitrary, Lemma 1 is proved.

3. The multinomial result. We have mentioned in Section 1 that the main results of this paper are obtained by approximating the original problem by an appropriate multinomial problem. In the present section we summarize the needed multinomial results which were obtained in [1], and sketch the ideas of the proofs, unencumbered by the tedious details of [1]. Actually, we do not need the full strength of the results of [1], which are broader than those of Lemma 8 below in that, in the derivation of Section 3 of [1], the calculations were carried out in fine detail in order to obtain an error term which can be used to calculate an upper bound on the departure of ϕ_n^* from minimax character (in view of the lack of knowledge about the distribution of D_n, it seems more difficult to obtain a useful bound of this kind when $m > 1$). In fact, if one does not bother to obtain an error term, it is obvious how to shorten considerably the proof of the multinomial result in Section 3 of [1], and we shall see that this simple multinomial result without error term rests mainly on a result of v. Mises ([7], especially pages 84–86) which is almost forty years old.

We now introduce the needed notation. Let h be a positive integer and let B_h be the family of $(h + 1)$-vectors $\pi = \{p_i , 1 \leqq i \leqq h + 1\}$ with real components satisfying $p_i \geqq 0$, $\sum p_i = 1$. Let B_h' be a specified subset of $B_h \cdot B_h'$ can actually be fairly arbitrary in structure; to avoid trivial circumlocutions, we shall suppose in this section that B_h' is the closure of an h-dimensional open subset of B_h, although it will be obvious that Lemmas 8, 9, and 10 hold much more generally. Let $T^{(n)} = \{T_i^{(n)}, 1 \leqq i \leqq h+1\}$, a vector of $h + 1$ chance variables, have a multinomial probability function arising from n observations with $h + 1$ possible outcomes, according to some π in B_h'; i.e., for integers $x_i \geqq 0$ with $\Sigma_1^{h+1} x_i = n$,

$$(3.1) \qquad P_\pi\{T_i^{(n)} = x_i, 1 \leqq i \leqq h+1\} = \frac{n!}{x_1! \cdots x_{h+1}!} p_1^{x_1} \cdots p_{h+1}^{x_{h+1}}.$$

Let L be a positive integer, let γ_i be an $(h + 1)$-vector, $1 \leqq i \leqq L$, and let $\rho_i = \gamma_i'\pi$ (scalar product) be corresponding linear functions of π, $1 \leqq i \leqq L$. To avoid trivialities, we assume at least one ρ_i is not constant on B_h'. Let $\mathcal{E}_n$ be the class of all (possibly randomized) vector estimators of $\rho = \{\rho_i, 1 \leqq i \leqq L\}$, the weight function (which depends on n) being the simple one for which the risk function of a procedure ψ_n in $\mathcal{E}_n$ is

$$(3.2) \qquad 1 - P_{\pi,\psi_n}\{|d_i - \rho_i| \leqq r/n^{1/2}, 1 \leqq i \leqq L\},$$

where r is a positive value and we have written $d = \{d_i, 1 \leqq i \leqq L\}$ for the vector of decisions. Let ψ_n^* be the nonrandomized estimator whose ith component is $\gamma_i'T^{(n)}/n$ (the allowable decisions may be restricted to $\gamma'\pi$ for π in B_h' with only trivial modifications in what follows). Finally, a point π in B_h' is called an interior point if all its components p_i are positive, and if it has a neighborhood (in B_h) which is a subset of B_h'. The required multinomial result is:

LEMMA 8. *For any interior point π^* of B_h' there is a sequence $\{\xi_n\}$ of a priori distributions on B_h' converging in distribution to the distribution which gives probability one to π^* and such that $\{\psi_n^*\}$ is asymptotically Bayes relative to $\{\xi_n\}$ as $n \to \infty$, uniformly for $0 \leqq r \leqq R$ for any $R < \infty$; i.e., such that, uniformly in such r,*

$$(3.3) \qquad \lim_{n\to\infty} \frac{\int P_\pi\{|\rho_i - \gamma_i'T^{(n)}/n| > r/n^{1/2}, 1 \leqq i \leqq L\}\, \xi_n(d\pi)}{\inf_{\psi_n \in \mathcal{E}_n} \int P_{\pi,\psi_n}\{|d_i - \rho_i| > r/n^{1/2}, 1 \leqq i \leqq L\}\, \xi_n(d\pi)} = 1.$$

Of course, continuity considerations show that the positive (since not all ρ_i are constant) limit of the numerator of (3.3) is obtained by putting $\pi = \pi^*$ instead of integrating with respect to ξ_n, and then using the multivariate central limit theorem to compute the limiting probability.

The idea of the proof of Lemma 8 is very simple. Let Γ^* be the (nonsingular) covariance matrix of the limiting h-variate normal distribution of $n^{1/2}(n^{-1}T_i^{(n)} - p_i^*)$, $1 \leqq i \leqq h$, when $\pi = \pi^*$. Let ϵ be a small positive value and let ξ be the uniform *a priori* distribution in the (solid) sphere of radius ϵ about π^* in B_h'. (ϵ is small enough that this sphere consists entirely of interior points.) According to the result [7] of v. Mises, for any π'' in this sphere, with probability one when $T^{(n)}$ is distributed according to π'', the *a posteriori* density function of $n^{1/2}(p_i - T_i^{(n)}/n)$, $1 \leqq i \leqq h$ (calculated assuming ξ to be the *a priori* distribution) will tend to the h-variate normal density with means 0 and covarance matrix Γ'' (corresponding to π'') as $n \to \infty$. If the *a posteriori* density were really normal with the stated parameters, it would follow at once from a result [8] of Anderson that the *a posteriori* probability of the event

$$\{n^{1/2}|\rho_i - d_i| \leqq r, 1 \leqq i \leqq L\} \tag{3.4}$$

(this probability is unity minus the *a posteriori* risk) is a maximum for $d = \gamma' T^{(n)}/n$, since the region (3.4) is for each d a convex symmetric (about a point depending on d) subset in the space of the h variables $n^{1/2}(p_i - T_i^{(n)}/n)$ (considering the latter to be unrestricted in magnitude). Since the actual *a posteriori* density is almost normal (with high probability as $n \to \infty$), ψ_n^* will be asymptotically Bayes. Finally, let ξ_n be the ξ just described when $\epsilon = \epsilon_n$, where ϵ_n goes to zero slowly enough that the above result still holds for ψ_n^* as $n \to \infty$. (For example, $\epsilon_n = n^{-\alpha}$ with $0 < \alpha < \frac{1}{2}$. The crucial consideration is that the radius $n^{1/2}\epsilon_n$ of the set of possible values of $n^{1/2}(\pi - \pi^*)$ approach infinity with n, as will therefore the radius of the set of possible values of $n^{1/2}(\pi - T^{(n)}/n)$ w.p.1 under ξ_n. The asymptotic problem is thus approximately one of estimating the mean of a multivariate normal distribution with known constant covariance matrix, when the mean can take on any value in an appropriate Euclidean space).

The actual proof—the precise handling of the approximations mentioned above, the uniformity in r, etc.—may be handled as in [1] or by complementing with appropriate estimates the argument of [7], but the main idea is really the simple one of [7].

The reason for wanting Lemma 8 in its stated form with the sequence $\{\xi_n\}$ shrinking down on π^* has to do with the problem of multinomial minimax estimation for the risk function (3.2). Let π^0 be the value of π at which the positive limit b (as $n \to \infty$) of the continuous risk function of ψ_n^* is a maximum. Since the ρ_i are not all constant, for any $\delta > 0$ there will, by continuity, be an interior point π^* of B_h' at which the limit of the risk function of ψ_n^* is at most $(1 + \delta)b$. From Lemma 8 and the sentence following (3.3) we conclude:

LEMMA 9. *$\{\psi_n^*\}$ is asymptotically minimax relative to B_h' and the risk function* (3.2).

We next consider a generalization of this result to other weight functions which are nondecreasing functions of $\max_i |d_i - \rho_i|$. Of course, the risk function is defined in the usual way. (A Bayes result analogous to Lemma 8 can be proved in the course of the demonstration, but we shall not bother to state it.) Let C_0 and C be positive constants such that

$$P_\pi\{n^{1/2} \max_i \gamma_i' |T^{(n)}/n - \pi| \geqq r\} < C_0 e^{-Cr^2} \tag{3.5}$$

for all r, all n, and all π in B_h'. The existence of such positive constants (which depend on h and the structure of B_h') follows from well known results on the multinomial (or, in fact, the binomial) distribution; in Section 4 we shall actually refer to (1.8) for appropriate values of these constants.

LEMMA 10. *Let $W(r)$ be a nondecreasing real function of r for $r \geqq 0$, not identically zero, and satisfying*

$$\int_0^\infty W(r) r e^{-Cr^2} \, dr < \infty. \tag{3.6}$$

Then $\{\psi_n^*\}$ is asymptotically minimax relative to B_h' and the weight functions

$$W_n(\pi, d) = W(n^{1/2} \max_i |\rho_i - d_i|). \tag{3.7}$$

The proof of Lemma 10 can be carried out, starting from scratch, along lines like those of Lemma 8. An easier proof, which was given in [1], rests upon the idea of reducing the proof essentially to that for the simple weight function already considered in Lemma 8. Specifically, if the *a posteriori* distribution of the variables $n^{1/2}(p_i - T_i^{(n)}/n)$ were actually normal with means 0 and the appropriate covarance matrix, then $d_i = \gamma_i' T^{(n)}/n$, $1 \leqq i \leqq L$, would minimize the *a posteriori* risk; for, if this choice of the d_i did not minimize the *a posteriori* risk and if H_1 and H_2 were respectively, the d.f.'s of $n^{1/2} \max_i |\rho_i - d_i|$ for the above choice of d_i and for a better choice, we would have

$$\int_0^\infty W(r) \, d[H_1(r) - H_2(r)] > 0,$$

which is easily seen to imply that $H_1(r') < H_2(r')$ for some r', contradicting Anderson's result cited previously (i.e., when the error terms are included, this contradicts the result of Lemma 8). The details of the proof are contained in Section 4 of [1].

We note that Lemma 10 exemplifies a principle which is of more general use in statistics: If one can verify suitable (asymptotic) Bayes results for an appropriate class of simple weight functions, the results will automatically hold for a general class of monotone weight functions.

We remark that Anderson's result can be used to prove the result of Lemma 10 for a larger class of weight functions, namely, every function of $n^{1/2}(d_i - \rho_i)$, $1 \leqq i \leqq L$, which is symmetric about the origin and which for each real value c has a convex (or empty) set for the domain where the function is $\leqq c$.

4. Proof of Theorem 1 when $\mathfrak{F}$ is replaced by $\mathfrak{F}_\epsilon'$. Define $\mathfrak{F}_\epsilon'$ to consist of every d.f. in $\mathfrak{F}^c$ which gives probability one to I^m and which can be realized as the d.f. of Z_1' (say) when Z_1 has a d.f. in $\mathfrak{F}_\epsilon$ and Z_1' is obtained from Z_1 by continuous monotonic transformations on the individual coordinate functions. Thus, $\mathfrak{F}_\epsilon' \supset \mathfrak{F}_\epsilon$, but $\mathfrak{F}_\epsilon'$ includes d.f.'s which are not in $\mathfrak{F}^*$. Clearly, for any F' in $\mathfrak{F}_\epsilon'$ there is an F in $\mathfrak{F}_\epsilon$ such that $G_n(r; F) = G_n(r; F')$ for all n and r.

In this section we use the results of Sections 2 and 3 to prove the following

LEMMA 11. *For m a positive integer, suppose that*

$$W_n(F, g) = W(n^{1/2} \sup_z |F(z) - g(z)|), \tag{4.1}$$

where $W(r)$ for $r \geqq 0$ is continuous, nonnegative, nondecreasing in r, not identically *zero, and satisfies* (1.4). *Then, for each* ϵ *with* $0 < \epsilon < 1$, $\{\phi_n^*\}$ *is asymptotically minimax relative to* $\{W_n\}$ *and* $\mathfrak{F}_\epsilon'$.

PROOF. We divide the proof into three paragraphs; ϵ is fixed in what follows.

1. By (1.4), (1.8), the last sentence of the first paragraph of this Section, and Lemma 1, the function $r_n(F, \phi_n^*)$ approaches a bounded limit as $n \to \infty$, *uniformly* for F in $\mathfrak{F}_\epsilon'$. (This limit is positive, by the known results in the case $m = 1$.) Hence, for any $\delta > 0$, there is a d.f. F_δ in $\mathfrak{F}_\epsilon$ and an integer N_δ such that

$$\sup_{F \varepsilon \mathfrak{F}_\epsilon'} r_n(F, \phi_n^*) < (1 + \delta) r_n(F_\delta, \phi_n^*) \tag{4.2}$$

for $n > N_\delta$. Define

$$r_{nk}(F, \phi_n) = E_{F,\phi_n} W(n^{1/2} \sup_{z \varepsilon A_k^m} |F(z) - g(z)|), \tag{4.3}$$

so that

$$r_{nk}(F, \phi_n^*) = \int_0^\infty W(r) d_r G_{n,k}(r; F). \tag{4.4}$$

Since $r_{nk} \leqq r_n$, it follows from (4.2) and the arbitrariness of δ that Lemma 11 will be proved if we show that

$$\liminf_{k\to\infty} \liminf_{n\to\infty} \inf_{\phi_n \varepsilon \mathfrak{D}_n} \sup_{F \varepsilon \mathfrak{F}_\epsilon'} r_{nk}(F, \phi_n) \geqq \lim_{n\to\infty} r_n(F_\delta, \phi_n^*). \tag{4.5}$$

2. Define

$$r_{\infty k}^* = \int_0^\infty W(r) d_r G_{\infty,k}(r; F_\delta) \tag{4.6}$$

and

$$r^* = \int_0^\infty W(r) d_r G(r; F_\delta). \tag{4.7}$$

Let $\mathfrak{F}_{\epsilon,k}'$ be the subset of $\mathfrak{F}_\epsilon'$ consisting of every absolutely continuous d.f. in $\mathfrak{F}_\epsilon'$ which has a density function which is a constant on each of the k^m open m-cubes of side $1/k$ in I^m whose corners are points of A_k^m. From equations (1.4) through (1.8) and the fact that $F_\delta \varepsilon \mathfrak{F}^*$, we have

$$\lim_{n\to\infty} r_{nk}(F_\delta, \phi_n^*) = r_{\infty k}^* \tag{4.8}$$

and

$$\lim_{n\to\infty} r_n(F_\delta, \phi_n^*) = r^* = \lim_{k\to\infty} r_{\infty k}^*. \tag{4.9}$$

Let $F_{\delta k}$ be that member of $\mathfrak{F}_\epsilon'$ for which $F_{\delta k}(z) = F_\delta(z)$ whenever $z \varepsilon A_k^m$. Clearly, for each k and n,

$$r_{nk}(F_{\delta k}, \phi_n^*) = r_{nk}(F_\delta, \phi_n^*). \tag{4.10}$$

From equations (4.6) through (4.10) and the fact that $\mathfrak{F}_{\epsilon k}' \subset \mathfrak{F}_\epsilon'$, we see that

(4.5) will be proved if we show that, for each fixed $k > 1$,

$$\liminf_{n\to\infty} \inf_{\phi_n \in \mathfrak{D}_n} \sup_{F \in \mathfrak{F}'_{\epsilon k}} r_{nk}(F, \phi_n) \geqq \lim_{n\to\infty} r_{nk}(F_{\delta k}, \phi_n^*). \tag{4.11}$$

Since a sufficient statistic for $\mathfrak{F}'_{\epsilon k}$ based on $Z^{(n)}$ is the collection $T^{(n,k)}$ of k^m real random variables which are equal to the number of components of $Z^{(n)}$ taking on values in each of the k^m cubes just described, we may replace $\mathfrak{D}_n$ in (4.11) by the class $\mathfrak{D}_{n,k}$ of decision functions depending only on $T^{(n,k)}$. But the definition of r_{nk} then shows that the left side of (4.11) may be viewed as the limiting minimax risk associated with the problem of estimating certain linear combinations of multinomial probabilities. If we put $h + 1 = k^m$ in Section 3 and think of the p_i as being assigned to the k^m cubes and think of the $L = (k + 1)^m$ quantities ρ_i as being the values of the unknown d.f. at the $(k + 1)^m$ points in A_k^m, then the left side of (4.11) without the limit in n may be identified with the minimax risk for a multinomial problem with the setup of Lemma 10. (We shall discuss B'_h and the C of (3.5) in the next paragraph.)

3. Fix $k > 1$. For any F in $\mathfrak{F}'_{\epsilon,k}$, let π_F be the associated multinomial probability vector whose components are the p_i described in the previous paragraph. Let B'_h be the set of all such π_F in $\mathfrak{F}'_{\epsilon,k}$. From the definitions of $\mathfrak{F}_\epsilon$ and $\mathfrak{F}'_\epsilon$ it is clear that B'_h is a closed convex h-dimensional subset of the h-dimensional set B_h, and thus satisfies the requirements of Lemma 10. For the ρ_i defined in the previous paragraph, we can clearly take the C and C_0 of (3.5) to be the c'_m and c_m^* of (1.8). Hence, from Lemma 10, for each k, we have for the multinomial problem of Section 3 where B'_h and the ρ_i are as described above and the function W is that given in the statement of Lemma 11,

$$\lim_{n\to\infty} \inf_{\psi_n \in \mathcal{E}_n} \sup_{\pi \in B'_h} r'(\pi, \psi_n) = \lim_{n\to\infty} \sup_{\pi \in B'_h} r'(\pi, \psi_n^*), \tag{4.12}$$

where we have written r' for the risk function in the multinomial problem. Since $r'(\pi_F, \psi_n^*) = r_{nk}(F, \phi_n^*)$ and since the left sides of (4.11) and (4.12) are equal because of the correspondence of $\mathcal{E}_n$ to $\mathfrak{D}_{n,k}$, of r' to r, and of B'_h to $\mathfrak{F}'_{\epsilon,k}$, we see that (4.11) follows from (4.12). Thus, Lemma 11 is proved.

5. Completion of the proof of Theorem 1; passage to the limit with ϵ. We now complete the proof of Theorem 1 by showing that $\mathfrak{F}_\epsilon$ is suitably dense in $\mathfrak{F}^*$ (and hence that $\mathfrak{F}'_\epsilon$ is suitably dense in $\mathfrak{F}^c$) as $\epsilon \to 0$. We require two lemmas to do this.

As in Section 2, the proof of the next two lemmas is very similar for all $m > 1$, but is most briefly written out when $m = 2$. For simplicity of presentation, we shall therefore again write out the details only in the case $m = 2$, and shall state explicitly all modifications for the case $m > 2$ which are not completely obvious.

Let $\mathfrak{F}^I$ denote the class of all d.f.'s on I^2 (in the general case, on I^m). For F in $\mathfrak{F}^I$ and $0 < \epsilon < 1$, define

$$\begin{aligned} \bar{F}(x, y) &= (1 - \epsilon)F(x, y / (1 - \epsilon)), \quad y \leqq 1 - \epsilon; \\ \bar{F}(x, y) &= (1 - \epsilon)F(x, 1) + x(y - 1 + \epsilon), \quad y > 1 - \epsilon. \end{aligned} \tag{5.1}$$

We shall not display the dependence on ϵ of the bar operation defined by (5.1). If $F \varepsilon \mathfrak{F}^*$ and we perform the bar operation of (5.1) on F to obtain $\bar{F}$ and then, interchanging the roles of x and y, perform the bar operation on $\bar{F}$ to obtain F^* (say), we clearly have $F^* \varepsilon \mathfrak{F}_\epsilon$. (In the case of chance vectors with m components, F^* is obtained after m such steps.) Let $\bar{Z}_1, \cdots, \bar{Z}_n$ be independent chance vectors with the common d.f. $\bar{F}$, let $\bar{S}_n$ be their sample (empiric) d.f., and define

$$\bar{D}_n = \sup_{z \varepsilon I} |\bar{S}_n(z) - \bar{F}(z)|.$$

Also, define $m = m(n, \epsilon)$ to be the greatest integer $\leqq n(1 - \epsilon)$.

We now prove the following lemma:

LEMMA 12. *We have*

$$(5.2) \qquad P_{\bar{F}}\{\bar{D}_n < r/n^{1/2}\} \leqq P_F\{D_m < [r(1 + \epsilon) + 7\epsilon^{1/4}] / m^{1/2}\} + o(1 \mid \epsilon, N(\epsilon) \mid r, F \varepsilon \mathfrak{F}^I).$$

PROOF. Let C^* be the event

$$\{|\bar{S}_n(1, 1 - \epsilon) - (1 - \epsilon)| < \epsilon^{1/4} n^{-1/2}\}.$$

From Chebyshev's inequality we obtain

$$(5.3) \qquad P_F\{C^*\} = 1 + o(1 \mid \epsilon \mid n, F \varepsilon \mathfrak{F}^I).$$

For small ϵ we have

$$(5.4) \qquad \left|1 - \frac{n(1 - \epsilon)}{n(1 - \epsilon) + \theta n^{1/2}\epsilon^{1/4}}\right| < 4\epsilon^{1/4}/n^{1/2}.$$

Hence, when C^* occurs and ϵ is small,

$$(5.5) \qquad \left|1 - \frac{n(1 - \epsilon)}{n\bar{S}_n(1, 1 - \epsilon)}\right| < 4\epsilon^{1/4}/n^{1/2}.$$

Since

$$(5.6) \qquad \left|\frac{(1 - \epsilon)\bar{S}_n(z)}{\bar{S}_n(1, 1 - \epsilon)} - \bar{S}_n(z)\right| \leqq \left|\frac{(1 - \epsilon)}{\bar{S}_n(1, 1 - \epsilon)} - 1\right|,$$

we have

$$(5.7) \qquad \bar{D}_n \geqq \sup_{\substack{0 \leqq x \leqq 1 \\ 0 \leqq y \leqq 1 - \epsilon}} \left|\frac{(1 - \epsilon)\bar{S}_n(z)}{\bar{S}_n(1, 1 - \epsilon)} - \bar{F}(z)\right| - \left|1 - \frac{(1 - \epsilon)}{\bar{S}_n(1, 1 - \epsilon)}\right|.$$

Also we have, for $y \leqq 1 - \epsilon$,

$$(5.8) \qquad E_{\bar{F}}\left\{\frac{n\bar{S}_n(z)}{m'} \middle| n\bar{S}_n(1, 1 - \epsilon) = m'\right\} = \bar{F}(z) / (1 - \epsilon) = F(x, y / (1 - \epsilon))$$

Hence the conditional d.f. of the first term on the right side of (5.7), given that $n\bar{S}_n(1, 1 - \epsilon) = m'$, is the same as the d.f. of $(1 - \epsilon)D_{m'}$. In what follows define $M' = n\bar{S}_n(1, 1 - \epsilon)$.

If m_1 and m_2 are two positive integers with $m_1 < m_2$, we can think of S_{m_2} as being obtained by adjoining $(m_2 - m_1)$ random vectors Z_i to the set of m_1 random vectors Z_i which gave rise to a corresponding realization of S_{m_1}. Hence, θ' denoting a value with $0 \leqq \theta' \leqq 1$, the corresponding values of $S_{m_1}(z)$ and $S_{m_2}(z)$ differ for all z by no more than

$$|S_{m_2}(z) - S_{m_1}(z)| = \left|\frac{m_1 S_{m_1}(z) + \theta'(m_2 - m_1)}{m_2} - S_{m_1}(z)\right| \leqq \frac{(m_2 - m_1)}{m_2}.$$

Thus, in C^*, where $|M' - m| < n^{1/2}\epsilon^{1/4} + 1$, we have that for each possible value of D_m there is a corresponding set of values of $D_{M'}$ of the same probability (these sets corresponding to different values of D_m arising from disjoint sets in the space of sequences $\{Z_i\}$) with $D_m \leqq D_{M'} + 2\epsilon^{1/4} m^{-1/2}$, provided $m >$ some $M(\epsilon)$.

From (5.3), (5.5), (5.7), (5.8), and the discussion of the previous paragraph, we have

$$\begin{aligned} &P_{\bar{F}}\{\bar{D}_n < r/n^{1/2}\} \\ (5.9)\qquad &\leqq \sup_{|m'-m|<n^{1/2}\epsilon^{1/4}+1} P_F\{(1-\epsilon)D_{m'} < (r + 4\epsilon^{1/4})/n^{1/2}\} + P_{\bar{F}}\{C^*\} \\ &\leqq P_F\left\{D_m < \frac{r(1+\epsilon) + 7\epsilon^{1/4}}{m^{1/2}}\right\} + o(1 \mid \epsilon, N(\epsilon) \mid r, F \,\varepsilon\, \mathfrak{F}^I), \end{aligned}$$

which proves Lemma 12.

We now prove

LEMMA 13. *For W satisfying the assumptions of Theorem* 1, *we have*

$$(5.10)\qquad \sup_{F\varepsilon\mathfrak{F}^*} r_n(F, \phi_n^*) = \sup_{F\varepsilon\mathfrak{F}_\epsilon} r_n(F, \phi_n^*) + o(1 \mid \epsilon, N'(\epsilon)).$$

PROOF. Define $m' = m'(n, \epsilon)$ to be the greatest integer $\leqq (1 - \epsilon)^2 n$. Using Lemma 12 a second time (with a trivial modification since $m'(n, \epsilon)$ may differ by unity from $m[m(n, \epsilon), \epsilon]$) to go from $\bar{F}$ to F^*, we have at once, for any W satisfying (1.4) and the other assumptions of Theorem 1,

$$(5.11)\qquad r_n(F^*, \phi_n^*) \geqq r_{m'}(F, \phi_{m'}^*) + o(1 \mid \epsilon, N(\epsilon) \mid F \,\varepsilon\, \mathfrak{F}^I).$$

From (5.11) and the fact that $F^* \,\varepsilon\, \mathfrak{F}_\epsilon$ if $F \,\varepsilon\, \mathfrak{F}^*$, we have

$$(5.12)\qquad \sup_{F\varepsilon\mathfrak{F}_\epsilon} r_n(F, \phi_n^*) \geqq \sup_{F\varepsilon\mathfrak{F}^*} r_{m'}(F, \phi_{m'}^*) + o(1 \mid \epsilon, N(\epsilon)).$$

Now, as in the first part of the proof of Lemma 11, we have that $r_n(F, \phi_n^*)$ approaches a bounded limit as $n \to \infty$, uniformly for F in $\mathfrak{F}_\epsilon$. Hence,

$$(5.13)\qquad \sup_{F\varepsilon\mathfrak{F}_\epsilon} r_n(F, \phi_n^*) = \sup_{F\varepsilon\mathfrak{F}_\epsilon} r_{m'}(F, \phi_{m'}^*) + o_\epsilon(1 \mid n),$$

where $o_\epsilon(1 \mid n)$ denotes a term which, for each ϵ, goes to 0 as $n \to \infty$ (not necessarily uniformly in ϵ). From (5.12), (5.13), and the fact that $\mathfrak{F}_\epsilon \subset \mathfrak{F}^*$, we obtain

$$(5.14)\qquad \sup_{F\varepsilon\mathfrak{F}^*} r_{m'}(F, \phi_{m'}^*) = \sup_{F\varepsilon\mathfrak{F}_\epsilon} r_{m'}(F, \phi_{m'}^*) + o(1 \mid \epsilon, N''(\epsilon)).$$

Since the possible values of m' for $n > N''(\epsilon)$ include all integers $>N''(\epsilon)(1-\epsilon)^2 - 1 = N'(\epsilon)$ (say), Lemma 13 follows from (5.14).

LEMMA 14. *The statement of Theorem 1 holds with $\mathfrak{F}$ replaced by $\mathfrak{F}^c$.*

PROOF. We have previously alluded to the fact that, if Z_1 has a d.f. F in $\mathfrak{F}^c$, then by appropriate monotonic transformations on the individual coordinates of Z_1 we can obtain a random vector Z_1' (say) such that Z_1' has d.f. F' (say) in $\mathfrak{F}^*$ and $G_n(r; F') = G_n(r; F)$ for all r and n. Hence,

$$\sup_{F \epsilon \mathfrak{F}^c} r_n(F, \phi_n^*) = \sup_{F \epsilon \mathfrak{F}^*} r_n(F, \phi_n^*). \tag{5.15}$$

Moreover, in the same way we have

$$\sup_{F \epsilon \mathfrak{F}_\epsilon} r_n(F, \phi_n^*) = \sup_{F \epsilon \mathfrak{F}_\epsilon'} r_n(F, \phi_n^*). \tag{5.16}$$

Lemma 14 now follows at once from Lemma 11, (5.16), Lemma 13, (5.15), and the fact that $\mathfrak{F}_\epsilon' \subset \mathfrak{F}^c$.

PROOF OF THEOREM 1. Theorem 1 now follows immediately from Lemma 14 and (1.9).

We remark that the proof of Theorem 1 is clearly valid when $\mathfrak{F}$ is replaced by a suitably large subset.

It is not really necessary to prove Theorem 1 by using (1.9) and proving the result first for $\mathfrak{F}^c$ (in Lemma 14). For Lemma 13 clearly holds if in (5.10) we replace $\mathfrak{F}^*$ by $\mathfrak{F}^I$ and $\mathfrak{F}_\epsilon$ by the class of d.f.'s obtained by substituting $\mathfrak{F}^I$ for $\mathfrak{F}^*$ in the definition of $\mathfrak{F}_\epsilon$; one can carry through the arguments of Sections 2 and 4 with this altered definition of $\mathfrak{F}_\epsilon$ (appropriate results from [1] still hold), and obvious analogues of (5.15) and (5.16) then yield Theorem 1.

In Section 7 we shall discuss various modifications of Theorem 1 obtained by altering the way in which W depends on $F(z) - g(z)$.

6. Integral weight functions. Since for $m > 1$ the procedure ϕ_n^* does not have constant risk for F in $\mathfrak{F}^c$ and any common weight functions of the form given in equation (5.1) of [1], there is no longer any special reason for considering weight functions for which the dependence on F of the integrand is of the form considered there. Therefore, to make the proof of this section as simple as possible, we shall consider here the analogue of the special case of Section 5 of [1] wherein $W(y, z)$ does not depend on z, relegating the consideration of more complicated weight functions to Section 7. Our result is

THEOREM 2. *Let $W(r)$ be a monotonically nondecreasing nonnegative real function of r for $r \geqq 0$ which is not identically zero and which satisfies*

$$\int_0^\infty W(r) r e^{-2r^2}\, dr < \infty. \tag{6.1}$$

Then $\{\phi_n^*\}$ is asymptotically minimax relative to $\mathfrak{F}^c$ and the weight functions

$$W_n(F, g) = \int W(n^{1/2}|F(x) - g(x)|)\, dF(x). \tag{6.2}$$

PROOF. As in Section 5 of [1], the proof of this theorem is essentially easier than that of Theorem 1, since it is centered about the one-dimensional asymptotic result (6.12) (for each z). The analytic details are often like corresponding ones of Section 5 of [1], to which we shall consequently sometimes refer. The proof will be conducted in four numbered paragraphs.

1. From (6.1) and the uniformity of approach to its continuous limit of the d.f. of $n^{1/2}[S_n(z) - F(z)]$ for all z for which $\frac{1}{2} - |F(z) - \frac{1}{2}| > \delta > 0$ and all F in $\mathfrak{F}^c$ (the F-measure of this set of z approaches 1 as $\delta \to 0$, uniformly in F), we conclude at once from (6.1) that $r_n(F, \phi_n^*)$ has a bounded limit uniformly for F in $\mathfrak{F}^c$, and thus that (4.2) is satisfied with $\mathfrak{F}'_\epsilon$ replaced by $\mathfrak{F}^c$, for some F_δ in $\mathfrak{F}^c$ (of course, r_n is now to be computed using (6.2)). We can clearly suppose, and hereafter do, that F_δ is a d.f. on I^m. Let $\mathfrak{F}'_{0k}$ denote the class of d.f.'s defined in paragraph 2 of the proof of Lemma 11, with $\epsilon = 0$; thus, the B'_h of paragraph 3 of that proof now coincides with the B_h there.

2. As in Section 5 of [1], we shall let $\{\xi_{kn}\}$ be a sequence of *a priori* probability measures on B_h (we shall think of $\mathfrak{F}'_{0k}$ and B_h interchangeably), and we shall write $P_z^*\{A\}$ for the probability of an event expressed in terms of $T^{(n,k)} = T^{(n)}$ when the latter has probability function

$$
\begin{aligned}
&P\{T_i^{(n)} = t_i^{(n)}, 1 \leqq i \leqq h+1\} \\
(6.3)\qquad &= \frac{1}{d(k,n,z)} \int_{B_h} f(z,\pi) P_\pi\{T_i^{(n)} = t_i^{(n)}, 1 \leqq i \leqq h+1\}\, d\xi_{kn}(\pi);
\end{aligned}
$$

here P_π is defined in (3.1) and $f(z, \pi)$ is the Lebesgue density at z (in I^m) of the d.f. $F(\cdot, \pi)$ in $\mathfrak{F}'_{0k}$ corresponding to a given π in B_h; $d(k, n, z)$ is chosen to make (6.3) a probability function. We take $f(z, \pi)$ to be constant on the interior of each of the k^m cubes in I^m; this determines (6.3) for all z with all irrational components (hereafter called irrational z), to which such z we may limit all further discussion. For each such z and possible value $t^{(n)}$ of $T^{(n,k)}$, we define

$$
(6.4)\qquad r_{kn}(z, \phi, t^{(n)}) = \int_{B_h} E_\phi W(n^{1/2}|g(z) - F(z,\pi)|)\, d_\pi \xi_{kn}^*(\pi, z, t^{(n)}),
$$

where, for Borel subsets B of B_h,

$$
(6.5)\qquad \xi_{kn}^*(B, z, t^{(n)}) = \frac{\int_B f(z,\pi) P_\pi\{t^{(n)}\}\, d\xi_{kn}(\pi)}{\int_{B_h} f(z,\pi) P_\pi\{t^{(n)}\}\, d\xi_{kn}(\pi)};
$$

we have used $P_\pi\{t^{(n)}\}$ to denote the function of (3.1).

For each n and k, if F is restricted to be in $\mathfrak{F}'_{0k}$, we may, as in Section 4, restrict our consideration to procedures ϕ in $\mathfrak{D}_{n,k}$. Denoting expectation with respect to P_z^* by E_z^*, we have as in (5.10) of [1],

$$
(6.6)\qquad \int r_n(F, \phi)\, d\xi_{kn} = \int_{I^m} E_z^* r_{kn}(z, \phi, T^{(n,k)})\, d(k,n,z)\, dz,
$$

where dz denotes the differential element of Lebesgue measure on I^m. For fixed n, k, z, and $t^{(n)}$, let $r_{kn}^*(z, t^{(n)})$ denote the infimum of (6.4) over $\mathfrak{D}_{n,k}$. In order to prove Theorem 2, according to (6.6) and the discussion of paragraph 1 of this proof, it clearly suffices to show that, for some $\{\xi_{kn}\}$,

$$\lim_{k\to\infty} \lim_{n\to\infty} \int_{I^m} E_z^* r_{kn}(z, \phi, T^{(n,k)})\, d(k, n, z)\, dz \geqq \lim_{n\to\infty} r_n(F_\delta, \phi_n^*). \tag{6.7}$$

3. Fix k. Let $\pi_{\delta k}$ be such that $F(z, \pi_{\delta k}) = F_\delta(z)$ for z in A_k^m. We may assume $\pi_{\delta k}$ is an interior point of B_h; for, if $\pi_{\delta k}$ were not an interior point, letting $F_\delta' = (1 - \delta')F_\delta + \delta' U$ where U is the uniform d.f. on I^m, we see easily that the right side of (6.7) can be decreased by at most a quantity which approaches 0 as $\delta' \to 0$ if F_δ is replaced by F_δ' there; we could thus replace $\pi_{\delta k}$ by the interior point π corresponding to F_δ' (for δ' small but positive) in what follows. Let ξ_{kn}, $n = 1, 2, \cdots$, be a sequence of *a priori* measures on B_h which "shrink down" on $\pi_{\delta k}$ as the ξ_n of Lemma 8 shrink down on π^*; e.g., ξ_{kn} is uniform on a sphere of radius $n^{-1/4}$ about $\pi_{\delta k}$. It follows at once that

$$\lim_{n\to\infty} d(k, n, z) = f(z, \pi_{\delta k}) \tag{6.8}$$

at all irrational z. Suppose we show that, for any irrational z and any $\epsilon > 0$, there is an $N = N(\epsilon, z, k)$ such that, for $n > N$, P_z^* assigns probability at least $1 - \epsilon$ to a set of $T^{(n,k)}$ values for which

$$r_{kn}^*(z, T^{(n,k)}) + \epsilon > \int_{-\infty}^{\infty} W(y) q(y, \sigma(z, k))\, dy, \tag{6.9}$$

where $q(y, \sigma) = (2\pi\sigma^2)^{-1/2} \exp(-y^2/2\sigma^2)$ and where $\sigma(z, k)$ is continuous in z and

$$\sigma(z, k) = F(z, \pi_{\delta k})[1 - F(z, \pi_{\delta k})] + o(1 \mid k \mid z) \leqq \tfrac{1}{4}. \tag{6.10}$$

Then, writing $V(z, k)$ for the expression on the right side of (6.9), we will clearly have (from (6.10), (6.1), and the continuity of q)

$$\begin{aligned} \lim_{n\to\infty} r_n(F_\delta, \phi_n^*) &= \int_{I^m} \lim_{k\to\infty} V(z, k)\, dF_\delta(z) \\ &= \lim_{k\to\infty} \int_{I^m} V(z, k)\, dF_\delta(z) \\ &= \lim_{k\to\infty} \int_{I^m} V(z, k) f(z, \pi_{\delta k})\, dz. \end{aligned} \tag{6.11}$$

Thus, an application of Fatou's lemma to the left side of (6.7) shows that (6.11) and (6.8) will imply (6.7). Thus, it remains to prove (6.9) for the appropriate values of the arguments there.

4. The proof of (6.9) is similar to that of Lemma 8. For fixed z, the expression of (6.5) is like the *a posteriori* probability measure of π when ξ_{kn} is the *a priori*

measure, except for the factor $f(z, \pi)$. In fact, by the shrinking property of ξ_{kn} as $n \to \infty$ and the nature of $f(z, \pi)$, one obtains in the manner of [7] (see [1] for details) that, for any $\epsilon' > 0$ and for n suitably large, with probability $>1 - \epsilon'$ under P_z^*, the joint density according to ξ_{kn}^* of the quantities $\bar{\gamma}_i = n^{1/2}(p_i - t_i^{(n)}/n)$, $1 \leq i \leq h$ (where we have written $\pi = (p_1, \cdots, p_{h+1})$), in a spherical region of probability $>1 - \epsilon'$ under ξ_{kn}^*, is at least $(1 - \epsilon')$ times the appropriate normal density for which the $\bar{\gamma}_i$ have means 0, var $\bar{\gamma}_i = p_{\delta i}(1 - p_{\delta i})$, $\mathrm{cov}(\bar{\gamma}_i, \bar{\gamma}_j) = -p_{\delta i}p_{\delta j}$ (the $p_{\delta i}$ being the components of $\pi_{\delta k}$). For $\epsilon'' > 0$, an elementary computation (the details being like those of [1], p. 661, except that now $m > 1$) then shows that, for a fixed arbitrary irrational z, the corresponding distribution of $n^{1/2}[F(z, \pi) - J_{n,k}(z)]$, where $J_{n,k}(z)$ is the obvious best linear estimator in $\mathfrak{D}_{n,k}$ of $F(z, \pi)$ for π in $\mathfrak{F}_{0k}'$ (not in general $S_n(z)$, unless $z \in A_k^m$), has, with probability $>1 - \epsilon''$ under P_z^*, an absolutely continuous component the magnitude of whose Lebesgue density is at least

$$(1 - \epsilon'')\, q(y, \sigma(z, k)) \tag{6.12}$$

on the interval $-1/\epsilon'' < y < 1/\epsilon''$, where $\sigma(z, k)$ is continuous in z and satisfies (6.10). Since ϵ'' is arbitrary, (6.9) follows easily from (6.12) and the trivial one-dimensional case of [8] (see [1] for details; the argument here is easier, since we have not yet included the additional dependence of W on other quantities as in [1] and Section 7 below). Thus, Theorem 2 is proved.

It is clear that Theorem 2 remains valid if $\mathfrak{F}^c$ is replaced by a suitably large subset. Further generalizations will be discussed in the next section.

7. Other loss functions. We list a few of the extensions of Theorems 1 and 2 which may be proved by the same methods with only minor modifications and no essential new difficulties in the proof. In fact, our treatment of the case $m > 1$ (compared with the argument of [1]) has been concentrated on the difficulty engendered by the nonconstancy of $r_n(F, \phi_n^*)$, and that nonconstancy (in the counterpart of modification F, below) is the only real new difficulty in any of the corresponding generalizations of Section 6 of [1] (the difficulty is more trivial there, where $m = 1$ and the nonconstancy is easier to deal with than in Theorems 1 and 2 above).

A. In Theorem 2, the form of W may be extended. For $m = 1$, the more general form $W(n^{1/2}|F(z) - g(z)|, F(z))$ was considered in Section 5 of [1]. The same form can be considered here, but perhaps the dependence on the second variable is no longer so natural; it may be replaced or supplemented, for example, by a dependence on the value of the marginal d.f.'s at the point z. The regularity condition which must be imposed on W in order for our method of proof to hold is, in any event, exactly the obvious analogue of that of Section 5 of [1]. For example, continuity and an appropriate integrability condition (the analogue of (5.5) of [1]) is more than enough.

B. In Theorem 2, W can be replaced by a measure (rather than a density) in the second argument of the W of A above (or its replacements, just above).

For example, when $m = 2$, one might be interested only in the estimation of the deciles of the marginal d.f.'s F_1 and F_2 (say) and, at each decile r of F_1, the deciles of the d.f. $F(r, y) / F_1(r)$ (and its counterpart with x and y interchanged).

C. An analogue of Theorem 2 (with any of the modifications noted above) for $\mathfrak{F}$ rather than $\mathfrak{F}^c$ is perhaps not too natural (see [1] for further comments), but can be given under suitable assumptions. An analogue of Theorem 1 or Theorem 2 for the class of purely discrete d.f.'s (e.g., on R^m, or on the integral lattice points of R^m) can also be given; for example, the former essentially follows from the fact that there is a discrete d.f. at which ϕ_n^* has almost the same risk as at F_δ when n is large (see (1.4) through (1.8)).

D. In Theorem 1, one can replace D_n by $\sup_z[|g(z) - F(z)|h(F(z))]$, where h is a suitably regular nonnegative function whose dependence on $F(z)$ may be replaced, e.g., by a dependence on the marginal d.f.'s, as in A above; a linear combination of such functions can also be employed. If h takes on only the values 0 and 1, this modification amounts to taking the supremum of the deviation over a suitable subset of R^m whose description depends on F.

E. In Theorem 1, one could consider the *measures* P, Q_n, and g^* corresponding to F, S_n, and g, and could let W depend on $\sup_A|P(A) - g^*(A)|$ where the supremum is taken over a suitable family of sets, e.g., rectangles with sides parallel to the coordinate axes. This presents no new difficulties.

F. The function h of D above, the second argument of W in A above, and the integrating measure of (6.2), can all be changed so as to depend only on z and not on $F(z)$ (or they can depend on both). This requires no new arguments, only obvious regularity conditions as on p. 664 of [1]. It is again the existence of an F_δ which is the crucial point.

G. The remarks on the *sequential* asymptotic minimax character of ϕ_n^* for suitable weight functions, which are contained on pp. 664–665 of [1], hold here without change.

H. Obvious combinations of the types of dependence of W_n on F and z which occur in Theorems 1 and 2 and in the previous remarks can be considered with no essential new difficulty. In fact, the asymptotic minimax character of ϕ_n^* seems to hold for a very general class of weight functions. The discussion of p. 664 of [1] indicates the possible breadth of that class, but we are even further than we were in the case $m = 1$ of [1] from being able to give a single simple, unified proof.

REFERENCES

[1] A. Dvoretzky, J. Kiefer, and J. Wolfowitz, "Asymptotic minimax character of the sample distribution function and of the classical multinomial estimator," *Ann. Math. Stat.*, Vol. 27 (1956), pp. 642–669.

[2] J. Kiefer, and J. Wolfowitz, "On the deviations of the empiric distribution function of vector chance variables," *Trans. Amer. Math. Soc.*, Vol. 87, Jan. 1958, pp. 173–186.

[3] M. D. Donsker, "Justification and extension of Doob's heuristic approach to the Kolomogorov-Smirnov theorems," *Ann. Math. Stat.*, Vol. 23 (1952), pp. 277–281.

[4] J. L. Doob, "Heuristic approach to the Kolmogorov-Smirnov theorems," *Ann. Math. Stat.*, Vol. 20 (1949), pp. 393–402.

[5] C. G. ESSEEN, "Fourier analysis of distribution functions," *Acta. Math.*, Vol. 77 (1945), pp. 1–125.

[6] H. BERGSTROM, "On the central limit theorem in the space R^k, $k > 1$," *Skand. Aktuarietids*, (1949), pp. 106–127.

[7] R. v. MISES, "Fundamentalsätze der Wahrscheinlichkeitsrechnung," *Math. Zeit.*, Vol. 4 (1919), pp. 1–97.

[8] T. W. ANDERSON, "The integral of a symmetric unimodal function," *Proc. Amer. Math. Soc.*, Vol. 6 (1955), pp. 170–176.

THE EQUIVALENCE OF TWO EXTREMUM PROBLEMS

J. KIEFER AND J. WOLFOWITZ

1. Introduction. Let $f_1, \ldots, f_k$ be linearly independent real functions on a space X, such that the range R of $(f_1, \ldots, f_k)$ is a compact set in k-dimensional Euclidean space. (This will happen, for example, if the f_i are continuous and X is a compact topological space.) Let S be any Borel field of subsets of X which includes X and all sets which consist of a finite number of points, and let $C = \{\xi\}$ be any class of probability measures on S which includes all probability measures with finite support (that is, which assign probability one to a set consisting of a finite number of points), and which are such that

$$m_{ij}(\xi) = \int_X f_i(x) f_j(x) \xi(dx) \qquad i, j = 1, \ldots, k$$

is defined. In all that follows we consider only probability measures ξ which are in C. Write $M(\xi)$ for the $k \times k$ matrix $\|m_{ij}(\xi)\|$. When $M(\xi)$ is non-singular, write $[M(\xi)]^{-1} = \|m^{ij}\|$. (We shall not always exhibit dependence on ξ.) Letting $f(x)$ denote the column vector with components $f_i(x)$, and letting primes denote transposes, we define

$$d(x; \xi) = f(x)'[M(\xi)]^{-1} f(x)$$

whenever $M(\xi)$ is non-singular.

We consider two extremum problems. The first is to choose ξ so that

$$\xi \text{ maximizes } \det M(\xi). \tag{1}$$

The second is to choose ξ so that

$$\xi \text{ minimizes } \max_x d(x; \xi). \tag{2}$$

We also note that the integral with respect to ξ of $d(x; \xi)$ is k; hence, $\max_x d(x; \xi) \geqslant k$, and thus a sufficient condition for ξ to satisfy (2) is

$$\max_x d(x; \xi) = k. \tag{3}$$

The result of this note is that (1), (2), and (3) are equivalent. This result, which seems to have interest *per se*, also strengthens and extends results of the authors **(1)** on the optimum design of regression experiments. A brief description of the connection with the design of such experiments is given below. The proof of the theorem is elementary and brief.

Received March 30, 1959. Research of J. Kiefer was sponsored by the Office of Naval Research. Research of J. Wolfowitz was supported by the United States Air Force under Contract no. AF 18(600)–685 monitored by the Office of Scientific Research.

2. The theorem. For every ξ consider $M(\xi)$ as a point in Euclidean k^2-space, let T be the totality of such points for all ξ in C, and let $\bar{T}$ be the convex closure of T. It is clear that every extreme point of $\bar{T}$ can be achieved by a ξ which assigns probability one to a single point. Since C contains every ξ with finite support, it follows that $T = \bar{T}$. The class C need not, of course, be convex. However, since our argument will be concerned only with the $M(\xi)$, we may argue below as if C were convex. Thus, if ξ_1 and ξ_2 are in C and

$$\frac{\xi_1 + \xi_2}{2}$$

is not, we may still discuss

$$M\left(\frac{\xi_1 + \xi_2}{2}\right),$$

because there exists a ξ in C with finite support, say ξ_3, such that

$$M(\xi_3) = M\left(\frac{\xi_1 + \xi_2}{2}\right).$$

Moreover, if $H - 1$ is the dimension of the linear space spanned by the functions $f_i f_j$, $i \leqslant j$, any $M(\xi)$ is equal to an $M(\xi')$ where the support of ξ' consists of at most H points. This can often be impoved, as in the case where X is the unit interval and $f_i(x) = x^{i-1}$.

Call a subset D of C *linear* if the following condition holds: For every α, $0 \leqslant \alpha \leqslant 1$, and every pair ξ_1, ξ_2 in D, $\alpha\xi_1 + (1 - \alpha)\xi_2$ is in D whenever it is in C. Thus, if C is convex, D is also convex.

We shall prove the following:

THEOREM. *Conditions* (1), (2), *and* (3) *are equivalent. The set B of all ξ satisfying these conditions is linear, and $M(\xi)$ is the same for all ξ in B.*

This result has a function space corollary which may be of interest. Suppose ξ satisfies (3) and that Q is a real $k \times k$ matrix such that $QM(\xi)Q'$ is the identity. Then $g = Qf$ is a vector of orthonormal functions with respect to ξ, and $g(x)'g(x) = d(x; \xi)$. Thus we have

COROLLARY. *If $f_1, \ldots, f_k$ are linearly independent, continuous, real functions on a compact space X, then there is a probability measure ξ on X and a linear transformation $g_i = \sum_j a_{ij} f_j$ such that $g_1, \ldots, g_k$ are orthonormal with respect to ξ and*

$$\max_x \sum_{i=1}^{k} g_i^2(x) = k.$$

The set of all such ξ is the set B of the theorem.

Proof of the theorem. We shall say that ξ is a *local solution* of (1) if $\det M(\xi) > 0$ and if, for every ξ',

$$\frac{\partial}{\partial\alpha}\log\det M([1-\alpha]\xi+\alpha\xi')|_{\alpha=0+}\leqslant 0. \tag{4}$$

Now, if $\det M(\xi) > 0$, A is such that $AM(\xi)A'$ is the identity, and $AM(\xi')A'$ is diagonal with diagonal elements b_i, then $\det M([1-\alpha]\xi+\alpha\xi') = \det A^{-2} \Pi_i[1-\alpha+\alpha b_i]$, from which we easily compute that $-\log\det M([1-\alpha]\xi+\alpha\xi')$ is convex in $\alpha(0<\alpha<1)$ and is strictly convex unless all $b_i = 1$ (that is, unless $M(\xi) = M(\xi')$). Hence, if $\det M(\xi') > \det M(\xi)$, equation (4) cannot hold for that ξ'. We conlcude that local solutions of (1) are actual solutions of (1), and of course the converse is true. Moreover, if $\det M(\xi) = \det M(\xi') = h > 0$, we have $\det M(\xi/2+\xi'/2) > h$ unless $M(\xi) = M(\xi')$, so that ξ and ξ' cannot both satisfy (1) unless $M(\xi) = M(\xi')$. It follows from this and the linearity in ξ of $M(\xi)$ that, if ξ and ξ' both satisfy (1), then so does $\alpha\xi+(1-\alpha)\xi'$, whenever it is in C.

It now suffices to prove that $\det M(\xi) > 0$ and ξ satisfies (4) for all ξ', if and only if ξ satisfies (2), and only if it satisfies (3). First suppose ξ satisfies (4) and that $\det M(\xi) > 0$. Performing the differentiation in (4), and denoting by M_{ij} the cofactor of m_{ij}, we have

$$\begin{aligned}(5)\quad 0 &\geqslant [\det M(\xi)]^{-1}\sum_{i,j}\frac{\partial\det M}{\partial m_{ij}}\frac{\partial m_{ij}([1-\alpha]\xi+\alpha\xi')}{\partial\alpha}\bigg|_{\alpha=0}\\ &= [\det M(\xi)]^{-1}\sum_{i,j}\left(\frac{\partial}{\partial m_{ij}}\sum_q m_{iq}M_{iq}\right)[m_{ij}(\xi')-m_{ij}(\xi)]\\ &= [\det M(\xi)]^{-1}\sum_{i,j}M_{ij}(\xi)[m_{ij}(\xi')-m_{ij}(\xi)] = \sum_{i,j}m^{ij}(\xi)m_{ij}(\xi')-k.\end{aligned}$$

Letting ξ' give measure one to the point x, we obtain

$$[f(x)]'M(\xi)^{-1}f(x)\leqslant k \tag{6}$$

for all x. Thus, (3) is satisfied and, as we have remarked, this implies (2).

Finally, if (2) is satisfied, we must have (6) for all x, since we have just seen that there always exist ξ's satisfying (3). Hence, for any ξ' with finite support, we obtain $\sum_{i,j}m^{ij}(\xi)m_{ij}(\xi')\leqslant k$. Hence this inequality is valid for all ξ', and (5) is satisfied. This completes the proof of the theorem.

3. Extensions and applications. We remark that it is easy to see that, if R is bounded but not compact, and if $\{\xi_i\}$ is a sequence of measures on S, then $\lim_i\det M(\xi_i)$ is a maximum if and only if $\lim_i\sup_x d(x;\xi_i)$ is a minimum, and if and only if $\lim_i\sup_x d(x;\xi_i) = k$. Similarly, the first part of the corollary holds with the replacement $\sup_x\sum_i g_i^2(x) < k+\epsilon$, for any $\epsilon > 0$.

We now describe briefly the statistical applications of the results. An integer N is given, and the statistician must choose N points $x_1,\ldots,x_N$ (not necessarily distinct) corresponding to which he obtains observations on uncorrelated random variables Y_i $(1\leqslant i\leqslant N)$ with common variance σ^2 (perhaps unknown) and with expectation $\sum_{j=1}^k\theta_j f_j(x_i)$, where the θ_j are unknown real parameters. If $\xi(x)$ denotes the proportion of x_i's which are equal to x, we find that the covariance matrix of best linear estimators of $\theta_1, \ldots, \theta_k$ is $N^{-1}\sigma^2[M(\xi)]^{-1}$. The function ξ is called the experiment or the experimental design. A criterion often adopted for choosing a design is to minimize the determinant of the above covariance matrix (the "generalized variance"). Another possible criterion is to minimize the maximum over x of the variance $N^{-1}\sigma^2 d(x; \xi)$ of the "best linear estimator," given ξ, of the "regression function" $\sum_j \theta_j f_j(x)$. If we consider not merely the class C_N of probability measures ξ which take on only integral multiples of N^{-1} as values, but rather *all* probability measures ξ in C, then our result is that the two optimality criteria are equivalent. Moreover, for any ξ with support on H points which satisfies (1), (2), and (3), there is clearly a ξ' in C_N which achieves (1), (2), and (3) to within a multiplicative factor $1 + 0(N^{-1})$, and is easy to write down from ξ. Since the *exactly* optimum designs are often difficult to obtain, depend on N, and differ for the two criteria, we see the practical importance of our considerations.

It is very helpful to use the interplay of the two criteria (1) and (2) in obtaining a solution. For example, one can sometimes guess that a solution exists which is a member of a class of ξ which depend on several parameters. One may use (1) as the more convenient initial approach, maximize det $M(\xi)$ over the parametric class, and then verify whether the maximum just obtained is indeed a maximum over *all* ξ (which may be difficult in terms of (1)) by verifying (3). It is useful to note that, if ξ has a set consisting of k points as its support, then it gives equal measure to each of these points. (This is part of Theorem 5 of **(1)**.) Examples which make use of such methods will appear elsewhere, as will generalizations such as one concerned with the minimization of the determinant of a principal minor of $M(\xi)^{-1}$.

Reference

1. J. Kiefer and J. Wolfowitz, *Optimum designs in regression problems*, Ann. Math. Stat., *30* (1959).

Cornell University

Offprint from "Archive for Rational Mechanics and Analysis".
Volume 4, Number 4, 1960, P. 371—386

Springer-Verlag, Berlin · Göttingen · Heidelberg

Simultaneous Channels

J. Wolfowitz

Communicated by M. Kac

1. Simultaneous discrete channels without memory

We begin[1] with a description of a single discrete channel and assume that the transmitted and received alphabets consist of the two elements 0 and 1. (All these terms will be precisely defined shortly.) Extension of our results to the case where the alphabets contain any finite number of symbols is trivial, so that there is no essential loss of generality. The channel probability function (c.p.f.) $w(i|j)$, $i, j=0, 1$, is a non-negative function such that

$$w(0|j)+w(1|j)=1 \qquad j=0,1.$$

A word of length n or an n-sequence is a sequence of n zeros and ones. Let u be any such word which may be "sent" or "transmitted" over the channel. The "received" sequence $\bar{v}=(Y_1, \ldots, Y_n)$ is a sequence of chance variables which, when the sequence $u=(x_1, \ldots, x_n)$ is sent, have respective distributions given by[2]

$$(1.1) \qquad P\{Y_i=j|u, Y_1, \ldots, Y_{i-1}\}=w(j|x_i) \qquad i=1, \ldots, n; \quad j=0,1.$$

A code of length N and probability of error $\leqq \lambda$ is a set

$$(1.2) \qquad \{(u_1, A_1), \ldots, (u_N, A_N)\}$$

where each u_i is an n-sequence, each A_i is a set of n-sequences, the A_i are disjoint and[3]

$$(1.3) \qquad P\{\bar{v} \in A_i | u_i\} \geqq 1-\lambda \qquad i=1, \ldots, N.$$

The practical application of this is as follows: When one wishes to transmit the ith word of a dictionary which contains N words, one sends the sequence u_i. Whenever the receiver receives a sequence which is in A_j he always concludes that the jth word has been sent. When the receiver receives a sequence not in $A_1 \cup \cdots \cup A_N$, he may draw any conclusion at all about the word that has been sent. The probability that any word transmitted will be correctly received is $\geqq 1-\lambda$.

[1] This research was supported by the United States Air Force under Contract No. AF 18(600)−685 monitored by the Office of Scientific Research.

[2] This symbol stands for the conditional probability that $Y_i=j$, given that u has been sent and $Y_1, \ldots, Y_{i-1}$ received.

[3] This symbol stands for the conditional probability that $\bar{v}$ is in A_i, given that u_i is transmitted.

A simultaneous channel is a set S^*, finite or infinite, of c.p.f.'s $w(i|j|s)$, indexed by $s \in S$. When the probability P is computed according to $w(\cdot|\cdot|s)$ we shall write P_s. A code for a simultaneous channel is a code as above for which (1.3) is satisfied for *all* s in S when P is replaced by P_s. Such codes are the subject of the present paper.

The practical application of a simultaneous channel is the same as for a single channel. Neither sender nor receiver has to know what c.p.f. governs the noise (error) in any one word. The c.p.f. may change from word to word in an arbitrary manner. It should not change from letter to letter within a word. (See, however, a remark in Section 7.)

The capacity C_0 of the simultaneous channel is a number defined below in Section 3. In the first part of this paper we prove the following theorems:

Theorem 1 *(Coding theorem). Let λ, $0<\lambda\leqq 1$, be any number. There exists a constant $K>0$ such that, for every n, there exists a code for the simultaneous channel with probability of error $\leqq\lambda$ and length greater than*

$$2^{nC_0-K\sqrt{n}}. \tag{1.4}$$

(Our proof of Theorem 1 really proves more and may suggest a method of code construction. Suppose one chooses any pair (u_1, A_1) according to the specifications of the proof of Theorem 1 (relations (4.2) and (4.3) below), and then continues with (u_2, A_2), (u_3, A_3), *etc*., for as long as the resulting set is a code with proper probability of error. Suppose further that successive choices are made, subject to the specifications of the proof of Theorem 1, as ineptly or as maliciously as possible so as to hinder the prolongation of the code. Nevertheless, the length of the code after no prolongation is possible will be no less than (1.4).)

Theorem 2 *(Strong converse of the coding theorem). Let λ, $0\leqq\lambda<1$, be any number. There exists a constant $K'>0$ such that, for every n, there does not exist a code for the simultaneous channel with probability of error $\leqq\lambda$ and length*

$$2^{nC_0+K'\sqrt{n}}. \tag{1.5}$$

(The terms "weak converse" and "strong converse" do not refer to the difference between (1.5) and

$$2^{n(C_0+\varepsilon)}, \tag{1.6}$$

with arbitrary positive ε. The latter, for sufficiently large n, is the usual formulation of the strong converse, and, of course, the difference between (1.5) and (1.6) is an added improvement. The terms strong and weak refer to the difference between the theorem's being valid for any λ ($0\leqq\lambda<1$) (strong converse) and being valid only for a sufficiently small λ (weak converse). The basic difference between the two converses may be explained as follows: Essentially, the smaller λ the larger must the A_i be, and the larger λ the smaller may the A_i be. The weak converse says that, if λ is sufficiently small and hence the A_i are sufficiently large, $2^{n(C_0+\varepsilon)}$ of the latter cannot be packed into the space of all n-sequences. The strong converse says that, no matter how large λ is and hence how small the A_i may be, it is still not possible to pack $2^{n(C_0+\varepsilon)}$ (actually even $2^{nC_0+K'\sqrt{n}}$) of the latter into the space of all n-sequences.)

In Sections 6 and 7 these results will be extended to channels other than the one described in the present section. The full name of the latter channel is the simultaneous discrete memoryless channel; it is studied in [5], where the capacity C_0 is defined and a coding theorem and weak converse proved. Our methods are slight improvements of the arguments of [1] and [2]. No previous familiarity with information theory and no knowledge of probability theory is needed to read this paper except perhaps for Sections 6 and 7 (unless an acquaintance with Chebyshev's inequality can be considered as knowledge of probability theory). The problems involved are really of a purely combinatorial character. Theorems 3 and 4 of Section 6 are proved exactly as in [2]. The treatment of the semi-continuous channel will require a result of [3] (see Section 7).

In Section 8 we treat the problems where either the sender or the receiver (but not both) knows the c.p.f. which governs the channel for a particular word. We obtain the capacities of, and prove the analogues of Theorems 1 and 2 for, these channels.

2. Combinatorial preliminaries

When an n-sequence is used as a transmitted sequence we shall refer to it as a u-sequence, and when it is used as a received sequence we shall refer to it as a v-sequence. Suppose u and v are a u-sequence and a v-sequence, respectively. Define $N(i|u)$, $i=0, 1$, as the number of elements i in u. Define $N(i, j|u, v)$, $i, j=0, 1$, as the number of indices k, $k=1, \ldots, n$, such that the kth element of u is i, and the kth element of v is j.

Let δ and $\delta_2>2$ be fixed positive numbers to be determined later. Let $w(\cdot|\cdot|s)$ be any c.p.f. (not necessarily in S^*). We shall say that the v-sequence v is generated(s) by the u-sequence u if

$$|N(i,j|u,v)-N(i|u)\,w(j|i|s)|\leqq\delta_2[N(i|u)\,w(0|i|s)\,w(1|i|s)]^{\frac{1}{2}}\quad i,j=0,1. \tag{2.1}$$

A u-sequence u will be said to be a qu-sequence if

$$|N(1|u)-qn|\leqq\delta\sqrt{n}. \tag{2.2}$$

Let π always denote the pair (π_0, π_1), $0\leqq\pi_0, \pi_1\leqq 1$, $\pi_0+\pi_1=1$. Define the following[4]:

$$H(X|\pi)=-\sum_{i=0}^{1}\pi_i\log\pi_i, \tag{2.3}$$

$$H(Y|\pi|s)=-\sum_{i=0}^{1}\left\{\left[\sum_{j=0}^{1}\pi_j\,w(i|j|s)\right]\log\left[\sum_{j=0}^{1}\pi_j\,w(i|j|s)\right]\right\}, \tag{2.4}$$

$$H(Y|X|\pi|s)=-\sum_{i=0}^{1}\sum_{j=0}^{1}\pi_i\,w(j|i|s)\log w(j|i|s), \tag{2.5}$$

$$H(X|Y|\pi|s)=-\sum_{i=0}^{1}\sum_{j=0}^{1}\pi_j\,w(i|j|s)\log\frac{\pi_j\,w(i|j|s)}{\pi_0\,w(i|0|s)+\pi_1\,w(i|1|s)}. \tag{2.6}$$

One verifies easily that

$$H(X|\pi)+H(Y|X|\pi|s)=H(X|Y|\pi|s)+H(Y|\pi|s). \tag{2.7}$$

[4] All the logarithms in this paper are to the base 2; this is due to a convention of no importance. The symbol 0 log 0 is always to be understood as 0.

For purposes only of mathematical manipulation it will frequently be convenient below to assume that the x_i are independent, identically distributed chance variables, with

$$P\{x_i = 1\} = \pi_1 = 1 - P\{x_i = 0\} = 1 - \pi_0.$$

For brevity we shall describe this[5] as the "stochastic input π". When the stochastic input π is employed, then the chance sequence u will be written as $u(\pi)$ and the chance received sequence $\bar{v}$ as $\bar{v}(\pi)$.

In all that follows *in this section* the letter K with or without a subscript or an accent will always denote a suitable positive constant *which does not depend on π or on the c.p.f.* (The latter need not belong to S^*.) In Lemmas 2 and 5, it will be assumed, without further statement, that:

$$\min(\pi_0, \pi_1) \geqq d > 0, \tag{2.8}$$

$$|w(0|1|s) - w(0|0|s)| = |w(1|1|s) - w(1|0|s)| \geqq a > 0. \tag{2.9}$$

Lemma 1. *Let $\gamma < 1$ be any positive number, and δ be greater than a lower bound which is a function only of γ. Then*

$$P\{u(\pi) \text{ is a } \pi_1 u\text{-sequence}\} > \gamma. \tag{2.10}$$

Proof. This follows at once from Chebyshev's inequality.

Lemma 2. *The number $B_1(\pi_1)$ of $\pi_1 u$-sequences satisfies*

$$2^{nH(X|\pi) - K_1\sqrt{n}} < B_1(\pi_1) < 2^{nH(X|\pi) + K_1\sqrt{n}}. \tag{2.11}$$

Proof. If u is any $\pi_1 u$-sequence, then

$$\begin{aligned} 2^{-nH(X|\pi) - K_1'\sqrt{n}} &< \pi_0^{n\pi_0 + \delta\sqrt{n}}\, \pi_1^{n\pi_1 + \delta\sqrt{n}} < P\{u(\pi) = u\} < \\ &< \pi_0^{n\pi_0 - \delta\sqrt{n}}\, \pi_1^{n\pi_1 - \delta\sqrt{n}} < 2^{-nH(X|\pi) + K_1'\sqrt{n}}. \end{aligned} \tag{2.12}$$

From this the lemma follows at once.

Lemma 3. *Let ε' be any positive number, and δ_2 greater than a suitable lower bound which depends only on ε'. Then, for any u-sequence u whatever, we have*

$$P_s\{\bar{v} \text{ is generated}(s) \text{ by } u \mid u\} > 1 - \varepsilon'. \tag{2.13}$$

Proof. This follows at once from Chebyshev's inequality. Since $\delta_2 > 2$ always $\varepsilon' < \frac{1}{4}$.

Lemma 4. *The number $B_2(\pi_1|s)$ of v-sequences generated* (s) *by any $\pi_1 u$-sequence satisfies*

$$2^{nH(Y|X|\pi|s) - K_2\sqrt{n}} < B_2(\pi_1|s) < 2^{nH(Y|X|\pi|s) + K_2\sqrt{n}}. \tag{2.14}$$

Proof. If v is any v-sequence generated by a $\pi_1 u$-sequence u, then

$$\begin{aligned} 2^{-nH(Y|X|\pi|s) - K_2'\sqrt{n}} &< \prod_{i,j} w(j|i|s)^{(n\pi_i + \delta\sqrt{n})\,w(j|i|s) + \delta_2\sqrt{n\,w(j|i|s)}} < \\ &< P_s\{\bar{v} = v \mid u\} < \prod_{i,j} w(j|i|s)^{(n\pi_i - \delta\sqrt{n})\,w(j|i|s) - \delta_2\sqrt{n\,w(j|i|s)}} < 2^{-nH(Y|X|\pi|s) + K_2'\sqrt{n}}. \end{aligned} \tag{2.15}$$

---

[5] The reader is invited to verify that this is purely a technical device for expeditiously achieving the proof and in no way contradicts our earlier statement that the problems treated in this paper are of combinatorial character. The use of the stochastic input is convenient but not at all indispensable.

The lemma follows at once from (2.15) and Lemma 3.

Lemma 5. *Let v be any v-sequence which is generated(s) by some π_1u-sequence. Then*

(2.16)
$$2^{-nH(Y|\pi|s)-K_s\sqrt{n}} < P_s\{\bar{v}(\pi) = v\} < 2^{-nH(Y|\pi|s)+K_s\sqrt{n}}.$$

Proof. We have that

(2.17)
$$h_i = \pi_0 w(i|0|s) + \pi_1 w(i|1|s) \geqq ad, \qquad i = 0, 1,$$

is the probability that the kth element of $\bar{v}(\pi)$, $k = 1, \ldots, n$, is i. Define, for $i = 0, 1$,

(2.18)
$$V_{i0} = nh_i - 2\sqrt{n}(\delta + \delta_2),$$

(2.19)
$$V_{i1} = nh_i + 2\sqrt{n}(\delta + \delta_2).$$

The number of elements i, $i = 0, 1$, in v, is greater than V_{i0} and less than V_{i1}. Hence

(2.20)
$$2^{-nH(Y|\pi|s)-K_s\sqrt{n}} < h_0^{V_{01}} h_1^{V_{11}} < P_s\{\bar{v}(\pi) = v\} < h_0^{V_{00}} h_1^{V_{10}} < 2^{-nH(Y|\pi|s)+K_s\sqrt{n}}.$$

3. Preliminaries on the channel

Define the capacity C_0 of the simultaneous channel S^* as

(3.1)
$$C_0 = \sup_{\pi} \inf_{s \in S} [H(X|\pi) - H(X|Y|\pi|s)].$$

It is obvious that the quantity in brackets which occurs in (3.1) is continuous in π, uniformly in the w, so that the supremum in (3.1) is actually attained for some π, henceforth to be designated as $\bar{\pi}$. It is easy to show that $C_0 = 0$ if and only if[6]

(3.2)
$$\inf_{s \in S} |w(0|1|s) - w(0|0|s)| = \inf_{s \in S} |w(1|1|s) - w(1|0|s)| = 0.$$

Let any c.p.f. $w(\cdot|\cdot|s)$ in the simultaneous channel be approximated by the c.p.f. $w(\cdot|\cdot|s')$ such that $w(0|j|s')$, $j = 0, 1$, is that multiple of $2^{-\sqrt{n}}$ which is closest to $w(0|j|s)$. (If there are two or more such c.p.f. any may be taken at pleasure.) The totality of all such approximating c.p.f. will be called the simultaneous canonical channel S_0^*. The index set for the canonical channel will be called S_0.

Let u be any u-sequence, s be any index in S such that $\min_{i,j} w(i|j|s) \geqq n^{-2}$, s' be the index in S_0 of the c.p.f. which approximates $w(\cdot|\cdot|s)$, and v be any v-sequence. Then

$$P_s\{\bar{v} = v|u\} \geqq (n^{-2})^n.$$

Also

$$\frac{(n^{-2})^n}{(n^{-2} - 2^{-\sqrt{n}})^n} \to 1 \quad \text{as} \quad n \to \infty.$$

---

[6] (3.2) is easily shown to be equivalent to

$$I = \inf_{s \in S} \sup_{\pi} [H(X|\pi) - H(X|Y|\pi|s)] = 0.$$

Obviously $I \geqq C_0$ always, and it is easy to see that the inequality sign can hold. Still $C_0 = 0$ implies $I = 0$. The quantity I is the infimum of the capacities of the individual c.p.f.'s which compose S^*.

Moreover, if $w(j_0|i_0|s) \leq n^{-2}$, then, no matter what u is, we have

$$P_s\{N(i_0, j_0|u, \bar{v}) \geq 1|u\} \leq \frac{1}{n},$$

which of course approaches zero as $n \to \infty$. Thus we have proved

Lemma 6. *Let b be a positive constant, u be any u-sequence, A be any set of v-sequences, and $s \in S$ be any index such that*

$$P_s\{\bar{v} \in A|u\} > b. \tag{3.3}$$

Let $s' \in S_0$ be the index of the approximating c.p.f. Then we have, for all $s \in S$, A, and u which satisfy (3.3),

$$\left|\frac{P_{s'}\{\bar{v} \in A|u\}}{P_s\{\bar{v} \in A|u\}} - 1\right| < a_n, \tag{3.4}$$

where $a_n \to 0$ as $n \to \infty$. (Of course, b is a constant independent of n.)

The proof of the following lemma is obvious.

Lemma 7. *There exists a positive constant K_4 with the following property. Let $w(\cdot|\cdot|s)$ be any c.p.f. in S^*, and $w(\cdot|\cdot|s')$ its approximating c.p.f. in S_0^*. Then, for any stochastic input π,*

$$|H(X|Y|\pi|s) - H(X|Y|\pi|s')| < \frac{K_4}{\sqrt{n}}. \tag{3.5}$$

4. Proof of the coding theorem (Theorem 1)

It is clearly sufficient to prove the theorem for n sufficiently large. Let $\lambda' < \lambda$ be any positive number. It follows from Lemma 6 that, for n sufficiently large, a code with probability of error $\leq \lambda'$ for the simultaneous canonical channel S_0^* is a code with probability of error $\leq \lambda$ for the simultaneous channel S^*.

We may assume $C_0 > 0$, or there is nothing to prove. In the former case (2.9) holds with some a for all s in S and in S_0. Also $0 < \bar{\pi}_1 < 1$, so that Lemma 5 holds with $\pi = \bar{\pi}$.

We may, and do, take $\lambda' < \frac{1}{2}$. Let $\gamma < 1$ be any positive number, and δ be sufficiently large for Lemma 1 to hold. Let δ_2 be sufficiently large so that Lemma 3 holds with $\varepsilon' = \lambda'/2$. Let

$$\{(u_1, A_1), \ldots, (u_N, A_N)\} \tag{4.1}$$

be a code with probability of error $\leq \lambda'$ for the simultaneous canonical channel S_0^*, such that the following conditions are fulfilled:

(4.2) $u_i, i = 1, \ldots, N$, is a $\bar{\pi}_1$ u-sequence.

(4.3) $A_i, i = 1, \ldots, N$, contains only v-sequences generated(s) by u_i for $s \in S_0$.

(4.4) The code is maximal in the sense that it is impossible to add another element (u_{N+1}, A_{N+1}) such that $\{(u_1, A_1), \ldots, (u_{N+1}, A_{N+1})\}$ is a code with probability of error $\leq \lambda'$ for the simultaneous canonical channel S_0^*, and (4.2) and (4.3) are also satisfied for $i = N+1$.

From (4.4) we obtain the following conclusion: Let u_0 be *any* $\bar{\pi}_1$ u-sequence (if one exists) not in the set $u_1, \ldots, u_N$. For some s_0 in S_0 we have

$$(4.5)\quad P_{s_0}\{\bar{v} \text{ is generated } (s_0) \text{ by } u_0 \text{ and belongs to } A_1 \cup \cdots \cup A_N | u_0\} > \frac{\lambda'}{4}.$$

Of course s_0 is a function of u_0, and (4.5) also holds if u_0 is in the set $\{u_1, \ldots, u_N\}$. Since there are fewer than $2^{\frac{3}{2}(\sqrt{n})}$ indices in S_0 it follows from Lemma 1 that there is an index s_{00} in S_0 such that the $\bar{\pi}_1$ u-sequences u for which

$$(4.6)\quad P_{s_{00}}\{\bar{v} \text{ is generated } (s_{00}) \text{ by } u \text{ and belongs to } A_1 \cup \cdots \cup A_N | u\} > \frac{\lambda'}{4}$$

have probability greater than $\gamma \cdot 2^{-\frac{3}{2}\sqrt{n}}$ when $\bar{\pi}$ is the stochastic input. Hence

$$(4.7)\quad \begin{aligned} P_{s_{00}}\{\bar{v}(\bar{\pi}) \text{ is generated } (s_{00}) \text{ by some } \bar{\pi}_1\ u\text{-sequence} \\ \text{and belongs to } A_1 \cup \cdots \cup A_N\} > \frac{\gamma\lambda'}{2^{3\sqrt{n}}}. \end{aligned}$$

It follows from (4.7) and Lemma 5 that $A_1 \cup \cdots \cup A_N$ contains at least

$$(4.8)\quad \gamma\lambda' \cdot 2^{-3\sqrt{n}} \cdot 2^{nH(Y|\bar{\pi}|s_{00}) - K_3\sqrt{n}} > 2^{nH(Y|\bar{\pi}|s_{00}) - K_5\sqrt{n}}$$

v-sequences generated (s_{00}) by some $\pi_1 u$-sequence, where $K_5 > K_3$ is sufficiently large.

We now obtain an upper bound on the number of v-sequences in $A_1 \cup \cdots \cup A_N$ generated (s_{00}) by some $\bar{\pi}_1 u$-sequence. The number of v-sequences in A_i, $i = 1, \ldots, N$, which are generated (s_{00}) by u_i is, by Lemma 4, less than $2^{nH(Y|X|\bar{\pi}|s_{00}) + K_2\sqrt{n}}$. Suppose that, for $s' \in S_0$ and not s_{00}, a v-sequence generated (s') by u_i is also generated (s_{00}) by some other $\bar{\pi}_1$ u-sequence. It follows from (2.18) and (2.19) for $\pi = \bar{\pi}$ that, for n sufficiently large,

$$(4.9)\quad |[\bar{\pi}_0 w(0|0|s_{00}) + \bar{\pi}_1 w(0|1|s_{00})] - [\bar{\pi}_0 w(0|0|s') + \bar{\pi}_1 w(0|1|s')]| < \frac{K_6}{\sqrt{n}},$$

$$(4.10)\quad |[\bar{\pi}_0 w(1|0|s_{00}) + \bar{\pi}_1 w(1|1|s_{00})] - [\bar{\pi}_0 w(1|0|s') + \bar{\pi}_1 w(1|1|s')]| < \frac{K_6}{\sqrt{n}}$$

where K_6 is a constant which depends only upon δ and δ_2.

Call an index $s' \in S_0$ "associated" with s_{00} if its c.p.f. satisfies (4.9) and (4.10). It follows that, if s' is associated with s_{00}, then, for n sufficiently large,

$$(4.11)\quad |H(Y|\bar{\pi}|s_{00}) - H(Y|\bar{\pi}|s')| < \frac{K_7}{\sqrt{n}}$$

where K_7 is a constant.

Let $s_{00}^* \in S_0$ be that index associated with s_{00} for which

$$(4.12)\quad H(Y|X|\bar{\pi}|s_{00}^*) = \max_{s'} H(Y|X|\bar{\pi}|s')$$

where the maximum is taken over all indices s' associated with s_{00}. From Lemma 4 we conclude that, if n is sufficiently large, A_i, $i = 1, \ldots, N$, contains fewer than

$$(4.13)\quad 2^{3\sqrt{n}} \cdot 2^{nH(Y|X|\bar{\pi}|s_{00}^*) + K_2\sqrt{n}}$$

v-sequences generated (s_{00}) by some $\bar{\pi}_1$ u-sequence. Thus an upper bound on the number of v-sequences in $A_1 \cup \cdots \cup A_N$ which are generated (s_{00}) by some

$\bar{\pi}_1$ u-sequence is

$$N \cdot 2^{3\sqrt{n}} \cdot 2^{n H(Y|X|\bar{\pi}|s_{00}^*) + K_2\sqrt{n}} \tag{4.14}$$

when n is sufficiently large. Bearing in mind that then

$$2^{n H(Y|\bar{\pi}|s_{00})} > 2^{n H(Y|\bar{\pi}|s_{00}^*) - K_7\sqrt{n}}, \tag{4.15}$$

we obtain, from (4.14) and (4.8), that

$$N > 2^{n[H(Y|\bar{\pi}|s_{00}^*) - H(Y|X|\bar{\pi}|s_{00}^*)] - \sqrt{n}[3 + K_2 + K_5 + K_7]} \geqq 2^{n C_0 - \sqrt{n}(3 + K_2 + K_5 + K_7 + K_4)} \tag{4.16}$$

for n sufficiently large. This proves the theorem.

5. Proof of the strong converse (Theorem 2)

We begin by proving

Lemma 8. *Let $w(\cdot|\cdot|s)$ be any c.p.f. (it need not be in S^*) which satisfies* (2.8) *and* (2.9), *and let β be any number, $0 \leqq \beta < 1$. Let $\{(u_1 A_1), \ldots, (u_N, A_N)\}$ be any code for $w(\cdot|\cdot|s)$ with probability of error $\leqq \beta$ and such that u_i is a π_1 u-sequence, $i = 1, \ldots, N$. Then*

$$N < 2^{n[H(X|\pi) - H(X|Y|\pi|s)] + K_8\sqrt{n}} \tag{5.1}$$

where K_8 is a positive function only of β and does not depend on $w(\cdot|\cdot|s)$.

Proof. Let δ_2 be large enough so that the ε' of Lemma 3 is less than $\frac{1-\beta}{2}$. Then

$$P_s\{\bar{v} \text{ is generated } (s) \text{ by } u_i \text{ and belongs to } A_i | u_i\} > \frac{1-\beta}{2}. \tag{5.2}$$

Hence, by (2.15) each A_i contains more than

$$\frac{(1-\beta)}{2} 2^{n H(Y|X|\pi|s) - K_2'\sqrt{n}} > 2^{n H(Y|X|\pi|s) - K_2''\sqrt{n}} \tag{5.3}$$

v-sequences generated (s) by u_i. By Lemma 5 the total number of v-sequences generated (s) by some π_1 u-sequence is less than

$$2^{n H(Y|\pi|s) + K_3\sqrt{n}}. \tag{5.4}$$

Lemma 8 follows from (5.3) and (5.4), with $K_8 = K_3 + K_2''$.

We now prove the strong converse when $C_0 > 0$. We may then assume that (2.9) is satisfied. It is clearly sufficient to prove the theorem for n sufficiently large. Let λ' be any number such that $\lambda < \lambda' < 1$. Let d, $0 < d < \frac{1}{2}$, be so small that

$$-d \log d - (1-d)\log(1-d) < \frac{C_0}{4}. \tag{5.5}$$

Let $\{(u_1, A_1), \ldots, (u_N, A_N)\}$ be any code with probability of error $\leqq \lambda$ for the simultaneous channel S^*. It follows from Lemma 6 that, for n sufficiently large (which we henceforth assume to be the case) this code is also a code with probability of error $\leqq \lambda'$ for the simultaneous canonical channel S_0^*. Divide the interval $[0, 1]$ into $J = n^{\frac{1}{2}}/2\delta$ intervals of length $2\delta/n^{\frac{1}{2}}$, and let $t_1, \ldots, t_J$ be the midpoints of these intervals. Then the code in question can be divided into J codes $W_1, \ldots, W_J$, such that all the u-sequences in W_i are t_i u-sequences, $i = 1, \ldots, J$. The sum of the lengths of all codes W_i such that $\min[t_i, (1-t_i)] \leqq d$ is, by (5.5) and Lemma 2, $< J \cdot 2^{n C_0/2}$, for n sufficiently large.

Consider now any code W_i, with $\min[t_i, (1-t_i)] > d$. Let $s_i \in S_0$ be such that, writing $\bar{t}_i = (t_i, 1-t_i)$,

$$H(X|\bar{t}_i) - H(X|Y|\bar{t}_i|s_i) = \min_{s \in S_0} [H(X|\bar{t}_i) - H(X|Y|\bar{t}_i|s)]. \tag{5.6}$$

Since W_i is a code for the c.p.f. $w(\cdot|\cdot|s_i)$ with probability of error $\leqq \lambda'$, it follows from Lemma 8 that the length of W_i is less than

$$2^{n(H(X|\bar{t}_i) - H(X|Y|\bar{t}_i|s_i)) + K_8\sqrt{n}}, \tag{5.7}$$

which, by Lemma 7, is less than

$$2^{nC_0 + \sqrt{n}(K_8+K_4)}. \tag{5.8}$$

Thus we have that

$$N < J[2^{nC_0/2} + 2^{nC_0 + \sqrt{n}(K_8+K_4)}], \tag{5.9}$$

which proves the theorem when $C_0 > 0$.

Suppose now that $C_0 = 0$. Then, for some $s \in S_0$, say s_0,

$$w(0|0|s_0) = w(0|1|s_0), \tag{5.10}$$

$$w(1|0|s_0) = w(1|1|s_0). \tag{5.11}$$

Since a code for the simultaneous channel S^* with probability of error $\leqq \lambda$ is a code for the simultaneous canonical channel S_0^* with probability of error $\leqq \lambda'$, and since $s_0 \in S_0$, we have from (5.10) and (5.11) that

$$P_{s_0}\{\bar{v} \in A_j | u_i\} = P_{s_0}\{\bar{v} \in A_j | u_j\} \geqq 1 - \lambda', \qquad i, j = 1, \ldots, N. \tag{5.12}$$

Set $i = 1$, say, in (5.12). Then clearly

$$N \leqq \frac{1}{1-\lambda'} \tag{5.13}$$

which proves the theorem.

(It is easy to see that we could have given the above proof of Theorem 2 without employing the simultaneous canonical channel.)

6. The *m*-finite simultaneous channel

We begin with a description of an *m*-finite channel. A sequence of $(m+1)$ elements, each zero or one, with m a non-negative integer, will be called an $(m+1)$-sequence. A c.p.f. $w(i|\alpha)$ ($i = 0, 1$; α any $(m+1)$-sequence) is any non-negative function such that $w(0|\alpha) + w(1|\alpha) = 1$ for every α. If the word of length n (n-sequence) $u = (x_1, \ldots, x_n)$ is "sent" or "transmitted" over the channel, the "received" sequence $v = (Y_1, \ldots, Y_{n-m})$ is an $(n-m)$-sequence of chance variables with respective distributions given by

$$P\{Y_i = 0 | u, Y_1, \ldots, Y_{i-1}\} = w(0|x_i, \ldots, x_{i+m}), \tag{6.1}$$

$$P\{Y_i = 1 | u, Y_1, \ldots, Y_{i-1}\} = w(1|x_i, \ldots, x_{i+m}) \qquad i = 1, \ldots, (n-m). \tag{6.2}$$

The capacity C' of this channel was given (independently) in [2] and [4] as follows: Let $l > m$ be any integer. Consider the discrete memoryless channel (to be called the l-modified channel) in which the transmitted *alphabet* consists of

l-sequences, the received *alphabet* consists of $(l-m)$-sequences, and the c.p.f. w_1 is as follows: Let β_1 be any l-sequence and β_2 be any $(l-m)$-sequence. Then $w_1(\beta_2|\beta_1)$ is equal to the probability that, in the *original* m-finite channel, when $n=l$ and the sequence β_1 is transmitted, the sequence β_2 is received. Let $C_0(l)$ be the capacity of the l-modified channel. Then

$$C' = \sup_l \frac{C_0(l)}{l} = \lim_{l\to\infty} \frac{C_0(l)}{l}. \tag{6.3}$$

(Since the alphabets involved have more than two elements $C_0(l)$ has, strictly speaking, not yet been defined. However, it is defined in complete analogy with C_0, as follows: Number the 2^l l-sequences and the 2^{l-m} $(l-m)$-sequences in some arbitrary but fixed manner. Let $\pi = (\pi(1), \ldots, \pi(2^l))$ be a probability distribution on the 2^l l-sequences, *i.e.*, a vector with 2^l non-negative components which add to one. Define

$$H(Y|\pi) = -\sum_j \Big(\sum_i \pi(i)\, w_1(j|i)\Big) \log \Big(\sum_i \pi(i)\, w_1(j|i)\Big), \tag{6.4}$$

$$H(Y|X|\pi) = -\sum_j \sum_i \pi(i)\, w_1(j|i) \log w_1(j|i) \tag{6.5}$$

where the summation with respect to i is from 1 to 2^l, and the summation with respect to j is from 1 to 2^{l-m}. Then

$$C_0(l) = \max_\pi \big(H(Y|\pi) - H(Y|X|\pi)\big).) \tag{6.6}$$

In extension of the definition of Section 2 we shall find it convenient to speak of the stochastic input π as follows: Assume that n is an intégral multiple of l. Consider

$$(x_{il+1}, \ldots, x_{(i+1)l}), \qquad i = 0, 1, \ldots, \left(\frac{n}{l} - 1\right), \tag{6.7}$$

as independent, identically distributed, l-vector chance variables, with probabilities given by the 2^l components of π. We shall call such a stochastic input one of independent, identically distributed blocks of length l. (Each vector with l-components is a block.) When we use the stochastic input π, we write $u(\pi)$ for the then chance sequence u, and $\bar{v}(\pi)$ for the chance received sequence. The capacity $C_0(l)$ is the maximum of the difference $H(Y|\pi) - H(Y|X|\pi)$ over all stochastic inputs π of independent, identically distributed blocks of length l.

The simultaneous m-finite channel S^* bears the same relation to the m-finite channel as the simultaneous (memoryless) channel bears to the memoryless channel. As before, let S be the set of indices of the c.p.f.'s.

We define the capacity C of the simultaneous m-finite channel by

$$C = \sup_{l>m} \max_\pi \inf_{s\in S} \frac{1}{l} \left[H(Y|\pi|s) - H(Y|X|\pi|s)\right]$$

where max with respect to π (easily shown to exist) is the maximum with respect to all stochastic inputs of independent, identically distributed blocks of length l.

Theorem 3. *Let λ, $0<\lambda\leqq 1$, and $\varepsilon>0$ be any numbers. For n sufficiently large there exists a code for the simultaneous m-finite channel with probability of error $\leqq\lambda$ and length greater than $2^{n(C-\varepsilon)}$.*

Theorem 4. *Let λ, $0 \leqq \lambda < 1$, and $\varepsilon > 0$ be any numbers. For n sufficiently large any code for the simultaneous m-finite channel with probability of error $\leqq \lambda$ cannot have a length greater than $2^{n(C+\varepsilon)}$.*

Theorem 5. *We have*

$$C = \lim_{l\to\infty} \max_{\pi} \inf_{s\in S} \frac{1}{l} [H(Y|\pi|s) - H(Y|X|\pi|s)].$$

Also $\frac{m}{l} < \varepsilon_1$ *implies*

$$\max_{\pi} \inf_{s\in S} \frac{1}{l} [H(Y|\pi|s) - H(Y|X|\pi|s)] > C - \varepsilon_1.$$

These theorems are proved exactly as in [2], starting from Theorems 1 and[7] 2.

7. Miscellaneous results

The approximation by the canonical channel (of order $2^{-\sqrt{n}}$) was an act of supererogation; a far coarser approximation would have sufficed. Let us devote a few remarks[8] to this.

In our proof of Theorem 1 we employed the following three facts:

(7.1) The canonical channel contains fewer than $2^{K_3\sqrt{n}}$ c.p.f.'s for n sufficiently large.

(7.2) If $w(\cdot|\cdot|s)$ is a c.p.f. in S^* and $w(\cdot|\cdot|s')$ its approximating c.p.f. in S_0^*, then

$$|H(X|Y|\pi|s) - H(X|Y|\pi|s')| < \frac{K_4}{\sqrt{n}}.$$

(7.3) Lemma 6. Actually a weaker statement, like (7.6) below, would suffice.

A much coarser approximation than to $2^{-\sqrt{n}}$ would already have given us (7.2), and, of course, the coarser the approximation the fewer the c.p.f.'s in the canonical channel. To see what the order of the approximation can be, consider the differential $d(-z\log z) = (-1 - \log z)\,dz$. If the approximation is Δ', then the critical question is the value of $\Delta = \Delta' \log \frac{1}{\Delta'}$. Thus, if $\Delta' = n^{-3}$, Lemma 6 obviously holds, and $\Delta = o(n^{-\frac{1}{2}})$. Then (7.2) is satisfied, and *a fortiori* (7.1) is.

In order to use our proof for the weaker version of Theorem 1 where (1.4) is replaced by

(1.4′) $$2^{n(C_0-\varepsilon)}$$

with $\varepsilon > 0$ arbitrary (the theorem then being valid for all n greater than a lower bound which depends on ε and S^*), we could make do with the following:

(7.4) The canonical channel contains no more than $2^{n\varepsilon}$ c.p.f.'s, for n sufficiently large.

---

[7] In a paper to appear shortly, the author has proved the results of Section 6 for the simultaneous analogue of what FEINSTEIN [4] has called the general finite memory channel. These results therefore include those of Section 6.

[8] The remarks made just after the statement of Theorem 1 apply here as well.

(7.5) If $w(\cdot|\cdot|s)$ and $w(\cdot|\cdot|s')$ are, respectively, a c.p.f. in S^* and its approximating c.p.f. in S_0^*, then

$$|H(X|Y|\pi|s) - H(X|Y|\pi|s')| < \varepsilon.$$

(7.6) The a_n of (3.4) is less than $\lambda/4$ when $b = \frac{1}{2}$ and n is sufficiently large.

It is clear that (7.5) will be satisfied if Δ' is less than a suitable function of ε. The number of c.p.f.'s in the canonical channel is then fixed (and finite), so that (7.4) is satisfied. It can also be shown that, if Δ' is less than a suitable function of λ, condition (7.6) is satisfied. Of course, one could use a Δ' which varies with the c.p.f. being approximated.

In some problems the c.p.f.'s in the simultaneous channel might depend upon n. If the number of the c.p.f.'s fulfills condition (7.1) or (7.4), our method might be applicable.

For the proof of the strong converse conditions (7.1)—(7.6) play no role, because the simultaneous canonical channel is then completely unnecessary. This may be readily verified from the proof itself.

Similar remarks obtain for the discrete m-finite simultaneous channel.

All the channels discussed so far in the paper are discrete, even when this adjective was omitted. "Discrete" means that both the transmitted and received alphabets have only finitely many symbols. We shall now very briefly discuss "semi-continuous" channels (see [3]). We shall limit ourselves to channels without memory, the extension to m-finite channels being routine.

In a semi-continuous channel the symbols of the received alphabet need not be finite in number and may lie in any probability (measure) space. We shall take the latter to be the real line, because the treatment of the probability space is routine. To prove the coding theorem for a semi-continuous channel (with a single c.p.f.) one approximates the channel by a suitable discrete channel, so that conditions (7.5) and one like (7.6) are satisfied. In order to carry through our proof of the coding theorem for a simultaneous semi-continuous memoryless channel one will have to try to construct a simultaneous canonical discrete memoryless channel which will satisfy (7.5) and, say, (7.6), and still satisfy (7.4). (We are assuming that the "discretization" (*e.g.*, the sets J of [3]) does not depend on n, so that the number of symbols in the received alphabet is fixed (independent of n). If this is not the case, the situation is more complicated and has to be considered *per se*.) For example, if the original simultaneous channel contains only a finite number of c.p.f.'s conditions (7.4)—(7.6) are easily satisfied. A more interesting case occurs when the noise (error of transmission) is Gaussian, additive and independent for each symbol, with variance one and a mean which lies in a bounded interval of the line. (Thus the mean is the same for each symbol of a u-sequence, but may vary from u-sequence to u-sequence.) The variance could also be allowed to vary in a bounded interval at a positive distance from zero. Conditions (7.4)—(7.6) can then be satisfied.

Consider now the question of proving the strong converse in the form (1.6) for the simultaneous semi-continuous memoryless channel. As we have remarked earlier, the canonical channel is not needed. The strong converse would hold if we had the semi-continuous analogue of Lemma 8 (in the form (1.6)). Lemma 3 of [3] is precisely that. Thus the strong converse holds (in the form (1.6)).

The continuous operation of a simultaneous discrete memoryless channel, say, can be conceived of as follows: Various words of length n are sent consecutively, and the c.p.f. may change from word to word. Suppose now that we wish to be able to change the c.p.f. within a word, *i.e.*, while transmitting a word. If the c.p.f. can change only when the symbol transmitted has an index which is an integral multiple of n/t, with $t>1$ a constant, then our earlier results obviously apply. A slightly more elaborate analysis will give results even when more frequent changes in the c.p.f. are permitted[9].

8. Simultaneous channels where either the sender or the receiver knows the c.p.f.

In this section we return to the simultaneous discrete memoryless channel S^* and study two new methods of transmission, each different from that of Section 1.

In the first method of transmission, at the beginning of each word, the receiver, but not the sender, knows the c.p.f. in S^* which governs the channel for that particular word (i.e., the c.p.f. according to which the received symbols are distributed). We call this simultaneous channel S_1^*. In the second method of transmission, at the beginning of each word, the sender, but not the receiver, knows the c.p.f. in S^* which governs the channel for that particular word[10]. We call this simultaneous channel S_2^*.

We shall show that the capacity of S_1^* is C_0, and the capacity of S_2^* is

$$I = \inf_{s\in S} \sup_{\pi} [H(X|\pi) - H(X|Y|\pi|s)]. \tag{8.1}$$

Thus the capacity of S_1^* is the same as that of S^*, and the receiver's knowing the c.p.f. (when the sender does not know it) does not increase the capacity of the channel. On the other hand, obviously $I \geqq C_0$, and the inequality sign can hold. Thus $I - C_0$ is the increase in capacity due to the sender's knowing the channel (when the receiver does not know it).

When we say that "we shall show that the capacity is so-and-so", we mean that we shall prove a coding theorem (analogue of Theorem 1) and a strong converse (analogue of Theorem 2) involving the capacity.

Extensions to the channels discussed in Sections 6 and 7 can be made similarly.

We begin with S_1^*, which is very easy. An element of a code is now a complex

$$(u, \{A(s), s\in S\})$$

which we write for short as

$$(u, \{A(s)\}).$$

[9] A problem where the c.p.f. changes from symbol to symbol according to an independent random procedure is treated in the interesting paper [6]. For further results on this problem see the author's paper in the Proceedings of the Purdue University Symposium, 1959.

[10] The problem treated in the paper [6] (already cited in footnote 8) is completely different. In this problem not only are the c.p.f.'s chosen stochastically for each symbol, but the sender knows only the c.p.f.'s for the symbols previously sent and the symbol presently being sent.

Here u is a u-sequence, and the $A(s)$ are (not necessarily disjoint) sets of v-sequences. A code of length N and probability of error $\leqq \lambda$ is a set

$$\{(u_1, \{A_1(s)\}), \ldots, (u_N, \{A_N(s)\})\} \tag{8.2}$$

where, for each $s \in S$, the $A_i(s)$, $i=1, \ldots, N$, are disjoint, and, for every $s \in S$,

$$P_s\{\bar{v} \in A_i(s) | u_i\} \geqq 1 - \lambda. \tag{8.3}$$

In other words, for any $s \in S$,

$$\{(u_1, A_1(s)), \ldots, (u_N, A_N(s))\} \tag{8.4}$$

is a code of length N and probability of error $\leqq \lambda$ for the c.p.f. $w(\cdot|\cdot|s)$. The practical use of (8.2) is as follows: When the receiver knows that $w(\cdot|\cdot|s)$ is the c.p.f. which governs the channel for a particular word, he uses the sets $A_1(s), \ldots, A_N(s)$ for decoding purposes.

Any code for S^* is also a code for S_1^*. This corresponds to the special case of (8.2) where, for each i, $i=1, \ldots, N$, $A_i(s)$ is the same for all $s \in S$. Since we have already proved a coding theorem for S^* (Theorem 1), and since we wish to prove that the capacity of S_1^* is C_0, it follows that the required coding theorem for S_1^* is already proved. It remains only to prove the strong converse.

To do so, proceed exactly as in Section 5 and consider the W_i such that $\min[t_i, (1-t_i)] > d$. In view of (8.4) we have that (5.7) and (5.8) hold. But then (5.9) holds and the strong converse is proved. This completes the proof that C_0 is the capacity of S_1^*.

We turn now to S_2^*. The element of a code is now a complex

$$(\{u(s), s \in S\}, A)$$

which we write for short as

$$(\{u(s)\}, A).$$

Here $u(s)$, $s \in S$, is a u-sequence, and A is a set of v-sequences. A code of length N and probability of error $\leqq \lambda$ is a set

$$\{(\{u_1(s)\}, A_1), \ldots, (\{u_N(s)\}, A_N)\} \tag{8.5}$$

where the A_i are disjoint, and, for every $s \in S$,

$$P_s\{\bar{v} \in A_i | u_i(s)\} \geqq 1 - \lambda. \tag{8.6}$$

In other words, for any $s \in S$,

$$\{(u_1(s), A_1), \ldots, (u_N(s), A_N)\} \tag{8.7}$$

is a code of length N and probability of error $\leqq \lambda$ for the c.p.f. $w(\cdot|\cdot|s)$. The practical use of (8.5) is this: When the sender wants to send the ith word and knows that $w(\cdot|\cdot|s)$ is the governing c.p.f., he sends $u_i(s)$.

In view of Lemma 6 we may, and do, confine ourselves everywhere below to $s \in S_0$.

Let $s_0 \in S_0$ be such that

$$\sup_\pi [H(X|\pi) - H(X|Y|\pi|s_0)] = \inf_{s \in S_0} \sup_\pi [H(X|\pi) - H(X|Y|\pi|s)]. \tag{8.8}$$

For each $s \in S_0$ let $\pi(s) = (\pi_0(s), \pi_1(s))$ be that stochastic input for which

$$(8.9)\qquad [H(X|\pi(s)) - H(X|Y|\pi(s)|s)] = \sup_{\pi} [H(X|\pi) - H(X|Y|\pi|s_0)].$$

From Lemma 7 it follows that each member of (8.8) differs from I by less than $K_4/\sqrt{n}$.

We now prove the coding theorem. We proceed exactly as in the proof of Theorem 1, with $u_i(s)$, $i = 1, \ldots, N$, a $\pi_1(s)$ u-sequence, $s \in S_0$. Each A_i, $i = 1, \ldots, N$, contains only v-sequences generated (s) by $u_i(s)$, $s \in S_0$. The analogue of (4.8) is obtained even more simply than in the proof of Theorem 1. This analogue says that, for some $s_{00} \in S_0$, the set $A_1 \cup \cdots \cup A_N$ contains at least

$$(8.10)\qquad 2^{nH(Y|\pi(s_{00})|s_{00}) - K_5\sqrt{n}}$$

v-sequences generated (s_{00}) by some $\pi_1(s_{00})$ u-sequence.

We now turn to establishing the analogue of (4.13). We seek an upper bound on the number of v-sequences in A_i, $i = 1, \ldots, N$, which are generated (s_{00}) by some $\pi_1(s_{00})$ u-sequence. In doing this we have to consider the possibility that a v-sequence v which is generated (s') by $u_i(s')$, $s' \in S_0$, $s' \neq s_{00}$, is also generated (s_{00}) by some $\pi_1(s_{00})$ u-sequence u' (say). When this is the case then, reasoning as we did in the argument which led to (4.11), we conclude that

$$(8.11)\qquad |H(Y|\pi(s_{00})|s_{00}) - H(Y|\pi(s')|s')| < \frac{K_7'}{\sqrt{n}}.$$

From (8.9) we then have

$$(8.12)\qquad |H(Y|X|\pi(s_{00})|s_{00}) - H(Y|X|\pi(s')|s')| < \frac{K_7'}{\sqrt{n}}.$$

From (8.12) and Lemma 4 we conclude that A_i, $i = 1, \ldots, N$, contains fewer than

$$(8.13)\qquad 2^{3\sqrt{n}} \cdot 2^{nH(Y|X|\pi(s_{00})|s_{00}) + \sqrt{n}(K_3 + K_7')}$$

v-sequences generated (s_{00}) by some $\pi_1(s_{00})$ u-sequence. This is of course the analogue of (4.13), and the proof of the coding theorem is completed exactly as the proof of Theorem 1.

The strong converse for the channel S_2^* hardly requires proof. We know that (8.7), for $s = s_0$, cannot be longer than

$$(8.14)\qquad 2^{nI + K'\sqrt{n}},$$

by Theorem 1 of the present paper applied to the channel containing the *single* c.p.f. $w(\cdot|\cdot|s_0)$. (For a channel with a single c.p.f. this is Theorem 2 of [1].) This completes our proof for S_2^*.

Let $\bar{C}$ be the capacity of either S_1^* or S_2^* when one of these channels is employed. Let $\varepsilon > 0$ be arbitrary. By the application of relation (74) (or (75)) of page 280 of [7] to the proof of the coding theorems of the present section one easily obtains that there exist positive constants c_1 and c_2 with the following property: Let λ_n be the smallest probability of error among all codes of length $2^{n(\bar{C}-\varepsilon)}$ for either of the channels S_1^* or S_2^*. Then

$$(8.15)\qquad \lambda_n < c_1 e^{-nc_2}.$$

Of course this method will also work for the channel S^*.)

References

[1] WOLFOWITZ, J.: The coding of messages subject to chance errors. Ill. J. Math. **1**, No. 4 591–606 (Dec. 1957).
[2] WOLFOWITZ, J.: The maximum achievable length of an error correcting code. Ill. J. Math. **2**, No. 3, 454–458 (Sept. 1958).
[3] WOLFOWITZ, J.: Strong converse of the coding theorem for semi-continuous channels. Ill. J. Math. **3**, No. 4 (Dec. 1959).
[4] FEINSTEIN, A.: On the coding theorem and its converse for finite-memory channels. Information and Control **2**, No. 1 25–44 (Apr. 1959).
[5] BLACKWELL, D., L. BREIMAN & A. J. THOMASIAN: Ann. Math. Stat., **30**, No. 4 (Dec. 1959).
[6] SHANNON, C. E.: Channels with side information at the transmitter. IBM J. Research and Development **2**, No. 4, 289–293 (Oct. 1958).
[7] LÉVY, P.: Théorie de l'addition des variables aléatoires. Paris: Gauthier-Villars 1937.

Cornell University
Ithaca, New York

(Received July 25, 1959)

Druck der Universitätsdruckerei H. Stürtz AG., Würzburg

Reprinted from INFORMATION AND CONTROL, Volume 5, No. 1, March 1962

INFORMATION AND CONTROL 5, 44–54 (1962)

Channels with Arbitrarily Varying Channel Probability Functions

J. KIEFER* AND J. WOLFOWITZ†

Cornell University, Ithaca, N. Y.

I. INTRODUCTION

We begin by defining several terms whose significance will be apparent shortly. Let $D = \{1, 2, \cdots, d\}$ and $B = \{1, 2, \cdots, b\}$ be, respectively, the input and output alphabets. Let $S = \{w(\cdot \mid \cdot \mid i), i = 1, \cdots, c\}$ be c channel probability functions (c.p.f.'s). This means that, for $j = 1, \cdots, d$, $w(\cdot \mid j \mid i)$ is a nonnegative function with domain B such that $\sum_{k=1}^{b} w(k \mid j \mid i) = 1$.

Let n be an integer. Call any sequence of n elements, each a member of the set D (respectively, the set B) a transmitted (resp. received) n-sequence. Call any sequence of n elements, each one of $\{1, \cdots, c\}$, a channel n-sequence. Let

$$u_0 = (d_1, \cdots, d_n), \qquad v_0 = (b_1, \cdots, b_n), \tag{1.1}$$

and

$$\gamma_0 = (c_1, \cdots, c_n) \tag{1.2}$$

be, respectively, a transmitted n-sequence, a received n-sequence, and a channel n-sequence. Suppose u_0 is "sent (or transmitted) over the channel when the transmission is governed by γ_0." The chance received n-sequence

$$v(u_0) = (Y_1(u_0), \cdots, Y_n(u_0)) \tag{1.3}$$

is a sequence of independent chance variables such that

$$P\{v(u_0) = v_0 \mid \gamma_0\} = \prod_{s=1}^{n} w(b_s \mid d_s \mid c_s), \tag{1.4}$$

* The research of this author under contract with the Office of Naval Research.

† The research of this author was supported by the U. S. Air Force under contract No. AF 18 (600)-685, monitored by the Office of Scientific Research.

where the symbol on the left is the probability that $v(u_0) = v_0$ when the transmission is governed by γ_0. Thus the significance of $w(\cdot \mid \cdot \mid i)$, $i = 1, \cdots, c$, can be looked upon as follows: When the "letter" j is sent, and $w(\cdot \mid \cdot \mid i)$ governs its transmission, the probability that the letter k will be received is $w(k \mid j \mid i)$. The n letters of a received word are independently distributed. (For more detail about applications see, for example, Wolfowitz (1961).

A code (n, N, λ) for the present problem (when the c.p.f. varies arbitrarily) is a system

$$\{(u_1, A_1), \cdots, (u_N, A_N)\} \tag{1.5}$$

where $u_1, \cdots, u_N$ are transmitted n-sequences, $A_1, \cdots, A_N$ are disjoint sets of received n-sequences, and for *every* channel n-sequence γ_0 we have

$$P\{v(u_i) \in A_i \mid \gamma_0\} \geqq 1 - \lambda, i = 1, \cdots, N. \tag{1.6}$$

A number C is called the capacity (of the channel) if, for any $\epsilon > 0$ and λ, $0 < \lambda < 1$, there exists a code $(n, 2^{n(C-\epsilon)}, \lambda)$ for all sufficiently large n, and for all sufficiently large n there does not exist a code $(n, 2^{n(C+\epsilon)}, \lambda)$. (See also Wolfowitz (1961), Section 5.6.) A number $R \geqq 0$ is called a (possible) rate of transmission if, for any $\epsilon > 0$ and λ, $0 < \lambda < 1$, there exists, for all sufficiently large n, a code $(n, 2^{n(R-\epsilon)}, \lambda)$.

The code described in (1.5) can be more fully described as a code where neither the sender nor the receiver knows the channel sequence which governs the transmission of a word. Codes are described below which apply to the cases where the sender or the receiver or both know the channel sequence which governs the transmission of a word. For all four of these situations we give in the present paper necessary and sufficient conditions for the existence of a positive (possible) rate of transmission. In the case where both sender and receiver know the c.p.f. we actually determine the capacity.

The reader will find no difficulty in verifying that the existence of a positive rate for certain infinite collections of c.p.f.'s (of the nature of S of Section II) can be obtained from the methods and results of this paper.

The reader may also find it interesting to compare the results of Sections III and IV with those in the case of a compound channel (where the unknown c.p.f. is the same for each letter; see, for example, Wolfowitz (1961), Chapter 4).

The codes of the present paper, including the one described above, are all "nonrandomized." For the case where neither the sender nor the receiver knows the channel sequence which governs the transmission of a word, Blackwell, Breiman, and Thomasian (1960) studied the relation between the "nonrandomized" and "randomized" capacities. Their methods do not apply to the problems of the present paper, as they indicate in their discussion on page 566. This discussion gives an example where no positive rate of transmission exists and is included in the necessity condition of Theorem 1 below.

II. CASE WHERE NEITHER THE SENDER NOR THE RECEIVER KNOWS THE CHANNEL SEQUENCE

For each j in D consider the smallest convex body $T(j)$ which contains the c points of b-space

$$(w(1 \mid j \mid i), w(2 \mid j \mid i), \cdots, w(b \mid j \mid i)), i = 1, \cdots, c. \quad (2.1)$$

We shall now prove

THEOREM 1. *Necessary and sufficient for the existence of a rate of transmission greater than zero when neither sender nor receiver knows the channel sequence is that, among the convex bodies $T(1), \cdots, T(d)$, at least two be disjoint.*

PROOF OF NECESSITY. Assume that no two of $T(1), \cdots, T(d)$ are disjoint. Fix n and $\lambda < \frac{1}{2}$. We shall show that any code can contain only one member (i.e., $N = 1$). If δ is a distribution on channel n-sequences we define, for the sake of brevity,

$$P\{v(u_0) \in A \mid \delta\} = \sum_{\gamma_0} P\{v(u_0) \in A \mid \gamma_0\} \, \delta(\gamma_0).$$

The idea of the proof will be this: For any fixed n we will construct a distribution δ such that

$$P\{v(u_1) \in A_2 \mid \delta\} \geqq 1 - \lambda. \quad (2.2)$$

Of course, from (1.6),

$$P\{v(u_1) \in A_1 \mid \delta\} \geqq 1 - \lambda. \quad (2.3)$$

Since A_1 and A_2 are disjoint and $\lambda < \frac{1}{2}$, (2.2) and (2.3) yield a contradiction.

Suppose

$$u_1 = (x_1, x_2, \cdots, x_n)$$
$$u_2 = (y_1, y_2, \cdots, y_n).$$

Let g_i, $i = 1, \cdots, n$, be a point common to $T(x_i)$ and $T(y_i)$. Let $s^{(i)} = (s_1^{(i)}, s_2^{(i)}, \cdots, s_c^{(i)})$ and $t^{(i)} = (t_1^{(i)}, t_2^{(i)}, \cdots, t_c^{(i)})$ be the "barycentric" coordinates of g_i in the sets $T(x_i)$ and $T(y_i)$, respectively. (Of course, these sets have at most $\min(b, c)$ extreme points, and properly a point in one of them has as many barycentric coordinates as the number of extreme points. When, as above, we write the barycentric coordinates as being c in number, it is understood that a coordinate which corresponds to an inner (nonextreme) point is zero). Suppose that, when x_i (resp. y_i) is sent, the c.p.f. which governs its transmission were chosen at random, with probability $s_j^{(i)}$ (resp. $t_j^{(i)}$) that $w(\cdot \mid \cdot \mid j)$ would be chosen. It would follow then that, when either x_i or y_i is sent (under the above conditions), the probability that k ($k = 1, \cdots, b$) would be received is the same and equal to the kth Cartesian coordinate of g_i.

Now let δ (resp. δ^*) be the distribution on the channel n-sequences implied by the following: The elements of the channel n-sequence are independent chance variables, the distribution of the ith chance variable, $i = 1, \cdots, n$, being $s^{(i)}$ (resp. $t^{(i)}$). It follows that

$$P\{v(u_1) \in A_2 \mid \delta\} = P\{v(u_2) \in A_2 \mid \delta^*\}. \tag{2.4}$$

The right member of (2.4) is $\geq 1 - \lambda$, by (1.6); this proves (2.2) and hence the necessity condition.

Proof of Sufficiency. We may suppose, without loss of generality, that $T(1)$ and $T(2)$ are disjoint. Then there is a plane in b-space which separates $T(1)$ and $T(2)$ and is disjoint from $T(1)$ and $T(2)$. Let $(l_1, \cdots, l_b, m)$ be its coordinates. Suppose an h-sequence (say z_1) consisting exclusively of ones or an h-sequence (say z_2) consisting exclusively of twos is sent over the channel, and let $N_i(z_j)$, $i = 1, \cdots, b$, $j = 1, 2$, be the number of elements i in the chance sequence $v(z_j)$. Let η, $0 < \eta < \frac{1}{16}$, be chosen arbitrarily. Now, reversing if necessary the indices 1 and 2, we may conclude from the law of large numbers that, when h is sufficiently large, the probability exceeds $1 - \eta$ that

$$\sum_{i=1}^{b} l_i N_i(z_1) < hm \text{ and } \sum_{i=1}^{b} l_i N_i(z_2) \geq hm,$$

no matter what channel h-sequence governs the transmission of z_1 or z_2. From the above it follows that, if we construct the u's of the code (1.5) of consecutive blocks of h ones or h twos, each block can be "decoded correctly" with probability at least $1 - \eta$. The code whose existence we shall now demonstrate will have its sequences u so constructed. Since the result to be proved is one for large n there is no loss of generality in assuming that n is an integral multiple of h.

We now digress for a moment to describe a "t-error correcting" code. Take $d = b$. Then the code (1.5) is called t-error correcting if, for $i = 1, \cdots, N$, A_i consists of all n-sequences which differ from u_i in at most t places. (The condition (1.6) is no longer required. Since the A_i are disjoint it follows that any two u's of the code must differ in at least $(2t + 1)$ places.)

Now let $d = b = 2$. Suppose that S now contains a continuum of c.p.f.'s, each indexed by θ, where θ takes all values in the interval $[0, \eta]$. The c.p.f. $w(\cdot \mid \cdot \mid \theta)$ is defined as follows: $w(1 \mid 1 \mid \theta) = w(2 \mid 2 \mid \theta) = 1 - \theta$, $w(1 \mid 2 \mid \theta) = w(2 \mid 1 \mid \theta) = \theta$. It follows from the law of large numbers that, whatever be λ(fixed), $0 < \lambda < 1$, for n sufficiently large a $2\eta n$-error correcting code of length N is a code (n, N, λ) for the channel just described

$$\text{(i.e., } d = b = 2,\ S = \{w(\cdot \mid \cdot \mid \theta) \mid \theta \in [0, \eta]\}).$$

Since $\eta < \frac{1}{16}$ it follows from a result of Gilbert (1952, Theorem 1) that there is a positive r such that, for all n sufficiently large, there exists a $2\eta n$-error correcting code of length 2^{nr} for the channel described above.

To construct a code for the channel of our original problem when n is sufficiently large, we proceed as follows: A block of h ones (resp. twos) of the transmitted alphabet of the original problem corresponds to the symbol 1 (resp. 2) of the transmitted alphabet of the new problem for which an error correcting code will be used

$$(d = b = 2,\ S = \{w(\cdot \mid \cdot \mid \theta) \mid \theta \in [0, \eta]\}).$$

All blocks of h letters of the received alphabet of the original problem, which satisfy $l_i N_i < hm$ (resp. $l_i N_i \geqq hm$), are to correspond to the symbol 1 (resp. 2) of the received alphabet of the new problem; here N_i, $i = 1, \cdots, b$, is the number of elements i in the block of h letters. It follows that, whatever be λ, $0 < \lambda \leqq 1$, when n is sufficiently large there exists a code (n, N, λ) for our original problem with $\log N$ greater than the largest integer in nr/h. This completes the proof of sufficiency.

III. CASE WHERE THE CHANNEL SEQUENCE IS KNOWN TO THE RECEIVER BUT NOT TO THE SENDER

Let Γ_n be the totality of all channel n-sequences. For the case described in the title of this section a code (n, N, λ) is a system

$$(u_1, \{A_1(\gamma_0), \gamma_0 \in \Gamma_n\}), \cdots, (u_N, \{A_N(\gamma_0), \gamma_0 \in \Gamma_n\}) \qquad (3.1)$$

where $u_1, \cdots, u_N$ are transmitted n-sequences, for each $\gamma_0 \in \Gamma_n$ the sets $A_1(\gamma_0), \cdots, A_N(\gamma_0)$ are disjoint sets of received n-sequences, and

$$P\{v(u_i) \in A_i(\gamma_0) \mid \gamma_0\} \geqq 1 - \lambda, \gamma_0 \in \Gamma_n, \quad i = 1, \cdots, N. \qquad (3.2)$$

(For the application of such a code see, for example, Wolfowitz (1961, Chapters 3 and 4).) This section is devoted to a proof of the following:

THEOREM 2. *Necessary and sufficient for the existence of a positive rate of transmission when the receiver but not the sender knows the channel sequence is that, for some pair d_1, d_2 of elements of D,*

$$\sum_{i=1}^{b} \mid w(i \mid d_1 \mid j) - w(i \mid d_2 \mid j) \mid > 0, \quad j = 1, \cdots, c. \qquad (3.3)$$

PROOF OF SUFFICIENCY. The proof will be similar to the proof of sufficiency in Theorem 1. Suppose (3.3) holds. For typographical simplicity assume $d_1 = 1, d_2 = 2$. Let $\eta, 0 < \eta < \frac{1}{16}$, be chosen arbitrarily. From (3.3) and the law of large numbers it is not difficult to obtain the conclusion that there exists a positive integer h with the following property: Let $h^1 \geqq h$ be any integer. Suppose that a block of h^1 ones or a block of h^1 twos is sent over the channel with the transmission of every letter governed by the same c.p.f. $w(\cdot \mid \cdot \mid i)$, $i = 1, \cdots, c$. Then, no matter which is the c.p.f., known to the receiver, the latter can "correctly decode" which block (of ones or of twos) has been sent with probability at least $1 - \eta$. To put it more precisely: Let z_1 (resp. z_2) be the transmitted h^1-sequence which consists exclusively of ones (resp., of twos). Let $\gamma^{(i)}$, $i = 1, \cdots, c$, be the channel h^1-sequence which consists exclusively of elements i. There is a partition of the space of all received h^1-sequences into two disjoint sets $B_1^{(i)}$ and $B_2^{(i)}$ such that

$$P\{v(z_j) \in B_j^{(i)} \mid \gamma^{(i)}\} \geqq 1 - \eta, \ j = 1, 2; \ i = 1, \cdots, c.$$

We now proceed as in the last paragraph of the proof of sufficiency in Theorem 1. Instead of using blocks of h ones and h twos we use blocks of ch ones and ch twos. In any channel ch-sequence at least h elements must be the same, and the receiver knows which they are. He can therefore decode correctly the block of ch elements of the transmitted alphabet with probability at least $1 - \eta$. The remainder of the proof of sufficiency is as in Theorem 1.

PROOF OF NECESSITY. Assume that (3.3) does not hold. Then, for any pair a_1, a_2 of elements of D there exists an element of $1, \cdots, c$, say $c^*(a_1, a_2)$, such that

$$\sum_{i=1}^{b} | w(i \mid a_1 \mid c^*) - w(i \mid a_2 \mid c^*) | = 0.$$

Fix n and $\lambda < \frac{1}{2}$. We show that any code (3.1) can contain only one member (i.e., $N = 1$). Suppose

$$u_1 = (x_1, \cdots, x_n), \qquad u_2 = (y_1, \cdots, y_n).$$

Let γ^* be the channel n-sequence whose ith element, $i = 1, \cdots, n$ is $c^*(x_i, y_i)$. Then obviously

$$P\{v(u_1) \in A_2(\gamma^*) \mid \gamma^*\} = P\{v(u_2) \in A_2(\gamma^*) \mid \gamma^*\} \geqq 1 - \lambda.$$

Also, by (3.2),

$$P\{v(u_1) \in A_1(\gamma^*) \mid \gamma^*\} \geqq 1 - \lambda.$$

The last two statements are obviously in contradiction. Necessity is proved.

IV. CASE WHERE THE CHANNEL SEQUENCE IS KNOWN TO THE SENDER BUT NOT TO THE RECEIVER

We begin by describing a code (n, N, λ) for the case described in the title; our description will be, for the sake of brevity, a little informal but completely intelligible. The sets $A_1, \cdots, A_N$ are as in (1.5). When the sender wishes to send the ith word he no longer sends u_i, $i = 1, \cdots, N$. Instead he has a rule f_i which operates as follows: Let

$$\gamma_0 = (c_1, c_2, \cdots, c_n)$$

be the channel n-sequence which will govern the transmission of the word. This sequence γ_0 is known to the sender in the following way. When the sender is sending the jth letter, $j = 1, \cdots, n$, he knows

$(c_1, \cdots, c_j)$. The rule $f_i = (f_i^{(1)}, \cdots, f_i^{(n)})$ for sending the ith word tells the sender successively what each letter is to be. The jth letter, $j = 1, \cdots, n$, is given by the rule to be $f_i^{(j)}(c_1, \cdots, c_j)$ and is a function of the arguments exhibited. The place of $u_1, \cdots, u_N$ in the code (1.5) is now taken by $f_1, \cdots, f_N$. Of course, the analogue of (1.6) must hold.

For $k = 1, \cdots, c$, let $D(k)$ be the set of d points in b-space

$$\{D(i \mid k) = (w(1 \mid i \mid k), w(2 \mid i \mid k), \cdots, w(b \mid i \mid k)), \quad i = 1, \cdots, d\}.$$

Now consider the totality of d^c sets $B^1(1), \cdots, B^1(d^c)$, each of which contains c points, one from each of $D(1), \cdots, D(c)$. Let $B(i)$ be the smallest convex body which contains the points of $B^1(i)$.

We now prove

THEOREM 3. *Necessary and sufficient for the existence of a positive rate of transmission when the sender but not the receiver knows the channel sequence is that at least two of* $B(1), \cdots, B(d^c)$ *be disjoint.*

PROOF OF SUFFICIENCY. Suppose $B(1)$ and $B(2)$, say, are disjoint, and $B(1)$ (resp. $B(2)$) contains $D(a_1(k) \mid k)$, $k = 1, \cdots, c$ (resp. $D(a_2(k) \mid k)$, $k = 1, \cdots, c$). The proof of sufficiency of Theorem 1 now applies with one difference. Instead of the sender sending long blocks of ones and long blocks of twos, he proceeds as follows: When he would wish to send a one (resp. a two) as part of a long block he sends the letter $a_1(k)$ (resp. $a_2(k)$) when he knows that $w(\cdot \mid \cdot \mid k)$ will be the c.p.f. according to which the received letter will be distributed. The complete proof is easy to supply after the model of the proof of Theorem 1.

PROOF OF NECESSITY. Suppose no two of $B(1), \cdots, B(d^c)$ are disjoint. Fix n and $\lambda < \frac{1}{2}$. Then the randomization argument of the proof of necessity of Theorem 1 can easily be applied to f_1 and f_2 to obtain the same contradiction as before.

V. CASE WHERE BOTH SENDER AND RECEIVER KNOW THE CHANNEL SEQUENCE

We now consider the case described in the title of this section. The receiver's knowing the channel n-sequence γ_0 (say) means that the sets A in the code corresponding to (1.5) are functions of γ_0, thus:

$$A_1(\gamma_0), \cdots, A_N(\gamma_0). \tag{5.1}$$

These sets are of course disjoint (for the same γ_0). There is such a system for every channel n-sequence. The words transmitted are defined by rules $f_1, \cdots, f_N$ as in Section IV.

Let $C(i)$, $i = 1, \cdots, c$, be the capacity of the c.p.f. $w(\cdot \mid \cdot \mid i)$ (i.e., of the discrete memoryless channel with (single) c.p.f. $w(\cdot \mid \cdot \mid i)$; see, e.g., Wolfowitz (1961, Chapter 3)).We will now show that the *capacity* of the channel of this section is the smallest of $C(1), \cdots, C(c)$, say C^*.

Clearly the capacity could not be greater than C^*. For, if $C^* = C(1)$, say, it would be enough to consider the channel sequence which consists entirely of ones to see that, for all sufficiently large n, there does not exist a code of length $2^{n(C^*+\epsilon)}$. It is therefore sufficient to prove that C^* is a possible rate of transmission for the present channel.

(As the channel has been defined above, the sender does not know the entire channel sequence in advance of sending a word (transmitted n-sequence); he knows only the c.p.f. for the letter he is sending and for the letters already sent. Whether the receiver knows the entire channel sequence in advance or not does not matter, since he does not "decode" the word (n-sequence) received until he has received the whole word. Suppose however, that the sender does know the entire channel sequence in advance of transmission. The argument of the preceding paragraph shows that the capacity of the channel (assuming that there is a capacity) could not be increased by this knowledge of the sender. On the other hand, this knowledge could not, obviously, decrease the capacity. It follows that, in determining the capacity of the channel of the present section we also determine the capacity of the channel modified so that the sender knows the entire channel sequence in advance of transmission of a word.)

This section is devoted to a proof of:

THEOREM 4. *The capacity of the channel of the present section, where the channel sequence is known to both sender and receiver, is C^*, the smallest of the capacities of the individual c.p.f.'s.*

As the earlier argument has shown, it is sufficient to show that C^* is a possible rate of transmission, which we now proceed to do. We shall write $\sqrt{n}$ as if it were always an integer, and leave to the reader the easy task of approximating it by an integer when that is necessary. Consider first another channel, say V, which is the same as the channel of the present section except that there are only c possible channel sequences (for each n) each one consisting of the same element repeated n times. Let $\epsilon > 0$ be fixed arbitrarily. Then (e.g., Wolfowitz (1961, Section 7.5)) there is a positive number α such that, when n is sufficiently large, there exists a code $((\sqrt{n}, 2^{\sqrt{n}(C^*-\epsilon)}, e^{-\alpha\sqrt{n}})$ for channel V. Let $z_1, \cdots, z_t$ be the elements u of this code $(t = 2^{\sqrt{n}(C^*-\epsilon)})$. We now construct a code $(n + s, t^{\sqrt{n}}, \sqrt{n}\, e^{-\alpha\sqrt{n}})$ for the channel of the present section, with $s = (\sqrt{n} - 1)c(c - 1) + (c - 1)$. Since $\sqrt{n}\, e^{-\alpha\sqrt{n}} \to 0$ as $n \to \infty$ it is easy to see that this proves the theorem.

Suppose the channel n-sequences of our problem were always of the following type: the first $\sqrt{n}$ elements are all the same, the second $\sqrt{n}$ elements are all the same, etc. Then we could construct the desired code for our problem as follows: Each element u is a succession of elements from $z_1, \cdots, z_t$, to a total of $\sqrt{n}$ elements in all. Each z sent can be correctly decoded with a probability at least $1 - e^{-\alpha\sqrt{n}}$, by the property of the code V. Hence the probability of error in the code just constructed is at most $\sqrt{n}\, e^{-\alpha\sqrt{n}}$. However, the channel n-sequences are not all of the above type.

We therefore proceed as follows: Suppose that in the case of the preceding paragraph we would have sent the sequence $z^{(1)}, z^{(2)}, \cdots, z^{(\sqrt{n})}$. Let $\gamma_0 = (c_1, c_2, \cdots, c_n)$ be the channel n-sequence which will govern the transmission of the word in our problem. The sender begins by sending the first element of $z^{(1)}$. Then, if $c_2 = c_1$ he sends the second element of $z^{(1)}$, and if $c_2 \neq c_1$ he sends the first element of $z^{(2)}$. The procedure at the third step may best be described by Table I. The procedure at the fourth and subsequent steps is now clear. As soon as a z has been entirely sent its place is taken by the next z whose transmission has not yet begun. The number of elements z which will be sent in this manner depends upon the sequence γ_0, because there will be "waste" at the end. However, it is clear that at least $(\sqrt{n} - c + 1)$ elements z will always be sent. Since the receiver knows the sequence γ_0 he can make each of the received symbols correspond to its own z.

Suppose $(c - 1)z$'s have not been sent. These could be in various positions (have various serial numbers), and both sender and receiver know their serial numbers after n symbols have been sent and received. Let the serial numbers be $\alpha_1 < \alpha_2 < \cdots < \alpha_{c-1}$. Beginning with the

TABLE I

$c_1 = c_2 = c_3$	Send third element of $z^{(1)}$
c_1, c_2, c_3 all different	Send first element of $z^{(3)}$
$c_1 \neq c_2 = c_3$	Send second element of $z^{(2)}$
$c_2 \neq c_1 = c_3$	Send second element of $z^{(1)}$
$c_1 = c_2 \neq c_3$	Send first element of $z^{(2)}$

$(n + 1)$st letter the sender sends $z^{(\alpha_1)}$ so that all its letters will be transmitted (not necessarily consecutively) under the same c.p.f.; at most $(\sqrt{n} - 1)\, c + 1$ letters will suffice for this. Then he sends $z^{(\alpha_2)}$ so that all its letters will be transmitted (not necessarily consecutively) under the same c.p.f., etc. At most $s = (\sqrt{n} - 1)\, c\, (c - 1) + (c - 1)$ letters will suffice to send all the missing z's. The sender sends exactly s letters; if fewer letters are needed the remainder can be any prearranged letters which are "ignored" by the receiver.

If fewer than $(c - 1)$ z's originally remained unsent the sender sends these as in the preceding paragraph, and then sends enough prearranged letters (which the receiver will "ignore") to make up a total of s letters. The receiver, who knows the channel sequence, knows the order in which all the z's have been sent, and knows which $\sqrt{n}$ received letters are to be used to decode any one z. Each z thus sent can be correctly decoded with a probability at least $1 - e^{-\alpha\sqrt{n}}$, by the property of the code V. Hence the probability of error in the code for our problem is at most $\sqrt{n}\ e^{-\alpha\sqrt{n}}$. This completes the proof of Theorem 4.

RECEIVED: October 9, 1961.

REFERENCES

BLACKWELL, D., BREIMAN, L., AND THOMASIAN, A. J., (1960), The capacities of certain channel classes under random coding. *Ann. Math. Stat.* **31**, No. 3, 558–567.

GILBERT, E. N., (1952), A comparison of signaling alphabets. *Bell System Tech. J.* **31**, 504–522.

WOLFOWITZ, J., (1961), "Coding Theorems of Information Theory." Springer, Berlin, Göttingen; Prentice-Hall, Englewood Cliffs, N. J.

Econometrica, Vol. 30, No. 3 (July, 1962)

BAYESIAN INFERENCE AND AXIOMS OF CONSISTENT DECISION[1]

BY J. WOLFOWITZ

One school of statistical thought holds that statistical decisions, when "rationally" made, are (and must be) made as if there were an a priori distribution on the states of Nature. Here "rational" means according to some set of axioms of "rational" choice of a decision function. In the present paper one aspect of one such axiom system is examined. The system is one of the simplest which has been proposed and can be regarded as a prototype of the others. It is argued that one of its axioms does not, upon closer scrutiny, appear very plausible and reasonable. It is demonstrated that this axiom requires the experimenter to have a preference among the states of Nature *in advance* of the experiment. Some related questions are discussed.

1. INTRODUCTION

THE DEBATE between the protagonists of the subjective (personal) philosophical basis of probability theory and of the frequentist (objective) basis is already of long duration. So is the debate between those who insist that all statistical inference must be Bayesian and those who belong to the objective school. The two debates overlap but are not the same. While those who hold to the personal basis of probability theory will also be Bayesians, one of the founders and staunchest proponents of the frequentist theory, R. v. Mises, was also a staunch Bayesian. (For example, he says in [**10**], p. 186, "Es gibt keine andere Lösung als die Bayessche Theorie.") The members of the Bayesian school divide, sometimes very sharply, on how the a priori distribution is to be determined. For example, v. Mises thought it should be on the basis of prior experience, while Jeffreys uses, inter alia, certain canonical distributions.

The cause of Bayesian inference and personal probability has recently been taken up with great energy in a number of papers (e.g., [**4**], [**5**], [**6**]) by L. J. Savage from a different starting point. Typical of his position is his treatment of the finite decision problem. (The problem is described in Section 2 below and, for example, in [**8**].) Professor Savage holds that, in this case, a certain set of axioms, due to D. V. Lindley and himself (see [**6**, Chapter 3, Section 4]),* is such that "rational" statisticians should (and

[1] Work under contract with the Office of Scientific Research of the U.S. Air Force.

* I am obliged to Professor Savage for the information that this argument is also referred to in the following papers of his: "The foundations of statistics reconsidered," *Proceedings of the Fourth Berkeley Symposium on Mathematical Statistics and Probability*, Berkeley and Los Angeles, 1961, University of California Press, and "La probabilita soggetiva nei problemi practici della statistica," to appear in the book *Induzione e Statistica*, and mimeographed in 1959 by the Istituto Matematico of the University of Rome.

perhaps do) act according to these axioms. A logical consequence of these axioms is that the statistician chooses his decision function as if there were an a priori distribution on the states of Nature. Professor Savage has also vigorously espoused the cause of personal probability and is a severe critic of modern statistical theory, most of which he condemns.[2]

The purpose of the present note is to make some remarks about this axiom system and its consequences.[3] For the sake of brevity I shall confine my remarks, as far as possible, to those arguments which, to the best of my knowledge, have not been made elsewhere. For example, arguments for the frequentist basis of probability theory and against the subjective basis have been given in many places (e.g., [**10**]) and none of them will be repeated here.

The points which I should like to make in this note and for which I will argue in subsequent sections are these:

(1) The axioms of behavior which Professor Savage espouses do not appear nearly so plausible and incontrovertible when one examines carefully one important assumption and its consequences.

(2) Even full acceptance of these axioms and their consequences is

[2] As an example consider the peroration of [**5**]: "I hope I have made you feel that statistics is going places today. There is a lot of exciting work to do that is relatively easy. Any of you who are inclined to work in statistical theory can get in on the ground floor now, for there are not any experts any more. Those who only use statistics without intending to do research in it must also remember that there are not any experts any more. Don't take any wooden nickels; do your own work honestly and thoughtfully without much reliance on rules; and don't believe in magic or powerful new tools." [The American colloquialism "don't take any wooden nickels" means "don't be gullible"; "to get in on the ground floor" means "to participate in the founding (of an enterprise)" J. W.]

Also, *ibid.*, Section 4: "Though it is not necessary to their philosophy, they do have a way of making it clear that they are the statisticians and you are the fellow who pays; so if you make a mistake, it is not exactly the statistician's mistake." It is difficult to understand why the statistician should assume more responsibility for his expert advice than, for example, a physician.

[3] I have selected Professor Savage's views for discussion because I consider him to be one of the leaders of his school of thought. A copy of the manuscript of the present paper was, upon its completion, sent to Professor Savage, who took the time and trouble to subject it to long and detailed criticism which is hereby gratefully acknowledged. Footnotes 2 and 6 have been added because of his comments. Footnotes 8, 9, and, of course, the present footnote, were also not in my original manuscript.

The body of the present manuscript differs from that of the original manuscript only in the following: (a) a quotation from another writer was deleted; (b) a reference to the particular axioms discussed was added in the second paragraph of Section 1; and (c) in Section 3, third paragraph, the phrase "the argument" was replaced by "one argument." (Professor Savage had informed me that he had given more than one set of axioms and more than one argument.)

compatible with holding the frequentist (objective) view of the foundations of probability.

It follows from the second of these points that a large part of Professor Savage's final conclusions rests not on the plausibility of his axioms but on the validity of the subjective basis of probability theory. In particular, much of his criticism of modern statistical theory flows from his acceptance of the subjective point of view and not from his axioms of behavior. His criticism is thus really a continuation of the long standing debate between subjectivists and frequentists. His claim that the subjective theory can solve many statistical problems not solvable in the objective theory has only as much validity as the basis of the subjective theory.

2. THE DECISION PROBLEM WHEN THE NUMBER OF DECISIONS AND OF STATES OF NATURE IS FINITE

Let us consider a simple problem in order not to obscure the discussion with irrelevant complications. The problem of the design of the experiment will not and should not arise; it would unnecessarily complicate the discussion and the optimum design of an experiment will in general depend on its analysis, which is precisely the point at issue. The problem will be posed in mathematical terms in order to avoid possible irrelevant debate about its practical aspects. I expect that it will be generally regarded as the idealization of a meaningful statistical problem.

We will be given an observation x on (a value of) a chance variable X, whose unknown distribution function G is a member of the set $\Omega = \{F_1, \ldots, F_m\}$, which is known to us, the statistician(s). (The chance variable X may be multidimensional, so that an observation on X may represent many observations.) Knowing x we have to make one of a set of decisions $D = \{d_1, \ldots, d_l\}$. We are given a function W such that $W(i, j)$ is the loss to the statistician when G is actually F_i and decision d_j is made.[4] The determination of W may be a difficult practical problem whose discussion would be irrelevant here. W is called the weight or loss function, and $-W$ is called the utility function.

A decision function δ is defined for all possible values which X can take; if y is any such value then

$$\delta(y) = (\delta_1(y), \ldots, \delta_l(y))$$

and

$$\delta_1(y) + \ldots + \delta_l(y) = 1 .$$

The nonnegative functions $\delta_1, \ldots, \delta_l$ should be measurable, which is a techni-

[4] For a more complicated weight function see [8].

cal mathematical requirement which need not concern us further. The significance of δ is this: If $x = y$ the statistician will decide among the decisions of D by means of an additional chance experiment such that the probability of making decision d_j is $\delta_j(y)$. The problem of the statistician is to choose a decision function.

The reader may, at this point, object to making the decision by means of an additional chance experiment. Such an objection may have considerable validity but is irrelevant to our main point and will therefore be ignored. If this objection is considered valid then we must limit ourselves to such δ as have the property that, for every y, one of $\delta_1(y), \ldots, \delta_l(y)$ is one and all others are zero. (If, for example, $F_1, \ldots, F_m$ have density functions with respect to Lebesgue measure this is no limitation [1]).

We come now to the idea of risk (risk point) and to the totality of risk points, which will play an important role in our discussion. The risk $r(\delta)$ of a decision function δ is defined[5] by

$$r(\delta) = (r_1(\delta), \ldots, r_m(\delta))$$

where

$$r_i(\delta) = \sum_{j=1}^{l} W(i,j)\, \pi(i,j \mid \delta)$$

and $\pi(i, j \mid \delta)$ is the probability that, if decision function δ is employed when $G = F_i$, decision d_j will be made. The quantity $r_i(\delta)$ is called the risk of δ when F_i is true (i.e., when $G = F_i$). Let S be the totality of points $r(\delta)$ as δ ranges over all possible decision functions. It is trivial that S is convex (i.e., if two points belong to S so does the line joining them) and much less trivial to prove that S is closed ([7], [2]).

Suppose that $\delta^{(1)}$ and $\delta^{(2)}$ are two decision functions such that $r(\delta^{(1)}) = r(\delta^{(2)})$; we shall regard such decision functions as equivalent and equally desirable. Suppose that $\delta^{(1)}$ and $\delta^{(2)}$ are not equivalent and are such that

$$r_i(\delta^{(1)}) \leqslant r_i(\delta^{(2)}), \qquad i = 1, \ldots, m,$$

so that the inequality sign holds for at least one value of i; we shall then regard $\delta^{(1)}$ as "better" than $\delta^{(2)}$ and consider that $\delta^{(1)}$ is to be preferred to $\delta^{(2)}$. Unfortunately, it is seldom true, except in trivial instances, that there is a decision function which is best in the sense that it is better than all other decision functions in the sense just described, or even that there is a decision function which is either better than, or equivalent to, any other decision function. It is this fact which is responsible for the present discussion. In

[5] This is the definition of risk generally accepted for various reasons. Our discussion would be unaffected if the definition were different.

order to designate some decision function as "best" an additional "principle" or basis for comparison has to be introduced.

Since two decision functions with the same risk are looked upon as equivalent, the choice of a decision function is equivalent to the choice of a risk point, and vice versa. Our problem is then the selection of a risk point from among the points of S.

3. SPECIALIZATION OF THE PREVIOUS PROBLEM

For simplicity we now take $m = l = 2$, $W(1, 1) = W(2, 2) = 0$, $W(1, 2) = W(2, 1) = 1$, and suppose that F_i has the density function f_i (with respect to Lebesgue measure, of course). The set S will look approximately like this:

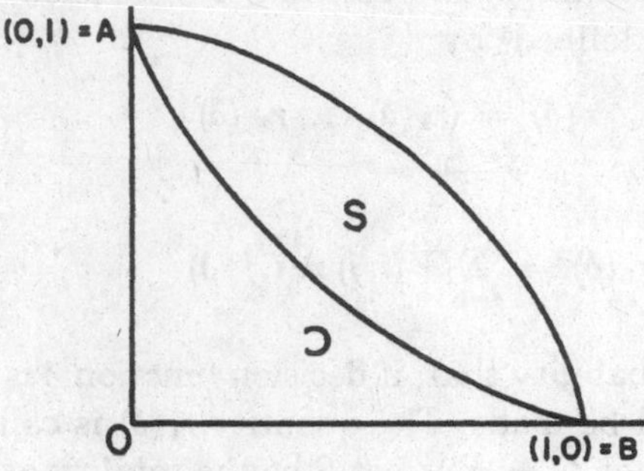

FIGURE 1

Suppose the point (a, b) lies in S. This means that there is a decision function whose risk is a when F_1 is true (when $G = F_1$) and whose risk is b when F_2 is true. In the present problem, because of the particular choice of W, a is the probability of making decision 2 when $G = F_1$ and b is the probability of making decision 1 when $G = F_2$. (We may consider that both of these situations correspond to error. If we borrow the terminology of testing hypotheses and call the statement that $G = F_1$ the "null hypothesis," then a and b are the probabilities of type 1 and type 2 error, respectively. Decision d_i is then the decision that $G = F_i$). It is trivially obvious that points not on the arc ACB correspond to decision functions which the statistician would not want to adopt.

(Parenthetically it may be remarked that, when the number of observations is large, the arc ACB will bulge into the angle AOB, and it is easy to choose a decision function which has uniformly small error).

I shall now describe briefly one argument by which Professor Savage obtains the decision function (point on the arc ACB) which is to be used, or, more properly, an argument which he prescribes for the statistician to follow in order to obtain the decision function which he (the statistician) should use. The statistician must subject himself to "self-interrogation" (Savage's phrase) in the following sense: He must be prepared to choose between any two points on the arc ACB (or perhaps any two points in S), i.e., given the two points, he must say which of the two points (decision functions) he prefers. (He may also prefer neither, i.e., be indifferent as to choice between them.) It is assumed that choices for different pairs of points will be "consistent," or, at least, should be "consistent," i.e., satisfy certain axioms (rules of choice) which are considered very reasonable and such that all or most statisticians will agree on the desirability of conforming to these rules. (A discussion of these axioms is here unnecessary, for reasons which will appear shortly.) It is then a consequence of these axioms that (except in certain uninteresting cases) the indifference contours (among the points of S) are parallel straight lines (with obviously negative slope). The slope of these lines is determined by the (consistent) answers to a number of requests to choose between two points of S. Suppose that these lines are parallel to $\gamma_1 a + \gamma_2 b = 1$, where we may take γ_1 and γ_2 to be positive. Then obviously the statistician will wish to choose a point (h_1, h_2) of S for which $\gamma_1 h_1 + \gamma_2 h_2$ is a minimum, and he will use a decision function which corresponds to this point. It is very easy to verify that such a decision function is given by the following procedure: Assume that the a priori probability that $G = F_1$ (resp. $G = F_2$) is $\gamma_1/(\gamma_1 + \gamma_2) = t_1$ (resp. $\gamma_2/(\gamma_1 + \gamma_2) = t_2$). After the observation x has been obtained the a posteriori probabilities are, respectively, $t_1 f_1(x)/(t_1 f_1(x) + t_2 f_2(x))$ and $t_2 f_2(x)/(t_1 f_1(x) + t_2 f_2(x))$. One then makes that decision whose a posteriori probability is the greater.

The conclusion of the above is that one should determine one's a priori probabilities by the "self-interrogation" process and then make one's decision on the basis of the a posteriori probabilities, since a rational or consistent procedure amounts to this anyhow. Although Professor Savage has more to say (which the reader will want to read) I believe that the above is a fair description of the nub of his argument.

The crux of the matter is that the indifference contours are parallel straight lines. From this fact alone the reader can easily deduce what choices would be regarded as consistent according to the axioms. It is also clear that the process of self-interrogation must be carried so far as to determine the slope of the indifference contours. The entire procedure thus rests upon the following: (1) the assumption that the points of S can be simply[6] ordered (i.e., that, whatever be two points of S, the statistician either prefers one to the other or is indifferent between them); (2) the

[6] In customary mathematical usage such orderings (among equivalence classes) are called simple or linear or complete or total. See, for example, [3, page 14].

assumption that the ordering obeys the axioms of consistency mentioned above; and (3) the ability of the statistician to determine the slope of the indifference contours by self-interrogation.

I should like to question the first of these assumptions, whose plausibility is in no way enhanced by whatever plausibility the axioms of consistency may possess. It seems to be difficult enough to get intelligent people to agree on what decision function should be preferred by the statistician; in fact, Professor Savage's whole effort is devoted to getting one to agree with his choice. Is it reasonable to demand of the statistician that he have a simple preference pattern over all possible decision functions? Is it not enough for him to select one decision function from S? True, one has to act and a choice must be made, but only one decision function need be chosen. Why, when there is so much debate over the choice of one, require the statistician to order all? We may illustrate the unreasonableness of this requirement by a homely example of the sort which Professor Savage uses frequently and effectively: When a man marries he presumably chooses, from among possible women, that one whom he likes best. Need he necessarily be able also to order the others in order of preference?

Under certain circumstances a conservative experimenter may place the greatest stress on avoiding error and may choose a minimax decision function. Granted that there is no a priori necessity[7] for one always to use such a decision function, how can one categorically and a priori brand this procedure as "inconsistent" and unreasonable? Surely no logical inconsistency is in any way involved and it is an arbitrary action to impute inconsistency to the minimax criterion. It is also easy to see that one can conceive of other logical criteria, not hitherto advocated, which lead to decision functions not in general attainable in the above manner which is held up to be the sole consistent one.

The statistician tempted to try to put the points of S in a simple order would do well to reflect on the following implication of assumption (1): In the process of ordering S he is also indicating a preference between the points A and B, whether explicitly during the self-interrogation process or implicitly through the order established by it. Now points A and B correspond to decision functions which require no observations at all. Therefore a preference for A over B, for example, corresponds to an a priori preference for the hypothesis that F_1 is the true distribution over the other hypothesis. Why need the experimenter or statistician be compelled to have such a preference a priori? The purpose of the experiment is to decide between the

[7] This seems a good place to correct a not infrequently made historical error, which pictures Wald as believing that one should always act according to the minimax criterion. Wald merely explored the consequences of this criterion without arguing for its necessity.

two hypotheses. Is it not perfectly reasonable to wish to make a decision solely on the basis of the experiment? May it not be that no other information is available, because the problem has not been studied before or because it is temporary and personal to the experimenter? Surely it is arbitrary and unreasonable to demand a preference for one or the other hypothesis in advance of the experiment.

Thus one who plays at this game of self-interrogation about his preferences (in a simple ordering) among the points S is also (if he accepts assumption 2) willy-nilly expressing an a priori preference between the two hypotheses. Moreover, it can now scarcely be considered surprising that a simple ordering of the points of S leads to an a priori distribution on the two hypotheses. For example, if the statistician is indifferent as between A and B the a priori distribution is obviously (1/2, 1/2).

Suppose, however, that in spite of the objections[8] (to the first assumption) one does make assumptions (1)–(3). At the conclusion of the self-interrogation one obtains a decision function which can perhaps best be described by reference to an a priori distribution. This is, however, completely compatible with a frequentist interpretation of the consequences of any decision function. The process of questioning the statistician (self-interrogation) has served to reveal the pair of errors which the statistician prefers among all points of S. His preference may even be based on his prior beliefs in the two hypotheses. At any rate, the statistician considers this the best method for selecting a decision function from the class of all decision functions. There is no contradiction, however, in interpreting the risk point as giving the long term relative frequency of error in the case of each hypothesis being true. Thus one can logically be a Bayesian as regards the choice of decision function, while holding to the frequentist interpretation of probability theory (this was v. Mises' position). One can hold such a position and still have reservations about drawing conclusions from results obtained by optional stopping, still prefer that one of two unbiased estimators whose variance is smaller, in short, still differ strongly with some of the criticisms made by Professor Savage and the methods advocated by him.

The latter's strictures against modern statistical theory really originate from an acceptance of the subjective theory of probability. The basic argument on this has been going on for a long time. Professor Savage gives

[8] Even Professor Savage appears to have some difficulties. Thus he says [6, Chapter 3, Section 4]): "It is, of course, inherent in this analysis that every pair of σ's [i.e., points of the quadrant (J.W.)] can be compared and that the relation 'is not preferred to', written '$\leqslant$', is transitive as usual. *These assumptions are not entirely above criticism* [emphasis mine, J.W.]. In particular they are beset by the difficulty of vagueness, but they seem interesting and the consequences of exploring them can be judged on their own merits."

it new form by directing his fire at most of the advances in statistics which have been made since the great burgeoning of the objective school of statistical inference (largely or wholly frequentist in its outlook on the foundations of probability) which began with Fisher and was continued by Neyman, Wald, and others. But it would be an error for the reader who merely accepts the axioms of choice (to which we have been referring) to conclude that he must also follow Professor Savage in rejecting modern statistical theory.[9]

Cornell University

REFERENCES

[1] Dvoretzky, A., A. Wald, and J. Wolfowitz: "Elimination of Randomization in Certain Statistical Decision Procedures and Zero-Sum Two Person Games," *Annals of Mathematical Statistics*, vol. 22, no. 1, March, 1951.

[2] ———: "Relations Among Certain Ranges of Vector Measures," *Pacific Journal of Mathematics*, vol. 1, 1951.

[3] Kelley, J. L.: *General Topology*, New York: Van Nostrand Co., 1955.

[4] Savage, L. J.: "Subjective Probability and Statistical Practice," to appear as part of a book, *The Foundations of Statistical Inference: A Symposium*, London: Methuen and Co., 1962.

[5] ———: "Bayesian Statistics," to appear in the proceedings of Purdue

[9] The present footnote is really a postscript to the manuscript, written as a footnote because it did not appear in my first version. (See footnote 3.) Since in many problems Bayes solutions form a complete class (see, for example, [8]) workers in decision theory are keenly interested in the structure of such solutions. Their study presents many problems; only the trivial ones are easy, and most are of great difficulty. An example of the latter is this: to determine completely the structure of Bayes solutions for sequential decision problems (see, for example, [9]). Such problems are a challenge to all statisticians, whether or not they adopt the Bayesian approach.

It should be noted that there is a great difference between (a) setting up a system of axioms, among which is the axiom of simple ordering, and deducing from these a Bayes criterion or minimax criterion or any other criterion, and (b) requiring the statistician to order simply the points of S and then to act according to a Bayes criterion implied by a set of axioms. The first of these procedures results in a mathematical theorem which, like all mathematical theorems, asserts that certain hypotheses (the axioms) imply certain conclusions. One can object to such a theorem only if its proof contains an error in logic (in which case it is not a proof and the "theorem" may not be a theorem). The second procedure deals with how the statistician should act when making statistical decisions. Professor Savage and I are concerned with the latter problem.

It should also be noted that, if the true (actual) distribution of X is chosen from $\{F_1, F_2\}$ by a chance process with (known) probabilities $(\pi, 1 - \pi)$, Professor Savage would, I am sure, concur with me in advising the statistician to use a Bayes solution with respect to $(\pi, 1 - \pi)$. However, this is not the point at issue.

University's Third Symposium on Decision and Information Processes, to be published by Macmillan.

[6] SAVAGE, L. J.: "The Subjective Basis of Statistical Practice," mimeographed notes, University of Michigan, July, 1961.

[7] WALD: A.: *Statistical Decision Functions*, New York: John Wiley and Sons, 1950.

[8] WALD, A., AND J. WOLFOWITZ: "Characterization of the Minimal Complete Class of Decision Functions When the Number of Distributions and Decisions Is Finite," *Proceedings of the 2nd Berkeley Symposium on Probability and Statistics*, University of California Press, 1951.

[9] ———: "Bayes Solutions for Sequential Decision Problems," *Annals Mathematical Statistics*, 21 (1950), pp. 82–99.

[10] v. MISES, R.: *Wahrscheinlichkeit, Statistik, und Wahrheit*, Vienna, 1951, Springer-Verlag, third edition.

[11] ———: *Probability, Statistics, and Truth* (English translation of an earlier edition of [10]), London: William Hodge and Co., Ltd., 1939.

PRODUCTS OF INDECOMPOSABLE, APERIODIC, STOCHASTIC MATRICES[1]

J. WOLFOWITZ

1. Introduction. A finite square matrix $P=\{p_{ij}\}$ is called stochastic if $p_{ij}\geqq 0$ for all i, j, and $\sum_j p_{ij}=1$ for all i. A stochastic matrix P is called indecomposable and aperiodic (SIA) if

$$Q = \lim_{n\to\infty} P^n$$

exists and all the rows of Q are the same. SIA matrices are defined differently in books on probability theory; see, for example, [1] or [2]. The latter definition is more intuitive, takes longer to state, is easier to verify, and explains why the probabilist is interested in SIA matrices. A theorem in probability theory or matrix theory then says that the customary definition is equivalent to the one we have given. The latter is brief and emphasizes the property which will interest us in this note.

We define $\delta(P)$ by

$$\delta(P) = \max_j \max_{i_1,i_2} | p_{i_1j} - p_{i_2j} | .$$

Thus $\delta(P)$ measures, in a certain sense, how different the rows of P are. If the rows of P are identical, $\delta(P)=0$ and conversely.

Let $A_1, \cdots, A_k$ be any square matrices of the same order. By a word (in the A's) of length t we mean the product of t A's (repetitions permitted).

The object of this note is to prove the following:

THEOREM. *Let* $A_1, \cdots, A_k$ *be square stochastic matrices of the same order such that any word in the* A's *is SIA. For any* $\epsilon>0$ *there exists an integer* $\nu(\epsilon)$ *such that any word* B *(in the* A's*) of length* $n\geqq\nu(\epsilon)$ *satisfies*

$$\delta(B) < \epsilon.$$

In words, the result is that any sufficiently long word in the A's has all its rows approximately the same.

It is sufficient to require that any word in the A's be indecomposable. (This means, in the terminology of [1], that it contains at most one closed set of states other than the set of all states, or, in the

Presented to the Society, June 6, 1962; received by the editors June 7, 1962.

[1] Research under contract with the Office of Scientific Research of the U. S. Air Force.

terminology of [2], that it contains only one ergodic class.) For, as pointed out by Thomasian [6], if C is a word which is indecomposable and has period $r>1$, C^r is decomposable.

The theorem has applications to coding (information) theory (see [3]) and obvious applications to the study of nonhomogeneous Markov chains.

For 2×2 matrices A the theorem is trivial. For then the determinant of each matrix is less than one in absolute value. The determinant of a long enough word has a value close to zero, so that the rows of the word are almost identical.

It would be desirable to have the hypothesis of the theorem require only that the A's be SIA, but it is easy to prove by a counter-example that this would be insufficient. Indeed (e.g., [4, Equation 4(b)]), one can construct two SIA matrices A_1 and A_2 such that A_1A_2 is decomposable. Hence for no n does the word $(A_1A_2)^n$ have all its rows approximately the same. Thus the condition of the theorem is not only sufficient but also necessary.

Thomasian [6] has proved a result which will be described at the end of §2 and which implies an algorithm such that, in a bounded number of arithmetical operations, starting with $A_1, \cdots, A_k$, one can determine whether *every* word in the A's (of whatever length) is SIA. The number of arithmetical operations required by his method may still be very large. At the end of §2 we make a suggestion which will usually substantially reduce the amount of calculation required.

A result related to ours has been proved by Sarymsakov [5]. Lemma 1 of his paper (which he considers its most important result) is Lemma 4 of the present paper, with our t replaced by two less than the order of the A's. Sarymsakov's conditions are only on the individual A's (and not on all the words in the A's) which is a pleasant feature. However, his conditions are rather strong and not obviously meaningful in a probabilistic sense. Essentially they are that each A remain SIA under permutation of rows and columns, the permutation of rows being *independent* of the permutation of the columns. As an example of the implications of this consider the matrix D, where

$$D = \begin{Bmatrix} 0 & 1 & 0 \\ 0 & 0 & 1 \\ x & x & x \end{Bmatrix}.$$

Here the x's stand for positive elements whose actual value is immaterial. D is obviously SIA. It does not fulfill the above described conditions of [5], as may be seen by putting the first column in third position.

2. **Proof of the theorem.** Henceforth, all matrices under discussion will be square and stochastic without any further mention of this fact. Capital Roman letters will always denote matrices. We will say that P_1 and P_2 are of the same type, $P_1 \sim P_2$, if they have zero elements and positive elements in the same places. Whether or not a matrix is SIA depends solely on its type.

Define

$$\lambda(P) = 1 - \min_{i_1, i_2} \sum_j \min (p_{i_1 j}, p_{i_2 j}).$$

If $\lambda(P) < 1$ we will call P a scrambling matrix. $\lambda(P) = 0$ if and only if $\delta(P) = 0$ and conversely. $\lambda(P) < 1$ implies that, for every pair of rows i_1 and i_2, there exists a column b (which may depend on i_1 and i_2) such that $p_{i_1 b} > 0$ and $p_{i_2 b} > 0$, and conversely. The introduction of λ is due to Hajnal [**4**]. Russian writers (e.g., [**5**]) use a related coefficient of "ergodicity."

We note the following for future reference:

LEMMA 1. *If one or more matrices in a product of matrices is scrambling, so is the product.*

This is Lemma 2 of [**4**].

LEMMA 2. *For any k*

$$\delta(P_1 P_2 \cdots P_k) \leq \prod_{i=1}^{k} \lambda(P_i).$$

This is Theorem 2 of [**4**].

We now prove

LEMMA 3. *If* P_2 *is an SIA matrix and*

$$P_1 P_2 \sim P_1, \tag{2.1}$$

then P_1 *is a scrambling matrix.*

PROOF. From (2.1) we obtain that, for every positive integral n,

$$P_1 P_2^n \sim P_1, \tag{2.2}$$

because

$$P_1 P_2^2 \sim P_1 P_2 \sim P_1.$$

From the definition of an SIA matrix it follows that, for n sufficiently large, P_2^n is a scrambling matrix. The desired result now follows from (2.2) and Lemma 1.

The conclusion of this lemma could easily be strengthened, but the present lemma will suffice for our purposes.

Let t be the number of different types of all SIA matrices of the same order as the A's.

LEMMA 4. *All words in the* A's *of length* $\geqq t+1$ *are scrambling matrices.*

PROOF. Consider any word in the A's of length $t+1$, say

$$B_1 \cdot B_2 \cdots B_{t+1}.$$

From the definition of t it follows that there exist integers a and b, $0 < a < b \leqq t+1$ such that

$$B_1 \cdot B_2 \cdots B_a \sim B_1 \cdot B_2 \cdots B_b. \tag{2.3}$$

It follows from Lemma 3 that $B_1 \cdot B_2 \cdots B_a$ is a scrambling matrix. The lemma now follows from Lemma 1.

LEMMA 5. *There exists a constant* d, $0 \leqq d < 1$, *with the following property: Let* C *be any word of length* $(t+1)$ *in the* A's. *Then* $\lambda(C) \leqq d$.

PROOF. There are a finite number of words of length $(t+1)$ in the A's. By Lemma 4 each of these words has a λ which is less than one. Let d be the largest of these values of λ.

PROOF OF THE THEOREM. Let h be such that

$$d^h < \epsilon.$$

The theorem now follows from Lemma 2, with

$$\nu(\epsilon) = h(t+1).$$

The result of Thomasian's referred to earlier is as follows: Let $\Gamma(k)$ be the set of types of all words in the A's of length $\leqq k$. If, for some k, $\Gamma(k) = \Gamma(k+1)$ and all the types in $\Gamma(k)$ are those of SIA matrices, then all words in the A's are SIA; the converse also holds. Let s be the order of the A's. If all words in the A's are SIA then the above property must hold for $k \leqq 2^{s^2}$. (Actually it must hold for $k \leqq t$.)

Since any word which has a scrambling matrix as a factor is SIA, we can at once disregard all A's which are scrambling and consider the classes Γ only for those A's which are not scrambling. Thus if all the A's are scrambling no computations are necessary. In computing the Γ's for the A's which are not scrambling we need not consider further any word, a factor of which is scrambling. This procedure will generally considerably reduce the amount of computation required. It is easy to verify whether or not a given matrix is scrambling.

The only place where we make use of the fact that there are only finitely many A's is in Lemma 5. If there are infinitely many A's but the conclusion of Lemma 5 holds then the theorem also holds. Some of the results on Markov chains in the literature are immediate consequences of this remark. In particular also, Lemma 4 always holds, even if there are infinitely many A's.

References

1. W. Feller, *An introduction to probability theory and its applications*, 2nd ed., Vol. 1, Wiley, New York, 1957.

2. J. L. Doob, *Stochastic processes*, Wiley, New York, 1953.

3. J. Wolfowitz, *Strong converse of the coding theorem for indecomposable channels*, Sankhya (1963). See also the abstract by the same title in Ann. Math. Statist. 33 (1962), 1212.

4. J. Hajnal, *Weak ergodicity in non-homogeneous Markov chains*, Proc. Cambridge Philos. Soc. 54 (1958), 233–246.

5. T. A. Sarymsakov, *Inhomogeneous Markov chains*, Teor. Verojatnost. i Primenen. 6 (1961), 194–201.

6. A. J. Thomasian, *A finite criterion for indecomposable channels*, Ann. Math. Statist. 34 (1963), 337–338.

Cornell University

Reprinted from THE ANNALS OF MATHEMATICAL STATISTICS
Vol. 36, No. 6, December, 1965

ON A THEOREM OF HOEL AND LEVINE ON EXTRAPOLATION DESIGNS

BY J. KIEFER[1] AND J. WOLFOWITZ[2]

Cornell University

0. Summary. Recent results [5] of Hoel and Levine (1964), which assert that designs on $[-1, 1]$ which are optimum for certain polynomial regression extrapolation problems are supported by the "Chebyshev points," are extended to cover other nonpolynomial regression problems involving Chebyshev systems. In addition, the large class of linear parametric functions which are optimally estimated by designs supported by these Chebyshev points is characterized.

1. Introduction. Let $f = (f_0, f_1, \cdots, f_m)$ be a row vector of $m + 1$ continuous real-valued functions on a compact set X, where $m > 0$. (Unprimed vectors will ordinarily denote row vectors, and the transpose of a vector or matrix will be denoted by a prime.) The f_i are assumed to be linearly independent on X. A design is a probability measure ξ (which can always be assumed discrete) on X. (For a discussion of this see [8].)

Write

$$m_{ij}(\xi) = \int f_i f_j \, d\xi, \qquad M(\xi) = \{m_{ij}(\xi), 0 \leqq i, j \leqq m\}. \tag{1.1}$$

If N uncorrelated observations with equal variance σ^2 (known or unknown) are made, taking $N\xi(x)$ observations at x for each x in X, and if the expected value of an observation at x is $\theta f(x)' = \sum_0^m \theta_i f_i(x)$ where $\theta = (\theta_0, \cdots, \theta_m)$ with the θ_i unknown real parameters, then, if $M(\xi)$ is nonsingular, $\sigma^2 N^{-1} M^{-1}(\xi)$ is the covariance matrix of best linear estimators of the vector θ. Moreover, setting

$$V(a, \xi) = aM^{-1}(\xi)a' \tag{1.2}$$

where $a = (a_0, a_1, \cdots, a_m)$ with the a_i real, $\sigma^2 N^{-1} V(a, \xi)$ is the variance of the best linear estimator of the linear parametric function $\theta a'$. The function V of (1.2) is defined to have the same meaning even if M is singular; in particular, $V(a, \xi) = \infty$ if $\theta a'$ is not estimatable under ξ.

As has been discussed in other papers, we do not restrict ξ to take on values which are integral multiples of N^{-1}. This allows us to obtain optimum design characterizations which cannot be obtained under that restriction, and at the same time yields designs which can be implemented in practice through the use of closely related ξ's which do take on only values which are integral multiples of N^{-1}.

Received 21 April 1965.

[1] Research supported by the Office of Naval Research under Contract No. Nonr 266(04) (NR 047-005).

[2] Research supported by the U. S. Air Force under Contract No. AF 18(600)-685, monitored by the Office of Scientific Research.

One problem of interest is to characterize, for each a, a design ξ, termed *a-optimum,* which minimizes $V(a, \xi)$; work on this problem has been done in [4] Elfving (1952), [3] Chernoff (1953), and [8] Kiefer and Wolfowitz (1959). A particular way in which this problem can arise, and which is of considerable practical importance, is that f is extended continuously so as to be defined on the set $Y \cup X$, and that for some point e in Y it is required to choose a design which estimates optimally the regression $\theta f(e)'$ at the point e, the design ξ still being restricted to be a probability measure on the set X of points at which we are permitted to take observations. For e in $Y - X$ (resp., $Y \cap X$) this may be called the problem of *extrapolation* (resp., *interpolation*) of the estimated regression to the point e of the set Y on which the regression function is of interest to us. This is clearly the problem of finding an a-optimum design when $a = f(e)$. (Other extrapolation and interpolation problems, such as that of minimizing $\max_{y \epsilon Y} V(f(y), \xi)$ for certain sets Y other than those of the type mentioned in the next two paragraphs, are considered in [9] Kiefer and Wolfowitz (1964a, b); when $Y = X$, this problem has been considered extensively in other papers.)

Hoel and Levine (1964) [5] have considered this extrapolation problem in the important case of univariate polynomial regression, where $X = [-1, 1]$, $Y = (-\infty, \infty)$, $f_i(y) = y^i$ for y in Y. Their elegant main result is that, for $|e| > 1$, there is an $f(e)$-optimum design which, for every e, is supported by the same set of $m + 1$ points (with weights which depend on e), namely, the "Chebyshev points" which were shown in [8] to be the support of the a-optimum design when $a = (0, 0, \cdots, 0, 1)$. (It will often be helpful to think of this a as $\lim_{e \to \pm\infty} [f_m(e)]^{-1} f(e)$.) It is well known (see Example 2 of Section 5) that this conclusion cannot be extended to the $f(x)$-optimum design for $|x| \leqq 1$.

(We are indebted to Dr. T. J. Rivlin for pointing out that part of the above-mentioned development on p. 1556 of [5] which shows that $\sum_0^m |L_i(e)|$ is maximized by the Chebyshev points, where the L_i are Lagrange interpolation polynomials, rediscovers a result proved by S. Bernstein on p. 186 of [2].)

A second result of Hoel and Levine is that, for e sufficiently large, their $f(e)$-optimum design also minimizes $\max_{-1 \leqq y \leqq e} V(f(y), \xi)$, which is proportional to the maximum over the interval $[-1, e]$ of the variance of the estimated regression. (In most of the literature to which we refer, $V(f(y), \xi)$ is denoted by $d(y, \xi)$.)

It is natural to ask then, whether in this polynomial case there are vectors a other than constant multiples of those in the one-dimensional set $\{f(e), |e| > 1\}$, such that there is an a-optimum design supported by this same set of Chebyshev points. One of the results of the present paper (Theorems 1 and 2 of Section 4 and Example 2 of Section 5) is that there are many such a, in fact, an $(m + 1)$-dimensional set of them in the $(m + 1)$-dimensional space of all a. This apparent anomaly (in view of the low dimensionality of these designs in the space of all admissible designs) is discussed after the introduction of necessary nomenclature, in the first paragraph of Section 4, and is illustrated in Example 2(b) of Section 5.

It is also natural to ask whether the two Hoel-Levine results and the result discussed in the previous paragraph can be extended to examples other than that of polynomial regression. Theorems 4, 5, 1 and 2 give such extensions under assumptions stated in the next section; the main assumptions (1 and 2) are related to the behavior of the f_i for a related Chebyshev approximation problem. (The proof of Theorem 4 reduces, in the polynomial case, to one which differs from that of [5].) It seems interesting to determine, under these assumptions, the set of vectors a for which there is an optimum design supported by (that is, which assigns positive probability to, and only to) the Chebyshev points, and Theorems 1 and 2 show that this is the set T^* defined in the next section. Also of interest is the set $\bar{T}^*$ of vectors for which there is an optimum design supported by some subset of the Chebyshev points. $\bar{T}^*$, which is not generally merely the closure of T^* (as will be seen in Example 2(b)), does not permit as simple an analysis as T^*. For the most part we are concerned with a subset R^* of T^*, where a related Chebyshev approximation problem has a solution of a particular form, and where the optimum design is unique (as it need not be in $T^* - R^*$). The sets R^* and $T^* - R^*$ are not difficult to characterize explicitly (see (2.20), Theorem 2, and (2.27)), but Theorem 3 describes the inclusion in R^* of a set A^* of vectors which is sometimes also easy to compute, namely, the set of vectors a for which $a'\theta$ is not estimatable for any design on fewer than $m + 1$ points. Theorem 3 is used in establishing Theorem 4.

Section 2 contains nomenclature, definitions, assumptions, and statements of results from previous papers which we shall use. Section 3 contains proofs of auxilliary lemmas which are used in the proofs of our main results (Theorems 1–5) in Section 4; Remark 5 in the latter section describes some further extensions of our results. Finally, Section 5 contains some examples which illustrate the relationship among T^*, R^*, A^*, and other sets we shall consider.

The preliminary propositions, examples, and theorems of Section 2 will be numbered decimally (2.x). All other theorems and examples, and all lemmas and remarks, will be numbered consecutively without indication of section.

2. Preliminaries. Our basic model of f, X, and Y will be as described in Section 1. Throughout this paper, except when explicitly stated to the contrary (as, in particular, in the extensions of Remark 5), X will be $[-1, 1]$ and Y will be $(-\infty, \infty)$. We shall usually denote points of X, Y and $Y - X$ by x, y, and e, respectively. As described in Section 1, we always assume, without further statement in the sequel,

ASSUMPTION 0. $f_0, f_1, \cdots, f_m$ are linearly independent on $[-1, 1]$ and continuous on $(-\infty, \infty)$.

The functions $f_0, \cdots, f_k$ are called a *Chebyshev system* on the set U if every linear combination $\sum_0^k c_i f_i$, with not all of the real constants c_i zero, has k or fewer zeros on U. This condition can be rephrased in any of several equivalent forms which will be useful in what follows: For distinct $x_0, x_1, \cdots, x_k$ in U,

(i) the vector $\sum_0^k c_i(f_i(x_0), \cdots, f_i(x_k))$ is not the zero vector unless all c_i are 0;

(ii) the matrix $\{f_i(x_j), 0 \leqq i, j \leqq k\}$ has rank $k + 1$;

(iii) the vector $(f_0(x_k), \cdots, f_k(x_k))$ cannot be represented as $\sum_0^{k-1} c_i(f_0(x_i), \cdots, f_k(x_i))$.

In the lemmas and theorems of Sections 3 and 4, we shall make use of the following five assumptions, stating explicitly where any of them is made. The first two assumptions are used in all five of the theorems.

Assumption 1. The functions $f_0, f_1, \cdots, f_{m-1}$ constitute a Chebyshev system on $[-1, 1]$.

If F is a continuous real-valued function on $[-1, 1]$, we shall say that F *changes direction* at x_0 if $-1 < x_0 < 1$ and if F has a local maximum or minimum at x_0. In particular, if F is constant on any open subinterval of $[-1, 1]$, it will be said to have infinitely many changes of direction.

Assumption 2. For each q in E^{m+1} (Euclidean $(m + 1)$-space), the function fq' on $[-1, 1]$ either has fewer than m changes of direction, or else is constant on $[-1, 1]$.

In proving our generalizations of the Hoel-Levine results we shall also use

Assumption 3. The continuous functions $f_0, f_1, \cdots, f_m$ are a Chebyshev system on $(-\infty, \infty)$;

Assumption 4. For $0 \leqq i < m$, we have $\lim_{e\to\pm\infty} f_i(e)/f_m(e) = 0$;

Assumption 5. $|f_m(e)|$ is strictly increasing (resp., decreasing) when e(resp., $-e$) is sufficiently large; $\lim_{e\to\pm\infty} |f_m(e)| = +\infty$; and, for $0 \leqq i < m$, the quantity

$$\sup_{0<|\Delta|\leqq 1} |f_i(e + \Delta) - f_i(e)|/|f_m(e + \Delta) - f_m(e)|$$

remains bounded as $e \to \pm\infty$.

Assumption 5 can of course be phrased in terms of derivatives, if they exist. We next remark briefly on Assumptions 1 and 2.

Remark 1(a). We shall prove in Lemma 2 that Assumption 2 implies that $f_0, f_1, \cdots, f_m$ constitute a Chebyshev system on $[-1, 1]$ with $\sum_0^m d_j f_j(x) \equiv 1$ for some $d_0, \cdots, d_m$. In the other direction we have the obvious

Proposition 2.1. *If* $f_0 \equiv 1$, *the* f_i *are differentiable, and* $\{df_i(x)/dx, 1 \leqq i \leqq m\}$ *constitute a Chebyshev system on* $[-1, 1]$, *then Assumption* 2 *holds.*

However, simple examples show that Assumption 2 is stronger than $\{f_i, 0 \leqq i \leqq m\}$ being a Chebyshev system. An illustration is

Example 2.1. If $\{f_i, 0 \leqq i \leqq m\}$ satisfies Assumption 2 and h is positive and continuous on $[-1, 1]$, then $\{hf_i, 0 \leqq i \leqq m\}$ is Chebyshev but need not satisfy Assumption 2 (e.g., let $f_i(x) = x^i$ and $h(x) = 2 + \sin 10x$).

Remark 1(b). The form of Assumptions 1 and 2 appears to single out f_m for special treatment. This asymmetric form of the first two assumptions can be replaced by a more symmetric form. For example, Assumptions 1 and 2 can be replaced by the hypotheses of the following:

Proposition 2.2 *Assumption* 2 *and the assumption that* $f_0, f_1, \cdots, f_m$ *is Chebyshev on a set* $[-1, 1] \cup \{v\}$ *for some point* $v \not\in [-1, 1]$ *imply that Assumptions* 1 *and* 2 *are satisfied for* $\{\bar{f}_i, 0 \leqq i \leqq m\}$ *where* $\bar{f}' = Qf'$ *for some nonsingular* Q. *In particular, this conclusion is implied by Assumptions* 2 *and* 3.

(The second assumption of Proposition 2.2, without the v, is the conclusion of Lemma 2 obtained under Assumptions 1 and 2. Example 2.1 above shows that Assumption 3 alone does not imply Assumption 2.)

PROOF OF PROPOSITION 2.2. We clearly need only prove that Assumption 1 is satisfied for $\bar{f}_0, \cdots, \bar{f}_{m-1}$ with some nonsingular Q. By linear independence there is a nonsingular Q such that $\bar{f}(v)' \equiv Qf(v)' = (0, 0, \cdots, 0, 1)'$. But then, writing $x_m = v$, if $0 \leqq x_0 < x_1 < \cdots < x_{m-1} \leqq 1$ we have

$$\det \{\bar{f}_i(x_j), 0 \leqq i, j < m\} = \det \{\bar{f}_i(x_j), 0 \leqq i, j \leqq m\}$$
$$= \det Q \det \{f_i(x_j), 0 \leqq i, j \leqq m\} \neq 0,$$

the last by the Chebyshev assumption for $[-1, 1] \cup \{v\}$. This proves the desired result.

The following example shows that the Chebyshev nature of $\{f_0, f_1, \cdots, f_m\}$ on $[-1, 1]$ does not by itself imply that $\{\bar{f}_0, \cdots, \bar{f}_{m-1}\}$ is Chebyshev for some Q, even if $f_0(x) \equiv 1$:

EXAMPLE 2.2. Suppose $f_0(x) \equiv 1$ on $[-1, 1]$, and let $f_1(x) = \cos \pi x$ and

$$f_2(x) = \sin \pi x \quad \text{if} \quad -1 \leqq x \leqq \tfrac{1}{2},$$
$$= (1 + \sin \pi x)/2 \quad \text{if} \quad \tfrac{1}{2} \leqq x \leqq 1.$$

This system $\{f_0, f_1, f_2\}$ was given in [11] Volkov (1958) as an example of a Chebyshev system on $[-1, 1]$ which cannot be extended to be Chebyshev on a larger interval (our example must of course be of this nature by Proposition 2.2). To see that this system is indeed Chebyshev, it is only necessary to graph the function $c_1f_1 + c_2f_2$ (with c_1 and c_2 not both zero) in each of the four cases $0 \leqq \pm c_1 \leqq c_2$ and $0 \leqq c_2 < \pm c_1$ and to note that this function assumes each value at most twice. If now $\{\bar{f}_0, \bar{f}_1\}$ were Chebyshev on $[-1, 1]$ with $\bar{f}_i = \sum_j q_{ij}f_j$, some linear combination of $\bar{f}_0$ and $\bar{f}_1$ would be of the form $g_1 = c_1f_1 + c_2f_2$ and (by the Chebyshev property) would have at most one zero on $[-1, 1]$. An examination of the four cases described above shows that any function of this form has two zeros on $[-1, 1]$, except for multiples of $f_1 + cf_2$ with $c > 2$. For g_1 of this last form we have $g_1'(r) = 0$ for a single r satisfying $-1 < r < -\frac{1}{2}$, where we use primes to denote derivatives in this paragraph. Let g_0 be a linear combination of $\bar{f}_0$ and $\bar{f}_1$ with g_0 positive throughout $[-1, 1]$; the existence of such a g_0 is shown in the first paragraph of Example 1 of Section 5. The development of that paragraph also shows (since $g_1(-1) < 0 < g_1(1)$) that g_1/g_0 is nondecreasing on $[-1, 1]$. Since $(g_1/g_0)' = g_1'/g_0 - g_1g_0'/g_0^2$, we see that $(g_1/g_0)'(r) \geqq 0$ if and only if $g_0'(r) \geqq 0$. If $g_0'(r) > 0$, then, since $g_0'(-1) < 0$ for g_1/g_0 to be increasing (because $g_1(-1) < 0, g_1'(-1) < 0$), there is an s with $-1 < s < r$ and $h'(s) = 0$; but then $(g_1/g_0)'(s) < 0$, a contradiction. Hence $g_0'(r) = 0$. But then, since $\{1, g_1, f_1\}$ span the same vector space as $\{f_0, f_1, f_2\}$ and $f_1'(r) \neq 0$, we conclude that $g_0 = k_0 + k_1g_1$ for some constants k_i with $k_0 \neq 0$. But then $g_1 + k_3(g_0 - k_1g_1)$ is a linear combination of $\bar{f}_0$ and $\bar{f}_1$ which, for a suitable choice of the constant k_3, has two zeros in a small neighborhood of r (since $g_1''(r) \neq 0$).

For reference in the proofs of Sections 3 and 4, we now summarize in Theorems 2.1–2.3 the known results of Chebyshev approximation theory and optimum design theory which will be used. Suppose $h_0, h_1, \cdots, h_m$ are continuous real-valued functions on $[-1, 1]$. Then $\sum_0^{m-1} c_i^* h_i$ is called a *best Chebyshev (uniform) approximation* to h_m on $[-1, 1]$ if $\max_{-1 \leqq x \leqq 1} |h_m(x) - \sum_0^{m-1} c_i h_i(x)|$ is minimized, over all choices of the real constants c_i, by the choice $c_i = c_i^*$ $(0 \leqq i < m)$. The vector $c^* = (c_0^*, \cdots, c_{m-1}^*)$ is then called a *Chebyshev vector*. A classical result of approximation theory ([1] Achieser (1956), p. 74) is

THEOREM 2.1. *If $h_0, \cdots, h_{m-1}$ form a Chebyshev system, then there is a unique Chebyshev vector c^*, and it is characterized by the fact that there are at least $m + 1$ points at which the residual $h_m - \sum_0^{m-1} c_i^* h_i$ attains its maximum in absolute value, this maximum being assumed with at least $m + 1$ successive alternations in sign.*

If c^* is a Chebyshev vector, we shall denote by $B(c^*)$ the set where $|h_m - \sum_0^{m-1} c_j^* h_j|$ attains its maximum on $[-1, 1]$.

We shall be concerned both with cases where $h_0, \cdots, h_{m-1}$ are a Chebyshev system, and with cases where they are not (see Lemma 8, etc.). It will be shown in Lemma 3 that, if Assumptions 1 and 2 hold, which in particular imply that Theorem 2.1 holds with $h_i = f_i$ $(0 \leqq i \leqq m)$, the set $B(c^*)$ consists of exactly $m + 1$ points $-1 = x_0^* < x_1^* < \cdots < x_m^* = 1$. We shall denote by X_m^* the set of these *Chebyshev points*.

If $h_0, h_1, \cdots, h_m$ are continuous real-valued linearly independent functions on $[-1, 1]$, and if $\sum_0^m \psi_i h_i(x)$ is the expected value of an observation at x, where the ψ_i are unknown real parameters and the observations are uncorrelated with common variance σ^2, we are led to consider the game with payoff function

$$K(\xi, q) = \int_{-1}^{1} [h_m(x) - \textstyle\sum_0^{m-1} q_j h_j(x)]^2 \xi(dx), \tag{2.1}$$

where the minimizing player may (by convexity of K in q) be restricted to pure strategies which are vectors $q = (q_0, \cdots, q_{m-1})$ of real components, while the maximizing player has probability measures ξ (which can be taken to be discrete) for his mixed strategies. From Section 2 of [8], we have the following results:

THEOREM 2.2. *The variance $\sigma^2 N^{-1} v_m(\xi)$ of the best linear estimator of ψ_m when the design ξ is used satisfies*

$$v_m(\xi) = 1/\min_q K(\xi, q). \tag{2.2}$$

Hence, ξ^* is optimum for estimating ψ_m if and only if it is maximin for the game with payoff K; i.e., *if and only if*

$$\min_q K(\xi^*, q) = \max_\xi \min_q K(\xi, q). \tag{2.3}$$

THEOREM 2.3. *The game with payoff K is determined, the minimizing player has a pure minimax strategy q^*, and the maximizing player has a maximin strategy*

ξ^* on $m + 1 - p$ points, where p is the dimensionality of the convex set of Chebyshev vectors for approximating h_m by $h_0, \cdots, h_{m-1}$ on $[-1, 1]$. The minimax strategies coincide with the Chebyshev vectors. (Hence, from standard game theory, ξ is maximal relative to q and q is minimal with respect to ξ, if and only if ξ is optimum and q is Chebyshev.) The design ξ^* is maximin if and only if, for any Chebyshev vector q^*, $\xi^*(B(q^*)) = 1$ and ξ^* satisfies the "orthogonality relations"

$$\int_{-1}^{1} [h_m(x) - \sum_0^{m-1} q_j^* h_j(x)] h_i(x) \xi^*(dx) = 0, \quad 0 \leq i < m. \tag{2.4}$$

In the setting of Section 1 and the first paragraph of the present section, we considered linear parametric functions $\theta a'$ for a in $B^* = E^{m+1} - \{0\}$ where E^k denotes Euclidean k-space and 0 denotes the origin in whatever Euclidean space is under consideration (or, where appropriate, a matrix of zeros). Clearly, for every a and every real $\lambda \neq 0$, ξ is a-optimum if and only if it is (λa)-optimum. Thus, from the point of view of characterizing a-optimum designs, one could replace B^* by the real projective m-space P^m of its equivalence classes under the equivalence $a \sim \lambda a$ for all $\lambda \neq 0$. Throughout most of the developments of Sections 3 and 4, it is not profitable to do this, and therefore *all starred sets, such as A^*, R^*, and T^* will be regarded as subsets of B^**. However, because of the special role of the last coördinate θ_m in Assumption 1 (where f_m, of the $m + 1$ functions f_i, is absent), it is sometimes more convenient, especially in explicit representations of the optimum designs for various parametric functions, to consider the a's in terms of two disjoint sets $B = E^m$ and $B_0 = P^{m-1}$, as follows: To each a with $a_m = 1$ we make correspond the vector $b = (b_0, \cdots, b_{m-1})$ in B defined by $b_i = a_i$ $(0 \leq i < m)$, and conversely; we shall thus think of the m-vector b as corresponding to the linear parametric function $\theta_m + \sum_0^{m-1} b_i\theta_i$ and shall call a design which is optimum for this linear parametric function *b-optimum*. Of course, for any a with $a_m \neq 0$ (whether or not $a_m = 1$), the a-optimum designs will then coincide with the b-optimum designs with

$$b = (a_0/a_m, \cdots, a_{m-1}/a_m).$$

Similarly, the a's with $a_m = 0$ but not all $a_i = 0$ correspond to points in B_0. We let Γ denote the mapping of B^* onto $B \cup B_0$ under this identification. Throughout this paper we shall use a to mean an element of B^* and b to mean an element of B. The disadvantages of working in terms of B and B_0 rather than B^* will be seen in Examples 1 and 2(c) of Section 5.

It will sometimes be useful in the developments which follow to work not with the space Ξ_m of probability $(m + 1)$-vectors with positive components (to be thought of as being on X_m^*), but rather with the set $\Xi_m^* = \{\eta: \eta = \lambda\xi$ for some ξ in Ξ_m and some real $\lambda \neq 0\}$, which can be regarded as the union of two congruent convex cones in E^{m+1}.

We shall apply Theorems 2.2–2.3 to the problem of determining a-optimum designs in the setting of the first paragraph of this section by using the following simple reduction of [7] Kiefer (1962), p. 795: For fixed a and i_0 with $a_{i_0} \neq 0$, write

$$
\begin{aligned}
&\varphi_{i_0} \equiv \varphi_{i_0,a} = \textstyle\sum_0^m a_i\theta_i/a_{i_0}, \\
(2.5)\qquad &\varphi_i = \theta_i, && i \neq i_0, \\
&g_{i_0} = f_{i_0}, \\
&g_i \equiv g_{i,a} = f_i - (a_i/a_{i_0})f_{i_0}, && i \neq i_0.
\end{aligned}
$$

Then $\sum_0^m \varphi_i g_i = \sum_0^m \theta_i f_i$, and the problem of estimating $a\theta'$ when the regression is $\theta f(x)'$ for $x \varepsilon [-1, 1]$ is the same as that of estimating φ_{i_0} when the regression is $\varphi g(x)'$ for $x \varepsilon [-1, 1]$. Thus, we can apply Theorems 2.2–2.3 to the problem of finding a-optimum designs by setting

$$
\begin{aligned}
(2.6)\qquad &\psi_m = \varphi_{i_0}, \qquad \{\psi_i, 0 \leqq i < m\} = \{\varphi_i, i \neq i_0\}, \\
&h_m = g_{i_0}, \qquad \{h_i, 0 \leqq i < m\} = \{g_i, i \neq i_0\}.
\end{aligned}
$$

In particular, the payoff function (2.1) becomes

$$
(2.7)\qquad K(\xi, q) = \int_{-1}^{1} [f_{i_0}(x) - \textstyle\sum_{i \neq i_0} q_i(f_i(x) - (a_i/a_{i_0})f_{i_0}(x))]^2 \xi(dx),
$$

the orthogonality relations (2.4) become

$$
\begin{aligned}
(2.8)\qquad &\int_{-1}^{1} [a_{i_0} f_{i_0}(x) - \textstyle\sum_{j \neq i_0} q_j^*(a_{i_0} f_j(x) - a_j f_{i_0}(x))] \\
&\qquad\qquad \cdot [a_{i_0} f_i(x) - a_i f_{i_0}(x)]\xi^*(dx) = 0, \quad i \neq i_0,
\end{aligned}
$$

and the related Chebyshev problem is that of approximating f_{i_0} on $[-1, 1]$ by $\{f_i - (a_i/a_{i_0})f_{i_0}, i \neq i_0\}$. All of these depend on the a and i_0 under consideration (although the latter dependence will be seen to be irrelevant).

The functions g_i of course satisfy Assumption 0 if the f_i do. However, the same is not true of Assumption 1 (as will be seen in Lemma 8). Thus, although Theorem 2.1 can be applied under Assumption 1 to help characterize a 0-optimum design (for estimating θ_m, where $b = 0$), and to find X_m^*, we cannot apply Theorem 2.1 in the same way to find other a-optimum designs.

We now define the sets whose study will be the chief concern of this paper. A parallel notation will be used throughout: A starred symbol D^* (say) will always be defined as a subset of B^* which is invariant under multiplication by a nonzero scalar, and we will then always write

$$
(2.9)\qquad D = \Gamma(D^*) \cap B, \qquad D_0 = \Gamma(D^*) \cap B_0;
$$

the starred sets D^* will be our main objects of interest in Sections 3 and 4, but it will be useful to consider the unstarred sets in the examples of Section 5. Under Assumptions 1 and 2, as we have mentioned, there is, corresponding to $\{f_0, f_1, \cdots, f_m\}$, the set of $m + 1$ Chebyshev points $\{x_0^*, x_1^*, \cdots, x_m^*\} = X_m^*$. We define, as used in Section 1,

$$
\begin{aligned}
(2.10)\qquad T^* = \{a: a \,\varepsilon\, B^* &\text{ and there is an } a\text{-optimum design} \\
&\text{supported by the entire set } X_m^*\}.
\end{aligned}
$$

One could instead study the set $\bar{T}^*$ (say) where there is an a-optimum design supported by a subset of $X_m{}^*$, but the set $\bar{T}^* - T^*$ is less susceptible to study by our methods. Let U be the subset of E^m defined by

$$(2.11)\qquad U = \{(x_0, x_1, \cdots, x_{m-1}): \text{ all } |x_i| \leq 1 \text{ and all } x_i \text{ different}\},$$

and write, for $\tilde{x} = (x_0, \cdots, x_{m-1})$ in E^m and a in B^*,

$$(2.12)\qquad P^*(\tilde{x}, a) = \det \begin{pmatrix} f_0(x_0) & \cdots & f_0(x_{m-1}) & a_0 \\ \vdots & \vdots\vdots\vdots & \vdots & \vdots \\ f_m(x_0) & \cdots & f_m(x_{m-1}) & a_m \end{pmatrix}.$$

We define

$$(2.13)\qquad A^* = \{a: a \,\varepsilon\, B^* \text{ and } P^*(\tilde{x}, a) \neq 0 \text{ for all } \tilde{x} \text{ in } U\}.$$

We also define

$$(2.14)\qquad N^* = \{a: a \,\varepsilon\, B^* \text{ and } a\theta' \text{ is only estimatable for designs supported by at least } m+1 \text{ points of } [-1, 1]\}.$$

It is easy to see (Lemma 1 below) that $N^* = A^*$. The usefulness of this set is of course that, for a in A^*, any a-optimum design has at least $m + 1$ points of support, so that $X_m{}^*$ is at least a possible candidate for this support.

For fixed a in B^*, suppose $a_{i_0} \neq 0$, and consider the system of m linear equations in the $m + 1$ unknowns η_j,

$$(2.15)\qquad \textstyle\sum_{j=0}^m (-1)^j \eta_j [a_{i_0} f_i(x_j{}^*) - a_i f_{i_0}(x_j{}^*)] = 0, \qquad 0 \leq i \leq m,\ i \neq i_0.$$

We shall also consider the related system

$$(2.16)\qquad \textstyle\sum_{j=0}^m (-1)^j \xi_j [a_{i_0} f_i(x_j{}^*) - a_i f_{i_0}(x_j{}^*)] = 0, \qquad i \neq i_0; \qquad \sum_{j=0}^m \xi_j = 1.$$

For $i_0 = m$, putting $b = \Gamma(a)$, (2.16) becomes

$$(2.17)\qquad \textstyle\sum_{j=0}^m (-1)^j \xi_j [f_i(x_j{}^*) - b_i f_m(x_j{}^*)] = 0, \qquad 0 \leq i < m; \qquad \sum_{j=0}^m \xi_j = 1.$$

(It will be clear from the derivation of (2.20) and (2.27) that for fixed a, an η in $\Xi_m{}^*$ is a solution of (2.15) or (2.23) for some i_0 for which $a_{i_0} \neq 0$, if and only if it is a solution for every such i_0.) The form (2.15) will be of chief concern in Sections 3 and 4. We shall use (2.17) extensively in Example 2(b) of Section 5. We now define

$$(2.18)\qquad \begin{aligned} R^* &= \{a: a \,\varepsilon\, B^* \text{ and for some } i_0 \text{ with } a_{i_0} \neq 0 \text{ (2.15) has a solution } \eta \text{ in } \Xi_m{}^*\} \\ &= \{a: a \,\varepsilon\, B^* \text{ and for some } i_0 \text{ with } a_{i_0} \neq 0 \text{ (2.16) has a solution } \xi \text{ in } \Xi_m\}. \end{aligned}$$

Write

$$F_{R^*} = \{(-1)^j f_i(x_j^*), \qquad 0 \leq i, \ j \leq m\},$$
$$F_{S^*} = \{f_i(x_j^*), \qquad 0 \leq i, \ j \leq m\}. \tag{2.19}$$

By Lemma 2, Assumption 2 implies that these matrices are nonsingular, a fact which we shall use repeatedly. Suppose $a = \eta F'_{R^*}$ for some η in Ξ_m^*. Since F_{R^*} is nonsingular by Assumption 2, $a_{i_0} \neq 0$ for some i_0, and a clearly satisfies (2.15), since the latter can be written as $a_{i_0}(F_{R^*}\eta')_i = a_i(F_{R^*}\eta')_{i_0}$. Also, this form of (2.15) shows that every a in R^* can be obtained in this way. Thus,

$$R^* = \{a: a = \eta F'_{R^*} \quad \text{for some} \quad \eta \quad \text{in} \quad \Xi_m^*\}$$
$$= (\Xi_m^*)F'_{R^*}. \tag{2.20}$$

We thus have an explicit representation of R^* as a pair of congruent open convex cones obtained from the linear mapping F_{R^*} acting on Ξ_m^*. One of these cones is spanned by the $m + 1$ half-lines consisting of the positive multiples of column vectors of F_{R^*}; the other, of the negative multiples.

From (2.18) we also have

$$R = \{b: b \, \varepsilon \, B \quad \text{and (2.17) has a solution in} \quad \Xi_m\}$$
$$= \{b: b_i = \textstyle\sum_{j=0}^m (-1)^j \xi_j f_i(x_j^*) / \sum_{j=0}^m (-1)^j \xi_j f_m(x_j^*), \tag{2.21}$$
$$0 \leq i < m, \quad \text{for some} \quad \xi \quad \text{in} \quad \Xi_m\}.$$

The importance of the set R^* is given in Theorem 1. For a in R^*, it will turn out that the residual of the best Chebyshev approximation of f_{i_0} by $\{f_i - (a_i/a_{i_0})f_{i_0}, i \neq i_0\}$, mentioned below (2.8), is *oscillatory* (that is, satisfies the condition of Theorem 2.1) even though (Lemma 8) $\{f_i - (a_i/a_{i_0})f_{i_0}, i \neq i_0\}$ is not a Chebyshev system for $a \, \varepsilon \, R^* - A^*$. It will also turn out that this residual attains its maximum in absolute value at the x_j^*. Hence, the residual at x_j^* (first factor of the integrand of (2.8)) is a multiple of $(-1)^j$ and thus, writing

$$\xi_j = \xi^*(x_j^*), \tag{2.22}$$

the orthogonality relations (2.8) will be seen to reduce to (2.16). The application of the game theory of Theorem 2.3 will be used, in the proof of Theorem 1, to show that $R^* \subset T^*$.

We have mentioned in the previous paragraph that $a \, \varepsilon \, R^*$ implies the oscillatory nature of a certain Chebyshev approximation problem. One could also study the designs in $T^* - R^*$, which are also supported by X_m^*, but for which (by Lemma 5 and Theorem 1) the solution to this Chebyshev approximation problem has *constant* nonzero residual. Paralleling the development indicated in the previous paragraph, we now consider, in place of (2.15), (2.16), (2.17), the systems

$$\textstyle\sum_{j=0}^m \eta_j[a_{i_0} f_i(x_j^*) - a_i f_{i_0}(x_j^*)] = 0, \qquad 0 \leq i \leq m, \quad i \neq i_0; \tag{2.23}$$

$$(2.24)\qquad \sum_{j=0}^{m} \xi_j[a_{i_0}f_i(x_j^*) - a_i f_{i_0}(x_j^*)] = 0, \qquad i \neq i_0,$$
$$\sum_{j=0}^{m} \xi_j = 1;$$

$$(2.25)\qquad \sum_{j=0}^{m} \xi_j[f_i(x_j^*) - b_i f_m(x_j^*)] = 0, \qquad 0 \leqq i < m,$$
$$\sum_{j=0}^{m} \xi_j = 1.$$

In place of (2.18) we now define

$$(2.26)\qquad \begin{aligned} S^* &= \{a: a \,\varepsilon\, B^* \text{ and for some } i_0 \text{ with } a_{i_0} \neq 0 \text{ (2.23) has a solution } \eta \text{ in } \Xi_m^*\} \\ &= \{a: a \,\varepsilon\, B^* \text{ and for some } i_0 \text{ with } a_{i_0} \neq 0 \text{ (2.24) has a solution } \xi \text{ in } \Xi_m\}. \end{aligned}$$

Using the second half of (2.19) we obtain in place of (2.20),

$$(2.27)\qquad \begin{aligned} S^* &= \{a: a = \eta F'_{S^*} \text{ for some } \eta \text{ in } \Xi_m^*\} \\ &= (\Xi_m^*)F'_{S^*}. \end{aligned}$$

In place of (2.21) we now have

$$(2.28)\qquad \begin{aligned} S &= \{b: b \,\varepsilon\, B \text{ and (2.25) has a solution in } \Xi_m\} \\ &= \{b: b_i = \textstyle\sum_{j=0}^{m} \xi_j f_i(x_j^*) / \sum_{j=0}^{m} \xi_j f_m(x_j^*), \quad 0 \leqq i < m, \\ &\qquad \text{for some } \xi \text{ in } \Xi_m\}. \end{aligned}$$

Using the results indicated at the outset of the present paragraph, we shall show in Theorem 2 that $T^* - R^* = S^*$. A major difference between R^* and S^*, which will be illustrated in Example 2(b) of Section 5, is that the a-optimum design is unique for $a \,\varepsilon\, R^*$, while for $a \,\varepsilon\, S^*$ we can have other a-optimum designs whose supports are not X_m^*.

While the set $\bar{T}^*$ defined just below (2.10) will be illustrated in Example 2(b), we shall not analyze $\bar{T}^*$ in general. Such an analysis would be more complicated than that of T^* because of the variety of forms the residual can now have and the necessity of determining when the orthogonality relations parallel to (2.15) or (2.23) do indeed correspond to an a for which the residual attains its maximum absolute value on the relevant subset of X_m^*. In particular, for $m > 1$ the set $\bar{T}^*$ is not generally the closure of T^*.

The closure of R^* or S^* (and, hence, of T^*) in $E^{m+1} - \{0\}$ is obviously obtained by replacing Ξ_m^* in (2.20) or (2.27) by the closure of Ξ_m^* in $E^{m+1} - \{0\}$; that is, by the set of all $(m + 1)$-vectors not all of whose components are zero, but whose nonzero components all have the same sign.

The definition of the generalization of the set considered by Hoel and Levine is

$$(2.29)\qquad H^* = \{a: a \,\varepsilon\, B^* \text{ and } a = \lambda f(e)' \text{ for some real } \lambda \neq 0 \text{ and some real } e \text{ with } |e| > 1\}.$$

(If $f(e) = 0$, $\theta f(e)'$ can of course be estimated without error; such e are excluded from H^*, and under Assumption 3 they obviously can not exist.) In particular, if $f_m(e) \neq 0$ for $|e| > 1$, H_0 is empty and

$$\Gamma(H^*) = H = \{b: b_i = f_i(e)/f_m(e), \quad 0 \leq i < m, \quad \text{for some} \quad e \quad \text{with} \quad |e| > 1\}. \tag{2.30}$$

The examples of Section 5 illustrate the sets defined in this section.

We add the definition of a concept which arises in the first paragraph of Section 4 and in Example 2(b) of Section 5, that of an *admissible design* ξ, which is a design such that for no ξ' is $M(\xi') - M(\xi)$ nonnegative definite and not the zero matrix. The meaning of this concept is discussed in [6] Kiefer (1959).

In Example 2(a) we shall introduce some additional material from the literature, which is used only there.

3. Auxiliary lemmas. The lemmas of this section will be used in proving our main results in the next section.

LEMMA 1. $N^* = A^*$.

PROOF. Since $\theta f(x_j)'$ is the expected value of an observation at x_j, any linear parametric function which is estimatable under a design supported at $\{x_0, x_1, \cdots, x_{m-1}\}$ or a subset thereof must be of the form $\sum_0^{m-1} \gamma_j \theta f(x_j)'$ for some real $\gamma_0, \cdots, \gamma_{m-1}$. Hence, $\theta a'$ is estimatable under a design on $\{x_0, x_1, \cdots, x_{m-1}\}$ if and only if there exist $\gamma_0, \cdots, \gamma_{m-1}$ such that $\sum_0^{m-1} \gamma_j \theta f(x_j)' = \theta a'$ for all θ; that is, such that $\sum_0^{m-1} \gamma_j f(x_j) = a$. This last is equivalent to $P^*(\bar{x}, a) = 0$. This completes the proof.

LEMMA 2. *Under Assumption* 2, $f_0, f_1, \cdots, f_m$ *form a Chebyshev system on* $[-1, 1]$ *and there are numbers* d_j *such that* $\sum_0^m d_j f_j(x) \equiv 1$ *for* x *in* $[-1, 1]$.

PROOF. If $f_0, f_1, \cdots, f_m$ are not a Chebyshev system, there is a vector q other than the zero vector such that $q'f$ has at least $m + 1$ zeros on $[-1, 1]$. Since $q'f$ clearly has at least one change of direction at some point strictly between any two successive zeros, it follows that $q'f$ has at least m changes of direction.

To prove the second assertion, write $h(x) = f(x) - f(0)$. Since $ch(0)' = 0$ for all c, Assumption 2 implies that, for each c, ch' either has fewer than m changes of direction or else is identically zero on $[-1, 1]$. If the latter holds for some c which is not the zero vector, we must have $cf(0)' \neq 0$ (since otherwise $cf(x)' \equiv 0$, contradicting Assumption 0), and $d_j = c_j/cf(0)'$ then yields the desired result. If $ch(x)'$ is not identically zero for all nonzero c, the h_i are linearly independent so that, by continuity, there are $m + 1$ points x_j $(0 \leq j \leq m)$ in $[-1, 1]$, none of them zero, such that $H = \{h_i(x_j), 0 \leq i, j \leq m\}$ is nonsingular. The $m + 1$ linear equations $cH = (0, \cdots, 0, 1)$ then have a solution $c = \bar{c}$ (say), and $\bar{c}h'$ then vanishes at the $m + 1$ points $0, x_0, x_1, \cdots, x_{m-1}$ and thus has at least m changes of direction, which is a contradiction.

LEMMA 3. *Under Assumptions* 1 *and* 2 *the residual* $f_m - \sum_0^{m-1} c_j^* f_j$ *of the unique best Chebyshev approximation to* f_m *on* $[-1, 1]$ *of the form* $\sum_0^{m-1} c_j f_j$, *attains its maximum in absolute value at exactly* $m + 1$ *points* $-1 = x_0^* < x_1^* \cdots < x_m^* = 1$, *the residual alternating in sign at successive* x_j^*.

PROOF. The alternating nature of the residual on $m + 1$ points follows from Assumption 1 and Theorem 2.1. It then follows from Assumption 2 that the residual cannot take on its maximum in absolute value at more than $m + 1$ points, and that $x_0^* = -1$, $x_m^* = 1$.

The reader is reminded of the definition which follows Theorem 2.1, according to which $\{x_0^*, x_1^*, \cdots, x_m^*\}$ will be called the *Chebyshev points of* $\{f_0, f_1, \cdots, f_m\}$.

LEMMA 4. *Assumption* 1 *implies that* $0 \varepsilon A$.

PROOF. The proof of Lemma 1, with $a = (0, 0, \cdots, 0, 1) = a^*$ (say), shows that θ_m is estimatable for a design on $\{x_0, x_1, \cdots, x_{m-1}\}$ if and only if $0 = P^*(\bar{x}, a^*) = \det\{f_i(x_j), 0 \leqq i, j < m\}$. The latter is not zero if the x_i are distinct, by Assumption 1.

LEMMA 5. *Under Assumptions* 1 *and* 2, *if an a-optimum design is supported by at least* $m + 1$ *points, then either* $a \varepsilon R^*$ *or else, for each* i_0 *for which* $a_{i_0} \neq 0$, *every best Chebyshev approximation of* f_{i_0} *by* $\{f_i - (a_i/a_{i_0})f_{i_0}, i \neq i_0\}$ *on* $[-1, 1]$ *has constant nonzero residual.*

PROOF. Suppose there is a best approximation $\sum_{i \neq i_0} c_i'[f_i - (a_i/a_{i_0})f_{i_0}]$ such that the residual $r(x) = f_{i_0}(x) - \sum_{i \neq i_0} c_i'[f_i(x) - (a_i/a_{i_0})f_{i_0}(x)]$ is not constant. By Theorem 2.3 and the hypothesis of the lemma, there are $m + 1$ points $x_0 < x_1 < \cdots < x_m$ in the support of ξ at which $|r(x)|$ attains its maximum on $[-1, 1]$. It follows easily from Assumption 2 that, if $r(x)$ is not constant, $r(x)$ alternates in sign at $x_0, x_1, \cdots, x_m$ and thus has m zeros on $[-1, 1]$. In that case the coefficient of f_m in r is not zero, since, if it were, r would be a linear combination of $f_0, \cdots, f_{m-1}$ which has m zeros but which is not identically zero, contradicting Assumption 1.

Writing $r(x) = q[f_m(x) - \sum_0^{m-1} h_i f_i(x)]$, it follows from the oscillation property of $q^{-1}r(x)$ at $x_0, x_1, \cdots, x_m$ and Theorem 2.1 that $\sum_0^{m-1} h_i f_i$ is the best Chebyshev approximation of f_m by $f_0, \cdots, f_{m-1}$. Hence $x_0, x_1, \cdots, x_m$ are the Chebyshev points and by Lemma 3 they in fact constitute the entire support of ξ. Thus, $a \varepsilon R^*$.

Finally, if $r(x) \equiv 0$, then by Theorem 2.2 the variance of the best linear estimator of $\theta a'$ is infinite, contradicting the fact that any design on $m + 1$ points yields an estimator with finite variance.

LEMMA 6. *For fixed* a *with* $a_{i_0} \neq 0$, *there are real constants* c_i *and* K *such that*

$$f_{i_0}(x) - \sum_{i \neq i_0} c_i[f_i(x) - (a_i/a_{i_0})f_{i_0}(x)] \equiv K \quad \text{for} \quad x \,\varepsilon\, [-1, 1], \tag{3.1}$$

if and only if $K \neq 0$ *and there are unique numbers* d_i *such that*

$$\sum_0^m d_i f_i(x) \equiv 1 \quad \text{for} \quad x \,\varepsilon\, [-1, 1], \tag{3.2}$$

and

$$K = a_{i_0}\left[\sum_{i=0}^m a_i d_i\right]^{-1}, \qquad c_i = -K d_i. \tag{3.3}$$

For fixed a *with* $a_{i_0} \neq 0$ *and fixed reals* c_i', *there are real constants* c_i *and* K *such that*

$$\begin{aligned} f_{i_0}(x) - \sum_{i \neq i_0} c_i[f_i(x) - (a_i/a_{i_0})f_{i_0}(x)] \\ \equiv K \sum_{i=0}^m c_i' f_i(x) \quad \text{for} \quad x \,\varepsilon\, [-1, 1] \end{aligned} \tag{3.4}$$

if and only if $\sum_0^m a_i c_i' \neq 0$ *and*

$$K = a_{i_0}[\sum_0^m a_i c_i']^{-1}, \qquad c_i = -Kc_i'. \tag{3.5}$$

Proof. First suppose (3.1) holds. Since it is impossible for all c_i to be 0 while $1 + \sum_{i \neq i_0} a_i c_i / a_{i_0}$ is also 0, the left side of (3.1) cannot be identically 0, by the linear independence of the f_i. Hence $K \neq 0$. The existence of numbers d_i satisfying (3.2) now follows, and their uniqueness is a consequence of the linear independence of the f_i. Substituting $K \sum_0^m d_i f_i(x)$ for K in (3.1), for each i $(0 \leqq i \leqq m)$ the coefficients of f_i on both sides must be the same, again by linear independence. This yields (3.3). The converse is obvious.

Finally, assuming (3.4), equality of the coefficients of f_i on both sides yields (3.5). The converse is again clear.

Lemma 7. *Under Assumptions* 1 *and* 2, *suppose that* $\sum_0^{m-1} c_j^* f_j$ *is the best Chebyshev approximation of* f_m *by* $\{f_0, f_1, \cdots, f_{m-1}\}$ *and write* $c_m^* = -1$; *furthermore, let* $d_0, \cdots, d_m$ *be the numbers whose existence is guaranteed by Lemma* 2. *Then* $a \,\varepsilon\, R^*$ *implies* $\sum_0^m a_i c_i^* \neq 0$, *and* $a \,\varepsilon\, S^*$ *implies* $\sum_0^m a_i d_i \neq 0$.

Proof. Suppose $a \,\varepsilon\, R^*$ but that $\sum_0^m a_i c_i^* = 0$. Multiply the ith orthogonality relation (2.15) by c_i^* and sum over $i \neq i_0$. We obtain

$$\sum_{j=0}^m (-1)^j \eta_j [\sum_{i=0}^m c_i^* f_i(x_j^*)] = 0. \tag{3.6}$$

Since the term in square brackets in (3.6) is some nonzero constant times $(-1)^j$, this leads to a contradiction. Similarly, if $a \,\varepsilon\, S^*$ but $\sum_0^m a_i d_i = 0$, multiplying the ith relation of (2.23) by d_i and summing over $i \neq i_0$ yields

$$\sum_{j=0}^m \eta_j [\sum_{i=0}^m d_i f_i(x_j^*)] = 0, \tag{3.7}$$

which yields a contradiction since $\sum_0^m d_i f_i \equiv 1$.

Lemma 8. *Suppose* $a_{i_0} \neq 0$. *Then* $\{a_{i_0} f_i - a_i f_{i_0}, i \neq i_0\}$ *is a Chebyshev system on* $[-1, 1]$ *if and only if* $a \,\varepsilon\, A^*$.

Proof. For $0 \leqq i \leqq m$ and $i \neq i_0$, subtract a_i/a_{i_0} times the i_0th row of the matrix of (2.12) from the ith row. We obtain

$$P^*(\bar{x}, a) = \pm a_{i_0} \det \{f_i(x_j) - (a_i/a_{i_0}) f_{i_0}(x_j), i \neq i_0, 0 \leqq j < m\},$$

which at once yields the conclusion.

Lemma 9. *Under Assumption* 3, $H^* \subset A^*$.

The proof is immediate.

Remark 2. Lemma 9 really uses something weaker than Assumption 3, namely, the nonvanishing of det $\{f_i(x_j), 0 \leqq i, j \leqq m\}$ when m different x_i's are in $[-1, 1]$ and one x_i is outside $[-1, 1]$.

4. Principal results. Our first result is that $a \,\varepsilon\, R^*$ implies that the unique a-optimum design is supported by the Chebyshev points, and that R is m-dimensional (and hence R^* is $(m + 1)$-dimensional, which we already knew from (2.20)). This last is perhaps surprising in view of the fact that, in such a simple example as that of polynomial regression, the designs on the Chebyshev points are only an m-parameter family out of the $(2m - 1)$-parameter family of designs on $m + 1$ points including ± 1, all of which are admissible (see [6]), and the Hoel-Levine set H is one-dimensional. We shall see why this is possible in Example 2(b), where it will be seen that infinitely many different supporting sets may yield a-optimum designs for the same a.

Theorem 1. *Under Assumptions 1 and 2, if $a \varepsilon R^*$ the orthogonality relations* (2.16) *have a unique solution* $(\xi_0, \cdots, \xi_m)$, *and the corresponding design* (2.22) *(which is supported by the Chebyshev points) is the unique a-optimum design. (Thus, $R^* \subset T^*$.) Furthermore, R contains a neighborhood of the origin in E^m (that is, R^* contains a neighborhood of* $(0, 0, \cdots, 0, 1)$ *in E^{m+1}).*

Proof. Suppose $a \varepsilon R^*$. Let $c_0^*, \cdots, c_m^*$ be as in Lemma 7. By Lemma 7, $\sum_0^m a_i c_i^* \neq 0$. Hence, by Lemma 6 with $c_i' = -c_i^*$ and i_0 such that $a_{i_0} \neq 0$, there are values $c_i = \bar{c}_i$ (say) and $K \neq 0$ (by (3.5)) such that, on $[-1, 1]$,

$$f_{i_0} - \textstyle\sum_{i \neq i_0} \bar{c}_i[f_i - (a_i/a_{i_0})f_{i_0}] \equiv K[f_m - \sum_0^{m-1} c_i^* f_i]. \tag{4.1}$$

By Lemma 3 the right member of (4.1) attains its maximum in absolute value on $[-1, 1]$ at, and only at, the Chebyshev points $x_0^*, \cdots, x_m^*$, and this is therefore true of the left member. Hence, ξ^* is maximal with respect to $\bar{c} = \{\bar{c}_i, i \neq i_0\}$ for the game with payoff (2.7) if and only if the support of ξ^* is a subset of X_m^*. On the other hand, since $f_{i_0}(x_j^*) - \sum_{i \neq i_0} \bar{c}_i[f_i(x_j^*) - (a_i/a_{i_0})f_{i_0}(x_j^*)]$ is some nonzero constant times $(-1)^j$, by (2.22) the orthogonality relations (2.8) reduce to (2.16), and thus $\bar{c}$ is minimal relative to any nonnegative solution to (2.16). One such strictly positive solution is of course guaranteed by the definition of R^*. Since ξ^* is maximal with respect to $\{\bar{c}_i, i \neq i_0\}$, and the latter is minimal with respect to ξ^*, standard game theory results, mentioned in Theorem 2.3, assert that ξ^* is maximin for the game with payoff (2.7). Hence ξ^* is a-optimum. Moreover, these same results assert that the left member of (4.1) is the residual of the Chebyshev approximation of f_{i_0} on $[-1, 1]$ by a linear combination of the functions $\{(f_i - (a_i/a_{i_0})f_{i_0}), i \neq i_0\}$, a fact of which we shall make use in a moment.

We now turn to the proof of uniqueness. According to Theorem 2.3, every a-optimum design ξ is maximal relative to the $\bar{c}$ of the previous paragraph, and hence, as deduced there, is supported by some subset of X_m^* and (by Theorem 2.3) satisfies (2.16) (with the interpretation (2.22)), with all ξ_i nonnegative (perhaps some zero). If there were more than one such solution, the $(m + 1) \times (m + 1)$ matrix L_a whose (i, j)th element is

$$\begin{aligned}(L_a)_{i,j} &= f_i(x_j^*) - (a_i/a_{i_0})f_{i_0}(x_j^*), \quad 0 \leqq i \leqq m, \quad i \neq i_0, \quad 0 \leqq j \leqq m, \\ &= (-1)^j, \qquad\qquad\qquad\qquad\qquad\qquad i = i_0, \quad 0 \leqq j \leqq m,\end{aligned} \tag{4.2}$$

would be singular, since (2.16) can be written as $L_a(\xi_0, -\xi_1, \xi_2, \cdots, (-1)^m \xi_m)' = (0, \cdots, 0, 1, 0, \cdots, 0)'$, with the 1 in i_0th place in the last vector. Now $(-1)^j$ can be written as $K_0 \sum_0^m c_i^* f_i(x_j^*)$ for some nonzero constant K_0. Hence (4.2) yields

$$L_a = \begin{pmatrix} I_{i_0} & \rho_1' & 0 \\ & K_0 c^* & \\ 0 & \rho_2' & I_{m-i_0} \end{pmatrix} F_{S^*} \tag{4.3}$$

where I_r is the $r \times r$ identity matrix, $c^* = (c_0^*, c_1^*, \cdots, c_m^*)$, and $\rho_1 = (-a_0/a_{i_0}, \cdots, -a_{i_0-1}/a_{i_0})$, $\rho_2 = (-a_{i_0+1}/a_{i_0}, \cdots, -a_m/a_{i_0})$. The second factor on the right side of (4.3) is nonsingular by Lemma 2. The determinant of the first factor, which can be computed by adding $-K_0 c_i^*$ times the ith row to the i_0th row for each $i \neq i_0$, is $K_0 a_{i_0}^{-1} \sum_0^m a_i c_i^*$, which is nonzero by Lemma 7. Hence, L_a is nonsingular and there is a unique a-optimum design.

It remains to show that R contains an open neighborhood of the origin. When $b = 0$, there is an optimum design ξ^* on the Chebyshev points by Lemma 3 and Theorem 2.3 with $h_i = f_i$ and $\psi_i = \theta_i$, and all ξ_i^* are then positive, since otherwise θ_m would be estimatable on fewer than $m + 1$ points, in violation of Lemmas 1 and 4. This shows that $0 \varepsilon R$. Moreover, if $a = (b, 1)$ in the first factor on the right side of (4.3) is varied by varying b in a small enough neighborhood of 0, $L_{(b,1)}$ remains nonsingular (since $\sum_0^{m-1} b_j c_j^* \neq -c_m^*$ for b near 0) and the coordinates ξ_j of the solution to (2.16), which will vary continuously with b, will remain positive as they were when $b = 0$. (Alternatively, this last sentence may be replaced by (2.20).) This completes the proof of Theorem 1.

Theorem 2. *Under Assumptions 1 and 2, $T^* - R^* = S^*$; and, if $a \varepsilon S^*$, the orthogonality relations (2.24) have a unique solution which corresponds to the design (2.22) on the entire set X_m^*. There is no other a-optimum design supported by X_m^* or a subset thereof. (There may be other a-optimum designs.)*

Proof. The proof parallels that of Theorem 1, so we merely outline the differences. Suppose $a \varepsilon S^*$. By Lemma 7, $\sum_0^m a_i d_i \neq 0$. By Lemmas 2 and 6 with $a_{i_0} \neq 0$, there are constants c_i^0 and $K \neq 0$ such that, on $[-1, 1]$,

$$f_{i_0}(x) - \sum_{i \neq i_0} c_i^0 [f_i(x) - (a_i/a_{i_0}) f_{i_0}(x)] \equiv K. \tag{4.4}$$

Hence, every ξ^* is maximal with respect to $c^0 = \{c_i^0, i \neq i_0\}$ for the game with payoff (2.7). By (2.22) and (4.4), the orthogonality relations (2.8) become (2.24). Therefore c^0 is minimal relative to any nonnegative solution of (2.24), while, as we have already seen, the latter is maximal relative to c^0. Hence, by the standard game theory results cited in the proof of Theorem 1, any nonnegative solution of (2.24) is a-optimum. One strictly positive solution of (2.24) is guaranteed by the definition of S^*, and this is surely a-optimum.

If there were two a-optimum designs with subsets of X_m^* for support, there would be more than one solution to (2.24), which can be written as $M_a(\xi_0, \xi_1, \cdots, \xi_m)' = (0, \cdots, 0, 1, 0, \cdots, 0)$, where M_a is obtained from the L_a of (4.2) by replacing $(-1)^j$ by 1 in the i_0th row. Since $\sum_0^m d_i f_i(x) \equiv 1$, the equation for M_a corresponding to (4.3) is obtained by replacing the i_0th row of the first factor on the right side by $(d_0, d_1, \cdots, d_m)$. Adding $-d_i$ times the ith row of this factor to the i_0th row for $i \neq i_0$, we obtain $a_{i_0}^{-1} \sum_0^m a_i d_i$ for the determinant of this factor, which is thus nonzero by Lemma 7. Hence there is only one a-optimum design supported by a subset of X_m^*.

By Lemma 5, $T^* = R^* \cup S^*$, so that it remains to show that R^* and S^* are disjoint. If, to the contrary, there were an a in $R^* \cap S^*$, then for $a_{i_0} \neq 0$ there would, by our previous development, be two different Chebyshev approximations to f_{i_0} by $\{f_i - (a_i/a_{i_0})f_{i_0}, i \neq i_0\}$, one with constant residual and one with oscillatory residual. The Chebyshev vectors are thus at least one-dimensional. Applying Theorem 2.3 with $p \geqq 1$, we conclude that there is an a-optimum design supported by m or fewer points. Since $a \varepsilon R^*$, this contradicts the conclusion of Theorem 1. The proof of Theorem 2 is now complete.

REMARK 3. Example 2(b) of Section 5 will illustrate the lack of uniqueness of a-optimum designs for $a \varepsilon S^*$, as well as the fact mentioned in Section 2 that the set $\bar{T}^*$, defined just below (2.10) and discussed above (2.29), has a more complicated structure than T^* (in particular, that $\bar{T}^*$ is not merely the closure of T^*).

THEOREM 3. *Under Assumptions* 1 *and* 2, $A^* \subset R^*$.

PROOF. Suppose $a \varepsilon A^*$. Let i_0 be such that $a_{i_0} \neq 0$. By Lemma 8, $\{(a_{i_0}f_i - a_i f_{i_0}), i \neq i_0\}$ is Chebyshev, and hence by Theorem 2.1 the best Chebyshev approximation of $a_{i_0}f_{i_0}$ by $\{(a_{i_0}f_i - a_i f_{i_0}), i \neq i_0\}$ has an oscillatory residual. Since $a \varepsilon A^*$, any a-optimum design is, by Lemma 1, supported by at least $m + 1$ points. Lemma 5 now yields $a \varepsilon R^*$.

The next (and last) two theorems of this section are direct generalizations of the Hoel-Levine results discussed in Section 1, since their example of polynomial regression satisfies the assumptions of these theorems.

THEOREM 4. *Under Assumptions* 2 *and* 3, $H^* \subset R^*$. *If also* $f_m(e) \neq 0$ *for* $|e| > 1$, *then* $\Gamma(H^*) = H \subset R$.

PROOF. By Proposition 2.2 of Remark 1(b), Assumptions 2 and 3 imply that Assumptions 1 and 2 are satisfied for $\{\bar{f}_i, 0 \leqq i \leqq m\}$ where $\bar{f}' = Qf'$, for some nonsingular Q. Let A_1^* and R_1^* be the sets defined by (2.13) and (2.18) if f is replaced by $\bar{f}$ and θ is replaced by $\bar{\theta} = \theta Q^{-1}$ (so that $\theta f' = \bar{\theta}\bar{f}'$). Then $a\bar{\theta}' = aQ'^{-1}\theta'$, so that the vector a in A_1^* must be multiplied by Q'^{-1} to give the corresponding vector in A^*; that is, $A^* = A_1^*Q'^{-1}$, and similarly $R^* = R_1^*Q'^{-1}$. Since $A_1^* \subset R_1^*$ by Theorem 3, we thus obtain $A^* \subset R^*$. Lemma 9 now completes the proof that $H^* \subset R^*$. The remainder of the theorem follows from (2.30).

REMARK 4. Assumptions 2 and 3 may be replaced in Theorem 4 by Assumptions 1 and 2 and the assumption indicated in Remark 2.

A consequence of our conclusion that $H^* \subset R^* \subset T^*$ under Assumptions 1–3 is the result of Hoel and Levine [5] mentioned in Section 1, that $H^* \subset T^*$ if $f_i(x) \equiv x^i$.

The last theorem of this section concerns $V(f(y), \xi) = f(y)M^{-1}(\xi)f(y)'$ which, we recall, is proportional to the variance of the best linear estimator of the regression function $\theta f(y)'$ at the point y of Y when the design ξ on $[-1, 1]$ is used.

THEOREM 5. *Under Assumptions* 1, 2, 4, *and* 5, *if k is a real function of e*, $1 < e < \infty$, *such that always* $k(e) \leqq e$ *and* $\liminf_{e \to +\infty} k(e) > -\infty$, *then for e (resp., $-e$) sufficiently large, the unique $f(e)$-optimum design* $\xi^{(e)}$ *minimizes* $\max_{k(e) \leqq y \leqq e} V(f(y), e)$ (*resp.*, $\max_{e \leqq y \leqq -k(-e)} V(f(y), e)$).

PROOF. We shall prove only the first conclusion (as $e \to +\infty$), the case $e \to -\infty$ being treated similarly.

By Assumption 4, $f(e) = f_m(e)(o(1), \cdots, o(1), 1)$ as $e \to +\infty$, so that for e

sufficiently large the second part of Theorem 1 shows that the $f(e)$-optimum design $\xi^{(e)}$ is unique and is supported by X_m^*. Moreover, the proof of Theorem 1 shows that

$$\lim_{e\to+\infty} \xi^{(e)}(x_i^*) = \xi^*(x_i^*), \qquad 0 \leqq i \leqq m, \tag{4.5}$$

where ξ^* is the unique optimum design for estimating θ_m.

Since $\xi^{(e)}$ minimizes $V(f(e), \xi)$, the theorem will follow if we prove that, for some real e_0,

$$\max_{k(e)\leqq y\leqq e} V(f(y), \xi^{(e)}) = V(f(e), \xi^{(e)}) \qquad \text{for } e > e_0. \tag{4.6}$$

It was shown in the proof of Theorem 1 that the unique optimum design ξ^* for estimating θ_m has $\xi^*(x_i^*) > 0$ for $0 \leqq i \leqq m$. It follows from Lemma 2 that $M(\xi^*)$ is nonsingular. Hence, from (4.5), $M(\xi^{(e)})$ is nonsingular for sufficiently large, and we can write

$$M^{-1}(\xi^{(e)}) = M^{-1}(\xi^*) + \{o(1)\} \qquad \text{as } e \to +\infty \tag{4.7}$$

where $\{o(1)\}$ is a matrix whose elements approach 0 as $e \to +\infty$. Since f is continuous and $V(f(y), \xi) = f(y)M^{-1}(\xi)f(y)'$, it follows from the nonsingularity of $M(\xi^*)$ and (4.7) that, for every compact set K,

$$\max_{y\epsilon K} |V(f(y), \xi^{(e)}) - V(f(y), \xi^*)| \to 0 \qquad \text{as } e \to +\infty. \tag{4.8}$$

We shall show below that there is an $\epsilon > 0$ and real k_0 and e_1, $k_0 < e_1$, such that $e > e_1$ implies

(4.9)
(a) $V(f(y), \xi^{(e)})$ is strictly increasing in y for $y \geqq e_1$;
(b) $V(f(e_1 + 1), \xi^{(e)}) - V(f(e_1), \xi^{(e)}) > \epsilon$;
(c) $k(e) \geqq k_0$;
(d) $V(f(e_1), \xi^*) = \max_{k_0\leqq y\leqq e_1} V(f(y), \xi^*)$.

If we let K be the interval $[k_0, e_1]$, we can by (4.8) find an e_2 such that the left side of (4.8) is $<\epsilon/2$ for $e > e_2$. Then (4.9) implies that $V(f(e_1 + 1), \xi^{(e)}) = \max_{k_0\leqq y\leqq e_1+1} V(f(y), \xi^{(e)})$ if $e > \max(e_1 + 1, e_2) = e_3$ (say); consequently, from (4.9) (a), we obtain (4.6) for $e_0 = e_3$.

We now prove (4.9). The hypothesis of the theorem on k implies (c) if e_1 is sufficiently large. (4.9) (d) follows from the validity of (4.9) (a) with $\xi^{(e)}$ replaced by ξ^*, which will be proved below, and from the fact that

$$\lim_{e\to+\infty} V(f(e), \xi^*) = \infty;$$

the latter follows from Assumption 4, according to which $V(f(e), \xi^*) = f_m^2(e)v_m(\xi^*)(1 + o(1))$ as $e \to +\infty$, where $v_m(\xi^*)$ is the lower right element of $M^{-1}(\xi^*)$, and from the fact (Assumption 5) that $f_m^2(e)$ approaches $+\infty$ with e. Next, we note that

$$\begin{aligned} V(f(y + \Delta), \xi) - V(f(y), \xi) \\ = [f(y + \Delta) + f(y)]M^{-1}(\xi)[f(y + \Delta) - f(y)]'. \end{aligned} \tag{4.10}$$

By Assumption 4, $f(y + \Delta) + f(y) = (f_m(y + \Delta) + f_m(y))(o(1), o(1), \cdots, o(1), 1)$ as $y \to +\infty$, with the $o(1)$ terms uniform for positive Δ. Similarly, by Assumption 5, $f(y + \Delta) - f(y) = (f_m(y + \Delta) - f_m(y))(O(1), O(1), \cdots, O(1), 1)$ as $y \to +\infty$, with the $O(1)$ terms uniform for $0 < \Delta \leqq 1$. We also note, from Assumption 5, that

$$f_m^2(y + \Delta) - f_m^2(y) > 0 \tag{4.11}$$

for y sufficiently large and all $\Delta > 0$. From these and (4.7), we have

$$\begin{aligned} V(f(y + \Delta), \xi^{(e)}) &- V(f(y), \xi^{(e)}) \\ &= [f_m^2(y + \Delta) - f_m^2(y)]v_m(\xi^*)(1 + o(1)) \end{aligned} \tag{4.12}$$

as $\min (y, e) \to +\infty$, uniformly for $0 < \Delta \leqq 1$. (4.11) and (4.12) yield (4.9) (a) and (b) for e_1 sufficiently large. (4.9) (a) with $\xi^{(e)}$ replaced by ξ^* is proved in the same way. This completes the proof of Theorem 5.

REMARK 5. *Extensions.* The results of this section can be extended by altering the nature of X and Y. For example, it is well known that much of the Chebyshev approximation theory, in particular Assumption 1 and Theorem 2.1, apply if X is a subset of the 1-sphere (boundary of the unit circle). Without going into further detail, we note that a case of practical importance which can be treated by our methods is that where there are open intervals in $[-1, 1]$ where observations are prohibited for technological reasons; $[-1, 1]$ is then replaced by a union of closed intervals. Similarly, Y can be altered from $(-\infty, \infty) - [-1, 1]$; for example, it may be that it only makes sense to define f on $[-1, \infty)$ because $x + 1$ is inherently nonnegative; for another example, if X is a union of disjoint intervals as mentioned just above, Y might be $(-\infty, \infty) - X$. For the required approximation theory results, see, e.g., [10a], section 2.3.3. These results apply, in particular, to the polynomial csae where X is two intervals, studied independently by [4a] Hoel (1965), some of whose arguments can be simplified by use of this theory.

5. Examples.

EXAMPLE 1. *The case $m = 1$.* For the sake of completeness (and for use in Example 2.2) we first determine the possible Chebyshev systems when $m = 1$, and then, in the next paragraph, show that under the stronger Assumptions 1 and 2 there is essentially only one example. We first show, then, that *if $\{f_0, f_1\}$ is a continuous Chebyshev system on $[-1, 1]$, then, for some nonsingular D and $g = fD$, we have $g_0(x) > 0$ and $h(x) = g_1(x)/g_0(x)$ strictly increasing for all x.* This result is probably known (although we did not succeed in finding a reference), and the proof is quite simple: We first show there is a linear combination $g_0 = fa'$ which is positive throughout $[-1, 1]$. By the Chebyshev assumption, there are linear combinations $G_i = fa^{(i)\prime}$, $i = \pm 1$, with $G_i(i) = 0$, $G_i(-i) = 1$. If $G_0 = G_1 + G_2$, then either (i) G_0 is such a g_0; or else (ii) G_0 has at least two zeros (contradicting the Chebyshev assumption); or else (iii) G_0 has a single zero at q (say) with $-1 < q < 1$, in which case $G_0 - (\text{sgn } G_1(q))G_1/2$ has at least two zeros. With the existence of a g_0 thus established, we need only observe that the Chebyshev nature of $\{1, h\}$ follows from that of $\{f_0, f_1\}$, and that h is hence strictly monotone and can be taken as increasing by making a change of sign if necessary.

If now we also impose Assumptions 1 and 2, since $G_0(\pm 1) = 1$ we see that $G_0(x) \equiv 1$, because otherwise G_0 would be a nonconstant function with at least one change of direction. Hence, in this case we can find D such that $g_0(x) \equiv 1$ on $[-1, 1]$ and $g_1(\pm 1) = \pm 1$, with g_1 strictly increasing on $[-1, 1]$. Write $f\theta' = g\psi'$. Now map X onto another copy Z of $[-1, 1]$ using the mapping g_1. The regression problem on Z with regression $\psi_0 + \psi_1 z$ then corresponds to that on X with regression $\psi_0 + \psi_1 g_1(x)$ in such a manner that if ξ' is a-optimum on Z, then an a-optimum design ξ on X is defined by $\xi(x) = \xi'(g_1(x))$.

For the linear regression problem on Z just described, it is easily verified that $A = \{b_0 : |b_0| < 1\}$ and that A_0 is empty (since ψ_0 can be estimated by an observation at $z = 0$). The Chebyshev points are $x_0^* = -1$, $x_0^* = 1$, so that

$$F_{R^*} = \begin{pmatrix} 1 & -1 \\ -1 & -1 \end{pmatrix}.$$

Hence, writing $\alpha = \eta_0 - \eta_1$ and $\sigma = -\eta_0 - \eta_1$, we have from (2.20) that a general element a of R^* has the form (α, σ). Since $\sigma \neq 0$ for $\eta \,\varepsilon\, \Xi_1^*$, we see that R_0 is empty. Since the range of α/σ for η in Ξ_1^* is the interval $(-1, 1)$, we conclude that $R = A$. The points $b_0 = \pm 1$ of B correspond to optimum designs on one point: $\xi(\pm 1) = 1$. All admissible designs in this problem are well known to be supported on X_1^* or a subset thereof, and from this or (2.27) we see that S^* consists of all points of B^* except R^* and $\Gamma^{-1}(b_0)$ for each of the two additional points $b_0 = \pm 1$ of B. (As Example 2 (b) and (c) shows, no such simple result holds when $m > 1$.) The point $S_0 = T_0$ corresponds to estimating ψ_0, which can be done optimally both by the design ξ' (say) for which $\xi'(1) = \xi'(-1) = \frac{1}{2}$ and also by any of an infinite number of inadmissible designs, the simplest of which is the design ξ'' (say) for which $\xi''(0) = 1$; it is easily verified that $M(\xi') - M(\xi'')$ is nonnegative definite, of rank 1.

The above characterizations also hold for the regression $g\psi'$ on X. The linear transformation which took f into g can then be used to characterize the corresponding sets for the original problem with regression $f\theta'$ on X. For example, as in the proof of Theorem 4, if $g' = Qf'$, then $(A^* \text{ for } g)Q'^{-1} = A^*$ for f. However, R need no longer be connected. For example, if $f_0(x) = 1 + x/2$, $f_1(x) \equiv 1$, so that

$$Q = \begin{pmatrix} 0 & 1 \\ 2 & -2 \end{pmatrix},$$

one obtains $(a_0, a_1) = (\bar{a}_0 + \bar{a}_1/2, \bar{a}_0)$, where (a_0, a_1) refers to f and $(\bar{a}_0, \bar{a}_1)$ refers to g (treated in the previous paragraph). Thus, for f we obtain

$$R = A = (-\infty, \tfrac{1}{2}) \cup (\tfrac{3}{2}, \infty),$$

with $R_0 = A_0 =$ "point at infinity." This unnecessary complication points up the advantage of working in terms of R^* (as described by (2.20)), whose geometric characteristics are unchanged by the linear transformation Q.

As for H, suppose we extend the map of $X \to Z$ to $(-\infty, \infty) \to (-\infty, \infty)$ by the identity map on $(-\infty, \infty) - [-1, 1]$. Write $\psi h(z)'$ for the regression function on $(-\infty, \infty)$ as extended from Z, so that $h_i(z) = z^i$ for $z \varepsilon Z$. Under Assumption 3 it is easy to see that the graph of h_1 crosses (and is not merely tangent to) that of h_0 at 1 and at no other point of $(-\infty, \infty)$, and that $h_1(z) = 0$ only at $z = 0$. Under Assumption 4 (for example, if $h_i(z) = z^i$ on $(-\infty, \infty)$) we obtain $H = (-1, 0) \cup (0, 1) = A - \{0\}$ (another result which does not hold if $m > 1$); if Assumption 4 does not hold (for example, if $h_0(z) = 1$ and $h_1(z) = 2z/[1 + |z|]$ for $|z| > 1$), H is a proper subset of $A - \{0\}$.

EXAMPLE 2. *Polynomial regression* ($f_i(x) = x^i$), $m > 1$.

(a) *General results.* The Chebyshev points in the polynomial case are well known (for example, see [1]) to be $x_j^* = -\cos(j\pi/m)$, $0 \leqq j \leqq m$. Thus, R^* and S^* can be described explicitly from (2.20) and (2.27), as we shall do in detail below for $m = 2, 3$.

The set A^* has been characterized in [10] as follows: Define real-valued functions S_j and Q_h on E^m (whose points we write as $x = (x_0, \cdots, x_{m-1})$) by

$$S_j(x) = (-1)^j \sum\nolimits^{(j)} x_{i_1} x_{i_2} \cdots x_{i_j}, \qquad 1 \leqq j \leqq m, \tag{5.1}$$

where $\sum^{(j)}$ denotes summation over the set $0 \leqq i_1 < i_2 < \cdots < i_j < m$, and

$$Q_h(x) = \sum\nolimits^{(h)} (1 - x_{i_1})(1 - x_{i_2}) \cdots (1 - x_{i_h})(1 + x_{i_{h+1}}) \cdots (1 + x_{i_m}) / \tbinom{m}{h}, \quad 0 \leqq h \leqq m, \tag{5.2}$$

where the m subscripts in the summand are distinct ($\sum^{(0)}$ consists of one term). Define the points $b^{(h)} = (b_0^{(h)}, \cdots, b_{m-1}^{(h)})$, $0 \leqq h \leqq m$, by

$$Q_h(x) = 1 + \sum\nolimits_{j=1}^m b_{m-j}^{(h)} S_j(x). \tag{5.3}$$

In particular, $b^{(m)} = (1, 1, \cdots, 1)$ and $b^{(0)} = ((-1)^m, (-1)^{m-1}, \cdots, (-1)^1)$. The points $b^{(0)}, \cdots, b^{(m)}$ can be shown not to lie in any hyperplane of E^m, so that they span an m-dimensional simplex. Let Δ_m denote this simplex minus the closed edge containing $b^{(0)}$ and $b^{(m)}$. The main result of [10] is

THEOREM 6. *For polynomial regression with $m > 1$, $A = \Delta_m$ and A_0 is empty.*

As we shall see in Example 2(c), R_0 is not generally empty.

We note that H is the twisted curve $\{(t^m, t^{m-1}, \cdots, t^1) : 0 < |t| < 1\}$, whose two open components have end-points $b^{(0)}$, $b^{(m)}$, and (in common) 0.

(b) *The case $m = 2$.* As in the case $m = 1$, a complete analysis of the a-optimum designs, for a in B^*, is possible here, but would be much more complicated as m increases, as will be seen in (c). We begin by describing the structure of b-optimum designs for all b in the (b_0, b_1)-plane B. From Theorem 6, we have

$A =$ triangle with vertices $(-1, 0)$, $(1, 1)$, $(1, -1)$,

minus closed segment joining the latter two.

Recalling the first paragraph of Example 2(a), we have $X_2^* = \{-1, 0, 1\}$ and

$$F_{R^*} = \begin{pmatrix} 1 & -1 & 1 \\ -1 & 0 & 1 \\ 1 & 0 & 1 \end{pmatrix}, \quad F_{S^*} = \begin{pmatrix} 1 & 1 & 1 \\ -1 & 0 & 1 \\ 1 & 0 & 1 \end{pmatrix}.$$

Setting $\alpha = \eta_0 + \eta_2$ and $\beta = \eta_2 - \eta_0$, we have from (2.20) that a general element a of R^*, obtained as a linear combination of the columns of F_{R^*}, is of the form $(\alpha - \eta_1, \beta, \alpha)$. Since $\alpha \neq 0$ for $\eta \, \varepsilon \, \Xi_m^*$, we obtain that R_0 is empty. Moreover, R is the set of points of the form $(b_0, b_1) = (1 - \eta_1/\alpha, \beta/\alpha)$. For $\eta \, \varepsilon \, \Xi_m^*$, the variables η_1/α and β/α can vary independently of each other over domains $(0, \infty)$ and $(-1, 1)$, respectively. Hence, $R = \{(b_0, b_1): b_0 < 1, |b_1| < 1\}$. Similarly, by (2.27), a point of S^* is of the form $(\alpha + \eta_1, \beta, \alpha)$, so that S_0 is empty, $S = \{(b_0, b_1): b_0 > 1, |b_1| < 1\}$. Thus, T_0 is empty and $\Gamma(T^*) = T = \{(b_0, b_1): b_0 \neq 1, |b_1| < 1\}$.

In subdividing the plane B into regions where the b-optimum designs are of various forms, we shall encounter repeatedly the parabola $b_0 = b_1^2$, which consists of 0, the set $H = \{(t^2, t): 0 < |t| < 1\}$, and the set J (say) where $b_0 = b_1^2 \geqq 1$. The point (t^2, t) of J with $|t| \geqq 1$ corresponds to the linear parametric function $f(t^{-1})\theta'$. Since $|t^{-1}| \leqq 1$, this linear parametric function can be estimated by the design $\xi^{(t)}$ (say) for which $\xi^{(t)}(t^{-1}) = 1$. It was shown in [6] that $\xi^{(t)}$ is admissible for $|t| \leqq 1$. Since $f(t^{-1})\theta'$ and its multiples are the only linear parametric functions estimatable under $\xi^{(t)}$, it follows that $\xi^{(t)}$ must be $f(t^{-1})$-optimum for $0 < |t| \leqq 1$. ($\xi^{(0)}$ will be discussed with B_0.) No two of these $\xi^{(t)}$'s allow estimation of the same linear parametric function. Hence, if there were a design other than $\xi^{(t)}$ which was also $f(t^{-1})$-optimum, it would have to be supported by at least two points, and thus it would allow estimation of some linear parametric function not estimatable under $\xi^{(t)}$, from which it follows easily that $\xi^{(t)}$ would be inadmissible. Thus, we have shown that

$$J \equiv \{(t^2, t): |t| \geqq 1\} = \{\text{points of } B \text{ where there is an optimum design supported by one point}\}.$$

(The analogue of this holds for general m, with $J = \{(t^m, t^{m-1}, \cdots, t): |t| \geqq 1\}$.)

In analyzing B further we shall use the fact that, for a design supported by *more than one point*, the residual to the best Chebyshev approximation of x^2 on $[-1, 1]$ by $\{1 - b_0x^2, x - b_1x^2\}$, being quadratic and attaining its maximum in absolute value at the points of support, must be of one of the following forms:

(i) a multiple of $x^2 - \frac{1}{2}$, with support a subset of X_2^*;

(ii) a constant;

(iii) a quadratic with derivative 0 at q, $0 < |q| \leqq 1$,

and with values of equal magnitude and opposite sign, at -1 and q if $q > 0$, and at 1 and q if $q < 0$, these two values being the support in the respective cases (the case $q = 0$ is covered by (i) above);

(iv) a multiple of $x^2 - L$ with $L < \frac{1}{2}$ and support $\{-1, 1\}$;

(v) a quadratic or linear function with nonzero derivatives of the same sign at ± 1, and with values of equal magnitude and opposite sign at the two points of support $-1, 1$.

Corresponding to these, there are three forms of the orthogonality relation (2.4) which we shall consider:

(I) the Equations (2.17), where we no longer demand that all ξ_j be positive, but only that two be positive and one nonnegative; this corresponds to (i) and, with $\xi(0) = 0$, to (iv) and the case of (ii) where the support is $\{-1, 1\}$;

(II) corresponding to (iii) and (v), the equations (a) and (b) for $q > 0$ and $q < 0$, respectively:

$$
(5.4)\quad
\begin{array}{ll}
\text{(a)} & \begin{aligned} \xi(-1)(1 - b_0) &- \xi(q)(1 - q^2 b_0) = 0, \\ \xi(-1)(-1 - b_1) &- \xi(q)(q - q^2 b_1) = 0, \\ \xi(-1) &+ \xi(q) = 1; \end{aligned} \\[2ex]
\text{(b)} & \begin{aligned} \xi(1)(1 - b_0) &- \xi(q)(1 - q^2 b_0) = 0, \\ \xi(1)(1 - b_1) &- \xi(q)(q - q^2 b_1) = 0, \\ \xi(1) &+ \xi(q) = 1; \end{aligned}
\end{array}
$$

(III) corresponding to the part of case (ii) not covered in (I), equations which will be discussed below, and which lead to (5.8).

It is trivial that for each fixed b there exists a vector $c = (c_0, c_1)$ which yields a residual with each of the possible sets of extrema and oscillations of sign represented by (I), (II), and (III). Hence, in each of these three cases, any ξ with the given support is maximal relative to any c yielding such a residual, and if the orthogonality relations are satisfied then c is minimal with respect to ξ. From Theorem 2.3 it then follows that ξ is b-optimum; it is unnecessary to go back to (i)–(v) and compare residuals to find which approximation is best, where the best approximation is not unique, etc.

The regions where these three forms hold can be described as follows: partition B into disjoint sets $B_{\text{I}}, B_{\text{II}}, B_{\text{III}}, J$, defined by

$$
(5.5)\quad
\begin{aligned}
B_{\text{I}} &= \{(b_0, b_1) : b_0 \leqq 1, |b_1| \leqq 1\} - \{(1, 1)\} - \{(1, -1)\}, \\
B_{\text{II}} &= \{(b_0, b_1) : |b_1| > 1, b_0 < b_1^2\}, \\
B_{\text{III}} &= \{(b_0, b_1) : b_0 > \max(b_1^2, 1)\}.
\end{aligned}
$$

We shall show that, for $L = \text{I, II, III}$, the orthogonality relations of case L have a solution on two or more points if and only if $b \,\varepsilon\, B_L$.

Case I is treated by the same computation which yields R; in fact, the Equations (2.17) have a nonnegative solution on the closure of R,

$$
(5.6)\quad
\begin{aligned}
cl(R) &= B_{\text{I}} \cup \{(1, 1) \cup \{(1, -1)\} \\
&= R \cup \{(b_0, 1) : b_0 < 1\} \cup \{(b_0, -1) : b_0 < 1\} \cup \{(1, b_1) : |b_1| < 1\} \\
&\quad \cup \{(1, 1)\} \cup \{(1, -1)\}.
\end{aligned}
$$

In this last partition of (5.6) we have, respectively, none of the three $\xi(x_j^*)$'s zero, only $\xi(-1) = 0$, only $\xi(1) = 0$, only $\xi(0) = 0$, $\xi(1) = 1$, and $\xi(-1) = 1$, the last two being points of J. (The point corresponding to $\xi(0) = 1$, which does not arise from (2.17), will be seen later to be in B_0.)

In describing Case II, we shall use the partition of B_{II} into disjoint sets L_s, $-\infty < s < \infty$, defined as follows:

$$(5.7) \qquad \begin{aligned} L_s &= \{(b_0, b_1): b_0 - sb_1 = 1 + s,\, b_0 < b_1^2\} && \text{if } s \geqq 0, \\ &= \{(b_0, b_1): b_0 - sb_1 = 1 - s,\, b_0 < b_1^2\} && \text{if } s \leqq 0. \end{aligned}$$

Thus, L_s is that portion not in $cl(B_{\text{III}})$ of a line passing through $(1, -1)$ if $s \geqq 0$ and through $(1, 1)$ if $s \leqq 0$; in particular, $L_0 = \{(1, b_1): |b_1| > 1\}$. Consider now the orthogonality relations (5.4) (a) in the case $0 < q < 1$. Equating the ratios $\xi(-1)/\xi(q)$ in the first two equations, one obtains $b_0 - sb_1 = s + 1$ where $s = (1 - q)/q > 0$; from the positivity condition $0 < \xi(-1)/\xi(q) < \infty$ one obtains $0 < (1 - b_0q^2)/(1 - b_0) < \infty$ or $\{b_0 > q^{-2}\} \cup \{b_0 < 1\}$, which with $b_0 - sb_1 = s + 1$ yields L_s as the subset of B for which the b-optimum design is supported by the two points -1, q and the residual has opposite signs at these two points. (The support $\{-1, q\}$ arises in case III with constant residual.) The case $-1 < q < 0$ of (5.4) (b) similarly yields L_s with $s = (1 + q)/q < 0$. Finally, the case $q = 1$ of (5.4) (a) coincides with $q = -1$ in (5.4) (b) and yields L_0 as the subset of B for which the b-optimum design is supported by the two points $1, -1$ with residual of opposite sign at the two points. (The set $\{(1, b_1): |b_1| < 1\}$ encountered in case I also has support $\{1, -1\}$. but with residual of the same sign at the two points.)

For any $b \varepsilon S$ every best Chebyshev approximation has constant residual (Theorem 2 and Lemma 5). Since for b in $B_{\text{I}} \cup B_{\text{II}}$ the residual is not constant, as we have seen, it follows that $S \subset B_{\text{III}}$. On $B_{\text{I}} \cup B_{\text{II}}$ the optimum design is unique because the orthogonality relations have a unique solution. For b in B_{III} there is no uniqueness of the b-optimum design. In fact, while the b-optimum design for each b in $B - B_{\text{III}}$ is unique and hence admissible, for each b in B_{III} there are infinitely many different supporting sets of admissible b-optimum designs, and also infinitely many different supporting sets (including supporting sets with an arbitrarily large finite number, or an infinite number, of points) of inadmissible b-optimum designs. Since the admissible b-optimum designs are of greater theoretical and practical interest, we shall exhibit only the totality of these, for each b in B_{III}. We shall then indicate by an example the existence of inadmissible b-optimum designs.

It was shown in [6] that the supports of admissible designs are of the form $\{-1, q, 1\}$ with $-1 < q < 1$, or subsets thereof, and conversely. The orthogonality relations (2.4) for the set $\{-1, q, 1\}$ in case III are

$$(5.8) \qquad \begin{aligned} &\xi(-1)(1 - b_0) &&+ \xi(q)(1 - b_0q^2) &&+ \xi(1)(1 - b_0) &&= 0, \\ &\xi(-1)(-1 - b_1) &&+ \xi(q)(q - b_1q^2) &&+ \xi(1)(1 - b_1) &&= 0, \\ &\xi(-1) &&+ \xi(q) &&+ \xi(1) &&= 1. \end{aligned}$$

We seek a nonnegative solution to these for which $0 < \xi(q) < 1$; this condition is equivalent to that of finding a solution for which at least two of the components are positive (to eliminate J) and for which $\xi(q) > 0$ (to eliminate the part of (ii) included in case I, namely, the interval $\{(1, b_1): |b_1| < 1\}$ denoted by L_0' below). In describing such solutions, it is convenient to write

$$
\begin{aligned}
L_s' &= \{(b_0, b_1): b_0 - sb_1 = 1 + s,\ b_0 > b_1^2\} && \text{if } s \geqq 0,\\
&= \{(b_0, b_1): b_0 - sb_1 = 1 - s,\ b_0 > b_1^2\} && \text{if } s \leqq 0;\\
M_r' &= \{(b_0, b_1): b_0 - rb_1 = 1 - r,\ b_0 > b_1^2\} && \text{if } r > 1,\\
&= \{(b_0, b_1): b_0 - rb_1 = 1 + r,\ b_0 > b_1^2\} && \text{if } r < -1,\\
&= \{(b_0, b_1): |b_1| = 1,\ b_0 > 1\} && \text{if } r = \infty.
\end{aligned}
\tag{5.9}
$$

Thus, L_s', L_s and the two points (q^{-2}, q^{-1}) and $(1, -\text{sign } q)$ of J (or $(1, 1)$ and $(1, -1)$ if $s = 0$) constitute a partition of the line encountered in conjunction with (5.7). For $r \neq \infty$, M_r' is the intersection with B_{III} of a line of slope $1/r$ through $(1, 1)$ if $r > 0$ and through $(1, -1)$ if $r < 0$, while M_∞' consists of two half-lines in B_{III}.

The Equations (5.8) have the formal solution

$$
\begin{aligned}
\xi(q) &= (b_0 - 1)/b_0(1 - q^2),\\
\xi(1) &= [1 + b_1(1 - q) - b_0q]/2b_0(1 - q),\\
\xi(-1) &= [1 - b_1(1 + q) + b_0q]/2b_0(1 + q).
\end{aligned}
\tag{5.10}
$$

The condition $0 < \xi(q) < 1$ is equivalent to $b_0 > 1$. We also require the nonnegativity of the numerators of $\xi(1)$ and $\xi(-1)$ in (5.10), with at least one being positive. It is easy to verify that $\xi(1) = 0$ on the line through $(1, -1)$ of slope $q/(1 - q)$, and that $\xi(-1) = 0$ on the line through $(1, 1)$ of slope $q/(1 + q)$. We conclude that, for $-1 < q < 1$, (5.8) has a solution for which $0 < \xi(q) < 1$ for b in the set V_q defined by

$$
\begin{aligned}
V_q &= \{\text{triangle with vertices } (1, 1)(1, -1), (q^{-2}, q^{-1})\} \cap B_{\text{III}} && \text{if } 0 < |q| < 1,\\
V_0 &= \{(b_0, b_1): b_0 > 1, |b_1| \leqq 1\}.
\end{aligned}
\tag{5.11}
$$

In each case, all three components of ξ are positive if b is in the interior of V_q, while two components are positive on that part of the boundary which is in V_q. The latter is M_∞' if $q = 0$ and, if $q \neq 0$, consists of the two open line segments L_s' and M_r', where $s = (1 - q)/q$ and $r = (1 + q)/q$ if $q > 0$, and $s = (1 + q)/q$ and $r = (1 - q)/q$ if $q < 0$. The rest of the boundary of V_q of course consists of the interval L_0' of B_{I} and the points $(1, 1)$, $(1, -1)$, and (if $q \neq 0$) (q^{-2}, q^{-1}) of J. Thus, for any point $b = (b_0, b_1)$ in B_{III} there is an admissible design supported by q and one or both of the points $1, -1$, provided that $b \in V_q$. From the condition of nonnegativity of $\xi(1)$ and $\xi(-1)$ in (5.10), this interval of q-values, always of positive length for b in B_{III}, is

$$\{q: (b_1 - 1)/(b_0 - b_1) \leqq q \leqq (b_1 + 1)/(b_1 + b_0)\},$$

the endpoints corresponding to designs for which ξ has only two nonzero components. Hence, for each b in B_{III}, there are infinitely many different supporting sets of admissible b-optimum designs.

As an illustration of inadmissible a-optimum designs for $\Gamma(a)$ in B_{III}, consider (2, 0, 1)-optimality, that is, optimality for estimating $2\theta_0 + \theta_2$. Among the admissible designs for this problem, obtained above, two examples are $q = 0$, $\xi(-1) = \xi(1) = \frac{1}{4}$, $\xi(0) = \frac{1}{2}$, for which

$$M(\xi) = \begin{pmatrix} 1 & 0 & \frac{1}{2} \\ 0 & \frac{1}{2} & 0 \\ \frac{1}{2} & 0 & \frac{1}{2} \end{pmatrix}, \quad M^{-1}(\xi) = \begin{pmatrix} 2 & 0 & -2 \\ 0 & 2 & 0 \\ -2 & 0 & 4 \end{pmatrix}, \tag{5.12}$$

for which $V(a, \xi) = (2, 0, 1)M^{-1}(\xi)(2, 0, 1)' = 4$, and the design with $q = \frac{1}{2}$, $\xi(1) = 0$, $\xi(-1) = \frac{1}{3}$, $\xi(\frac{1}{2}) = \frac{2}{3}$, for which $M(\xi)$ is singular, but for which $V(a, \xi)$ is again 4. Among the many inadmissible designs are symmetric designs supported by $\{(1 - \epsilon), -(1 - \epsilon), 0\}$ with $0 < \epsilon < 1 - 2^{-\frac{1}{2}}$ and with $\xi'(\pm(1 - \epsilon)) = \frac{1}{4}(1 - \epsilon)^2$. For such a design

$$M(\xi') = \begin{pmatrix} 1 & 0 & \frac{1}{2} \\ 0 & \frac{1}{2} & 0 \\ \frac{1}{2} & 0 & (1-\epsilon)^2/2 \end{pmatrix},$$

$$M^{-1}(\xi') = [4/(1 - 4\epsilon + 2\epsilon^2)] \begin{pmatrix} (1-\epsilon)^2/2 & 0 & -\frac{1}{2} \\ 0 & 1 - 4\epsilon + 2\epsilon^2)/2 & 0 \\ -\frac{1}{2} & 0 & 1 \end{pmatrix},$$

so that again $V(a, \xi) = 4$. The inadmissibility of such designs is exhibited in the fact that, for ξ given by (5.12), $M(\xi) - M(\xi')$ (or $M^{-1}(\xi') - M^{-1}(\xi)$) is nonnegative definite of rank one. It is not difficult to obtain inadmissible b-optimum ξ's here supported by any number of points $\geqq 2$ (a 2-point design being given by ξ' just above when $\epsilon = 1 - 2^{-\frac{1}{2}}$), or even with ξ absolutely continuous with positive Lebesgue density on $[-1, 1]$.

It remains to consider B_0. The unique optimum design for estimating θ_0 is, by the same argument used in discussing J, that for which $\xi(0) = 1$. For $-\infty < s < \infty$, the unique optimum design for estimating $s\theta_0 + \theta_1$, to which corresponds the problem of approximating x on $[-1, 1]$ by $\{x^2, 1 - sx\}$, is easily found by calculations parallel to those for B (solutions to the orthogonality relations now existing only in the case corresponding to II above). We obtain an optimum design supported by $\{-1, q\}$ if $q > 0$ and $s = (1 - q)/q$, and by $\{1, q\}$ if $q < 0$ and $s = (1 + q)/q$. In particular, for $s = 0$ we obtain the unique optimum design for estimating θ_1, for which $\xi(-1) = \xi(1) = \frac{1}{2}$. We note, then, that if we think of $B_0 = P^1 = \{b_{(s)}, -\infty < 1/s \leqq \infty\}$ in the usual manner, $b_{(s)}$ being the "point at infinity" of all lines in B of slope $1/s$, $-\infty < 1/s \leqq \infty$, then the optimum designs for these points $b_{(s)}$ of B_0 can be obtained, for

$-\infty < 1/s < \infty$, as limits of the corresponding designs for the family L_s; for $s = \infty$ we have the optimum design for estimating θ_0 considered at the outset of this paragraph, which can be thought of conveniently as the limit of designs as $b \to \infty$ in R or in $B_{\rm I}$ or in $B_{\rm III}$ or in J.

We note also that, in the notation of and the sentence following (2.10),

$$\bar{T} = \{(b_0, b_1): |b_1| \leqq 1\} \cup \{(1, b_1): -\infty < b_1 < \infty\},$$
$$\bar{T}^* = \Gamma^{-1}(\bar{T}) \cup \{b_{(0)}\} \cup \{b_{(\infty)}\}.$$

As an example of the explicit computation of how large the "sufficiently large" of Theorem 5 is, we consider the case $k(e) \equiv -1$, $e > 1$. For $b = \Gamma(f(e)) = (e^{-2}, e^{-1})$, writing $e^{-1} = t$, (2.17) yields

$$\xi^{(e)} = (\xi_0^{(e)}, \xi_1^{(e)}, \xi_2^{(e)}) = [2(2-t^2)]^{-1}(1-t, 2(1-t^2), 1+t),$$

so that

$$M(\xi^{(e)}) = [1/(2-t^2)]\begin{pmatrix} 2-t^2 & t & 1 \\ t & 1 & t \\ 1 & t & 1 \end{pmatrix},$$

$$M^{-1}(\xi^{(e)}) = [(2-t^2)/(1-t^2)]\begin{pmatrix} 1 & 0 & -1 \\ 0 & 1 & -t \\ -1 & -t & 2 \end{pmatrix},$$

and thus

$$V(f(y), \xi^{(e)})\cdot(1-t^2)/(1-2t^2) = 1 - y^2 - 2ty^3 + 2y^4 = p_t(y) \quad \text{(say)}.$$

The function p_t is easily seen to have local minima at $y = [3t \pm (9t^2+16)^{\frac{1}{2}}]/8$ and a local maximum at $y = 0$, all three of these points being in $[-1, 1]$. Thus, $\max_{-1 \leqq y \leqq e} p_t(y) = \max(p_t(-1), p_t(0), p_t(t^{-1}))$. Since $p_t(-1) > p_t(0)$, we seek t such that $p_t(-1) - p_t(t^{-1}) \leqq 0$, that is, such that $2t^4 - t^3 + t^2 + 2t - 2 \leqq 0$. This last is satisfied for $t \leqq .694$, or $t^{-1} \geqq 1.44$. Thus, for $e \geqq 1.44$ the design $\xi^{(e)}$ minimizes $\max_{-1 \leqq y \leqq e} V(f(y), \xi)$. We remark that it is even easier to conclude that, since $p_t(y)$ is increasing for $y \geqq 1$, the design $\xi^{(e)}$ minimizes $\max_{1 \leqq y \leqq e} V(f(y), \xi)$ for $e \geqq 1$.

(c) *The case* $m = 3$. The set $A = \Delta_3$ of Theorem 6 is determined by the points

$$b^{(0)} = (-1, 1, -1),$$
$$b^{(1)} = (1, -\tfrac{1}{3}, -\tfrac{1}{3}),$$
$$b^{(2)} = (-1, -\tfrac{1}{3}, \tfrac{1}{3}),$$
$$b^{(3)} = (1, 1, 1).$$

The set X_3^* is $\{-1, -\frac{1}{2}, \frac{1}{2}, 1\}$, and

$$F_{R^*} = \begin{pmatrix} 1 & -1 & 1 & -1 \\ -1 & \frac{1}{2} & \frac{1}{2} & -1 \\ 1 & -\frac{1}{4} & \frac{1}{4} & -1 \\ -1 & \frac{1}{8} & \frac{1}{8} & -1 \end{pmatrix}.$$

Setting $\alpha = \eta_0 + \eta_3$, $\beta = \eta_0 - \eta_3$, $\gamma = \eta_1 - \eta_2$, $\sigma = \eta_0 + \eta_1 + \eta_2 + \eta_3$, we obtain from (2.20) that a general element of R^* has the form

$$a = (\beta - \gamma, (\sigma - 3\alpha)/2, (4\beta - \gamma)/4, (\sigma - 9\alpha)/8). \tag{5.13}$$

Thus, R_0 is no longer empty as it was when $m = 2$. To obtain R, we consider η to be in Ξ_m and thus $\sigma = 1$, and find

$$\begin{aligned} R = \{(8[\beta - \gamma]/[1 - 9\alpha], 4[1 - 3\alpha]/[1 - 9\alpha],\\ 2[4\beta - \gamma]/[1 - 9\alpha]) : (\alpha, \beta, \gamma) \varepsilon \{0 < \alpha < 1,\\ \alpha \neq \tfrac{1}{9}\} \cap \{-1 < \beta/\alpha < 1\} \cap \{-1 < \gamma/(1 - \alpha) < 1\}\}; \end{aligned} \tag{5.14}$$

the value $\alpha = \frac{1}{9}$ yields points in R_0, discussed below. The variables α, β/α $\gamma/(1 - \alpha)$ vary independently in (5.14), so that for b in R the range of b_1 is $(-\infty, 1) \cup (4, \infty)$. For each fixed value k of b_1 (that is, for each fixed value of α) the range of (b_0, b_2) in (5.14) is an open parallelogram $R(k)$ (say) in the plane $b_1 = k$, symmetric about $(0, k, 0)$, but whose dimensions and angles depend on k. Thus,

$$R = \bigcup_{b_1<1 \text{ or } >4} R(b_1)$$

is no longer connected as it was when $m = 2$.

The set R_0 can be obtained as the set of elements of (5.13) with $\sigma = 9\alpha = 1$ and $b_1 = (\sigma - 3\alpha)/2 = \frac{1}{3}$; this is, by an analysis similar to that of (5.14), the set of ratios $(b_0/b_1, b_2/b_1) = (3(\beta - \gamma), 3(\beta - \gamma/4))$ in the region $|\beta| < \frac{1}{9}$, $|\gamma| < \frac{8}{9}$. This can be thought of as a "parallelogram at infinity" corresponding to the ratios $(b_0/b_1, b_2/b_1)$ of (5.14) as $\alpha \to \frac{1}{9}$.

The set S^* can be analyzed similarly. As was the case with R, the sets S and T no longer have the simple structure of the case $m = 2$. As in the next to last paragraph of Example 1, this again points up the greater simplicity of working with R^*, S^*, and T^*. The convexity of the cones which constitute half of R^* and S^* can of course be carried over to R and S in a different parametrization, one in which R_0 and S_0 are empty so that $R = \Gamma(R^*)$ can be thought of as a base (section) of a cone which constitutes half of R^*, and similarly for S. Thus, in place of x^3 we seek a function $\bar{f}_3(x) = \sum_0^3 \lambda_i x^i$ such that, if we work in terms of $\bar{f} = (\bar{f}_0, \bar{f}_1, \bar{f}_2, \bar{f}_3)$ instead of f, the quantity $\sum_0^3 (-1)^j \eta_j \bar{f}_3(x_j^*)$, which corresponds to the last element of (5.13), and the quantity $\sum_0^3 \eta_j \bar{f}_3(x_j^*)$ for the corresponding development for S^*, are never 0 for $\eta \varepsilon \Xi_3^*$. Writing $\lambda = (\lambda_0, \lambda_1, \lambda_2, \lambda_3)$, this says that all non-zero elements (there is at least one such) of $(\zeta_0, \zeta_1, \zeta_2, \zeta_3) = \lambda F_{R^*}$ must be of the same sign, and all non-zero elements of $(\zeta_0, -\zeta_1, \zeta_2, -\zeta_3) = \lambda F_{S^*}$ must be of the same sign. Hence, either (i) ζ_0 or $\zeta_2 \neq 0$, $\zeta_0\zeta_2 \geqq 0$, $\zeta_1 = \zeta_3 = 0$, or else (ii) ζ_1 or $\zeta_3 \neq 0$, $\zeta_1\zeta_3 \geqq 0$, $\zeta_2 = \zeta_4 = 0$. Since

$$\begin{pmatrix} -1 & 1 & 4 & -4 \\ -1 & 2 & 1 & -2 \\ 1 & 2 & -1 & -2 \\ 1 & 1 & -4 & -4 \end{pmatrix} F_{R^*} = \begin{pmatrix} 6 & 0 & 0 & 0 \\ 0 & \frac{3}{2} & 0 & 0 \\ 0 & 0 & \frac{3}{2} & 0 \\ 0 & 0 & 0 & 6 \end{pmatrix},$$

the solutions in case (i) are easily seen to be $\lambda = \pm[k_0(-1, 1, 4, -4) + k_2(1, 2, -1, -2)]$ with $k_0 \geqq 0, k_2 \geqq 0, k_0 + k_2 > 0$, and in case (ii) they are $\lambda = \pm[k_1(-1, 2, 1, -2) + k_3(1, 1, -4, -4)]$ with $k_1 \geqq 0, k_3 \geqq 0, k_1 + k_3 > 0$. For any such λ and $\bar{f}_3$ we can, for example, take $\bar{f}_i = x^i$ for $0 \leqq i \leqq 2$ and the transformation from f to $\bar{f}$ will be nonsingular.

A development analogous to that of the previous paragraph can be carried out for general m.

We shall not analyze B^* further in the manner of Example 2(b). The number of cases to be treated and the complexity of the resulting regions increase with m, as is evident even from the above characterization of R.

EXAMPLE 3. Other Chebyshev systems are discussed in the literature of approximation theory. As illustrated in Example 2.1 of Remark 1, Assumption 2 is somewhat stronger than the assumption that $\{f_i, 0 \leqq i \leqq m\}$ is Chebyshev. The sufficient condition for Assumption 2 which is given in Proposition 2.1 of Remark 1(a) is useful in applications, as is the condition of Proposition 2.2.

REFERENCES

[1] ACHIESER, N. I. (1956). *Theory of Approximation*. Ungar, New York.

[2] BERNSTEIN, S. (1926). *Lecons sur les Propriétés Extrémales et la Meilleure Approximation des Fonctions Analytiques d'une Variable Réele*. Gauthier-Villars, Paris.

[3] CHERNOFF, H. (1953). Locally optimum designs for estimating parameters. *Ann. Math. Statist.* **24** 586–602.

[4] ELFVING, G. (1952). Optimum allocation in linear regression theory. *Ann. Math. Statist.* **23** 255–262.

[4a] HOEL, P. G. (1965). Optimum designs for polynomial extrapolation. *Ann. Math. Statist.* **36** 1483–1493.

[5] HOEL, P. G. and LEVINE, A. (1964). Optimal spacing and weighting in polynomial prediction. *Ann. Math. Statist.* **35** 1553–1560.

[6] KIEFER, J. (1959) Optimum experimental designs. *J. Roy. Statist. Soc. Ser. B* **21** 272–319.

[7] KIEFER, J. (1962). Two more criteria equivalent to D-optimality of designs. *Ann. Math. Statist.* **33** 792–796.

[8] KIEFER, J. and WOLFOWITZ, J. (1959). Optimum designs in regression problems. *Ann. Math. Statist.* **30** 271–294.

[9] KIEFER, J. and WOLFOWITZ, J. (1964 a, b). Optimum extrapolation and interpolation designs, I and II. *Ann. Inst. Statist. Math.* **16** 79–108 and 295–303.

[10] KIEFER, J. and WOLFOWITZ, J. (1965). On a problem connected with the Vandermonde determinant. *Proc. Amer. Math. Soc.* **16**.

[10a] TIMAN, A. F. (1963). *Theory of Approximation of Functions of a Real Variable*, Macmillan, N.Y.

[11] VOLKOV, V. I. (1958). Some properties of Chebyshev systems. *Kalinin Gos. Ped. Inst. Uč. Zap.* **26** 41–48.

Reprinted from THE ANNALS OF MATHEMATICAL STATISTICS
Vol. 37, No. 3, June, 1966

REMARK ON THE OPTIMUM CHARACTER OF THE SEQUENTIAL PROBABILITY RATIO TEST[1]

BY J. WOLFOWITZ

Cornell University

There is a small lacuna in the proof ([1]) of the property stated in the title of this note. In some recent papers many pages are devoted to correcting it. Since all "other" proofs of the optimum character of the sequential probability ratio test follow all the principal ideas of [1] and differ from the latter only in very minor details, it seems appropriate to show, as we will, that the lacuna can be filled in a very simple and obvious way. The present note assumes familiarity only with Lemmas 1, 2, and 3 of [1]. The gap in [1] is in Lemma 1, where it is claimed that the test S^* there constructed minimizes the average risk.

We shall replace Lemmas 1, 2, and 3 of [1] by Lemma A whose statement is that of Lemma 1 plus that of Lemma 2 plus that of Lemma 3. The proof of Lemma A will be that given for Lemma 1, followed by that given for Lemma 2, followed by that given for Lemma 3, followed by the remarks which we now make.

At the end of the proof of Lemma 2 we already have that S^* is the sequential probability ratio test.

We now prove that S^* minimizes the average risk. We adopt the notation and terminology of [1]. Suppose there were a test S such that

$$R(S) = R(S^*) - \delta, \qquad \delta > 0. \tag{1}$$

We shall construct a sequence of tests $S_0(=S)$, S_1, S_2, $\cdots$, such that, for $i = 0, 1, 2, \cdots$,

$$R(S_{i+1}) \leqq R(S_i) + \delta(2^{-i-2}) \tag{2}$$

and

$$\lim_{i\to\infty} R(S_i) = R(S^*). \tag{3}$$

From this it follows that

$$R(S^*) \leqq R(S) + \delta/2. \tag{4}$$

The contradiction between (1) and (4) proves the desired result.

If t is any sequential test let $n(t)$ be its associated stopping variable; the value of $n(t)$ at the point $\omega = x_1, x_2, \cdots$ will be denoted by $n(\omega, t)$. Let $r_j(\omega) = r_j(x_1, \cdots, x_j) = p_{1j}/p_{0j}$ as in [1]. Let T be the totality of all sequential tests t such that $E_i[n(t)] < \infty$, $i = 0, 1$. Define

$$T_0 = \{t \,\varepsilon\, T \mid r_j(\omega) \geqq A \quad \text{or} \quad \leqq B \Rightarrow n(\omega, t) \leqq j, j \geqq 1\}$$

Received 4 November 1965.

[1] Research under contract with the Office of Naval Research.

and

$$T_j = \{t \, \varepsilon \, T_0 \mid P_i\{\omega \mid n(\omega, t) = j, B < r_j(\omega) < A\} = 0, i = 0, 1\},$$

where P_i indicates probability under the hypothesis H_i. We note that, if S is not in T_0, we may, by Lemma 3, (c), replace it by an obvious modification $S' \, \varepsilon \, T_0$ such that $R(S') \leqq R(S)$. We therefore assume that S is in T_0. Let α be the integer such that $S \, \varepsilon \, \bar{T}_\alpha \bigcap_{i=1}^{\alpha-1} T_i$.

Let $\epsilon > 0$ be arbitrary. It follows from Lemma 3 and the argument of Lemma 2 that there exists a sequential test $Z(\epsilon)$ (resp. $Z'(\epsilon)$) which requires at least one additional observation, and is such that

$$l(x_1, \cdots, x_\alpha) - E(L_n \mid (x_1, \cdots, x_\alpha), Z(\epsilon)) > -\epsilon$$

for any $(x_1, \cdots, x_\alpha)$ such that $B \leqq r_\alpha(x_1, \cdots, x_\alpha) \leqq W_0 g_0 / W_1 g_1$ (resp., is such that

$$l(x_1, \cdots, x_\alpha) - E(L_n \mid (x_1, \cdots, x_\alpha), Z'(\epsilon)) > -\epsilon$$

for any $(x_1, \cdots, x_\alpha)$ such that $W_0 g_0/W_1 g_1 < r_\alpha(x_1, \cdots, x_\alpha) \leqq A$.) Now modify S as follows: when $B < r_\alpha(x_1, \cdots, x_\alpha) \leqq W_0 g_0/W_1 g_1$ (resp., when $W_0 g_0/W_1 g_1 < r_\alpha(x_1, \cdots, x_\alpha) < A$) and S calls for stopping taking observations, replace S by the test $Z(\delta/4)$ (resp., $Z'(\delta/4)$). Call the resulting test S_1. Since S, Z, and Z' fulfill the required measurability and integrability conditions, so does S_1, which also satisfies (2) for $i = 0$.

We repeat the above procedure in an obvious way on S_1, replacing $\delta/4$ by $\delta/8$, and obtaining S_2. We repeat the same procedure on $S_2, S_3, \cdots$, using $\delta/16, \delta/32, \cdots$. Since the index α increases by at least one each time, (3) follows. this completes the proof.

REFERENCE

[1] Wald, A. and Wolfowitz, J. (1948). Optimum character of the sequential probability ratio test. *Ann. Math. Statist.* **19** 326–339.

THE MOMENTS OF RECURRENCE TIME

J. WOLFOWITZ[1]

In connection with Poincaré's recurrence theorem Kac [1] obtained the mean of the recurrence time (formula (3) below) and the author [2] gave a very simple proof of this result. Recently Blum and Rosenblatt [3] obtained[2] the higher moments (formula (2) below). In the present note we obtain both results by an exceedingly simple and perspicuous argument. This note is entirely self-contained.

Let Ω be a point set, m a probability measure on Ω, and T a one-to-one ergodic measure-preserving transformation of Ω into itself. Let $A \subset \Omega$ be such that $m(A)>0$. For any point a in Ω let $n(a)$ be the smallest positive integer such that $T^n a \in A$; if no such integer exists let $n(a)=\infty$. Define $A_k = \{a \in A \mid n(a)=k\}$, $\bar{A}=\Omega-A$, and $\Gamma_k = \{a \in \bar{A} \mid n(a)=k\}$. Borrowing the notation of [3] we will define

$$p_n = m\{\Gamma_n \cup \Gamma_{n+1} \cup \cdots\}, \tag{1}$$

for $n \geqq 1$. We will also make use of the usual combinatorial symbol $(k)_j = k(k-1)\cdots(k-j+1)$ for k and j positive integers, with $(k)_0=1$.

Our object will be to prove that

$$D_j = \int_A [n(a)]_j dm = j(j-1)\sum_{k=j-2}^{\infty} (K)_{(j-2)} p_{k+1} \tag{2}$$

Received by the editors August 1, 1966.

[1] Fellow of the John Simon Guggenheim Memorial Foundation. Research supported in part by the U. S. Air Force under Contract AF 18(600)-685 with Cornell University.

[2] These moments have also been obtained by F. H. Simons, Notice #40 of the Eindhoven Technical School, December 23, 1966.

for $j \geqq 2$, the result of [3]. The result of [1] (also proved in [2]) is

$$D_1 = 1. \tag{3}$$

By Poincaré's recurrence theorem (e.g., [2]; ergodicity of T is not required) one has that $m(A_\infty)=0$. The ergodicity of T implies that $m(\Gamma_\infty)=0$.

The basic formula of our argument will be

$$T(A_k \cup \Gamma_k) = \Gamma_{k-1} \tag{4}$$

for $k \geqq 2$; it is so obvious as not to require proof. Using (4) repeatedly for $k=n+1, n+2, \cdots$ we obtain that

$$m(\Gamma_n) = \sum_{k=n+1}^{\infty} m(A_k), \qquad n \geqq 1; \tag{5}$$

$$p_n = \sum_{k=n+1}^{\infty} (k-n)m(A_k), \qquad n \geqq 1. \tag{6}$$

Thus

$$\begin{aligned} p_1 &= m(A_2) + 2m(A_3) + 3m(A_4) + \cdots \\ &= D_1 - \sum_{k=1}^{\infty} m(A_k) = D_1 - m(A). \end{aligned} \tag{7}$$

Obviously

$$p_1 = m(\overline{A}) = 1 - m(A), \tag{8}$$

so that (7) and (8) prove (3).

Using (6) in the right member of (2) we obtain that the coefficient of $m(A_k)$, $k \geqq j$, in the right member of (2) is

$$j(j-1)\left[\sum_{i=1}^{k-j+1} i(k-i-1)_{(j-2)}\right], \tag{9}$$

which is easily shown (e.g., by induction) to equal $(k)_j$. This proves (2).

References

1. M. Kac, *On the notion of recurrence in discrete stochastic processes*, Bull. Amer. Math. Soc. **53** (1947), 1002–1010.

2. J. Wolfowitz, *Remarks on the notion of recurrence*, Bull. Amer. Math. Soc. **55** (1949), 394–395.

3. J. R. Blum and J. I. Rosenblatt, *On the moments of recurrence time*, University of New Mexico mimeographed report, 1966; J. Math. Sci. (Delhi) **2** (1967), 1–6.

Cornell University

MAXIMUM PROBABILITY ESTIMATORS

L. WEISS[1] AND J. WOLFOWITZ[2]

(Received Jan. 28, 1967; revised March 17, 1967)

1. The program of [1] and [2]

The present paper improves the results of [1], which in turn improved the result of [2]. Some familiarity with the results of these papers would help understand the rationale of the present paper. However, we shall describe in very great brevity some of the ideas of [1] and [2] so as to make the present paper as self-contained as possible. Readers familiar with [1] may at this point proceed at once to section 2. Example III of section 5 briefly states the results of [1] in the case of a one-dimensional unknown parameter and shows them to be a special case of the present result.

The purpose (of [1] and [2]) was to develop a unified theory of asymptotically efficient estimators, and in the present paper we extend this theory further. The most important existing theory is that of maximum likelihood, which goes back to the 1920's and is due to R. A. Fisher; an enormous literature has grown up around it. This theory has the following inadequacies:

1.1) It has not been applied with any success to any case other than the so-called "regular" case (see [3], for example). The latter is that of independent, identically distributed chance variables with strong regularity conditions on their common density. There is nothing irregular about most of the non-regular cases, which include some of the most common and important problems of statistics, examples of which are: estimation of the parameters of stochastic difference equations and of Markov chains in general, and estimation of the parameters of uniform and exponential distributions.

1.2) In the regular case the problem was to cope, in a non-artificial and statistically meaningful way, with the difficulties presented by Hodges' superefficient estimators (for an example of the latter see, e.g., [2], p. 268).

---

1) Research supported by NSF Grant GP 3783.

2) Fellow of the John Simon Guggenheim Memorial Foundation. Research supported by the U.S. Air Force under Grant AF 18(600)-685 to Cornell University.

1.3) Even in the regular case the competition (to be the asymptotically efficient estimator) is limited to estimators which are asymptotically normally distributed, an ad hoc restriction not justifiable on statistically operational grounds.

1.4) Comparisons between estimators are made on the basis of the variances of their limiting distributions. This comparison is irrelevant for many distributions other than the normal.

In [2] the present line of investigation was initiated and applied to the regular case; the limitation to asymptotically normal estimators and comparison by variances (1.3) and 1.4) above) was removed. It was argued that only statistically operational restrictions can justifiably be imposed on the totality of estimators admitted to competition, and it was concluded that the restriction to be imposed is the uniform convergence of the distribution of the properly normalized estimators ([2], p. 270, third paragraph). This position was also taken in [1] (i.e., property U in the first paragraph on page 68 et seq.), and is taken in the present paper.

It was pointed out in [1] that the full force of the uniformity condition was never employed, and that certain special consequences of the uniformity condition (e.g., (3.7) or (3.17) of [1]) were sufficient to give the results. In the present paper the consequence of the uniformity condition which assures the desired result is (3.4) below. It is our position that (3.7) of [1] and (3.4) below are statistically artificial restrictions, and the uniformity condition is the statistically justifiable one. However, we state the theorem below with (3.4) purely for *mathematical* convenience. Simple regularity conditions on the distributions of the chance variables, together with the uniformity assumption, will more than suffice to imply (3.4).

Throughout our previous work and in the present work we regard the assumption that competing estimators need have a particular limiting distribution, e.g., normal, as statistically unjustifiable and unreasonably restrictive. Hence the variance cannot serve as a basis for comparison among different estimators. In [1] for the one-dimensional case we used the limit of the probability assigned to any interval centered at the value of the parameter, and in [2] the interval did not need to be centered at the parameter. More about this in section 2.

2. Comparison of the results of [1] with those of the present paper

In all that follows it is assumed that the uniformity condition or its appropriate consequence holds on the estimators, but no other condition whatever. Let $X(n)$ be the chance variables observed at the nth stage and $k(n)$ be the normalizing factor of the estimators (of course $k(n) \to$

∞ as $n \to \infty$). Let $r > 0$ be arbitrary. In [1] the present authors described an estimator (really a sequence of estimators) Z_n (which depends on r), called a generalized maximum likelihood (m.l.) estimator such that, if T_n is any competing estimator,

$$(2.1) \quad \lim_{n\to\infty} \mathrm{P}\{-r < k(n)(Z_n - \theta) < r \mid \theta\} \geqq \overline{\lim_{n\to\infty}}\, \mathrm{P}\{-r < k(n)(T_n - \theta) < r \mid \theta\}$$

for every scalar value θ in the parameter space Θ. (For definitions of this standard notation see section 3). Similar results held in the multi-dimensional parameter case. In the regular case the generalized m.l. estimator became the classical m.l. estimator, but the generalized m.l. estimator applied in a host of problems which hitherto could not be treated in the framework of the classical theory.

The following problems still remained:

2.2a) There was no "mechanical" method of finding generalized m.l. estimators like that of finding m.l. estimators.

2.2b) The estimator in the multi-parameter case was even more difficult to find and had other drawbacks.

2.2c) Would (2.1) hold for intervals not centered at the origin or for regions other than intervals? (For example, in the regular case Kaufman ([5], see also [1], section 5) had more general regions than intervals. For a comparison of Kaufman's result with the application of the result of the present paper to this case see section 5, II, and section 6, both below.)

2.2d) There were important problems for which the generalized m.l. estimator does not exist, for example, the problem of [4], discussed in section 7 below.

The maximum probability estimator of the present paper answers the problems raised in (2.2 a–c). It includes the generalized m.l. estimator as a special case, and may even exist where the latter does not *even* for intervals centered at the origin. (section 7 below.) Its optimal property need not be limited to intervals centered at the origin, as is the case with the generalized m.l. estimator.

3. Maximum probability estimators

For each positive integer n let $X(n)$ denote the (finite) vector of (observed) chance variables of which the estimator is to be a function. $X(n)$ need not have n components, nor need its components be independent or identically distributed. Let $K_n(x \mid \theta)$ be the density, with respect to a σ-finite measure μ_n, of $X(n)$ at the point x (of some appropriate space) when θ is the value of the (unknown to the statistician) parameter. The latter is known to be a point of the "parameter space" Θ.

An estimator T_n is a Borel measurable function of $X(n)$ with values in $\bar{\Theta}$. The set Θ is a closed region of m-dimensional Euclidean space and is contained in a closed region $\bar{\Theta}$ so that every (finite) boundary point of Θ is an inner point of $\bar{\Theta}$. Let $P\{\quad|\theta\}$ denote the probability of the relation in braces when θ is the parameter of the density of $X(n)$. We assume that, for almost every (μ_n) point x, $K_n(x\mid\theta)$ is a measurable function of θ in $\bar{\Theta}$.

Let R be some measurable set region in m-space. An estimator (chance variable) Z_n will be called a maximum probability estimator with respect to R if, for the sequence $\{k(n)\}\to\infty$ of normalizing factors (precise definition in (3.2) below), and for almost every (μ_n) value x of $X(n)$, $Z_n(x)$ is equal to a value of d in $\bar{\Theta}$ for which

$$\int K_n(x\mid\theta)d\theta\,, \tag{3.1}$$

the integral over the region $\{d-[k(n)]^{-1}R\}$, is a maximum (with respect to d); the set $\{d-[k(n)]^{-1}R\}$ will now be described. For each n,

$$k(n)=(k_1(n),\cdots,k_m(n))\,,$$

a vector with m components. Write $d=(d_1,\cdots,d_m)$. Then

$$d-[k(n)]^{-1}R=\{(z_1,\cdots,z_m)\in\bar{\Theta}\mid d_i-[k_i(n)]^{-1}y_i=z_i,\\ i=1,\cdots,m,(y_1,\cdots,y_m)\in R\}.$$

We assume that such an estimator exists for the problem under consideration, whatever be the value of θ in Θ. (See, however, the remarks at the end of this section. Also, it is actually sufficient that the probability, that such an estimator exists, should approach one.)

In all that follows the statement $k(n)\to\infty$ is to mean that each component of $k(n)$ approaches infinity. We define the vector

$$k(n)(Z_n-\theta)=k_1(n)(Z_{n1}-\theta_1),\cdots,k_m(n)(Z_{nm}-\theta_m)$$

and similarly for $k(n)(T_n-\theta)$. Also we define

$$|k(n)(Z_n-\theta)|=\max_i|k_i(n)(Z_{ni}-\theta_i)|$$

$$|k(n)(\theta-\theta_0)|=\max_i|k_i(n)(\theta_i-\theta_{0i})|\,.$$

THEOREM. *Let $\{Z_n\}$ be a maximum probability estimator which satisfies the following conditions for some sequence $k(n)\to\infty$ $(n=1,2,\cdots)$, and any $h>0$. As $n\to\infty$ we have*

$$\lim P\{k(n)(Z_n-\theta)\in R\mid\theta\}=\beta\,, \tag{3.2}$$

say, uniformly for all $\theta \in H$, *where*[3)]

$$H=\{\theta \,\big|\, |k(n)(\theta-\theta_0)| \leqq h\}, \qquad \theta_0 \in \Theta .$$

As $n \to \infty$ *and* $M \to \infty$ *we have*

$$\lim \mathrm{P}\{|k(n)(Z_n-\theta)|<M \,|\, \theta\}=1, \tag{3.3}$$

uniformly for all θ *in some neighborhood*[4)] *of* θ_0. *Let* $\{T_n\}$ *be any estimator such that, as* $n \to \infty$,

$$\lim [\mathrm{P}\{k(n)(T_n-\theta) \in R \,|\, \theta\} - \mathrm{P}\{k(n)(T_n-\theta_0) \in R \,|\, \theta_0\}]=0 \tag{3.4}$$

uniformly for all $\theta \in H$. *Then*

$$\beta \geqq \overline{\lim}\, \mathrm{P}\{k(n)(T_n-\theta_0) \in R \,|\, \theta_0\}. \tag{3.5}$$

The proof will be quite easy. Suppose, to the contrary, that the right member of (3.5) exceeds the left member by $3\gamma>0$. Later we will choose h (large) as a function of γ. For each n consider the following Bayes problem: The parameter θ is a chance variable with a distribution such that the vector $k(n)(\theta-\theta_0)$ is distributed uniformly in H. The value of the loss function $L(d, \theta)$ when d is the value of the estimator (the estimate) and θ is the "true" value of the parameter is -1 if d lies in the set $\theta+[k(n)]^{-1}R$ and zero otherwise. Let z be the radius of a sphere with center at the origin which contains R. Let B_n be a Bayes estimator. (Conceivably a Bayes estimator may fail to exist for some points in the sample space. The argument below will show that this is of no consequence.) Obviously $Z_n=B_n$ whenever

$$|k(n)(Z_n-\theta_0)| \leqq (h-z). \tag{3.6}$$

Let A be the set of points θ such that

$$|k(n)(\theta-\theta_0)| \leqq (1-\gamma)^{1/m}h. \tag{3.7}$$

For convenience write $\gamma_0=1-(1-\gamma)^{1/m}$. Suppose that $(\gamma_0 h-z)>0$ and

$$|k(n)(Z_n-\theta_0)| \leqq (\gamma_0 h-z). \tag{3.8}$$

Then (3.7) and (3.8) imply (3.6). Now choose h so large that, using (3.3), we have, for all n sufficiently large,

$$\mathrm{P}\{\text{inequality (3.8)} \,|\, \theta\}>1-\gamma \tag{3.9}$$

---

3) This is a brief way of saying the following: The convergence in (3.2) is uniform for all sequences $\{\theta_n\}$ such that θ_n satisfies $|k(n)(\theta_n-\theta_0)| \leqq h$ for $n=1, 2, \cdots$.

4) It is actually enough if (3.3) holds uniformly in a neighborhood

$$|k_0(n)(\theta-\theta_0)|< \text{a constant},$$

where $k_0(n)=(k_{01}(n), \cdots, k_{0m}(n))$ is such that $k_{0i}(n)/k_i(n) \to 0$, $k_{0i}(n) \to \infty$, as $n \to \infty$, for $i=1, \cdots, m$.

uniformly for all θ in H. Consequently, letting h^* be the reciprocal of the volume of H, we have from (3.2),

$$\overline{\lim_{n\to\infty}}\, h^* \int_H [\mathrm{P}\{k(n)(B_n-\theta) \in R \mid \theta\}]d\theta \tag{3.10}$$

$$\leqq \lim_{n\to\infty} h^* \int_A [\mathrm{P}\{k(n)(Z_n-\theta) \in R \mid \theta\}]d\theta + 2\gamma < \beta + 2\gamma .$$

Since B_n is a Bayes estimator we have

$$h^* \int_H [\mathrm{P}\{k(n)(B_n-\theta) \in R \mid \theta\}]d\theta \geqq h^* \int_H [\mathrm{P}\{k(n)(T_n-\theta) \in R \mid \theta\}]d\theta . \tag{3.11}$$

Hence

$$\overline{\lim_{n\to\infty}}\, h^* \int_H [\mathrm{P}\{k(n)(B_n-\theta) \in R \mid \theta\}]d\theta \tag{3.12}$$

$$\geqq \overline{\lim}\, \mathrm{P}\{k(n)(T_n-\theta_0) \in R \mid \theta_0\}.$$

From (3.10) and (3.12) we obtain

$$\beta + 2\gamma > \overline{\lim}\, \mathrm{P}\{k(n)(T_n-\theta_0) \in R \mid \theta_0\}. \tag{3.13}$$

This contradiction proves the theorem.

When this theorem is invoked to justify the use of a maximum probability estimator the hypotheses of the theorem must of course be satisfied for all possible θ_0, since θ is unknown. Hence they should be satisfied for all points in Θ. When $n \geqq N_0$ it will be possible to choose Θ suitably small (depending on N_0) by making use of a consistent estimator of θ, as is done in [1]. (See also the next two paragraphs.)

We note that any estimator which differs from Z_n by a chance variable T_n', say, such that, for all $\theta \in \Theta$,

$$\lim \mathrm{P}\{k(n)(Z_n-\theta) \in R \mid \theta\} = \lim \mathrm{P}\{k(n)(Z_n + T_n' - \theta) \in R \mid \theta\} ,$$

has the same asymptotic efficiency properties as Z_n. In all that follows we shall simply call $Z_n + T_n'$ a maximum probability estimator without any further qualification. Thus we shall not bother to distinguish between different elements of what is for our purposes (asymptotic efficiency) an equivalence class. This procedure has an obvious justification.

Sometimes it may happen that the function which maximizes (3.1) for all d in $\bar{\Theta}$ is useless for our purposes, because it does not satisfy the hypotheses (3.2) and (3.3) of the theorem. For example, in [1], section 4, example VIII, this maximizing function would be the m.l. estimator which there is a manifestly absurd estimator. However, the device used in [1], especially in the example cited, and referred to in the present paper two paragraphs earlier, can often be used with success. We describe the use of this device in the case $m=1$, where it is simplest to describe; the general case is the same. Suppose ϕ_n is a consistent estimator of θ, with normalizing factor $k(n)$. Then we maximize (3.1) with respect to d in the interval

$$[\phi_n(x)-w(n)[k(n)]^{-1}, \phi_n(x)+w(n)[k(n)]^{-1}].$$

Here $w(n)$ is any sequence such that $w(n)\to\infty$, $w(n)[k(n)]^{-1}\to 0$. Thus the domain of d is a small interval which depends on x. The maximizing value of d in this interval often fulfills the hypotheses of the theorem and is then obviously a member of the equivalence class of maximum probability estimators. This device is also used below to simplify obtaining a maximum probability estimator; see the argument of section 7, especially after (7.7).

Suppose Z'_n satisfies the regularity requirements imposed above on Z_n, and is such that the integral (3.1) over the set

$$\{Z'_n(x)-[k(n)]^{-1}R\}$$

is greater than the supremum of (3.1) with respect to d, minus l_n. Here $\{l_n\}$ is a sequence of positive constants which approaches zero. Then obviously Z'_n is asymptotically efficient in the sense of the theorem, by the same proof.

4. Extension to more general regions

In all the foregoing R was assumed to be a fixed region, but this is not always necessary or desirable. For example, when estimating a scale parameter like the standard deviation (σ, say) of a normal distribution, one is more likely to be concerned with relative error than with absolute error, i.e., with the probability that the estimator lies in the interval $(\sigma_0 d_2, \sigma_0 d_1)$, $0<d_2<1<d_1$, rather than with the probability that it lies in the interval $(\sigma_0-d_3, \sigma_0+d_4)$, $0<d_3, d_4$. We will show that this problem, too, can often (when the simple hypotheses are satisfied) be asymptotically efficiently solved in the same easy way as in the preceding section.

Let $R(\theta)$, $\theta\in\bar{\Theta}$, be a set in the parameter space which depends upon θ. We consider the same Bayesian problem as in the proof of the theorem, except that R is replaced by $R(\theta)$. Let V_n be the Cartesian product of $\bar{\Theta}$ with the interval

(4.1) $$[\theta_0-h[k(n)]^{-1}, \theta_0+h[k(n)]^{-1}].$$

For any point $d\in\bar{\Theta}$ define the set $W_n(d)$ as follows:

(4.2) $$W_n(d)=\{(d,\theta)\mid(d,\theta)\in V_n,\ d\in[\theta+[k(n)]^{-1}R(\theta)]\}.$$

Then a Bayesian estimator B_n of θ can be described as follows: $B_n(x)$ is a value of d for which the integral

(4.3) $$\int K_n(x\mid\theta)d\theta$$

over $W_n(d)$ is a maximum with respect to d in $\bar{\Theta}$. If there is a (chance variable) Borel measurable function Z_n of $X(n)$ which is *not* a function of θ_0 and is such that

(4.4) $$P\{Z_n=B_n\mid\theta\}\to 1$$

as $n\to\infty$, uniformly for θ in H, and if the other conditions (or similar conditions) of the theorem are fulfilled, then the conclusion and proof of the theorem hold as in section 3. Indeed, the idea of the maximum probability estimator is that, while it does not depend on θ_0, it is asymptotically equivalent to a Bayes "estimator" which maximizes the desired probability. The latter "estimator" cannot really serve as an estimator because it depends on the value of θ_0. Define

$$V_n(d)=\left\{\theta\mid d\in[\theta+[k(n)]^{-1}R(\theta)]\right\}.$$

Then the maximum probability estimator Z_n is defined as follows for almost every (μ_n) value x of $X(n)$: $Z_n(x)$ is equal to a value of d in $\bar{\Theta}$ for which

$$\int K_n(x\mid\theta)d\theta,$$

the integral over $V_n(d)$, is a maximum.

When $R(\theta)$ is at all unusual or atypical it is best to go back to first principles to give reasonable sufficient conditions for (4.4) to hold. It would probably be easier to give such conditions than to verify the hypotheses of a general theorem. We shall therefore content ourselves with giving such conditions for the problem of estimating a (positive) scale $(m=1)$ parameter θ, where we would like to maximize the probability that the estimator lies within the interval $\left(\theta_0\left(1-\frac{b_1}{k(n)}\right),\ \theta_0\left(1+\frac{b_2}{k(n)}\right)\right)$, b_1 and b_2 given positive constants. In practice b_1 and b_2 are usually equal, but this is not necessary for our purposes.

Proceeding as in the proof of the theorem, we define $L(d,\theta)=-1$ if $\theta\left(1-\frac{b_1}{k(n)}\right)<d<\theta\left(1+\frac{b_2}{k(n)}\right)$ and 0 otherwise. Then $Z_n=B_n$ whenever

(4.5) $$\frac{\theta_0 b_2}{k(n)}-\frac{h}{k(n)}\left(1+\frac{b_2}{k(n)}\right)\leqq Z_n-\theta_0\leqq\frac{-\theta_0 b_1}{k(n)}+\frac{h}{k(n)}\left(1-\frac{b_1}{k(n)}\right).$$

Let $b=\max(b_1, b_2)$. Then (4.5) is satisfied when

(4.6) $$|k(n)(Z_n-\theta_0)|\leqq h\left(1-\frac{b}{k(n)}\right)-\theta_0 b.$$

When n is large (4.6) is satisfied if

(4.7) $$|k(n)(Z_n-\theta_0)|\leqq(h-2\theta_0 b).$$

Then (3.7) and

(4.8) $$|k(n)(Z_n-\theta)|\leqq(\gamma_0 h-2\theta_0 b)$$

imply (4.7). As in the proof of the theorem in section 3, it will now be sufficient if we can choose h so large that, for all n large enough,

(4.9) $$\mathrm{P}\{\text{inequality (4.8)}\,|\,\theta\}>1-\gamma$$

uniformly for all θ in H. The only essential difference now is the presence of the unknown θ_0 in the right member of (4.8). A sufficient condition for the existence of an h such that (4.9) holds for large n is therefore, for example, that $\bar{\Theta}$ be bounded.

5. Applications and examples

I) The regular case, $m=1$ (m is the dimensionality of the space which contains θ). The asymptotic behavior of $K_n(\cdot|\theta)$ is given, for example, in (2.3) of [1]. From this one can determine Z_n for a given R. When R is the interval (a, b) we may set

$$Z_n=\hat{\theta}_n+(2\sqrt{n})^{-1}(a+b).$$

When $a=-b$ this is the result of [1] in the regular case. Thus even in this special case (the regular case, $m=1$) our present results are more general than those of [1]. When the competing estimator T_n is asymptotically normally distributed the latter result of [1] implies the comparison between variances which is the classical efficiency result announced by Fisher.

II) The regular case, $m>1$. Essentially the same remarks apply here as in example I. Hence, when R is bounded, convex, and symmetric about the origin, it follows from the theorem of Anderson [6] that $Z_n=\hat{\theta}_n$. (Anderson's theorem implies the following: Let A^* be a convex, symmetric set. Then the integral of a normal density whose means are zero, over the set $d+A^*$, is a maximum with respect to d

at $d=0$.) This proves the result of Kaufman [5] (or [1], section 5) under the limitation (not made in [5]) that R is bounded. Our result can also be extended to unbounded R; this will be described in section 6. On the other hand, our results are not limited to convex and symmetric R.

III) Generalized maximum likelihood estimators. The results of [1] for $m=1$.

An estimator W_n is called a generalized maximum likelihood estimator if, for some $r>0$, the following hold[5]: There is a set S_n in the space of $X(n)$ such that

$$\lim P\{X(n) \in S_n \mid \theta_0\} = 1,$$

$$\lim P\left\{X(n) \in S_n \,\middle|\, \theta_0 + \frac{r}{k(n)}\right\} = 1,$$

and there exist two sequences

$$\{a_n(X(n), \theta, r)\}, \qquad \{a'_n(X(n), \theta, r)\}$$

of Borel measurable functions of the arguments exhibited such that $a_n(X(n), \theta, r)$ (resp. $a'_n(X(n), \theta, r)$) converges stochastically to zero as $n \to \infty$ when θ (resp. $\theta + r/k(n)$) is the parameter of the density of $X(n)$, and, whenever $X(n) \in S_n$ we have

$$\text{(5.1)} \qquad \frac{K_n\big(X(n) \mid \theta + 2r/k(n)\big)}{K_n(X(n) \mid \theta)} \leqq 1 \Longrightarrow W_n < \theta + \frac{r}{k(n)} + \frac{a_n}{k(n)},$$

and

$$\text{(5.2)} \qquad \frac{K_n\big(X(n) \mid \theta + 2r/k(n)\big)}{K_n(X(n) \mid \theta)} > 1 \Longrightarrow W_n > \theta + \frac{r}{k(n)} + \frac{a'_n}{k(n)},$$

and (3.2) holds with Z_n replaced by W_n and R by the interval $(-r, r)$. It was shown in [1] that (3.5) of the present paper holds with R the interval $(-r, r)$.

Consider $K_n(X(n) \mid \theta)$, for each value of $X(n)$, as a function of θ. It follows easily that, whenever (5.1) and (5.2) hold, the integral

$$\int_{d - r[k(n)]^{-1}}^{d + r[k(n)]^{-1}} K_n(x \mid \theta) d\theta$$

[5] The language of ([1], p. 76), implies that there (3.5a) and (3.6a) imply (3.5) and (3.6). This is wrong, but the basic idea is correct, namely (what should actually have been said in [1]), that (3.5a) and (3.6a) can replace (3.5) and (3.6) in the hypothesis and proof of theorem 3.1 of [1] without any change in the conclusion. Actually (3.5a) and (3.6a) are consequences of (3.5) and (3.6). The latter are (5.1) and (5.2) of the present paper.

is a maximum with respect to d at a point $\hat{d}_n$, say, such that

$$\lim P\{|k(n)(\hat{d}_n(X(n))-W_n(X(n)))|<\delta\,|\,\theta_0\}=1$$

for any arbitrary (fixed) $\delta>0$. Thus[6] W_n is asymptotically equivalent to Z_n, and a generalized m.l. estimator is, in this case, a maximum probability estimator. We shall show in section 7 by an example that *even for this* R the converse is not true. Thus the idea of maximum probability estimators enlarges the idea of generalized maximum likelihood estimators, which in turn was a generalization of the idea of maximum likelihood estimators, even in the special case when R is an interval centered at the origin.

IV) Translation and scale parameter cases. Let $X(n)=(X_1,\cdots,X_n)$ be n independent, identically distributed chance variables with common density (Lebesgue measure) $f(x\,|\,\theta)=f(x-\theta)$ at the point x of the (Euclidean line) sample space when θ $(m=1)$ is the value of the parameter. Let $\Theta=(-\infty,\infty)$. Obviously, for any R

$$Z_n(x_1,\cdots,x_n)+h=Z_n(x_1+h,\cdots,x_n+h),$$

i.e., Z_n is an invariant estimator. Similar remarks apply to the scale parameter case and appropriate $R(\theta)$.

V) Let S_n be any statistic sufficient for θ. Then Z_n is a function of S_n.

6. Extension to unbounded R; transformation of parameters

The limitation to bounded R was made simply for convenience. By taking a suitable sequence $R_1, R_2,\cdots$ of increasing, bounded regions such that $\cup R_i=R$, and employing the theorem (of section 3), one can obtain the result for unbounded R. The details can best be handled in each special case by going back to first principles.

Consider the regular case for $m\geqq 2$. Let R (independent of θ) be convex, symmetric about the origin and unbounded. Let R_i be the intersection of R with the sphere of radius i and center at the origin. Then (3.5) holds for R_i and $Z_n=\hat{\theta}_n$. Obviously therefore (3.5) holds for R with $Z_n=\hat{\theta}_n$. Thus the full result of [5] is a special case of that of the present paper.

If Z_n is a maximum probability estimator of θ with respect to R, and if g is a transformation from m-space to m'-space $(m'\leqq m)$, then it is clear that, under mild conditions on g, $g(Z_n)$ is a maximum probability estimator of $g(\theta)$ with respect to $g(R)$.

6) This follows from A) on page 75 of [1].

7. Estimation of the location of a discontinuity of a density function

The generalized maximum likelihood estimator exists in most problems involving a one-dimensional parameter and solves the asymptotic efficiency problem for intervals R centered at the origin. However, even in those cases the formula (3.1) offers the great advantage of a mechanical method of finding the generalized m.l. estimator. In order to study an interesting example of a problem where $R=(-r, r)$ and the generalized m.l. estimator does *not* exist, we consider a slightly simpler version of a problem studied in [4].

Let $X(n)=X_1, \cdots, X_n$, where the X's are independent chance variables with a common density $f(\cdot \mid \theta)$ defined as follows: Let $B>1$ be a constant known to the statistician. Then

$$f(x \mid \theta)=\begin{cases} B, & 0 \leqq x \leqq \theta \\ \dfrac{1-B\theta}{1-\theta}, & \theta < x \leqq 1 \\ 0, & \text{otherwise.} \end{cases}$$

The set $\bar{\Theta}=[0, 1/B]$.

Let $H_n(\cdot)$ be the empiric distribution function, i.e., $nH_n(x)$ is the number of X's whose values are $\leqq x$. Then

$$K_n(x \mid \theta)=B^{nH_n(\theta)}\left[\frac{1-B\theta}{1-\theta}\right]^{n(1-H_n(\theta))}.$$

Let $Y_1 \leqq Y_2 \leqq \cdots \leqq Y_n$ be the n X's arranged in order of size. It is easy to see that, for any i, $K_n(x \mid \theta)$ decreases monotonically as θ moves from Y_i to Y_{i+1}, and at Y_{i+1} the function $K_n(x \mid \theta)$ jumps upward. It is also easy to see from this behavior of the likelihood function that no generalized m.l. estimator exists, with probability approaching one as n increases, no matter what the true value θ_0 is ($\theta_0>0$).

It was proved in [4] that $\hat{\theta}_n=Y_a$ for some integer a, and that $n(\hat{\theta}_n-\theta_0)$ has a limiting distribution which is continuous in θ_0. It follows easily that a/n converges stochastically to $B\theta_0$. Since $K_n(x \mid Y_a) \geqq K_n(x \mid Y_{a+1})$ we have, writing $w=n(Y_{a+1}-Y_a)$,

$$\left[\frac{1-\dfrac{w}{n(1-Y_a)}}{1-\dfrac{Bw}{n(1-BY_a)}}\right]^{n-a} \geqq B\left[\frac{1-Y_a-\dfrac{w}{n}}{1-BY_a-\dfrac{Bw}{n}}\right]. \tag{7.1}$$

The left member, is for all large n, with probability approaching one, close to

$$\exp\left\{\frac{w(B-1)}{(1-\theta_0)}\right\}. \tag{7.2}$$

The right member converges stochastically to

$$B\left(\frac{1-\theta_0}{1-B\theta_0}\right). \tag{7.3}$$

Let $s(\theta_0)$ be the root in s of the equation

$$\exp\left\{\frac{s(B-1)}{(1-\theta_0)}\right\}=B\left(\frac{1-\theta_0}{1-B\theta_0}\right). \tag{7.4}$$

It follows that, for any $\delta>0$,

$$\lim \mathrm{P}\left\{w>s(\theta_0)-\frac{\delta}{2}\,\middle|\,\theta\right\}=1 \tag{7.5}$$

uniformly for all θ in a small neighborhood of θ_0. We shall now obtain a maximum probability estimator for $R=(-r, r)$, with

$$r<s(\theta_0)-\delta\,. \tag{7.6}$$

Let $\hat{\theta}_n=Y_a$, and let b be an integer such that

$$a-\sqrt{n}<b<a+\sqrt{n}\,.$$

Let $I(b)$ be an interval

$$\left[Y_b,\, Y_b+\frac{z}{n}\right],$$

with $Y_b+z/n\leqq Y_{b+1}$. Then

$$\begin{aligned} n\int_{I(b)} & K_n(x\mid\theta)d\theta \\ &=zB^b\left[\frac{1-BY_b}{1-Y_b}\right]^{n-b}\left[1+\frac{z(1-b/n)(1-B)}{2(1-Y_b)(1-BY_b)}+t_n(x)\right] \\ &=zB^b\left[\frac{1-BY_b}{1-Y_b}\right]^{n-b}\left[1+\frac{z(1-a/n)(1-B)}{2(1-Y_a)(1-BY_a)}+t'_n(x)\right] \end{aligned} \tag{7.7}$$

where $t_n(X(n))$ and $t'_n(X(n))$ converge stochastically to zero uniformly for θ in a suitable neighborhood of θ_0.

Let Z'_n be a maximum probability estimator of θ (i.e., it maximizes (3.1)) subject to the condition

$$|Z'_n-\hat{\theta}_n|\leqq\frac{1}{\sqrt{n}}\,.$$

Obviously

$$\lim P\{Z_n \neq Z'_n \mid \theta\} = 0 \tag{7.8}$$

uniformly for all θ in

$$H' = (\theta_0 - hn^{-1}, \theta_0 + hn^{-1}),$$

where h is a fixed large number as in the proof of the theorem in section 3. It follows from (7.5) and (7.7) that

$$\begin{aligned} &\lim\left[n\int_{H'} P\{|\hat{\theta}_n + rn^{-1} - \theta| < rn^{-1} \mid \theta\}d\theta\right] \\ &= \lim\left[n\int_{H'} P\{|Z'_n - \theta| < rn^{-1} \mid \theta\}d\theta\right] \end{aligned} \tag{7.9}$$

Hence

$$\begin{aligned} &\lim\left[n\int_{H'} P\{|Z_n - \theta| < rn^{-1} \mid \theta\}d\theta\right] \\ &= \lim\left[n\int_{H'} P\{|\hat{\theta}_n + rn^{-1} - \theta| < rn^{-1} \mid \theta\}d\theta\right] \end{aligned} \tag{7.10}$$

It now follows from the argument of the theorem that the estimator

$$\hat{\theta}_n + rn^{-1}$$

has the asymptotic optimum property of Z_n relative to the interval $R = (-r, r)$ under the restriction (7.6).

If r does not satisfy (7.6), the argument above does not apply. However, it seems a reasonable conjecture that Z_n will be as given above. To prove this would require a detailed investigation of the joint asymptotic distribution theory of neighboring Y's. Since the example has served its purpose, we forego this.

CORNELL UNIVERSITY

REFERENCES

[1] L. Weiss and J. Wolfowitz, "Generalized maximum likelihood estimators," *Teoriya Vyeroyatnostey*, 11, No. 1 (1966), 68-93.

[2] J. Wolfowitz, "Asymptotic efficiency of the maximum likelihood estimator," *Teoriya Vyeroyatnostey*, 10, No. 2 (1965), 267-281.

[3] H. Cramér, *Mathematical Methods of Statistics*, Princeton University Press, 1946.

[4] H. Chernoff and H. Rubin, "The estimation of the location of a discontinuity in density," *Proc. 3rd Berkeley Symp.*, University of California Press, 1946.

[5] S. Kaufman, "Asymptotic efficiency of the maximum likelihood estimator," *Ann. Inst. Statist. Math.*, 18, No. 2 (1966), 155-178.

[6] T. W. Anderson, "The integral of a symmetric unimodal function," *Proc. Amer. Math. Soc.*, 6, No. 2 (1955), 170-176.

MARCH 1967 VOL. 18, NO. 7

THE NEW YORK STATISTICIAN

Official Publication of the New York Area Chapter of the American Statistical Association

REMARKS ON THE THEORY OF TESTING HYPOTHESES*

An Invited Paper . . . . *By J. Wolfowitz***

Elsewhere ("Reflections on the future of mathematical statistics," in the S. N. Roy Memorial Volume, edited by Professor R. C. Bose of the University of North Carolina) I have written of the divorcement from reality of much of current research in mathematical statistics. Since good work in applications is impossible without a good theory, this fact has grave implications for statistical practice. The blame for this situation must, in large part, be shared by the practical workers in statistics (a group of whom I am now addressing in their journal). It is their role to formulate the problems which they want the theoreticians to solve, to press actively for the solution of these problems, and to refuse to accept as answers unrealistic solutions or "solutions" which do not answer the real questions. However, the history of the theory of testing hypotheses is an example of collaboration between theoreticians and practical statisticians which has resulted in the greater obfuscation of important statistical problems and the sidetracking of much statistical effort.

No textbook on statistics but devotes a large fraction of its contents to tests of hypotheses (or tests of significance). Usually many tests are described, and the distribution theory of the test statistics developed at great length. Often detailed algorithms for the computation of the test statistics are given. Theoretical books discuss the partition of the sample space into two regions, etc., etc. Yet how many of these books discuss what it means, in any real problem, to test a hypothesis, which hypotheses should be tested, what one does if the hypothesis is accepted, what one does if the hypothesis is rejected? Surely it is these questions that an intelligent practical statistician would want to be able to answer, and it is the distribution theory he would be willing to take on faith. Of what avail is it to know an efficient method of computation unless one understands what to do with the results of the computations?

Surely the most influential book in the entire history of statistics has been R. A. Fisher's "Statistical Methods for Research Workers." A typical edition devotes a major fraction of the book (actually most of it) to details of various tests of significance and either one sentence (!) (in my opinion) or one paragraph (according to the most extreme estimate) to the rationale of tests of significance. We can afford the luxury of space to quote the one sentence in full: "Critical tests of this kind may be called tests of significance, and when such tests are available we may discover whether a second sample is or is not significantly different from the first." The early teaching of Wilks at Princeton, coming as it did at a time when there were few centers of instruction in mathematical statistics and even fewer good books, had a profound influence on the development of mathematical statistics. The 1946 edition of the influential notes of his Princeton lectures devotes one *sentence* to the rationale of tests of significance, so that once again we can afford the space to quote: "As a general rule one sets up a test with the hope of rejecting the hypothesis, and for this reason the hypothesis is often called a NULL HYPOTHESIS in such cases." (Emphasis by Wilks.) The same author's exhaustive and impressive 1962 book, of approximately 650 pages, omits even this one sentence explanation! The rationale of tests of hypotheses fares no better in Snedecor's influential exegesis of Fisher's methods, which has served as a bible to very large numbers of experimental workers.

Even more important is the fact that many actual statistical problems are not really problems in testing hypotheses, and that statisticians perform a Procrustean operation in fitting such problems into the testing hypotheses framework. Consider the statistician at an agricultural experiment station who tests (by the analysis of variance) whether eight varieties of corn, say, have the same mean yield. Surely no reasonable person will doubt (even without performing an experiment) that the answer to this question is "no," that eight different varieties could not possibly have the same mean yield. (Even if, by some miracle, they did have the same yield, what great disaster would be likely to occur if one wrongly concluded that their yields were not the same?) But, it will be argued, one is really trying to decide whether the variation among the eight mean yields is small. Is this really the question which the statistician most often wants to answer? I suggest to you that in such problems the statistician is most often interested in knowing which are the better varieties and by how much they are better. Always the problem is to *compare* the different varieties. The possible conclusion that they are not very dif-

*This note was written while the author was a fellow of the John Simon Guggenheim Memorial Foundation.

**Professor of Mathematics, Cornell University.

ferent is a very special one and only one of very many possible meaningful conclusions. Why make the entire problem hinge on this one special question?

Consider now what happens after the statistician has tested the (null) hypothesis that all the mean yields are equal. Suppose first, he concludes that this hypothesis is not rejected. In acting on this conclusion he will a) incur no loss if the null hypothesis is really true and b) incur a probably great loss even if seven of the varieties are essentially equivalent and the eighth distinctly superior. (When the latter situation is the case the probability of not rejecting the null hypothesis can be very substantial.) The fact is, though, that even a casual search of the literature will show that almost no experimenter acts afterward as if he *really believed* that there is no difference among the mean yields, so as to carry out the logical consequences of such a belief. Suppose, on the other hand, that the conclusion is that the null hypothesis is rejected, that there is a significant difference among the mean yields of the varieties. The theory is now silent about what he should do next, although what has been reached is simply a perfectly reasonable and foreseeable conclusion, the second of only two possible ones (a conclusion which, as we have said earlier, could without great loss have been reached in advance of the experiment). Thus, if the conclusion is to accept (i.e., not to reject) the null hypothesis, the statistician acts as if he did not really believe the conclusion, and, if the conclusion is to reject the null hypothesis, the statistician does not know what to do next. Does this make sense? Is this really what the statistician wants?

It is interesting to note how this question is treated in a book ("Mathematical Methods of Statistics," 1946, Princeton, N.J., Princeton University Press) by a very distinguished mathematician, Harald Cramér. I believe that he senses these difficulties but, in an attempt to cope with them in one or two examples (though not systematically), he is led into error. Thus, on page 422, in an attempt to make sure of the correctness of a conclusion reached from a test of hypothesis, he performs different tests of hypotheses on the same data, without saying anything about how many tests to perform or about the grave consequences to the level of significance. On page 457, in trying to decide what hypothesis to test, he ends up by testing an hypothesis on the data which suggested it.

The use of the conventional levels of significance (.05, .01, .001) without any regard for the power of the test is an absurdity which has been pointed out in many places. However, it also has an effect to which not enough attention has been paid and which we will describe in connection with the following illustration. I was once consulted by research workers in the cancer laboratory of a distinguished medical school. Their method of testing a proposed anti-cancer drug was this: Fifty patients suffering from terminal cancer would be divided at random into two equal-sized groups. One group was given the drug, the other not. The mean lengths of life for the two groups were then tested for equality (by Student's t), using one of the conventional levels of significance. When one considers the sample size and the small level of significance it is hardly surprising that almost all the results were non-significant. Even a drug which can achieve an appreciable prolongation of life would have a sizeable probability of producing a non-significant result. The research workers must have sensed this (they certainly did not understand it; see the second paragraph below) because, at the end of each report on a drug, they concluded that "further study" was "necessary."

Now certainly serious scientists should not go off half-cocked and hail some drug on the basis of one experiment (particularly one with a limited amount of data). But neither is it their function always so to qualify and hedge their conclusions that they never make an error because they never really say anything. Whether one wishes it or not, the fact is that only some of the already available drugs can be adequately tested, if only because new drugs are constantly being created and will also require testing. To say, of virtually all drugs tested, that they require further testing, is a useless conclusion, about as valuable as the recommendation that interest rates or wages should be set at the *right* level or that children should receive the *proper* amount of discipline. Let the reader search his soul, his publications, and the publications of others, to see how often conclusions are profoundly qualified because of the improper statistical formulation and the desire to avoid being burdened with responsibility for error.

The conclusion of my venture into assisting cancer research can be told briefly. I tried to explain to the doctors the meaning of the power of a test, and the advisability of testing fewer drugs but by larger samples whose size would be computed on the basis of the power of the test. The reactions were bewilderment, boredom, and an end to the consultations, in that order. The doctors thought that I was hedging (presumably they felt that this should be left to them). They expected from the statistician a simple and unequivocal answer in the form of a magic number. Incidentally, this raises another interesting question which also arises in many other scientific investigations. If the drug is created by a chemist and the definitive number produced by a statistician, what then is the scientific role of the scientist (e.g., doctor) in the investigation?

The above problem, properly interpreted, really could give an instance of correct application of the theory of testing hypotheses. Generally speaking, the latter can occur whenever one has to decide on the basis of statistical data, between (only) *two* courses of action whose desirability depends upon which of two statistical hypotheses is true. If the possible decisions are to adopt one hypothesis, or to adopt the other, or a vague "further experimentation is necessary," then this problem does not fall within the framework of the theory of testing hypotheses. On the other hand, a prescribed routine of testing which terminates only when one of two decisions is made, e.g., the Wald sequential method of deciding between two values of an unknown parameter, does fall under the theory of testing hypotheses. The above problem in cancer research could very reasonably be formulated as a testing hypotheses problem. Let the experimenter decide on an amount μ, say, of prolongation of life, which he would consider as indicating importance for this drug and perhaps related drugs. One would then compute the sample size for an experiment to decide whether the mean prolongation of life due to the drug is zero or (at least) μ, which would produce reasonable and desired levels of errors of types I and II. It is very likely that such a sample size would make it impossible to test all the drugs which have been presented. (We have already seen that all were not really effectively tested under the procedure actually adopted.) The experimenter must then choose, on the basis of his theory of cancer and cancer chemotherapy, which drugs he will choose to test. (His scientific judgment plays the crucial role here and may distinguish the great and/or the lucky scientist.) The conclusion of each experiment will then be either "this drug does not merit further testing" or "this drug is sufficiently promising to warrant further investigation" (and "we really mean it and are not just passing the scientific buck").

The advice implied in Wilks' notes, that the null hypothesis is the hypothesis one hopes to be able to reject on the basis of the experiment, may often be sound. Sound practice requires due attention to sample size and levels of type I and II error, and a specific plan for a course of action in the event that the null hypothesis is *not* rejected. Otherwise one's actions may resemble those of a manufacturer who keeps putting the same lot through sample inspection until it finally passes and is whisked away to the consumer.

To sum up the present, already overlong note, which is to be an editorial and not a full and detailed analysis of the theory of testing hypotheses: Many, many statistical problems which are treated as problems in testing hypotheses are not really such problems at all. Many other problems could properly be treated as problems in testing hypotheses, but only after correct and drastic reformulation. The result would then be better science and less scientific evasion. The practical statisticians who now accept useless theory should rebel and not do so any more. It is not necessary for them to question or challenge mathematical arguments and conclusions which are usually impeccable and which the practical statistician is often not competent to criticize. But, when mathematical conclusions are to be translated into recommendations for action, the practical statistician should understand and question and be critical. The literature shows that he has frequently failed to exercise critical judgment.

What about the mathematical statisticians, especially the talented and mathematically productive among them? Most of the papers they write on testing hypotheses are, *at their best,* mathematically ingenious and difficult, of no general mathematical interest, and of no lasting statistical importance. They rehash the old problems and prove obscure and difficult-to-prove properties of tests, which themselves are of limited utility and applicability. What a waste of talent and energy! Why not tap this reservoir of talent for productive purposes by posing real problems and showing that you will be satisfied only with real answers?

Reflections on the Future of Mathematical Statistics.

J. WOLFOWITZ[1], *Cornell University*

1. INTRODUCTION

During his stay at Columbia University in 1948-9, S. N. Roy often discussed with me the directions in which, we thought, (mathematical) statistics would or should develop. It is therefore not unfitting to discuss this subject in a volume dedicated to his memory. Nor is it entirely superfluous to do so, for the future of statistics, both as a field for research and as a discipline to be studied, is far from assured. Mathematical fashions change, and fields once regarded as of burning research interest, (e.g., projective geometry) are now considered uninteresting or played out. Some subjects, for example, differential equations, we will always have with us, at least as subjects to be studied; the needs of science and technology alone will suffice for this. However, one needs to examine whether present day statistics really

[1] The writing of this paper was supported by the Air Force Office of Scientific Research, Office of Aerospace Research, U.S. Air Force, under AFOSR Grant No. 396–63 to Cornell University.

meets scientific and technological needs. Statistics arose in response to such needs, and these are, or should be, intimately connected with the directions of research. It seems to me obvious that, were it not for these needs, the future of statistics as a discipline for research and its appeal to pure mathematicians would be limited indeed.

However, even at this point we must already introduce an obvious caution. Suppose a mathematical study does have its *raison d' être* in the solution of certain mathematical problems which arise in science and technology. It would be a grave mistake to construe the content of the science too narrowly. Even an applied subject needs a theory which can often facilitate the solution of "practical" problems. Rigorous justification of results obtained has many advantages: protection from error, a sense of psychological security, aesthetic grounds, the possibility of suggesting new solutions and interesting new problems. A colleague of mine once said that there are two kinds of applied mathematics, pure applied mathematics and applied applied mathematics. A branch of applied mathematics which is to endure needs to contain both.

2. THE NEO-BAYESIANS

The period after World War II saw the emergence of a school of neo-Bayesians, deeply critical of previous developments and filled with a proselytizing missionary zeal. Not their philosophical doctrines but their role in the future of statistics will be discussed here.

One of the leaders of this school has written (Savage [1]): "I hope that I have made you feel that statistics is going places today (i.e., is achieving considerable sucess, J. W.). There is a lot of exciting work to do that is relatively easy. Any of you who are inclined to work in statistical theory can get in on the ground floor now (i.e., be in at the founding of the subject, J. W.), for there are not any experts any more. Those who only use statistics without intending to do research in it must also remember that there are not any experts any more. Don't take any wooden nickels (i.e., don't be gullible, J. W.); do your own work honestly and thoughtfully without much reliance on rules; and don't believe in magic or powerful new tools". Even accepting this position for the sake of argument one cannot but raise certain obvious points. How can any scientific endeavor which is "relatively easy" be "exciting" or hold out much prospect as a research discipline? The role of the scientist is to push back the frontiers of knowledge in important directions. This is never easy, or the frontiers would not be where they are. If a certain field or certain directions have reached the state where they are sufficiently explored, the scientist worth his salt and not yet ready to retire, looks for challenging problems in other fields. According to the Bayesian point of view one has only to determine the a priori distribution and then compute the a posteriori distribution. Can the study of this engage for long the serious efforts of first-rate minds? What challenges can this offer to brilliant young students? It is obvious that, if one accepts the Bayesian view as the answer to the problems of statistics, then the answer is so sweeping as to deny the existence of enough challenging problems to constitute a research discipline with much attraction to first-rate minds.

Since the Bayesian answer is so sweeping and relatively so easy to give, and since (the Bayesians say) there are so few (if any) difficult problems to solve (when one adopts their point of view), it is a continual source of wonderment to me to see the growth of some Bayesian university statistics departments. What can one member of such a department teach that is so different from what his colleague is teaching at the same time?

3. DECISION THEORY

The rest of this essay is addressed only to those who believe that many, if not most, of the problems which actually rise in statistics can and should be solved in the framework of decision theory.

Decision theory arose to replace the theory of testing hypotheses. In the latter theory the formulation of the problems is inadequate for most of the problems which arise in actual statistical investigations. These inadequacies have often been discussed and need not be considered here.[2] (It is surely an irony of history that decision theory was founded by a theoretician like Wald, and that it was not those who use the results of statistical research, the experimental scientists, who first rebelled against fitting their problems into the Procrustean bed of testing hypotheses. No one is so impractical as the so-called practical men.) It is obvious that decision theory has failed the high hopes once held for it, and it is pertinent briefly to discuss here the reasons for this.

Overzealous converts to decision theory probably helped, by exaggerated claims, to launch the subject and Wald's book to a bad start. Even the jacket of the book contained advertising matter which claimed that the book explained a new theory which would enable the experimenter properly to design experiments and draw conclusions from them. It should have added "eventually, after the theory is properly developed". The practical statistician looked at the formidable notation and terminology, and at the theorems on the closure of classes of decision functions in various topologies, and drew the only conclusion he could draw from these, namely, that a statistical theory which needed such a formidable mathematical apparatus merely to explain its fundamental ideas was forever beyond his reach and useless to him[3]. Consequently it was certainly impossible for him to formulate his actual problems in decision theoretic terms and to ask the mathematicians for solutions which would be meaningful to him.

Thus the mathematician working in statistics was not presented with actual problems to which to apply the theory. Per-

[2] For a recent discussion see the paper listed as [7] below.

[3] The fundamental ideas can be explained, even rigorously, in much simpler and more readily accessible settings. See, for example, [2], Chapter 6, or [3], Section 1.

haps he had no feeling for, or much interest in, "applied" problems, except when the latter could be posed in attractive mathematical form. All his professional instincts always pushed him to generalize and deepen. When even the fundamentals of the theory were presented in such abstract terms, was he not justified in pushing further in the same direction? Also, Wald's proofs of his closure theorems and other developments could be made "cleaner" and much more elegant by use of the new abstract formulations which are so characteristic of the post-World War II mathematics. This was a challenge to the technician who, within varying limits, must be present in every professional mathematician. To meet this challenge required no new statistical ideas and really no basically new mathematical ideas, but simply technical mathematical knowledge and facility.

For these and perhaps other reasons the relatively few papers on decision theory written after Wald's death in 1950 are usually attempts at mathematical "cleanups", often *à la* Bourbaki, of his closure theorems. Such work obviously has a limited future, either on the basis of its mathematical interest or for its applicability to actual statistical problems. Only the emergence of a group of workers who will concern themselves with problems of greater mathematical novelty and greater statistical applicability will revivify the theory. On the basis of past history, it is highly unlikely that any stimulus in the latter direction will come from the applied statisticians. If there is to be such a revival it will have to come from the mathematical statisticians themselves.

4. IDEALIZATION

The role played by mathematical idealization in the "development" of decision theory and other branches of mathematical statistics is so important and so peculiarly different from its role in other sciences as to deserve special discussion. Mathematical idealization is a method of simplification and approximation and is employed in all sciences where mathematics is used, e.g., idealized gases in physics, etc. In mathematical statistics a crucial application occurs as follows: In actuality the observations which occur in any statistical experiment take values which belong to a (finite) set of integral multiples of some unit. Thus, the chance variables in the mathematical formulation of the problem should all take their values on a (finite) lattice. When the unit of measurement is small the problem can often be simplified and a good approximation obtained by replacing the distributions of the "actual" chance variables by continuous approximations. This is the idealization which occurs most frequently and which is of greatest importance in statistics.

In some problems involving chance variables with continuous distributions there arise measure theoretical difficulties not present when the chance variables are all discrete. (It is precisely to avoid these difficulties that the book [2] on decision theory deals almost entirely with discrete chance variables. Presumably its writers considered the measure theoretic problems as extraneous to the essence of decision theory.) For example, many of the difficult and delicate problems in the study of sufficient statistics arise *solely* for this reason. Some subtle and difficult papers have been written to solve these problems. The trouble is that most, if not all, of these problems are of no particular interest to the measure theorist and really have no enduring place in mathematics except possibly in mathematical statistics. In mathematical statistics they owe their existence to the idealization described above. Is there not something basically wrong with an idealization which *creates* difficult problems rather than serves to avoid them? And is there not something basically wrong with a subject if difficult problems arise from a supposedly simplifying idealization?

The reasons for studying such problems are easy to explain. When the problems are difficult they present a challenge to first-rate research workers which the latter find hard to resist. Once such a paper is written it is a challenge to improve it or carry it further. It would be wrong to say of any one paper that it should not have been written. But, unless a subject can produce many interesting and difficult papers which are not of this artificial character, its future would seem to be far from assured.

Of course our discussion has been of the difficult papers which it is an achievement to write. Trivial papers are written in every subject but do not determine its future, except when only trivial papers are being written.

It is also unnecessary to say that nothing in the above is to be construed as an argument against the need for rigorous proofs of statistical theorems. It seems difficult to believe that the value of, and need for, rigorous proof can seriously be questioned, but this has been done by Barnard [5]. In an earlier paper [4] this author had omitted to justify an inversion in the order of two limit processes, the most difficult part of an otherwise easy argument. This omission having been called to his attention, he supplied the missing argument in [5]. In the course of the latter he inveighed against pettifogging pedants who think that, even in mathematical statistics, one needs to justify the inversion of two limit processes. (What remarkable "theorems" one could prove without this vulgar constraint!) It is amusing to note that this author, in the very same paper [4], expects his reader to be familiar with Weill's *Integration dans les groupes topologiques* and seriously discusses the case where the statistical parameter to be estimated is an element of a general topological group (presumably not Euclidean space!)

5. THE THEORY OF TESTING HYPOTHESES

It has been said earlier that the theory of decision functions arose because of the inadequacies of the theory of testing hypotheses, and that the remainder of this essay would be addressed to those who believe that many, if not most, statistical problems are to be solved in the context of decision theory. Nevertheless, it is in order to make some remarks about work in testing hypotheses because a) many workers who teach the desirability of decision theory still write about testing hypotheses, and b) one of the most distinguished books on statistics is the book by Lehmann [6] on testing hypotheses.

Probably the most telling criticism of the theory of testing hypotheses is that it poses the wrong problems. For example, an experimenter is rarely, if ever, really interested in testing the hypothesis that eight varieties of wheat all have the same yield. (It is almost incredible that they should. What is really being tested is that the variation among their yields is "not large".) Yet this is what several famous tests are designed to test. In fact, as the tests are universally applied, it is concluded that all eight varieties have the same, or essentially the same, yield, unless the contrary is clearly demonstrated, i.e., unless a "significant" result is obtained! (Nor do these tests say what one is to do when a significant result is obtained.) Yet, first-rate statisticians, who freely acknowledge the validity of these criticisms, write papers about the admissibility, stringency, etc., of these tests. At their best, these papers deal with difficult problems whose solution requires considerable skill and ingenuity. However, they do not contribute to the essential development of statistics or to the solution of realistic problems, nor are they of permanent mathematical interest per se. What a pity to waste this talent and energy!

This is not the place to review Lehmann's book, but some comments on it are in order because of the great influence this book has and is bound to have. (It also contains material which does not fall under the theory of testing hypotheses and is not included in our comments.) The exercises in the book are an essential and important part of it. They carry the theory further, and many of them are extremely clever and ingenious. In fact, they are much more interesting than many of the formally stated theorems, which are relatively obvious or have proofs which are rather easy.

One comes away with a general impression of relatively few deep and difficult theorems, and of many clever and ingenious examples, mostly involving the binomial, normal, Poisson, and other distributions of the exponential family. So many ingenious tests about the latter have been studied, and so few problems of practical interest solved. Is this material likely to survive into the future, either because it meets the needs of experimental science or because of its enduring mathematical interest per se?

The author of [6], in the preface to his book and elsewhere, acknowledges the validity of some of the criticisms of the theory of testing hypotheses. In fact, from his remarks and other evidence, one concludes that he published the book because he had already expended so much ingenuity and effort on its contents. I agree enthusiastically that this book should have been published and that it is a distinguished book. I am delighted to have at hand in permanent form so many clever and ingenious examples. But I would consider it a disaster for statistics if this book should determine the direction of research for any appreciable period of time[4].

---

[4] Perhaps this is the place to raise a different and minor point, relevant to a discussion of what is one of the best and most influential books on statistical theory. A student of pure mathematics learns Poincaré's recurrence theorem or Hilbert's Nullstellensatz or about Finsler spaces or Lie groups. After all, how better to identify theorems than by the names of their discoverers? Incidentally, then, one also gets a little feeling for the grandeur of the subject and of its great men. My reading of Lehmann's work which, in this respect, is admittedly only casual, turns up only two theorems named after their discoverers. One is the "Neyman-Pearson fundamental lemma" which, no matter how "fundamental" it may be, is pretty trivial to prove and not difficult to discover. The other is the "Hunt-Stein" theorem, also not difficult to prove. Surely these distinguished writers have discovered deeper theorems to which their names could, with more luster, properly be attached. And what about other eminent writers, in particular such colossi as Fisher and Wald? Are there no theorems which should bear their names? The proof of what is perhaps the deepest theorem in the book displays most conspicuously not the name of him who brilliantly conjectured it, but of someone who later gave an alternate proof of one of its lemmas. What kind of feeling for the history and development of the subject is the student likely to get from such a treatment? Just as a taste for good literature is developed by reading good books, so good taste in science comes from knowing the great achievements of the past and how they fitted into the structure of the knowledge of their time.

6. CONCLUDING REMARKS

Just as it was necessary to hope that Lehmann's book does not determine the course of future research, although the book is so distinguished and eminently worth publishing, so it is necessary to understand the criticisms which have been made in this paper in the light of their context. Let us, therefore, recapitulate and summarize the principal of these.

Except perhaps for a few of the deepest theorems, and perhaps not even these, most of the theorems of statistics would not survive in mathematics if the subject of statistics itself were to die out. In order to survive the subject must be more responsive to the needs of application. On the basis of past experience it is too much to expect that the formulation of actual problems will come from "practical" experimenters; the change will have to come from within the ranks of the mathematical statisticians themselves. Some current work, e.g., in the design of experiments, does meet these needs as applied applied statistics and/or pure applied statistics.

Even when a branch of, or tendency in, statistics represents a sterile direction, it is a serious mistake to say of any one paper that, for this reason alone, it should not have been written. Rather one can only say of a series of papers that they do not develop the subject in a fruitful direction. It is very important to stress this. Obviously one does not want to dissuade anyone who is challenged by the difficulty of a problem from attempting to solve it. Nor should one dictate within strict bounds what should or should not be published. Dictatorship in science is as dangerous and repugnant as it is in other fields. (Of course, some choice and selection are absolutely unavoidable. Obviously everything cannot and should not be published. This has as an immediate consequence that some papers will be dismissed as "not interesting". Probably the best hope of avoiding error lies in the multiplicity of journals in many different countries.) What is to be hoped for is that the scientific concensus (to borrow a word now in vogue in another context) will be such that gifted workers will be induced to work in directions fruitful for statistics. It is for this reason that I have avoided discussing specific books or papers except in one or two egregious instances, e.g., Lehmann's book because of its great importance.

It goes without saying that "responsiveness to the needs of science and technology" should never be construed in any narrow sense. The subject obviously needs a theory (pure applied statistics) which should forge ahead and also interact with problems in application. What we must guard against is the development of a theory which, on the one hand, bears little or no relation to the actual problems of statistics, and which, on the other hand, when viewed as pure mathematics, is not interesting per se nor likely to survive.

Finally, let me anticipate the *tu quoque* criticism by pleading guilty to it in advance. Obviously my views on, say, testing hypotheses have changed and developed over the years. Criticism can be cogent even if the critic himself is not without blemish.

For any science, as in nuclear reactions, there is a critical mass, this time of scientific workers, who should be numerous enough to stimulate each other, gifted enough to see and to solve problems, and, by the esteem in which they hold some work and the disesteem in which they hold other work, successful in guiding the science into fertile and fruitful directions. Perhaps this essay may evoke a sympathetic response among others of similar but hitherto unvoiced views, who will be numerous enough and gifted enough to constitute such a critical mass.

References

[1] Savage, L. J. "The Subjective Basis of Statistical Practice," mimeographed notes, University of Michigan, July, 1961.

[2] Blackwell, D. and M. A. Girshick. *Theory of Games and Statistical Decisions*, John Wiley and Sons, New York, 1954.

[3] Wald, A. and J. Wolfowitz. "Characterization of the Minimal Complete Class of Decision Functions when the Number of Distributions and Decisions is Finite," *Proceedings of the Second Berkeley Symposium on Probability and Statistics*, pp. 149–158, University of California Press, 1951.

[4] Barnard, G. A. "The Frequency Justification of Sequential Tests," *Biometrika,* **39** (1952), 144–150.

[5] Barnard, G. A. "The Frequency Justification of Sequential Tests—Addendum," *Biometrika,* **40** (1953), 468–469.

[6] Lehmann, E. L. *Testing Statistical Hypotheses,* John Wiley and Sons, New York, 1959.

[7] Wolfowitz, J. "Remarks on the theory of testing hypotheses" *The New York Statistician,* **18** (1967), No. 7, 1–3.

(*Received Jan. 1, 1966.*)

Maximum Probability Estimators with a General Loss Function[1)]

L. Weiss[2)] and J. Wolfowitz[3)]
Cornell University, Ithaca, New York

1. Introduction.

The present paper is an extension of [1], but familiarity with the latter is not essential for an understanding of the present paper. The purpose of [1] was to solve the problem of asymptotic estimation in the general case, thus extending the results of [2] which in turn considerably generalized the method of maximum likelihood. Among the inadequacies of the classical maximum likelihood theory are the following: 1) The theory applies only under very onerous regularity conditions (the so-called "regular" case of Cramer [5] and others) which exclude many of the most frequent problems of statistics. For example, the case where the density, at a point x, of a chance variable whose distribution depends upon a parameter θ, is $e^{-(x-\theta)}$ when $x \geq \theta$ and zero otherwise, is not "regular"! 2) Only estimators which are asymptotically normally distributed are allowed to enter into competition with the maximum likelihood estimator. This is convenient for the theory and allows comparison on the basis of variances, but does not correspond to practical application or necessity. This requirement begs the question whether estimators which are not asymptotically normally distributed may not sometimes actually be more efficient. 3) The classical results are largely limited to the case m = 1, where m is the dimension of the unknown parameter. 4) The theory applies mainly to the case of independent, identically distributed chance variables.

1) Presented by the second author at the International Symposium on Probability and Information Theory, held April 4 and 5, 1968 at McMaster University, Hamilton, Ontario, Canada.

2) Research supported by NSF Grant GP 7798.

3) Fellow of the John Simon Guggenheim Memorial Foundation. Research supported by the U. S. Air Force under Grant AF 18(600)-685 to Cornell University.

The theory of maximum probability estimators is not subject to these limitations. Let $X(n)$ be the observed chance variable whose density with respect to a σ-finite measure μ_n depends upon an unknown parameter θ and at the point x is given by $K_n(x|\theta)$. Thus, in the regular case, $X(n) = (X_1,\cdots,X_n)$, where the X_i's are independently and identically distributed with common density function $f(\cdot|\theta)$, say, μ_n is Lebesgue measure on n-space, and $K_n(x|\theta)$, where $x = (x_1,\cdots,x_n)$, is $\prod_{i=1}^{n} f(x_i|\theta)$. (For precise definitions and notation in the case $m = 1$ see Section 2 below and for general m see [1].) Let R be a bounded measurable set in m-space (the space of the unknown parameter), and let $k(n)$ be a normalizing factor. The maximum probability estimator Z_n of θ (with respect to R) is equal to a value of d for which the integral

$$\int K_n(X(n)|\theta)d\theta$$

over the set $\{d - [k(n)]^{-1}R\}$, is a maximum. Under certain conditions (see [1] and Section 6 below) the inequality

$$\begin{aligned} &\lim P\{k(n)(Z_n-\theta) \in R|\theta\} \\ &\geq \overline{\lim}\, P\{k(n)(T_n-\theta) \in R|\theta\} \end{aligned} \tag{1.1}$$

holds for any competing estimator T_n which satisfies the reasonable condition (3.4) of [1], which is the analogue of (2.8) below. (For a discussion of practical requirements which imply this condition see [2], [3], and Section 5 below).

The many examples of [1] and [2], which range from the classical "regular" case through non-"regular" cases frequently encountered, e.g., in reliability theory, to complex and perhaps artificial examples, show the wide applicability of the method. In the very special "regular" case, when R is convex and symmetric (with respect to the origin), Z_n is (equivalent to) the maximum likelihood estimator, and the inequality (1.1) then already implies the classical efficiency result of Fisher when m = 1. The general theory of [1] also allows R to depend on θ.

The results of the present paper apply for a general loss function, and not just for the special loss function implied in the above problems. Consequently our results include those of [1] (hence those of [2]) and [6] as special cases.

Sections 2 and 3 contain the statements and proofs of the theorems in the case m = 1, the case chosen, in the interests of simplicity, for explicit statement. In Section 4 we show how the results can be extended at once to the case m > 1. Section 5 contains a discussion of the assumptions under which the present theorems and the results of [1] are proved. Some applications are given in Section 6.

2. Statement of the theorem for $m = 1$.

For each positive integer n let $X(n)$ denote the (finite) vector of (observed) chance variables of which the estimator is to be a function. $X(n)$ need not have n components (although the number of components will approach infinity), nor need its components be independently or identically distributed. Let $K_n(x|\theta)$ be the density, with respect to a σ-finite (positive) measure μ_n, of $X(n)$ at the point x (of the appropriate space) when θ is the value of the (unknown to the statistician) parameter. The latter is known to be a point of the "parameter space" Θ. Any estimator T_n is a Borel measurable function of $X(n)$ with values in $\overline{\Theta}$; the set Θ is a closed region of m-dimensional Euclidean space and is contained in a closed region $\overline{\Theta}$ such that every (finite) boundary point of Θ is an inner point of $\overline{\Theta}$. Let $P\{\ |\theta\}$ denote the probability of the relation in braces when θ is the parameter of the density of $X(n)$. We assume that $K_n(x|\theta)$ is a Borel measurable function of both arguments jointly. Although our results are valid for any (finite) m, in this section we shall proceed for $m = 1$ in order to keep down the complexity of the notation.

Let $L_n(z,\theta)$ be a non-negative loss function, i.e., when the value of the estimator (function of $X(n)$) is z, and the value of the parameter which determines the density of $X(n)$ is θ, the loss incurred by the statistician is $L_n(z,\theta)$. In many problems one will have

$$L_n(z,\theta) = k(n)\ L(z,\theta),$$

where $k(n)$ ($\to \infty$) is the normalizing factor of [1] and the theorems below. For any $y > 0$ define

$$s_n^*(y) = \sup L_n(z,\theta),$$

the supremum being taken over all z and θ such that $|z-\theta| \leq y$.

Let $\{k(n)\}$, $\{k_1(n)\}$, $\{k_2(n)\}$ be sequences of positive numbers such that, as $n \to \infty$,

$$(2.1) \quad k_2(n) \to \infty, \ \frac{k_2(n)}{k_1(n)} \to 0, \ \frac{k_1(n)}{k(n)} \to 0.$$

Write for brevity

$$h_1(n) = \frac{k_1(n)}{k(n)}, \ h_2(n) = \frac{k_2(n)}{k(n)}$$

and

$$(2.2) \qquad s(n) = s_n^*(h_2(n)).$$

We assume that $s(n) < \infty$ for all n. Let Y_n be an estimator defined as a value of d which maximizes

$$(2.3) \quad \int_{d-h_2(n)}^{d+h_2(n)} [s(n) - L_n(d,\theta)]K_n(X(n)|\theta)d\theta.$$

We assume that such an estimator exists. This assumption is, however, not necessary. It is enough if the integral is maximized to within ℓ_n, where $\{\ell_n\}$ is a sequence of positive numbers which approach zero. Also, remarks corresponding to those made in [1], Section 3, third paragraph from the end, obviously apply, so that what has really been defined is an <u>equivalence class</u> of estimators.

We shall phrase our theorem in terms of a particular value of θ, say θ_o, as was done in [1] and [4] for convenience. Of course the hypotheses of the theorem should be satisfied for all possible θ_o(in Θ).

Define

$$H_n = \{\theta \,\big|\, |(\theta-\theta_o)| \leq h_1(n)\}$$

The phrase "converges uniformly in H_n" will mean uniform convergence for all sequences $\{\theta_n\}$ such that $\theta_n \in H_n$.

We shall prove the following theorem for a general loss function L_n:

Theorem 1. Let the estimator Y_n satisfy the following three conditions uniformly in H_n:

(2.4) $\lim_{n\to\infty} E\{L_n(Y_n,\theta)|\theta\} = \beta$, say,

(2.5) $\lim_{n\to\infty} [s(n)P\{|k(n)(Y_n - \theta)| > k_2(n)|\theta\}] = 0$,

and

(2.6) $\lim_{n\to\infty} \int_{B_n(\theta)} L_n(Y_n(x),\theta)K_n(x|\theta)d\mu_n(x) = 0$,

where

(2.7) $B_n(\theta) = \{x \,\big|\, |k(n)(Y_n-\theta)| > k_2(n)\}$.

Let $\{T_n\}$ be any estimator for which the following two conditions hold uniformly in H_n:

(2.8) $\lim_{n\to\infty} [E\{L_n(T_n,\theta)|\theta\} - E\{L_n(T_n,\theta_o)|\theta_o\}] = 0$

and

(2.9) $\lim_{n\to\infty} [s(n)P\{|k(n)(T_n-\theta)| > k_2(n)|\theta\}] = 0$.

(For a discussion of these conditions see Section 5.) Then

(2.10) $\beta \leq \underline{\lim}\, E\{L_n(T_n,\theta_o)|\theta_o\}$,

so that Y_n is asymptotically efficient in this sense.

If $s(n) < V$ (say) for all n sufficiently large, conditions (2.5) and (2.9) take an especially simple form. For more special but very important loss functions we shall also prove the following:

Theorem 2. If, for all n sufficiently large, $L_n(z,\theta)$ is a monotonically non-decreasing function of $|z-\theta|$, Theorem 1 holds even without the condition (2.9).

Theorem 3. If, for all n sufficiently large, $L_n(z,\theta) = s(n)$ for $|z-\theta| > h_2(n)$, Theorem 1 holds even without the condition (2.9).

3. Proofs of the theorems.

We first prove Theorem 1. Suppose that

$$(3.1)\quad \beta - \underline{\lim}\, E\{L_n(T_n,\theta_o)|\theta_o\} = 4\gamma > 0$$

Define $L_n^*(d,\theta)$ as follows:

$$L_n^*(d,\theta) = L_n(d,\theta),\ k(n)|\theta-d| \leq k_2(n),$$

$$L_n^*(d,\theta) = s(n),\ k(n)|\theta-d| > k_2(n).$$

Let $\varphi_n^*(X(n))$ be an estimator which minimizes

$$(3.2)\quad \int_{H_n}\int L_n^*(\varphi_n(x),\theta)K_n(x|\theta)d\mu_n(x)d\theta$$

with respect to φ_n. (If a minimum does not exist it is sufficient to minimize (3.2) to within ℓ_n, where the sequence of positive numbers $\{\ell_n\}$ approaches zero.) Obviously, whenever

$$(3.3)\quad |(Y_n-\theta_o)| \leq h_1(n) - h_2(n),$$

we can, and will, set $\varphi_n^* = Y_n$. The inequalities

$$(3.4)\quad |(\theta-\theta_o)| \leq h_1(n) - 2h_2(n)$$

and

$$(3.5)\quad |(Y_n-\theta)| \leq h_2(n)$$

imply (3.3). For n sufficiently large it follows from (2.1), (2.4), and (2.5) that

$$(3.6)\quad \lim_{n\to\infty} [h_1(n)]^{-1} \int E\{L_n^*(Y_n,\theta)|\theta\}d\theta = 0,$$

the integral in (3.6) being over the set

$$(3.7)\quad \{\theta \,\big|\, (h_1(n)-2h_2(n)) < |\theta-\theta_o| \leq h_1(n)\}.$$

Hence, for n sufficiently large, we obtain from (3.4), (3.5), (3.6), and (2.5) that

$$(3.8)\quad [h_1(n)]^{-1} \int_{H_n} E\{L_n^*(Y_n,\theta)|\theta\}d\theta$$

$$\leq [h_1(n)]^{-1} \int_{H_n} E\{L_n^*(\varphi_n^*,\theta)|\theta\}d\theta + \gamma$$

$$\leq [h_1(n)]^{-1} \int_{H_n} E\{L_n^*(T_n,\theta)|\theta\}d\theta + \gamma.$$

The last member of (3.8) is not greater than

$$(3.9)\quad [h_1(n)]^{-1} \int_{H_n} E\{L_n(T_n,\theta)|\theta\}d\theta + \gamma$$

$$+ s(n)[h_1(n)]^{-1} \int_{H_n} P\{|T_n-\theta| > h_2(n)|\theta\}d\theta.$$

From (3.8), (3.9), (2.6), and (2.9) we obtain that, for n sufficiently large,

$$(3.10)\quad [h_1(n)]^{-1} \int_{H_n} E\{L_n(Y_n,\theta)|\theta\}d\theta$$

$$\leq [h_1(n)]^{-1} \int_{H_n} E\{L_n(T_n,\theta)|\theta\}d\theta + 2\gamma.$$

From (3.10), (2.4), and (2.8) we obtain

$$(3.11)\quad \beta < \underline{\lim}\, E\{L_n(T_n,\theta_o)|\theta_o\} + 3\gamma.$$

This contradicts (3.1) and proves Theorem 1.

We now prove Theorem 2. When $L_n(z,\theta)$ is a monotonically non-decreasing function of $|z-\theta|$ we obtain (3.10) immediately from (3.8) and (2.6), without any use of (2.9). Theorem 2 follows.

Theorem 3 follows from (3.8) and the definition of L_n^*.

Theorems 2 and 3 have been stated so as to make Theorem 1 directly applicable to several important problems without the need for condition (2.9). It is obvious that other reasonable conditions on the loss functions $\{L_n\}$ will enable us to dispense with (2.9).

4. Extension to $m > 1$.

In [1] the results were given for general m. The extension to general m of the present results follows in the same straightforward manner as in [1]. No further comment is necessary except for the remarks of the following two paragraphs.

In the statements and proofs of Theorems 1, 2, and 3 for $m = 1$ we "truncated" the integration at a distance $h_2(n)$ on either side of θ. This is not necessary. One can integrate for different distances on either side of θ, provided only that these distances be such that they converge to zero, and that, after normalization (i.e., multiplication by $k(n)$), they converge to infinity. (Of course, the hypotheses and the definition of L_n^* have to be changed in a corresponding manner.) Not only does this freedom exist in the case $m > 1$, but also the "shape" of the set over which one integrates need not be that of an interval. It is sufficient that the normalized (for a description of the normalization see [1]; the various components $k(n)$ can be different, and even of a different order) set of integration expand steadily so as to include the entire m-space, and that the unnormalized set shrink steadily to the null set. This additional freedom will make it easier to apply the theory for some loss functions. The application of these remarks to the m-dimensional extension of Theorem 3 will be obvious.

As a consequence of the above remarks it is clear that, for the m-dimensional extension of Theorem 2, many definitions could be used to generalize the notion of a "loss function which is a monotonically non-decreasing function of $|z-\theta|$." In the proof of

Theorem 2 this was used only to assert that

$$(4.1)\quad E\{L_n^*(T_n,\theta)|\theta\} \leq E\{L_n(T_n,\theta)|\theta\},$$

and this can be achieved in many ways (when $m = 1$ and when $m > 1$). For example, if L_n is a monotonically non-decreasing function of $|z-\theta|$ along every <u>line</u> passing through θ, (4.1) will obviously hold for any region of integration of the sort described in the preceding paragraph. The function L_n^* will be constant along that part of any line through θ which is outside the set of integration. The constant may depend on the line.

5. Discussion of the assumptions.

The hypotheses of the theorems of [1] and of the present paper are phrased in terms of the behavior of the maximum probability estimator. These hypotheses will generally be easier to verify than the elaborate conditions on $K_n(\cdot|\cdot)$ imposed in many studies of maximum likelihood in the regular case. The abundance and range of examples for the application of maximum probability estimators are the best evidence of the applicability of the theory.

Much more important is the justification of the restrictions on the competing estimators, namely (2.8) and (2.9). Clearly, if these were artificially constructed restrictions designed to eliminate from competition the really efficient estimators, the results about the efficiency of the maximum probability estimator would have no value. We now recapitulate briefly the views expressed elsewhere ([3] and [2], Section 3). The justification of any restrictions must rest on the use to which the estimators will be put in statistical practice. In such practice, an asymptotic theory will be used as follows: When n is large the statistician will act as if the limiting results apply. Thus, if the estimator (properly normalized) has a limiting distribution, the statistician will act as if this is actually the distribution. Since the value of θ is unknown the approach to the limit distribution must be uniform in order for the errors due to this approximation to be uniformly bounded. This was the position taken in [3], where no attempt was made to weaken this assumption mathematically, since it was considered to correspond to the pragmatic situation. However, full mathematical use was not made of this assumption, and only a certain consequence of it was utilized in [3]. For various reasons we have since decided to formulate our results using only the mathematical consequence which we actually utilize. This is (3.7) of [2] (see the first two paragraphs of Section 3 of [2]), (3.4) of [1], and (2.8) of the present paper.

The assumption (2.9) of the present paper is a bound on the rapidity of approach to θ of the competing estimator. While the assumption seems not unreasonable the case for it is far from being as compelling as that for (2.8). The condition (2.9) may well be due to our particular method of proof and/or the desire to include all loss functions which are bounded below. For many loss functions natural in practical applications it will be possible to dispense entirely with (2.9), as Theorems 2 and 3 already show.

6. Some applications.

Let R be a bounded measurable set in m-space. Let

$$L_n(z,\theta) = 0 \text{ if } z \in \{\theta + [k(n)]^{-1}R\}$$
$$L_n(z,\theta) = 1 \text{ otherwise.}$$

Since R is bounded we have that, for large enough n, $s(n) = 1$ and $L_n(z,\theta) = s(n)$ for $|z-\theta| > h_2(n)$. Applying the m-dimensional analogue of Theorem 3 we essentially obtain the theorem of [1] (for fixed R). To verify this for $m = 1$ we shall compare the hypotheses of Theorem 3 above with those of the theorem of [1]. (The conclusions are obviously the same.) The reason for the word "essentially" is that there is actually a slight difference, due largely to an accident of formulation, between the definition of H_n of the present paper and that of H of [1], and (3.3) of [1] and (2.6) of the present paper. Our comparison is to be understood as modulo this difference.

condition of Theorem 3	condition of theorem of [1]
(2.4)	(3.2)
(2.5) $\Longleftrightarrow$ (2.6)	(3.3)
(2.8)	(3.4)

Now let $R(\theta)$ be, for each θ in Θ, a measurable set in the space of θ, and let these sets be uniformly bounded, i.e., the distance from the origin to any point in any $R(\theta)$ is less than a constant. Then the result of [1] about maximum probability estimators relative to $R(\theta)$ follows from the m-dimensional analogue of the present Theorem 3.

It is not necessary for the application of Theorem 3 that R or $R(\theta)$ be bounded. They may actually depend upon n, e.g., R_n or $R_n(\theta)$, and expand or move out with n, provided only that they not, say, move out faster than $k_2(n)$.

For $m > 1$ let R be a closed convex set, symmetric with respect to the origin. We may assume R unbounded, or our discussion would be contained in that above. Let

$$L_n(z,\theta) = 0 \text{ if } z \in \{\theta + [k(n)]^{-1}R\}$$

$$L_n(z,\theta) = 1 \text{ otherwise.}$$

This loss function satisfies the requirements of the m-dimensional analogue of Theorem 2. Suppose the hypotheses of the problem fit it into the regular case (e.g., [5], Sec. 32.3, or [2], Sec. 2.) Then the classical maximum likelihood estimator $\hat{\theta}_n$ is a maximum probability estimator relative to R (the proof of this, given in [1] for a finite R, is valid verbatim for the application of Theorem 2 to our present R), and is asymptotically efficient relative to R by Theorem 2, since the conditions of Theorem 2 are satisfied in the regular case by $\hat{\theta}_n$. This proves the result of Kaufman [6].

Let μ_n be Lebesgue measure, $X(n) = (X_1,\cdots,X_n)$, $m = 1$, and

$$K_n(x_1,\ldots,x_n|\theta) = (2\pi)^{-\frac{n}{2}} \exp\{-\tfrac{1}{2}\sum_1^n (x_i - \theta)^2\}.$$

Let $L_n(z,\theta) = n(z-\theta)^2$, $k(n) = n^{1/2}$, $k_1(n) = n^{1/3}$, $k_2(n) = n^{1/4}$. Then $s(n) = n^{1/2}$. We conclude that $\overline{X}(n) = n^{-1}\sum_1^n X_i$ is the maximum probability estimator Y_n. Hence any estimator T_n which satisfies (2.8) satisfies

(6.1) $\underline{\lim_{n\to\infty}} E\{(T_n-\theta)^2|\theta\} \geq 1.$

Such an estimator need not be unbiased or asymptotically normally distributed. Of course, if T_n is asymptotically normally distributed about θ it would be more intelligent to concern one's self with the variance of the limiting distribution; this can be done using an R (as in [1]) which is symmetric about the origin. If the limiting distribution of T_n is not normal about θ the second moment may truly not be the appropriate measure of loss. However, this does not affect the validity of (6.1) or its value as an illustration.

In [2] we computed a maximum probability estimator with respect to $R = (-r,r)$ for a class of densities which includes the exponential. We now consider the following illustration: Let $X(n) = (X_1,\cdots,X_n)$, where the X_i's are independently distributed with the common density function

$$f(x|\theta) = e^{-(x-\theta)}, \quad x \geq \theta$$

$$f(x|\theta) = 0, \quad x < \theta$$

and μ_n is Lebesgue measure. The loss function L_n is defined by

$$L_n(z,\theta) = A \text{ when } z - \theta < -\frac{c_1}{n}$$

$$= 0 \text{ when } \frac{-c_1}{n} \leq z - \theta \leq \frac{c_2}{n}$$

$$= B \text{ when } z - \theta > \frac{c_2}{n}$$

Here A, B, c_1, and c_2 are positive constants.

Let $J = \max(A,B)$. Since $n\, h_2(n) \to \infty$ we have that $s(n) = J$ for all n sufficiently large, which we henceforth assume to be the case.

We seek a value of d which maximizes

$$(6.2)\quad \int_{d-h_2(n)}^{d+h_2(n)} [J-L_n(d,\theta)]K_n(X(n)|\theta)d\theta.$$

Let $q = \theta - d$, and $v = \min(X_1,\ldots,X_n)$. We note that $L_n(d,q+d)$ depends only on q, and we denote $J - L_n(d,q+d)$ by $M_n(q)$. Hence

$$M_n(q) = J - A \text{ if } q > \frac{c_1}{n}$$

$$M_n(q) = J \text{ if } \frac{-c_2}{n} \leq q \leq \frac{c_1}{n}$$

$$M_n(q) = J - B \text{ if } q < \frac{-c_2}{n} .$$

Also

$$K_n(X(n)|q+d) = 0 \quad \text{if } v < q + d,$$

$$K_n(X(n)|q+d) = \exp\{n(q+d) - \sum_1^n X_i\} \quad \text{if } v \geq q + d.$$

Maximizing (6.2) therefore is equivalent to maximizing

$$(6.3)\quad e^{nd} \int_{-h_2(n)}^{\min(h_2(n),v-d)} M_n(q)e^{nq}\,dq.$$

Suppose first that

$$(6.4)\quad -h_2(n) \leq v - d \leq \frac{-c_2}{n} .$$

Then (6.3) is

$$\frac{J-B}{n}\,[e^{nv} - e^{-nh_2(n) + nd}],$$

whose maximum is attained at the minimum possible value of d, i.e., at

(6.5) $$d = v + \frac{c_2}{n}.$$

Suppose now that

(6.6) $$\frac{-c_2}{n} \leq v - d \leq \frac{c_1}{n}.$$

Then (6.3) is

(6.7) $$\frac{Je^{nv}}{n} + \frac{e^{nd}}{n}[- Be^{-c_2} - (J-B)e^{-nh_2(n)}]$$

whose maximum is attained at the minimum value of d, i.e., at

(6.8) $$d = v - \frac{c_1}{n}$$

Suppose now that

(6.9) $$\frac{c_1}{n} \leq v - d \leq h_2(n)$$

Then (6.3) is

(6.10) $$\frac{J-A}{n} e^{nv} + \frac{e^{nd}}{n}[-Be^{-c_2} + Ae^{c_1} - (J-B)e^{-nh_2(n)}].$$

If

(6.11) $$Ae^{c_1} - Be^{-c_2} > 0$$

then, for n sufficiently large, since $n\, h_2(n) \to \infty$, the maximum of (6.10) is attained at the maximum value of d, i.e., at

(6.12) $$d = v - \frac{c_1}{n}$$

If

(6.13) $$Ae^{c_1} - Be^{-c_2} \leq 0$$

then the maximum value of (6.10) is, for n sufficiently large, attained at the minimum value of d, i.e., at

$$d = v - h_2(n). \tag{6.14}$$

Finally, suppose that

$$v - d \geq h_2(n) \tag{6.15}$$

Then the upper limit of integration in (6.3) is $h_2(n)$, and (6.3) is maximized at the maximum value of d, i.e., at

$$d = v - h_2(n) \tag{6.16}$$

Putting together all of the above computations we conclude that we may set, when n is sufficiently large,

$$Y_n = v - \frac{c_1}{n} \text{ when } Ae^{c_1} - Be^{-c_2} > 0,$$

$$Y_n = v - h_2(n) \text{ when } Ae^{c_1} - Be^{-c_2} \leq 0.$$

When $Ae^{c_1} - Be^{-c_2} \leq 0$ (and hence $A < B$) the above estimator Y_n depends upon $h_2(n)$, which is at least partly arbitrary. For any n the expected loss of Y_n is

$$A[1-e^{c_1 - nh_2(n)}] + B[e^{-c_2 - nh_2(n)}]$$

$$\text{when } nh_2(n) - c_1 \geq 0,$$

$$B[e^{-c_2 - nh_2(n)}] \text{ when } nh_2(n) - c_1 < 0.$$

Thus this expected loss approaches A no matter how $nh_2(n) \to \infty$.

7. Asymptotic minimax solutions of sequential point estimation problems.

In his paper [7] with the above name Wald considers the following problem: Let $X(n) = (X_1,\ldots,X_n)$ be independent, identically distributed chance variables with a common density function $f(\cdot|\theta_o)$ which satisfies the requirements of the regular case and certain other restrictions. Let T be any sequential stopping variable (rule), a the cost of a single observation, and φ_T the estimator of θ after the termination of sampling. Let the risk r of this procedure be defined by

$$(7.1)\quad r(\theta_o,T,a) = aE\{T|\theta_o\} + E\{\varphi_T-\theta_o)^2 \mid \theta_o\}$$

The stopping variables $\{T^*(a)\}$ are said to be asymptotically minimax if

$$(7.2)\quad \lim_{a\to 0} \frac{\sup_{\theta} r\,(\theta,T^*(a),a)}{\inf_{T(a)} \sup_{\theta} r\,(\theta,T(a),a)} = 1.$$

Let

$$(7.3)\quad g(\theta) = E\{(\frac{\partial \log f(X_1|\theta)}{\partial\theta})^2 |\theta\}$$

and

$$(7.4)\quad 0 < g = \inf_{\theta\in\Theta} g(\theta)$$

Wald proves that the following (non-sequential) procedures are asymptotically minimax: Take

$$(7.5)\quad N(a) = \text{smallest integer} \geq \frac{1}{\sqrt{ag}}$$

observations, and then estimate θ_o by the maximum likelihood estimator $\hat{\theta}$ (N(a)). The basic ideas of the proof are the following: When a is small a large number of observations will obviously be required. When the number of observations is large, say n, the maximum likelihood estimator is approximately the most efficient estimator and the best risk is uniformly close (because of his conditions) to

(7.6) $$an + \frac{1}{ng(\theta)}.$$

At the value of θ least favorable to the statistician this risk is

(7.7) $$an + \frac{1}{ng} .$$

Minimizing (7.7) with respect to n (for fixed a) we obtain (7.5).

Now consider a similar problem for a non-regular case. Let $m(z,\theta)$ be a non-negative function of z and θ, and let the risk r of a sequential stopping variable T now be defined by

(7.8) $$r(\theta_o,T,a) = a\, E\{T|\theta_o\} + E\{m(\varphi_T,\theta_o)|\theta_o\},$$

where φ_T is the estimator of θ_o after the sampling stops according to rule T. Let Y_n be a maximum probability estimator for the loss function $L_n(z,\theta) = n^{\alpha} m(z,\theta)$, and suppose that the conditions of Theorem 1 are fulfilled for every point θ_o in Θ, and that

(7.9) $$\lim_{n\to\infty} n^{\alpha} E\{m(Y_n,\theta_o)|\theta_o\} = \gamma(\theta_o), \text{ say, uniformly in } \theta_o \in \Theta.$$

Then, for large n, the smallest risk is uniformly close to

(7.10) $an + \frac{\gamma(\theta_o)}{n^\alpha}$.

Let

(7.11) $\gamma = \sup_{\theta\epsilon\Theta} \gamma\ (\theta)$.

Then the following procedures are asymptotically minimax:

Take

(7.12) $N(a)$ = smallest integer $\geq (\frac{\alpha\gamma}{a})^{\frac{1}{\alpha+1}}$

observations, and then estimate θ_o by the maximum probability estimator Y_n. The argument is the same as before, the maximum probability estimator Y_n being uniformly asymptotically efficient in the present problem, just as the maximum likelihood estimator was uniformly asymptotically efficient in the regular case (under Wald's additional restrictions) for the loss function $L_n(z,\theta) = n(z-\theta)^2$.

It is obvious that the above argument will apply for normalizing factors other than powers of n and "cost" functions other than an. Wald (ibid.) also gives an actually sequential procedure T in which the set $\{T=n\}$ of sequences of observations $(x_1,x_2,\ldots.)$ is determined by the sequence of estimates of the contribution to the risk (to be made if one stops and estimates by $\hat{\theta}_T$) based on the sequence $\hat{\theta}_1(x_1)$, $\hat{\theta}_2(x_1,x_2)$, ..., $\hat{\theta}_n(x_1,\ldots,x_n)$. Here the sequence of estimators $\hat{\theta}_1,\hat{\theta}_2,\ldots$ is required by Wald to converge strongly to θ_o uniformly in $\theta_o\epsilon\Theta$. What we have done above with Y_n can easily be extended to this second procedure of Wald's. As Wald points out, the second procedure will require fewer observations except at the least favorable values of θ_o, and this is also the case with our generalization.

References

[1] L. Weiss and J. Wolfowitz - "Maximum probability estimators" Ann. Inst. Stat. Math., 19, No. 2 (1967), 193-206.

[2] L. Weiss and J. Wolfowitz - "Generalized maximum likelihood estimators" Teoriya Vyeroyatnostey, 11, No. 1 (1966), 68-93.

[3] J. Wolfowitz - "Asymptotic efficiency of the maximum likelihood estimator" Teoriya Vyeroyatnostey, 10, No. 2 (1965), 267-281.

[4] A. Wald - "Note on the consistency of the maximum likelihood estimate" Ann. Math. Stat., 20, No. 4 (1949), 595-600.

[5] H. Cramér - "Mathematical methods of statistics" Princeton University Press, 1946, Princeton, N. J.

[6] S. Kaufman - "Asymptotic efficiency of the maximum likelihood estimator" Ann. Inst. Stat. Math. 18, No. 2 (1966), 155-178.

[7] A. Wald "Asymptotic minimax solutions of sequential point estimation problems" Proc. Second Berkeley Symposium on Mathematical Statistics and Probability, 1950. Berkeley and Los Angeles, University of California Press, 1951.

Z. Wahrscheinlichkeitstheorie verw. Geb. 15, 186–194 (1970)

The Capacity of a Channel with Arbitrarily Varying Channel Probability Functions and Binary Output Alphabet*

R. AHLSWEDE and J. WOLFOWITZ

Summary. Let $X=\{1,\dots,a\}$ be the "input alphabet" and $Y=\{1,2\}$ be the "output alphabet". Let $X^t=X$ and $Y^t=Y$ for $t=1,2,\dots$, $X_n=\prod_{t=1}^{n} X^t$ and $Y_n=\prod_{t=1}^{n} Y^t$. Let S be any set, $\mathscr{C}=\{w(\cdot|\cdot|s)|s\in S\}$ be a set of $(a\times 2)$ stochastic matrices $w(\cdot|\cdot|s)$, and $S^t=S$, $t=1,\dots,n$. For every $s_n=(s^1,\dots,s^n)\in\prod_{t=1}^{n} S^t$ define $P(\cdot|\cdot|s_n)$ by $P(y_n|x_n|s_n)=\prod_{t=1}^{n} w(y^t|x^t|s^t)$ for every $x_n=(x^1,\dots,x^n)\in X_n$ and every $y_n=(y^1,\dots,y^n)\in Y_n$. Consider the channel $\mathscr{C}_n=\{P(\cdot|\cdot|s_n)|s_n\in S_n\}$ with matrices $w(\cdot|\cdot|s)$ varying arbitrarily from letter to letter. The authors determine the capacity of this channel when a) neither sender nor receiver knows s_n, b) the sender knows s_n but the receiver does not, and c) the receiver knows s_n but the sender does not.

1. Introduction

Let $X=\{1,\dots,a\}$ be the "input alphabet" and $Y=\{1,2\}$ be the "output alphabet" of the channels we shall study below. Results for $a>2$ will not appear until later sections. Hence, to simplify matters, we assume henceforth that $a=2$ unless the contrary is explicitly stated. (The case $a=1$ is trivial.) Let $X^t=X$ and $Y^t=Y$ for $t=1,2,\dots$. By $X_n=\prod_{t=1}^{n} X^t$ we denote the set of input n-sequences (words of length n) and by $Y_n=\prod_{t=1}^{n} Y^t$ we denote the set of output n-sequences.

Let S be any set, and let $\mathscr{C}=\{w(\cdot|\cdot|s)|s\in S\}$ be a set of $(a\times 2)$ stochastic matrices $w(\cdot|\cdot|s)$. We shall refer to a $w(\cdot|\cdot|s)$ in the sequel either as a "matrix" or as a "channel probability function" (c.p.f.). Let $S^t=S$, $t=1,\dots,n$. For every n-sequence $s_n=(s^1,\dots,s^n)\in\prod_{t=1}^{n} S^t$ we define $P(\cdot|\cdot|s_n)$ by

(1.1) $$P(y_n|x_n|s_n)=\prod_{t=1}^{n} w(y^t|x^t|s^t)$$

for every $x_n=(x^1,\dots,x^n)\in X_n$ and every $y_n=(y^1,\dots,y^n)\in Y_n$.

Now consider the channel

(1.2) $$\mathscr{C}_n=\{P(\cdot|\cdot|s_n)|s_n\in S_n\}.$$

Suppose that sender and receiver want to communicate over the channel $\mathscr{C}_n$ without knowing which channel n-sequence s_n will govern the transmission of any

* Research of both authors supported by the U.S. Air Force under Grant AF-AFOSR-68-1472 to Cornell University.

word (input n-sequence). A code (n, N, λ) is a system

(1.3) $$\{(u_1, A_1), \ldots, (u_N, A_N)\}$$

where the message sequence $u_i \in X_n$, $A_i \subset Y_n$, $i=1, \ldots, N$, $A_i \cap A_j = \emptyset$ for $i \neq j$, and

(1.4) $$P(A_i | u_i | s_n) \geqq 1-\lambda, \qquad i=1, \ldots, N, \text{ and all } s_n \in S_n.$$

A number C is called the capacity of the channel if, for any $\varepsilon > 0$ and any λ, $0<\lambda<1$, the following is true for all n sufficiently large: There exists a code $(n, 2^{n(C-\varepsilon)}, \lambda)$ and there does not exist a code $(n, 2^{n(C+\varepsilon)}, \lambda)$.

The channel described above has been called (see [2]) a channel with arbitrarily varying c.p.f.'s, which we abbreviate thus: a.v.ch. The study of a.v.ch. was initiated in [2]. The authors of [2] did not limit themselves to the case where Y has only two elements, and obtained various partial results. We limit ourselves here to the case $|Y|=2$, but for the problems we treat our results go considerably beyond [2] and give the capacity of the particular channels studied. The problems where $|Y|>2$ seem to be unamenable to our methods.

Theorem 1 can easily be improved by using sharper estimates of the maximal code length for discrete memoryless channels.

2. Preliminary Lemmas

Once again we remind the reader that $a=2$ until the contrary is explicitly stated. In particular, $a=2$ in Sections 2 and 3.

We shall consider first a special case of a.v.ch. from which the general case can be easily derived. Consider two matrices w, w'. We denote the i-th row vector in w by i and the i-th row vector in w' by i'. We represent these vectors as points in E^2. Let the matrices w, w' be such that their representation is given by the following Fig. 1, in which the abscissa is the first coordinate of a vector:

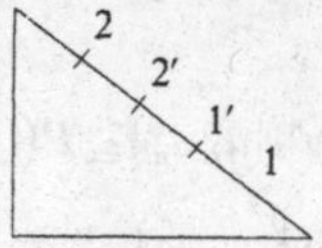

Fig. 1

If we define $w(\cdot|\cdot|1) = w(\cdot|\cdot)$ and $w(\cdot|\cdot|2) = w'(\cdot|\cdot)$, then $P(\cdot|\cdot|s_n)$ can be defined as in (1.1).

We say that the code (1.3) is a *strict* maximum likelihood code (s.m.l.c.) with respect to $P(\cdot|\cdot|s_n^*)$, where $s_n^* = (2, \ldots, 2)$, if, for given $u_1, \ldots, u_N$,

(2.1) $$A_i = \{y_n | y_n \in Y_n \text{ and } P(y_n|u_i|s_n^*) > P(y_n|u_j|s_n^*) \text{ for } j \neq i\}$$

for $i=1, \ldots, N$. (Cf. [4], 7.3.1.)

Define

$$_1A_i^t = \{y_n | y_n \in A_i \text{ and } y^t = u_i^t\},$$

$$_2A_i^t = \{y_n | y_n \in A_i \text{ and } y^t \neq u_i^t\},$$

$$_1A_i^{*t} = \{y^1, \ldots, y^{t-1}, y^{t+1}, \ldots, y^n)| \text{ there exists } y^t \text{ such that } (y^1, \ldots, y^n) \in {}_1A_i^t\},$$

$$_2A_i^{*t} = \{y^1, \ldots, y^{t-1}, y^{t+1}, \ldots, y^n)| \text{ there exists } y^t \text{ such that } (y^1, \ldots, y^n) \in {}_2A_i^t\}.$$

Lemma 1. *With w and w' as in the figure, if $\{(u_i, A_i) | i=1, \ldots, N\}$ is a s.m.l.c. with respect to $P(\cdot|\cdot|s_n^*)$, then*

$$ {}_1A_i^{*t} \supset {}_2A_i^{*t}, \quad i=1, \ldots, N; \ t=1, \ldots, n, \tag{2.2}$$

and

$$P(A_i | u_i | s_n) \geqq P(A_i | u_i | s_n^*) \tag{2.3}$$

for $i=1, \ldots, N$ and all $s_n \in S_n$.

Proof. Suppose first that $u_i^t = 1$.

Let $(y^1, \ldots, y^{t-1}, y^{t+1}, \ldots, y^n) \in {}_2A_i^{*t}$ and $(y^1, \ldots, y^{t-1}, u_i^t, y^{t+1}, \ldots, y^n) \notin {}_1A_i^t$. This could occur for only one of two reasons:

(a) there exists u_j, $j \neq i$, such that

$$P((y^1, \ldots, y^{t-1}, u_i^t, y^{t+1}, \ldots, y^n) | u_j | s_n^*) > P((y^1, \ldots, y^{t-1}, u_i^t, y^{t+1}, \ldots, y^n) | u_i | s_n^*) \tag{2.4}$$

or

(b) $(y^1, \ldots, y^{t-1}, u_i^t, y^{t+1}, \ldots, y^n) \notin \bigcup_{i=1}^{n} A_i$, and there exists a $k \neq i$ such that

$$P((y^1, \ldots, y^{t-1}, u_i^t, y^{t+1}, \ldots, y^n) | u_k | s_n^*) = P((y^1, \ldots, y^{t-1}, u_i^t, y^{t+1}, \ldots, y^n) | u_i | s_n^*). \tag{2.5}$$

Whatever the situation may be, we can find a $j \neq i$, such that

$$P((y^1, \ldots, y^{t-1}, u_i^t, y^{t+1}, \ldots, y^n) | u_j | s_n^*) \geqq P((y^1, \ldots, y^{t-1}, u_i^t, y^{t+1}, \ldots, y^n) | u_i | s_n^*). \tag{2.6}$$

In case $w(\cdot|2|2) \equiv w(\cdot|1|2)$, $N=1$ and the lemma holds. We can therefore assume that $w(1|2|2) \neq w(1|1|2)$. This implies that $w(1|1|2) > 0$ (see Fig. 1).

Suppose first that $u_j^t = 1$. Multiplying both sides of (2.6) by $\dfrac{w(2|1|2)}{w(1|1|2)} \geqq 0$ we obtain

$$P((y^1, \ldots, y^{t-1}, 2, y^{t+1}, \ldots, y^n) | u_j | s_n^*) \geqq P((y^1, \ldots, y^{t-1}, 2, y^{t+1}, \ldots, y^n) | u_i | s_n^*). \tag{2.7}$$

This contradicts the fact that $(y^1, \ldots, y^{t-1}, y^{t+1}, \ldots, y^n) \in {}_2A_i^{*t}$ and proves (2.2) in this case.

Suppose now that $u_j^t = 2$. It follows from $w(1|2|2) \leqq w(1|1|2)$, $w(1|1|2) > 0$, and (2.6) that

$$\begin{aligned} &P((y^1, \ldots, y^{t-1}, y^{t+1}, \ldots, y^n) | (u_j^1, \ldots, u_j^{t-1}, u_j^{t+1}, \ldots, u_j^n) | s_n^*) \\ &\geqq P((y^1, \ldots, y^{t-1}, y^{t+1}, \ldots, y^n) | (u_i^1, \ldots, u_i^{t-1}, u_i^{t+1}, \ldots, u_i^n) | s_n^*). \end{aligned} \tag{2.8}$$

However, $w(2|2|2) \geqq w(2|1|2)$ and (2.8) imply that

$$P((y^1, \ldots, y^{t-1}, 2, y^{t+1}, \ldots, y^n) | u_j | s_n^*) \geqq P((y^1, \ldots, y^{t-1}, 2, y^{t+1}, \ldots, y^n) | u_i | s_n^*), \tag{2.9}$$

which also contradicts the fact that $(y^1, \ldots, y^{t-1}, y^{t+1}, \ldots, y^n) \in {}_2A_i^{*t}$. This proves (2.2) when $u_i^t = 1$. The proof when $u_i^t = 2$ is (symmetrically) the same.

We now prove (2.3) inductively. Assume that (2.3) holds for $s_n' \in S_n$. We shall show that (2.3) then holds for s_n, where s_n is obtained from s_n' by changing the element 2 in the k-th component of s_n' to a 1.

Define $B(i,k)$ by

$$ {}_1A_i^{**k} = {}_2A_i^{**k} \cup B(i,k), \qquad {}_2A_i^{**k} \cap B(i,k) = \emptyset. $$

This definition is possible because of (2.2). Let

$$ P\big(B(i,k)|(u_i^1, \ldots, u_i^{k-1}, u_i^{k+1}, \ldots, u_i^n)|s_n'\big) = a_1', $$
$$ P\big({}_2A_i^{**k}|(u_i^1, \ldots, u_i^{k-1}, u_i^{k+1}, \ldots, u_i^n)|s_n'\big) = a_2'. $$

If $u_i^k = 1$, then

$$ P(A_i|u_i|s_n') = w(1|1|2)\, a_1' + a_2' $$

and if $u_i^k = 2$, then

$$ P(A_i|u_i|s_n') = w(2|2|2)\, a_1' + a_2'. $$

If now in the k-th component of s_n' we replace $w(\cdot|\cdot|2)$ by $w(\cdot|\cdot|1)$, then in both cases we get $P(A_i|u_i|s_n) \geqq P(A_i|u_i|s_n')$, because $w(1|1|1) \geqq w(1|1|2)$ and $w(2|2|1) \geqq w(2|2|2)$. This completes the proof of the lemma.

We now need the following definitions:

(2.10) The entropy of a probability vector $\pi = (\pi_1, \ldots, \pi_c)$ is defined to be

$$ H(\pi) = -\sum_{i=1}^{c} \pi_i \log \pi_i. $$

(2.11) The rate for the probability vector π on X and c.p.f. $w(\cdot|\cdot|s)$ is $R(\pi, w(\cdot|\cdot|s)) = H(\pi'(s)) - \sum_i \pi_i H(w(\cdot|\cdot|s))$, where $\pi'(s) = \pi \cdot w(\cdot|\cdot|s)$.

(2.12) $N(n, \lambda)$ is the maximal length of a (N, n, λ)-code for $\mathscr{C}_n$.

(2.13) For every fixed $i \in X$, $T(i)$ denotes the minimal closed convex system of probability distributions on Y which contains all distributions $\{w(\cdot|i|s)|s \in S\}$.

(2.14) The set of $(a \times 2)$ stochastic matrices

$$ \overline{\mathscr{C}} = \{(w(j|i))_{\substack{i=1,\ldots,a \\ j=1,2}} | w(\cdot|i) \in T(i),\ i = 1, \ldots, a\} $$

is called the row convex closure of the set $\mathscr{C}$.

We shall need

Lemma 2. *Let $\{(u_i, A_i)|i = 1, \ldots, N\}$ be a code with average error $\bar{\lambda}$ for a single channel n-sequence. There exists a subcode of length $N/2$ with maximal error $\lambda = 2\bar{\lambda}$.* (See [4], Lemma 3.1.1.)

Lemma 3. *An (n, N, λ) code for $\mathscr{C}_n$ is an (n, N, λ) code for $\overline{\mathscr{C}}_n$, and conversely.*

Proof. Denote by Σ the σ-field of all subsets of S, and by Σ_n the σ-field of all subsets of S_n.

$$ P(A_i|u_i|s_n) \geqq 1 - \lambda \qquad \text{for all } s_n \in S_n $$

implies that

$$ \int_{S_n} d\, q_{u_i}(s_n)\, P(A_i|u_i|s_n) \geqq 1 - \lambda $$

for all probability distributions q_{u_i} on (S_n, Σ_n) and all $i=1, \ldots, N$. Any element $P(\cdot|\cdot)$ of $\bar{\bar{\mathscr{C}}}$ can be approximated row-wise arbitrarily closely by expressions of the form

$$\int_S d\, q_x(s)\, P(\cdot|x|s), \qquad x\in X$$

where q_x is concentrated on finitely many points. Any element $P_n(\cdot|\cdot)$ of $\bar{\bar{\mathscr{C}}}_n$ can be approximated row-wise arbitrarily closely by expressions of the form

$$\int_{S_n} d\, q^*_{u_i}(s_n)\, P(\cdot|u_i|s_n), \qquad i=1, \ldots, N,$$

where $q^*_{u_i}$ is a distribution on S_n which is concentrated on finitely many points and which is a product of suitable distributions q_{a_j}, where $j=1, \ldots, n$ and $u_i=(a_1, \ldots, a_n)$. This proves the first part of the lemma. The converse is obvious.

Lemma 4.

$$\max_\pi \min_{w\in\bar{\bar{\mathscr{C}}}} R(\pi, w) = \min_{w\in\bar{\bar{\mathscr{C}}}} \max_\pi R(\pi, w).$$

Proof. It is known that $R(\pi, w)$ is concave in π for each w and convex in w for each π. $\bar{\bar{\mathscr{C}}}$ and $\{\pi\}$ are normcompact convex sets and $R(\pi, w)$ is normcontinuous in both variables. Therefore the minimax theorem ([6]) is applicable and the desired result follows.

This lemma is due to Stiglitz [7]. His proof was given here because it is so brief.

3. The Capacity when $a=2$

We shall now prove

Theorem 1. *Define*

$$C=\max_\pi \inf_{w\in\bar{\bar{\mathscr{C}}}} R(\pi, w).$$

For every λ, $0<\lambda<1$, the following estimates hold:

a) $N(n, \lambda) > e^{Cn-k(\lambda)\sqrt{n}}$,

b) $N(n, \lambda) < e^{Cn+k(\lambda)\sqrt{n}}$

where $k(\lambda)$ is a known function of λ and $n=1, 2, \ldots$.

Proof. Let w' be such that $\max_\pi R(\pi, w') = \inf_{w\in\bar{\bar{\mathscr{C}}}} \max_\pi R(\pi, w)$. It follows from Lemma 3 that a λ-code for $\mathscr{C}_n$ is also a λ-code for the d.m.c. determined by w'. Therefore statement b) is a consequence of the strong converse for the d.m.c. ([3], [4]). We can assume without loss of generality (w.l.o.g.) that w' has a representation

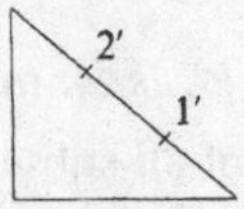

Fig. 2

Choose any $\bar{w}\in\mathscr{C}$. Then $\bar{w}(1|1)\geqq w'(1|1)$, $\bar{w}(2|2)\geqq w'(2|2)$, because otherwise we could, by convex combinations, produce a matrix $w^*\in\bar{\bar{\mathscr{C}}}$ with $\max_\pi R(\pi, w^*)<$

$\max\limits_{\pi} R(\pi, w')$. We therefore have the representation

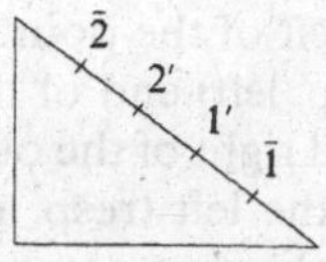

Fig. 3

Now let $w \in \overline{\mathscr{C}}$ be such that

$$w(1|1) \geqq w^{**}(1|1), \quad w(2|2) \geqq w^{**}(2|2)$$

for all $w^{**} \in \overline{\mathscr{C}}$. Obviously $\overline{\mathscr{C}}$ = row convex closure of $\{w, w'\}$. Again by Lemma 3 it is sufficient to prove a) for $\mathscr{C} = \{w, w'\}$.

It follows from Shannon's random coding theorem ([5], [4]) that we can find a s.m.l.c. for the d.m.c. w' with average error $\bar{\lambda} = \lambda/2$ and length $N(n, \lambda) > e^{Cn - k(\lambda)\sqrt{n}}$. By Lemma 2 there exists a subcode with length $N(n, \lambda)/2$ and maximal error λ. Application of Lemma 1 completes the proof.

An examination of the proof of Theorem 1 shows the following:

(3.1) $$T(1) \cap T(2) \neq \emptyset \Leftrightarrow C = 0.$$

This was proved in [2], Theorem 1. When $C = 0$ then, in the next to the last diagram, 1′ and 2′ coincide, and conversely.

(3.2) C is the capacity of the channel with $|S| = 1$ whose single matrix has, as its i-th row, $i = 1, 2$, the point of $T(i)$ closest to $T(i')$, $i' \neq i$.

4. Extension of Theorem 1 to $a > 2$

Theorem 2. *Theorem* 1 *holds verbatim when* $a > 2$.

For each pair $i \neq i'$ let $C(i, i')$ be the capacity of the matrix whose rows are the point of $T(i)$ nearest to $T(i')$ and the point of $T(i')$ nearest to $T(i)$. (When $T(i) \cap T(i') \neq \emptyset$ these points may not be uniquely defined, but $C(i, i') = 0$ anyhow.)

Theorem 3. *Under the conditions of Theorem* 2 *we have*

(4.1) $$C = \max_{i \neq i'} C(i, i').$$

From (4.1) we easily obtain

(4.2) $$C > 0 \Leftrightarrow T(i) \cap T(i') = \emptyset \quad \text{for some pair } i, i'.$$

((4.2) is a special case for $b = 2$ of Theorem 1 of [2].)

Before proving Theorems 2 and 3 we shall need

Lemma 5. *Let M be an $(a \times b)$ stochastic matrix. Suppose all rows are convex linear combinations of two (extreme) rows. Let M' be the $(2 \times b)$ matrix of these rows. Then the capacity of M equals the capacity of M'.*

This lemma must be present, explicitly or implicitly, in one of Shannon's papers. It is easily proved from the expression for the capacity of an individual channel given, e.g., in Theorem 3.1.1 of [4].

We now prove Theorems 2 and 3. In order to define a convenient terminology let us say that, in the next to the last diagram (which appears in the proof of Theorem 1), the point $2'$ is to the left of the point $1'$, the point $1'$ is to the right of $2'$, and $1'$ (resp. $2'$) is the right (resp. left) end of the interval $[1', 2']$. If two points coincide, each is to be to the left and right of the other. Let z (resp. z') be the farthest to the right (resp., to the left) of the left (resp. right) ends of the intervals $T(i)$, $i=1, \ldots, a$. Without loss of generality we assume that z is the left end of $T(1)$ and z' is the right end of $T(2)$.

Suppose first that z is to the left of z'. Then obviously $C=0$. According to Theorem 1 of [2], the capacity of the channel is zero. Hence Theorems 2 and 3 are valid in this case.

Suppose now that z is to the right of z'. From Lemma 5 it follows that C is the capacity of the (2×2) matrix with rows z and z'. Using only the letters 1 and 2 of the input alphabet we see that the capacity of the channel is at least C. Thus the proof of Theorems 2 and 3 will be complete when we prove the converse part.

Suppose that the message sequences u_i, $i=1, \ldots, N$, consist only of 1's and 2's. Then the converse is obvious (or follows from Theorem 1). Suppose now all the other input letters are also used in the u_i. Since we require the error of decoding to be no greater than λ for *every* word u_i and *every* channel n-sequence, we can picture the situation as if some malevolent being, to be called, say, the "jammer", could choose the c.p.f. for each letter *after* he knows the letter being sent. It follows from Lemma 3 that he can achieve that the point on the diagram which corresponds to the letter being sent lies in the interval $[z, z']$. The desired converse now follows from Lemma 5. This completes the proof of Theorems 2 and 3.

Let $m_1, \ldots, m_a$ (resp., $m'_1, \ldots, m'_a$) be the left (resp., the right) end points of $T(1), \ldots, T(a)$, respectively. Let μ (resp., μ') be that one of m_i (resp., m'_i), $i=1, \ldots, a$, which is farthest to the right (resp., to the left). A convenient way of computing C is given by the following, which we state as a theorem for ease of reference:

Theorem 3'. *If μ' is to the right of μ, then $C=0$. Otherwise C is the capacity of the matrix with rows μ and μ'.*

5. The Case where the Sender but not the Receiver Knows the c.p.f. for Each Letter, and $a \geqq 2$

We now study the case described in the title of this section. We assume that the sender knows the c.p.f. for each letter in advance of sending that letter. We also assume that he knows all the preceding c.p.f.'s, but does *not* know any future c.p.f.'s.

Using the method of proof of Theorem 4.9.1 of [4] (see also the proof of Theorem 4.8.1 of [4]) one can show that the capacity of our channel is unaltered if we limit ourselves to codes where the sender chooses the next letter to be sent solely on the basis of the c.p.f. which will govern the transmission of *this* letter and not on the basis of preceding c.p.f.'s. (The fundamental reason for this is that the channel is memoryless.) Henceforth we limit ourselves to such codes.

To make the proof easier to follow we start with the case $|S|=2$, and then remove this limitation. Denote the two matrices in $\mathscr{C}$ by A and B, say. We can describe the codes for the present channel by the following device: The sender's (input) alphabet is to consist of a^2 pairs (i, j), $i, j = 1, \ldots, a$. The "letter" (i, j) means that, when the sender knows that A (resp. B) will govern the transmission of the letter, he sends the letter i (resp., j). By this simple device we have reduced the problem to that treated in Theorem 2.

We have already seen that Theorem 3 or Theorem 3′ implies that the expression C of Theorem 1 is a function only of the closed convex sets $T(i)$, $i = 1, \ldots, a$. We may therefore write

$$C = C(T(1), \ldots, T(a)). \tag{5.1}$$

It is clear (e.g., from Theorem 3′) that the right member of (5.1) is well defined even if the number of sets T is infinite, provided that, in the computations implied by Theorems 3 and 3′, we replace the operation "max" by "sup".

Now let

$$A = \begin{pmatrix} g_1 \\ \vdots \\ g_a \end{pmatrix}, \quad B = \begin{pmatrix} h_1 \\ \vdots \\ h_a \end{pmatrix}.$$

Consider the following a^2 convex bodies, each determined by the two points exhibited:

$$T(g_i, h_j), \quad i, j = 1, \ldots, a.$$

It follows from Theorem 2 that we have proved that the capacity of the channel being discussed is

$$C(\{T(g_i, h_j), i, j = 1, \ldots, a\}). \tag{5.2}$$

We now drop the restriction that $|S| = 2$. Consider the totality $\mathscr{B}' = \{B'\}$ of sets B' such that each B' consists of *exactly* one row from each matrix $w(\cdot|\cdot|s)$, $s \in S$. Let B be the convex hull of B' and $\mathscr{B} = \{B\}$ be the totality of sets (intervals) B. The general case follows from the preceding remarks and the argument of Theorem 2 of [8]. We have thus proved

Theorem 4. *When the sender, but not the receiver, knows the c.p.f. being used for each letter of an a.v.ch., the capacity of the channel is* $C(\mathscr{B})$.

6. The Case where the Receiver but not the Sender Knows the c.p.f. for Each Letter, and $a \geq 2$

We now study the case described in the title of this section. We introduce one change from the preceding channels: We assume that the jammer is allowed to choose each c.p.f. (for each letter) by a random process, i.e., the i-th c.p.f., $i = 1, \ldots, n$, is chosen according to a probability distribution q_i on (S, Σ). The receiver knows the sequence $(q_1, q_2, \ldots, q_n)$ when he decodes the received n-sequence (i.e., decides which transmitted n-sequence u_i was sent). This assumption is very realistic in this case and in the case treated in Theorems 1 and 2. It was not made explicitly there because it is unnecessary; Lemma 3 essentially involves it.

We shall now prove

Theorem 5. *When the receiver (but not the sender) knows the sequence* $(q_1, \ldots, q_n)$, *the capacity of the channel is the same as that given in Theorem 2, i.e.,* *the capacity is the same as it would be if the receiver did not know the sequence* $(q_1, \ldots, q_n)$.

Proof. Since the capacity cannot be less than that in Theorem 2, it remains only to prove the converse. The capacity of Theorem 2 is, by Lemma 4, the smallest of the capacities of the matrices in $\bar{\bar{\mathscr{C}}}$. Clearly, the jammer can choose a q for each letter which will produce (row-wise) the matrix in $\bar{\bar{\mathscr{C}}}$ whose capacity is smallest. This proves the desired result.

7. Miscellaneous Remarks

Theorems 2, 4, and 5 hold, with essentially the same proofs, when a is not finite. One uses the argument of Theorem 2 in Chapter II of [8].

Theorems 1 and 3 of [2] were proved for arbitrary but finite a and b and $|S| < \infty$. It is a consequence of Theorems 2 and 4 of the present paper and the argument of Theorem 2 in Chapter II of [8] that Theorems 1 and 3 of [2] hold for arbitrary (not necessarily finite) a and S, and $b=2$. An examination of the proofs of Theorems 1 and 3 of [2] in the light of these latter results shows that the restriction (made in [2]) to $|S| < \infty$ was unnecessary, and that the proofs of [2] carry over verbatim to arbitrary S.

References

1. Ahlswede, R., Wolfowitz, J.: Correlated decoding for channels with arbitrarily varying channel probability functions. Inform. and Control **14**, 457–473 (1969).
2. Kiefer, J., Wolfowitz, J.: Channels with arbitrarily varying channel probability functions. Inform. and Control **5**, 44–54 (1962).
3. Wolfowitz, J.: The coding of messages subject to chance errors. Illinois Jour. Math. **1**, 591–606 (1957).
4. – Coding theorems of information theory. Berlin-Heidelberg-New York: Springer, first edition, 1961; second edition, 1964.
5. Shannon, C. E.: Certain results in coding theory for noisy channels. Inform. and Control **1**, 6–25 (1957).
6. Kakutani, S.: A generalization of Brouwer's fixed point theorem. Duke math. J. **8**, 457–458 (1941).
7. Stiglitz, I. G.: Coding for a class of unknown channels. IEEE Trans. Inform. Theory **IT-12**, 189–195 (1966).
8. Ahlswede, R.: Beiträge zur Shannonschen Informationstheorie im Falle nichtstationärer Kanäle. Z. Wahrscheinlichkeitstheorie verw. Geb. **10**, 1–42 (1968).

Professor R. Ahlswede
Ohio State University
Columbus, Ohio, USA

Professor J. Wolfowitz
University of Illinois
Dept. of Mathematics
Urbana, Ill. 61801, USA

(Received January 7, 1969)

MAXIMUM PROBABILITY ESTIMATORS AND ASYMPTOTIC SUFFICIENCY

L. WEISS[1] AND J. WOLFOWITZ[2]

(Received June 25, 1969)

1. Introduction

In two recent papers ([1], [2]; see also [3] and [4]) we developed a theory of estimators which we called "maximum probability" estimators. These include the classical maximum likelihood estimator as a very special case, and are asymptotically efficient in all point estimation problems of the statistical literature. The hypotheses of the theorems which imply the asymptotic efficiency of the m. p. (maximum probability) estimators were formulated in terms of certain properties of the m. p. estimators ((2.4)–(2.6) below). These hypotheses are satisfied in all cases of statistical interest, and in most cases are easy to verify. To derive these hypotheses for *all* problems to which the theory applies, from one set of conditions imposed on the families of distributions whose parameters are being estimated, would be very complicated and unproductive, precisely because of the wide applicability of the results. To cover even a substantial number of applications the conditions would have to be so involved that it would be more difficult to verify them than directly to verify the needed hypotheses on the m. p. estimator.

In the present paper we discuss these hypotheses on the m. p. estimator in a number of important and typical cases. Each of the cases discussed includes many, many parametric families, and these cases already include many problems of the statistical literature. It will be easy for the reader to extend the method to other problems. Indeed, the method is so simple and perspicuous that we give it in all detail only in the first case, and content ourselves with sketching it in the other cases.

Our arguments make it clear that the reason why these hypotheses are satisfied by the m. p. estimator is the asymptotic behavior of the likelihood function in a neighborhood of the true value of the parameter.

---

1) Research supported by NSF Grant GP 7798.
2) Research supported by the U. S. Air Force under Grant AF-AFOSR-68-1472.

(Clearly, in an asymptotic theory only behavior in such a neighborhood is of interest.) In the cases studied below and in all cases of statistical interest, this local behavior of the likelihood function depends on a statistic which is therefore asymptotically sufficient for the parameter being estimated in the essential meaning of sufficiency. The m. p. estimator is (asymptotically) a function of this statistic, and the hypotheses on the m. p. estimator which we now desire to verify follow from this.

The reason for the asymptotic efficiency of m. p. estimators is that, asymptotically, they are Bayes solutions with respect to a uniform (or other smooth) a priori distribution on a neighborhood of the " true " value of the parameter. (The neighborhood must shrink with sufficient speed.) To find such a Bayes solution explicitly is, on the face of it, absurd, because it is precisely the true value of the parameter which we wish to estimate. Nevertheless, the m. p. estimator coincides with such a solution with a probability which approaches one; this was proved in [1] and [2]. This fact provides an intuitive explanation of the reason why the classical maximum likelihood estimator is efficient in the regular case and confirms the outline of argument given in [5]. This becomes even clearer as a result of the argument of the present paper. Since the classical case has been treated in the literature in many, many papers, we include it below among the cases where we content ourselves with sketching the argument.

The next section of this paper is essentially a reproduction of Section 2 of [2], since we have anyway to introduce definitions and notation. The paragraph which includes (2.3) and the one following are slightly expanded for greater readability. Discussions of how the assumptions of the theorems correspond to statistical applications are given in [1] and [2], Section 5 (see also [7]).

2. Description of the m. p. estimator

For each positive integer n let $X(n)$ denote the (finite) vector of (observed) chance variables of which the estimator is to be a function. $X(n)$ need not have n components (although the number of components will approach infinity), nor need its components be independently or identically distributed. Let $K_n(x|\theta)$ be the density, with respect to a σ-finite (positive) measure μ_n, of $X(n)$ at the point x (of the appropriate space) when θ is the value of the (unknown to the statistician) parameter. The latter is known to be a point of the " parameter space " Θ. Any estimator T_n is a Borel measurable function of $X(n)$ with values in $\bar{\Theta}$; the set Θ is a closed region of m-dimensional Euclidean space and is contained in a closed region $\bar{\Theta}$ such that every (finite) boundary point of Θ is an inner point of $\bar{\Theta}$. Let $P\{\ |\theta\}$ denote the probability of the relation in braces when θ is the parameter of the density of $X(n)$. We assume that $K_n(x|\theta)$ is a Borel measurable function of both arguments jointly. Although our results are valid for any (finite) m, in this section we shall proceed for $m=1$ in order to keep down the complexity of the notation.

Let $L_n(z, \theta)$ be a non-negative loss function, i.e., when the value of the estimator (function of $X(n)$) is z, and the value of the parameter which determines the density of $X(n)$ is θ, the loss incurred by the statistician is $L_n(z, \theta)$. In many problems one will have

$$L_n(z, \theta)=k(n)L(z, \theta),$$

where $k(n)$ $(\to\infty)$ is the normalizing factor of [1] and the theorems below. For any $y>0$ define

$$s_n^*(y)=\sup L_n(z, \theta),$$

the supremum being taken over all z and θ such that $|z-\theta|\leqq y$.

Let $\{k(n)\}$, $\{k_1(n)\}$, $\{k_2(n)\}$ be sequences of positive numbers such that, as $n\to\infty$,

$$k_2(n)\to\infty\ , \qquad \frac{k_2(n)}{k_1(n)}\to 0\ , \qquad \frac{k_1(n)}{k(n)}\to 0\ . \tag{2.1}$$

Write for brevity

$$h_1(n)=\frac{k_1(n)}{k(n)}\ , \qquad h_2(n)=\frac{k_2(n)}{k(n)}$$

and

$$s(n)=s_n^*(h_2(n))\ . \tag{2.2}$$

We assume that $s(n)<\infty$ for all n. Let Y_n be an estimator defined as a value of d which maximizes

$$\int_{d-h_2(n)}^{d+h_2(n)}[s(n)-L_n(d, \theta)]\,K_n(X(n)|\theta)\,d\theta\ . \tag{2.3}$$

We assume that such an estimator exists. (It need not be unique.) This assumption is, however, not necessary. It is enough if the integral is maximized to within l_n, where $\{l_n\}$ is a sequence of positive numbers which approach zero. This means that

$$\frac{\displaystyle\int_{Y_n-h_2(n)}^{Y_n+h_2(n)}[s(n)-L_n(Y_n, \theta)]K_n(X(n)|\theta)\,d\theta}{\displaystyle\sup_d\int_{d-h_2(n)}^{d+h_2(n)}[s(n)-L_n(d, \theta)]K_n(X(n)|\theta)\,d\theta}\geqq 1-l_n\ .$$

(Since Y_n is an estimator it is Borel measurable. This means that the maximizing values of d for different values of $X(n)$ "link up" in a measurable way.) This shows that Y_n is not uniquely defined. Our theorems apply to any estimator Y_n which satisfies the hypotheses.

In [1], Section 3, third paragraph from the end, it was pointed out that, from any estimator Y_n to which the theorems apply, one could obtain infinitely many others by adding suitable chance variables. Analogous remarks apply here as well.

Although it is obvious that the m. p. estimator is far from unique (indeed, no estimator which is only asymptotically efficient can ever be unique), and that we are really dealing with a class of estimators, in order to avoid tediousness we shall often speak below of *the* m. p. estimator. No error or misunderstanding will be caused by this.

We shall phrase our theorem in terms of a particular value of θ, say θ_0, as was done in [1] and [6] for convenience. Of course the hypotheses of the theorem should be satisfied for all possible θ_0 (in Θ).

Define

$$H_n = \{\theta \mid |(\theta - \theta_0)| \leqq h_1(n)\} .$$

The phrase "converges uniformly in H_n" will mean uniform convergence for all sequences $\{\theta_n\}$ such that $\theta_n \in H_n$.

In [2] we have proved the following theorem for a general loss function L_n:

THEOREM 2.1. *Let the estimator Y_n satisfy the following three conditions uniformly in H_n:*

$$\lim_{n\to\infty} E\{L_n(Y_n, \theta)|\theta\} = \beta , \qquad \text{say,} \tag{2.4}$$

$$\lim_{n\to\infty} [s(n)P\{|k(n)(Y_n - \theta)| > k_2(n)|\theta\}] = 0 , \tag{2.5}$$

and

$$\lim_{n\to\infty} \int_{B_n(\theta)} L_n(Y_n(x), \theta) K_n(x|\theta)\, d\mu_n(x) = 0 , \tag{2.6}$$

where

$$B_n(\theta) = \{x \mid |k(n)(Y_n - \theta)| > k_2(n)\} . \tag{2.7}$$

Let $\{T_n\}$ be any estimator for which the following two conditions hold uniformly in H_n:

$$\lim_{n\to\infty} [E\{L_n(T_n, \theta)|\theta\} - E\{L_n(T_n, \theta_0)|\theta_0\}] = 0 \tag{2.8}$$

and

$$\lim_{n\to\infty} [s(n)P\{|k(n)(T_n - \theta)| > k_2(n)|\theta\}] = 0 . \tag{2.9}$$

(For a discussion of these conditions see Section 5 *of* [2].) *Then*

$$\beta \leqq \underline{\lim}\, E\{L_n(T_n, \theta_0)|\theta_0\}\,, \tag{2.10}$$

so that Y_n *is asymptotically efficient in this sense.*

If $s(n) < V$ (say) for all n sufficiently large, conditions (2.5) and (2.9) take an especially simple form. For more special but very important loss functions we have also proved in [2] the following:

THEOREM 2.2. *If, for all* n *sufficiently large,* $L_n(z, \theta)$ *is a monotonically non-decreasing function of* $|z-\theta|$, *Theorem* 2.1 *holds even without the condition* (2.9).

THEOREM 2.3. *If, for all* n *sufficiently large,* $L_n(z, \theta) = s(n)$ *for* $|z-\theta| > h_2(n)$, *Theorem* 2.1 *holds even without the condition* (2.9).

3. A slight modification of the m. p. estimator. Specialization of the loss function

In [1], Section 3, second paragraph from the end, we discuss a modification of the m. p. estimator to be used in the event that the hypotheses on Y_n as defined in (2.3) are not satisfied. We shall now describe this modification and prove our results for it. We do this in order to avoid having to study the global behavior of the likelihood function, with the consequent need for onerous and unnecessary restrictions on the density functions. In many, if not most, cases of statistical interest the estimator defined in (2.3) will already satisfy the desired conditions. In any case our results do not lose in generality from this modification.

We shall use the usual notation O_p, o_p. To say that A is $O_p(a)$ means that, for arbitrary positive $\alpha < 1$, there is a positive constant $K(\alpha)$ such that

$$P\{|A| < |a| K(\alpha)\,|\,\theta_0\} \geqq \alpha$$

for all n sufficiently large. The definition of $o_p(a)$ is analogous.

Let $\varphi_n(X(n))$ be any estimator such that $\varphi_n - \theta_0 = O_p([k(n)]^{-1})$. This means that φ_n is an estimator to within the same order as the best estimator, but, of course, φ_n need not be efficient. The idea is that, starting from an inefficient estimator, we will obtain an efficient estimator. As an example, let $X(n) = (X_1, \cdots, X_n)$, where the X_i's are independently distributed with the common density (Lebesgue measure) $(2\pi)^{-1/2}\cdot \exp\left\{-\frac{1}{2}(x-\theta)^2\right\}$. Let $\bar{X}_n = n^{-1}\sum_1^n X_i$, and $M_n = \text{median}(X_1, \cdots, X_n)$. Then M_n is an inefficient estimator which can serve as φ_n. On the other hand,

$\bar{X}_n+(\log n)^{-1}$ is a consistent but truly absurd and inefficient estimator which no statistician would dream of using. The purpose of our restriction on φ_n is to eliminate such absurd estimators. The order $[k(n)]^{-1}$ of the efficient estimator is easy to determine in most problems of statistical interest.

Suppose an m. p. estimator Y_n satisfies the hypotheses of Theorem 2.1, say, and a competing estimator T_n, which also satisfies the hypotheses, had a normalizing coefficient $k'(n)$ such that $\frac{k'(n)}{k(n)}\to\infty$. Then T_n would be superior to Y_n by an order of magnitude. However, because of the conclusion of Theorem 2.1, this is impossible if the loss function puts a premium on the estimator's being as close as possible to the value of the parameter being estimated, that is, for any reasonable loss function. This shows that something must be said about the loss function, and we shall do this in a moment.

First we define Y_n as a value of d which maximizes (2.3) subject to the restriction

$$\varphi_n(X(n))-h_1(n)\leqq d\leqq\varphi_n(X(n))+h_1(n) . \tag{3.1}$$

This definition of Y_n will apply throughout the remainder of the paper.

We have seen that reasonable restrictions should be placed on L_n if one is to obtain reasonable results. Define

$$\begin{aligned} L_n(z,\theta)&=0\ , & |z-\theta|&\leqq\frac{r}{k(n)} \\ L_n(z,\theta)&=1\ , & |z-\theta|&>\frac{r}{k(n)} \end{aligned} \tag{3.2}$$

where $r>0$ is a constant. This is a very reasonable, important, and natural loss function. In the classical regular case the m. p. estimator for this loss function is the maximum likelihood estimator. Thus one is able, from Theorem 2.3 or earlier results, to obtain the classical results about the minimality of the variance of the limiting normal distribution of the maximum likelihood estimator. (See [1], [7], and [3], particularly the latter, where many other applications are made.) It might be thought that for this purpose one would use $L_n(z,\theta)\equiv n(z-\theta)^2$, but this gives the limit of the second moments rather than the second moment of the *limiting distribution*. The latter quantity is the statistically meaningful one.

Unless the contrary is explicitly stated below, we shall take L_n to be defined by (3.2). We do this in order to avoid tediousness and because it will be easy to treat each new case ab initio. The crucial point, which we shall always discuss, is the asymptotic behavior of the likelihood function in the neighborhood of the true value θ_0, and the resultant asymptotically sufficient statistic. In a few instances we shall, by way of examples, discuss other loss functions.

4. The range Θ of θ a half-line. Generalization of Example I of [3]

Let $X(n)=(X_1, \cdots, X_n)$ and the X's be independently and identically distributed. Their common density function $f(\cdot|\theta)$ will now be described. In later sections only the common density function $f(\cdot|\theta)$ will be given explicitly, unless $X(n)$ is not as defined above. Also μ_n will always be Lebesgue measure unless the contrary is explicitly stated.

The exponential density is $e^{-(x-\theta)}$ when $x\geqq\theta$, and zero when $x<\theta$. The maximum likelihood estimator of θ is then

$$w_n=\min(X_1, \cdots, X_n)\,.$$

The efficient (m. p.) estimator is $w_n-\frac{r}{n}$. (We will shortly see the essential reason for this. For a proof by another method see [3].) The densities we shall now describe will also be positive only when $x\geqq\theta$. (For additional remarks about the densities see the paragraph which follows Theorem 4.2.) To avoid the tedious we make no attempt to give the most general conditions. For the same reason only the argument of the present section will be given in detail, and the argument of other sections will be sketched.

Let Θ be the real line. Let $f(x|\theta)$ be continuous in (x, θ) for all (x, θ) such that $x\geqq\theta$. Let

$$\begin{aligned} f(x|\theta)&=0\,, \quad x<\theta\,, \\ f(x|\theta)&>0\,, \quad x\geqq\theta\,, \end{aligned} \tag{4.1}$$

and write

$$f(\theta+|\theta)=h(\theta)>0\,. \tag{4.2}$$

We assume that $\frac{\partial f(x|\theta)}{\partial\theta}$ exists and is continuous in (x, θ) for all (x, θ) such that $x>\theta$. Hence

$$\int_\theta^\infty \frac{\delta f(x|\theta)}{\partial\theta}\,dx=h(\theta)\,.$$

We also assume that

$$(4.3) \qquad \int_\theta^\infty \left(\frac{\partial \log f(x|\theta)}{\partial \theta} \right)^2 f(x|\theta)\, dx < \infty$$

for all θ in Θ, and one of the following conditions:

(4.4) The second derivative $\frac{\partial^2 \log f(x|\theta)}{\partial \theta^2}$ exists for $x > \theta$ and is bounded by D, say,

or

$$(4.5) \qquad E\left\{ \left| \frac{\partial^2 \log f(X_1|\theta_0)}{\partial \theta^2} \right| \theta_0 \right\} = g(\theta_0) < \infty$$

and

$$(4.6) \qquad \left| \frac{\partial^2 \log f(x|\theta_0)}{\partial \theta^2} - \frac{\partial^2 \log f(x|\theta)}{\partial \theta^2} \right| < G|\theta - \theta_0|$$

where G is a constant.

It will be apparent from the argument below that even in many cases where some of the above conditions are not fulfilled the same conclusion can be obtained by a not very different argument.

We now take $\phi_n = w_n$, $k(n) = n$, $k_1(n) = n^{1/4}$, $k_2(n) = n^{1/8}$. We can draw the desired conclusion from a Taylor expansion. We proceed under condition (4.4), in such a way as to make the argument under the other condition as similar as possible. With probability approaching one (as $n \to \infty$).

$$(4.7) \qquad 0 < w_n - \theta_0 \leqq h_1(n) .$$

Consequently, with probability approaching one, we have, for all θ such that

$$(4.8) \qquad w_n - 2h_1(n) \leqq \theta < w_n ,$$

that the relation

$$(4.9) \qquad \left| \log K_n(X(n)|\theta) - \log K_n(X(n)|\theta_0) - \frac{n(\theta-\theta_0)}{n} \sum_1^n \frac{\partial \log f(X_i|\theta_0)}{\partial \theta} \right| < \frac{n(\theta-\theta_0)^2 D}{2}$$

holds. The right member of this inequality approaches zero as $n \to \infty$. Write

$$\frac{\log K_n(X(n)|\theta_0)}{n} = M_1 \qquad \text{(say)}$$

$$\frac{\log K_n(X(n)|\theta)}{n} = M_2(\theta) \quad \text{(say)}.$$

By the central limit theorem,

$$\text{(4.10)} \qquad \frac{1}{n}\sum_{1}^{n}\frac{\partial \log f(X_i|\theta_0)}{\partial\theta} = h(\theta_0) + O_p(n^{-1/2}).$$

Hence, when θ satisfies (4.8), we have, with probability approaching one,

$$\text{(4.11)} \quad \exp\{nM_2(\theta)\} = \exp\{n[M_1 + (\theta-\theta_0)(h(\theta_0)+O_p(n^{-1/2}))] + O(n^{-1/2})\}.$$

Thus, with probability approaching one, when n is sufficiently large $M_2(\theta)$ is monotonically increasing in the interval (4.8), which includes the essential part of the interval (3.1) of the present problem. Hence $Y_n = w_n - \frac{r}{n}$ (for the loss function (3.2)). This includes the result of [3] cited earlier in this section.

We verify easily that the estimator Y_n satisfies the conditions (2.4)–(2.6) of Theorem 2.1 (for the loss function (3.2)).

We now consider a general loss function. Since

$$w_n = \theta_0 + O_p\left(\frac{1}{n}\right)$$

we have

$$\text{(4.12)} \qquad h(w_n) = h(\theta_0) + O_p\left(\frac{1}{n}\right).$$

Hence

$$\text{(4.13)} \qquad \frac{\exp\{nM_2(\theta)\}}{\exp\{nM_2'(\theta)\}} = \exp\{O_p(n^{-1/4})\}$$

uniformly for all θ which satisfy (4.8), where

$$\text{(4.14)} \qquad M_2'(\theta) = M_1 + (\theta-\theta_0)h(w_n), \qquad \theta < w_n.$$

Let

$$\chi(\theta, w_n) = 1, \qquad \theta < w_n,$$

$$\chi(\theta, w_n) = 0, \qquad \theta \geqq w_n.$$

Just for a moment and for typographical simplicity let d_1 be such that

$$c_1 = \int_{d_1-h_2(n)}^{d_1+h_2(n)} \chi(\theta, w_n)[s(n) - L_n(d_1, \theta)]\exp\{nM_2(\theta)\}\, d\theta$$

is a maximum, d_2 be such that

$$c_2=\int_{d_2-h_2(n)}^{d_2+h_2(n)} \chi(\theta, w_n)[s(n)-L_n(d_2, \theta)] \exp\{nM_2'(\theta)\}\, d\theta$$

is a maximum, and let

$$c_3=\int_{d_1-h_2(n)}^{d_1+h_2(n)} \chi(\theta, w_n)[s(n)-L_n(d_1, \theta)] \exp\{nM_2'(\theta)\}\, d\theta\ .$$

Then we have

$$c_2 \geqq c_3 \geqq c_1(1+O_p(n^{-1/4}))$$

by (4.13). Bearing in mind the remarks made after (2.3) we see that we have proved

THEOREM 4.1. *For the densities which satisfy the conditions of this section, the estimator Y_n^* which maximizes*

$$\int_{d-h_2(n)}^{d+h_2(n)} \chi(\theta, w_n)[s(n)-L_n(d, \theta)] \exp\{n\theta h(w_n)\}\, d\theta \tag{4.15}$$

with respect to d in the interval

$$[w_n-h_1(n),\ w_n+h_1(n)] \tag{4.16}$$

is an m.p. estimator.

Y_n^* is a function only of w_n. Hence, for any loss function for which Y_n^* is asymptotically efficient, the statistic w_n is asymptotically sufficient. The statistician need not know the entire sample $X(n)$; it is sufficient if he knows only w_n. The asymptotic Bayes solution which Y_n^* really is depends only upon w_n.

As an example of the application of Theorem 4.1 we consider the problem treated in Section 6 of [2]. Here A, B, c_1, and c_2 are positive constants, and

$$\begin{aligned} L_n(z, \theta) &= A \quad \text{when} \quad z-\theta < \frac{-c_1}{n} \\ &= 0 \quad \text{when} \quad \frac{-c_1}{n} \leqq z-\theta \leqq \frac{c_2}{n} \\ &= B \quad \text{when} \quad z-\theta > \frac{c_2}{n}\ . \end{aligned} \tag{4.17}$$

In [2] $f(\cdot|\theta)$ is the exponential distribution:

$$\begin{aligned} f(x|\theta) &= e^{-(x-\theta)}\ , \quad x \geqq \theta \\ &= 0 \qquad\quad\ , \quad x < \theta\ . \end{aligned} \tag{4.18}$$

Now we take $f(\cdot|\theta)$ to be any density of the present section.

Following [2] and just for this example, we define

$$J=\max(A, B)$$

$$(4.19)\qquad \begin{aligned} M_n(q)&=J-A &&\text{when } q>\frac{c_1}{n} \\ M_n(q)&=J &&\text{when } \frac{-c_2}{n}\leqq q\leqq\frac{c_1}{n} \\ M_n(q)&=J-B &&\text{when } q<\frac{-c_2}{n}\,. \end{aligned}$$

In [2] the m. p. estimator Y_n was obtained by maximizing

$$(4.20)\qquad \int_{-h_2(n)}^{\min(h_2(n),\, w_n-d)} M_n(q)\, e^{n(q+d)}\, dq\,.$$

In view of Theorem 4.1 the estimator Y_n^* for the generalized exponential distribution and the loss function (4.17) is to be obtained by maximizing an expression obtained from (4.20) by replacing $n(q+d)$ by $n(q+d)\,h(w_n)$.

To perform the maximization most expeditiously we leave unaltered w_n, $h_2(n)$, q and d in (4.20), and replace n, c_1, and c_2 in (4.19) and (4.20) by $nh(w_n)$, $c_1h(w_n)$, and $c_2h(w_n)$, respectively. The maximizing value of d can then be obtained without difficulty from the argument of [2], and is the value of d in the interval $\left[w_n-h_2(n),\ w_n-\frac{c_1}{n}\right]$ which maximizes

$$(4.21)\qquad \frac{J-A}{nh(w_n)}\, e^{nw_nh(w_n)} + \frac{e^{ndh(w_n)}}{nh(w_n)}\left[-Be^{-c_2h(w_n)}+Ae^{c_1h(w_n)}-(J-B)\,e^{-nh(w_n)h_2(n)}\right].$$

Whether Y_n^* or Y_n satisfies conditions (2.4)–(2.6) depends, of course, on the loss function L_n. It is clear that this will happen for reasonable loss functions. It does not seem to us that giving sufficient conditions on L_n is likely to be productive, because verifying the conditions (2.4)–(2.6) directly will usually be easier than verifying the sufficient conditions. However, we will give one simple set of sufficient conditions for (2.5) and (2.6) to be satisfied.

Suppose L_n satisfies

$$(4.22)\qquad L_n(z, \theta)=L_n(z+z', \theta+z')$$

whenever the arguments are in Θ, and

$$(4.23)\qquad L_n(z, \theta_2)\geqq L_n(z, \theta_1) \qquad \text{whenever}\quad z\geqq\theta_1\geqq\theta_2\,.$$

The loss functions (3.2), (4.17), and

$$L_n(z, \theta) = \text{constant}\ n^{\beta}|z-\theta|^{\gamma} + \delta \tag{4.24}$$

where β, δ, and $\gamma > 0$ are constants, satisfy (4.22) and (4.23). Now (4.22) implies that $Y_n^* \geqq w_n - h_2(n)$. Also (4.23) implies that $Y_n^* \leqq w_n$. Thus

$$w_n - h_2(n) \leqq Y_n^* \leqq w_n\ . \tag{4.25}$$

We have

$$P\{|k(n)(w_n - \theta)| > k_2(n)|\theta\} < \exp\left\{-\frac{1}{2}h(\theta_0)k_2(n)\right\} \tag{4.26}$$

for all θ in H_n and n sufficiently large. Hence also

$$P\{|k(n)(Y_n^* - \theta)| > k_2(n)|\theta\} < \exp\left\{-\frac{1}{2}h(\theta_0)k_2(n)\right\} \tag{4.27}$$

for all θ in H_n and n sufficiently large.

If $s(n) \leqq n^{\alpha}$, as is the case under (3.2), (4.17), or (4.24), it follows from (4.27) that (2.5) is satisfied by Y_n^*. If L_n is bounded by the right member of (4.24), (2.6) is satisfied by Y_n^*.

We may summarize these remarks in

THEOREM 4.2. *If L_n satisfies* (4.22) *and* (4.23), *then* (4.25) *holds. If L_n is such that $s(n) \leqq n^{\alpha}$ for large enough n,* (2.5) *is satisfied by Y_n^*. If L_n is bounded by the right member of* (4.24), (2.5) *and* (2.6) *are satisfied by Y_n^*.*

It is apparent from our arguments that essential use was made *only* of the behavior of $f(x|\theta)$ in a θ-neighborhood of θ_0 for (fixed) x such that $f(x|\theta_0) > 0$. The convenient but inessential assumption, that $\dfrac{\partial f(x|\theta)}{\partial\theta}$ exists and is continuous in (x, θ) for all (x, θ) such that $x > \theta$, has the consequence that

$$A(\theta_0) = E\left\{\frac{\partial \log f(X_i|\theta_0)}{\partial\theta}\,\middle|\,\theta_0\right\} > 0\ , \tag{4.28}$$

which played an important role in our conclusions above. It is not, however, essential that (4.28) be true. The reader can easily change the argument and conclusion in the cases when (4.28) is not true. In Section 7 below, for example, $A(\theta_0)$ can be positive, negative, or zero.

In Sections 6 and 7 more information about $f(\cdot|\cdot)$ is given than was given in the present section. The m. p. estimator is therefore correspondingly modified.

5. The regular case

The regular case is described in [8] and discussed, inter alia, in [1] and [3]. It is the case almost always treated in the literature and the only one for which the maximum likelihood (m. l.) estimator has been shown to be asymptotically efficient in any sense.

In this case let $\phi_n(X(n))=\hat{\theta}_n$, the m.l. estimator, and $k(n)=n^{1/2}$. Also let $k_1(n)=n^{1/32}$, $k_2(n)=n^{1/64}$. The interval (3.1) is now

$$[\hat{\theta}_n-n^{-15/32},\ \hat{\theta}_n+n^{-15/32}]=U(n)\,, \qquad \text{say.} \tag{5.1}$$

Define the interval

$$[\theta_0-n^{-7/16},\ \hat{\theta}_n+n^{-7/16}]=V(n)\,, \qquad \text{say.} \tag{5.2}$$

It is well known that, in the regular case,

$$\hat{\theta}_n=\theta_0+O_p(n^{-1/2})\,. \tag{5.3}$$

Hence, with probability approaching one as $n\to\infty$,

$$U(n)\subset V(n)\,. \tag{5.4}$$

It is well known that, with probability approaching one as $n\to\infty$, for all θ within $V(n)$ (and hence for all θ within $U(n)$), we have

$$\log K_n(X(n)|\theta)-\log K_n(X(n)|\hat{\theta}_n)=-\frac{n(\theta-\hat{\theta}_n)^2}{2}[I(\hat{\theta}_n)+z(\theta)] \tag{5.5}$$

where

$$I(\theta_*)=E\left\{\left(\frac{\partial \log f(X_i|\theta_*)}{\partial\theta}\right)^2\middle|\theta_*\right\}>0\,, \tag{5.6}$$

and the chance variables $\{z(\theta)\}$ satisfy

$$\max_{\theta\in V(n)}|z(\theta)|=O_p(n^{-7/16})\,.$$

Define

$$nM_2'(\theta)=\log K_n(X(n)|\hat{\theta}_n)-\frac{n(\theta-\hat{\theta}_n)^2I(\hat{\theta}_n)}{2}\,. \tag{5.7}$$

Then

$$\frac{K_n(X(n)|\theta)}{\exp\{nM_2'(\theta)\}}=\exp\left\{\frac{-n(\theta-\hat{\theta}_n)^2z(\theta)}{2}\right\}=\exp\{O_p(n^{-3/8})\} \tag{5.8}$$

uniformly for all θ in $U(n)$. Let Y_n^* be the estimator which maximizes

$$(5.9) \qquad \int_{d-h_2(n)}^{d+h_2(n)} [s(n)-L_n(d,\theta)] \exp\left\{\frac{-n(\theta-\hat{\theta}_n)^2 I(\hat{\theta}_n)}{2}\right\} d\theta$$

with respect to d in $U(n)$. The argument after (4.14) now applies with obvious changes and we obtain

THEOREM 5.1. *In the regular case the estimator Y_n^* defined in connection with* (5.9) *is an m. p. estimator.*

We see at once that, for the loss function (3.2), $Y_n^* = \hat{\theta}_n$.

Consider the loss function so often used in the literature,

$$(5.10) \qquad L_n(z,\theta) = n(z-\theta)^2 .$$

By differentiation or other methods one proves that $Y_n^* = \hat{\theta}_n$ also for the loss function (5.10).

One verifies easily that, in the regular case, (2.4)-(2.6) are satisfied by $Y_n^* = \hat{\theta}_n$ in the cases of loss functions (3.2) and (5.10).

We have already shown in [1] and [3] that the classical result, due to R. A. Fisher, about the asymptotic efficiency of the *m. l.* estimator, in the sense of minimal variance of the limiting normal distribution, can be obtained as a consequence of our results for the regular case and the loss function (3.2). The loss function (5.10) does not seem as well suited for this purpose, as might at first glance appear, because in the case (5.10) our theorem reaches a conclusion about the limit of the normalized second moment of $\hat{\theta}_n$ about θ_0, which would then have to be shown to be equal to the second moment about zero of the limiting distribution of $\sqrt{n}(\hat{\theta}_n - \theta_0)$.

6. The range of X an interval. Generalization of Example II of [3]

Suppose always that $\theta < B(\theta)$, $\dfrac{dB}{d\theta} < 0$ and continuous, and

$$(6.1) \qquad f(x|\theta) = 0 , \qquad x < \theta \quad \text{or} \quad x > B(\theta) .$$

We leave it to the reader to fill in the remaining regularity conditions, as in Section 4, so that the present argument, essentially like that of Section 4, goes through. (See also [3], example II.)

Now $k(n) = n$. Take $k_1(n) = n^{1/4}$, $k_2 = n^{1/8}$. In order that $K_n(X(n)|\theta)$ be positive we must have

$$(6.2) \qquad \theta < w_n \quad \textit{and} \quad \theta < B^{-1}(v_n)$$

where $v_n = \max(X_1, \cdots, X_n)$. To the left of $\min\{w_n, B^{-1}(v_n)\} = Z_n$, say, and close to Z_n, we can represent $\log K_n(X(n)|\theta)$ in the form (4.9). We consider the sign of

$$E\left\{\frac{\partial \log f(X_i|\theta_0)}{\partial\theta}\bigg|\,\theta_0\right\}=A(\theta_0)\,, \qquad \text{say.} \tag{6.3}$$

Differentiating both sides of

$$1\equiv\int_{\theta}^{B(\theta)} f(x|\theta)\,dx \tag{6.4}$$

with respect to θ, we obtain that

$$A(\theta_0)=f(\theta_0|\theta_0)-f(B(\theta_0)|\theta_0)\,B'(\theta_0)>0\,, \tag{6.5}$$

since $B'<0$. Hence, as in Section 4, $K_n(X(n)|\theta)$ is monotonically increasing in a suitable neighborhood to the left of Z_n. For the loss function (3.2) the m. p. estimator Y_n^* is therefore $Z_n-\frac{r}{n}$, as was shown in [3]. It is easy to show that (2.4)–(2.6) are then satisfied by Y_n^*.

For other loss functions one can also proceed as in Section 4.

7. The range of X an interval. Generalization of Example III of [3]

This is the same as the example of Section 6, except that now always $B'>0$.

We know that, with probability one,

$$w_n>\theta_0 \qquad \text{and} \qquad v_n<B(\theta_0)\,. \tag{7.1}$$

Hence, with probability one,

$$B^{-1}(v_n)<w_n\,. \tag{7.2}$$

Outside the θ-interval

$$[B^{-1}(v_n),\, w_n]\,, \tag{7.3}$$

$K_n(X(n)|\theta)$ is zero. Inside the interval we represent $\log K_n(X(n)|\theta)$ in the form (4.9). As in Section 6, when the loss function is that of (3.2), which we now consider, the crucial question is the sign of $A(\theta_0)$, whose value is given by (6.5).

Case 1: $A(\theta_0)<0$. Then $\log K_n(X(n)|\theta)$ is monotonically decreasing in the essential interval, and

$$Y_n^*=B^{-1}(v_n)+\frac{r}{n}\,. \tag{7.4}$$

Case 2: $A(\theta_0)>0$. Then $\log K_n(X(n)|\theta)$ is monotonically increasing in the essential interval, and

$$Y_n^* = w_n - \frac{r}{n}. \tag{7.5}$$

Case 3 : $A(\theta_0)=0$. Then $\log K_n(X(n)|\theta)$ is essentially constant in the essential interval (essentially constant means modulo the argument of Section 4), and we obtain the following conclusion: $Y_n^*(X(n))$ is any point in the interval

$$\left[B^{-1}(v_n)+\frac{r}{n},\ w_n-\frac{r}{n}\right] \tag{7.6}$$

or

$$\left[w_n-\frac{r}{n},\ B^{-1}(v_n)+\frac{r}{n}\right], \tag{7.7}$$

whichever applies (i.e., according as $B^{-1}(v_n)+\frac{r}{n} \lesseqgtr w_n-\frac{r}{n}$), provided Y_n^* is Borel measurable. This extends the result of [4].

For other loss functions one can again proceed as in Section 4.

8. The double exponential distribution

Let Θ be the real line, and

$$f(x|\theta)=\frac{1}{2}e^{-|x-\theta|}. \tag{8.1}$$

Suppose $n=2m+1$, m an integer. Let $\phi_n=Z_n$, the median of the X's, $k(n)=n^{1/2}$, $k_1(n)=n^{1/32}$, say, and $k_2(n)=n^{1/64}$, say. Suppose L_n is given by (3.2). Then Y_n maximizes

$$\int_{d-rn^{-1/2}}^{d+rn^{-1/2}} K_n(X(n)|\theta)\,d\theta. \tag{8.2}$$

The likelihood function $K_n(X(n)|\theta)$ (of θ) is always continuous, strictly increasing when $\theta<Z_n$, and strictly decreasing when $\theta>Z_n$. Obviously, therefore $Y_n=Y_n^{**}$, where Y_n^{**} is determined by the relation

$$K_n(X(n)|Y_n^{**}-rn^{-1/2})=K_n(X(n)|Y_n^{**}+rn^{-1/2}). \tag{8.3}$$

It is essentially proved in [3] that the chance variables $\sqrt{n}(Y_n^{**}-\theta_0)$, $\sqrt{n}(Z_n-\theta_0)$ have the same limiting distribution. Bearing in mind the definition of L_n (in (3.2)) we conclude that we may take $Y_n=Z_n$. It is easy to verify that (2.4)–(2.6) are satisfied.

9. The two-dimensional generalizations of the densities of Section 4

Let $\theta=(\theta_1, \theta_2)$, $\theta_0=(\theta_{10}, \theta_{20})$, and Θ be the real plane. We consider the two-dimensional (in both X and θ) generalizations of the densities of Section 4, with such regularity conditions that

$$A_1(\theta_0)=E\left\{\frac{\partial \log f(X_i|\theta_0)}{\partial\theta_1}\bigg|\theta_0\right\}>0\ , \tag{9.1}$$

and

$$A_2(\theta_0)=E\left\{\frac{\partial \log f(X_i|\theta_0)}{\partial\theta_2}\bigg|\theta_0\right\}>0\ . \tag{9.2}$$

Carrying through an argument analogous to that of Section 4, we obtain that, in the region

$$\{\theta_1\leqq w_{1n} \text{ and } \theta_2\leqq w_{2n}\} \tag{9.3}$$

the likelihood function (of θ) $K_n(X(n)|\theta)$ is, with probability approaching one, monotonically increasing in the intersection of a suitable neighborhood of $w_n=(w_{1n}, w_{2n})$ with the region (9.3).

We now define $L_n^{(2)}$, the two-dimensional analogue of the loss function (3.2), as follows: Let $z=(z_1, z_2)$. Then

$$L_n^{(2)}(z, \theta)=1-(1-L_n'(z_1, \theta_1))(1-L_n''(z_2, \theta_2))\ , \tag{9.4}$$

where $L_n'(z_1, \theta_1)=L_n(z_1, \theta_1)$ with $r=r_1$ and $L_n''(z_2, \theta_2)=L_n(z_2, \theta_2)$ with $r=r_2$. For the loss function (9.4) the m. p. estimator Y_n is therefore given by

$$Y_n=\left(w_{1n}-\frac{r_1}{n},\ w_{2n}-\frac{r_2}{n}\right)\ . \tag{9.5}$$

Theorem 4.1 can be extended to these densities in an analogous manner. Sections 7 and 8 can be similarly extended to multi-dimensional θ.

10. Examples where the likelihood estimator is inconsistent

One simple and natural such example is given in [9], as follows. Let Θ be the real half-plane $\theta_2>0$, where $\theta=(\theta_1, \theta_2)$, $\theta_0=(\theta_{10}, \theta_{20})$. Let

$$f(x|\theta)=\frac{1}{2\sqrt{2\pi}}\,e^{(-1/2)(x-\theta_1)^2}+\frac{1}{2\sqrt{2\pi}\theta_2}\,e^{(-1/2)((x-\theta_1)^2/\theta_2)} \tag{10.1}$$

be the density of the chance variable X_i at the point x. Now write

$$\phi_n=(\phi_{1n}, \phi_{2n}) \tag{10.2}$$

and choose

$$k(n)=(\sqrt{n}\,,\ \sqrt{n}\,) \tag{10.3}$$

$$\phi_{1n}=n^{-1}\sum_{i=1}^{n} X_i\ . \tag{10.4}$$

We note that $E\{X_1\}=\theta_1$ and

$$E\{X_1^2\}=\theta_1^2+\frac{1}{2}(1+\theta_2)\ . \tag{10.5}$$

We therefore choose

$$\phi_{2n}=2\left[n^{-1}\sum_{i=1}^{n} X_i^2-(\phi_{1n})^2-\frac{1}{2}\right]. \tag{10.6}$$

In a suitable neighborhood of ϕ_n the likelihood function $K_n(X(n)|\theta)$ is regular in both θ_1 and θ_2. We conclude: For the loss function (9.4) the m. p. estimator Y_n is the value of θ which maximizes $K_n(X(n)|\theta)$ with respect to θ in the region

$$\{|\theta_1-\phi_{1n}|<n^{-15/32},\ |\theta_2-\phi_{2n}|<n^{-15/32}\}\ . \tag{10.7}$$

Other loss functions can be treated similarly.

Now let Θ be as before, and

$$\begin{aligned} f(x|\theta)&=\frac{1}{2}e^{-(x-\theta_1)}+\frac{1}{2\theta_2}e^{-(x-\theta_1)/\theta_2}, \qquad x\geqq\theta_1 \\ &=0\ , \quad x<\theta_1\ . \end{aligned} \tag{10.8}$$

The density is studied in Example VIII of [3]. Now let

$$k(n)=(n,\sqrt{n}\,)\ , \qquad k_1(n)=(n^{1/4},\ n^{1/32})\ , \qquad \phi_{1n}=w_n\ . \tag{10.9}$$

Obviously $K_n(X(n)|\theta_1,\theta_2)=0$ whenever $\theta_1>w_n$. For $\theta_1\leqq w_n$, $K_n(X(n)|\theta)$ is monotonically increasing in θ_1, when θ_2 is fixed. For the loss function (9.4) we then obviously have

$$Y_{1n}=w_n-\frac{r_1}{n}\ . \tag{10.10}$$

Let $Z_1,\cdots,Z_{n-1}$ be the non-zero chance variables among X_1-w_n, $\cdots, X_n-w_n$; with probability one there are $(n-1)$ of the former. Let $K_n'(\cdot|\theta)$ be the density of $(w_n, Z_1,\cdots,Z_{n-1})$, and $K_n''(\cdot|\theta)$ be the density of w_n. It is proved in [3] that the ratio

(10.11) $$\frac{K_n'(w_n, Z_1, \cdots, Z_{n-1}|\theta)}{K_n''(w_n|\theta) \prod_{i=1}^{n-1} \left(\frac{1}{2} e^{-Z_i} + \frac{1}{2\theta_2} e^{-Z_i/\theta_2}\right)}$$

converges stochastically to 1, uniformly in the region

(10.12) $$\{\theta \mid |\theta_1 - \theta_{10}| < n^{-1/2},\ |\theta_2 - \theta_{20}| < n^{-1/4}\} .$$

Now

$$\int_0^\infty z\left(\frac{1}{2} e^{-z} + \frac{1}{2\theta_2} e^{-z/\theta_2}\right) dz = \frac{1+\theta_2}{2} .$$

We therefore let

(10.13) $$\phi_{2n} = 2 \sum_{i=1}^{n-1} Z_i - 1 .$$

In the θ_2-interval

(10.14) $$(\phi_{2n} - n^{-15/32},\ \phi_{2n} + n^{-15/32})$$

the estimation of θ_2 is the same as in the regular case. Hence in the case of the loss function (9.4), we can define $Y_{2n} = m_{2n}$, where m_{2n} is the value of θ_2 which maximizes

(10.15) $$\prod_{i=1}^{n-1} \left(\frac{1}{2} e^{-Z_i} + \frac{1}{2\theta_2} e^{-Z_i/\theta_2}\right)$$

with respect to θ_2 in the interval (10.14).

To summarize; In the case of the density (10.8) and the loss function (9.4),

$$Y_n = \left(w_n - \frac{r_1}{n},\ m_{2n}\right) .$$

This result was first proved in Example VIII of [3].

REFERENCES

[1] Weiss, L. and Wolfowitz, J. (1967). Maximum probability estimators, *Ann. Inst. Stat. Math.*, **19**, 193-206.

[2] Weiss, L. and Wolfowitz, J. (1968). Maximum probability estimators with a general loss function, Proceedings of the International Symposium on Probability and Information Theory held at McMaster University. Heidelberg and New York, Springer Verlag.

[3] Weiss, L. and Wolfowitz, J. (1966). Generalized maximum likelihood estimators, *Teoriya Vyeroyatnostey*, **11**, 68-93.

[4] Weiss, L. and Wolfowitz, J. (1968). Generalized maximum likelihood estimators in a particular case, *Teoriya Vyeroyatnostey*, **13**, 657-662.

[5] Wolfowitz, J. (1953). The method of maximum likelihood and the Wald theory of decision functions, *Indagationes Mathematicae*, **15**, 114-119.

[6] Wald, A. (1949). Note on the consistency of the maximum likelihood estimate, *Ann. Math. Stat.*, **20**, 595-600.

[7] Wolfowitz, J. (1965). Asyptotic efficiency of the maximum likelihood estimator, *Teoriya Vyeroyatnostey*, **10**, 267-281.

[8] Cramér, H. (1946). Mathematical methods of statistics, *Princeton University Press*, Princeton, N. J.

[9] Kiefer, J. and Wolfowitz, J. (1956). Consistency of the maximum likelihood estimator in the presence of infinitely many incidental parameters, *Ann. Math. Stat.*, **27**, **4**, 887-906.

Z. Wahrscheinlichkeitstheorie verw. Geb. 16, 134–150 (1970)

Asymptotically Efficient Non-Parametric Estimators of Location and Scale Parameters

L. WEISS* and J. WOLFOWITZ**

* Research supported by National Science Foundation Grant No. GP-7798.
** Research supported by the U.S. Air Force under Grant AF-AFOSR-68-1472, monitored by the Office of Scientific Research.

1. Introduction. Statement of the Problem

Let $G(\cdot)$ be a (cumulative) distribution function (c.d.f.), and $\mu, \mu_1, \mu_2, \sigma, \sigma_1, \sigma_2$ parameters unknown to the statistician. Of these the μ's are real and the σ's positive. Let $U_1, \ldots, U_n$ be independent chance variables with c.d.f. $G(x-\mu)$ at the point x, and $V_1, \ldots, V_n$ be independent chance variables with c.d.f. $G(x/\sigma)$ at the point x. Let

$$X_1^{(n)} \leqq X_2^{(n)} \leqq \cdots \leqq X_n^{(n)} \tag{1.1}$$

be the ordered U's, and

$$Y_1^{(n)} \leqq Y_2^{(n)} \leqq \cdots \leqq Y_n^{(n)} \tag{1.2}$$

be the ordered V's.

Bennett [1], Jung [3], and Blom [2] proved that

$$\mu_n(X) = \frac{\sum_{j=1}^{n} A_j^{(n)} \left(X_j^{(n)} - G^{-1}\left(\frac{j}{n}\right)\right)}{\sum_{j=1}^{n} A_j^{(n)}} \tag{1.3}$$

and

$$\sigma_n(Y) = \frac{\sum_{j=1}^{n} B_j^{(n)} Y_j^{(n)}}{\sum_{j=1}^{n} B_j^{(n)} G^{-1}\left(\frac{j}{n}\right)} \tag{1.4}$$

are, respectively, asymptotically efficient estimators of μ and σ, under certain regularity conditions on G. The coefficients $A_j^{(n)}$ and $B_j^{(n)}$ are constants given in Section 2 below. Here "asymptotically efficient" is meant in the classical sense of minimal variance of the limiting normal distribution.

Now let p and q be constants such that $0<p<q<1$. Throughout this paper, whenever we write np and nq we always mean the largest integer in np and nq, respectively. (We do this for typographical simplicity.) If, in the previous paragraph, one limits one's self to estimators which are functions only of

$$X_{np}^{(n)}, X_{np+1}^{(n)}, \ldots, X_{nq}^{(n)} \tag{1.5}$$

or of

$$Y_{np}^{(n)}, Y_{np+1}^{(n)}, \ldots, Y_{nq}^{(n)}, \tag{1.6}$$

respectively, then weaker regularity conditions are needed. By a more expeditious argument which uses results of [7], Weiss [6] proved that then the estimators $\hat{\mu}_n(X)$

$$= \frac{\sum_{j=np+1}^{nq-1} \frac{1}{n} A_j^{(n)}\left(X_j^{(n)} - G^{-1}\left(\frac{j}{n}\right)\right) + A_{np}'^{(n)}(X_{np}^{(n)} - G^{-1}(p)) + A_{nq}'^{(n)}(X_{nq}^{(n)} - G^{-1}(q))}{\sum_{j=np+1}^{nq-1} \frac{1}{n} A_j^{(n)} + A_{np}'^{(n)} + A_{nq}'^{(n)}} \tag{1.7}$$

and

$$\hat{\sigma}_n(Y) = \frac{\sum_{j=np+1}^{nq-1} \frac{1}{n} B_j^{(n)} Y_j^{(n)} + B_{np}'^{(n)} Y_{np}^{(n)} + B_{nq}'^{(n)} Y_{nq}^{(n)}}{\sum_{j=np+1}^{nq-1} \frac{1}{n} B_j^{(n)} G^{-1}\left(\frac{j}{n}\right) + B_{np}'^{(n)} G^{-1}(p) + B_{nq}'^{(n)} G^{-1}(q)} \tag{1.8}$$

where $A_{np}'^{(n)}$, $A_{nq}'^{(n)}$, $B_{np}'^{(n)}$, and $B_{nq}'^{(n)}$ are given in Section 2 below, are asymptotically efficient in the very general sense of [9] against all reasonable competitors, for the loss function (see Section 6 of [9]) which assigns loss zero when the estimator differs from the parameter being estimated by at most $r/\sqrt{n}$, $r>0$ arbitrary and fixed, and assigns a constant positive loss otherwise. (See also [8] and [10].)

Now let $U_1^{(1)}, \ldots, U_{n_1}^{(1)}$ be independent chance variables with the common c.d.f. $G(x-\mu_1)$, $U_1^{(2)}, \ldots, U_{n_2}^{(2)}$ be independent chance variables with the common c.d.f. $G(x-\mu_2)$, $V_1^{(1)}, \ldots, V_{n_1}^{(1)}$ be independent chance variables with the common c.d.f. $G(x/\sigma_1)$, and finally $V_1^{(2)}, \ldots, V_{n_2}^{(2)}$ be independent chance variables with the common c.d.f. $G(x/\sigma_2)$. These four sets of chance variables are also to be independent of each other. Then obviously

$$\hat{\mu}_{n_1}(X^{(1)}) - \hat{\mu}_{n_2}(X^{(2)}) \tag{1.9}$$

and

$$\frac{\hat{\sigma}_{n_1}(Y^{(1)})}{\hat{\sigma}_{n_2}(Y^{(2)})} \tag{1.10}$$

are, respectively, asymptotically efficient estimators of $(\mu_1-\mu_2)$ and σ_1/σ_2 under the condition

$$\underline{\lim}\frac{n_1}{n_2} > 0, \quad \underline{\lim}\frac{n_2}{n_1} > 0. \tag{1.11}$$

(Condition (1.11) is necessary to assure that the two samples are comparable in size and that one can normalize the estimators for comparing efficiencies.)

In this paper we consider the problem of estimating $(\mu_1-\mu_2)$ and σ_1/σ_2 *when* $G(\cdot)$ *is unknown to the statistician.* Let the coefficients A^* and B^* be as defined in Section 5. They are simply estimators of all the coefficients A and B such that the maximum, over all the coefficients, of the absolute error of estimation of each coefficient, is $O_p(n^{-\alpha})$, $\alpha>0$. (A definition of this conventionally employed symbol is given in Section 2. Actually even worse estimators would suffice.) Now let $\mu_n^*(X)$ and $\sigma_n^*(Y)$ be defined just as $\hat{\mu}_n(X)$ and $\hat{\sigma}_n(Y)$ were, except that the A's, A''s, B's and B''s are replaced by the respective starred (*) elements. We prove, under mild regularity conditions on $G(\cdot)$ and, of course, the essential condition (1.11), that, among all estimators of $(\mu_1-\mu_2)$ and σ_1/σ_2, respectively, which are functions only of

$$X_{n_1 p}^{(n_1,1)}, \ldots, X_{n_1 q}^{(n_1,1)}, X_{n_2 p}^{(n_2,2)}, \ldots, X_{n_2 q}^{(n_2,2)} \tag{1.12}$$

or of

$$Y_{n_1 p}^{(n_1,1)}, \ldots, Y_{n_1 q}^{(n_1,1)}, Y_{n_2 p}^{(n_2,2)}, \ldots, Y_{n_2 q}^{(n_2,2)} \tag{1.13}$$

respectively, the estimators

$$\mu_{n_1}^*(X^{(1)}) - \mu_{n_2}^*(X^{(2)}) \tag{1.14}$$

and

$$\frac{\sigma_{n_1}^*(Y^{(1)})}{\sigma_{n_2}^*(Y^{(2)})} \tag{1.15}$$

are asymptotically efficient in the sense of [9].

This is one of the very rare instances in the statistical literature where an asymptotically *efficient* estimator has been achieved for a non-parametric problem.

The *only* reason for limiting one's self to functions of (1.12) and (1.13) is to help assure that the regularity conditions on G will be satisfied. One can move p towards zero and q towards one provided the regularity conditions on G continue to be fulfilled. The regularity conditions are described in the next section.

We consider it likely that our estimators will be relatively good for many distributions even in middle-size samples, especially when the A^*'s and B^*'s used below are fairly good estimators.

2. Derivation of the Coefficients A, A', B, and B'

These coefficients were given in [6] with a rather terse explanation of their derivation. We shall now explain in more detail. We limit ourselves to the derivation of the B's since the derivation of the A's is similar.

Let g be the derivative (density) of G, and g' and g'' its first two derivatives. The density of V_i at x is therefore $\sigma^{-1}g(x/\sigma)$, and the density of U_i at x is $g(x-\mu)$. Sufficient regularity conditions for the argument below are the following: The function $G^{-1}(\cdot)$ is single-valued in some open interval which contains the closed interval $[p,q]$. In this open interval the derivatives g, g', g'' exist, g is bounded below by a positive number, and g'' is uniformly continuous. These conditions suffice for both the A's and the B's, and may be replaced by similar conditions.

From now and for the remainder of the paper, we shall always write $X_j^{(1)}$ for $X_j^{(n_1,1)}$, $Y_j^{(2)}$ for $Y_j^{(n_2,2)}$, etc. This will make for typographical simplicity without danger of error.

The logarithm $L(s)$ of the likelihood function of $Y_{np}^{(n)}, \ldots, Y_{nq}^{(n)}$ is given by

$$L(s) = (np-1)\log G\left(\frac{Y_{np}^{(n)}}{s}\right) + \sum_{j=np}^{nq} \log g\left(\frac{Y_j^{(n)}}{s}\right) - [n(q-p)+1]\log s + n(1-q)\log G\left(\frac{Y_{nq}^{(n)}}{s}\right) \tag{2.1}$$

This is a regular case and the maximum likelihood (m.l.) estimator is asymptotically efficient with respect to the loss function described in Section 1. We use the method of [7] to obtain an estimator asymptotically equivalent to the m.l. estimator.

Let B and s be arbitrary but fixed positive numbers. We show that

$$L\left(\frac{s}{1+\frac{B}{\sqrt{n}}}\right) - L\left(\frac{s}{1-\frac{B}{\sqrt{n}}}\right) \leqq 0 \Rightarrow s \leqq \hat{\sigma}_n(Y) + \frac{a_n(s, Y)}{\sqrt{n}} \tag{2.2}$$

and

$$L\left(\frac{s}{1+\frac{B}{\sqrt{n}}}\right) - L\left(\frac{s}{1-\frac{B}{\sqrt{n}}}\right) > 0 \Rightarrow s > \hat{\sigma}_n(Y) + \frac{a_n(s, Y)}{\sqrt{n}} \tag{2.3}$$

where $a_n(s, Y)$ converges stochastically to zero when either

$$G\left(\frac{x}{\frac{s}{1+\frac{B}{\sqrt{n}}}}\right) \text{ is the c.d.f. of } V_i^{(n)} \tag{2.4}$$

or

$$G\left(\frac{x}{\frac{s}{1-\frac{B}{\sqrt{n}}}}\right) \text{ is the c.d.f. of } V_i^{(n)}. \tag{2.5}$$

It then follows from the results of [7] that $\hat{\sigma}_n(Y)$ is asymptotically efficient.

The notation $t_n = O_p(n^\alpha | \sigma_n)$ is to mean that the chance variable t_n which is a function of the $V^{(n)}$'s is of probability order n^α when the c.d.f. of $V_i^{(n)}$ is $G(x/\sigma_n)$, i.e., that, for any $\varepsilon > 0$ there exists $K(\varepsilon) > 0$ such that, for all n sufficiently large,

$$P\{n^{-\alpha}|t_n| < K(\varepsilon)\} > 1 - \varepsilon.$$

Define $W_j^{(n)}$, for $np \leqq j \leqq nq$, by

$$Y_j^{(n)} = s\left[G^{-1}\left(\frac{j}{n}\right) + n^{-\frac{1}{2}} W_j^{(n)}\right]. \tag{2.6}$$

It follows from the theorem of Kolmogorov-Smirnov that

$$\max_j W_j^{(n)} = O_p(1|s). \tag{2.7}$$

Hence

$$\max_j W_j^{(n)} = O_p\left(1 \middle| \frac{s}{1 \pm \frac{B}{\sqrt{n}}}\right). \tag{2.8}$$

Now, using (2.6), expand the various terms in

$$L\left(\frac{s}{1+\frac{B}{\sqrt{n}}}\right) - L\left(\frac{s}{1-\frac{B}{\sqrt{n}}}\right) \tag{2.9}$$

in finite Taylor series and collect terms. We give typical expressions from which the entire expansion can be reconstituted.

$$\log\left(1+\frac{B}{\sqrt{n}}\right) = \frac{B}{\sqrt{n}} - \frac{B^2}{2n} + O(n^{-\frac{3}{2}}) \tag{2.10}$$

$$\log G\left(\frac{Y_{np}^{(n)}}{\frac{s}{1+\frac{B}{\sqrt{n}}}}\right)$$

$$= \log G\left(G^{-1}(p) + n^{-\frac{1}{2}}[BG^{-1}(p) + W_{np}^{(n)}] + n^{-1} B W_{np}^{(n)}\right)$$

$$= \log\left(p + [n^{-\frac{1}{2}}(BG^{-1}(p) + W_{np}^{(n)}) + n^{-1} B W_{np}^{(n)}]\, g(G^{-1}(p))\right.$$

$$\left. + \tfrac{1}{2} n^{-1}(BG^{-1}(p) + W_{np}^{(n)})^2\, g'(G^{-1}(p))\right) \tag{2.11}$$

$$= \log p + \frac{1}{p}\left\{[n^{-\frac{1}{2}}(BG^{-1}(p) + W_{np}^{(n)}) + n^{-1} B W_{np}^{(n)}]\, g(G^{-1}(p))\right.$$

$$\left. + \frac{n^{-1}}{2}(BG^{-1}(p) + W_{np}^{(n)})^2\, g'(G^{-1}(p))\right\}$$

$$- \frac{1}{2p^2}\{[n^{-\frac{1}{2}}(BG^{-1}(p) + W_{np}^{(n)})]\, g(G^{-1}(p))\}^2$$

$$\log g\left(\frac{\frac{Y_j^{(n)}}{s}}{1+\frac{B}{\sqrt{n}}}\right)$$

$$=\log g\left(G^{-1}\left(\frac{j}{n}\right)+n^{-\frac{1}{2}}\left[BG^{-1}\left(\frac{j}{n}\right)+W_j^{(n)}\right]+n^{-1}BW_j^{(n)}\right)$$

$$=\log\left(g\left(G^{-1}\left(\frac{j}{n}\right)\right)+\left[n^{-\frac{1}{2}}\left(BG^{-1}\left(\frac{j}{n}\right)+W_j^{(n)}\right)\right.\right.$$

$$\left.+n^{-1}BW_j^{(n)}\right]g'\left(G^{-1}\left(\frac{j}{n}\right)\right)$$

$$\left.+\frac{n^{-1}}{2}\left(BG^{-1}\left(\frac{j}{n}\right)+W_j^{(n)}\right)^2 g''\left(G^{-1}\left(\frac{j}{n}\right)\right)\right)=\log g\left(G^{-1}\left(\frac{j}{n}\right)\right)$$

$$\text{(2.12)}\qquad +\frac{\left[n^{-\frac{1}{2}}\left(BG^{-1}\left(\frac{j}{n}\right)+W_j^{(n)}\right)+n^{-1}BW_j^{(n)}\right]g'\left(G^{-1}\left(\frac{j}{n}\right)\right)}{g\left(G^{-1}\left(\frac{j}{n}\right)\right)}$$

$$+\frac{n^{-1}\left(BG^{-1}\left(\frac{j}{n}\right)+W_j^{(n)}\right)^2 g''\left(G^{-1}\left(\frac{j}{n}\right)\right)}{2g\left(G^{-1}\left(\frac{j}{n}\right)\right)}$$

$$-\frac{n^{-1}\left(BG^{-1}\left(\frac{j}{n}\right)+W_j^{(n)}\right)^2\cdot g'^2\left(G^{-1}\left(\frac{j}{n}\right)\right)}{2g^2\left(G^{-1}\left(\frac{j}{n}\right)\right)}.$$

We also make use of the fact that

$$\frac{1}{n}\sum_{j=np}^{nq}\left(\frac{g'\left(G^{-1}\left(\frac{j}{n}\right)\right)}{g\left(G^{-1}\left(\frac{j}{n}\right)\right)}\right)\cdot G^{-1}\left(\frac{j}{n}\right)$$

$$\text{(2.13)}\qquad =\int_p^q \frac{g'(G^{-1}(x))}{g(G^{-1}(x))}G^{-1}(x)\,dx+O(n^{-1})$$

$$=\int_{G^{-1}(p)}^{G^{-1}(q)} x\,g'(x)\,dx+O(n^{-1})$$

$$=g(G^{-1}(q))\,G^{-1}(q)-g(G^{-1}(p))\,G^{-1}(p)-(q-p)+O(n^{-1}).$$

We now collect terms in the expansion of (2.9), and then replace the W's in terms of the Y's according to (2.6). It was shown in [6] that the denominator of the right member of (1.8) is positive. Except for positive factors we obtain that

(2.9) equals

$$B\left(s-\hat{\sigma}_n(Y)-\frac{a_n(s,Y)}{\sqrt{n}}\right)$$

where $a_n(s, Y)$ has the property described above and the coefficients $B_j^{(n)}$ are given below.

In [6] it was shown that

$$(2.14)\quad \begin{aligned} B_j^{(n)} &= \left[\left(\frac{g'\left(G^{-1}\left(\frac{j}{n}\right)\right)}{g\left(G^{-1}\left(\frac{j}{n}\right)\right)}\right)^2 G^{-1}\left(\frac{j}{n}\right)\right. \\ &\quad \left.-\frac{g''\left(G^{-1}\left(\frac{j}{n}\right)\right)}{g\left(G^{-1}\left(\frac{j}{n}\right)\right)} G^{-1}\left(\frac{j}{n}\right)-\frac{g'\left(G^{-1}\left(\frac{j}{n}\right)\right)}{g\left(G^{-1}\left(\frac{j}{n}\right)\right)}\right], \\ B_{np}'^{(n)} &= -\left[g(G^{-1}(p))-\frac{1}{p}g^2(G^{-1}(p))\,G^{-1}(p)+g'(G^{-1}(p))\,G^{-1}(p)\right], \\ B_{nq}'^{(n)} &= g(G^{-1}(q))+\frac{1}{1-q}g^2(G^{-1}(q))\,G^{-1}(q)+g'(G^{-1}(q))\,G^{-1}(q). \end{aligned}$$

$$(2.15)\quad \begin{aligned} A_j^{(n)} &= \left[\left(\frac{g'\left(G^{-1}\left(\frac{j}{n}\right)\right)}{g\left(G^{-1}\left(\frac{j}{n}\right)\right)}\right)^2-\frac{g''\left(G^{-1}\left(\frac{j}{n}\right)\right)}{g\left(G^{-1}\left(\frac{j}{n}\right)\right)}\right], \\ A_{np}'^{(n)} &= -\left[g'(G^{-1}(p))-\frac{1}{p}g^2(G^{-1}(p))\right], \\ A_{nq}'^{(n)} &= g'(G^{-1}(q))+\frac{1}{1-q}g^2(G^{-1}(q)). \end{aligned}$$

3. Asymptotically Efficient Estimators of σ_1/σ_2

For $i=1,2$ define

$$(3.1)\quad c(n,i,j)=G^{-1}\left(\frac{j}{n}\right)\cdot\sigma_i,$$

$$(3.2)\quad d(n,i,j)=\frac{1}{\sigma_i}g\left(G^{-1}\left(\frac{j}{n}\right)\right),$$

$$(3.3)\quad e(n,i,j)=\frac{1}{\sigma_i^2}g'\left(G^{-1}\left(\frac{j}{n}\right)\right),$$

$$(3.4)\quad f(n,i,j)=\frac{1}{\sigma_i^3}g''\left(G^{-1}\left(\frac{j}{n}\right)\right),$$

and

(3.5) $$h(n,i)=\sigma_i\left[g\left(G^{-1}\left(\tfrac{1}{2}\right)\right)\right]^{-1}.$$

Also define, for $i, k=1,2$,

$$S(n,i,k)=-Y_{np}^{(k)}\left[d(n,i,np)\,h(n,i)-\frac{1}{p}\,d^2(n,i,np)\,h(n,i)\,c(n,i,np)\right.$$
$$\left.+e(n,i,np)\,h(n,i)\,c(n,i,np)\right]+Y_{nq}^{(k)}\left[d(n,i,nq)\,h(n,i)\right.$$
$$\left.+\frac{1}{1-q}\,d^2(n,i,nq)\,h(n,i)\,c(n,i,nq)+e(n,i,nq)\,h(n,i)\,c(n,i,nq)\right]$$

(3.6)
$$+\frac{1}{n}\sum_{j=np+1}^{nq-1} Y_j^{(k)}\left[e^2(n,i,j)\,d^{-2}(n,i,j)\,h(n,i)\,c(n,i,j)\right.$$
$$\left.-f(n,i,j)\,d^{-1}(n,i,j)\,h(n,i)\,c(n,i,j)-e(n,i,j)\,d^{-1}(n,i,j)\,h(n,i)\right]$$
$$=(\text{say})\;Y_{np}^{(k)}\,C(n,i,np)+Y_{nq}^{(k)}\,C(n,i,nq)+\frac{1}{n}\sum_{j=np+1}^{nq-1} Y_j^{(k)}\,C(n,i,j).$$

Then obviously

(3.7) $$\frac{\hat{\sigma}_{n_1}(Y^{(1)})}{\hat{\sigma}_{n_2}(Y^{(2)})}=\frac{S(n_1,1,1)}{S(n_2,1,2)}=\frac{S(n_1,2,1)}{S(n_2,2,2)}.$$

In Section 5 below we will estimate the elements of (3.1)–(3.5); the corresponding estimators will be the same quantities starred (*). When these estimators are inserted into the middle member of (3.6) we call the left member $S^*(n,i,k)$ and the corresponding C's are then written $C^*(n,i,j)$. We will prove below that

(3.8) $$\frac{S^*(n_1,1,1)}{S^*(n_2,1,2)}=W_1\ (\text{say})$$

and

(3.9) $$\frac{S^*(n_1,2,1)}{S^*(n_2,2,2)}=W_2\ (\text{say})$$

are both asymptotically efficient estimators of σ_1/σ_2 by showing that

(3.10) $$\sqrt{n}\left(W_1-\frac{\hat{\sigma}_{n_1}(Y^{(1)})}{\hat{\sigma}_{n_2}(Y^{(2)})}\right)=o_p(1)$$

and

(3.11) $$\sqrt{n}\left(W_2-\frac{\hat{\sigma}_{n_1}(Y^{(1)})}{\hat{\sigma}_{n_2}(Y^{(2)})}\right)=o_p(1).$$

(The statement "$=o_p(1)$" means "converges stochastically to zero".) Thus W_1 and W_2 are asymptotically as efficient as the best estimator which can be computed with full knowledge of G, although W_1 and W_2 are non-parametric and do not require any knowledge of G!! From (3.10) and (3.11) it follows that W_1 and W_2

are asymptotically equivalent, and the estimator

$$W=\frac{W_1+W_2}{2}, \tag{3.12}$$

which one might use because it is symmetric, asymptotically efficient. Of course, one could also weight W_1 and W_2 unequally.

The estimators of (3.1)–(3.5) to be given in Section 5 are *uniformly* $O_p(n^{-\alpha})$, $\alpha>0$, and thus uniformly consistent. It is even possible to estimate all the c's uniformly to within $O_p(n^{-\frac{1}{2}})$, but certainly not the derivatives of G. Thus $\alpha<\frac{1}{2}$. Now obviously (under condition (1.11))

$$\frac{\hat{\sigma}_{n_1}(Y^{(1)})}{\hat{\sigma}_{n_2}(Y^{(2)})}=\frac{\sigma_1}{\sigma_2}+\frac{\gamma}{\sqrt{n_1}} \tag{3.13}$$

where

$$\gamma=O_p(1). \tag{3.14}$$

The efficiency of the estimator is determined by the distribution of γ. Hence, even if α were $\frac{1}{2}$, the introduction of the estimators $B_j^{n(*)}$ into the right member of (3.6) could be expected to affect γ and hence the efficiency of the estimators W_1 and W_2. Since $\alpha<\frac{1}{2}$ the errors in the coefficients are enormous compared with the total error permitted in W_1 and W_2 if the latter are to be efficient.

First let $n_1=n_2=n$. Write

$$\begin{aligned} S^*(n,i,k)=&\,Y_{np}^{(k)}\big(C(n,i,np)+n^{-\alpha}\Delta(n,i,np)\big)\\ &+Y_{nq}^{(k)}\big(C(n,i,nq)+n^{-\alpha}\Delta(n,i,nq)\big)\\ &+\frac{1}{n}\sum_{j=np+1}^{nq-1}Y_j^{(k)}\big(C(n,i,j)+n^{-\alpha}\Delta(n,i,j)\big), \end{aligned} \tag{3.15}$$

where

$$\max_{np\leqq j\leqq nq}|\Delta(n,i,j)|=O_p(1). \tag{3.16}$$

Let

$$\begin{aligned} D(n,i)=&\,C(n,i,np)\,G^{-1}(p)+C(n,i,nq)\,G^{-1}(q)\\ &+\frac{1}{n}\sum_{j=np+1}^{nq-1}C(n,i,j)\,G^{-1}\left(\frac{j}{n}\right), \end{aligned} \tag{3.17}$$

$$\begin{aligned} H(n,i)=&\,G^{-1}(p)\cdot\Delta(n,i,np)+G^{-1}(q)\cdot\Delta(n,i,nq)\\ &+\frac{1}{n}\sum_{j=np+1}^{nq-1}G^{-1}\left(\frac{j}{n}\right)\cdot\Delta(n,i,j), \end{aligned} \tag{3.18}$$

$$Y_j^{(k)}=\sigma_k\left(G^{-1}\left(\frac{j}{n}\right)+\delta_j^{(k)}\cdot n^{-\frac{1}{2}}\right), \tag{3.19}$$

$$\begin{aligned} D'(n,i,k)=&\,C(n,i,np)\cdot\delta_{np}^{(k)}+C(n,i,nq)\cdot\delta_{nq}^{(k)}\\ &+\frac{1}{n}\sum_{j=np+1}^{nq-1}C(n,i,j)\cdot\delta_j^{(k)} \end{aligned} \tag{3.20}$$

and

$$D''(n, i, k) = \delta_{np}^{(k)} \cdot \Delta(n, i, np) + \delta_{nq}^{(k)} \cdot \Delta(n, i, nq)$$

(3.21)
$$+ \frac{1}{n} \sum_{j=np+1}^{nq-1} \delta_j^{(k)} \cdot \Delta(n, i, j).$$

Then, clearly, the left members of (3.17), (3.18), (3.20), and (3.21) are all $O_p(1)$.

We then have

(3.22)
$$\frac{S(n, 1, 1)}{S(n, 1, 2)} = \frac{\sigma_1}{\sigma_2} \cdot \frac{D(n, 1) + n^{-\frac{1}{2}} D'(n, 1, 1)}{D(n, 1) + n^{-\frac{1}{2}} D'(n, 1, 2)}$$

and

(3.23)
$$W_1 = \frac{\sigma_1}{\sigma_2} \cdot \frac{D(n, 1) + n^{-\frac{1}{2}} \cdot D'(n, 1, 1) + n^{-\alpha} H(n, 1) + n^{-\alpha-\frac{1}{2}} \cdot D''(n, 1, 1)}{D(n, 1) + n^{-\frac{1}{2}} \cdot D'(n, 1, 2) + n^{-\alpha} H(n, 1) + n^{-\alpha-\frac{1}{2}} \cdot D''(n, 1, 2)}.$$

From results of [6] we conclude easily that

$$\lim_{n \to \infty} D(n, i) > 0.$$

An easy computation now verifies (3.10). The proof of (3.11) is practically the same.

It remains to consider the general situation, where not necessarily $n_1 = n_2$, but (1.11) holds. Without loss of generality we assume that $n_2/n_1 = \beta > 1$. Write $n = n_1$. $S(n, i, 1)$ and $S^*(n, i, 1)$ are defined exactly as before. In the definition of $S(n_2, i, 2)$ and hence that of $S^*(n_2, i, 2)$ we now change only the definitions of the C's in the following way: For any integer j, $n_2 p \leqq j \leqq n_2 q$, define an integer j^* by the requirement that

$$\left| \frac{j^*}{n} - \frac{j}{n_2} \right|$$

is a minimum. Then, in the definition of $S(n_2, i, 2)$ we replace $C(n_2, i, j)$ by $C(n, i, j^*)$, and in the definition of $S^*(n_2, i, 2)$ we replace $C^*(n_2, i, j)$ by $C^*(n, i, j^*)$. This has the effect of changing $S^*(n_2, i, 2)$ by $O_p(n^{-1})$. We may write

$$Y_j^{(2)} = \sigma_2 \left(G^{-1}\left(\frac{j^*}{n}\right) + \delta_j'^{(2)} \cdot n^{-\frac{1}{2}} \right)$$

where

$$\max_j \delta_j'^{(2)} = O_p(1).$$

The proof of efficiency is now essentially the same as in the case $n_1 = n_2$, after allowance is made for terms of $O_p(n^{-1})$. This completes the proof.

We notice that the above proof requires only that $\alpha > 0$. Thus even poor estimators of the quantities (3.1)–(3.5) will still produce asymptotically efficient W_1 and W_2. An advantage of efficient estimators of (3.1)–(3.5) is that their use makes the distributions of W_1 and W_2 approach their limiting distributions faster. Also estimators which use efficient A^* and B^* are probably better for small samples. In Section 5 we content ourselves with giving estimators sufficient for the purposes of the present paper.

10*

4. Asymptotically Efficient Estimators of $\mu_1 - \mu_2$

We have, when $n_1 = n_2 = n$,

$$\hat{\mu}_n(X^{(1)}) - \hat{\mu}_n(X^{(2)}) = A'^{(n)}_{np}(X^{(1)}_{np} - X^{(2)}_{np}) + A'^{(n)}_{nq}(X^{(1)}_{nq} - X^{(2)}_{nq}) \tag{4.1}$$
$$+ \frac{\displaystyle\sum_{j=np+1}^{nq-1} \frac{1}{n} A^{(n)}_j (X^{(1)}_j - X^{(2)}_j)}{A'^{(n)}_{np} + A'^{(n)}_{nq} + \displaystyle\sum_{j=np+1}^{nq-1} \frac{1}{n} A^{(n)}_j}.$$

To obtain the estimator $\mu^*_n(X^{(1)}) - \mu^*_n(X^{(2)})$ one replaces the A's by their estimators A^*'s in the right member of (4.1). Define

$$v(n, j) = (A^{(n)*}_j - A^{(n)}_j) \cdot n^{\alpha'}. \tag{4.2}$$

The A^*'s to be defined in Section 5 will have the property that $\alpha' > 0$ and

$$\max_j |v(n, j)| = O_p(1). \tag{4.3}$$

Obviously

$$X^{(1)}_j - X^{(2)}_j = n^{-\frac{1}{2}} \cdot \lambda(n, j) + \mu_1 - \mu_2, \tag{4.4}$$

where

$$\max_j |\lambda(n, j)| = O_p(1). \tag{4.5}$$

Define

$$F_1(n) = A'^{(n)}_{np}\, \lambda(n, np) + A'^{(n)}_{nq}\, \lambda(n, nq) + \frac{1}{n} \sum_{j=np+1}^{nq-1} A^{(n)}_j\, \lambda(n, j), \tag{4.6}$$

$$F_2(n) = \lambda(n, np)\, v(n, np) + \lambda(n, nq)\, v(n, nq) + \frac{1}{n} \sum_{j=np+1}^{nq-1} \lambda(n, j)\, v(n, j), \tag{4.7}$$

$$F_3(n) = A'^{(n)}_{np} + A'^{(n)}_{nq} + \frac{1}{n} \sum_{j=np+1}^{nq-1} A^{(n)}_j \tag{4.8}$$

and

$$F_4(n) = v(n, np) + v(n, nq) + \frac{1}{n} \sum_{j=np+1}^{nq-1} v(n, j). \tag{4.9}$$

We then have

$$\hat{\mu}_n(X^{(1)}) - \hat{\mu}_n(X^{(2)}) = (\mu_1 - \mu_2) + \frac{F_1(n)}{\sqrt{n}\, F_3(n)} \tag{4.10}$$

and

$$\mu^*_n(X^{(1)}) - \mu^*_n(X^{(2)}) = (\mu_1 - \mu_2) + \frac{F_1(n) + n^{-\alpha'} F_2(n)}{\sqrt{n}\,(F_3(n) + n^{-\alpha'} F_4(n))}. \tag{4.11}$$

It was proved in [6] that

$$\lim_{n \to \infty} F_3(n) > 0. \tag{4.12}$$

A trivial calculation now verifies that

$$(4.13)\qquad \sqrt{n}\,[(\hat{\mu}_n(X^{(1)})-\hat{\mu}_n(X^{(2)}))-(\mu_n^*(X^{(1)})-\mu_n^*(X^{(2)}))]=o_p(1).$$

This completes the proof of the asymptotic efficiency of our estimator when $n_1=n_2$.

The modification of the proof when $n_1 \neq n_2$ and condition (1.11) is satisfied is essentially the same as that at the end of Section 3.

5. Estimation of the Coefficients *A* and *B*

For typographical simplicity we shall, in what follows, write $n^{\frac{3}{4}}$, np, etc., when we mean the largest integer in $n^{\frac{3}{4}}$, np, etc.; no error will be caused thereby. We assume that $n_1=n_2=n$. When this is not the case the changes needed below are obvious. We set

$$(5.1)\qquad c^*(n,i,j)=Y_j^{(i)}.$$

We shall use a notation which is illustrated by the statement

$$(5.2)\qquad c^*(n,i,j)\sim c(n,i,j)[O_p(n^{-\frac{1}{2}})].$$

This means that

$$(5.3)\qquad \max_j |c^*(n,i,j)-c(n,i,j)|=O_p(n^{-\frac{1}{2}}).$$

We now write $\beta_1=n^{\frac{3}{4}}$. For every integer j such that

$$(5.4)\qquad \beta_1(j'-1)<j\leqq\beta_1 j',\qquad j' \text{ an integer}$$

we set

$$(5.5)\qquad d^*(n,i,j)=\frac{n^{-\frac{1}{4}}}{c^*(n,i,\beta_1 j')-c^*(n,i,\beta_1(j'-1))}.$$

There are two sources of error in this estimator: 1) the error in c^* at the ends of the interval (5.4); 2) the error caused by setting d^* constant throughout the interval (5.4). The sources of error in the estimators below correspond to these two. For d^* as in (5.5) the first of these errors is $O_p(n^{-\frac{1}{2}}/n^{-\frac{1}{4}})=O_p(n^{-\frac{1}{4}})$. The second error is always of the same order as the "length" of the interval (5.4), $n^{-\frac{1}{4}}$. Hence

$$(5.6)\qquad d^*(n,i,j)\sim d(n,i,j)[O_p(n^{-\frac{1}{4}})].$$

Now let $\beta_2=n^{\frac{7}{8}}$. For every integer j such that

$$(5.7)\qquad \beta_2(j'-1)<j\leqq\beta_2 j',\qquad j' \text{ an integer}$$

we set

$$(5.8)\qquad e^*(n,i,j)=\frac{d^*(n,i,\beta_2 j')-d^*(n,i,\beta_2(j'-1))}{c^*(n,i,\beta_2 j')-c^*(n,i,\beta_2(j'-1))}.$$

The error of first type is here, by (5.6), $O_p\left(\frac{n^{-\frac{1}{4}}}{n^{-\frac{1}{8}}}\right)=O_p(n^{-\frac{1}{8}})$. The error of second type is of order $n^{-\frac{1}{8}}$, the "length" of the interval (5.7). Hence

$$(5.9)\qquad e^*(n,i,j)\sim e(n,i,j)[O_p(n^{-\frac{1}{8}})].$$

Now let $\beta_3 = n^{\frac{15}{16}}$. For every integer j such that

(5.10) $$\beta_3(j'-1) < j \leqq \beta_3\, j', \qquad j' \text{ an integer}$$

we set

(5.11) $$f^*(n, i, j) = \frac{e^*(n, i, \beta_3\, j') - e^*(n, i, \beta_3(j'-1))}{c^*(n, i, \beta_3\, j') - c^*(n, i, \beta_3(j'-1))}.$$

From (5.9) it follows that the error of first type is

$$O_p\left(\frac{n^{-\frac{1}{8}}}{n^{-\frac{1}{16}}}\right) = O_p(n^{-\frac{1}{16}}).$$

The "length" of the interval is of probability order $n^{-\frac{1}{16}}$. Hence

(5.12) $$f^*(n, i, j) \sim f(n, i, j)[O_p(n^{-\frac{1}{16}})].$$

Finally we simply set

(5.13) $$h^*(n, i) = \left[d^*\left(n, i, \frac{n}{2}\right)\right]^{-1}.$$

Obviously

(5.14) $$h^*(n, i) \sim h(n, i)[O_p(n^{-\frac{1}{4}})].$$

It is clear that, with the use of c^*, d^*, e^*, f^*, and h^*, the estimators $B_j^{(n)*}$ can be constructed to within $O_p(n^{-\frac{1}{16}})$.

We now estimate the coefficients $A_j^{(n)}$, which will be very easy. Suppose first that $n_1 = n_2 = n$. With the above definitions of d^*, e^*, and f^*, we set

(5.15) $$d^*(n, j) = \tfrac{1}{2}[d^*(n, 1, j) + d^*(n, 2, j)],$$

(5.16) $$e^*(n, j) = \tfrac{1}{2}[e^*(n, 1, j) + e^*(n, 2, j)],$$

(5.17) $$f^*(n, j) = \tfrac{1}{2}[f^*(n, 1, j) + f^*(n, 2, j)],$$

(5.18) $$A_{np}'^{(n)*} = -\left[e^*(n, n\,p) - \frac{1}{p}(d^*(n, n\,p))^2\right],$$

(5.19) $$A_{nq}'^{(n)*} = \left[e^*(n, n\,q) + \frac{1}{1-q}(d^*(n, n\,q))^2\right],$$

(5.20) $$A_j^{(n)*} = \left(\frac{e^*(n, j)}{d^*(n, j)}\right)^2 - \frac{f^*(n, j)}{d^*(n, j)}.$$

Again, the quantity α' of Section 4 is $\frac{1}{16}$.

When $n_1 \neq n_2$ the necessary modifications of the above procedure are obvious. For example, one uses together $d^*(n_1, 1, j_1)$ and $d^*(n_2, 2, j_2)$, where j_1/n_1 and j_2/n_2 are essentially equal. The weighting can be equal, i.e., $\frac{1}{2}, \frac{1}{2}$, or perhaps proportional to sample size. Such estimators will make $\mu_{n_1}^*(X^{(1)}) - \mu_{n_2}^*(X^{(2)})$ asymptotically efficient.

6. Asymptotically Efficient Estimators of $(\mu_1-\mu_2)$ and σ_1/σ_2 Jointly

This problem now presents no difficulty and the methods used previously carry over in toto. For this reason we omit the details. The basic estimators are given in the middle of p. 133 of [6]. They have a structure which permits the straightforward application of our methods.

In the definitions of r_1, r_2, t_1, and t_2 of [6] there occur integrals. It may help the reader in his calculations to know that these integrals approximate the following sums, respectively:

(6.1)
$$\frac{1}{n}\sum_{j=np+1}^{nq-1}\left\{\left[\frac{g'\left(G^{-1}\left(\frac{j}{n}\right)\right)}{g\left(G^{-1}\left(\frac{j}{n}\right)\right)}\right]^2 G^{-1}\left(\frac{j}{n}\right) - \frac{g''\left(G^{-1}\left(\frac{j}{n}\right)\right)}{g\left(G^{-1}\left(\frac{j}{n}\right)\right)}G^{-1}\left(\frac{j}{n}\right) - \frac{g'\left(G^{-1}\left(\frac{j}{n}\right)\right)}{g\left(G^{-1}\left(\frac{j}{n}\right)\right)}\right\},$$

(6.2) the same sum as (6.1), except that the bracket { } is followed by the factor $G^{-1}(j/n)$

(6.3)
$$\frac{1}{n}\sum_{j=np+1}^{nq-1}\left\{\left[\frac{g'\left(G^{-1}\left(\frac{j}{n}\right)\right)}{g\left(G^{-1}\left(\frac{j}{n}\right)\right)}\right]^2 - \frac{g''\left(G^{-1}\left(\frac{j}{n}\right)\right)}{g\left(G^{-1}\left(\frac{j}{n}\right)\right)}\right\},$$

and

(6.4) the same sum as (6.3), except that the bracket { } is followed by the factor $G^{-1}(j/n)$.

7. The Problem of Takeuchi [4]

Consider the special case where g is symmetric about zero. Then $\mu_n(X)$ of (1.3) becomes

(7.1)
$$t_n(X) = \frac{\sum_{j=1}^{n}\frac{1}{n}A_j^{(n)}X_j^{(n)}}{\sum_{j=1}^{n}\frac{1}{n}A_j^{(n)}}$$

and $\hat{\mu}_n(X)$ of (1.7) becomes, for $p=1-q<\frac{1}{2}$,

(7.2)
$$\hat{t}_n(X) = \frac{A'^{(n)}_{np}(X^{(n)}_{np}+X^{(n)}_{nq}) + \frac{1}{n}\sum_{j=np+1}^{nq-1}A_j^{(n)}X_j^{(n)}}{2A'^{(n)}_{np} + \frac{1}{n}\sum_{j=np+1}^{nq-1}A_j^{(n)}}.$$

Depending upon which regularity conditions are fulfilled, the statistician may use

$$t_n^{**}(X) = \frac{\sum_{j=1}^{n} \frac{1}{n} A_j^{(n)*} X_j^{(n)}}{\sum_{j=1}^{n} \frac{1}{n} A_j^{(n)*}} \tag{7.3}$$

or

$$t_n^{*}(X) = \frac{A_{np}^{\prime(n)*}(X_{np}^{(n)} + X_{nq}^{(n)}) + \frac{1}{n} \sum_{j=np+1}^{nq-1} A_j^{(n)*} X_j^{(n)}}{2A_{np}^{\prime(n)*} + \frac{1}{n} \sum_{j=np+1}^{nq-1} A_j^{(n)*}}. \tag{7.4}$$

If the regularity assumptions of Section 2 are satisfied we can estimate the A's almost exactly as in Section 5, with only two slight differences: a) there is now only one sample, and d^*, e^*, and f^* are calculated from it, and b) we make use of the present symmetry properties of the A's. For $j = 1, 2, \ldots, [n/2]$, we have

$$-G^{-1}\left(\frac{j}{n}\right) = G^{-1}\left(\frac{n-j}{n}\right) \tag{7.5}$$

and hence

$$A_j^{(n)} = A_{n-j}^{(n)}. \tag{7.6}$$

We set

$$\begin{aligned} A_j^{(n)*} = A_{n-j}^{(n)*} = {} & \frac{1}{2}\left(\frac{e^*(n,j)}{d^*(n,j)}\right)^2 + \frac{1}{2}\left(\frac{e^*(n,n-j)}{d^*(n,n-j)}\right)^2 \\ & - \frac{1}{2}\,\frac{f^*(n,j)}{d^*(n,j)} - \frac{1}{2}\,\frac{f^*(n,n-j)}{d^*(n,n-j)} \end{aligned} \tag{7.7}$$

and

$$A_p^{\prime(n)*} = A_{n-p}^{\prime(n)*}. \tag{7.8}$$

It is now easy to prove that $t_n^*(X)$ is an asymptotically efficient estimator among all estimators which are functions of $X_{np}^{(n)}, \ldots, X_{n(1-p)}^{(n)}$ only. We have, for $np \leqq j \leqq n(1-p)$,

$$X_j^{(n)} = \mu + G^{-1}(j/n) + n^{-\frac{1}{2}} \lambda_j, \qquad \max_j \lambda_j = O_p(1).$$

Hence

$$\hat{t}_n(X) = \mu + n^{-\frac{1}{2}} \cdot \frac{A_{np}^{\prime(n)}(\lambda_{np} + \lambda_{nq}) + \frac{1}{n} \sum_{np+1}^{nq-1} A_j^{(n)} \lambda_j}{2A_{np}^{\prime(n)} + \frac{1}{n} \sum_{np+1}^{nq-1} A_j^{(n)}} + O\left(\frac{1}{n}\right) \tag{7.9}$$

and

$$t_n^*(X) = \mu + n^{-\frac{1}{2}} \cdot \frac{A_{np}^{\prime(n)*}(\lambda_{np} + \lambda_{nq}) + \frac{1}{n} \sum_{np+1}^{nq-1} A_j^{(n)*} \lambda_j}{2A_{np}^{\prime(n)*} + \frac{1}{n} \sum_{np+1}^{nq-1} A_j^{(n)*}} + O\left(\frac{1}{n}\right). \tag{7.10}$$

Also

(7.11) $$A_j^{(n)*}=A_j^{(n)}+\Delta(n,j)\cdot n^{-\alpha},\quad np\leqq j\leqq nq$$

and

(7.12) $$\max_j \Delta(n,j)=O_p(1).$$

It is proved in [6] that the denominator of (7.9) approaches a positive number as $n\to\infty$. It is now trivial to verify that

$$\sqrt{n}\,(t_n^*(X)-\hat{t}_n(X))=o_p(1),$$

so that $t_n^*(X)$ is asymptotically efficient.

The problem of estimating μ by $t_n^{**}(X)$ has been studied by Takeuchi [4], who used $A_j^{(n)*}$ different from ours. We do not understand his argument.

8. Asymptotic Confidence Intervals for the Problem of Section 7

We shall give asymptotic confidence intervals (with confidence coefficient ρ, say) based on $t_n^*(X)$. We know that, for large n, $\sqrt{n}\,(t_n^*(X)-\mu)$ has a distribution which is close to the normal distribution with mean zero and a variance V which we shall give below. An estimator $V^*(n)$, which can be found by straightforward application of our methods of Section 7 (which make use of the symmetry properties of G), will have the property that

(8.1) $$|V^*(n)-V\,|\xrightarrow{P}o.$$

From this asymptotic confidence intervals can be obtained at once. For example, let $z(\rho)$ be defined by

$$\frac{1}{\sqrt{2\pi}}\int_{-z(\rho)}^{z(\rho)}\exp\left\{-\frac{t^2}{2}\right\}dt=\rho.$$

Then

$$\left[t_n^*(X)-\frac{z(\rho)[V^*(n)]^{\frac{1}{2}}}{\sqrt{n}},\ t_n^*(X)+\frac{z(\rho)[V^*(n)]^{\frac{1}{2}}}{\sqrt{n}}\right]$$

is such an asymptotic confidence interval.

To facilitate the computation of V we use the results of [5]. Let a be the denominator of the right member of (7.2). Define

$$a_1=\sqrt{p(1-p)}\cdot\frac{g'(G^{-1}(p))}{g(G^{-1}(p))},$$

$$a_2=a+\frac{2a_1^2}{1-p}.$$

Consider $a\hat{t}_n(X)$ as the sum of three terms:

$$A_{np}'^{(n)}X_{np}^{(n)},\ A_{np}'^{(n)}X_{nq}^{(n)},\ \frac{1}{n}\sum_{j=np+1}^{nq-1}A_j^{(n)}X_j^{(n)}.$$

It follows immediately from the theorem of [6] that $a^2 V$ is given by

$$a_2 + 2p(1-p)[A'^{(n)}_{np}]^2 + 4\sqrt{\frac{p}{1-p}}\,\frac{A'^{(n)}_{np}\cdot a_1}{g(G^{-1}(p))} + \frac{2p^2[A'^{(n)}_{np}]^2}{[g(G^{-1}(p))]^2}.$$

Note Added in Proof. In the paper "Efficiency-Robust Estimation of Location" by C. van Eeden, Annals of Mathematical Statistics, 1970, similar results for location parameters, using a different method, are given.

References

1. Bennett, C. A.: Asymptotic properties of ideal linear estimators. Ph. D. dissertation, Univ. of Michigan, 1952.
2. Blom, G.: Statistical estimates and transformed beta-variables. Thesis, University of Stockholm, Sweden, 1958. Uppsala: Almqvist and Wiksells 1958.
3. Jung, J.: On linear estimates defined by a continuous weight function. Ark. Mat. 3, **15**, 199-209 (1955).
4. Takeuchi, K.: A uniformly asymptotically efficient robust estimator of a location parameter. Report IMM 375, Courant Institute of Mathematical Sciences, New York University, May, 1969.
5. Weiss, L.: On the asymptotic distribution of an estimate of a scale parameter. Naval Res. Logist. Quart. **10**, 1-9 (1963).
6. – On estimating location and scale parameters from truncated samples. Naval Res. Logist. Quart. **11**, 125-134 (1964).
7. – Wolfowitz, J.: Generalized maximum likelihood estimators. Teor. Verojatn. Primen **11**, 68-93 (1966).
8. – – Maximum probability estimators. Ann. Inst. statist. Math. **19**, 193-206 (1967).
9. – – Maximum probability estimators with a general loss function. Proc. McMaster Symposium. Berlin-Heidelberg-New York: Springer; Lecture Notes in Mathematics, **89**, 232-256 (1969).
10. – – Maximum probability estimators and asymptotic sufficiency. Submitted to the Ann. Inst. statist. Math.

Professor L. Weiss
Dept. of Operations Research
Cornell University
College of Engineering
Upson Hall
Ithaca, New York 14850, U.S.A.

Professor J. Wolfowitz
Dept. of Mathematics
University of Illinois
Urbana, Ill. 61801, U.S.A.

(Received October 17, 1969)

Z. Wahrscheinlichkeitstheorie verw. Geb. 24, 203–209 (1972)

Optimal, Fixed Length, Nonparametric Sequential Confidence Limits for a Translation Parameter

L. Weiss * and J. Wolfowitz **

1. Statement of Results

Let $X_1, X_2, \ldots$ be independent random variables with the common density function $g(x-\theta)$ at the point x, where θ, which is to be estimated, is a real parameter, unknown to the statistician. (Precise regularity conditions and a precise description of the procedure will be given in Section 2.) The function g is completely unknown to the statistician (except for the regularity conditions given below) and certainly does not belong to any given parametric family.

Let n_0 be a convenient positive integer greater than 6, and let ρ, $0<\rho<1$, be an arbitrary number (ρ will be the confidence coefficient), and let $z(\rho)$ be the function of ρ to be defined below. Let $V_0^*(n)$ and t_n^* be certain functions of $X_1, \ldots, X_n$. Define the stopping variable $N(s)$ for $s>0$ as

$$N=N(s)=\text{smallest } n \geqq n_0 \quad \text{such that} \quad z(\rho)[V_0^*(n)]^{\frac{1}{2}} \leqq s\,n. \tag{1.1}$$

Now observe the next $N^2(s)$ random variables $X_{N(s)+1}, X_{N(s)+2}, \ldots, X_{N(s)+N^2(s)}$ and define the function $\bar{t}_{N(s)}^*$ by $\bar{t}_{N(s)}^* = t_{N^2(s)}^*(X_{N(s)+1}, \ldots, X_{N(s)+N^2(s)})$. Define the random inverval $I_{N(s)}$ by

$$(\bar{t}_{N(s)}^* - s,\ \bar{t}_{N(s)}^* + s). \tag{1.2}$$

We shall prove the following: With probability one,

$$\lim_{n\to\infty} V_0^*(n) = \sigma^2, \tag{1.3}$$

say, where σ^2 is a constant defined below in Lemma 8, and

$$\lim_{s\to 0} \frac{s^2[N(s)+N^2(s)]}{\sigma^2[z(\rho)]^2} = 1. \tag{1.4}$$

We shall also prove that

$$\lim_{s\to 0} P\{\theta \in I_{N(s)}\} = \rho \tag{1.5}$$

and

$$\lim_{s\to 0} \frac{s^2\, E[N(s)+N^2(s)]}{\sigma^2[z(\rho)]^2} = 1. \tag{1.6}$$

It will be seen that the quantity σ^2 of (1.3), which depends on the function g only, is the smallest possible. This is asymptotic efficiency.

* Research supported by National Science Foundation Grant No. GP 21184.

** Research supported by the U.S. Air Force under Grant AF-AFOSR-70-1947, monitored by the Office of Scientific Research.

2. Introduction

The present line of study began with Wald [5], but this fact is usually overlooked in the literature. Chow and Robbins [6] began the nonparametric version of the investigation. Recently, Sen and Ghosh [7] returned to the nonparametric problem, and, using ranks, obtained a smaller σ^2 than that of [6]. They assumed, inter alia, that g is symmetric about zero.

We, too, make this last assumption[1], and postulate the regularity conditions of [1] (see Sections 7 and 8). We then prove (1.3)–(1.6), with the smallest possible σ^2. Indeed, the σ^2 we obtain is such that, even if it were known to the statistician that g is a member of a given parametric family, and the method of [5] were then applied to obtain the smallest possible σ^2, no smaller σ^2 would be obtained. This statement is modulo the interval $[p, 1-p]$ (see [1]) in which the postulated regularity conditions are fulfilled.

The present paper is a sequel to [1], whose notation is adopted below unless the contrary is explicitly stated. Familiarity with [1] will be assumed in order to avoid needless repetition.

The functions t_n^*, $V^*(n)$, and $z(\rho)$ are defined in Sections 7 and 8 of [1][2]. Let p be the quantity of Section 7 of [1]. It will be assumed below that the regularity conditions of [1] are satisfied in the closed interval $[p, 1-p]$. Of course, p should be taken as small as possible.

Whenever, in what follows, there appear expressions like $\sup_j$, it is always to be understood that the supremum is taken over all j such that $np<j<n(1-p)$.

Let M be a large constant; what "large" means here will be explained below. We define

$$V_0^*(n)=\min\{V^*(n), M\}.$$

3. Auxiliary Results

Let $S_n(\cdot)$ be the empiric distribution function of n independent random variables with the common distribution function $G_0(\cdot)$. The following result is part of Lemma 2 of [2]:

Lemma 1. *There exists a positive constant C such that, for all positive integers n and all* $r \geqq 0$,

$$P\left\{\sup_x |S_n(x)-G_0(x)| \leqq \frac{r}{\sqrt{n}}\right\} > 1 - C e^{-2r^2} \tag{3.1}$$

The next five lemmas are concerned with the behavior of certain estimators described in (5.15)–(5.20) of [1]. Below, $Y_1^{(n)}<\cdots<Y_n^{(n)}$ denote the ordered values of n independent and identically distributed random variables, each with distribution function $G(\cdot)$. In (5.15)–(5.20) of [1], the X's defined in [1] should have been used, rather than the Y's defined in [1].

---

[1] Our results are valid under more general conditions, for example the following: There exists a positive constant k (not necessarily known to the statistician) such that, in the notation of [1], $G^{-1}(x) = -k\, G^{-1}\left(\frac{1-x}{k}\right)$ for all x such that $\frac{1}{k+1} \leqq x \leqq 1-p$. When $k=1$, g is symmetric about zero. A study of the most general conditions is in progress.

[2] In the definition of t_n^* in (7.10) of [1], the expression $(\lambda_{np}+\lambda_{nq})$ should be deleted from the denominator.

We remark that Lemmas 3–6 below can easily be strengthened, for example by increasing the upper bound on δ, or by bringing in $\log n$, or in similar ways. This is so because the estimators of Section 5 of [1] are merely sufficient for their purposes ([1], p. 143, last paragraph) and can easily be improved.

Lemma 2. *For a given δ in the open interval $(0, \frac{1}{2})$, with probability one*

$$\limsup_{n\to\infty} \sup_j n^{\frac{1}{2}-\delta}\left|Y_j^{(n)} - G^{-1}\left(\frac{j}{n}\right)\right| = 0. \tag{3.2}$$

Proof. From Lemma 1 and the fact ([1]) that g is bounded below by a positive constant in $[G^{-1}(p), G^{-1}(1-p)]$,

$$P\left\{\sup_j n^{\frac{1}{2}-\delta}\left|Y_j^{(n)} - G^{-1}\left(\frac{j}{n}\right)\right| > n^{-\delta/2}\right\} < C\, e^{-\tau n^{\delta}} \tag{3.3}$$

where τ is a positive constant. The desired result follows from the Borel-Cantelli lemma.

Lemma 3. *For a given δ in the open interval $(0, \frac{1}{4})$, with probability one*

$$\limsup_{n\to\infty} \sup_j n^{\frac{1}{4}-\delta}\,|d^*(n,j) - d(n,j)| = 0. \tag{3.4}$$

Proof. Apply Lemma 2 above to (5.5) of [1].

Lemma 4. *For a given δ in the open interval $(0, \frac{1}{8})$, with probability one*

$$\limsup_{n\to\infty} \sup_j n^{\frac{1}{8}-\delta}\,|e^*(n,j) - e(n,j)| = 0. \tag{3.5}$$

Proof. Apply Lemmas 2 and 3 above to (5.8) of [1].

Lemma 5. *For a given δ in the open interval $(0, \frac{1}{16})$, with probability one*

$$\limsup_{n\to\infty} \sup_j n^{\frac{1}{16}-\delta}\,|f^*(n,j) - f(n,j)| = 0. \tag{3.6}$$

Proof. Apply Lemmas 2 and 4 above to (5.11) of [1].

Lemma 6. *For a given δ in the open interval $(0, \frac{1}{16})$, with probability one*

$$\limsup_{n\to\infty} \sup_j n^{\frac{1}{16}-\delta}\,|A_j^{(n)*} - A_j^{(n)}| = 0 \tag{3.7}$$

and

$$\lim_{n\to\infty} n^{\frac{1}{16}-\delta}\,|A_{np}^{\prime(n)*} - A_{np}^{\prime(n)}| = 0. \tag{3.8}$$

Proof. This follows from Lemmas 3, 4, and 5 above applied to (5.18)–(5.20) of [1].

Lemma 7. *With probability one,*

$$\lim_{n\to\infty} |V^*(n) - V| = 0. \tag{3.9}$$

(Compare (8.1) of [1].)

Proof. Denote by $a(n)$ the quantity called a in [1], Section 8. Recall that $q = 1-p$, and

$$a(n) = 2A_{np}^{\prime(n)} + \frac{1}{n}\sum_{np+1}^{nq-1} A_j^{(n)} \tag{3.10}$$

and that it is proved in [4] that

$$\lim_{n\to\infty} a(n) = g'(G^{-1}(q)) + \frac{1}{1-q} g^2(G^{-1}(q)) - g'(G^{-1}(p)) + \frac{1}{p} g^2(G^{-1}(p)) + \int_{G^{-1}(p)}^{G^{-1}(q)} \left\{ \left[\frac{g'(y)}{g(y)} \right]^2 - \frac{g''(y)}{g(y)} \right\} g(y)\, dy = t_1, \quad \text{say.} \tag{3.11}$$

Let $a^*(n)$ be the expression in the right member of (3.10) with the A's replaced by A^*'s. Then it follows from Lemma 6 above that with probability one

$$\lim_{n\to\infty} a^*(n) = t_1. \tag{3.12}$$

It was proved in [3] that $t_1 > 0$. The desired result now follows from Lemma 6.

It is easy to verify that we can replace (3.9) by

$$\lim_{n\to\infty} n^{\frac{1}{16}-\delta} |V^*(n) - V| = 0 \tag{3.13}$$

with probability one, for $0 < \delta < \frac{1}{16}$. (See also the paragraph preceding Lemma 2.) However, in the present paper we do not need this stronger result.

Lemma 8. *With probability one,*

$$\lim_{n\to\infty} V^*(n) = \sigma^2 = V \tag{3.14}$$

where

$$t_1^2 \left[\sigma^2 - \frac{1}{t_1} \right] = 2p \left(\frac{g'(G^{-1}(p))}{g(G^{-1}(p))} \right)^2 + 2p(1-p) \left\{ \frac{1}{p} g^2(G^{-1}(p)) - g'(G^{-1}(p)) \right\}^2 + \frac{4p \left\{ \frac{1}{p} g^2(G^{-1}(p)) - g'(G^{-1}(p)) \right\} g'(G^{-1}(p))}{(g(G^{-1}(p)))^2} + \frac{2p^2 \left\{ \frac{1}{p} g^2(G^{-1}(p)) - g'(G^{-1}(p)) \right\}^2}{(g(G^{-1}(p)))^2}. \tag{3.15}$$

Proof. It follows from the last displayed expression in Section 8 of [1], or from [4], where it is proved, that

$$V = \sigma^2. \tag{3.16}$$

The lemma therefore follows from Lemma 7.

4. Proof of the Results

Lemma 8 is (1.3). The statements (1.4) and (1.6) are immediate consequences of Lemmas 1 and 2 of [6], the proof of [6] is a modification of one in [8]. The relation (1.5) is a consequence of (1.4), the asymptotic distribution of $\bar{t}^*_{N(s)}$ (proved in [1], Section 7, with use made of the results of [4] about the asymptotic distribution of $\hat{t}_n$), and the independence of $\bar{t}^*_{N(s)}$ and $V_0^*(N(s))$.

It is to achieve this latter independence that we "waste" a fraction

$$\frac{N(s)}{N(s)+N^2(s)}$$

of observations. Of course this fraction approaches zero as s approaches zero. We could just as easily have wasted a fraction

$$\frac{N(s)}{N(s)+N^k(s)}$$

of observations ($k>1$), or other such fractions.

The independence of $\bar{t}^*(N(s))$ and $V_0^*(N(s))$ enables us to give at once a proof of (1.5). The method usually employed in the literature is that of [9]. Its application in [6] is immediate, its application in [7] is, and in the present paper would also be, difficult and involved. By "wasting" a fraction of the observations which approaches zero we completely bypass this difficulty.

In place of the interval $I_{N(s)}$ we could, of course, have employed the interval

$$\left(\bar{t}^*_{N(s)}-\frac{z(\rho)\,[V_0^*(N(s))]^{\frac{1}{2}}}{N(s)},\ \bar{t}^*_{N(s)}+\frac{z(\rho)\,[V_0^*(N(s))]^{\frac{1}{2}}}{N(s)}\right)$$

with the same result (1.5).

The definition of $V_0^*(n)$ was designed to insure that

$$E\{\sup_n V_0^*(n)\}<\infty, \tag{4.1}$$

which is needed to apply Lemma 2 of [6] so as to obtain (1.6) of [6]. In the latter paper a random variable v_n corresponds to our $V^*(n)$, and the device used by us to achieve (4.1) could have been used in [6]. The authors of [6] prefer to prove essentially (4.1), no doubt for aesthetic reasons. It seems likely that one could prove

$$E\{\sup_n V^*(n)\}<\infty. \tag{4.2}$$

However, $V^*(n)$ is a more complicated function than v_n by several orders of magnitude, and the proof of (4.2) is likely to be so laborious that its character would outweigh all aesthetic considerations. We have therefore proceeded as above.

The constant M need only be a number which exceeds any possible value of σ^2. Such a number should be available in any problem arising in practice.

Denote the σ^2 of the present paper by $\sigma_1^2(p)$, that of [6] by σ_6^2, and that of [7] for their "normal scores" statistic by σ_7^2. The three schemes are not strictly comparable because they postulate different regularity conditions. However, a comparison of the σ^2 may still be of interest. The significance of the σ^2 lies in the fact that (1.6) and its analogues in [6] and [7] hold. (To obtain σ_7^2, therefore, employ (3.6) and (3.7) of [7]. In the notation of [7], $\sigma_7^2=A^2/4B^2(F)$.) Thus, the smaller σ^2 the better. Sen and Ghosh ([7], Section 6) state that

$$\sigma_7^2\leqq\sigma_6^2 \tag{4.3}$$

for all distributions with a density and a finite second moment. Modulo the different regularity conditions, we therefore have from the result of [1], Section 7, and (4.3) that

$$\sigma_1^2(0) \leqq \sigma_7^2 \leqq \sigma_6^2. \tag{4.4}$$

Obviously, $\sigma_1^2(p)$ is monotonically increasing in p. It would be interesting to make numerical comparisons between $\sigma_1^2(p)$ and σ_7^2 for various p and various densities g. The quantity σ_7^2 depends also on the "score-function" of [7]. Computations performed by us, for several important g and score-functions, give $\sigma_1^2(p)$ substantially smaller than σ_7^2 for small p.

Gleser ([10]), using methods similar to those of [6], obtained sequential confidence intervals for regression coefficients in the nonparametric case. Sen and Ghosh ([11]) extended their method, based on ranks, to this problem, and state that a comparison like that of (4.3) holds. Koul ([14]) gave other confidence regions, also based on ranks. The present authors, in [12] [3], obtained asymptotically optimal nonsequential point estimators of nonparametric regression coefficients. This method can also be extended to optimal sequential estimation by confidence intervals of fixed length. Details will appear elsewhere.

Finally, we remark that the nonsequential methods of [1] and [12] are truly "robust" in the full statistical meaning of this word, as used in the papers of Huber and others (see [13] for some recent references), and should be considered among the solutions of the problem of robustly estimating a translation parameter. The papers on robust estimation mentioned above, and many others, all judge the efficiency of a procedure by its asymptotic variance, just as is done in [1] and [12]. All assume that the parametric form of the density g is known up to a small "contaminating" component; this strong assumption is not made in either [1] or [12]. Of course, the sequential procedures of [6, 7, 10, 11] and the present paper are all fully nonparametric and truly robust.

References

1. Weiss, L., and Wolfowitz, J.: Asymptotically efficient non-parametric estimators of location and scale parameters. Z. Wahrscheinlichkeitstheorie verw. Gebiete **16**, 134–150 (1970).
2. Dvoretzky, A., Kiefer, J., and Wolfowitz, J.: Asymptotic minimax character of the sample distribution function and of the classical multinomial estimator. Ann. Math. Statistics **27**, 642–669 (1956).
3. Weiss, L.: On estimating location and scale parameters from truncated samples. Naval Res. Logist. Quart. **11**, 125–134 (1964).
4. Weiss, L.: On the asymptotic distribution of an estimate of a scale parameter. Naval Res. Logist. Quart. **11**, 1–9 (1963).
5. Wald, A.: Asymptotic minimax solutions of sequential point estimation problems. Proc. Second Berkeley Sympos. Math. Statist. Probab., Univ. Calif. 1950/1951, 1–12.
6. Chow, Y.S., and Robbins, H.: On the asymptotic theory of fixed-width sequential confidence intervals for the mean. Ann. Math. Statistics **36**, 457–462 (1965).
7. Sen, P.K., and Ghosh, M.: On bounded length sequential confidence intervals based on one-sample rank order statistics. Ann. Math. Statistics **42**, 189–203 (1971).

[3] In the notation of [12], define T_n as $\frac{1}{n}\sum_{j=1}^{k(n)} N_j(n)\, m_j(n)$, S_n as $\frac{1}{n}\sum_{j=1}^{k(n)} N_j(n)(m_j(n)-T_n)^2$. Then $T=\lim_{n\to\infty} T_n$, $S=\lim_{n\to\infty} S_n$. The following changes should be made in [12]: In the definitions of $\bar{Q}_1(n)$, $\bar{Q}_2(n)$ in Sections 4 and 5, and of $\alpha_j(n)$, $\beta_j(n)$ in Section 5, replace T, S by T_n, S_n respectively.

8. Doob, J.L.: Renewal theory from the point of view of the theory of probability. Trans. Amer. Math. Soc. **63**, 422–438 (1948).
9. Anscombe, F.J.: Large sample theory of sequential estimation. Proc. Cambridge Philos. Soc. **48**, 600–607 (1952).
10. Gleser, L.J.: On the asymptotic theory of fixed-size sequential confidence bounds for a linear regression parameter. Ann. Math. Statistics **36**, 463–467 (1965).
11. Sen, P.K., and Ghosh, M.: On bounded length confidence intervals for a regression coefficient based on a class of rank statistics. Institute of Statistics, University of North Carolina, Chapel Hill, N.C. Mimeo series #680 (1970).
12. Weiss, L., and Wolfowitz, J.: Asymptotically efficient estimation of non-parametric regression coefficients. In: Statistical Decision Theory and Related Topics. Proc. of a Symposium held at Purdue University, November 1970. Shanti S. Gupta and James Yackel, *eds.* New York-London: Academic Press 1971.
13. Jaeckel, L.A.: Robust estimates of location: Symmetry and asymmetric contamination. Ann. Math. Statistics **42**, 1020–1034 (1971).
14. Koul, H.L.: Asymptotic behavior of a class of confidence regions based on ranks in regression. Ann. Math. Statistics **42**, 466–476 (1971).

L. Weiss
Department of Operations Research
Cornell University
Upson Hall
Ithaca, N.Y. 14850
USA

J. Wolfowitz
Department of Mathematics
University of Illinois
Urbana, Ill. 61801
USA

(Received September 10, 1971)

An Asymptotically Efficient, Sequential Equivalent of the *t*-Test

By L. Weiss and J. Wolfowitz

Cornell University *University of Illinois*

[Received December 1971. Revised March 1972]

Summary

Let $X_1, X_2, \ldots,$ be independent, normally distributed chance variables, each with mean μ and variance σ^2, both unknown to the statistician. Let it be required to test sequentially the hypothesis that (make a decision between) $\mu = \mu_1$ against the alternative $\mu = \mu_2$, $\mu_1 < \mu_2$. When the sample size is fixed the Student *t*-test is usually used for this purpose. In the present paper we give a sequential procedure for performing this test which is consistent and asymptotically efficient as $(\mu_2 - \mu_1)$ approaches zero.

Keywords: SEQUENTIAL *t*-TEST

1. Introduction

Let $X_1, X_2, \ldots,$ be independent, normally distributed chance variables, each with mean μ and variance σ^2, both unknown to the statistician. Let it be required to test sequentially the hypothesis that (make a decision between) $\mu = \mu_1$ against the alternative $\mu = \mu_2, \mu_1 < \mu_2$. When the sample size is fixed the Student *t*-test is usually used for this purpose.

In the present paper we give a sequential procedure for performing this test which is consistent and asymptotically efficient as $\mu_2 - \mu_1 = 1/t$, say, approaches zero. We will now explain what is meant by "consistent" and by "asymptotically efficient".

Let α and β be fixed small positive numbers; their significance is that the probabilities of the wrong decision when $\mu = \mu_1$ and $\mu = \mu_2$, respectively, will be very close to α and β. Write $A = (1-\beta)/\alpha$, $B = \beta/(1-\alpha)$. Suppose that the statistician knew the (positive) value of σ^2, and used the Wald sequential probability ratio test (WSPRT) with bounds A and B (Wald, 1947) to test the hypothesis $\mu = \mu_1$ against the alternative $\mu = \mu_2$. Let $N_W(\theta|t)$ be the number of observations required by this test to reach a decision when $\mu = \theta$. It is known (Wald and Wolfowitz, 1948) that for any t, $EN_W(\mu_1|t)$ and $EN_W(\mu_2|t)$ are both minimal, among all tests where the statistician knows σ^2 and the respective probabilities of error do not exceed those of the WSPRT. Let $L_W(\theta|t)$ be the probability of making the decision that $\mu = \mu_1$ by the procedure just described, when $\mu = \theta$. Let $L(\theta|t)$ be the probability, when $\mu = \theta$, of making the decision that $\mu = \mu_1$ by the procedure to be described below, which does not require a knowledge of σ^2. Let $N(\theta|t)$ be the number of observations required to reach a decision by this latter procedure, when $\mu = \theta$. Without loss of generality we set $\mu_1 = -1/2t$, $\mu_2 = 1/2t$. We shall show that, as $t \to \infty$, uniformly with respect to θ in any bounded interval,

$$\lim_{t\to\infty} L(\theta/t|t) = \lim_{t\to\infty} L_W(\theta/t|t) \quad \text{(consistency)}, \tag{1.1}$$

$$\lim_{t\to\infty} \frac{EN(\theta/t|t)}{EN_W(\theta/t|t)} = 1 \quad \text{(asymptotic efficiency)}. \tag{1.2}$$

Consequently our procedure is suitable for use when the statistician wishes to decide whether $\mu = \mu_1$ or $\mu = \mu_2$, with σ^2 unknown and $(\mu_2 - \mu_1)$ small. The procedure is then almost as efficient as the best procedure possible when σ^2 is known to the statistician.

Our procedure will be described precisely in the next section. Roughly speaking, it consists of taking a preliminary sample of random size to obtain an estimate $s^2(t)$ of σ^2, and then proceeding according to a WSPRT with bounds A and B, as if $\sigma^2 = s^2(t)$.

A procedure for our problem was indicated by Wald (1947). It involves integrating the densities with respect to a distribution on σ^2. No optimal properties have yet been proved for it. On the other hand, our procedure is intended only for small $(\mu_2 - \mu_1)$, which is not the case for Wald's procedure. Wald's procedure in the asymptotic case was studied by Schwarz (1971) by his method of asymptotic shapes.

2. PRECISE DESCRIPTION OF PROCEDURE

It will simplify the notation if we introduce the sequence of independent chance variables $Z_1, Z_2, \ldots$ with the same distribution as the X's. Define $\bar{Z}(k)$ as $(Z_1 + \ldots + Z_k)/k$, and $Y_k = (k\sigma^2)^{-1}[\{Z_1 - \bar{Z}(k)\}^2 + \ldots + \{Z_k - \bar{Z}(k)\}^2]$ for $k = 2, 3, \ldots$. Let c be a constant greater than unity. Define the chance variable $N_1(t)$ as the smallest $k \geqslant 2$ such that $Y_k \leqslant k^c/(t\sigma^2)$. Clearly $N_1(t)$ is defined with probability one ($N_1(t) = \infty$ with probability zero). The distribution of $N_1(t)$ does not depend on μ.

Suppose σ^2 were known to the statistician and the latter used a WSPRT on the X's. Then the statistician would continue sampling as long as

$$\log B < -\frac{1}{2\sigma^2}\left\{\sum_1^k (X_i - \mu_2)^2 - \sum_1^k (X_i - \mu_1)^2\right\} < \log A. \tag{2.1}$$

In our procedure the statistician does not know σ^2, and replaces σ^2 by $s^2(t) = \{N_1(t)\}^c/t$. Thus sampling continues as long as

$$(\log B)\frac{s^2(t)}{\sigma^2} < -\frac{1}{2\sigma^2}\left\{\sum_1^k (X_i - \mu_2)^2 - \sum_1^k (X_i - \mu_1)^2\right\} < (\log A)\frac{s^2(t)}{\sigma^2}. \tag{2.2}$$

When σ^2 multiplied by the expression in the middle of (2.2) exceeds $s^2(t)\log A$, the statistician decides that $\mu = \mu_2$. When σ^2 multiplied by the expression in the middle of (2.2) is less than $s^2(t)\log B$, the statistician decides that $\mu = \mu_1$. Let $N_2(\theta\,|\,t)$ be the number of observations required to make a decision according to the procedure based on (2.2), when $\mu = \theta$. Obviously

$$N(\theta\,|\,t) = N_1(t) + N_2(\theta\,|\,t). \tag{2.3}$$

3. PROOF OF CONSISTENCY

From (7:18), (A:31), (A:51) and (A:64) of Wald (1947), we obtain the following for any $\theta < 0$:

$$\frac{A^{-2\theta} - 1}{A^{-2\theta} - \{B^{-2\theta}/\delta(\theta/t)\}} \leqslant L_W\left(\frac{\theta}{t}\,\middle|\,t\right) \leqslant \frac{\delta(\theta/t)A^{-2\theta} - 1}{\delta(\theta/t)A^{-2\theta} - B^{-2\theta}}, \tag{3.1}$$

where

$$\lim_{t\to\infty} \delta(\theta/t) = 1. \tag{3.2}$$

We note that $\delta(\theta|t)$ is defined in an Appendix of (Wald, 1947) devoted to refining the approximations given by assuming the test statistic does not overshoot the decision bounds. The approach in (3.2) is uniform for θ bounded in absolute value. Hence, for $\theta<0$,

$$\lim_{t\to\infty} L_W(\theta/t\,|\,t) = \frac{A^{-2\theta}-1}{A^{-2\theta}-B^{-2\theta}}, \tag{3.3}$$

and the approach is uniform for θ bounded in absolute value.

Let $L_W\{\theta\,|\,t, s^2(t)\}$ be the conditional probability, given the first sample, of making the decision $\mu=\mu_1$ under the procedure (2.2) when $\mu=\theta$. Obviously, for any θ and t,

$$L(\theta/t\,|\,t) = EL_W(\theta/t\,|\,t, s^2(t)\}. \tag{3.4}$$

Define $A\{s(t)\}$ and $B\{s(t)\}$ by

$$\log A\{s(t)\} = \{s^2(t)/\sigma^2\}\log A$$
$$\log B\{s(t)\} = \{s^2(t)/\sigma^2\}\log B.$$

Now let $\theta<0$. By exactly the same argument which yielded (3.1) we obtain that

$$\frac{[A\{s(t)\}]^{-2\theta}-1}{[A\{s(t)\}]^{-2\theta}-[B\{s(t)\}]^{-2\theta}/\delta(\theta/t)} \leqslant L_W\{\theta/t\,|\,t, s^2(t)\}$$
$$\leqslant \frac{\delta(\theta/t)\,[A\{s(t)\}]^{-2\theta}-1}{\delta(\theta/t)\,[A\{s(t)\}]^{-2\theta}-[B\{s(t)\}]^{-2\theta}}. \tag{3.5}$$

Since $s^2(t)$ converges in probability to σ^2 as t increases, it follows from (3.2) and (3.5) that the chance variable $L_W\{\theta/t\,|\,t, s^2(t)\}$ converges in probability to the right member of (3.3), as t increases.

This convergence of $L_W\{\theta/t\,|\,t, s^2(t)\}$ is uniform for $\theta<0$ and bounded in absolute value, because of (3.2) and the fact that the distribution of $s^2(t)$ does not depend on μ. It follows from (3.4) that $L(\theta/t\,|\,t)$ converges uniformly, as $t\to\infty$, to the right member of (3.3). Thus we have

$$\lim_{t\to\infty} L(\theta/t\,|\,t) = \lim_{t\to\infty} L_W(\theta/t\,|\,t), \tag{3.6}$$

and the convergence of both members is uniform for $\theta<0$ and bounded in absolute value. This proves (1.1) for $\theta<0$. The proof for $\theta>0$ is essentially the same. The proof for $\theta=0$ is by a continuity argument as in Wald (1947). This completes the proof of consistency.

4. PROOF OF ASYMPTOTIC EFFICIENCY

In a later paragraph we will show that $EN_2(\theta/t\,|\,t)$ is of the order t^2. We will now show that $EN_1(t)$ is of a lower order in t (so that $EN_1(t)$ can be neglected in the limit when $EN(\theta/t\,|\,t)$ is computed). To prove the latter, we note that, since the Z's are normally distributed and hence have moments of all orders, it follows that

$$E\{\sup_k Y_k\} < \infty, \tag{4.1}$$

and hence that

$$E\{\sup_k Y_k^{1/c}\} < \infty. \tag{4.2}$$

Now $N_1(t)$ is also the smallest k such that $Y_k^{1/c} \leqslant k/\{(t\sigma^2)^{1/c}\}$. From Lemma 2 of Chow and Robbins (1965) and (4.2) we therefore obtain

$$\lim_{t\to\infty} \frac{EN_1(t)}{(t\sigma^2)^{1/c}} = 1. \tag{4.3}$$

Since $c>1$ this is the desired result.

Exactly the same argument which yielded (4.3) also gives

$$\lim_{t\to\infty} \frac{E[\{N_1(t)\}^c]}{t\sigma^2} = \lim_{t\to\infty} \frac{Es^2(t)}{\sigma^2} = 1. \tag{4.4}$$

Since $\{s^2(t)\}/\sigma^2$ converges in probability to one as $t\to\infty$, we draw from (4.4) the following conclusion: Let $h>0$ be arbitrary. Define the set

$$\bar{R}(t) = [(x_1, x_2, \ldots, x_{N_1(t)}) \,|\, 1-h < \{s^2(t)\}/\sigma^2 < 1+h],$$

and let $\{Es^2(t)\}/\sigma^2 = R_1(t)+R_2(t)$, where $R_1(t)$ is the integral over the set $\bar{R}(t)$ and $R_2(t)$ is the integral over the complement of $\bar{R}(t)$. Then

$$\lim_{t\to\infty} R_2(t) = 0. \tag{4.5}$$

Now let $\theta\neq 0$ be any point in a bounded interval. For any such θ we have from (A:77), (A:78), (A:125) and (A:126) of Wald (1947) that the value of $M_W\{\theta/t\,|\,t, s^2(t)\}$, the expected number of observations required by the procedure (2.2), is given by

$$M_W\{\theta/t\,|\,t,s^2(t)\}$$

$$= \frac{\sigma^2 t^2}{\theta}\left(L_W\left\{\frac{\theta}{t}\middle| t, s^2(t)\right\}\left\{\frac{s^2(t)}{\sigma^2}\log B + \frac{\gamma_1}{t}\right\} + \left[1 - L_W\left\{\frac{\theta}{t}\middle| t, s^2(t)\right\}\right]\left\{\frac{s^2(t)}{\sigma^2}\log A + \frac{\gamma_2}{t}\right\}\right). \tag{4.6}$$

Here γ_1 and γ_2 are functions of $\{\theta, t, s^2(t)\}$ which, for every value of $\{t, s^2(t)\}$, and every θ in a bounded interval, are bounded above in absolute value by Γ_1 and Γ_2 (say), respectively. (See (A:125) and (A:126) of Wald (1947).) Obviously

$$EN_2(\theta/t\,|\,t) = EM_W(\theta/t\,|\,t, s^2(t)). \tag{4.7}$$

The same argument which led to (4.6) also yields

$$EN_W\left(\frac{\theta}{t}\middle| t\right) = \frac{\sigma^2 t^2}{\theta}\left[L_W\left(\frac{\theta}{t}\middle| t\right)\left(\log B + \frac{\gamma_1'}{t}\right) + \left\{1 - L_W\left(\frac{\theta}{t}\middle| t\right)\right\}\left(\log A + \frac{\gamma_2'}{t}\right)\right], \tag{4.8}$$

where $|\gamma_1'| \leqslant \Gamma_1$, $|\gamma_2' \leqslant \Gamma_2$. This is valid for every t and every θ $(\neq 0)$ in a bounded interval.

From (4.5) to (4.7) and the remarks which follow (3.5), we obtain that, uniformly for all θ in a bounded interval and $\theta\neq 0$,

$$\lim_{t\to\infty} \frac{\theta EN_2(\theta/t\,|\,t)}{\sigma^2 t^2} = (\log B)\frac{A^{-2\theta}-1}{A^{-2\theta}-B^{-2\theta}} + (\log A)\frac{1-B^{-2\theta}}{A^{-2\theta}-B^{-2\theta}}$$

$$= T(\theta), \quad \text{say.} \tag{4.9}$$

From (4.8) we obtain that, uniformly for such θ,

$$\lim_{t\to\infty} \frac{\theta EN_W(\theta/t\,|\,t)}{\sigma^2 t^2} = T(\theta). \tag{4.10}$$

19

Let θ^* be the (unique) root of $T(\theta) = 0$. From (4.9) and (4.10) we obtain the desired result (1.2) except for $\theta = 0$ and $\theta = \theta^*$. The result (1.2) is proved for the latter values by a continuity argument as in Wald (1947).

5. CONCLUDING REMARKS

The test constructed above depended on a value c, which could be any arbitrary value above unity. Whatever the choice of c, the resulting test is consistent and asymptotically efficient. Some choices of c may be better than others in the sense that the resulting test has characteristics for fixed $\mu_2 - \mu_1$ closer to the asymptotic characteristics (as $\mu_2 - \mu_1$ approaches zero), but no information about this is available.

ACKNOWLEDGEMENTS

This research was supported by National Science Foundation Grant No. GP 21184 (L. Weiss) and by the U.S. Air Force under Grant No. AF-AFOSR-70-1947, monitored by the Office of Scientific Research (J. Wolfowitz).

REFERENCES

CHOW, Y. S. and ROBBINS, H. (1965). On the asymptotic theory of fixed-width sequential confidence intervals for the mean. *Ann. Math. Statist.*, **36**, 457–462.

SCHWARZ, G. (1971). A sequential Student test. *Ann. Math. Statist.*, **42**, 1003–1009.

WALD, A. (1947). *Sequential Analysis*. New York: Wiley.

WALD, A. and WOLFOWITZ, J. (1948). Optimum character of the sequential probability ratio test. *Ann. Math. Statist.*, **19**, 326–339.

The Annals of Statistics
1973, Vol. 1, No. 5, 944-947

MAXIMUM LIKELIHOOD ESTIMATION OF A TRANSLATION PARAMETER OF A TRUNCATED DISTRIBUTION

BY L. WEISS[1] AND J. WOLFOWITZ[2]

Cornell University and University of Illinois

$f(x)$ is a uniformly continuous density which equals zero for negative values of x, has a right-hand derivative equal to α at $x = 0$, where $0 < \alpha < \infty$, and satisfies certain regularity conditions. $X_1, \cdots, X_n$ are independent random variables with the common density $f(x - \theta)$, θ an unknown parameter. Let $\hat{\theta}_n$ denote the maximum likelihood estimator of θ, and define α_n by the equation $2\alpha_n^2 = \alpha n \log n$. It was shown by Woodroofe that the asymptotic distribution of $\alpha_n(\hat{\theta}_n - \theta)$ is standard normal. It is shown in the present paper that $\hat{\theta}_n$ is an asymptotically efficient estimator of θ.

Let f be a uniformly continuous density which vanishes on $(-\infty, 0]$ and is subject to regularity conditions to be described. Among these conditions is one which requires that $\alpha = \lim f'(x)$ exists as $x \to 0$ from the right, with $0 < \alpha < \infty$. Let $\Theta = (-\infty, \infty)$ be the parameter space of the unknown parameter θ. Let $X_1, \cdots, X_n$ be independent chance variables with the common density $f(x - \theta)$, at the point x of the real line. Let $\{\hat{\theta}_n\}$ be a consistent sequence of roots of the likelihood equation. It was proved by Woodroofe ([4]) under two different sets of regularity conditions (either of which we henceforth adopt) that, for any θ,

$$(1) \qquad \lim_{n\to\infty} P_\theta\{\alpha_n(\hat{\theta}_n - \theta) < y\} = \frac{1}{(2\pi)^{\frac{1}{2}}} \int_{-\infty}^{y} e^{-y^2/2}\, dy$$

for any y, where $2\alpha_n^2 = \alpha n \log n$.

It is obvious that we are here dealing with what is called a "non-regular" case (see, for example, [1]) since the normalizing factor is not $n^{\frac{1}{2}}$. Consequently the question remains open whether the maximum likelihood (m.l.) estimator $\hat{\theta}_n$ is asymptotically efficient. If a proof of this is not available then only faith in the eventual appearance of such a proof would justify the statistician's use of the m.l. estimator. In this note we prove the asymptotic efficiency of the m.l. estimator by proving that it is asymptotically equivalent to a maximum probability (m.p.) estimator ([2], [3]). The precise statement of efficiency is given in the theorem below. Our proof will be brief and will utilize some results of [4].

Let θ_0 now be any fixed point in Θ. We shall say that a sequence of functions $\{l_n(\cdot)\}$ converges in $H(h)$ (to a constant) if the following is true: Let $\{y_n, n = 1, 2, \cdots\}$ be any sequence of real numbers such that $|\alpha_n(y_n - \theta_0)| \leq h$ for $n = 1, 2, \cdots$. Then the sequence of real numbers $\{l_n(y_n)\}$ converges (to the constant).

Received August 1972; revised January 1973.

[1] Research supported by NSF Grant GP 31430X.

[2] Research supported by the U. S. Air Force under Grant AF-AFOSR 70-1947, monitored by the Office of Scientific Research.

Let $R = (-r, r)$ be any interval centered at the origin. We now state and prove the following.

THEOREM. *Let $\{T_n\}$ be any competing sequence of estimators such that, for any $h > 0$, we have, for θ in $H(h)$,*

$$\lim_{n\to\infty} [P_\theta\{\alpha_n(T_n - \theta) \text{ in } R\} - P_{\theta_0}\{\alpha_n(T_n - \theta_0) \text{ in } R\}] = 0 . \tag{2}$$

Then

$$\limsup_{n\to\infty} P_{\theta_0}\{\alpha_n(T_n - \theta_0) \text{ in } R\} \leqq \frac{1}{(2\pi)^{\frac{1}{2}}} \int_{-r}^{r} e^{-y^2/2}\, dy . \tag{3}$$

This is the statement of efficiency for $\hat{\theta}_n$. If, as is usually required in the literature (but not necessary for us), $\alpha_n(T_n - \theta)$ is also asymptotically normally distributed (in P_θ-probability for every θ) with mean 0 and variance $\sigma_\theta^2(T)$, from (1) and the Theorem we immediately obtain that

$$\sigma_\theta^2(T) \geqq 1 , \qquad \theta \text{ in } \Theta . \tag{4}$$

This is the classical statement of asymptotic efficiency.

PROOF OF THE THEOREM. An m.p. estimator Z_n can be obtained as follows. Let $\{\lambda_n\}$ be a sequence of positive numbers which approach zero. Then Z_n satisfies

$$\alpha_n \int_{Z_n - r/\alpha_n}^{Z_n + r/\alpha_n} \prod_{i=1}^n f(X_i - \theta)\, d\theta \geqq \alpha_n(1 - \lambda_n) \sup_d \int_{d - r/\alpha_n}^{d + r/\alpha_n} \prod_{i=1}^n f(X_i - \theta)\, d\theta . \tag{5}$$

Define $t = \alpha_n(\theta - \hat{\theta}_n)$ and

$$V_n(t) = [\prod_{i=1}^n f(X_i - \theta)][\prod_{i=1}^n f(X_i - \hat{\theta}_n)]^{-1} . \tag{6}$$

Since the second factor of $V_n(t)$ does not depend on θ, we may rewrite (5) as

$$\int_{\alpha_n(Z_n - \hat{\theta}_n) - r}^{\alpha_n(Z_n - \hat{\theta}_n) + r} V_n(t)\, dt \geqq (1 - \lambda_n) \sup_d \int_{\alpha_n(d - \hat{\theta}_n) - r}^{\alpha_n(d - \hat{\theta}_n) + r} V_n(t)\, dt . \tag{7}$$

Let $k(\cdot)$ be any positive function defined on the positive integers such that, as $n \to \infty$,

$$k(n) \to \infty , \qquad \alpha_n^{-1} k(n) \to 0 . \tag{8}$$

It is a consequence of a remark made in [2] (proved in greater detail in [3]) that, in the present problem, Z_n remains asymptotically efficient if the supremum operation in the right member of (5) is performed with respect to d in the θ-interval

$$\left[\Psi_n(X_1, \cdots, X_n) - \frac{k(n)}{2}\alpha_n^{-1}, \Psi_n(X_1, \cdots, X_n) + \frac{k(n)}{2}\alpha_n^{-1}\right] \tag{9}$$

where Ψ_n is any estimator of θ_0 such that

$$|\Psi_n - \theta_0| = O_p(\alpha_n^{-1}) . \tag{10}$$

Of course, the length of the interval (9) approaches zero. As Ψ_n we choose $\hat{\theta}_n$, which, it follows from (1), satisfies (10). For $k(\cdot)$ we have a choice among many functions, and we choose

$$k(n) = (\log n)^{\frac{1}{2}} , \quad \forall n . \tag{11}$$

It will be shown in the Appendix that Lemmas 3.4 and 3.5 of [4] hold with their constant k replaced by our $k(n)$; we assume this for the moment. It then follows from this version of these lemmas that

$$\sup_t \frac{|\log V_n(t) + \frac{1}{2}t^2|}{t^2} \tag{12}$$

converges to 0 in P_{θ_0}-probability as $n \to \infty$; the supremum in (12) is with respect to t in the t-interval

$$[\hat{\theta}_n - (\log n)^{\frac{1}{4}}, \hat{\theta}_n + (\log n)^{\frac{1}{4}}] . \tag{13}$$

(When $t = 0$ the expression being maximized in (12) becomes 0/0, and we define it as 0.) The t-interval (13) contains the t-interval into which the θ-interval (9) with the present choice of Ψ_n and $k(n)$ is transformed by the relation $t = \alpha_n(\theta - \hat{\theta}_n)$. From the above we conclude that, if we set $Z_n = \hat{\theta}_n$, there exists a sequence $\{\lambda_n\}$ such that (7) is satisfied with P_{θ_0}-probability approaching one. This proves that $\hat{\theta}_n$ is asymptotically equivalent to the m.p. estimator Z_n.

It remains only to prove that $\hat{\theta}_n$ satisfies the conditions (3.2) and (3.3) of [2] (or the conditions (3.5) and (3.6) of [3]). This follows immediately from (1) and the fact that θ is a translation parameter. This proves the Theorem, so $\hat{\theta}_n$ is asymptotically efficient.

As mentioned earlier, the $\{\hat{\theta}_n\}$ of [4] is a *consistent* sequence of roots of the likelihood equation. Since the statistician solves the likelihood equation for a particular value of n, how is he to recognize which roots belong to a consistent sequence and which do not? (As pointed out to one of us by the late Professor Abraham Wald, the same question can be raised about Cramér's proof of the consistency of the maximum likelihood estimator [1, Section 33.3].) The following procedure may be helpful; it is an application to the present case of the remark made in [2] and used by us earlier in this paper. Let $k_1(\cdot)$ be any function on the positive integers such that $k_1(n) \uparrow \infty$, $n^{-\frac{1}{2}}k_1(n) \to 0$. Then any consistent sequence will eventually lie in an interval of length $n^{-\frac{1}{2}}k_1(n)$ centered at $\min(X_1, \cdots, X_n)$.

APPENDIX

Define

$$M_n = \min(X_1, \cdots, X_n), \qquad N_n = \max(X_1, \cdots, X_n).$$

An examination of the proofs of Lemmas 3.4 and 3.5 of [4] shows that the proofs would remain unchanged if our $k(n)$ were substituted for their k, excepting only that we must now show for Lemma 3.4 that

$$P_{\theta_0}\left[M_n - \theta_0 \geqq \frac{\delta_n}{\varepsilon} = \frac{k(n)}{\varepsilon\alpha_n}\right] \to 1 \tag{14}$$

and for Lemma 3.5 that, in addition,

$$P_{\theta_0}\left[b - N_n \geqq \frac{\delta_n}{1-\beta}\right] \to 1 . \tag{15}$$

For large n the left member of (14) is greater than

$$\left(1 - \frac{\alpha\delta_n^2}{\varepsilon^2}\right)^n = \left(1 - \frac{2(\log n)^{\frac{1}{2}}}{\varepsilon^2 n \log n}\right)^n \to 1 .$$

This proves (14). As for (15), this statement is weaker than the conclusion of Lemma 2.2 of [4], so it certainly holds.

The above argument shows that $k(\cdot)$ could have been any positive function which satisfies (8), (14), and (15).

REFERENCES

[1] Cramér, H. (1961). *Mathematical Methods of Statistics*. Princeton Univ. Press.

[2] Weiss, L. and Wolfowitz, J. (1967). Maximum probability estimators. *Ann. Inst. Statist. Math.* **19** 193-206.

[3] Weiss, L. and Wolfowitz, J. *Asymptotic Methods in Statistics*, Chapter 3. In preparation.

[4] Woodroofe, M. (1972). Maximum likelihood estimation of a translation parameter of a truncated distribution. *Ann. Math. Statist.* **43** 113-122.

Department of Operations Research
Cornell University
Ithaca, New York 14850

Department of Mathematics
University of Illinois
Urbana, Illinois 61801

Z. Wahrscheinlichkeitstheorie verw. Gebiete 30, 117–128 (1974)

Asymptotically Efficient Non-parametric Estimators of Location and Scale Parameters. II

J. Wolfowitz*

9. Introduction

The present paper is a continuation of the paper [1] of the same name. In [1] the authors showed how to construct (asymptotically) efficient estimators of scale and location parameters and of the two jointly, when the form of the density function is unknown to the statistician (i.e., in the non-parametric case). Their estimators are functions of the "middle" $n(q-p)$ observations, where n is the total number of observations and $0<p<q<1$. The estimators are efficient modulo this fact. Since p can be chosen close to zero and q close to 1, the demands of statistical applications would probably be better served by improving this estimator rather than by eliminating the restriction to the middle $n(q-p)$ observations. However, for the purposes of statistical theory and the eventual development of a theory of non-parametric estimation, it seems of some interest to eliminate this "waste" of observations.

In the present paper we construct an estimator of the scale parameter σ which is asymptotically as efficient as the best estimator which can be constructed when the form of the density function is known to the statistician and all the observations are used. (Actually, we estimate the ratio of two σ's, because the assumptions we make are not sufficient to identify σ; see [1] and a remark in Section 15 below. If the parameter σ is identified then the method given below gives an efficient estimator of it.) It will be readily seen that the same method is applicable to estimating a location parameter μ, and μ and σ jointly. The parameter σ was chosen because a choice had to be made (it is not necessary to do both) and because it is perhaps slightly the more difficult of the two[1]. We believe that the method developed is of general interest and that it will be applicable in the development of a general theory of non-parametric estimation which has begun to emerge only recently.

In the present paper we assume familiarity with [1], whose notation and definitions are assumed herewith. Other notation will be added in Section 10, where the assumptions are stated. The numbering of the sections follows that of [1] consecutively. The assumptions will be discussed in Section 15, where the relation of this paper to work by other authors will be discussed.

---

* Research supported by the U.S. Air Force under Grant AF-AFOSR-70-1947, monitored by the Office of Scientific Research.

[1] By our method, that is. It has not as yet been estimated by any other method.

After the formulae (1.3) and (1.4) (see [1]) of Bennett, Jung, and Blom were discovered, and even more after the formulae (1.7) and (1.8) of Weiss, it was trivial to conjecture that these formulae could be used to obtain non-parametric estimators of μ and σ. The difficulty was to carry out this program, since the error in estimating just one coefficient exceeds, by an order of magnitude, the error permitted the entire estimator (see [1]). In the same way, it is trivial to conjecture that full efficiency can be obtained by pushing the "cut-off" points p_n and q_n to 0 and 1, respectively, as $n \to \infty$. Carrying out this program, as is done in the papers cited in Section 15 below and in the present paper, is not at all a trivial matter, and encounters a number of difficulties.

We now give an extremely brief outline of the present paper. In Sections 11 and 12 we assume that appropriate cut-off points for every n, and all the coefficients $B_j^{(n)}$, are known to the statistician, and that, between these cut-off points, the positive lower bound on the derivative g is known to the statistician. (Of course, in the nature of the problem, this is impossible, but we assume it temporarily.) In these sections we then obtain bounds (which approach zero as $p \to 0$ and $q \to 1$) on the difference between the distribution of the normalized estimator and the desired limiting normal distribution. Thus, if these (impossible) conditions were to be fulfilled, an efficient estimator would be at hand. Throughout this paper, the form of the estimator is always that of [1], but the crucial question always is what the suitable cut-off points are. The final decision is made only in Section 14, after a number of changes in different steps.

In the first half of Section 13 we prove that, if the coefficients $B_j^{(n)}$ are not known to the statistician, but estimated as in [1], then an estimator which is a function of certain of the observed variables is still efficient. In the second half of this section we estimate the cut-off points, so that, with probability approaching one, a satisfactory lower bound on g between the cut-off points, can be given.

This would seem to remove the assumptions with which the argument of Sections 11 and 12 was carried out, assumptions which involve knowledge by the statistician which he cannot possibly possess. Three obstacles still remain:

(9.1) The cut-off points determined are chance variables, not the constants which occur in the proofs of normality, like that of [2], for example.

(9.2) Since the chance cut-off points are functions of the middle observations, the latter are now not necessarily independently and identically distributed with the common truncated distribution, as required by the proofs of normality, e.g., that of [2]. Every method of moving out the cut-off points, as functions of the n observations, so that $p \to 0$ and $q \to 1$, must reckon with this difficulty.

(9.3) One must take into account how many of the $n(q-p)$ observations lie in the prescribed interval. For fixed p and q, as in [1], this was of no consequence for the limiting distribution.

In Section 14 these difficulties are resolved and the final estimator is given. This estimator has in the limit, after normalization, the same distribution as the most efficient estimator which can be constructed as a function of all the observations and with full knowledge of the form of the distribution function.

10. Assumptions

Before giving the assumptions we add some more notations, necessary or useful facts.

Define

$$V(p,q)=\Bigg\{\int_{G^{-1}(p)}^{G^{-1}(q)}\left(y\frac{g'(y)}{g(y)}\right)^2 g(y)\,dy-(q-p)$$

$$+\frac{(G^{-1}(p)\,g(G^{-1}(p)))^2}{p}+\frac{(G^{-1}(q)\,g(G^{-1}(q)))^2}{1-q}$$

$$+2G^{-1}(q)\,g(G^{-1}(q))-2G^{-1}(p)\,g(G^{-1}(p))\Bigg\}^{-1},$$

$$V^*=\left\{\int_{-\infty}^{\infty}\left(y\frac{g'(y)}{g(y)}\right)^2 g(y)\,dy-1\right\}^{-1},$$

$$V'(p,q)=\left\{\int_{G^{-1}(p)}^{G^{-1}(q)}\left(\frac{g'(y)}{g(y)}\right)^2 g(y)\,dy+\frac{g^2(G^{-1}(p))}{p}+\frac{g^2(G^{-1}(q))}{1-q}\right\}^{-1},$$

$$V^{**}=\left\{\int_{-\infty}^{\infty}\left(\frac{g'(y)}{g(y)}\right)^2 g(y)\,dy\right\}^{-1}.$$

The significance of these quantities is as follows: $\sigma^2 V(p,q)$ is the variance of the normal distribution with mean zero which is the limit of the distribution of

$$\sqrt{n}\,(\hat{\sigma}_n(Y)-\sigma),$$

where $\hat{\sigma}_n(Y)$ is the estimator of (1.8). $V'(p,q)$ is the variance of the normal distribution with mean zero which is the limit of the distribution of

$$\sqrt{n}\,(\hat{\mu}_n(X)-\mu),$$

where $\hat{\mu}_n(X)$ is the estimator of (1.7). Both of these results are derived in [2], and, in a simpler way, in [5]. When g is symmetric about zero, the value of $V'(p,p)$ is supposedly given at the top of p. 150 of [1], but the expression given there involves an algebraic error. The correct expression for $V'(p,q)$ in the general case is given above in this section. V^* and V^{**}, respectively, are of course the variance of the limiting normal distributions of the maximum likelihood estimators of σ and μ, respectively, when the statistician knows the form of g and the latter satisfies the conditions of the "regular" case. These are then the best variances (of the limiting distribution) which the statistician can achieve.

(We take this opportunity to correct a few minor errors in [1]. In Section 8, $V(n)$ should be replaced by V throughout. That V is now called $V'(p,p)$ for the symmetric case, because we now need a finer differentiation. In (7.12) there should be absolute values about $\Delta(n,j)$. In the denominator of (7.10) the factor $(\lambda_{np}+\lambda_{nq})$ should be deleted. A non-trivial error made in [1] was to omit the requirement that g'' satisfy a Lipschitz condition.)

(The error made in $V(n)$ of [1] is carried over into [12], (3.9), (3.13), (3.16), and Lemma 8.)

In the present paper we make the following assumptions:

Assumption 1. $0 < V^* < \infty$. Let T_n' be any estimator of σ which is such that, for every σ, $\sqrt{n}(T_n' - \sigma)$ is asymptotically normally distributed with mean zero and variance $V'(\sigma)$. Then, for (Lebesgue) almost every σ, $V'(\sigma) \geqq \sigma^2 V^*$.

Assumption 2. The set where g is positive is the entire line. The derivatives g', g'', and g''' exist and are bounded above in absolute value.

Assumption 3. As $p \to 0$ and $q \to 1$, $V(p, q) \to V^*$.

Assumption 4. $E|V_i|^3 < \infty$. (For the definition of V_i see the first paragraph of [1].)

Assumption 5. There exists an s, $0 < s < 1$, such that

a) for $x < 0$,

$$G\left(\frac{x}{\sigma}\right) < \left[\frac{1}{\sigma} g\left(\frac{x}{\sigma}\right)\right]^s,$$

and

b) for $x > 0$,

$$1 - G\left(\frac{x}{\sigma}\right) < \left[\frac{1}{\sigma} g\left(\frac{x}{\sigma}\right)\right]^s.$$

The quantity s may depend on σ.

The assumptions will be discussed in Section 15. We will carry out the proof in the next four sections under the following additional assumptions, for the sake of a little simplicity. These additional assumptions will be eliminated in an almost trivial manner in Section 14.

Assumption 2'. The statistician knows K_1, the largest of the bounds in Assumption 2.

Assumption 4'. The statistician knows an upper bound on

$$\frac{E|V_i|^3}{[EV_i^2 - (EV_i)^2]^{\frac{3}{2}}}.$$

Assumption 5'. The statistician knows the number s.

11. Heuristic Introduction to the Proof

In this section we describe some of the basic ideas of the proof in a non-rigorous manner for the sake of easier understanding. These ideas are carried out rigorously in Section 12. The proof is then completed in Sections 13 and 14, which depend on ideas not discussed in this section. Only a superficial familiarity with [1] is needed for Sections 11 and 12.

Let α_i, $i = 1, 2, \ldots$ be a descending sequence of positive numbers such that $\alpha_1 < 1$ and $\alpha_i \to 0$. Assume now that the statistician knows sequences $\{p_i\}$ and $\{q_i\}$ such that

(11.1) always $0 < p_i < q_i < 1$, and $(q_i - p_i) \to 1$, and

(11.2) whenever $\sigma G^{-1}(p_i) < x < \sigma G^{-1}(q_i)$,

$$\alpha_i < \frac{1}{\sigma} g\left(\frac{x}{\sigma}\right) = d(x), \quad \text{say.}$$

Suppose also that the statistician knows the coefficients $B_j^{(n)}$ in the estimator (1.8) of [1], i.e.,

$$(11.3)\quad \hat{\sigma}_n^{(i)}(Y) = \frac{\sum_{j=np_i+1}^{nq_i-1} \frac{1}{n} B_j^{(n)} Y_j^{(n)} + B'^{(n)}_{np_i} Y^{(n)}_{np_i} + B'^{(n)}_{nq_i} Y^{(n)}_{nq_i}}{\sum_{j=np_i+1}^{nq_i-1} \frac{1}{n} B_j^{(n)} G^{-1}\left(\frac{j}{n}\right) + B'^{(n)}_{np_i} G^{-1}(p_i) + B'^{(n)}_{nq_i} G^{-1}(q_i)}.$$

(Of course, the problem is such that the statistician cannot possibly know these things. In Sections 13 and 14 we will remove these assumptions.)

We will then find an integer N_i such that, when $n \geqq N_i$, the distribution of

$$(11.4)\quad \sqrt{n}\,(\hat{\sigma}_n^{(i)}(Y) - \sigma)$$

differs from the normal distribution with mean zero and variance $\sigma^2 V(p_i, q_i)$, by at most α_i, uniformly in the argument of the distribution function and all G which satisfy out assumptions.

From the determination of N_i it will follow that:

(11.5) The above conclusion about the uniform approach of the distribution of (11.4) to its limit is a fortiori true if, in $\hat{\sigma}_n^{(i)}(Y)$, we replace p_i and q_i, respectively, by p and q such that $p_i < p < q < q_i$.

(11.6) We may choose the N_i strictly increasing. Then $N_i \uparrow \infty$.

For any n let $i(n)$ be the largest i such that $N_i \leqq n$. It follows from the above that the distribution of

$$(11.7)\quad \sqrt{n}(\hat{\sigma}_n^{(i(n))}(Y) - \sigma)$$

approaches the normal distribution with mean zero and variance $\sigma^2 V^*$. Hence $\hat{\sigma}_n^{i(n)}(Y)$ is an (asymptotically) efficient estimator of σ.

The coefficients $B_j^{(n)}$ are, of course, unknown to the statistician. We will estimate them as in [1]. We will have to prove, and will do so in Section 13, that

(11.8) The ratio of the estimators of σ_1 and σ_2, with the coefficients estimated as in [1], is an asymptotically efficient estimator of $\frac{\sigma_1}{\sigma_2}$. (This was also proved in [1], but there α_i did not approach zero.) The distribution function of this ratio differs from the distribution function of the ratio of estimators with coefficients known, by at most a multiple of α_i.

(11.9) We will determine $\{p_i\}$ and $\{q_i\}$ so that (11.1) and (11.2) are satisfied, not literally, but in a probabilistic sense. This will cause some difficulties which will have to be eliminated by additional arguments. This will be done in the second half of Section 13 and in Section 14.

12. Proof of the Theorem when the Statistician Knows $\{p_i\}$, $\{q_i\}$, and the Coefficients $B_j^{(n)}$

Our proof in this case leans heavily on the clever proof of normality in [2], p. 4–6, and is actually a refinement of this proof for purposes not present in [2]. For this reason and to avoid needless repetition we assume familiarity with these pages of [2] and will indicate where modifications are to be made. Our problem, not present in [2], is to obtain an N_i large enough so that the distribution of (11.7) is, uniformly in the argument of the distribution function, within α_i of the normal distribution with mean zero and variance $\sigma^2 V(p_i, q_i)$.

Let p and q of [2] now be p_i and q_i, respectively. We now concern ourselves with the conditional distribution of S_n of [2], given that $V_n = v$ and $W_n = w$, with $K\left(\frac{j}{n}\right)$ of [2] equal our $B_j^{(n)}$. From the definition of V_n and W_n of [2] it follows that there exists an absolute (i.e., independent of G) constant H_i and an N_i such that, for $n \geqq N_i$,

$$P\{|V_n| < H_i, |W_n| < H_i\} \geqq 1 - \frac{\alpha_i}{6}. \tag{12.1}$$

The inequality in (12.1) provides the bounded region in the (v, w) plane discussed in [2], p. 6. The expression $Q''(\theta_j, n)$ of [2] comes from the derivative of $B_j^{(n)}$ of [1]. The latter (derivative) is a sum of terms, each of which consists of a product of non-negative powers of g', g'', g''', and G^{-1}, divided by a positive power of g. Now g', g'', g''' are, by Assumption 2, bounded above in absolute value. $\left|G^{-1}\left(\frac{j}{n}\right)\right|$ is bounded above by $|G^{-1}(\frac{1}{2})| + \alpha_i^{-1}\sigma^{-1}$ which, for i sufficiently large, is less than $i\alpha_i^{-1}$. Consequently, in the interval (p_i, q_i), $Q''(\theta_j, n)$ of [2] is bounded above in absolute value by a known multiple of a negative power of α_i. Increasing N_i, if necessary, this bound can be made less than N_i^γ for a $\gamma < \frac{1}{2}$.

It follows from [3], Lemma 2, that, for i sufficiently large,

$$P\left\{\max_{np_i \leqq j \leqq nq_i} \left[Y_j - G^{-1}\left(\frac{j}{n}\right)\right]^2 < (n\alpha_i^2)^{-1}\right\} > 1 - \frac{\alpha_i}{6}. \tag{12.2}$$

In the notation of [2], therefore, we have then

$$|\delta_n(v, w)| < \frac{n \cdot n^\gamma \cdot \max_{np_i \leqq j \leqq nq_i} \left[Y_j - G^{-1}\left(\frac{j}{n}\right)\right]^2}{2n^{\frac{1}{2}}}. \tag{12.3}$$

Hence, increasing N_i if necessary, we have

$$P\left\{|\delta_n(v, w)| < \frac{\alpha_i}{6}\right\} > 1 - \frac{\alpha_i}{6}. \tag{12.4}$$

We can now directly consider the conditional distribution of S_n of [2], given that $V_n = v$ and $W_n = w$. From the argument of [2], especially p. 6, line 6, we see that everything now depends on the distribution of T_n of [2]. The function Q ([2], p. 3, line 9) is still at our disposal, subject to the condition given in [2]. We define $Q(z)$ as

$$\int_{G^{-1}(\frac{1}{2})}^{z} K(G(y))\, dy. \tag{12.5}$$

The argument by which we proved Q'' bounded by N_i^{γ} applies a fortiori to Q', and we obtain that

$$|Q(z)| < |z - G^{-1}(\tfrac{1}{2})|\, N_i^{\gamma}. \tag{12.6}$$

From (12.6), the definition of the chance variables Z_i of [2], the definition of T_n, Assumption 4′, and the Berry-Esseen theorem, we conclude that, for N_i sufficiently large, the distribution of T_n of [2] differs from the normal distribution with mean zero and variance one by less than $\frac{\alpha_i}{6}$, uniformly for v and w in the set $|v| < H_i$, $|w| < H_i$, for i large.

The chance variable (11.7) is a linear function of T_n, V_n, and W_n of [2]. Following the proof of [6], p. 367–370, it is not difficult to prove that, when N_i is sufficiently large, because of Assumptions 1 and 2, the distribution of the normalized chance variable V_n differs from its limiting normal distribution by less than $\frac{\alpha_i}{6}$, and the conditional distribution of the normalized chance variable W_n, given that $|V_n| < H_i$, differs from its limiting normal distribution by less than $\frac{\alpha_i}{6}$. We have already proved the corresponding result for the conditional distribution of T_n, given the event in (12.1). Now, it is easy to see that, if the distribution of a chance variable Z' differs from a distribution G', say, by less than β, then, for any non-zero constant c, the distribution of cZ' differs from the distribution $G_0' \left(G_0'(x) \equiv G' \left(\frac{x}{c} \right) \text{ when } c > 0, \text{ with a corresponding definition when } c < 0 \right)$ by less than β. From these facts it is not difficult to see that, when N_i is sufficiently large, the distribution of (11.4) differs from the normal distribution with mean zero and variance $\sigma^2 V(p_i, q_i)$ by at most α_i. Since $V(p_i, q_i) \to V$ it follows that $\hat{\sigma}_n^{i(n)}(Y)$ is an asymptotically efficient estimator of σ.

If G, $\{p_i\}$, and $\{q_i\}$ were known to the statistician our work would now be finished. Of course, they are not known, and, in the nature of the problem, cannot be known.

13. Proof of (11.8). Determination of p_i and q_i

We begin by proving (11.8). It follows from Lemma 2 of [3] that, for large i, a "belt" of constant half-thickness $(\alpha_i n^{\frac{1}{2}})^{-1}$, about the graph of the distribution function $G\left(\frac{x}{\sigma}\right)$, will include the graph of the empiric distribution function of $Y_1, \ldots, Y_n$ with probability $\geqq 1 - \alpha_i$. We now retrace the argument of [1], Section 5, conditioned upon the event in the last sentence. In the interval (p_i, q_i), $G^{-1}\left(\frac{j}{n}\right)$

is bounded above in absolute value by $|G^{-1}(\frac{1}{2})| + \alpha_i^{-1}\sigma^{-1}$, which is less than $i\alpha_i^{-1}$ for i sufficiently large. We will now estimate the errors incurred in estimating

$$\sigma G^{-1}\left(\frac{x}{\sigma}\right), \quad \frac{1}{\sigma} g\left(\frac{x}{\sigma}\right), \quad \frac{1}{\sigma^2} g'\left(\frac{x}{\sigma}\right), \quad \text{and} \quad \frac{1}{\sigma^3} g''\left(\frac{x}{\sigma}\right)$$

as in [1], Section 5. This will give us simultaneous bounds on the errors of all the $C(n, i, j)$ of Section 3.

The error in estimating $c(n, i, j)$ of [1] is, from the construction of the "belt", not greater than $(\alpha_i^2 n^{\frac{1}{3}})^{-1}$.

Proceeding as in (5.5), the first of the errors in estimating $d(n, i, j)$ of [1] is not greater than $2(\alpha_i^2 n^{\frac{1}{4}})^{-1}$. The second of the errors is not greater than $K_1(\alpha_i n^{\frac{1}{4}})^{-1}$, where K_1 is the largest of the bounds on $|g'|$, $|g''|$, $|g'''|$. For large enough i, the larger of these two bounds is $2(\alpha_i^2 n^{\frac{1}{4}})^{-1}$.

We continue in this manner as in Section 5. The details are not important, because it is obvious what the conclusion will be, and the actual computations are tedious. The conclusion is that, for large i, the error in estimating $C(n, i, j)$ of [1] is, with probability $\geqq 1-\alpha_i$, simultaneously for all j such that Y_j lies in the interval $(\sigma G^{-1}(p_i), \sigma G^{-1}(q_i))$, bounded in absolute value by a multiple of $n^{-\frac{1}{16}}$ multiplied by a negative power of α_i. (In [1] there was no need to take into account this negative power of α_i, because α_i did not approach zero.) Increasing N_i, if necessary, we can make this bound less than $n^{-\frac{1}{32}}$.

Let $K_i' = \alpha_i^{-\frac{1}{2}}$. Then

$$P\{\sigma G^{-1}(p_i) < Y_i < \sigma G^{-1}(q_i) \text{ when } np_i + K_i'\sqrt{np_i} < j < nq_i - K_i'\sqrt{n(1-q_i)}\} > 1 - 2\alpha_i. \tag{13.1}$$

From now on in this section we shall use the estimator (11.3) with np_i replaced by $np_i + K_i'\sqrt{np_i}$ and nq_i replaced by $nq_i - K_i'\sqrt{n(1-q_i)}$.

We now proceed almost exactly as in Section 3 of [1]. Let $\hat{\sigma}_n^{i(n)}(Y_1)$ and $\hat{\sigma}_n^{i(n)}(Y_2)$ be, respectively, estimators of σ_1 and of σ_2, computed as though G (and hence the coefficients B_j) were known. Let W_1 be the ratio of the corresponding estimators with the coefficients estimated as in [1], Section 5. As in [1], Section 3, using the bounds obtained above, we conclude that, with probability $\geqq 1 - 6\alpha_{i(n)}$,

$$\sqrt{n}\left(W_1 - \frac{\hat{\sigma}_n^{i(n)}(Y_1)}{\hat{\sigma}_n^{i(n)}(Y_2)}\right) < 4(\alpha_{i(n)}^2 n^{\frac{1}{32}})^{-1}. \tag{13.2}$$

Increasing N_i, if necessary, the bound in (13.2) can be made less than α_i. This proves (11.8).

We now turn to (11.9). Increase N_i, if necessary, so that $N_i > \alpha_i^{-3}$. Let $H_n(\cdot)$ be the empiric distribution function of $Y_1, \ldots, Y_n$. Using the method of [4] we construct a confidence belt of constant thickness with confidence coefficient $\geqq 1-\alpha_i$. Increasing N_i, if necessary, we make the half-thickness of this belt less than α_i^2. From this we obtain a maximal x-interval, say J_0, such that, with confidence coefficient $\geqq 1-\alpha_i$, for every $x \in J_0$,

$$\min[H(x), 1 - H(x)] \geqq \alpha_i^s. \tag{13.3}$$

It follows from Assumption 5 that $d(x) \geqq \alpha_i$ for $x \in J_0$, with probability $\geqq 1-\alpha_i$. Let J^* be the set of indices j such that $Y_j \in J_0$. Let

(13.4) $$j_1 = \min\{j | j \in J^*\}, \quad j_2 = \max\{j | j \in J^*\}$$

and

(13.5) $$p_i = \frac{j_1 + K_i' n^{\frac{1}{2}}}{n}, \quad q_i = \frac{j_2 - K_i' n^{\frac{1}{2}}}{n}.$$

Then

(13.6) $$P\{d(x) \geqq \alpha_i \text{ whenever } \sigma G^{-1}(p_i) < x < \sigma G^{-1}(q_i)\} > 1 - 3\alpha_i.$$

It follows, from the above construction of p_i and q_i, by a tedious but obvious argument (for which, in the paragraph which follows (13.2), we made the half-thickness of the confidence belt less than α_i^2), that

(13.7) $$(q_i - p_i) \quad \text{converges stochastically to one.}$$

However, this is not (11.9), because the p_i and q_i we have just constructed are chance variables. Moreover, these p_i and q_i are not necessarily independent of the Y_j with $j_1 \leqq j \leqq j_2$, so that the latter Y_j cannot be used as in the theory developed in this section. Section 14 is devoted to overcoming these difficulties. When this will have been done the proof of the theorem will be complete.

14. Completion of the Proof

Let $\{N_i\}$ be the sequence hitherto obtained, and replace each N_i by $2N_i^2$. This is to be the final sequence $\{N_i\}$. Always, as before, $i(n)$ is the largest i such that $N_{i(n)} \leqq n$. Write $n' = n - \sqrt{n}$. We have, from the construction of the final sequence $\{N_i\}$, that always

(14.1) $$n' p_i + K_i' \sqrt{n' p_i} < n' p_i + n' \sqrt{p_i} < 2n' \sqrt{p_i}$$

and

(14.2) $$n'(1-q_i) + K_i' \sqrt{n'(1-q_i)} < n'(1-q_i) + n' \sqrt{1-q_i} < 2n' \sqrt{1-q_i}.$$

Define

(14.3) $$P\{p_i \geqq \tfrac{1}{64}\} = \beta_n, \quad P\{q_i \leqq 1 - \tfrac{1}{64}\} = \beta_n'.$$

Since p_i and q_i converge stochastically to zero and one, respectively, $\beta_n + \beta_n' \to 0$.

We are now ready to give the definition of our final estimator and to prove the desired result.

The original observed (i.i.d.) chance variables are $V_1, \ldots, V_n$. Let $Y_1, \ldots, Y_{n'}$ *now* be the chance variables $V_{\sqrt{n}+1}, \ldots, V_n$ ordered in ascending size. We determine p_i and q_i in the manner described in Section 13 as functions of $V_1, \ldots, V_{\sqrt{n}}$. From the construction of Section 13, Chebyshev's inequality, (14.1), and (14.2), it follows that the probability exceeds $1 - 4\alpha_i - \beta_n - \beta_n'$ that $p_i < \frac{1}{64}$, $1 - q_i < \frac{1}{64}$, and

(14.4) $$Y_{2n'\sqrt{p_i}}, \ldots, Y_{n'(1-2\sqrt{1-q_i})}$$

lie in an interval in which $d(\cdot) > \alpha_i$. Our final estimator is defined to be the estimator of [1] in terms of the variables (14.4), with, of course, the coefficients B_j estimated as in [1]. It follows from the final definition of N_i and Section 13 that the distribution of our normalized estimator of σ, i.e., of $\sqrt{n}$(estimator $-\sigma$), when the chance variables p_i and q_i are fixed, differs from the normal distribution with mean zero and variance $\sigma^2\, V(2\sqrt{p_i}, 1-2\sqrt{1-q_i})$ by at most ε_i, on a set of values of p_i and q_i whose probability exceeds $1-4\alpha_i-\beta_n-\beta'_n$ for large i. From Assumption 3 it follows that $V(2\sqrt{p_i}, 1-2\sqrt{1-q_i})$ converges stochastically to V^*. Since $\alpha_i \subset 0$, the limiting distribution of our normalized estimator of σ is normal, with mean zero and variance $\sigma^2\, V^*$, the smallest (for (Lebesgue) almost every σ) variance which can be attained by an asymptotically normal estimator even when the statistician knows G. The corresponding conclusion holds for our estimator of $\dfrac{\sigma_1}{\sigma_2}$. This proves the desired result.

15. Miscellaneous Remarks

First we eliminate Assumptions 2′ and 4′. To do this, compute N_i as if K_1 were i, and i were also an upper bound on the expression in Assumption 4′. For all i sufficiently large this procedure will be correct. To eliminate Assumption 5′, compute N_i as if $s=(-\log \alpha_i)^{-\frac{1}{2}}$. For all i sufficiently large this, too, will be correct. Thus we have eliminated Assumptions 2′, 4′ and 5′ in a trivial manner. Another method of doing this is to estimate the several quantities from a "wasted" subsample of size $\sqrt{n}$.

Our program for proving the (asymptotic) efficiency of our non-parametric estimator of σ is to show that its limiting distribution is the same as that of the best estimator which can be constructed even when the statistician knows g. Assumption 1 says that the maximum likelihood estimator is efficient. Sufficient conditions for the latter are known (e.g., [10, 11]), but necessary conditions are not known and, in the nature of things, are not likely to be known. It is clear, however, that this assumption implies conditions which are not necessarily independent of the conditions in the other assumptions. The first part of Assumption 2 is usually made in proofs of properties of the maximum likelihood estimator.

Assumption 3 is perhaps essential for the problem. Because of the obvious difficulties in the tails of the distribution, where g is small, it would seem that one is forced to omit the end observations and move out with p and q at the proper rate only.

Assumption 4 could be weakened by requiring only the finiteness of a $(2+\delta)$-th moment. This assumption is probably not an essential one and may be necessary only because of the particular method of proof employed.

Assumption 5 says in effect that the tails of the distribution must not approach zero too slowly. It is probable that it is needed because of the particular method of proof. An assumption about the monotonicity of g would render it unnecessary, as would other assumptions.

In [7–9] the authors give non-parametric estimators of μ whose limiting distribution after normalization has variance V^{**}. The first result of this kind was in [7], which is a brilliant tour de force not likely to be applicable to other problems. Nor has it yet been demonstrated that the methods of the interesting papers [8] and [9] will solve the problem of estimating σ, or of estimating μ and σ jointly, or of estimating the appropriate functions of these parameters when they are not identified. Incidentally, it is not easy to give natural conditions under which σ will be identified. The assumptions of these papers and those of the present paper are not directly comparable, since no set is uniformly stronger or weaker than another.

We devote a few words to Assumption 2. The actual estimator contains no derivatives, but only difference quotients. Preliminary work by the author suggests that one may be able to dispense with at least some of these derivatives. Assumption 1 may well have implicit consequences about derivatives. Also, in a non-parametric problem, where the statistician does not know the function g, is it likely that he would know that g is symmetric[2]? How shall we compare such an assumption with Assumption 2?

Until recently, the theory of non-parametric estimation consisted of a number of ad hoc procedures for which no optimality properties had been proved and, most probably, could not be proved because the procedures were not really optimal. With the publication of the two pioneering papers of this series and then of the subsequent papers, asymptotically optimal estimation procedures have been obtained in all papers for μ, and in one paper for μ, σ, and μ and σ jointly. Now it is obvious that no statistician dealing with a practical problem will ever employ the estimators of [7, 1, 8, 9], or the present paper. They are too complicated to compute, and the convergence rate of their distributions to their limits is too slow. Their value lies in their being existence theorems, as it were, that optimal estimators do exist and can be obtained in these different ways. It seems to the author that a next big step in the theory would be to get away from the limitation to scale and location parameters, in the direction of the general parameters treated in the parametric theory. The method of proof of the present paper may lend itself to this purpose.

When a comprehensive non-parametric theory of estimation becomes available, it will be desirable, on the one hand, to reduce the regularity assumptions, and, on the other, to make the theory accessible for practical purposes, e.g., by speeding up convergence and making compromises in the interest of computational simplicity. It may then happen that, by using estimators which are functions of the middle $n(q-p)$ observations with $(q-p)$ close to unity, one will gain much practical convenience and applicability, at the expense of a small loss of efficiency.

Acknowledgement. The author gratefully acknowledges valuable communications with Professor Lionel Weiss. The formulae for the V's given in Section 10 were furnished by him. Interesting conversations were held with Miss Abigail Sachs.

[2] Except in [1] (and the present paper), μ has always been estimated under the assumption that g is symmetric. This is also the case in the manuscript by C.J. Stone just received by us.

References

1. Weiss, L., Wolfowitz, J.: Asymptotically efficient non-parametric estimators of location and scale parameters. Z. Wahrscheinlichkeitstheorie verw. Gebiete **16**, 134–150 (1970)
2. Weiss, L.: On the asymptotic distribution of an estimate of a scale parameter. Naval Res. Logist. Quart. **10**, 1–9 (1963)
3. Dvoretzky, A., Kiefer, J., Wolfowitz, J.: Asymptotic minimax character of the sample distribution function and of the classical multinomial estimator. Ann. Math. Statist. **27**, 642–669 (1956)
4. Wald, A., Wolfowitz, J.: Confidence limits for continuous distribution functions. Ann. Math. Statist. **10**, 105–118 (1939)
5. Weiss, L.: Asymptotic properties of maximum likelihood estimators in some non-standard cases. J. Amer. Statist. Assoc. **66**, 345–350 (1971)
6. Cramer, H.: Mathematical methods of statistics. Princeton: Princeton University Press 1946
7. Van Eeden, C.: Efficiency-robust estimation of location. Ann. Math. Statist. **41**. 172–181 (1970)
8. Fabian, V.: Asymptotically efficient stochastic approximation; the RM case. Ann. Statist. **1**, 486–495 (1973)
9. Sacks, J.: An asymptotically efficient sequence of estimators of a location parameter. (To appear in Ann. Statist.)
10. LeCam, L.: Les proprietes asymptotiques des solutions de Bayes. Publ. Inst. Statist. Univ. Paris 7, fascicule 3–4, 17–35
11. Bahadur, R.R.: On Fisher's bound for asymptotic variances Ann. Math. Statist. **35**, 1545–1552 (1964)
12. Weiss, L., Wolfowitz, J.: Optimal, fixed length, non-parametric sequential confidence limits for a translation parameter. Z. Wahrscheinlichkeitstheorie verw. Gebiete **24**, 203–209 (1972)

J. Wolfowitz
Department of Mathematics
University of Illinois
Urbana, Ill. 61801 USA

(Received October 12, 1973)

ТЕОРИЯ ВЕРОЯТНОСТЕЙ И ЕЕ ПРИМЕНЕНИЯ
Том XX 1975 Выпуск 2

MAXIMUM PROBABILITY ESTIMATORS IN THE CLASSICAL CASE AND IN THE «ALMOST SMOOTH» CASE

J. WOLFOWITZ *

1. Introduction. For each positive integer n let $X(n)$ denote the (finite) vector of (observed) chance variables of which the estimator is to be a function. $X(n)$ need not have n components, nor need its components be independently or identically distributed. Let $K_n(x \mid \theta)$ be the density, with respect to a σ-finite measure μ_n, of $X(n)$ at the point x (of the appropriate space) when θ is the value of the (unknown to the statistician) parameter. The latter is a point of the known open set Θ. An estimator (of θ) is a Borel measurable function of $X(n)$ with values in Θ. Although it is not essential, in this paper we assume, for the sake of simplicity, that Θ is a subset of the real line.

For each n let $k(n)$ be a normalizing factor for the family $K_n(\cdot \mid \cdot)$. Let R be a bounded, Borel measurable subset of the real line. A maximum probability (m. p.) estimator (with respect to R) is one which maximizes, with respect to d,

$$\int K_n(X(n) \mid \theta)\, d\theta, \tag{1.1}$$

the integral being over the set $\left\{d - \frac{R}{k(n)}\right\}$. For simplicity we assume that there is a unique maximum. In [2] and [3] the authors discuss a) what to do when a maximum in Θ does not exist, b) how to choose the estimator when there is more than one maximum (of course, any estimator which maximizes (1.1) has the property of asymptotic efficiency described below), c) the case where R is not bounded and d) the case where R is a function of θ (this is especially interesting when one is interested in relative, rather than absolute, errors of estimation).

Let $h > 0$ be any number. We shall say that a sequence $\{\theta_n\}$ is in $H(h)$ (for a special point which below will always be θ_{00}), if $|k(n)(\theta_n - \theta_{00})| \leqslant h$ for $n = 1, 2, \ldots$. The symbols $\mathbf{P}_{\theta'}$ and $\mathbf{E}_{\theta'}$ are to denote, respectively, the probability (of an event) and expected value when θ' is the actual («true») value of the parameter.

In 1967 (see [2] and [3]) L. Weiss and the present author proved the following

Theorem. *Let M_n be an m. p. estimator with respect to R such that:*

* Research partly supported by the U. S. Air Force under Grant AF-AFOSR-70-1947, monitored by the Office of Scientific Research.

For any $h > 0$ *and any sequence* $\{\theta_n\}$ *in* $H(h)$ *we have*

$$\lim_{n\to\infty} \mathbf{P}_{\theta_n} \{k(n)(M_n - \theta_n) \in R\} = \beta(\theta_{00}), \text{ say.} \tag{1.2}$$

Let ε *and* δ *be arbitrary but positive. For* h *sufficiently large we have, for any sequence* $\{\theta_n\}$ *in* $H(h)$,

$$\underline{\lim}_{n\to\infty} \mathbf{P}_{\theta_n} \{|k(n)(M_n - \theta_n)| < \delta h\} \geqslant 1 - \varepsilon. \tag{1.3}$$

Let T_n *be any (competing) estimator such that for any* $h > 0$ *and any sequence* $\{\theta_n\}$ *in* $H(h)$ *we have*

$$\lim_{n\to\infty} [\mathbf{P}_{\theta_n} \{k(n)(T_n - \theta_n) \in R\} - \mathbf{P}_{\theta_{00}} \{k(n)(T_n - \theta_{00}) \in R\}] = 0. \tag{1.4}$$

Then

$$\overline{\lim}_{n\to\infty} \mathbf{P}_{\theta_{00}} \{k(n)(T_n - \theta_{00}) \in R\} \leqslant \beta(\theta_{00}). \tag{1.5}$$

The inequality (1.5) justifies the claim that the m. p. estimator is asymptotically efficient for R. A more general result was proved in [5] for a general loss function, not just the special 0—1 loss function implied by R.

The equation (1.4) is a perfectly reasonable restriction on competing estimators. It is difficult to conceive of a statistician's using, in an actual problem, an estimator which does not satisfy (1.4). The latter restriction eliminates pathological constructs like the superefficient estimator of Hodges. (Of course, the latter estimator was deliberately constructed for a specific purpose.) This restriction is much, much weaker and more reasonable than the usual one that the competing estimator, after proper normalization, be asymptotically normally distributed. The latter restriction is artificial and is made solely to permit comparison of variances for the convenience of the statistician. It does not correspond to any actual need on the part of the researcher who applies statistical theory to actual problems.

The conditions (1.2) and (1.3), on the other hand, are really regularity conditions on the system $\{K_n\}$. In most of their papers L. Weiss and the present author took the position that it is easiest to verify the regularity conditions by directly verifying (1.2) and (1.3). In their view a list of conditions on $\{K_n\}$ which would cover all or most cases of theoretical and practical interest would have to be of very great length, and the verification of these conditions would be more difficult than direct verification of conditions (1.2) and (1.3). We still adhere to this view. For the special cases: 1) the so-called «regular» case and 2) a large class of densities whose ranges depend on θ, sets of conditions have been given in [3] and [4], respectively.

In a series of interesting papers of which [1] is the first, Ibragimov and Hasminski have developed a new approach to asymptotic distribution theory by means of the study of a stochastic process naturally associated with the problem. In [1] they study the classical «regular» case (see also [6]). In the present paper we assume some familiarity with [1]. To the conditions of [1] on the common density function of independently distributed chance variables (on K_n in this special case, therefore) we add two conditions and then prove that, as a consequence, the conditions (1.2) and (1.3) are satisfied (in the regular case). Our proof utilizes the ideas and order of proof of [1] to obtain (1.2) and (1.3). The theorems of [1] are examined and, where needed, slightly strengthened, but the ideas are those of [1].

In Section 4 we show (see also [2] and [3]) that the theorem stated at the beginning of this section, together with (1.2) and (1.3) which will be proved in the present paper for the regular case, imply the classical result on the asymptotic efficiency of the maximum likelihood (m. l.) estimator.

We remark in passing that, when one is dealing with a translation parameter, (1.2) and (1.3) take an especially simple form; one simply replaces θ_n by θ_{00} in both. If the distribution of $k(n)(M_n - \theta_{00})$ approaches a limit (in $\mathbf{P}_{\theta_{00}}$-probability) as $n \to \infty$, (1.3) is obviously satisfied, and (1.2) is satisfied for many R, especially if the limit is continuous.

In Section 5 we show that the considerations of Section 4 carry over immediately to the «almost smooth» case of [8].

2. Regularity conditions. We postulate the following regularity conditions, which are those of [1] except where otherwise stated.

I_1. The chance variables $X_1, X_2, \ldots$ are independently distributed. Their common distribution $\mathbf{P}_\theta$ is absolutely continuous with respect to some σ-finite measure ν defined on a σ-field $\mathscr{X}$ of subsets of Euclidean k-space. We write $\frac{d\mathbf{P}_\theta}{d\nu}$ as $f(\cdot, \theta)$.

I_2. Θ is an open subset of the real line.

I_3. The density $f(x, \theta)$ is measurable $\mathscr{X} \times \mathscr{E}$, where $\mathscr{E}$ is the σ-algebra of measurable subsets of the line.

I_4. If $\theta \neq \theta'$, then

$$\int |f(x, \theta) - f(x, \theta')| \, \nu(dx) > 0.$$

II_1. For any fixed x the function $f(x, \theta)$ is defined and absolutely continuous on the closure Θ^c of Θ, and, for ν-almost every x, every point $\theta \in \Theta$ is a Lebesgue point of the function $f'_\theta(x, \theta)$.

II_2. For every $\theta \in \Theta$,

$$I(\theta) = \int \frac{|f'_\theta(x, \theta)|^2}{f(x, \theta)} \nu(dx) < \infty.$$

Here the integrand is set equal to zero whenever the denominator is zero. We write $I_0 = I(\theta_0)$, $I_{00} = I(\theta_{00})$.

II_3. $I(\cdot)$ is a continuous function of θ on Θ^c.

II_4. There exists a $p \geqslant 0$ such that

$$\sup_{\theta \in \Theta} (1 + |\theta|)^{-p} I(\theta) < \infty.$$

II_5. For some $\delta' > 0$, the function

$$I_{2+\delta'}(\theta) = \int \frac{|f'_\theta(x,\theta)|^{2+\delta'}}{|f(x,\theta)|^{1+\delta'}} \nu(dx)$$

is bounded on every compact subset of Θ. (See [1], Theorem 2.6. In the case of a translation parameter this assumption can be omitted; see [1], Theorem 2.7.)

III_1. There exists a $\delta > 0$ such that

$$\sup_{\theta \in \Theta} |\theta - \theta_0|^\delta \int \sqrt{f(x,\theta) f(x,\theta_0)}\, \nu(dx) < \infty$$

for every $\theta_0 \in \Theta$.

It is proved in [1] (Lemma 2.3) that the above conditions imply that

$$\int \sqrt{f(x,\theta+\varepsilon) f(x,\theta)}\, \nu(dx) = 1 - \frac{1}{8} I(\theta)\varepsilon^2 + o(\varepsilon^2) \quad (2.1)$$

as $\varepsilon \to 0$. For our purposes we need the stronger condition

III_2. Let $h > 0$ be any number and $\{\theta_n\}$ be any sequence in $H(h)$ (henceforth $k(n) = \sqrt{n}$). Then, for every θ_{00},

$$\int \sqrt{f(x,\theta_n) f(x,\theta_n+\varepsilon)}\, \nu(dx) = 1 - \frac{1}{8} I_{00}\varepsilon^2 + \bar{o}(\varepsilon^2), \quad (2.2)$$

where $\bar{o}(\varepsilon^2)$ means that $\frac{\bar{o}(\varepsilon^2)}{\varepsilon^2} \to 0$ as $\varepsilon \to 0$ and $n \to \infty$, independently. The symbol $\bar{o}$ will always be used in this sense hereafter. It is not difficult to give conditions on f which imply (2.2), but, since only (2.2) is needed, we content ourselves with postulating it. The same is true of (2.3) which follows and which also is not postulated in [1].

III_3. Let $\{a_n\}$ be any sequence with limit a, say. For any $h > 0$ and any sequence $\{\theta_n\}$ in $H(h)$, we have

$$\int \sqrt{f(x,\theta_n) f(x,\theta_n + a_n)}\, \nu(dx) \to \int \sqrt{f(x,\theta_{00}) f(x,\theta_{00} + a)}\, \nu(dx). \quad (2.3)$$

3. Examination of the proofs of [1]. We now begin an examination of the proofs of [1] so as to strengthen their conclusions for our purposes. We can expect slightly stronger conclusions because of our Assumptions II_5, III_2, and III_3, especially the latter two. We will use the following notation: Lemma «n» or Theorem «n» will always be from [1]. The new and corresponding lemmas and theorems will be numbered «n'». Our great indebtedness to the methods of [1] will be apparent throughout. The strengthening of the results of [1] will be very slight, just enough to produce the desired result. The introduction of the stochastic processes $\{Y_n(\theta)\}$ and the line of argument associated with them is entirely due to [1].

The present paper again confirms our opinion that, in general, it is easier to verify (1.2) and (1.3) directly. It is only the fact that the authors of [1] have carried through their long argument that makes it worthwhile to add the slight additional argument of the present paper.

In [1] θ_0 denotes the «true» value of θ, i. e. the common density of the X's is $f(x, \theta_0)$. Now the true value will be θ_{00}. The reader will find it helpful to identify the θ_0 of [1] successively with the elements θ_n of the sequence $\{\theta_n\}$ which lies in $H(h)$. The reason for this is to be found in the conditions (1.2) and (1.3) of the theorem on m. p. estimators. We now define

$$Y_n(\theta) = \sum_{j=1}^{n} \log \frac{f\left(X_j, \theta_n + \frac{\theta}{\sqrt{n}}\right)}{f(X_j, \theta_n)}. \tag{3.1}$$

(Comparing this definition with that in [1] will help to understand the roles of θ_0 and θ_{00}.)

Define

$$A_{\varepsilon\delta}^{(n)} = \left\{x : \left| \log \frac{f(x, \theta_n + \varepsilon)}{f(x, \theta_n)} \right| > \delta \right\}, \tag{3.2}$$

with the convention that $\frac{0}{0} = 1$.

Lemma 2.1′. *For any* $\delta > 0$

$$\sup_{|\theta - \theta_n| < \varepsilon} \int f(x, \theta) \nu(dx) = \bar{o}(\varepsilon^2), \tag{3.3}$$

where the domain of integration is $A_{\varepsilon\delta}^{(n)}$.

The proof is almost the same as that of Lemma 2.1. Condition II_5 and Theorem 2.6 enable us to replace the right member of (2.6) of [1] by $\bar{o}(\varepsilon^2)$. After that the proof is that of Lemma 2.1 with $o(\varepsilon^2)$ replaced by $\bar{o}(\varepsilon^2)$.

Lemma 2.4′. *For any* $\delta > 0$, *with integrals extended over* $\bar{A}_{\varepsilon\delta}^{(n)}$, *we have*

$$\int \log \frac{f(x, \theta_n + \varepsilon)}{f(x, \theta_n)} f(x, \theta_n) \nu(dx) = -\frac{\varepsilon^2}{2} I_{00} + \bar{o}(\varepsilon^2), \tag{3.4}$$

$$\int \log^2 \frac{f(x, \theta_n + \varepsilon)}{f(x, \theta_n)} f(x, \theta_n) \nu(dx) = \varepsilon^2 I_{00} + \bar{o}(\varepsilon^2). \tag{3.5}$$

The proof of this is now the same as that of Lemma 2.4, with $o(\varepsilon^2)$ replaced by $\bar{o}(\varepsilon^2)$ throughout.

Now write

$$y_j^{(n)}(\theta) = y_j^{(n)} = \log \frac{f\left(X_j, \theta_n + \frac{\theta}{\sqrt{n}}\right)}{f(X_j, \theta_n)}.$$

Then, by Lemma 2.1′,

$$\sum_{j=1}^{n} \mathbf{P}_{\theta_n} \{|y_j^{(n)}| \geq \delta\} = o(1) \quad \text{as} \quad n \to \infty. \tag{3.6}$$

By Lemma 2.4′, as $n \to \infty$,

$$\sum_{j=1}^{n} \left[\int_{|x|<\delta} x^2 d\mathbf{P}_{\theta_n} \{y_j^{(n)} < x\} - \left(\int_{|x|<\delta} x d\mathbf{P}_{\theta_n} \{y_j^{(n)} < x\} \right)^2 \right] = I_{00}\theta^2 + o(1), \tag{3.7}$$

$$\sum_{j=1}^{n} \int_{|x|<\delta} x d\mathbf{P}_{\theta_n} \{y_j^{(n)} < x\} = -\frac{1}{2} I_{00}\theta^2 + o(1). \tag{3.8}$$

It follows from the normal convergence criterion (e. g. [7], p. 316) that the distribution of $Y_n(\theta)$ approaches the normal distribution with mean $-\frac{1}{2}I_{00}\theta^2$ and variance $I_{00}\theta^2$. This proves part A of Theorem 2.1'. Part B is proved as in [1]. Thus we have proved

Theorem 2.1'. *Let the finite-dimensional chance variables of the process $Y_n(\cdot)$ be distributed according to $\mathbf{P}_{\theta_n}$-probability. Then their distributions approach, as $n \to \infty$, the distributions of the corresponding chance variables of the process*

$$Y(\cdot):\ Y(\theta) = \theta\sqrt{I_{00}}\xi - \frac{\theta^2}{2}I_{00},$$

where ξ is normally distributed with mean zero and variance one.

Write $Z_n(\theta) = \exp\{Y_n(\theta)\}$. In [1] the authors prove

Theorem 2.2. *If $\theta_1 \leqslant \theta_2$ then*

$$\mathbf{E}\,|Z_n^{1/2}(\theta_2) - Z_n^{1/2}(\theta_1)| \leqslant \frac{(\theta_2-\theta_1)^2}{4}\max I(\theta), \tag{3.9}$$

where the maximum is with respect to θ in the interval $\frac{\theta_1}{\sqrt{n}} \leqslant \theta - \theta_{00} \leqslant \frac{\theta_2}{\sqrt{n}}$. The expected value is with respect to $\mathbf{P}_{\theta_{00}}$-probability, i. e., the common distribution of $X_1, \ldots, X_n$ is $\mathbf{P}_{\theta_{00}}$.

Obviously we immediately have

Theorem 2.2'. *If $\theta_1 \leqslant \theta_2$, then*

$$\mathbf{E}\,|Z_n^{1/2}(\theta_2) - Z_n^{1/2}(\theta_1)| \leqslant \frac{(\theta_2-\theta_1)^2}{4}\max I(\theta), \tag{3.10}$$

where the maximum is with respect to θ in the interval $\frac{\theta_1-h}{\sqrt{n}} \leqslant \theta - \theta_{00} < \frac{\theta_2+h}{\sqrt{n}}$. The expected value is with respect to $\mathbf{P}_{\theta_n}$-probability, which is now the common distribution of $X_1, \ldots, X_n$.

In [1] the authors prove

Theorem 2.3. *Whatever be $N > 0$, there exists an n_0 and a constant c_N, which depends only on N, such that, for $n > n_0$,*

$$\mathbf{P}_{\theta_{00}}\left\{\sup_{|\theta|>A} Z_n(\theta) > \frac{1}{A^N}\right\} \leqslant \frac{c_N}{A^N}, \tag{3.11}$$

$$\mathbf{P}_{\theta_{00}}\left\{\sup_{l\leqslant|\theta|\leqslant l+1} Z_n(\theta) \geqslant \frac{1}{l^N}\right\} \leqslant \frac{c_N}{l^N}, \quad l \geqslant 1. \tag{3.12}$$

We would like to carry this theorem immediately over into Theorem 2.3' as was done with Theorem 2.2. However, it is not immediately obvious that then c_N would not also have do depend on n, which cannot be allowed to happen. We therefore have to examine the proof of Theorem 2.3.

We recall Assumptions III_2 and III_3 and prove

Lemma 2.7'. *For any $K > 0$ there exists a positive constant c such that, for all θ with $|\theta| < Kn^{1/2}$,*

$$\mathbf{P}_{\theta_n}\{Z_n(\theta) \geqslant \exp(-c\,\theta^2)\} \leqslant \exp(-c\,\theta^2) \tag{3.13}$$

The proof of the first part of the lemma, i. e., for $0 \leqslant \theta \leqslant Kn^{1/2}$, is as in Lemma 2.6, because of Assumption III_2. The proof of the second part would go through as in Lemma 2.6, if we could assert that $\varphi_0 < 1$. We now have, in place of $\varphi(\varepsilon)$,

$$\varphi_n(\varepsilon) = \int \sqrt{f(x, \theta_n) f(x, \theta_n + \varepsilon)}\, \nu(dx)$$

and

$$\varphi_0 = \lim_{n\to\infty} \sup_{k \leqslant \varepsilon \leqslant K} \varphi_n(\varepsilon).$$

From Assumption III_3 we easily obtain that $\varphi_0 < 1$.

The remainder of the proof of Theorem 2.3′ is the same as that of Theorem 2.3; one has only to replace $\mathbf{P}$ by $\mathbf{P}_{\theta_n}$.

Theorems 2.4 and 2.5 of [1] are now valid with the same proofs, if θ_0 is replaced by θ_n and $\mathbf{P}$ by $\mathbf{P}_{\theta_n}$.

4. Efficiency of the m. l. estimator. Let $R = (-r, r)$. Exactly as in the argument of [1] which follows Theorem 2.5, but making use of our Section 3, we obtain that, for any $h > 0$ and any $\{\theta_n\}$ in $H(h)$,

$$\lim \mathbf{P}_{\theta_n} \{\sqrt{n}(M_n - \theta_n) < y\} = \sqrt{\frac{I_{00}}{2\pi}} \int_{-\infty}^{y} \exp\left\{-\frac{1}{2} I_{00} x^2\right\} dx \tag{4.1}$$

for every y. From (4.1) we obtain (1.2) and (1.3) at once. Hence M_n is asymptotically efficient in the sense of (1.5) for $R = (-r, r)$.

A special case of (4.1) needed for our next purposes is

$$\lim \mathbf{P}_{\theta_{00}} \{\sqrt{n}(M_n - \theta_{00}) < y\} = \sqrt{\frac{I_{00}}{2\pi}} \int_{-\infty}^{y} \exp\left\{-\frac{1}{2} I_{00} x^2\right\} dx. \tag{4.2}$$

Now let $\hat{\theta}_n$ be the m. l. estimator. Under the assumptions of [1] it is there proved that

$$\lim \mathbf{P}_{\theta_{00}} \{\sqrt{n}(\hat{\theta}_n - \theta_{00}) < y\} = \sqrt{\frac{I_{00}}{2\pi}} \int_{-\infty}^{y} \exp\left\{-\frac{1}{2} I_{00} x^2\right\} dx. \tag{4.3}$$

Hence $\hat{\theta}_n$ has the same asymptotic distribution as M_n. Under our assumptions, therefore, $\hat{\theta}_n$ is asymptotically efficient in the sense of (1.5) for $R = (-r, r)$.

Now let T_n be any competing estimator such that

$$\lim \mathbf{P}_{\theta_{00}} \{\sqrt{n}(T_n - \theta_{00}) < y\} = \frac{1}{\sqrt{2\pi}\,\sigma(T \mid \theta_{00})} \int_{-\infty}^{y} \exp\left\{-\frac{1}{2} \frac{x^2}{\sigma^2(T \mid \theta_{00})}\right\} dx. \tag{4.4}$$

(This condition is required by the classical theory, but not by the theory of m. p. estimators (e. g., [2], [3]).) Since $\hat{\theta}_n$ is efficient in the sense of (1.5) for $R = (-r, r)$, it follows that, if $\{T_n\}$ satisfies the regularity condition (1.4) for this R, we have, for every θ_{00},

$$(I_{00})^{-1} \leqslant \sigma^2(T \mid \theta_{00}), \tag{4.5}$$

the classical statement of efficiency for the m. l. estimator.

Under additional conditions it is proved in [3] that

$$\lim \mathbf{P}_{\theta_{00}} \{M_n \neq \hat{\theta}_n\} = 0, \tag{4.6}$$

i. e., the m. p. and m. l. estimators are asymptotically the same (for $R = (-r, r)$). These conditions of [3] can certainly be weakened.

Suppose now that (1.2) and (1.3) are satisfied for some bounded, Borel measurable subset R of the line. Suppose also that (4.6) holds for this R. Let T_n be any competing estimator which need not be asymptotically normally distributed, but which satisfies (1.4) for this R. Then $\hat{\theta}_n$ is more efficient than T_n in the sense of (1.5) for this R and every $\theta_{00} \in \Theta$.

For unbounded R see also [5], Section 6.

5. The «almost smooth» case. Our discussion of this case, which is treated in [8], will be very simple and brief. A first discussion of this case is in [9], and in [10] the authors proved that the m. l. estimator is, in the case of [9], efficient in the sense of (1.5) for $R = (-r, r)$.

Now, in the case treated in [8], it is there proved (under the assumptions of [8]) that, for every y,

$$\lim \mathbf{P}_{\theta_{00}} \left\{ \sqrt{n \log \sqrt{n}}\, (\hat{\theta}_n - \theta_{00}) < y \right\} = \frac{\sqrt{B}}{\sqrt{2\pi}} \int_{-\infty}^{y} \exp\left\{ -\frac{1}{2} B x^2 \right\} dx, \quad (5.1)$$

where B is given in [8], (1.7). Exactly as from [1], it follows from the results of [8] that also

$$\lim \mathbf{P}_{\theta_{00}} \left\{ \sqrt{n \log \sqrt{n}}\, (M_n - \theta_{00}) < y \right\} = \frac{\sqrt{B}}{\sqrt{2\pi}} \int_{-\infty}^{y} \exp\left\{ -\frac{1}{2} B x^2 \right\} dx \quad (5.2)$$

for every y. Thus $\hat{\theta}_n$ and M_n have the same limiting distribution. As remarked in [10] and elsewhere, and in the last but one paragraph of Section 1 of the present paper, it follows from (5.2) that the regularity conditions (1.2) and (1.3) are automatically satisfied when θ (as in [8]) is a translation parameter and $R = (-r, r)$. Hence $\hat{\theta}_n$ and M_n are both asymptotically efficient in the sense of (1.5) for $R = (-r, r)$. If the competing estimator T_n, which satisfies (1.4), is also asymptotically normally distributed with parameters $(\theta_{00}, \sigma^2 (T \mid \theta_{00}))$ (after normalization by $k(n) = \sqrt{n \log \sqrt{n}}$), we have

$$B^{-1} \leqslant \sigma^2 (T \mid \theta_{00})$$

for every $\theta_{00} \in \Theta$.

Department of Mathematics
University of Illinois

Поступила в редакцию
30.5.74

REFERENCES

[1] **И. А. Ибрагимов, Р. З. Хасьминский,** Асимптотическое поведение некоторых статистических оценок в гладком случае, Теория вероят. и ее примен., **XVII**, 3 (1972), 469—486.

[2] **L. Weiss, J. Wolfowitz,** Maximum probability estimators, Ann. Inst. Statist. Math., Tokyo, 19, 2 (1967), 193—206.

[3] **L. Weiss, J. Wolfowitz,** Maximum probability estimators, Lecture Notes in Mathematics, v. 424, Springer Verlag, Berlin-Heidelberg—New York, 1974.

[4] **L. Weiss, J. Wolfowitz,** Maximum probability estimators and asymptotic sufficiency, Ann. Inst. Statist. Math., Tokyo, **22**, 2 (1970), 225—244.

[5] **L. Weiss, J. Wolfowitz,** Maximum probability estimators with a general loss function, Springer-Verlag, Berlin—Heidelberg—New York, Lecture Notes in Mathematics, vol. 89, 1969.

[6] **H. Cramér,** Mathematical methods of statistics, Princeton University Press, Princeton, N. Y., U. S. A., 1946.

[7] **M. Loève,** Probability theory, 3rd edition, D. Van Nostrand Co., Princeton, N.Y., U.S.A., 1963.

[8] **И. А. Ибрагимов, Р. З. Хасьминский,** Асимптотический анализ статистических оценок для «почти гладкого» случая, Теория вероят. и ее примен., **XVIII**, 2 (1973), 250—260.

[9] **M. Woodroofe,** Maximum likelihood estimation of a translation parameter of a truncated distribution, Ann. Math. Statist., 43 (1972), 113—122.

[10] **L. Weiss, J. Wolfowitz,** Maximum likelihood estimation of a translation parameter of a truncated distribution, Ann. Statist., 1, 5 (1973), 944—947.

ОЦЕНКИ МАКСИМАЛЬНОЙ ВЕРОЯТНОСТИ В КЛАССИЧЕСКОМ СЛУЧАЕ И В «ПОЧТИ ГЛАДКОМ» СЛУЧАЕ

Дж. ВОЛФОВИТЦ (США)

(*Резюме*)

Путем применения теоремы из [2] и небольшой модификации предположений и метода [1] и [8] устанавливается асимптотическая эффективность оценки максимального правдоподобия.

J. Appl. Prob. **12**, 713–723 (1975)

SIGNALLING OVER A GAUSSIAN CHANNEL WITH FEEDBACK AND AUTOREGRESSIVE NOISE

J. WOLFOWITZ, *University of Illinois*

Abstract

We study in detail the case of first-order regression, but our results can be extended to the general regression in a straightforward manner. An average energy constraint ((1.2) below) is imposed on each signal. In Section 2 we give an optimal linear signalling scheme (definition and proof in Section 4) for this channel. We conjecture that this scheme is optimal among all signalling schemes. Then the capacity C of the channel is (see Section 5) $-\log b$, where b is the unique positive root (in x) of the equation $x^2 = (1 + g^2(1 + |\alpha| x)^2)^{-1}$. Here α is the regression coefficient, and g^2 is the ratio of the average energy per signal to the variance of the noise. An equivalent expression is $C = \frac{1}{2}\log(1 + g^2(1 + |\alpha| b)^2)$.

GAUSSIAN CHANNEL; FEEDBACK; ENERGY CONSTRAINT FOR EACH SIGNAL; OPTIMAL LINEAR FUNCTION OF SIGNALS; ESTIMATION; LINEAR CAPACITY; REGRESSION; OPTIMAL LINEAR SIGNALLING SCHEME

1. Introduction

In this paper we shall deal explicitly with a regression of order one (described in (1.1) below). For the description of a general regression see [1]. Our results apply equally well to a general regression.

Let α, $|\alpha| < 1$, be the regression coefficient. Let $w_1, w_2, \cdots$ be independent Gaussian chance variables, all with mean zero and variance σ^2. Let z_0 be the state of the channel at the beginning of transmission; z_0 is known to both sender and receiver. Let $x_1, x_2, \cdots$ be the consecutive signals sent by the sender. The signals received by the receiver will be denoted by $y'_1, y'_2, \cdots$. We have

$$\begin{aligned} y'_1 &= x_1 + \alpha z_0 + w_1, \\ y'_2 &= x_2 + \alpha^2 z_0 + \alpha w_1 + w_2, \\ &\cdots \quad \cdots \quad \cdots \quad \cdots \\ y'_i &= x_i + \alpha^i z_0 + \alpha^{i-1} w_1 + \cdots + \alpha w_{i-1} + w_i. \end{aligned} \tag{1.1}$$

After sending $x_1, \cdots, x_i$, the sender (because of the feedback) knows $y'_1, \cdots, y'_i$

Received 3 April 1975.

Research supported by the U.S. Air Force under contract AFOSR 70–1947, monitored by the Office of Scientific Research.

$(i = 1, 2, \cdots)$, and can choose x_{i+1} as a function of the latter. The (energy) constraint on the x's is

$$Ex_i^2 \leqq g^2\sigma^2, \tag{1.2}$$

where $g > 0$ is a given number.

The sender wishes to transmit some one of e^{nR} given messages over the channel. The message to be transmitted can be chosen by him in any manner, arbitrarily or by a random mechanism or in any other way. (The statistics of the message play no role in all that follows.) After the n signals $y_1', \cdots, y_n'$ have been received by the receiver, transmission of this particular message will stop and the receiver will decide (decode) what message was sent. Given n and R, the sender will choose $x_1, \cdots, x_n$ so as to minimise the probability of an incorrect decision. Alternatively, his object can be described in the following manner: Let λ, $0 < \lambda < 1$, be a given upper bound on the probability of an incorrect decoding of the message after n transmissions. The sender chooses his functions $x_1, \cdots, x_n$ so as to make R as large as possible.

In [7] Butman formulated the general mth-order regression problem and derived the optimal linear signal structure for the feedback case. For the first-order channel he showed that the capacity for the channel with feedback can exceed that for the channel without feedback. The general mth-order regression was also considered by Tiernan and Schalkwijk [1], who gave upper bounds on the capacity of the channel. Other interesting bounds were given in the thesis of Tiernan [2]. For the interesting special case $m = 1$, $\alpha = 0$, Schalkwijk [3] gave optimal functions $x_1, \cdots, x_n$ (see also Wolfowitz [4]). The capacity of the channel in this case had already been found by Shannon [5] (or see Wolfowitz [6], Section 9.2, and Wolfowitz [4]). All these authors impose the more general energy constraint

$$n^{-1} \sum_1^n Ex_i^2 \leqq g^2\sigma^2.$$

After the present paper was written Dr. J. C. Tiernan kindly made availabel to the author an as yet unpublished paper [8] in which he studies, *inter alia*, the problem of the present paper, by means of the center-of-gravity feedback method of Schalkwijk [9]. He gives such a signal scheme whose rate is C (C is given explicitly in Section 5 below).

The approach of the present paper is new and entirely different. We set ourselves (and solve) the problem of obtaining an optimal linear signalling scheme (for a definition see Section 4), and obtain the maximum possible rate of transmission for a linear signalling scheme. This rate is shown in Section 5 to be C. We conjecture that C is the capacity of the channel. The signalling scheme of [8] is linear, so that it follows from our results that it, too, is optimal in this class.

The author thanks Dr. J. C. Tiernan and Miss Rachel Sachs for interesting conversations.

2. The method of signalling

Until Section 6 is reached we assume that $\alpha > 0$ except for occasional remarks which will leave no ambiguity.

Let the e^{nR} messages to be transmitted be numbered consecutively, beginning with 1. When k is the number of the message to be transmitted we send

$$x_1 = g\sigma k e^{-nR}. \tag{2.1}$$

Then the message received is

$$y'_1 = x_1 + \alpha z_0 + w_1. \tag{2.2}$$

Our method will be to send, for $i = 2, \cdots, n$

$$x_i = u_1^{(i)} w_1 + u_2^{(i)} w_2 + \cdots + u_{i-1}^{(i)} w_{i-1}, \tag{2.3}$$

and the method of calculating the u's will be described. It will simplify the notation to set $u_j^{(i)} = 0$ for all $j \geqq i$. Then we have

$$\begin{aligned} y'_i = \alpha^i z_0 &+ w_1(\alpha^{i-1} + u_1^{(i)}) + w_2(\alpha^{i-2} + u_2^{(i)}) \\ &+ \cdots + w_i(\alpha^0 + u_i^{(i)}). \end{aligned} \tag{2.4}$$

Since z_0, the initial state, is known to both sender and receiver, they both know

$$y_i = y'_i - \alpha^i z_0, \qquad i = 1, \cdots, n. \tag{2.5}$$

It is the y's which will be used in our calculations.

We set $u_1^{(2)} = g$, i.e.,

$$x_2 = g w_1. \tag{2.6}$$

Then

$$y_2 = w_1(\alpha + g) + w_2. \tag{2.7}$$

To decode k it is sufficient to estimate w_1; more about this later. Let $L(i)$ be the estimator of w_1 which we shall construct; $L(i)$ is a function of $y_2, \cdots, y_i$. First we let

$$L(2) = \frac{y_2}{\alpha + g}. \tag{2.8}$$

Then $L(2)$ is unbiased, normally distributed, and has variance

$$\frac{\sigma^2}{(\alpha+g)^2}. \tag{2.9}$$

It is now clear why we chose x_2 as in (2.6). The variance (2.9) is minimised subject to the energy constraint on the signal. (If α were negative we would have set $x_2 = -gw_1$.)

We shall now determine the u's in

$$x_3 = u_1^{(3)}w_1 + u_2^{(3)}w_2. \tag{2.10}$$

Then

$$y_3 = w_1(\alpha^2 + u_1^{(3)}) + w_2(\alpha + u_2^{(3)}) + w_3. \tag{2.11}$$

$$y_3 - y_2(\alpha + u_2^{(3)}) = w_1(\alpha^2 + u_1^{(3)} - [\alpha+g][\alpha + u_2^{(3)}]) + w_3. \tag{2.12}$$

We set

$$L(3) = (y_3 - y_2[\alpha + u_2^{(3)}])(\alpha^2 + u_1^{(3)} - [\alpha+g][\alpha + u_2^{(3)}])^{-1}. \tag{2.13}$$

Then $L(3)$ is unbiased, normally distributed, and has variance

$$\sigma^2(\alpha^2 + u_1^{(3)} - [\alpha+g][\alpha + u_2^{(3)}])^{-2}. \tag{2.14}$$

We now minimise (2.14) with respect to the u's, subject to the energy constraint

$$\sum_{j=1}^{s-1} (u_j^{(s)})^2 = g^2, \qquad s \geqq 2 \tag{2.15}$$

for the present particular value $s = 3$. We obtain the values

$$u_1^{(3)} = -g[1 + (g+\alpha)^2]^{-1/2}$$

$$u_2^{(3)} = g(g+\alpha)[1 + (g+\alpha)^2]^{-1/2}.$$

Inserting these values into (2.14) we obtain that the minimum variance of $L(3)$ is σ^2 divided by

$$g^2(\alpha + \sqrt{(1 + (g+\alpha)^2)})^2 = M(3), \quad \text{say}. \tag{2.16}$$

We now describe how, for any $i \geqq 3$, we can obtain $u_1^{(i)}, \cdots, u_{i-1}^{(i)}$, when all u's with superscripts smaller than i have been determined. We have

$$\sum_{j=1}^{s} w_j(\alpha^{s-j} + u_j^{(s)}) = y_s, \quad s = 2, \cdots, i-1. \tag{2.17}$$

$$\sum_{j=1}^{i-1} w_j(\alpha^{i-j} + u_j^{(i)}) = y_i - w_i. \tag{2.18}$$

To find $L(i)$, set $w_i = 0$ in (2.18) and solve for w_1 in terms of $y_2, \cdots, y_i$ in (2.17)–(2.18); this expression is $L(i)$. Denote the square of the coefficient of y_i in

$L(i)$ by $[M(i)]^{-1}$. This definition is consistent with that of $M(3)$ given earlier. $M(i)$ is therefore the square of the determinant of the equations (2.17)–(2.18) (with $w_i = 0$). Hence $[M(i)]^{1/2}$ is linear (but not necessarily homogeneous) in $u_1^{(i)}, \cdots, u_{i-1}^{(i)}$, with coefficients which are polynomials in the u's with superscripts smaller than i, i.e., with known coefficients. To determine $u_1^{(i)}, \cdots, u_{i-1}^{(i)}$ we maximise $M(i)$ subject to the constraint (2.15). We now verify the properties of $L(i)$.

$L(i)$ is linear in $y_2, \cdots, y_i$, say $a_2y_2 + \cdots + a_iy_i$. Then w_1 is equal to $a_2y_2 + \cdots + a_{i-1}y_{i-1} + a_i(y_i - w_i)$, identically in $y_2, \cdots, y_i$, and w_i. Hence $L(i)$ is an unbiased estimator of w_1, normally distributed with variance $a_i^2\sigma^2$. The quantity a_i^2 is $[M(i)]^{-1}$.

After n signals have been sent the receiver decides that k (the number of the message sent) is that one of the integers $1, 2, \cdots, e^{nR}$ nearest to

$$\frac{((y_1 - L(n))e^{nR})}{g\sigma}. \tag{2.19}$$

If there are two such integers the receiver chooses either at pleasure.

We shall show later that

$$\lim_{n\to\infty} \frac{1}{2n} \log M(n) = C, \tag{2.20}$$

where C is given in Section 5. Let $R < C$. In order that the receiver make the correct decision according to our rule (2.19) it is sufficient (and necessary) that

$$\frac{|w_n|\, e^{nR}}{g\sigma\sqrt{(M(n))}} < \frac{1}{2}. \tag{2.21}$$

Let $2\varepsilon = C - R$, and N be so large that $n \geqq N$ implies

$$\sqrt{(M(n))} > e^{n(C-\varepsilon)}.$$

Then condition (2.21) is satisfied when

$$|w_n| < \frac{g\sigma}{2} e^{n\varepsilon}. \tag{2.22}$$

Since w_n is normal with mean zero and variance σ^2, the probability that (2.22) will not occur for $n \geqq N$ is less than

$$c_1 \exp\{-c_2 e^{nc_3}\}, \tag{2.23}$$

where c_1, c_2 and c_3 are positive constants which depend on $g\sigma$, ε and σ^2. Using the same argument we can replace $\varepsilon = (C - R)/2$ in the above by a multiple of $(C - R)$ different from $\frac{1}{2}$; this may give a better bound (2.23) for large n.

3. The recursion formulae

For notational convenience write $D(i)$ for the value of the determinant of the equations (2.17)–(2.18) with $w_i = 0$, and let $D(1) = 1$. We already know that $D(2) = (g + \alpha)$. Expand the determinant in terms of the elements of its last row. After some obvious reduction we obtain, for $i \geqq 2$

$$D(i) = \sum_{j=1}^{i-1} D(j)(u_j^{(i)} + \alpha^{i-j})(-1)^{i-1+j}. \tag{3.1}$$

Maximising $[D(i)]^2 = M(i)$ subject to the constraint (2.15) by using a Lagrange multiplier and differentiating, we obtain, for $i \geqq 2$,

$$u_j^{(i)} = \lambda_i D(j)(-1)^{i-1+j}, \quad j = 1, \cdots, i-1, \tag{3.2}$$

where λ_i satisfies

$$g^2 = \lambda_i^2 \sum_{j=1}^{i-1} M(j). \tag{3.3}$$

Thus (3.2) gives two roots, depending upon the sign of λ_i. One root corresponds to the maximum of $M(i)$ and the other root to the minimum. Clearly $\lambda_2 = g$.

From (3.1) and (3.2) we obtain, for $i \geqq 2$,

$$D(i) = \lambda_i \sum_{j=1}^{i-1} M(j) + \sum_{j=1}^{i-1} D(j)\alpha^{i-j}(-1)^{i-1+j}. \tag{3.4}$$

Now, with $i \geqq 3$, replace $D(i-1)$ in the right member of (3.4) by its value obtained by putting $(i-1)$ into the argument of the left member of (3.4). We obtain

$$\begin{aligned} D(i) &= \lambda_i \sum_{j=1}^{i-1} M(j) + \sum_{j=1}^{i-2} D(j)\alpha^{i-j}(-1)^{i-1+j} \\ &\quad + \alpha\left[\lambda_{i-1} \sum_{j=1}^{i-2} M(j) + \sum_{j=1}^{i-2} D(j)\alpha^{i-1-j}(-1)^{i+j}\right] \\ &= \lambda_i \sum_{j=1}^{i-1} M(j) + \alpha\lambda_{i-1} \sum_{j=1}^{i-2} M(j). \end{aligned} \tag{3.5}$$

This formula is valid for $i \geqq 3$. Since $D(i)$ has maximum absolute value and α has been assumed > 0, λ_i and λ_{i-1} must have the same sign, for all $i \geqq 3$. Since $\lambda_2 = g > 0$, all λ_i are positive. From (3.3) and (3.5) we obtain

$$D(i) = g\left(\sum_{j=1}^{i-1} (D(j))^2\right)^{1/2} + \alpha g\left(\sum_{j=1}^{i-2} (D(j))^2\right)^{1/2}.$$

This formula is valid for $i \geqq 3$; the square roots in it are of course positive. This is the fundamental recursion formula for $D(i)$. We see from (3.6) that always $D(i)>0$.

The formulae (3.2) for the signalling coefficients $u_j^{(i)}$ are

$$u_j^{(i)} = (-1)^{i-1+j} g D(j) \left(\sum_{j=1}^{i-1} (D(j))^2 \right)^{-1/2} \tag{3.7}$$

$$i \geqq 2, \quad j = 1, \cdots, i-1.$$

The square root is positive.

4. Optimality of our signalling scheme

By a linear signalling scheme we shall mean a system of functions $x_2, x_3, \cdots$ which are linear in the w's and can be used for signalling. Then obviously the x's should be homogeneously linear and $x_2 = \pm g w_1$. Also x_i is a function only of $w_1, \cdots, w_{i-1}$. Consequently, for any linear signalling scheme the determinant of the equations which correspond to (2.17)–(2.18) must be the same form as the determinant of the latter, i.e., it can differ only in the $u_j^{(i)}$. For a competing linear scheme call $\bar{u}_j^{(i)}$ the corresponding values and let $\bar{D}(i)$ be the value of the determinant. The $\bar{u}_j^{(i)}$ must, of course, also satisfy the constraint (2.15).

We shall now prove that

$$D(i) \geqq |\bar{D}(i)|, \qquad i \geqq 2 \tag{4.1}$$

and that, unless the $\bar{u}_j^{(i)}$ are all the same as the $u_j^{(i)}$, the equality sign cannot hold for all $i \geqq 2$. In view of the significance of $M(i)$ in the estimation of w_1, this means that our scheme is optimal among all linear signalling schemes.

The inequality (4.1) requires proof because we obtained the maximising $u_j^{(i)}$ only consecutively with i. Thus, while (4.1) is obvious for $i = 2$ and 3, it is conceivable that there might exist a linear signalling scheme such that for some $n \geqq 4, D(i) \geqq |\bar{D}(i)|$ for $i = 2, \cdots, n-1$, and then $D(n) < |\bar{D}(n)|$.

Let i be any fixed integer $\geqq 4$. Our method of proof will be to consider the determinant of the equations (2.17)–(2.18) with $w_i = 0$, and maximise the absolute value of the determinant *simultaneously* with respect to all the $u_j^{(s)}$, which are now regarded as undetermined variables subject to (2.15). Thus, $2 \leqq s \leqq i$, $1 \leqq j \leqq s-1$. Of course, if the maximising values $\bar{u}_j^{(s)}$ for $s < i$, depended on i, the result would not be a possible signalling scheme. We will see that the $\bar{u}_j^{(s)}$ are precisely the $u_j^{(s)}$ we found earlier when we maximised consecutively.

Still with i fixed, let $\bar{D}(s)$, $s \leqq i$, be the value of the determinant of order $(s-1)$ in the upper left corner of the determinant we are maximising, after the $\bar{u}_j^{(s)}$ have been inserted. Let $\bar{M}(s) = (\bar{D}(s))^2$ and $\bar{M}(1) = \bar{D}(1) = 1$. Differentiating with a Lagrange multiplier gives us the equations

(4.2) $$\bar{u}_j^{(s)} = \bar{\lambda}_s \bar{D}(j)(-1)^{s-1+j} \qquad 2 \leqq s \leqq i, \quad 1 \leqq j \leqq s-1.$$

The equations with superscript s are obtained from the expansion of the determinant in terms of the elements of the $(s-1)$st row, which is the one which contains the variables $\bar{u}_j^{(s)}$. Here $\bar{\lambda}_s$ satisfies

(4.3) $$g^2 = \bar{\lambda}_s^2 \sum_{j=1}^{s-1} \bar{M}(j), \qquad 2 \leqq s \leqq i.$$

From the form of the determinant we obtain, for $2 \leqq s \leqq i$,

(4.4) $$\bar{D}(s) = \sum_{j=1}^{s-1} \bar{D}(j)(\bar{u}_j^{(s)} + \alpha^{s-j})(-1)^{s-1+j}.$$

This is the analogue of (3.1). From (4.2) and (4.4) we obtain, exactly as (3.5) was obtained,

(4.5) $$\bar{D}(s) = \bar{\lambda}_s \sum_{j=1}^{s-1} \bar{M}(j) + \alpha \bar{\lambda}_{s-1} \sum_{j=1}^{s-2} \bar{M}(j).$$

This is valid for $3 \leqq s \leqq i$. From (4.3) and (4.5) we obtain, for $3 \leqq s \leqq i$,

(4.6) $$\bar{M}(s) = \left[\pm g \left(\sum_{j=1}^{s-1} \bar{M}(j) \right)^{1/2} \pm \alpha g \left(\sum_{j=1}^{s-2} \bar{M}(j) \right)^{1/2} \right]^2.$$

The sign $\pm$ means that we have not determined whether the sign should be $+$ or $-$.

Let a, $2 \leqq a \leqq i$, be the smallest superscript for which $u_j^{(a)} \neq \bar{u}_j^{(a)}$ for some j. There must be such an a, or there would be nothing to prove. (The following remark, which is not needed for the argument, may help the reader to understand the situation. $D(t) = \bar{D}(t)$, $1 \leqq t < a$. We have $\lambda_a = -\bar{\lambda}_a$, i.e., $\bar{u}_j^{(a)} = -u_j^{(a)}$, $1 \leqq j \leqq a-1$. $D(a) > |\bar{D}(a)|$.) If $a = i$ our result is proved. Suppose therefore that $a < i$. Then $M(a) > \bar{M}(a)$. From (3.6), (4.6) and the fact that $M(s) = \bar{M}(s)$ for $1 \leqq s \leqq a-1$, we obtain that $M(a+1) > \bar{M}(a+1)$. Repeating the argument we obtain that $M(a+2) > \bar{M}(a+2), \cdots, M(i) > \bar{M}(i)$. The last inequality violates the fact that $\bar{M}(i)$ was maximised by the $\bar{u}$'s. The optimality property is proved.

5. The capacity of the channel

Define

(5.1) $$a_n = \frac{\sqrt{(M(1) + M(2) + \cdots + M(n-2))}}{\sqrt{(M(1) + M(2) + \cdots + M(n-1))}}.$$

Hence, from (3.6),

(5.2) $$D(3) = g\sqrt{(M(1) + M(2))}(1 + \alpha a_3),$$

so that

$$M(1)+M(2)+M(3)=(M(1)+M(2))(1+g^2(1+\alpha a_3)^2). \tag{5.3}$$

In general, for $n \geqq 3$,

$$\sum_{i=1}^{n} M(i)=(M(1)+M(2))\prod_{i=3}^{n}(1+g^2(1+\alpha a_i)^2). \tag{5.4}$$

From (5.1) and (5.4) we obtain

$$a_{n+1}^2=(1+g^2(1+\alpha a_n)^2)^{-1}. \tag{5.5}$$

Let b be the unique positive root of

$$x^2=\frac{1}{1+g^2(1+\alpha x)^2}, \tag{5.6}$$

then $b<1$. Consider the function

$$y=f(x)=\frac{1}{\sqrt{(1+g^2(1+\alpha x)^2)}} \tag{5.7}$$

for $0 \leqq x \leqq 1$. Now $0<y<1$, and

$$\frac{dy}{dx}=\frac{-g^2\alpha(1+\alpha x)}{(1+g^2(1+\alpha x)^2)^{3/2}}, \tag{5.8}$$

so that dy/dx is always negative. Hence, if $b<x<1$, then $0<f(x)<b$, and if $0<x<b$, then $b<f(x)<1$. Thus the elements of $\{a_n\}$ fall alternately on either side of b.

We have

$$\left|\frac{dy}{dx}\right|=\frac{\alpha}{(1+\alpha x)(1+g^2(1+\alpha x)^2)^{1/2}}\cdot\frac{g^2(1+\alpha x)^2}{(1+g^2(1+\alpha x)^2)} \tag{5.9}$$

so that

$$\left|\frac{dy}{dx}\right|<\alpha<1. \tag{5.10}$$

Hence, for all $n \geqq 3$,

$$|a_{n+1}-b|<\alpha\,|a_n-b|. \tag{5.11}$$

We have therefore proved that

$$\lim a_n=b. \tag{5.12}$$

From (3.6) and (5.4) we obtain

$$D(n)=(g+\alpha g a_n)(M(1)+M(2))^{1/2}\cdot\left(\prod_{i=3}^{n-1}(1+g^2(1+\alpha a_i)^2)\right)^{1/2}. \tag{5.13}$$

From (5.12) and (5.13) we obtain that

$$\text{(5.14)} \quad \lim (1/n) \log D(n) = \tfrac{1}{2} \log (1 + g^2(1 + \alpha b)^2)$$

$$= -\log b = C, \text{ the capacity of the channel.}$$

6. The case $\alpha < 0$

The method described in the previous sections clearly applies when $\alpha < 0$ and also when the regression is of order higher than one. We content ourselves with giving the result for $\alpha < 0$.

As we have already seen, now

$$\text{(6.1)} \quad x_2 = -g w_1$$

$$\text{(6.2)} \quad y_2 = (\alpha - g) w_1 + w_2$$

$$\text{(6.3)} \quad D(2) = \alpha - g.$$

Write $D(1) = 1$. We verify that the relations (3.1)–(3.5) hold, by the same argument as before. Since $\alpha < 0$ we conclude from (3.5) that λ_{i-1} and λ_i are of different signs, i.e., the λ's alternate in sign. The value $\lambda_2 = -g$ is compatible with (3.2), (3.3) and (3.5). Hence, from (3.3),

$$\text{(6.4)} \quad \lambda_i = g \left[\sum_{j=1}^{i-1} M(j) \right]^{-1/2} (-1)^{i+1}$$

where the square root is, of course, positive. The $M(2)$ of this section equals the $M(2)$ of Section 2. Hence our present $M(j)$ are the same as the $M(j)$ of Section 3, but our present $D(j)$ alternate in sign. The quantity $(1/n) \log |D(n)|$ depends only upon $|\alpha|$.

It was noted already in [1] that the capacity of the channel depends only upon $|\alpha|$.

7. Generalisations

I. Let the message to be sent now be a continuous variable θ. It is convenient to assume $|\theta| \leqq g\sigma$, as otherwise a reduction factor will have to be used for the first signal to meet the energy constraint. The statistics of the message play no role in what follows. The message to be sent can be selected arbitrarily, at random, or in any other way. The message is sent over the channel studied in the preceding sections. Let $x_1 = \theta$. After n signals in all according to the linear signalling scheme of Section 2, we have

$$\text{(7.1)} \quad E(y_1 - \theta - L(n))^2 \sim \sigma^2 e^{-2nC}.$$

No linear signalling scheme (and no $L(n)$) can make the right member of (7.1) essentially smaller, uniformly in θ. For, if it could, we could have used that linear signalling scheme in the earlier sections. Our conjecture implies that no signalling scheme can make the right member (7.1) essentially smaller, uniformly in θ.

II. The previous sections did not make full use of the assumptions that the w's are independent, identically distributed, with the normal distribution with mean zero and variance σ^2. We did not use the full implications of independence; it is enough if the w's are uncorrelated. They need not have a common distribution, let alone a normal one. It is enough if all have mean zero and variance σ^2. The means need not be zero if each mean is known to both sender and receiver. Even the assumption of common variance can be modified. No linear signalling scheme can then have a greater rate for this channel than that of the scheme described in the present paper. For, if it did, this linear scheme could have been used earlier. This means that the capacity of the channel is at *least* that given in our paper. It may be larger, but, in that case, rates close to this (larger) capacity can be achieved only by non-linear signalling schemes. These remarks already appeared in [4].

In the case of the channel studied in the previous sections of the present paper (the w's independent, Gaussian, with a common distribution with mean zero) we conjectured that the greatest rates can already be achieved by linear schemes. For the more general channels described in the preceding paragraph this is most probably not the case.

References

[1] Tiernan, J. C. and Schalkwijk, J. P. M. (1974) An upper bound to the capacity of the band-limited Gaussian autoregressive channel with noiseless feedback. *IEEE Trans. Inf. Theory* IT–**20**, 311–316.

[2] Tiernan, J. C. (1972) Autoregressive Gaussian channels with noiseless feedback. Ph.D. thesis, University of California at San Diego.

[3] Schalkwijk, J. P. M. (1966) A coding scheme for additive noise channels with feedback. *IEEE Trans. Inf. Theory* IT–**12**, 183–189.

[4] Wolfowitz, J. (1968) Note on the Gaussian channel with feedback and a power constraint. *Information and Control* **12**, 71–78.

[5] Shannon, C. E. (1948) A mathematical theory of communication. *Bell System Tech. Journal* **27**, 379–424 and 623–657.

[6] Wolfowitz, J. (1964) *Coding Theorems of Information Theory*, 2nd ed., Springer-Verlag, Heidelberg and New York.

[7] Butman, S. (1969) A general formulation of linear feedback communication systems with solutions. *IEEE Trans. Inf. Theory* IT–**15**, 392–400.

[8] Tiernan, J. C. (1975) Analysis of the center-of-gravity processor for the autoregressive forward channel with noiseless feedback. To appear.

[9] Schalkwijk, J. P. M. (1966) Center-of-gravity information feedback. Project No. 502–3, May 1966, Applied Research Laboratory, Sylvania Electronic Systems, Waltham, Mass.

Z. Wahrscheinlichkeitstheorie verw. Gebiete
34, 73–85 (1976)

Zeitschrift für
Wahrscheinlichkeitstheorie
und verwandte Gebiete

Asymptotically Minimax Estimation of Concave and Convex Distribution Functions

J. Kiefer* and J. Wolfowitz**
Department of Mathematics, Cornell University, Ithaca, N.Y. 14853, USA and
Department of Mathematics, University of Illinois, Urbana, Ill. 61801, USA

1. Introduction

Roughly speaking, this paper deals with the problem of efficiently estimating a distribution function when essentially nothing is known about it except that it is concave (or convex). A precise formulation of the problem is given in Section 3. The entire paper is written for concave distributions, but, mutatis mutandis, all results hold for convex distributions. The problem is one of a class with a large literature.

To the best of our knowledge our problem was first treated by Grenander [5]. He proved that the maximum likelihood estimator of a concave distribution F is the least concave majorant C_n of the empiric distribution function F_n of the n independent observations. It follows immediately from Marshall's lemma (Lemma 3 below, which asserts that $\sup_x |C_n(x)-F(x)| \leqq \sup_x |F_n(x)-F(x)|$) that C_n is a uniformly consistent estimator of F. (Throughout we use "estimator" as an abbreviation for "sequence of estimators". Also we omit the phrase "on its interval of support" when describing a d.f. as "concave" or "convex".)

As a consequence of studies of reliability theory there arose an interest in related problems, e.g., estimating increasing (decreasing) failure rate distributions, unimodal distributions, and others. There is now a very large literature, much of which is cited in [6]. The latter book is a good introduction to the subject and a guide to further reading. It would be invidious for us to cite a limited number of papers and ignore others equally worthy, and any attempt at completeness in our references would be out of the question for practical reasons. The authors take this opportunity to express their gratitude to Professor Frank Proschan, for guidance to the literature to which he himself is a distinguished contributor.

In many of the papers of this large literature, some of them very ingenious, the authors obtain estimators by application of the maximum likelihood method

* Research under NSF Grant MPS72-04998 A02.
** Research supported by the U.S. Air Force under Grant AFOSR-70-1947, monitored by the Office of Scientific Research.

or some other method (often quite difficult to do) and prove that their estimators are consistent. Obviously, efficiency is the important property, and proving consistency is only a first step in shrinking the class of possible estimators among which the asymptotically efficient estimators are to be found. The only reason for using maximum likelihood estimators is the hope that the success of this method in the classical case may carry over to these difficult non-parametric problems. No successful attempt has ever been made to show that the maximum likelihood method or any other method will produce asymptotically efficient estimators in these non-parametric problems.

The essential reason (and intuitive basis) for the asymptotic efficiency of the maximum likelihood estimator in the classical case is now well understood ([7], pp. 3.11–12). It is that the maximum likelihood estimator is asymptotically equivalent to the Bayes estimator with respect to an a priori distribution which is uniform on a small interval (of length $O(n^{-1/2})$) centered at the true value of the parameter. One verifies easily that the proof of this fact does not carry over to the present problem. The fact that, even when estimating a single parameter, any serious departure from the classical conditions has as a consequence that the maximum likelihood estimator is no longer efficient, does not bode well for the efficiency of the maximum likelihood estimator for our non-parametric problem.

In [1] and [2] the present authors (in [1] in collaboration with A. Dvoretzky), proved that F_n is asymptotically minimax, in several reasonable senses there precisely defined, as an estimator of F when F is known only to belong to the class of all distribution functions (d.f.'s) or to the class of all continuous d.f.'s. In the present paper we prove (in Section 4) that this is also true when F is known only to belong to the class of all concave (convex) d.f.'s. Now F_n need not be concave (convex), and may therefore be considered by the statistician as an unsuitable estimator of F. However, it follows immediately from Marshall's lemma cited earlier that C_n, which is concave (convex) by definition (in the convex case C_n is defined to be the greatest convex minorant of F_n), and hence suitable to be used as an estimator, is also asymptotically minimax for estimating F in the senses defined in [1, 2], and below in the present paper.

In Section 3 we prove, under certain additional restrictions, that

$$\sup_x |C_n(x) - F_n(x)| = o_P(n^{-1/2}).$$

Consequently, in this case, C_n is essentially no better than F_n for estimating F, in spite of the Marshall lemma, except of course for the fact that C_n is concave (convex).

2. Preliminaries

In this section we give some definitions and probability estimates that may be useful in a variety of applications.

The rv's $X_1, X_2, \ldots$ are i.i.d. according to some d.f. F on the reals. We define B_F to be the smallest closed interval to which F assigns probability one, and define

$$\alpha_0(F)=\sup\{x\colon F(x)=0\},$$
$$\alpha_1(F)=\inf\{x\colon F(x)=1\}, \tag{2.1}$$

with the convention that $\alpha_0=-\infty$ or $\alpha_1=+\infty$ if there is no x satisfying the appropriate condition in braces. By f we denote a derivative of the absolutely continuous part of F. For any real function g on B_F, we define

$$\|g\|=\sup_{x\in B_F}|g(x)|. \tag{2.2}$$

The words "convex" and "concave" are always used with the phrase "and continuous in the interior of B_F" being understood. The empiric d.f. based on n observations is denoted by F_n. The functions F and F_n are defined so as to be right continuous.

Although the normal approximation can be verified to hold for the binomial tail probabilities that arise in the sequel, we shall use the more elementary estimates obtained from Markov's inequality, since the best power of n obtainable in Theorem 1 by the present methods is the same for the two estimates. (The logarithms of these probabilities are asymptotically the same, in the domain we encounter.) Also, as in [8], Lemma 1, it suffices to use below an algebraically simple approximation to the best choice of the coefficient t, rather than the more complex best choice.

Lemma 1. *For positive $p_n\to 0$ and positive $\delta_n\to 0$, if F is the uniform* d.f. *on* [0, 1],

$$P\{|F_n(p_n)-p_n|\geqq\delta_n p_n\}\leqq 2e^{-np_n\delta_n^2[1-o(1)]/2}, \tag{2.3}$$

where the $o(1)$ term depends only on δ_n and p_n.

Proof. For $t>0$, we have

$$\begin{aligned}P\{F_n(p)\geqq(1+\delta)p\}&=P\{e^{tnF_n(p)}\geqq e^{nt(1-\delta)p}\}\\&\leqq e^{-nt(1-\delta)p}Ee^{tnF_n(p)}\\&=[e^{-t(1-\delta)p}(pe^t+1-p)]^n.\end{aligned} \tag{2.4}$$

The substitution $t=\delta$ and expansion of each exponential in the last expression to terms in t^2 plus remainder, yields the result for positive deviations. The result for negative deviations is obtained, similarly, by substituting $t=-\delta$ into

$$P\{F_n(p)\leqq(1-\delta)p\}\leqq[e^{-t(1-\delta)p}(pe^t+1-p)]^n, \tag{2.5}$$

valid for $t<0$.

We now define and study an interpolating process for any fixed, continuous F on R^1 and, correspondingly, for F_n. For each positive integer k, let $a_j^{(k)}$ be any values satisfying

$$\begin{aligned}&F(a_j^{(k)})=j/k \quad \text{if } 0<j<k,\\&a_0^{(k)}=\alpha_0(F),\\&a_k^{(k)}=\alpha_1(F).\end{aligned} \tag{2.6}$$

Each interval $[a_j^{(k)}, a_{j+1}^{(k)}]$ is assigned the same probability $1/k$ in (2.6); we do not require the obvious generalization to unequal probabilities in our applications.

Let $L^{(k)}$ be any nondecreasing function on R^1 satisfying

$$L^{(k)}(a_j^{(k)}) = F(a_j^{(k)}), \quad 0 \leqq j \leqq k. \tag{2.7}$$

The choice of $L^{(k)}$ just above Lemma 4 in Section 3 is piecewise linear, but this is not assumed in Lemma 2, and other choices may prove convenient in other applications. We define $L_n^{(k)}$ by

$$\begin{aligned} L_n^{(k)}(x) = & F_n(a_j^{(k)}) + k[F_n(a_{j+1}^{(k)}) - F_n(a_j^{(k)})][L^{(k)}(x) - F(a_j^{(k)})] \\ & \text{for } a_j^{(k)} \leqq x \leqq a_{j+1}^{(k)}, \ 0 \leqq j \leqq k. \end{aligned} \tag{2.8}$$

Thus, $L_n^{(k)}$ is also nondecreasing, and

$$L_n^{(k)}(a_j^{(k)}) = F_n(a_j^{(k)}), \quad 0 < j < k. \tag{2.9}$$

In the application where $L^{(k)}$ is piecewise linear, so is $L_n^{(k)}$.

In the sequel, $\{k_n, n \geqq 1\}$ will always denote a sequence of positive integers satisfying, for large n,

$$n^{b_1} < k_n < n^{b_2} \quad \text{for some } b_i > 0 \text{ with } b_2 < 1. \tag{2.10}$$

We shall write $p_n = 1/k_n$. The condition (2.10) can be weakened in Lemma 2, but that would not improve the power of n on the left side of (3.18). This lemma sharpens some estimates of [9].

Lemma 2. *Under* (2.10) *there is a positive value C_0 such that, for sufficiently large n, for all continuous F,*

$$\begin{aligned} P_F\{n^{1/2} & \sup_x |F_n(x) - F(x) - L_n^{(k_n)}(x) + L^{(k_n)}(x)| \\ & > [C_0 k_n^{-1} \log k_n]^{1/2}\} < n^{-2}. \end{aligned} \tag{2.11}$$

Proof. For typographic simplicity, we drop the superscript k_n throughout. On the interval $[a_j, a_{j+1}]$, we obtain from (2.7)-(2.8)

$$\begin{aligned} & F_n(x) - F(x) - L_n(x) + L(x) \\ & \quad = \{F_n(x) - F_n(a_j) - p_n^{-1}[F(x) - F(a_j)][F_n(a_{j+1}) - F_n(a_j)]\} \\ & \qquad + \{[F(x) - L(x)](p_n^{-1}[F_n(a_{j+1}) - F_n(a_j)] - 1)\} \\ & \quad = G_n(x) + H_n(x) \quad \text{(say)}. \end{aligned} \tag{2.12}$$

We shall prove that, for a suitable C_0, and for n sufficiently large, we have, uniformly in j $(0 \leqq j \leqq k_n)$,

$$P_F\{n^{1/2} \sup_{a_j \leqq x \leqq a_{j+1}} |G_n(x)| > \tfrac{1}{2}[C_0 k_n^{-1} \log k_n]^{1/2}\} < 4^{-1} k_n^{-1} n^{-2}, \tag{2.13}$$

and also (2.13) with H_n replacing G_n. Summing over these two sources of deviation and the $k_n + 1$ intervals (a_j, a_{j+1}) then yields (2.11).

Given the event $n[F_n(a_{j+1}) - F_n(a_j)] = n_0 > 0$, the process $\{n n_0^{-1} G_n(x), a_j \leqq x \leqq a_{j+1}\}$ has the law of "empiric minus true continuous d.f.", on an interval to which the latter assigns probability one. Consequently ([1]), for all positive n_0, all $d_n > 0$, and some constant C,

$$P\{n^{1/2} \sup_{a_j \leqq x \leqq a_{j+1}} |G_n(x)| > d_n | n[F_n(a_{j+1}) - F_n(a_j)] = n_0\} \leqq C e^{-2d_n^2 n/n_0}. \tag{2.14}$$

Now fix $\delta_n < 1$ at a small enough positive value so that the $o(1)$ term in (2.3) is $> -1/2$ for k_n sufficiently large. The estimate (2.3), in terms of the present setting, yields

$$P\{n[F_n(a_{j+1}) - F_n(a_j)] \geqq 2np_n\} \leqq 2e^{-C'np_n} \tag{2.15}$$

for some positive C'. Consequently, from (2.15) and (2.14),

$$P\{n^{1/2} \sup_{a_j \leqq x \leqq a_{j+1}} |G_n(x)| \geqq d_n\} \leqq C e^{-d_n^2/p_n} + 2e^{-C'np_n}. \tag{2.16}$$

From the fact that $\sup_{a_j \leqq x \leqq a_{j+1}} |F(x) - L(x)| \leqq p_n$, and from Lemma 1, we have

$$\begin{aligned} P\{n^{1/2} \sup_{a_j \leqq x \leqq a_{j+1}} |H_n(x)| \geqq d_n\} &\leqq P\{n^{1/2} |F_n(a_{j+1}) - F_n(a_j) - p_n| \geqq d_n\} \\ &\leqq 2e^{-d_n^2[1+o(1)]/2p_n}, \end{aligned} \tag{2.17}$$

provided that $\delta_n = d_n/n^{1/2} p_n \to 0$. Because of (2.10), the $o(1)$ term depends on δ_n and n, since p_n is a function of n.

Now substitude $d_n^2 = 4^{-1} C_0^2 p_n \log p_n^{-1}$ into (2.16) and (2.17). From (2.10), for C_0 sufficiently large, each of the right sides of (2.16) and (2.17) is seen to be $< 1/4n^2 k_n$, and the condition below (2.17) is satisfied. This completes the proof of (2.13), and thus of Lemma 2.

3. Estimating Concave F: Closeness of Two Estimators

Throughout this section we assume $X_1, X_2, \ldots$ are i.i.d. rv's with common d.f. F, concave on the interval $[\alpha_0(F), \infty)$; note that $\alpha_0(F)$ is finite, and F may have a jump at $\alpha_0(F)$. Denote the class of all such F by $\mathscr{F}$. Many results of this subject, in the literature or the present paper, are valid only for a subset of $\mathscr{F}$. Thus, F might also be assumed continuous, $\alpha_0(F)$ might be assumed known, and/or $\alpha_1(F)$ might be assumed finite, or an upper bound on $\alpha_1(F)$ might be assumed known. We shall state such restrictions where they are used. When $\alpha_0(F)$ is known, we shall take it to be zero, without loss of generality.

Assuming F continuous and $\alpha_0(F) = 0$, the maximum likelihood estimator of F, based on n observations, was shown by Grenander [4] to be C_n, the smallest concave majorant of F_n satisfying $C_n(0) = 0$. In the present section we show that, under certain assumptions, F_n and C_n are quite close for large n. A theoretical consequence is that the two estimators enjoy similar optimum properties (treated in the next section). A practical consequence is that, instead of computing C_n exactly, the statistician may often find it satisfactory to use any concave function sufficiently close to F_n; if he does not care whether his estimator is concave, he can even use F_n itself.

The ML estimator remains the same if $\alpha_1(F)$ is assumed finite, as is done in (3.2).

For simplicity of presentation, we shall prove the next four lemmas and Theorem 1 under a single set of assumptions, and thereafter discuss the assumptions and remark on which of the conclusions can be modified to hold under other assumptions. We write, for twice differentiable F with $\alpha_0(F)=0$,

$$\begin{aligned}\gamma(F) &= \sup_{0<x<\alpha_1(F)} |-f'(x)| / \inf_{0<x<\alpha_1(F)} f^2(x),\\ \beta(F) &= \inf_{0<x<\alpha_1(F)} |-f'(x)/f^2(x)|.\end{aligned} \tag{3.1}$$

Note that γ and β are invariant under changes of scale and location. We hereafter assume

$$\begin{aligned}&F \text{ continuous with } \alpha_0(F) \text{ known to be } 0 \text{ and } \alpha_1(F) \text{ unknown but known to be } <\infty;\\ &F \text{ concave and twice continuously differentiable on } (0, \alpha_1(F));\ \beta(F)>0,\ \gamma(F)<\infty.\end{aligned} \tag{3.2}$$

We remind the reader that $\|g\| = \sup_{0\leqq x\leqq \alpha_1(F)} |g(x)|$.

We now give an outline of our method of estimating the magnitude of $C_n - F_n$:

(A) For every concave function h on $[0, a_1(F)]$, it is proved ([5]) that

$$\|C_n - h\| \leqq \|F_n - h\|. \tag{3.3}$$

(B) Under the assumption (3.2), Lemma 1 is used to show that, for suitable k_n and functions $L^{(k)}$ and $L_n^{(k)}$ of (2.7)–(2.8), the event

$$A_n = \{L_n^{(k_n)} \text{ is concave on } [0, \infty)\} \tag{3.4}$$

has probability near one.

(C) Under A_n, we can use (A) with $h = L_n^{(k_n)}$ to show that

$$\|C_n - F_n\| \leqq 2\|F_n - L_n^{(k_n)} + L^{(k_n)} - F\| + 2\|L^{(k_n)} - F\|. \tag{3.5}$$

(D) The first term on the right side of (3.5) is estimated by using Lemma 2; the second term is estimated by a simple analytic argument.

Lemma 3 (Marshall [5]). *If h is concave, (3.3) holds.*

We now define $L^{(k)}$ to be linear on each of the intervals $[a_j^{(k)}, a_{j+1}^{(k)}]$, for $0 \leqq j \leqq k-1$.

Lemma 4. *If (3.2) holds, for k_n sufficiently large (depending only on $\beta(F)$),*

$$1 - P\{A_n\} \leqq 2k_n e^{-n\beta^2(F)/80k_n^3}. \tag{3.6}$$

Proof. For $0 \leqq j \leqq k_n - 1$, write

$$\begin{aligned}T_{n,j} &= F_n(a_{j+1}^{(k_n)}) - F_n(a_j^{(k_n)})\\ \Delta_{n,j} &= a_{j+1}^{(k_n)} - a_j^{(k_n)}.\end{aligned} \tag{3.7}$$

Since $L^{(k_n)}$ is linear on each of the k_n successive intervals of length $\Delta_{n,j}$ $(0 \leqq j \leqq k_n - 1)$, it follows from (2.8) and the computation of the derivative of $L_n^{(k_n)}$ in each interval that

$$A_n = \bigcap_{j=0}^{k_n-2} \{T_{n,j}/\Delta_{n,j} \geqq T_{n,j+1}/\Delta_{n,j+1}\}$$
$$= \bigcap_{j=0}^{k_n-2} B_{n,j} \quad \text{(say)}. \tag{3.8}$$

For $0 \leqq j \leqq k_n - 2$, and $0 < \delta_n < 1/3$, the event $B_{n,j}$ is a consequence of

$$|T_{n,i} - k_n^{-1}| \leqq \delta_n/k_n \quad \text{for } i = j, j+1 \tag{3.9}$$

and (since $1 + 3\delta_n > (1+\delta_n)/(1-\delta_n)$)

$$\Delta_{n,j+1}/\Delta_{n,j} \geqq 1 + 3\delta_n. \tag{3.10}$$

We next verify that (3.10) holds for $0 \leqq j \leqq k_n - 2$, provided that $\delta_n \leqq \beta(F)/6k_n < 1/3$. Since $dF^{-1}(t)/dt = 1/f(F^{-1}(t))$ and $d^2F^{-1}(t)/dt^2 = (-f'/f^3)(F^{-1}(t))$, a second order Taylor expansion about $(j+1)/k_n$ yields

$$\Delta_{n,j+1} = F^{-1}((j+2)/k_n) - F^{-1}((j+1)/k_n)$$
$$= k_n^{-1}/f(a_{j+1}^{(k_n)}) + (2k_n^2)^{-1}(-f'(\xi)/f^3(\xi)) \tag{3.11}$$

for some ξ between $a_{j+1}^{(k_n)}$ and $a_{j+2}^{(k_n)}$. Since f is decreasing, we obtain

$$\Delta_{n,j} \leqq k_n^{-1}/f(a_{j+1}^{(k_n)})$$

and also

$$\Delta_{n,j+1}/\Delta_{n,j} \geqq 1 + (2k_n)^{-1} f(a_{j+1}^{(k_n)})(-f'(\xi)/f^3(\xi))$$
$$\geqq 1 + (2k_n)^{-1}(-f'(\xi)/f^2(\xi))$$
$$\geqq 1 + (2k_n)^{-1}\beta(F). \tag{3.12}$$

Hence (3.10) holds if $\delta_n = \beta(F)/6k_n$. With this choice of δ_n we consequently obtain (3.6) from (3.9), (3.10), and Lemma 1, since the conditions $\delta_n \to 0$ and $p_n \to 0$ of that lemma are automatically satisfied as $k_n \to \infty$.

Lemma 5. *If A_n occurs* (3.5) *holds.*

Proof. From Lemma 3 and (3.3) with $h = L_n^{(k_n)}$,

$$0 \leqq C_n - F_n \leqq \|C_n - L_n^{(k_n)}\| + \|L_n^{(k_n)} - F_n\|$$
$$\leqq 2\|L_n^{(k_n)} - F_n\|$$
$$\leqq 2\|L_n^{(k_n)} - F_n + F - L^{(k_n)}\| + 2\|L^{(k_n)} - F\|. \tag{3.13}$$

Lemma 6. *If F satisfies* (3.2),

$$\|F - L^{(k)}\| \leqq \gamma(F)/2k^2. \tag{3.14}$$

Proof. Fix j, $0 \leqq j \leqq k-1$, and define $g(x) = F(x + a_j^{(k)}) - L(x + a_j^{(k)})$ on the interval $[0, \Delta_j]$, where $\Delta_j = a_{j+1}^{(k)} - a_j^{(k)}$. Then $g \geqq 0$, and

$$g(0) = g(\Delta_j) = 0. \tag{3.15}$$

Hence, a Taylor expansion about 0 gives, on $[0, \Delta_j]$,

$$g(x) = g'(0+)x + g''(\xi_x)x^2/2, \tag{3.16}$$

where $0 < \xi_x < x$. Evaluating (3.16) at $x = \Delta_j$ and again using (3.15) yields $g'(0+) = -\Delta_j g''(\xi_{\Delta_j})/2$. Since $g''(\xi_x) \leqq 0$, (3.16) becomes

$$\begin{aligned} g(x) &= -g''(\xi_{\Delta_j})\Delta_j x/2 + g''(\xi_x)x^2/2 \\ &\leqq -g''(\xi_{\Delta_j})\Delta_j^2/2. \end{aligned} \tag{3.17}$$

Since $g''(x) = f'(x + a_j^{(k)})$, and since $\Delta_j = 1/kf(a_j^{(k)} + \xi')$ for some ξ' in $(0, \Delta_j)$ by Rolle's theorem, the result (3.14) follows from (3.17).

Theorem 1. *If F satisfies* (3.2), *for all sufficiently large n (depending only on $\beta(F)$ and $\gamma(F)$)*

$$P_F\{\|C_n - F_n\| > n^{-2/3}(\log n)^{5/6}\} < 2n^{-2}, \tag{3.18}$$

so that

$$P_F\{\lim_{n\to\infty} [n^{2/3}/\log n]\,\|C_n - F_n\| = 0\} = 1. \tag{3.19}$$

Proof. Let $k_n = [\beta^2(F)n/200 \log n]^{1/3}$. By Lemma 4, for n sufficiently large (depending on $\beta(F)$) $1 - P\{A_n\} \leqq n^{-2}$. If A_n occurs, by Lemma 5, (3.5) holds. Note that (2.10) is satisfied. By Lemma 2, with probability at least $1 - n^{-2}$, the first term on the right side of (3.5) is at most $2C_0^{1/2}(200/\beta^2(F))^{1/6} n^{-2/3}(\log n)^{2/3}$, for all sufficiently large n (independent of F). By Lemma 6, the second term on the right side of (3.5) is at most $\gamma(F)(200/\beta^2(F))^{2/3} n^{-2/3}(\log n)^{2/3}$. The combination of these estimates yields (3.18).

Under (3.18) the limiting laws of $n^{1/2}\|C_n - F\|$ and $n^{1/2}\|F_n - F\|$ are of course the same.

Remarks. 1. In order to discuss the extent to which our results hold under assumptions other than (3.2), we must first describe the way in which the definition of C_n is altered under the various possible restrictions mentioned in the first paragraph of the present section. If $\alpha_0(F) = 0$ but the assumption of continuity of F is dropped, C_n is defined to be the smallest concave majorant of F_n, subject to $C_n(0-) = 0$; thus, $C_n(0) = F_n(0)$. This is the ML estimator in an extended sense [3]. If also $\alpha_0(F)$ is unknown, C_n is the smallest concave majorant of F_n on the interval $[\min_{1 \leqq i \leqq n} x_i, \infty)$, subject to being 0 to the left of this interval. This last estimator, although discontinuous, may also be used when $\alpha_0(F)$ is unknown but F is assumed continuous, in which case an ML estimator does not exist; alternatively, it may be modified to be a "neighborhood ML estimator" [3].

The above definitions are unchanged if $\alpha_1(F)$ is assumed finite or if an upper bound on $\alpha_1(F)$ is assumed known. If $\alpha_1(F)$ is known exactly, wp 1 the ML estimator does not exist, but these estimators, or slight modifications seem appropriate.

2. It is not difficult to see that Lemmas 3 and 5 remain valid for any of the alterations of Remark 1.

3. The proofs of Lemmas 4 and 6, and thus of Theorem 1, can be seen to hold under any or all of the following modifications of (3.2): F can be permitted to

have a jump at $\alpha_0(F)$; $\alpha_0(F)$ can be unknown; $\alpha_1(F)$ can be known. In addition, if f' fails to exist on a finite set H (where f can even have jumps) and (3.2) holds when the set H is removed in the definition of β and γ, Theorem 1 remains valid; it is only necessary to adjoin H to the set of $a_j^{(k)}$, for each k, removing the original $a_j^{(k)}$ closest to each member of H in order to make sure the analogue of Lemma 4 still holds.

4. The proofs of Lemmas 1, 2, and 4 have been kept short by using simple rather than sharp estimates; however, such estimates, or the weakening of (3.18) to require only that the probability approach zero, would not improve the power of n attainable in (3.18)–(3.19) by the present approach.

Of more interest is an extension of Theorem 1 obtainable when $\beta(F)=0$, due to f' vanishing on a finite set. We indicate the construction for the simple case where that set consists of the single point $\alpha_1(F)$. Thus, we assume that $f'(\alpha_1(F)-)=0$ and that, for $0<\varepsilon<1$,

$$0<\beta(F,\varepsilon)\overset{\text{def}}{=}\inf_{0<x<F^{-1}(1-\varepsilon)}[-f'(x)/f^2(x)]. \tag{3.20}$$

In order that $\gamma(F)$ still be finite, we assume also that $0<f(\alpha_1(F))=\bar{C}(F)$ (say).

The idea is to use $\beta(F,\varepsilon)$ in place of β in the last line of (3.12), which entails having ξ bounded away from $\alpha_1(F)$ for each k; this last necessitates changing the definition of $L^{(k)}$ slightly. Writing $a_*^{(k)}=F^{-1}(1-1/2k)$, we alter the definition of $L^{(k)}$ just above Lemma 4 *only* on the interval $[a_{k-1}^{(k)},\alpha_1(F)]$, where we define

$$L^{(k)}(x)=\begin{cases}1-1/2k & \text{at } x=a_*^{(k)},\\ \text{linear} & \text{on } [a_{k-1}^{(k)},a_*^{(k)}],\\ F(x) & \text{on } [a_*^{(k)},\alpha_1(F)].\end{cases} \tag{3.21}$$

It is clear that $L^{(k)}$ is concave, and we examine the proof of Lemma 4. For $j\leqq k_n-3$, the development of (3.11) and (3.12) still holds, with $\xi<a_{k-1}^{(k)}$. The event B_{n,k_n-2} is now seen to occur if, in analogy to (3.9)–(3.10),

$$\begin{aligned}&|T_{n,k_n-2}-k_n^{-1}|\leqq\delta_n/k_n,\\ &|F_n(a_*^{(k_n)})-F_n(a_{k_n-1}^{(k_n)})-(2k_n)^{-1}|<\delta_n/2k_n,\\ &2\Delta_{k_n-1}/\Delta_{k_n-2}\geqq 1+3\delta_n.\end{aligned} \tag{3.22}$$

The last of these is seen to hold (upon replacing k_n by $2k_n$ in the last expression of (3.11), etc.) if $\delta_n=\beta(F,1/2k_n)/12k_n$, which choice of δ_n can of course also be used for $j<k_n-2$. We obtain (3.6) with $\beta^2(F)/80$ replaced by $\beta^2(F,1/2k_n)/300$.

The conclusion of Lemma 6 is unaltered, and the proof of Theorem 1 is changed by replacing $200/\beta^2(F)$ everywhere in the proof by approximately $800/\beta_n^2(F)$. This entails solving approximately, for each large n, the relationship

$$\frac{k_n^3}{\beta^2(F,1/2k_n)}=n/800\log n \tag{3.23}$$

for k_n; since the denominator of the left side is non-increasing in k_n, there is a unique positive solution, and k_n may be taken to be the closest integer to it. This determination makes $1-P\{A_n\}<n^{-2}$ for n large. Assuming (2.10) holds, we apply Lemma 2 as before. The first term on the right side of (3.5), of order $(\log n/n)^{2/3}\beta_n^{-1/6}$, is of smaller order than the second term, of order $(\log n/\beta_n^2 n)^{2/3}$.

Thus, (3.18) is replaced by

$$P_F\{\|C_n - F_n\| \geqq 2\gamma(F)k_n^{-2}\} < n^{-2}, \tag{3.24}$$

where k_n is determined by (3.23), with the obvious analogue for (3.19).

To indicate the domain of this extension, we consider an example. Suppose $-f'(x) \sim A(F)(\alpha_1(F)-x)^q$ as $x \uparrow \alpha_1(F)$, for some $q>0$. Then $\beta(F,\varepsilon) \sim -f'(\alpha_1(F) - \varepsilon/\bar{C}(F))/\bar{C}^2(F) \sim A(F)\bar{C}^{q-2}(F)\varepsilon^q$ as $\varepsilon \downarrow 0$. Thus, (3.23) yields k_n of order $(n/\log n)^{1/(3+2q)}$, so that (2.10) holds, and consequently the term k_n^{-2} in (3.24) is of order $(n/\log n)^{-2/(3+2q)}$. Since the conclusion of Theorem 1, which gives only an upper bound on $\|C_n - F_n\|$, is of interest only when that bound is $o(n^{-1/2}) < O_p(\|F_n - F\|)$, we see that $q < 1/2$ is the domain of interest.

5. Although the device just described works equally well for modifiying Lemma 4 when $\alpha_1(F) = +\infty$, neither our original proof nor our modification yields a useful extension of Theorem 1 in that case because of the failure of Lemma 6. In fact, whether or not $\alpha_1(F) < \infty$, if $f(x) \downarrow 0$ as $x \uparrow \alpha_1(F)$ it is not difficult to show that $-f'/f^2$ is unbounded and hence $\gamma(F) = +\infty$. Much more is true, as we now show in demonstrating why the modification of Remark 4 does not work for Lemma 6. Supposing $\alpha_1(F) = +\infty$, we shall show that it is impossible that

$$-f'(x)/f^2(x) < \delta/[1-F(x)] \tag{3.25}$$

for any $\delta < 1$ and all large x; putting $x = F^{-1}(1-1/2k)$, this shows that Lemma 6 would yield the inadequate estimate $\sup_{0<x<F^{-1}(1-1/2k)} [F(x) - L(x)] = O(k^{-1})$. If (3.25) held, we would have $-f'/f < \delta f/[1-F]$ on an interval $[t_\delta, \infty)$. Integration yields $-\log[f(x)f(t_\delta)] < -\delta \log\{[1-F(x)]/[1-F(t_\delta)]\}$, or $f(x)/[1-F(x)]^\delta$ bounded below by a value $C_\delta > 0$. A second integration yields $(1-\delta)^{-1}\{[1-F(t_\delta)]^{1-\delta} - [1-F(x)]^{1-\delta}\} \geqq C_\delta(x - x_\delta)$, presenting a contradiction as $x \to \infty$. In similar fashion, if $\alpha_1(F) < \infty$ one can show, in place of (3.25), that it is impossible for $-f'[1-F]^\delta/f^2$ to be bounded for any $\delta < 1$, and this in turn implies the inadequacy of Lemma 6, in that it cannot provide an estimate of order k^{-r} for $F - L$, for any $r > 1$. An analogous result holds if f is unbounded at 0 and truncation at $F^{-1}(1/2k)$ is attempted.

6. If there is an interval $[r_1, r_2]$ of positive length on which f is a positive constant, and if $\bar{L}_n$ is the chord between $(r_1, F_n(r_1))$ and $(r_2, F_n(r_2))$, the probability that $F_n(x) < \bar{L}_n(x) - n^{-1/2}$ for some x in $[r_1, r_2]$ is bounded away from 0 as $n \to \infty$. Thus, there is no useful extension of Theorem 1 if F is not *strictly* concave on $(0, \alpha_1(F))$. However, if F is concave and (3.2) is satisfied when $(0, \alpha_1(F))$ is replaced by the complement therein of such an interval $[r_1, r_2]$, the limiting law of $n^{1/2}\|C_n - F_n\|$ can be expressed as the law of the deviation of a Brownian bridge *below* a chord between points of its graph on a corresponding interval. The limiting law of $\|C_n - F\|$, and the case where there is more than one such interval, can be treated similarly.

4. Asymtotic Minimax Character of F_n and C_n for Estimating a Concave d.f.

Throughout this section we consider

$$\mathscr{F} = \{F\colon F \text{ is continuous and concave on } [0, \infty)\}. \tag{4.1}$$

The conclusions stated below, regarding asymptotic optimality of F_n for estimating F in $\mathscr{F}$, will easily be seen to hold for the ~~more restricted~~ classes of concave d.f.'s discussed in Sections 2 and 3. Moreover, for every concave d.f. F there is a sequence of strictly concave d.f.'s $\{H_j\}$ such that $P_F\{\lim_{j\to\infty} dH_j(x)/dF(x)=1\}=1$, from which it follows that for each sample size n, each estimator of F has the same supremum risk over $\mathscr{F}$ as over

$$\mathscr{F}^* = \{F\colon F \text{ is continuous and strictly concave on } [0, \infty)\}. \tag{4.2}$$

Consequently, our conclusions hold if (4.1) or any of its variants is reduced by requiring strict concavity.

The space D of decisions (possible estimates of F) is any collection of real functions on R^1 that includes the functions F_n. If D is restricted by demanding that the estimate of F be continuous, our optimality conclusions still hold with F_n replaced by any of several possible continuous modifications F_n^* that have been suggested and for which $\sup_x |F_n(x)-F_n^*(x)| \leq c_n$ wp 1 under every F, where $c_n = o(n^{-1/2})$. If D is further restricted by demanding that the estimate be concave, no such F_n or F_n^* is usable. However, the simple fact (Lemma 3) that $\|C_n - F\| \leq \|F_n - F\|$, together with the monotone form of W below and the optimality conclusion when use of F_n *is* permitted, imply that C_n is still asymptotically minimax for D restricted to concave estimates.

Let W be any nonnegative nondecreasing function on the nonnegative reals for which

$$\int_0^\infty W(r) r e^{-2r^2}\, dr < \infty. \tag{4.3}$$

We assume W is not identically zero. A nonrandomized estimator g_n of F, based on n observations, takes on values in D and has risk function

$$r_n(F; g_n) = E_F W(n^{1/2} \|g_n - F\|). \tag{4.4}$$

We will not discuss the routine measure-theoretic background and consideration of randomized estimators which are treated in detail in Section 1 of [1]. In the statements that follow, g_n is permitted to be randomized. Our main result is

Theorem 2. *Under the above assumptions,*

$$\lim_{n\to\infty} \frac{\sup_{F\in\mathscr{F}} r_n(F; F_n)}{\inf_{g_n} \sup_{F\in\mathscr{F}} r_n(F; g_n)} = 1. \tag{4.5}$$

We remark that it is much simpler to prove asymptotic "optimality" results by replacing r_n by its limit, restricting one's self to regular sequences $\{g_n\}$ for which such limits exist. This is a weaker type of result than ours, and does not imply uniformity in F as $n \to \infty$ of the type exhibited in (4.5). From a practical point of view, such uniformity is important.

Before outlining the proof of Theorem 2, we mention that many other loss and risk functions could be considered in place of those of (4.4). For example,

$$\int_0^\infty E_F W(n^{1/2} |g_n(x) - F(x)|)\, dF(x) \tag{4.6}$$

gives a risk function of integrated form rather than a function of the maximum deviation. More general forms are discussed in [1] and [2], and the optimality results for such loss functions in our setting of concave F are proved by using the developments of those earlier papers in the same way that we now use them to prove (4.5) under (4.4). The reader who is unacquainted with those papers may want to consult pp. 649–650 of [1] and 477–478 of [2] for some explanatory intuitive comments.

Proof of Theorem 2. The required mathematical developments are contained in [1] and [2], and it is only necessary to fit them into the present context. It will be simpler to use the second of these references for part of the results (even though a multivariate d.f. is the ultimate interest there), because of the usable form in which the required preliminary results of [2] are set forth, and because the corresponding results of [1], although more refined in their explicit presentation of an error estimate, are based on Bayes procedure calculations for a particular form of prior law which must be modified (as it is in [2]) to fit the present problem.

For each positive integer h, let $\mathscr{F}^{(h)}$ be the class of all absolutely continuous d.f.'s F of the following form: $F(0)=0$, $F(1)=1$, and the density F' is some constant p_i (say) on each interval $([i-1]/[h+1], i/[h+1])$ for $1 \leqq i \leqq h+1$; furthermore,

$$p_1 \geqq p_2 \geqq \cdots \geqq p_{h+1}. \tag{4.7}$$

This last implies that $\mathscr{F}^{(h)} \subset \mathscr{F}$.

Let U denote the uniform d.f. on [0, 1], and write $\|\Psi\|_h = \max_{1 \leqq i \leqq h+1} |\Psi(i/[h+1])|$. Working in the spirit of [1] and [2], we shall prove (4.5) by showing (a) that the (constant) risk function of F_n for the original problem of this section is close to the value $E_U W(n^{1/2}\|F_n - U\|_h)$ when n and h are large, and that (b) for fixed h and large n, the procedure F_n is approximately minimax for the problem obtained by replacing $\mathscr{F}$ by $\mathscr{F}^{(h)}$ and $\|\cdot\|$ in (4.4) by $\|\cdot\|_h$. More precisely, writing

$$r_n^{(h)}(F; g_n) = E_F W(n^{1/2}\|g_n - F\|_h) \tag{4.8}$$

for F in $\mathscr{F}^{(h)}$, and noting that $r_n^{(h)} \leqq r_n$, and that $U \in \mathscr{F}^{(h)}$, we see that (4.5) is implied by

$$\lim_{h\to\infty} \lim_{n\to\infty} r_n^{(h)}(U; F_n) = \lim_{n\to\infty} r_n(U; F_n) \tag{4.9}$$

and

$$\lim_{n\to\infty} \frac{\sup_{F\in\mathscr{F}^{(h)}} r_n^{(h)}(F; F_n)}{\inf_{g_n} \sup_{F\in\mathscr{F}^{(h)}} r_n^{(h)}(F; g_n)} = 1. \tag{4.10}$$

The result (4.9) is precisely the result (4.5) of [1], and we turn to (4.10).

For fixed h, the problem with $\mathscr{F}$ replaced by $\mathscr{F}^{(h)}$ and r_n replaced by $r_n^{(h)}$ is treated in Section 3 of [2] as a multinomial problem; the multinomial $(h+1)$-vector with i-th-component $nF_n(i/[h+1]) - nF_n([i-1]/[h+1])$ is sufficient for $\mathscr{F}^{(h)}$, and this allows (4.10) to be proved by analyzing the multinomial problem. The result (4.10) is then obtained from Lemma 10 of [2], once one notes two aspects of the developments there. First, the integrability assumption (3.6) of [2], used there because of the consideration of multivariate d.f.'s, can be replaced by our (4.3) in the present univariate setting. Second, the set B'_h of [2] is our (4.7), and thus satisfies the assumption, made there, requiring it to be the closure of its interior. This completes the proof of Theorem 2.

We have not attempted here to obtain a more precise estimate of the departure from minimaxity (departure from 1 of the ratio in (4.5)) as was done in [1]. Indeed, such an estimate seems more difficult to obtain here because the concavity restriction does not permit us to use a sequence of prior laws centered at the uniform law (all $p_i = 1/[h+1]$) for the multinomial problem as we did in [1]; the uniform law is now an extreme point of the set (4.7) on which the prior laws must be supported.

References

1. Dvoretzky, A., Kiefer, J., Wolfowitz, J.: Asymptotic minimax character of the sample distribution function and of the classical multinomial estimator. Ann. Math. Statist. **27**, 642–669 (1956)
2. Kiefer, J., Wolfowitz, J.: Asymptotic minimax character of the sample distribution function for vector chance variables. Ann. Math. Statist. **30**, 463–489 (1959)
3. Kiefer, J., Wolfowitz, J.: Consistency of the maximum likelihood estimator in the presence of infinitely many incidental parameters. Ann. Math. Statist. **27**, 887–906 (1956)
4. Grenander, U.: On the theory of mortality measurement. Part II. Skand. Akt. Tid. **39**, 125–153 (1956)
5. Marshall, A. W.: Discussion of Barlow and van Zwet's papers in M. L. Puri (Ed.). Nonparametric techniques in statistical inference. p. 175–176. Cambridge University Press, 1970
6. Barlow, R. E., Bartholomew, D. J., Bremner, J., M., Brunk. H. D.: Statistical inference under order restrictions. New York: Wiley 1972
7. Weiss, L., Wolfowitz, J.: Maximum probability estimators and related topics. Lecture Notes in Math. **424**. Berlin, Heidelberg, New York: Springer 1974
8. Kiefer, J.: Skorokhod embedding of multivariate r.v.'s, and the sample d.f. Z. Wahrscheinlichkeitstheorie verw. Gebiete **24**, 1–35 (1972)
9. Kiefer, J., Wolfowitz, J.: On the deviations of the empiric distribution function of vector chance variables. Trans. Amer. Math. Soc. **87**, 173–186 (1958)

Received October 13, 1975

Ann. Inst. Statist. Math.
28 (1976), Part A, 359-370

ASYMPTOTICALLY EFFICIENT ESTIMATORS WHEN THE DENSITIES OF THE OBSERVATIONS HAVE DISCONTINUITIES

J. WOLFOWITZ*

(Received Dec. 18, 1975)

1. Introduction

Let $X_1, X_2, \cdots$ be independent chance variables with a common density function $f(x, \theta_{00})$ at the point x; the density function (d.f.) depends on the value θ_{00} of the real parameter θ, which is unknown to the statistician. In [1] Ibragimov and Hasminski obtained the asymptotic distribution of a stochastic process (similar to that in (4.1) below) naturally connected with the sequence $X_1, X_2, \cdots$, and of certain functionals of this process. This was done under more or less classical "regular" conditions on $f(\cdot, \cdot)$. A special case of this result is the asymptotic normality of the maximum likelihood (m.l.) estimator. (Throughout this paper we do not distinguish between "estimator" and "sequence of estimators." No confusion will ever be caused thereby.)

Since Fisher's famous paper on the m.l. estimator, asymptotic efficiency has usually been defined as the property of having minimum variance of the limiting normal distribution, among all estimators with limiting normal distribution (with the standard normalization). This definition is already inadequate in the classical "regular" case, because it limits the competition to be an efficient estimator to estimators which have a limiting normal distribution. It fails completely outside the regular case, because then asymptotic normality is also very exceptional.

In [11] and [12] (see also [7]) L. Weiss and the present author introduced a notion of asymptotic efficiency (with respect to a set R) which does not have these inadequacies (or other inadequacies of the classical theory). It also permits the statistician to choose his own loss function, via the set R or directly (see [12]). If, now, one requires f to satisfy the conditions of the (classical) regular case, and, if *in addition* (what this theory does not require), one admits to competition *only* asymptotically normal estimators, and chooses $R=(-r, r)$ to be any interval centered at the origin, this notion of efficiency specializes to

* Research supported by the U.S. Air Force under Grant AFOSR-76-2877, monitored by the Office of Scientific Research.

Fisher's notion.

The asymptotic efficiency of the m.l. estimator in the regular case and in Fisher's sense was first proved by LeCam in a series of papers of which [3] is the latest. Later Bahadur gave another proof in [4]. In [2] the present author, by making a slight change in the argument of [1] and then applying the theory of maximum probability (m.p) estimators (a basic theorem of which is cited in Section 2 below), was able to prove the asymptotic efficiency of the m.l. estimator with respect to $R=(-r, r)$ (and other R) in the regular case and in the non-regular "almost smooth" case (see also [5]). If, *in addition*, as is done in the classical Fisher theory, one limits the competition to asymptotically normal estimators, this specializes to efficiency in the Fisher sense.

In a large number of cases where the d.f. is discontinuous and the normalizing coefficient is n, the number of observations, Weiss and the present author (e.g., [6]-[9]) obtained m.p. estimators, in particularly simple and explicit form, principally for $R=(-r, r)$, but also, in some cases, for other R (loss functions). They verified that the conditions (2.2) and (2.3) below are satisfied, so that these estimators are asymptotically efficient (for their respective loss functions). In almost all of these cases the m.l. estimator is different from the m.p. estimator for $R=(-r, r)$, is not asymptotically efficient for that R, and is not asymptotically efficient in the Fisher sense.

In [10] Ibragimov and Hasminsky obtained the asymptotic distribution of the stochastic process (4.1) below and of functionals of it, for f with a finite number of discontinuities and satisfying certain regularity conditions. Their results which we need are given in Section 3 below. The remainder of this paper (i.e., its new contribution) is devoted to a further development of their results. Our further work proceeds for $R=(-r, r)$. We shall slightly modify the argument of [10], so that we will be able to verify conditions (2.2) and (2.3) below for our present R. *Hence the m.p. estimator is, under the conditions of* [10] *and the additional conditions of Section* 5 *below, asymptotically efficient with respect to* $R=(-r, r)$.

This R was chosen because a) it is extremely important b) the reader can readily see how to proceed for many other R or even a continuous loss function c) treating a general R would make the paper considerably more complicated. One should not, however, conclude that the theory can treat only this particular R.

The present paper requires no familiarity with m.p. estimators beyond what can be learned from Section 2. The results of [10] which we need are given in Section 3. The new results of the present paper are given in Section 4, with an outline of the proof. The reader interested only in these need read no further. In Section 5 are given the additional assumptions (beyond those of [10]) which are needed for Section 6. It is only for Sections 5 and 6 that familiarity with the details of [10] is required; Sections 1–4 can be read without any familiarity with [10]. The argument of the present paper is given in Section 6, and for its understanding the reader has to be acquainted with the details of [10]. It consists of a detailed examination of the proofs of [10], so as to obtain slightly stronger conclusions. These then suffice for the application of m.p. theory to obtain the stronger result about asymptotic efficiency. Our great debt to [10] is thus apparent.

In [10], Theorem 4.3, a comparison is made of the asymptotic efficiencies of a certain Bayes estimator, a certain m.p. estimator, and the m.l. estimator. The authors conclude that this Bayes estimator is efficient, and, by implication, that this m.p. estimator is not. The comparison and its conclusion are not correct. We shall return to it at the end of Section 2.

The author is grateful to Professor I. A. Ibragimov for valuable correspondence about his paper [10].

2. Introduction to m.p. estimators

For each positive integer n let $X(n)$ denote the (finite) vector of (observed) chance variables of which the estimator is to be a function. $X(n)$ need not have n components, nor need its components be independently or identically distributed. Let $K_n(x|\theta)$ be the density, with respect to a σ-finite measure μ_n, of $X(n)$ at the point x (of the appropriate space) when θ is the value of the (unknown to the statistician) parameter. The latter is a point of the known open set Θ. An estimator (of θ) is a Borel measurable function of $X(n)$ with values in Θ. Although it is not essential, in this paper we assume, for the sake of simplicity, that Θ is a subset of the real line.

For each n let $k(n)$ be a normalizing factor for the family $K_n(\cdot|\cdot)$. Let R be a bounded, Borel measurable subset of the real line. A maximum probability (m.p.) estimator (with respect to R) is one which maximizes, with respect to d,

$$\int K_n(X(n)|\theta)d\theta \,, \tag{2.1}$$

the integral being over the set $\{d-R/k(n)\}$. For simplicity we assume that there is a unique maximum. In [7] the authors discuss a) what to do when a maximum in Θ does not exist b) how to choose the estimator when there is more than one maximum (of course, any estimator which maximizes (2.1) has the property of asymptotic efficiency described below*) c) the case where R is not bounded and d) the case where

* Of course, when the conditions of the Theorem below are satisfied.

R is a function of θ (this is especially interesting when one is interested in relative, rather than absolute, errors of estimation).

Let $h>0$ be any number. We shall say that a sequence $\{\theta_n\}$ is in $H(h)$ (for a special point which below will always be θ_{00}) if $|k(n)(\theta_n-\theta_{00})| \leqq h$ for $n=1, 2, \cdots$. The symbols $P_{\theta'}$ and $E_{\theta'}$ are to denote, respectively, the probability (of an event) and expected value when θ' is the actual ("true") value of the parameter.

In 1967 (see [11] and [7]) L. Weiss and the present author proved the following

THEOREM. *Let M_n be an m.p. estimator with respect to R such that:*

(2.2) *For any $h>0$ and any sequence $\{\theta_n\}$ in $H(h)$, we have*

$$\lim_{n\to\infty} P_{\theta_n}\{k(n)(M_n-\theta_n)\in R\}=\beta(\theta_{00}) , \quad \textit{say} .$$

(2.3) *Let ε and δ be arbitrary but positive. For h sufficiently large we have, for any sequence $\{\theta_n\}$ in $H(h)$,*

$$\varliminf_{n\to\infty} P_{\theta_n}\{|k(n)(M_n-\theta_n)|<\delta h\}\geqq 1-\varepsilon .$$

Let T_n be any (competing) estimator such that

(2.4) *for any $h>0$ and any sequence $\{\theta_n\}$ in $H(h)$, we have*

$$\lim_{n\to\infty}[P_{\theta_n}\{k(n)(T_n-\theta_n)\in R\}-P_{\theta_{00}}\{k(n)(T_n-\theta_{00})\in R\}]=0 .$$

Then

$$\lim_{n\to\infty} P_{\theta_{00}}\{k(n)(T_n-\theta_{00})\in R\}\leqq \beta(\theta_{00}) . \tag{2.5}$$

The inequality (2.5) justifies the claim that the m.p. estimator is asymptotically efficient (with respect to R). A more general result was proved in [12] for a general loss function, not just the special 0-1 loss function implied by R.

Thus we see that an m.p. estimator is always defined with respect to a given R or loss function. It is meaningless to speak of an m.p. estimator unless R or the loss function are at least implicitly understood. It is for the statistician to choose R or the loss function, because different statisticians in general have different loss functions. Of course, the freedom to choose R is a very valuable one from the statistician's point of view. The m.p. estimator is asymptotically efficient with respect to the R (or loss function) for which it was constructed.

In [10], Theorem 4.3, the authors compare three classes of estimators:

1) Bayes estimators with respect to a given class of "smooth" a

priori distributions and loss functions W which are essentially as follows for small errors: when g is the "true" value of the parameter θ and h is the estimated value, the loss $W(h, g)=|h-g|^a$, for some given $a\geqq 1$.

2) The m.p. estimator with respect to $R=(-r, r)$.

3) The m.l. estimator.

The comparison is made as follows: Let T_n be any of the above estimators. Consider

$$\lim_{n\to\infty} n^a \mathrm{E}_\theta |T_n-\theta|^a . \tag{2.6}$$

Among the three classes of estimators that one is considered asymptotically efficient for which the quantity (2.6) is least (for all θ). It can scarcely come as a surprise to the reader that the Bayes estimator emerges as efficient.

The above comparison is very unreasonable. The Bayes estimator is constructed for the very loss function on the basis of which comparison is made, while the m.p. estimator is constructed for a completely different loss function. If the m.p. estimator were constructed for the loss function W then its value of (2.6) would be the least possible such value.

As for the m.l. estimator, we have remarked in Section 1 that, anyway, it is not, in general, efficient in the present case (roughly speaking, densities with discontinuities). Even so, extrapolating from the regular case (Fisher's theory) it would appear that the proper loss function for comparing it is the one implied by $R=(-r, r)$, not the one in (2.6). Although we have no intention of arguing for the efficiency of the m.l. estimator in this case (a lost cause to begin with), even so this particular comparison is unfair.

3. The needed results of Ibragimov-Hasminski

Let θ_{00} be the "true" value of the parameter θ. Write, for any real θ,

$$Z_n(\theta)=\prod_{j=1}^{n} \frac{f(X_j, \theta_{00}+\theta/n)}{f(X_j, \theta_{00})} , \tag{3.1}$$

where $X_1, X_2, \cdots$ are independent chance variables with the common density $f(\cdot\,|\theta_{00})$. Under certain conditions on f it is shown in [10] that the finite-dimensional distributions of Z_n and the distributions of certain functionals of Z_n approach the corresponding distributions generated by the stochastic process (θ real)

$$Z(\theta)=\exp\left\{\alpha_0\theta+\sum_{k=1}^{r} \alpha_k \nu_k(\theta)\right\} \tag{3.2}$$

where the α's are suitable functions of θ_{00}, and the $\nu_k(\cdot)$ are independent Poisson processes. We have

$$\nu_k(\theta)=\nu_k^+(\theta)+\nu_k^-(\theta) ;$$

ν_k^+ and ν_k^- are independent Poisson processes, $\nu_k^+(\theta)=0$, $\theta\leqq 0$, $\nu_k^-(\theta)=0$, $\theta\geqq 0$ (the process $\nu_k^-(\cdot)$ is monotonically non-increasing in θ), and

$$(3.3)\qquad \begin{aligned} \lambda_k^+ &= \mathrm{E}\,\nu_k^+(1)=p_k x_k'(\theta_{00}) &&\text{when } x_k'(\theta_{00})>0 ,\\ \lambda_k^+ &= \mathrm{E}\,\nu_k^+(1)=-q_k x_k'(\theta_{00}) &&\text{when } x_k'(\theta_{00})<0 ,\\ \lambda_k^- &= \mathrm{E}\,\nu_k^-(-1)=q_k x_k'(\theta_{00}) &&\text{when } x_k'(\theta_{00})>0 ,\\ \lambda_k^- &= \mathrm{E}\,\nu_k^-(-1)=-p_k x_k'(\theta_{00}) &&\text{when } x_k'(\theta_{00})<0 . \end{aligned}$$

The $\{p_k, q_k\}$ are suitable non-negative functions of θ_{00}, and r is the number of discontinuities x_k. Define

$$(3.4)\qquad Y^k(\theta)=\theta(p_k-q_k)x_k'(\theta_{00})+\mathrm{sign}\,(\theta\cdot x_k'(\theta_{00}))\log\frac{q_k}{p_k}[\nu_k^+(\theta)+\nu_k^-(\theta)] ,$$

with the stipulation that $0\cdot\infty=0$. Then

$$(3.5)\qquad Z(\theta)=\exp\left\{\sum_{k=1}^{r} Y^k(\theta)\right\} .$$

As we have remarked in Section 2, all the results of this paper will be with respect to $R=(-r, r)$. Let M_n be the m.p. estimator with respect to this R. Then the distribution of $n(M_n-\theta_{00})$ approaches the distribution of M, where M is the point at which the chance function φ,

$$(3.6)\qquad \varphi(x)=\int_{x-r}^{x+r} Z(\theta)d\theta ,$$

attains its maximum with respect to x.

Let $\hat{\theta}_n$ be the m.l. estimator of θ ($\hat{\theta}_n$ is a function of $X_1,\cdots,X_n$). Define $\hat{\theta}$ as the chance variable which satisfies

$$(3.7)\qquad \max\{Z(\hat{\theta}-0), Z(\hat{\theta}+0)\}=\sup_\theta Z(\theta) .$$

The joint distribution of $(n(M_n-\theta_{00}), n(\hat{\theta}_n-\theta_{00}))$ approaches the distribution of $(M, \hat{\theta})$.

All the results of this section are due to Ibragimov and Hasminski [10].

There are important cases for which, with probability one

$$(3.8)\qquad M_n=\hat{\theta}_n-\frac{r}{n}+o\left(\frac{1}{n}\right)$$

and consequently

$$M=\hat{\theta}-r\,. \tag{3.9}$$

If

$$2r<\frac{\min_k \log (p_k(\theta_{00})/q_k(\theta_{00}))}{\sum (p_k(\theta_{00})-q_k(\theta_{00}))}\,, \tag{3.10}$$

then (3.8) and (3.9) hold. If (3.10) does not hold then, with positive probability, (3.9) does not hold. If the numbers

$$\log \frac{p_k(\theta_{00})}{q_k(\theta_{00})}\,, \qquad k=1,\cdots,r,$$

are of the same sign or their absolute values are commensurable, there exists a positive number $S(\theta_{00})$ such that $2r<S(\theta_{00})$ implies that (3.8) holds with probability one.

4. The results of this paper. Preliminaries for the proofs

Let $h>0$ be any number which is fixed for the remainder of the argument. Let θ_{00} be the true value of θ. Let $\{\theta_n\}$ be any sequence such that $n|\theta_n-\theta_{00}|\leqq h$. (The normalizing factor $k(n)=n$.) In [10] the true value of θ is designated by θ_0. The reader will find it helpful to identify the θ_0 of [10] successively with the elements θ_n of the sequence $\{\theta_n\}$ which lies in $H(h)$. Henceforth we define $Z_n(\theta)$ not as in (3.1) but as

$$Z_n(\theta)=\prod_{j=1}^{n} \frac{f(X_j\,,\, \theta_n+\theta/n)}{f(X_j\,,\, \theta_n)}\,. \tag{4.1}$$

The distribution of $Z_n(\theta)$ for all θ is now determined not by θ_{00} but by θ_n; i.e., $X_1,\cdots,X_n$ are independent chance variables with the common d.f. $f(\cdot\,|\theta_n)$. Our object is to show that the results cited in the preceding section hold also with this new definition of Z_n, i.e., that the finite-dimensional distributions of Z_n and the distributions of certain functionals of Z_n approach the corresponding distributions generated by the stochastic processes $Z(\theta)$ as defined in (3.2).

Consequently it will follow from these results that, for our R, (2.2) and (2.3) are satisfied. *Hence M_n is asymptotically efficient with respect to R, according to the theorem of Section* 2.

A number of special results are proved in [6] by a different and simpler method. However, the results of [10] are general and elegant, and it seems to us interesting and worthwhile to show that one can obtain by this method not only conclusions about asymptotic distributions, but also about asymptotic efficiency. This is especially so because the needed modification of the argument is so small. The results on efficiency are obtained through application of the theory of m.p. estimators. In [6], the m.p. estimators are obtained in a particularly simple and explicit form. This is not possible for a general theory, except in special cases such as those described in the last paragraph of Section 4.

Up to this point no familiarity with the results of [10] was required. From now on this will no longer be the case. In the next section we examine the proofs of [10] so as to strengthen their conclusions for our purposes. We will use the following notation: Lemma "*n*" or Theorem "*n*" will always be from [10]. The new and corresponding lemmas and theorems will be numbered "*n*′." Our indebtedness to the methods of [10] is thus apparent. We will strengthen the discussed conclusions of [10] only slightly, but this will be sufficient for our purposes.

Our argument thus consists of 1) examining the proofs of [10] to see whether they yield the slightly stronger conclusions we need and 2) modifying the proofs when this is necessary. To avoid tediousness we will not always restate the lemmas and theorems of [10], which can be there consulted, and mostly confine ourselves to indicating the necessary changes in the argument, where this is needed and not obvious. The reader not interested in the proofs need read no further.

5. Assumptions

We postulate conditions I–V of [10], with the following additions:

(5.1) We require that the derivatives $x'_j(\cdot)$, $j=1,\cdots,k$, also be continuous.

(5.2) We require that the limit in III_2 of [10] be approached uniformly in every compact θ-set.

We have no need of condition VI of [10], because we make no use of Bayes solutions.

Henceforth, whenever we say that conditions I–V are fulfilled, it will always be understood that conditions (5.1) and (5.2) are included. Condition (5.1) is added to those in II and condition (5.2) is added to those in III.

For Theorem 4.2′ below and our efficiency result we also need that

$$c=\sum_{k=1}^{r}[p_k(\theta_{00})-q_k(\theta_{00})]x'_k(\theta_{00})\neq 0\ . \tag{5.3}$$

6. Analysis of the proofs of [10]

Throughout [10] the authors write p_k and q_k, short for $p_k(\theta_0)$ and $q_k(\theta_0)$. Now p_k and q_k are to be understood as follows: When we are dealing with Z_n, p_k and q_k are $p_k(\theta_n)$ and $q_k(\theta_n)$. When we are dealing with Z, p_k and q_k are $p_k(\theta_{00})$ and $q_k(\theta_{00})$. The same applies to λ^+, λ^-, and the x'_k.

The symbol $\bar{o}(\varepsilon)$ in what follows is to mean the following: $\bar{o}(\varepsilon)/\varepsilon \to 0$ as $\varepsilon \to 0$ and $n \to \infty$, independently of each other.

Lemma 2.1′ is exactly the same as Lemma 2.1, except for the trivial replacement of θ_0 by θ_{00}. (This is simply a change of notation.) Lemma 2.1 really deals with the process $Z(\cdot)$, which is the same for us as it is in [10].

Define $a_n(\sigma, \zeta)$ and $b_n(\tau, z)$ as $a(\sigma, \zeta)$ and $b(\tau, z)$ of [10], except that θ_0 is replaced by θ_n (a_n and b_n are the corresponding functions for the $Z_n(\cdot)$ process). Define $\psi_{\varepsilon,n}(s, t)$ as $\psi_\varepsilon(s, t)$ with θ_0 replaced by θ_n. Lemma 2.2′ states that, under conditions I–III, as $\varepsilon \to 0$,

$$\psi_{\varepsilon,n}(s, t) = 1 + \varepsilon[a_n(\sigma, s) + b_n(\tau, t)] + \bar{o}(\varepsilon) . \tag{6.1}$$

The proof proceeds as in [10] after the following lemmas are proved. We write $I_{1,n}$ and $I_{2,n}$ for I_1 and I_2 with θ_0 replaced by θ_n.

Lemma 2.3′ states that $I_{1,n}$ equals the expression for I_1 of Lemma 2.3, except that θ_0 is replaced by θ_n and $o(\varepsilon)$ by $\bar{o}(\varepsilon)$. The proof is essentially the same as that of Lemma 2.3, except that in line 1^-, page 561, one uses the continuity of $x'_j(\cdot)$. By "essentially the same" in this and subsequent arguments we do not preclude that minor (but not major) changes may be needed, and we leave these to the reader.

Lemma 2.4′, is the same as Lemma 2.4, except that θ_0 is replaced by θ_n and $o(\varepsilon)$ by $\bar{o}(\varepsilon)$. The proof is essentially the same as that of Lemma 2.4. Lemma 2.5′ has the same relationship to Lemma 2.5 as the previous Lemma 2.4′ to 2.4, and will be a consequence of Lemma 2.6′.

Lemma 2.6′ is obtained from Lemma 2.5 by replacing θ_0 by θ_n and $o(\varepsilon)$ by $\bar{o}(\varepsilon)$. Its proof follows that of Lemma 2.6 until line 5 on page 563. For the last equality on this line we use our Assumption III_2. In line 7 replace $|\theta - \theta_0| < \varepsilon$ by $|\theta - \theta_0| < \varepsilon + h/n$.

Lemma 2.7′ is also obtained from Lemma 2.7 by replacing θ_0 by θ_n and $o(\varepsilon)$ by $\bar{o}(\varepsilon)$. The following changes have to be made in the argument: On page 564, in lines 10^- and 9^- we have $\bar{o}(\varepsilon)$ from Lemma 2.5′. In lines 7^- and 6^- we have $\bar{o}(\varepsilon)$. In line 4^- factors s_j and t_i have been omitted in [10] and the constant C may have to be enlarged. Elsewhere in the argument we have $\bar{o}(\varepsilon)$ for $o(\varepsilon)$, and this is true in

(2.15) because R is bounded. This proves Lemma 2.7′, and hence Lemma 2.2′.

Theorem 2.1′, which is the same as Theorem 2.1 (bearing in mind the new definition of $Z_n(\cdot)$), is now proved essentially as Theorem 2.1 was proved.

Theorem 2.2′, which is the same as Theorem 2.2 except for replacing θ_0 by θ_n, is now proved in essentially the same way. The constant C of (2.19) may have to be enlarged.

The corollary to Theorem 2.2 now holds with the following changes: In the left member of its inequality, $Z_n(\cdot)$ has the definition of the present paper. In the right member, θ_0 is replaced by θ_{00} and C is enlarged.

The equation

$$\int \sqrt{f(x, \theta_n) f(x, \theta_n+\varepsilon)}\, dx = 1 - a\varepsilon + \bar{o}(\varepsilon), \qquad a>0, \tag{6.2}$$

can be verified without much difficulty. Using it and proceeding as in Lemma 2.7′ of [2] we obtain

LEMMA 2.8′. *Let $K>0$ be arbitrary, and postulate conditions* I–III. *Then there exists a number $C>0$ such that for $|\theta|<Kn$,*

$$P_{\theta_n}\{Z_n(\theta) > e^{-C|\theta|}\} \leqq e^{-C|\theta|}. \tag{6.3}$$

The following lemma is proved essentially as Lemma 2.8 of [1], with the obvious changes required by the replacement of θ_0 by θ_n:

LEMMA 2.9′. *Suppose conditions* I–V *to be fulfilled. Then for every $N>0$ there exists a number $K>0$, such that, for all $n>n_0(N)$ and $|\theta|>Kn$,*

$$P_{\theta_n}\{Z_n(\theta) > |\theta|^{-N}\} \leqq |\theta|^{-N}. \tag{6.4}$$

From Lemmas 2.8′ and 2.9′ we obtain

LEMMA 2.10′. *Suppose conditions* I–V *hold. For each $N>0$ there exists a constant C_N such that, for all $n>n_0(N)$,*

$$P_{\theta_n}\{Z_n(\theta) \geqq |\theta|^{-N}\} \leqq C_N |\theta|^{-N}. \tag{6.5}$$

The remainder of the complicated proof of Theorem 2.3 in [10] now applies with the obvious modifications, and we have

THEOREM 2.3′. *Suppose conditions* I–V *are fulfilled. For any $N>0$ there exists a constant C_N, which depends only on N, such that for $n>n_0(N)$ and for all $A>0$,*

(6.6) $$P_{\theta_n}\{\sup_{|\theta|>A} Z_n(\theta)>A^{-N}\}\leqq C_N A^{-N}.$$

Let $D=\{f\}$ denote the set of all functions which a) are defined on the entire real line, b) have limits from the right and left at every point, c) are right-continuous, and d) satisfy $\lim f(x)=0$ as $|x|\to\infty$. On D define the Skorokhod metric:

$$\rho(f, g)=\inf_{\lambda}\,[\sup_x |f(x)-g(\lambda(x))|+\sup_x |x-\lambda(x)|]\,,$$

where the infimum is taken over all monotonic, continuous, one-to-one functions λ from the real line to the real line. Let $\mathscr{D}$ denote the set of all functionals, continuous on D.

THEOREM 2.4′. *Suppose conditions* I–V *are fulfilled. For any functional* $\varphi\in\mathscr{D}$ *the distribution of the chance variable* $\varphi(Z_n)$ *approaches the distribution of the chance variable* $\varphi(Z)$.

The proof of this theorem is essentially the same as that of Theorem 2.4, making use of the primed lemmas above.

Finally, the only other result we need is the following, which corresponds to a part of Theorem 4.2:

THEOREM 4.2′. *Suppose that conditions* I–V *and* (5.3) *are fulfilled. Then*

$$\lim_{n\to\infty} P_{\theta_n}\{n(M_n-\theta_n)<y\}=P_{\theta_{00}}\{M<y\}$$

for any real y.

This result is proved in essentially the same way as the corresponding part of Theorem 4.2.

UNIVERSITY OF ILLINOIS

REFERENCES

[1] Ibragimov, I. A. and Hasminski, R. A. (1972). Asymptotic behavior of some statistical estimators in the smooth case, *Teor. Vyeroyat. Primen.*, XVII, 469-486.

[2] Wolfowitz, J. (1975). Maximum probability estimators in the classical case and in the "almost smooth" case, *Teor. Vyeroyat. Primen.*, XX, 371-379.

[3] LeCam, L. (1960). Locally asymptotically normal families of distributions, *Univ. Calif. Pub. Statist.*, **3**, 37-98.

[4] Bahadur, R. R. (1964). On Fisher's bound for asymptotic variances, *Ann. Math. Statist.*, **35**, 1545-1552.

[5] Ibragimov, I. A. and Hasminski, R. Z. (1973). Asymptotic analysis of statistical estimators for the "almost smooth" case, *Teor. Vyeroyat. Primen.*, XVIII, 250-260.

[6] Weiss, L. and Wolfowitz, J. (1970). Maximum probability estimators and asymptotic sufficiency, *Ann. Inst. Statist. Math.*, **22**, 225-244.

[7] Weiss, L. and Wolfowitz, J. (1974). Maximum probability estimators and related topics, *Lecture Notes in Mathematics*, No. 424, Springer-Verlag, Berlin-Heidelberg-New York.

[8] Weiss, L. and Wolfowitz, J. (1966). Generalized maximum likelihood estimators, *Teor. Vyeroyat. Primen.*, XI, 68-93.

[9] Weiss, L. and Wolfowitz, J. (1968). Generalized maximum likelihood estimators in a particular case, *Teor. Vyeroyat. Primen.*, XIII, 657-662.

[10] Ibragimov, I. A. and Hasminski, R. Z. (1972). Asymptotic behavior of statistical estimators for samples with discontinuous density, *Mat. Sbornik*, 187(129), 554-586.

[11] Weiss, L. and Wolfowitz, J. (1967). Maximum probability estimators, *Ann. Inst. Statist. Math.*, **19**, 193-206.

[12] Weiss, L. and Wolfowitz, J. (1969). Maximum probability estimators with a general loss function, Proc. Int. Symp. on Prob. and Information Theory, held at McMaster University, Hamilton, Ontario, Canada, 1968, *Lecture Notes in Mathematics*, No. 89, Springer-Verlag, Berlin-Heidelberg-New York, 232-256.

Reprinted from JOURNAL OF COMBINATORICS, INFORMATION & SYSTEM SCIENCES Vol. 4, No. 2, 117–122 (1979)

On List Codes

J. WOLFOWITZ*

Department of Mathematics, University of South Florida, Tampa, FL 33620 USA

1. INTRODUCTION

In the present paper we prove a new coding theorem and strong converse for list codes for a discrete memoryless channel (d.m.c.). When properly specialized these imply as consequences, inter alia, the Shannon coding theorem and the strong converse for a d.m.c., both in the stronger form ([4], [2], [3]) with $0\left(\frac{1}{\sqrt{n}}\right)$ in the exponent. The results can be extended to other channels. The idea of list codes is due to Elias [1].

A rough statement of both results is this: Let N be the code length and L the maximum decoding list length which is permitted to the decoder. Let $I(X, Y)$ denote, as usual, the amount of information between the input and output variables. Let n be the word length (block length). Then

$$\frac{1}{n}[\log N - \log L] \sim I(X, Y).$$

Precise definitions and statements are given below.

The two principal tools used are the idea of π-sequences and generated sequences, and the Shannon method of random codes ([2] or [3], Section 7.3), applied not in the usual way but on the π-sequences. The first idea was introduced and developed in [4], and is the principal tool in [2] and [3]. The notion of π-sequences has been regularly rediscovered (e.g., [6], [7], [8]) by various engineering writers, who call them "typical" sequences. (This is a more felicitous name when there is no doubt about the distribution of which they are typical. This need not always be the case and then one needs to call the sequences π-typical, etc.) Some Soviet writers use the term "fan" for the set of generated sequences. The reader unfamiliar with π-sequences and generated sequences can learn more than he needs to know for understanding the present paper by reading the short Section 2.1 of [2] or [3]. Familiarity with this section is henceforth assumed, and its notation and terminology are adopted henceforth.

Research under grant No. MCS 78–02148 from the National Science Foundation (USA).

Except where otherwise stated, the results below are intended to apply to a d.m.c., whose channel probability function (c.p.f.) will be denoted by $w(\cdot|\cdot)$. The probability is $w(j \mid i)$ that, when the letter i is sent over the channel, the letter j will be received.

2. Preliminaries

Let $A = \{1, \ldots, a\}$ and $B = \{1, \ldots, b\}$ be, respectively, the sender's and receiver's alphabets. Let A_n^* denote the Cartesian product of n A's, and similarly for B_n^*. The symbol n will denote the number of letters in the message sequences of the sender and the received sequences of the receiver (the "block length"). X is a chance variable with values in A and distribution π. Y is a chance variable with values in B. The conditional distribution of Y, given that $X = i$, is $w(\cdot \mid i)$. $H(X)$ and $H(Y)$, as usual, denote the respective entropies, and $H(X \mid Y)$ and $H(Y \mid X)$ the respective conditional entropies. Also as usual,

$$I(X, Y) = H(X) - H(X \mid Y) = H(Y) - H(Y \mid X) = I(Y, X).$$

K will always denote a suitable positive constant, and K's in different places will, in general, not be the same. In the definitions of π-sequences and generated sequences there occur certain multiplicative constants. In the present paper we shall not be limited to the particular values of these constants given in [2] or [3], Section 2.1, but will always choose the values of these constants in a way suitable for the result desired. The choice will always be obvious to the reader. For example, one can always choose the constant in the definition of generation so that the probability of being generated is as close to one as desired. Finally, when the sequence u of n letters is sent over the channel, the chance sequence received will be denoted by $v(u)$.

A list code (n, d, l, λ) is a system $\{(u_1, A_1), \ldots, (u_N, A_N)\}$ such that the "message sequences" u_i are points in A_n^*, $N = \exp_2 \{nd\}$, and each $A_i \subset B_n^*$. Suppose the sender sends u_i, so that the receiver receives $v(u_i)$. The "decoding list" of $v(u_i)$ is the set of all u_j such that $v(u_i) \in A_j$. The "length" of the list is the number of sequences in the set; denote it by $L(v(u_i))$. A "decoding error" is said to occur if either of the following takes place:

(a) u_i is not on the decoding list.

(b) $L(v(u_i)) > \exp_2 \{nl\}$.

For every i, $i = 1, \ldots, N$, the probability of a decoding error is less than λ. Of course, $0 < \lambda < 1$. N is called the code length, and $\exp_2 \{nl\}$ is the maximum decoding list length permitted to the decoder.

A list code $(n, d, 0, \lambda)$ is essentially an ordinary code. We say "essentially" because the A_i need not be disjoint. When the received sequence falls into more than one decoding set, an error occurs. However, the operation of the code is really little different from that of an ordinary code. (See Section 4, Part 1).

3. The Coding Theorem for List Codes

Let $d > 0$ and λ, $0 < \lambda < 1$, be arbitrary. For n sufficiently large there exists a list code (n, d, l, λ), say $\{(u_1', A_1'), \ldots, (u_N', A_N')\}$, all of whose message sequences u_i' are π-sequences, such that

$$l = \max\left\{\left[d - I(X, Y) + \frac{m}{\sqrt{n}}\right], 0\right\},$$

and $m > 0$ is a number which can be determined from the proof. Each A_i' consists of all sequences generated by u_i'.

Proof. We construct a random code of length $2N = \exp_2\{nd + 1\}$ as follows: The message sequences $u_1, \ldots, u_{2N}$ are independent chance variables, each uniformly distributed on the π-sequences, of which there are at least $\exp_2\{nH(X) - K\sqrt{n}\}$ (Lemma 2.1.7 of [2] or [3]). Each decoding set A_i is to consist of all sequences generated by u_i.

Let $y \in B_n^*$ be a member of A_i. Let $M(y)$ be the number of A_j to which y belongs. Then

$EM(y) \leqslant 1 +$ (upper bound on the number of π-sequences which generate y) $\times$ (upper bound on the probability that u_1 is any one particular π-sequence) $\times \exp_2\{nd + 1\}$

$$< 1 + \exp_2\{nH(X \mid Y) + K\sqrt{n}\} \times \exp_2(-nH(X) + K\sqrt{n}\} \times \exp_2\{nd + 1\}$$

The first factor of the second term of the right member comes from Lemma 6.2 of [4], and the second factor from Lemma 2.1.7 of [2] or [3]. Hence

$$EM(y) < 1 + \exp_2\{n[d - I(X, Y)] + K\sqrt{n}\}.$$

Suppose first that $d \geqslant I(X, Y)$. Then, by Markov's inequality, we can easily choose an $m > 0$ such that, for all n large enough, the probability that

$$M(y) > \exp_2\{n[d - I(X, Y)] + m\sqrt{n}\}$$

is as small as desired. Also, by proper choice of the constant in the definition of generation we can make the probability that $v(u_i)$ is not in A_i as small as desired.

Let λ_i be the probability of a decoding error when u_i is sent. Since u_i is a chance variable so is λ_i. The argument of the last paragraph shows that

$$\frac{1}{2N} \sum_{i=1}^{2N} E\lambda_i$$

can be made as small as desired, in particular, less than $\frac{\lambda}{2}$. Hence there exists a set $\{u_1^0, \ldots, u_{2N}^0\}$ of points in A_n^* with the following property: Let $\lambda(u_i^0)$ be the decoding error when u_i^0 is sent and the message sequences of the code are $u_1^0, \ldots, u_{2N}^0$. Then

$$\frac{1}{2N} \sum_{i=1}^{2N} \lambda(u_i^0) < \frac{\lambda}{2}.$$

(Recall that the corresponding decoding sets A_i^0 consist of all sequences generated by u_i^0.) Hence there exists a subset of N points of $u_1^0, \ldots, u_{2N}^0$, say $u_1', \ldots, u_N$, such that

$$\lambda(u_i') < \lambda, \; i = 1, \ldots, N.$$

This proves the coding theorem when $d \geqslant I(X, Y)$.

Suppose now that $d < I(X, Y)$. Then, for all n large enough, the probability that y, which belongs to some A_i, should also belong to another A_i, can be made as small as desired. The remainder of the proof is now as before. The proof of the theorem is complete.

4. Some Consequences of the Coding Theorem

(1) Let $d < I(X, Y)$. We obtain the Shannon coding theorem. Indeed, let $d = I(X, Y) - \frac{K}{\sqrt{n}}$, where $K > 0$ is properly chosen. We obtain the Shannon coding theorem with $0\left(\frac{1}{\sqrt{n}}\right)$ in the exponent of the length, as in [4], Theorem 1, or [2] or [3], Theorem 3.2.1.

The A_i obtained from Theorem 1 are not disjoint. If the received sequence lies in more than one A_i the receiver declares a decoding error. It is well known ([4], Theorem 1, or [2] or [3], Theorem 3.2.1) that a code satisfying the Shannon coding theorem, even with $0\left(\frac{1}{\sqrt{n}}\right)$ in the exponent of the length, can be achieved with the u_i all π-sequences and the disjoint A_i containing only sequences generated by u_i. To make our present A_i disjoint simply delete from all A_i any sequence which belongs to more than one A_i. This shows that the idea of list codes is the more general and includes that of ordinary codes.

(2) Let $d = H(X)$. Our coding theorem implies the result of Ahlswede ([5], Lemma 1). If, instead of using Lemma 6.2 of [4], we use the corresponding upper bound for arbitrary varying channels ([3], Chapter 6) and compound channels ([2] or [3], Chapter 4) respectively, we obtain Lemma 2 and (2.1) of [5].

5. Strong Converse of the Coding Theorem for List Codes

Let $\{(u_1, A_1), \ldots, (u_N, A_N)\}$ be a list code (n, d, l, λ), such that each u_i is a π-sequence. Then, if n is sufficiently large,

$$l > d - I(X, Y) - \frac{K}{\sqrt{n}}.$$

The limitation to sequences u_i which are π-sequences is not a drawback, because any code can be decomposed into subcodes of "constant composition". (See proof of Theorem 3.3.1 in [2] or [3], or proof of Theorem 2 in [4].)

Proof. Let $s > 1$ be such that $(s - 1)$ is sufficiently small. From each A_i delete the sequences not generated by u_i; call the result A_i'. If the multiplicative constant in the definition of generation is properly chosen, which we henceforth assume to be the case, then the system $\{(u_1, A_1'), \ldots, (u_N, A_N')\}$ is a list code $(n, d, l, s\lambda)$. Now, from each A_i', delete any sequence z such that $L(z) > \exp_2 \{nl\}$, and call the result A_i''. Since the probability of any such sequence z has already been counted in the probability of error of the given code when u_i is sent, the system $\{(u_1, A_1''), \ldots, (u_N, A_N'')\}$ is a list code $(n, d, l, s\lambda)$.

By Lemma 2.1.5 of [2] or [3], the total number of sequences in $\bigcup A_i''$ is greater than

$$\exp_2 \{n[d + H(Y \mid X)] - K\sqrt{n}\}.$$

(Each sequence in $\bigcup A_i''$ is counted as many times as it occurs). Also, by Lemma 2.1.4 of [2] or [3], the total number of sequences in $\bigcup A_i''$ is less than

$$\exp_2 \{n[H(Y) + l] + K\sqrt{n}\}.$$

Hence, when n is sufficiently large,

$$H(Y) + l > d + H(Y \mid X) - \frac{K}{\sqrt{n}},$$

so that then

$$l > d - I(X, Y) - \frac{K}{\sqrt{n}},$$

as was to be proved.

If we set $l = 0$ we immediately obtain the strong converse, indeed the strong converse with $0\left(\frac{1}{\sqrt{n}}\right)$ in the exponent (Lemma 3.3.1 and Theorem 3.3.1 of [2] or [3]). Our theorem also shows that the result in Lemma 1 of [5], cited in our Section 4 above, is, to within terms of order $0\left(\frac{1}{\sqrt{n}}\right)$ in the exponent, the best possible.

The author had interesting conversations with Mr. N. B. Sachs in the course of preparation of this paper.

References

[1] P. Elias (1955), "List decoding for noisy channels", Technical Report 335, Research Laboratory of Electronics, Mass. Inst. Tech., Cambridge, Mass.

[2] J. Wolfowitz, "Coding theorems of information theory", Springer-Verlag. New York and Heidelberg. (First edition, 1960, second edition, 1964).

[3] J. Wolfowitz, Third, revised and enlarged edition of [2], 1978.

[4] J. Wolfowitz (1957), "The coding of messages subject to chance errors", *Illinois Jour. Math.*, **1**(4), 591–606.

[5] R. Ahlswede (1973), "Channel capacities for list codes", *Jour. App. Prob.*, **10** 824–836.

[6] J. K. Wolf (1977), "The AEP property of random sequences and applications to information theory" Parts I, II and III, in "*Information Theory: New Trends and Open Problems*", ed. G. Longo, Springer-Verlag, Heidelberg and New York, pp. 125–171.

[7] G. D. Forney (Jr.) (Winter 1972), "Information theory", Stanford University Course Notes.

[8] A. D. Wyner and J. Ziv (January, 1976), "The rate distortion function for source coding with side information at the decoder", *IEEE Trans. on Information Theory*, IT-22, No. 1, 1–10.

Received : May, 1978

Z. Wahrscheinlichkeitstheorie verw. Gebiete
46, 307 - 315 (1979)

Zeitschrift für
Wahrscheinlichkeitstheorie
und verwandte Gebiete

Codes within Codes

J. Wolfowitz*

Department of Mathematics, University of South Florida, Tampa, Florida 33620, USA

Summary. We prove coding theorems for list codes for compound channels and for codes within codes. The theorems imply corresponding results for what are usually called simply "codes", which are list codes where one is the maximum decoding list length permitted to the decoder. The Bergmans coding theorem for degraded channels ([4], Theorem 15.2.1) and the positive part of the Wyner-Ziv theorem ([4], Theorem 13.2.1) are easy consequences of our results.

1. Introduction

Let X_1, X_2, X_3 be chance variables as follows: X_i takes values in the set $M_i = \{1, 2, \ldots, a_i\}$. Write M_{in} for the Cartesian product of M_i by itself n times. The distribution of X_i is $\pi_i(\cdot)$, and the conditional distribution of X_i, given $X_j = x$, is $w_{i|j}(\cdot|x)$. The entropy symbol H is used in the standard way, and so is the symbol I for the amount of information. Some familiarity with Section 2.1 of [4] is assumed. The notation will be somewhat different, but the differences will be explained below. As in [3] and the latter parts of [4], the constants used in the definition of π-sequences and of generation are always to be chosen suitably, and are not limited to those given in Chapter 2 of [4]. The choice of these constants will always be obvious.

In all that follows K, with or without a subscript, will always denote a suitable positive constant. The same K in different places need not be the same. All our results will be valid only for sufficiently large n; this is always to be understood even if not explicitly stated. Our channels will always be discrete memoryless channels (d.m.c.'s), even if this is not explicitly stated. The channel probability function (c.p.f.) will usually be obvious, but it may be given anyway. For example, for a code $X_a \to X_b$, the c.p.f. is $w_{b|a}(\cdot|\cdot)$.

In Section 3 the same notation is used for more than three variables X. Its use will be obvious.

---

* Research supported by the National Science Foundation under Grant No. MCS 78-02148

2. List Codes

All codes in this paper (unless the contrary is explicitly stated) will be list codes, which are a generalization of, and include, the codes defined and studied in [4]. We now describe list codes and state the essential results from [3].

A list code $X_a \to X_b$, say, with parameters

$$(\pi_a, n, N, \lambda, l) \tag{2.1}$$

(the notation differs slightly from that of [3]) is a set of π_a-sequences in M_{an}, say

$$\{u_1, \dots, u_N\}. \tag{2.2}$$

These are the message sequences of the code. Let G_i be the set of sequences in M_{bn} generated by u_i; G_i is the "decoding set" for u_i. Let v_0 be any sequence in M_{bn}. The "decoding list" of v_0 is the set of all u_j's such that v_0 lies in G_j. The "length" of the list is the number of sequences in the set; denote it by $L(v_0)$. When u_i is sent and v_0 is received, a "decoding error" is said to occur if either of the following takes place:

a) u_i is not on the decoding list

b) $L(v_0) > \exp_2\{nl\}$. (2.3)

For every i, $i=1, \dots, N$, the probability of a decoding error is less than λ. Of course, $0<\lambda<1$. N is called the code length, and $\exp_2\{nl\}$ is the maximum decoding list length permitted to the decoder.

When $l=0$ we could try to use the above code as if it were an ordinary code as defined in [4]. The obstacle would be that the G_i are not disjoint. If, from each G_i, we delete all sequences which belong to any G_j, $j \neq i$, and call the result G'_i, the G'_i are disjoint. Let $v(u_i)$ be the chance received sequence when u_i is sent over the channel. Then

$$P\{v(u_i) \in G'_i\} > 1-\lambda \tag{2.4}$$

by definition of a list code for $l=0$. Thus $\{(u_1, G'_1), \dots, (u_N, G'_N)\}$ is a code like those of [4], with parameters (n, N, λ).

This last code is still somewhat special because of the composition of the G'_i and the character of the u_i. Nevertheless, the Shannon coding theorem is proved in [3] even for this special code.

The following coding theorem for list codes is proved in [3] and will be used by us as a tool: Let $d>0$ and λ, $0<\lambda<1$, be arbitrary. For n sufficiently large there exists a list code $X_a \to X_b$ with parameters

$$(\pi_a, n, \exp_2\{nd\}, \lambda, l) \tag{2.5}$$

such that

$$l = \max\left\{\left[d - I(X_a, X_b) + \frac{m}{\sqrt{n}}\right], 0\right\}, \tag{2.6}$$

and $m>0$ is a number which can be determined from the proof.

If, in the above formula for l, we set $d<I(X_a, X_b)-\frac{m}{\sqrt{n}}$, we obtain the Shannon coding theorem in a slightly stronger form.

The following strong converse is proved in [3]: Consider a list code $X_a \to X_b$ with parameters (2.5). This time the decoding set for u_i, $i=1, \ldots, \exp_2\{nd\}$, need *not* consist of all sequences generated by u_i. (This makes our result more general, just as requiring the decoding sets in the coding theorem to consist of all sequences generated by u_i made that result stronger.) Then, for n sufficiently large, we have

$$l>d-I(X_a, X_b)-\frac{K}{\sqrt{n}}. \tag{2.7}$$

Whenever we discuss list codes it will always be assumed, unless the contrary is explicitly stated, that each decoding set consists of all sequences generated by the message.

3. Compound List Codes

Compound channels are studied in Chapter 4 of [4]. Here we want to find the capacity of a compound channel under list coding, with the list codes as described in Section 2. Let d always be $\frac{1}{n}\log_2 N$, where N is the length of the code. Let $X_1, X_2, \ldots, X_c$ be the set of chance variables X. The number of c.p.f.'s in the compound channel will be $c-1$. This is a different case from the compound channels of Chapter 4 of [4], which may contain infinitely many c.p.f.'s. Applying the method of proof of the coding theorem in [3], we shall obtain the following result.

Theorem 1. *Let n be sufficiently large and λ, $0<\lambda<1$, be arbitrary. Call the channel with c.p.f. $w_{i|1}(\cdot|\cdot)$ the ith channel, $i=2, \ldots, c$. There exists a set of π_1-sequences*

$$\{u_1, \ldots, u_N\} \tag{3.1}$$

such that they, together with the decoding sets, the jth of which consists of all the sequences generated $X_1 \to X_i$ by u_j, constitute a list code for the ith channel with parameters

$$(\pi_1, n, N, \lambda, l_i), \tag{3.2}$$

where

$$l_i=\max\left\{\left[d-I(X_1, X_i)+\frac{m}{\sqrt{n}}\right], 0\right\}, \tag{3.3}$$

and $m>0$ is a constant which may be determined from the proof.

The receiver (decoder) does not know which c.p.f. is being employed by the sender or by whatever mechanism governs the transmission of a word. Since n is sufficiently large, however, we can "waste" the first $\sqrt[4]{n}$ letters by always using them to send a prearranged sequence. The first $\sqrt[4]{n}$ letters of the received sequence can then tell the receiver, with probability as close to one as desired for n sufficiently large, which c.p.f. is being used for transmission. The remaining $(n-\sqrt[4]{n})$ letters are used to send and decode the message being sent. The decoding list is made up according to the decision on which c.p.f. is being used. (In this connection see Theorems 4.5.1 and 4.5.2 of [4], and the heuristic explanation of them in the last paragraph on page 41 of [4].)

To prove the theorem we have to show that the probability of error, when any of the $(c-1)$ c.p.f.'s is used and any message is sent, can be made arbitrarily small. This is done by a method very similar to the method used in the proof of the coding theorem of [3]. Let $M_i(y)$, $i=2, \ldots, c$, correspond to $M(y)$ of [3] for the ith c.p.f. Suppose the channel to be used is chosen at random with probability $\frac{1}{c-1}$ for each. Then, by an argument almost the same as that of the coding theorem of [3], one obtains that, when the constant m is properly chosen,

$$\frac{1}{c-1} \sum_{i=2}^{c} P\{M_i(y) > \exp_2\{n[d-I(X_1, X_i)] + m\sqrt{n}]\} < \varepsilon \tag{3.4}$$

where $\varepsilon > 0$ is arbitrarily small. As in [3], one ends up with a (deterministic) code of length N with the following property: Let u_j be a message sequence of the code, and $\lambda_i(u_j)$ the error of decoding when u_j is sent and channel i is used for transmission. Then

$$\frac{1}{c-1} \sum_{i=2}^{c} \lambda_i(u_j) < 2\varepsilon, \qquad j=1, \ldots, N. \tag{3.5}$$

But then

$$\lambda_i(u_j) < 2(c-1)\,\varepsilon, \qquad i=2, \ldots, c; \qquad j=1, \ldots, N. \tag{3.6}$$

This proves Theorem 1.

Corollary 1. *Let X_1, X_2, X_3, in this order, form a Markov chain. Let $0 < \lambda < 1$. For all n sufficiently large there exists a set of message sequences*

$$\{u_1, \ldots, u_N\}$$

with

$$\frac{1}{n} \log_2 N = I(X_1, X_3) - \frac{m}{\sqrt{n}},$$

which are the message sequences of a compound channel with two c.p.f.'s, $w_{2|1}$ and $w_{3|1}$. The parameters of the code are, for each channel,

$(\pi_1, n, N, \lambda, 0)$,

so that this code acts essentially like the codes of Chapter 4 of [4].

Proof. From the inequality at the bottom of page 128 of [4] we have that

$$I(X_1, X_3) \leqq I(X_1, X_2).$$

The corollary follows from this and Theorem 1.

The corollary can be immediately extended to more than three chance variables. This is true of the next corollary as well.

Corollary 2. *Let X_1, X_2, X_3, in this order, form a Markov chain. If the message sequence is sent to a first receiver using c.p.f. $w_{2|1}$, and the sequence received is then sent to a second receiver using the c.p.f. $w_{3|2}$, Theorem 1 holds with $c=3$.*

This is so because the c.p.f. $w_{3|1}$ is the c.p.f. of the sequence received by the second receiver, when transmission proceeds as described in the hypothesis of the corollary.

4. Codes within Codes

Theorem 2. *Let X_1, X_2, X_3, in this order, form a Markov chain. Let n and K_1 (below) be sufficiently large. Let*

$$\{u_1, \ldots, u_N\} \tag{4.1}$$

be the message sequences of a code with parameters

$$(\pi_1, n, N, \lambda_1, 0) \tag{4.2}$$

for the c.p.f. $w_{3|1}$, with

$$N = \exp_2 \{nI(X_1, X_3) - K_1 \sqrt{n}\}. \tag{4.3}$$

Then, for any $\lambda_2 > \lambda_1$ and for every u_i, $i=1, \ldots, N$, there exists a code with parameters

$$(w_{2|1}(\cdot \mid u_i), n, T, \lambda_2, l) \tag{4.4}$$

for the c.p.f. $w_{3|2}$. Here the first parameter in (4.4) *means that the message sequences of the code are generated by u_i. Denote these message sequences by*

$$\{s_1(u_i), \ldots, s_T(u_i)\}, \qquad i=1, \ldots, N. \tag{4.5}$$

The relation between l and T is given by

$$l = \max\left\{\left[\frac{1}{n}\log_2 T - I(X_2, X_3 \mid X_1) + \frac{K_2}{\sqrt{n}}\right], 0\right\}. \tag{4.6}$$

The decoding set for the message sequence $s_j(u_i)$ consists of all sequences generated by $s_j(u_i)$ with respect to the c.p.f. $w_{3|2}$. The receiver does not know to which u_i the message $s_j(u_i)$ "belongs".

Proof. For each u_i the existence of a code (4.5) which satisfies (4.4) and (4.6) is proved almost exactly as in the proof of the coding theorem of [3]. Thus, *when the receiver knows* u_i, he can decode among the sequences (4.5) with suitably small probability of error and suitable decoding list length. However, the receiver does not know u_i, and thus the union, over all i, of the sequences (4.5) enters into the decoding competition. This is the problem that must be resolved by proper choice of the sequences (4.5).

Let (X_1^i, X_2^i, X_3^i), $i=1, \ldots, n$, be independent chance variables with the same distribution as (X_1, X_2, X_3). Write $X_{jn}=(X_j^1, X_j^2, \ldots, X_j^n)$, $j=1, 2, 3$. We also choose the multiplicative constants in the definitions of generation so that each of the following is true:

$$P\{X_{2n} \text{ is generated by } x_1 \,|\, X_{1n}=x_1\}>1-\frac{\varepsilon}{4}, \tag{4.7}$$

$$P\{X_{3n} \text{ is generated by } x_1 \,|\, X_{1n}=x_1\}>1-\frac{\varepsilon}{4}, \tag{4.8}$$

$$P\{X_{3n} \text{ is generated by } x_2 \,|\, X_{2n}=x_2\}>1-\frac{\varepsilon}{4}. \tag{4.9}$$

The numbers ε and δ are small positive numbers to be chosen suitably in a manner which will be obvious below.

For each u_i let $G_i \subset M_{3n}$ be the set of sequences generated $X_1 \to X_3$ by u_i which correctly decode u_i. Thus the various G_i are disjoint. The "random walk" from u_i to $x_3 \in G_i$ is "equivalent" to a random walk from u_i to some $x_2 \in M_{2n}$ and then from x_2 to x_3. Let t_i be the probability, according to $w_{2|1}(\cdot|u_i)$, of the set $\bar{S}_i$ of points $x_2 \in M_{2n}$ such that the probability $\gamma_i(x_2)$, according to $w_{3|2}(\cdot|x_2)$, of moving from x_2 into G_i is less than $1-s\lambda_1$, where $s=1+\delta$. Then

$$(1-t_i)+t_i(1-s\lambda_1)>1-\lambda_1, \tag{4.10}$$

so that

$$1-t_i>\frac{\delta}{1+\delta}. \tag{4.11}$$

Let $S_i \subset M_{2n}$ consist of all points not in $\bar{S}_i$ which are generated $X_1 \to X_2$ by u_i. Then

$$P\{X_{2n} \in S_i | X_{1n}=u_i\}>\frac{\delta}{1+\delta}-\frac{\varepsilon}{4}>0 \tag{4.12}$$

for small enough ε.

We now go back to the beginning of this proof. For each u_i the existence of a code (4.5) with elements in S_i which satisfies (4.4) and (4.6) except possibly for the bound on the maximum error is proved almost exactly as in the proof of the coding theorem of [3]. If the receiver of the messages of this code knows u_i, he can decode among the sequences (4.5) with suitably small probability of error and suitably small decoding list length. It remains to show that he can also decode the message u_i with suitably large probability of being correct.

The message sequence actually sent was a member of the union of (4.5) over all i. Suppose it was actually a member of (4.5) for u_i. The probability that X_{3n} be in G_i is $\geqq 1-s\lambda_1$, since, for given i all $s_i(u_i)$ are in S_i. Thus the receiver decodes u_i correctly with a probability only slightly less than $1-\lambda_1$. Once u_i has been correctly decoded $s_j(u_i)$ can be decoded with probability of error as small as desired and suitable decoding list length, by the construction of the code (4.5). This proves the theorem.

This proof shows the great advantage of using list codes in proofs, even for $l=0$. There is no need to prove that the decoding sets for the different $s_j(u_i)$ are disjoint. (See Section 5 of the present paper and the proof of Theorem 15.2.1 of [4].)

Corollary. *Let the l of* (4.6) *be zero. The message sequences*

$$\{s_j(u_i),\ i=1,\ldots,N;\ j=1,\ldots,T\} \tag{4.13}$$

form a code $X_2 \to X_3$ (i.e., *with respect to the c.p.f.* $w_{3|2}$) *with parameters*

$$(\pi_2, n, \exp_2\{nI(X_2,X_3)-K_3\sqrt{n}\},\ \lambda_2, 0). \tag{4.14}$$

This follows immediately from Theorem 2 and the fact that

$$I(X_1,X_3)+I(X_2,X_3|X_1)=I([X_1,X_2],X_3)=I(X_2,X_3). \tag{4.15}$$

The last equality follows from the fact that X_1, X_2, X_3 form a Markov chain.

5. Application to Degraded Channels

Degraded channels were introduced by Cover [5] and the coding theorem for them was proved by Bergmans [1]. We also refer the reader to Chapter 15 of [4]. It is assumed that the reader is familiar with the statement of the coding theorem (Theorem 15.2.1 of [4]), which we will now show to be an immediate consequence of Theorem 2.

Let X_1, X_2, X_3, X_4 form a Markov chain, in this order. From the conclusion at the bottom of page 128 of [4] we have that

$$I(X_1,X_3) \geqq I(X_1,X_4). \tag{5.1}$$

Let (4.1) be a code $X_1 \to X_4$ with

$$N=\exp_2\{nI(X_1,X_4)-K_3\sqrt{n}\}. \tag{5.2}$$

As in Theorem 2 there exists a code $X_2 \to X_3$ as in (4.4) with $l=0$ and

$$T = \exp_2 \{nI(X_2, X_3 | X_1) - K_4 \sqrt{n}\}. \tag{5.3}$$

The only thing left to verify is that, as in the proof of Theorem 2, the receiver who corresponds to X_3 can decode which u_i was sent. Since the u's form a code $X_1 \to X_4$, a receiver who would correspond to X_4 could decode u_i. Then the receiver who corresponds to X_3 surely can do so, since X_1, X_2, X_3, X_4 form a Markov chain. A graphic way to see this is as follows: The receiver X_3 performs a chance experiment with c.p.f. $w_{4|3}$ on the sequence he receives. (This is really what the channel does.) The resulting sequence can be used to decode u_i as in the code $X_1 \to X_4$.

We have just proved Bergmans coding theorem [1]. The proof is simpler than that of [4], Theorem 15.2.1, because one does not have to worry about decoding sets being disjoint.

6. Application to Determining the Rate Distortion Function for Source Coding with Side Information at the Decoder

The subject of rate distortion was introduced by Shannon ([6], see also [4], Sections 11.1–11.4) and that of source coding by Slepian and Wolf ([9], see also [4], Section 12.1). In [7] the present author gave a "geometric" proof of Shannon's result which makes the latter more intuitive. The problem described in the title of this section was solved by Wyner and Ziv ([2], see also [8]). In [4], Theorem 13.2.1, the present author gave a somewhat simpler proof of the positive half of the Wyner-Ziv theorem. In the present section we use the results on compound codes of Section 3 to give an even simpler proof of this result. Without more ado we adopt the notation of Theorem 13.2.1 of [4]. Further references are to its proof.

In the proof of the theorem we now construct a compound code for the channels $Z \to X$ and $Z \to Y$, with the parameters of Theorem 1 as follows: Let π be the distribution of Z, which is some chance variable which satisfies the conditions of the theorem. Choose the N of Theorem 1 to be

$$\exp_2 \{nI(Z, X) - K_1 \sqrt{n}\}. \tag{6.1}$$

Let l_1 correspond to the $Z \to X$ code and l_2 to the $Z \to Y$ code. Then, by Theorem 1, we can have $l_1 = 0$ and

$$l_2 < I(Z, X) - I(Z, Y) + \frac{K_2}{\sqrt{n}}. \tag{6.2}$$

From (6.2), Lemma 6.2 of [10], and the Slepian-Wolf theorem ([4], Theorem 12.1.1), Theorem 13.2.1 follows easily.

During the writing of this paper the author had interesting conversations with Miss S.E. Wolfowitz.

References

1. Bergmans, P.P.: Random coding theorem for broadcast channels with degraded components. IEEE Trans. Information Theory, IT-**19**, 197–207 (1973)
2. Wyner, A.D., Ziv, J.: The rate distortion function for source coding with side information at the decoder. IEEE Trans. Information Theory, IT-**21**, 294–300 (1975)
3. Wolfowitz, J.: On list codes. [To appear in the Journal of Combinatorics, Information and System Sciences]
4. Wolfowitz, J.: Coding theorems of information theory. Third edition. Berlin-Heidelberg-New York: Springer 1978
5. Cover, T.M.: Broadcast channels. IEEE Trans. Information Theory, IT-**18**, 2–14 (1972)
6. Shannon, C.E.: Coding theorems for a discrete source with a fidelity criterion. Information and decision processes; R.E. Machol, editor. New York: McGraw-Hill 1960
7. Wolfowitz, J.: Approximation with a fidelity criterion. Proc. Fifth Berkeley Sympos. Math. Statist. Probab.: Univ. Calif. 1965/1966, 565–573
8. Wolfowitz, J.: An upper bound on the rate distortion function for source coding with partial side information at the decoder. [To appear]
9. Slepian, D., Wolf, J.K.: Noiseless coding of correlated information sources. IEEE Trans. Information Theory, IT-**19**, 471–480 (1973)
10. Wolfowitz, J.: The coding of messages subject to chance errors. Illinois J. Math., **1**, 591–606 (1957)

Received August 25, 1978

Z. Wahrscheinlichkeitstheorie verw. Gebiete
50, 245–255 (1979)

Zeitschrift für
Wahrscheinlichkeitstheorie
und verwandte Gebiete

The Rate Distortion Function for Source Coding with Side Information at the Decoder

J. Wolfowitz*

Department of Mathematics, University of South Florida
Tampa, Fl 33620, USA

To Professor Leopold Schmetterer, on the occasion of his 60th birthday, with high esteem

1. Description of the Problem

The description which follows is intended to be self-contained. If anything is unintentionally omitted or a symbol unspecified, what is omitted will be found in [1], especially Sections 11.5 and 13.2. The notation of [1] is adopted in toto. In addition, we adopt the following convention: If, for example, Z, X, Y, U, are chance variables, $Z \to X \to Y \to U$ means that they form a Markov chain in this order.

Let (X, Y) be a chance variable such that X takes values in $A = \{1, \ldots, a\}$ and Y takes values in $B = \{1, \ldots, b\}$. Write A_n^* for the Cartesian product of nA's, and similarly for B_n^* and C_n^*, where $C = \{1, \ldots, c\}$ is called the "reproduction alphabet". Let d be a given "distortion function", defined on $(A \times C)$, with of course finite values. We extend the definition of d to $(A_n^* \times C_n^*)$ as follows: Let $a_n = (a^1, \ldots, a^n)$, $c_n = (c^1, \ldots, c^n)$. Then

$$d(a_n, c_n) = \frac{1}{n} \sum_{i=1}^{n} d(a^i, c^i). \tag{1.1}$$

No confusion will be caused by the use of d in these two senses.

Let (X^i, Y^i), $i = 1, \ldots, n$, be independent chance variables with the same distribution as (X, Y). Write $X_n = (X^1, \ldots, X^n)$, and similarly for Y_n. The "side information" function g_n is defined on B_n^* and takes values in $\{1, \ldots, \|g_n\|\}$. Let R_0 be a given constant, $0 \leqq R_0 \leqq H(Y)$. It is required that

$$\overline{\lim_{n \to \infty}} \frac{1}{n} \log \|g_n\| \leqq R_0. \tag{1.2}$$

---

* Research supported by the National Science Foundation under Grant No. MCS 79-05813

The "encoding" function f_n is defined on A_n^* and takes values in $\{1, \ldots, \|f_n\|\}$. Let z be a given number in the appropriate interval and $\varepsilon > 0$ be arbitrary. The "reproducing" function γ_n:

$$(1, \ldots\ldots, \|f_n\|) \times (1, \ldots\ldots, \|g_n\|) \to C_n^*,$$

is to satisfy, for n large enough,

$$Ed(X_n, \gamma_n(f_n(X_n), g_n(Y_n))) < z + \varepsilon. \tag{1.3}$$

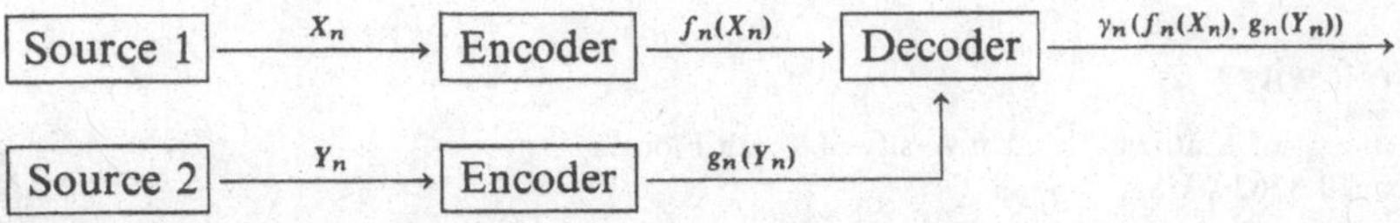

Thus, for z fixed throughout our argument, f_n, g_n, and γ_n depend upon ε. This dependence will not be exhibited, but it is to be understood implicitly. The objective of this paper is to discuss the rate distortion function

$$D(z) = \lim_{\varepsilon \downarrow 0} \lim_{n \to \infty} \frac{1}{n} \log(\text{minimal } \|f_n\|) \tag{1.4}$$

which is written as $D(z|\text{dec})$ in [1].

If $R_0 = H(Y)$ one can conveniently take $g_n(Y_n) \equiv Y_n$, as it is easy to verify. The case $R_0 = H(Y)$ is called the case of complete information, and any case where $R_0 < H(Y)$ is called "a" or "the" case of partial information. The case $R_0 = 0$ was solved by Shannon ([5], also [1], Sect. 11.1–11.4), and the case $R_0 = H(Y)$ was solved by Wyner and Ziv ([2], Sect. 11.5 and 13.2; henceforth written as W-Z). An upper bound on $D(z)$ in the case of partial information was given in [3].

W-Z prove their result in [2] by obtaining upper and lower bounds on $D(z)$, bounds which coincide. Their method of obtaining the upper bound has been streamlined in [1] and [3]. Even more difficult and clever is their method of obtaining the lower bound. Their method has hitherto not been successfully extended to the case of partial information and, in the opinion of the present writer, it is unlikely that it can ever be automatically extended. The difficulty is to find an analogue to (13.2.17) of [1] and the argument which follows it.

In the opinion of the present writer, it is important to find a method for obtaining lower bounds in source coding problems which will operate in an almost automatic way. The Slepian-Wolf (S-W) theorem has made it possible to find upper bounds in many source coding problems. If an almost automatic way of obtaining lower bounds is not found, the subject of source coding will not be unified and will become a collection of special problems, each solved by a clever ad hoc method.

We believe that the chance variables Z_n determined by the parent function (see Sect. 2) will provide the tool for a general method. They uncover what the S-W method so cleverly conceals. As an illustration, in Sect. 3, using the Z_n, we obtain the lower bound of W-Z, which is their more difficult result, by a simple method. In Sect. 6 we do the same by a trivial method.

In Sect. 4 we reduce the problem of obtaining the lower bound in the case of partial information to a problem in analysis. Although we have not been able to solve this problem as of this writing, we have little doubt that the problem will shortly be solved, and that the solution will be as we have conjectured it, the upper bound obtained in [3]. In Sect. 6 we obtain the result of Sect. 4 by a trivial method.

In Sect. 7 we give an even more streamlined method, almost a trivial one, of obtaining the upper bounds of [2] and [3].

The conditions (1.2)–(1.4) of the present paper are all limiting conditions. One should therefore always speak of n large enough or write $\{f_n\}$, for example. In order not to weary the reader we do not always bother to do this. It will be evident that no error will be caused thereby.

2. The Parent Function of f_n

Let f_n be the function defined in Sect. 1, whether for the problem with complete or with partial side information at the decoder. The function f_n is obtained by applying the S-W theorem to a certain function of X_n, say $F_n(X_n)$, which we shall call a "parent" function of f_n. The function F_n takes the values $1, 2, \ldots, \|F_n\|$. Application of the S-W theorem permits a drastic reduction (in general) of $\|F_n\|$, i.e., $\|f_n\| \ll \|F_n\|$. The process by which F_n is obtained from f_n is thus the reverse of the S-W process. We may also call F_n a parent function of the problem if only one parent function is involved.

The values of a parent function are $1, 2, \ldots, \|F_n\|$. Only its contours are relevant, i.e., those sets of values of X_n (sets of sequences) where F_n is a constant. In our problem the contours of F_n are determined by the pairs ($f_n(X_n)$=constant, $g_n(Y_n)$=constant).[2] The set $\{f_n(X_n)=\text{constant}\}$ is the union of a number of such contours, which are all represented by one value of $f_n(X_n)$; this reflects the economy achieved by the use of the S-W theorem. Adjoining the value of $g_n(Y_n)$ determines the actual contour among the union of contours.

We have spoken of "a" parent function for two reasons: a) The enumeration of the contours is obviously not uniquely defined. b) Even the function f_n is not uniquely defined. The sequence $\{f_n\}$ has to satisfy certain limit conditions, but any finite number of f_n may be arbitrarily defined. Moreover, infinitely many f_n may be changed if the changes are so small that their effect vanishes in the limit. These remarks apply equally well to $\{F_n\}$.

Thus there are really only equivalence classes of $\{f_n\}$ and $\{F_n\}$. Nevertheless, it will often be convenient to speak of "the" parent function when there is no possibility of the reader's being misled.

It will be easier first to discuss F_n in detail in the W-Z case of complete side information. We shall take $g_n(Y_n) \equiv Y_n$. One could just as well take $\log \|g_n(Y_n)\| = nH(Y) + \sqrt{n} \log n$, for example.

For any i, $1 \leqq i \leqq n$, the set of X_n-sequences, which is the contour of $F_n(X_n) = a$ constant, has a (conditional) probability distribution on X^i which can be

[2] A contour of F_n is as follows: $\{f_n(X_n) = c_1$, a constant$\}$ determines a set of values of X_n. Now adjoin a value of Y_n, say c_2. (In the general case, a value of $g_n(Y_n)$). The contour $\{f_n(X_n) = c_1, Y_n = c_2\}$ is a subset of $\{f_n(X_n) = c_1\}$ for which $\gamma_n(c_1, c_2)$ is constant.

If the construction of f_n of Section 7 below is used, the contours of F_n are the decoding sets of the codes constructed there. In [1], Theorem 13.2.1, the contours of F_n are the G's.

represented by a probability a-vector. Let s_n, say, be the number of different such vectors for all i and all values of F_n. Number these vectors arbitrarily from 1 to s_n. Let Z^i be a chance variable which takes the values $1, \ldots, s_n$ with the same probabilities as those of the contours of F_n which give rise to the corresponding vectors. Z^i is defined for $i=1, \ldots, n$. Then, for each i, $Z^i \to X^i \to Y^i$. Moreover, the value of $(Z^1, \ldots, Z^n) = Z_n$ is determined by the value of F_n.[3] Also $Z_n \to X \to Y_n$.

Suppose $Z^i = z_0$, $Y^i = y$. Write the resulting conditional distribution of X^i as

$$(\alpha_1(z_0, y), \ldots, \alpha_a(z_0, y)). \tag{2.1}$$

Write the c a-vectors

$$(d(1, i), d(2, i), \ldots, d(a, i)), \quad i=1, \ldots, c. \tag{2.2}$$

Let c_0 be a value of i such that

$$\sum_{j=1}^{a} \alpha_j(z_0, y)\, d(j, c_0) \tag{2.3}$$

is least. Denote c_0 by $\psi_0(z_0, y)$ and the value of (2.3) by $\delta(z_0, y)$.

From the previous paragraphs and the definition (1.1) it follows that, for our problem, we can replace $F_n(X_n)$ by the sequence

$$Z_n = (Z^1, \ldots, Z^n) \tag{2.4}$$

and

$$Ed(X_n, \gamma_n(f_n(X_n), Y_n)) \tag{2.5}$$

by

$$\frac{1}{n} \sum_{i=1}^{n} E\delta(Z^i, Y^i). \tag{2.6}$$

Obviously, the quantity in (2.6) is the smallest value which the quantity in (2.5) can take, given f_n. Moreover, if two values of F_n produce the same value of Z_n, we can "consolidate" the two values of F_n without increasing (2.6). Hence

$$\|F_n(X_n)\| \geqq \|Z_n\|, \tag{2.7}$$

where the right member of (2.7) denotes the number of different values (sequences) which Z_n can take with positive probability.

Suppose now that V is a given, finite-valued chance variable such that $V \to X \to Y$. We shall show that one can construct a function $F_n(X_n)$ such that the corresponding quantity (2.6) differs from $E\delta(V, Y)$ by less than an arbitrarily given positive ε for n sufficiently large. To do this let V^i, $i=1, \ldots, n$, be independent chance variables, each with the same distribution as V. Using Lemma 11.2.1 of [1], one constructs a code $V \to X$ just as in [1], p. 150. The values of F_n (sequences V_n) are the message sequences of the code. The decoding sets of the code contain only sequences generated by the message sequences. By

---

3 A simple illustration may help you to understand these definitions. Suppose $n = 2$, $a = 3$, and a contour of F_n is $\{(1, 1), (1, 2), (2, 3)\}$, with probability $\frac{1}{9}$ for each pair. Then Z^1 takes that value, say α, which corresponds to $(\frac{2}{3}, \frac{1}{3}, 0)$, and Z^2 takes that value, say β, which corresponds to $(\frac{1}{3}, \frac{1}{3}, \frac{1}{3})$. The probability is $\frac{1}{3}$ that $(Z^1, Z^2) = (\alpha, \beta)$.

proper choice of constants in the definition of generation, the Y-sequences generated by the sequences in a decoding set are generated by its message sequence with a probability as close to 1 as desired. One readily verifies, as in [1], p. 150, that the quantity (2.6) differs from $E\delta(V, Y)$ by less than ε for n sufficiently large. Let Z_n be as before for the function F_n just constructed. Then

$$\frac{1}{n}\log\|Z_n\| < \bar{I}(V, X) + \frac{K}{\sqrt{n}}, \tag{2.8}$$

where K will be the generic designation for a suitable positive constant.

Suppose that in the previous paragraph V is a sequence of m chance variables, which need not be independent or identically distributed, but should be finite valued, and suppose that $V \to X_m \to Y_m$. Then one can construct, as in the previous paragraph, a function F_n, $n = mr + s$, $0 \leqq s < m$, r sufficiently large, so that the resulting Z_n does in blocks of length m what was previously done by single Z^i's. The proof is essentially the same.

Write the V of the preceding paragraph as $V_m = V^1, \ldots, V^m$. The V^i are, of course, not necessarily independently distributed. Let $V_0^1, \ldots, V_0^m (= V_{0m})$ be independent chance variables with the same (marginal) distributions as $V^1, \ldots, V^m$, respectively. Then one can construct a function F_n as in the last paragraph, again in blocks of length m, through a code $V_{0m} \to X_m \to Y_m$.

The idea of the paragraph which follows (2.7) is used in [1] and [3] to obtain an upper bound on $\|f_n\|$, and also in the new, very simple method of Sect. 7 below.

3. The Lower Bound on $\|f_n\|$ in the Case of Complete Information

We may, and do, assume $g_n(Y_n) \equiv Y_n$. Then

$$\begin{aligned}
\log\|f_n\| &\geqq H(f_n) = \bar{I}(X_n, f_n) = H(X_n) - H(X_n \mid f_n) \\
&= H(X_n) - H(X_n \mid Y_n, f_n) - \bar{I}(X_n, Y_n \mid f_n) \\
&\geqq H(X_n) - H(X_n \mid Y_n, Z_n) - H(Y_n) + H(Y_n \mid X_n) \\
&= \bar{I}(X_n, Z_n \mid Y_n).
\end{aligned} \tag{3.1}$$

The sixth member of the above inequality is due to the following:

a) $H(Y_n) \geqq H(Y_n \mid f_n)$.

b) $H(Y_n \mid X_n, f_n) = H(Y_n \mid X_n)$.

c) The pair (Y_n, f_n) determines the pair (Y_n, Z_n).

Since $X^1, \ldots, X^n$ are independent, by a well known inequality ([1], (11.3.9)),

$$\bar{I}(X_n, Z_n) \geqq \sum_{i=1}^{n} \bar{I}(X^i, Z^i).$$

The same proof gives

$$\bar{I}(X_n, Z_n \mid Y_n) \geqq \sum_{i=1}^{n} \bar{I}(X^i, Z^i \mid Y_n). \tag{3.2}$$

Now $Z^i \to X^i \to Y^i$, and, when Y^i is given, $Y_{i-1}, Y^{i+1}, \ldots, Y^n$ give no additional information about X^i. Hence the right member of (3.2) equals

$$\sum_{i=1}^{n} \bar{I}(X^i, Z^i \mid Y^i), \tag{3.3}$$

the lower bound obtained by W-Z by another method ([1], (13.2.21)). They then employ an argument based on the Carathéodory theorem to prove that there exists a chance variable V such that $V \to X \to Y$, V takes the values $1, \ldots, (a+1)$,

$$nE\delta(V, Y) = \sum_{i=1}^{n} E\delta(Z^i, Y^i), \tag{3.4}$$

and

$$n\bar{I}(V, X \mid Y) \leqq \sum_{i=1}^{n} \bar{I}(Z^i, X^i \mid Y^i). \tag{3.5}$$

From (3.3) and (3.5) we therefore obtain that, for n sufficiently large,

$$\frac{1}{n} \log \|f_n\| \geqq \min I(V, X \mid Y), \tag{3.6}$$

where the minimum is with respect to all chance variables V such that $V \to X \to Y$, V takes the values $1, \ldots, (a+1)$, and $\delta(V, Y) \leqq z + \varepsilon$. This is the final W-Z lower bound. This is also essentially the upper bound obtained by W-Z ([1], Theorem 13.2.1, $\varepsilon \to 0$), and consequently is the value of the rate distortion $D(z)$.

4. The Lower Bound on $\|f_n\|$ in the Case of Partial Information

Let $g_n(Y_n)$ be given. We first proceed as in Sect. 2, not to obtain a parent function but to obtain chance variables $(U^1, \ldots, U^n) = U_n$ which will be equivalent to $g_n(Y_n)$ for all our purposes. Let $\{g_n(Y_n) = \text{constant}\}$ be a set of Y_n-sequences which form a contour of g_n. For any i, $1 \leqq i \leqq n$, this set defines a (conditional) probability distribution on Y^i, which can be represented by a probability b-vector. Let s'_n say, be the number of different such vectors for all i and all values of g_n. Number these vectors arbitrarily from 1 to s'_n. Let U^i be a chance variable which takes the values $1, \ldots, s'_n$ with the same probabilities as those of the contours of g_n which give rise to the corresponding vectors. U^i is defined for $i = 1, \ldots, n$. Then, for each i, $U^i \to Y^i \to X^i$. Moreover, the value of $(U^1, \ldots, U^n) = U_n$ is determined by $g_n(Y_n)$. Also $U_n \to Y_n \to X_n$. Let $f_n(X_n)$ be an optimal f_n for the given $g_n(Y_n)$. This means that, subject to the condition (1.3) (of course g_n must satisfy condition (1.2)), $\|f_n\|$ is minimal. The pair $(f_n(X_n), g_n(Y_n))$ determines the pair (Z_n, U_n). Then $U_n \to Y_n \to X_n \to Z_n$ and $U^i \to Y^i \to X^i \to Z^i$ for each i. The number $\|f_n\|$ is determined for each value of g_n by the S-W theorem. It follows from Sect. 7 below or from an argument like that of [3] that two values of g_n which give the same value of U_n can be consolidated in the process of constructing an optimal f_n for a given g_n. Hence the pair (U_n, f_n) determines the pair (U_n, Z_n).

Suppose $Z^i = z_0$, $U^i = u$. These determine the (conditional) distribution of X^i in a manner similar to that of Sect. 2, since $U^i \to Y^i \to X^i \to Z^i$. Consequently $\delta(z_0, u)$ can be determined in a manner similar to that of Sect. 2. Again,

$$\frac{1}{n} \sum_{i=1}^{n} E\delta(Z^i, U^i) \tag{4.1}$$

is the minimal value of

$$Ed(X_n, \gamma_n(f_n(X_n), g_n(Y_n))) \tag{4.2}$$

for the given g_n and a function f_n optimal for g_n.

As in Sect. 3, we have

$$\begin{aligned}\log \|f_n\| &\geqq H(X_n) - H(X_n \mid f_n)\\ &= H(X_n) - H(X_n \mid U_n, f_n) - \bar{I}(X_n, U_n \mid f_n)\\ &\geqq H(X_n) - H(X_n \mid U_n, Z_n) - H(U_n) + H(U_n \mid X_n)\\ &= \bar{I}(X_n, Z_n \mid U_n) \geqq \sum_{i=1}^{n} \bar{I}(X^i, Z^i \mid U^i).\end{aligned} \tag{4.3}$$

Returning to condition (1.2), we have

$$nR_0 \geqq \log \|g_n\| \geqq \log \|U_n\| \geqq \bar{I}(U_n, Y_n) \geqq \sum_{i=1}^{n} \bar{I}(U^i, Y^i).$$

Thus the condition

$$nR_0 \geqq \sum_{i=1}^{n} \bar{I}(U^i, Y^i) \tag{4.4}$$

is, if anything, weaker than condition (1.2), and therefore must hold.

We now return to inequality (4.3). Since $Z^i \to X^i \to Y^i \to U^i$, it follows, by the same argument as that of W-Z ([1], last two paragraphs of Chap. 13) that there exists a chance variable V_{00}^i such that V_{00}^i takes the values $1, \ldots, (a+1)$, $V_{00}^i \to X^i \to Y^i \to U^i$,

$$E\delta(V_{00}^i, U^i) = E\delta(Z^i, U^i), \tag{4.5}$$

and

$$\bar{I}(V_{00}^i, X^i \mid U^i) = \bar{I}(Z^i, X^i \mid U^i). \tag{4.6}$$

In the next section we prove that there exists a chance variable U_0^i, which takes the values $1, \ldots, (b+2)$, such that

$$V_{00}^i \to X^i \to Y^i \to U_0^i, \tag{4.7}$$

$$E\delta(V_{00}^i, U_0^i) = E\delta(V_{00}^i, U^i), \tag{4.8}$$

$$\bar{I}(V_{00}^i, X^i \mid U_0^i) = \bar{I}(V_{00}^i, X^i \mid U^i), \tag{4.9}$$

and

$$\bar{I}(Y^i, U_0^i) = \bar{I}(Y^i, U^i). \tag{4.10}$$

From (4.3) we therefore have

$$\log \|f_n\| \geqq \sum_{i=1}^{n} \bar{I}(X^i, V_{00}^i \mid U_0^i). \tag{4.11}$$

All of the above was proved under the assumption that $g_n(Y_n)$ is given.

We conjecture that there exist chance variables V, U such that $V \to X \to Y \to U$, V takes the values $1, \ldots, (a+1)$, U takes the values $1, \ldots, (b+2)$,

$$E\delta(V, U) = \frac{1}{n} \sum_{i=1}^{n} E\delta(V_{00}^i, U_0^i), \tag{4.12}$$

$$\bar{I}(Y, U) = \frac{1}{n} \sum_{i=1}^{n} \bar{I}(Y^i, U_0^i) \tag{4.13}$$

and

$$\bar{I}(V, X \mid U) \leqq \frac{1}{n} \sum_{i=1}^{n} \bar{I}(V_{00}^i, X^i \mid U_0^i). \tag{4.14}$$

(Compare (3.4) and (3.5)). We have not yet been able to prove this conjecture. Suppose it to be true. Then it follows from (4.11) that

$$\frac{1}{n} \log \|f_n\| \geqq \min I(V, X \mid U), \tag{4.15}$$

where the minimum is over all chance variables V, U such that

$$V \to X \to Y \to U, \tag{4.16}$$

V takes the values $1, \ldots, (a+1)$, U takes the values $1, \ldots, (b+2)$,

$$E\delta(V, U) \leqq z, \tag{4.17}$$

and

$$\bar{I}(Y, U) \leqq R_0. \tag{4.18}$$

This is also the upper bound obtained in [3], and would therefore be the value of $D(z)$. This last conclusion of course rests on the validity of the conjecture.

5. Application of the Carathéodory Theorem

Consider the totality T of all vectors with $(b+3)$ components, as follows: The first b components are a probability b-vector; it can be regarded as the conditional distribution of Y^i, given $U^i = u$, say. The next component is $H(Y^i) - H(Y^i \mid U^i = u)$, which is determined by the first b components. The next component is $E\delta(u, V_{00}^i) = E[\delta(U^i, V_{00}^i) \mid U^i = u]$, which is also determined by the first b components. The last component is $H(V_{00}^i \mid U^i = u) - H(V_{00}^i \mid X^i)$, which, like the others, is also determined by the first b components.

The set T is the map of the compact, connected set of the vectors which are its first b components, under a continuous transformation. It is therefore compact and connected. Its dimension is (at most) $b+2$, since the first b components sum to one. Let $\bar{T}$ be its convex hull. Suppose U^i takes t values. Then a certain convex combination of t vectors with weights equal to the respective probabilities of the values of U^i lies in $\bar{T}$. The components of this (combination) vector are as follows: The first b components are the distribution of Y^i, which is of course the same as that of Y. The next component is $\bar{I}(Y^i, U^i)$. The next component is $E\delta(U^i, V_{00}^i)$. The last component is $\bar{I}(V_{00}^i, X^i \mid U^i)$. By the generalized Carathéodory theorem ([4], Theorem 18) every point of $\bar{T}$ can be represented as a convex combination of $b+2$ points of T. Hence we have proved the following: Suppose given $V_{00}^i \to X^i \to Y^i \to U^i$, as in Sect. 4. (We recall that V_{00}^i takes $(a+1)$ values.) There exists a chance variable U_0^i, which takes $(b+2)$ values, such that

$$\begin{aligned}
&V_{00}^i \to X^i \to Y^i \to U_0^i,\\
&E\delta(V_{00}^i, U_0^i) = E\delta(V_{00}^i, U^i),\\
&\bar{I}(Y^i, U_0^i) = \bar{I}(Y^i, U^i),\\
&\bar{I}(V_{00}^i, X^i \mid U_0^i) = \bar{I}(V_{00}^i, X^i \mid U^i).
\end{aligned}$$

6. The Lower Bound Revisited

Consider first the problem where $g_n(Y_n) \equiv Y_n$. The pair (f_n, Y_n) determines (Z_n, Y_n). Hence

$$H(f_n(X_n), Y_n) \geqq H(Z_n, Y_n),$$

so that

$$H(f_n(X_n) \mid Y_n) \geqq H(Z_n \mid Y_n).$$

Hence

$$\begin{aligned}
\log \|f_n\| &\geqq H(f_n) \geqq H(Z_n \mid Y_n)\\
&\geqq \bar{I}(Z_n, X_n \mid Y_n) \geqq \sum_{i=1}^{n} \bar{I}(Z^i, X^i \mid Y_n)\\
&= \sum_{i=1}^{n} \bar{I}(Z^i, X^i \mid Y^i).
\end{aligned}$$

Now consider a general $g_n(Y_n)$. The pair (f_n, U_n) determines the pair (Z_n, U_n). Hence

$$H(f_n(X_n) \mid U_n) \geqq H(Z_n \mid U_n)$$

and, as before

$$\log \|f_n\| \geqq \sum_{i=1}^{n} \bar{I}(Z^i, X^i \mid U^i).$$

7. A General Slepian-Wolf Theorem. The Upper Bound on $\|f_n\|$

Suppose given N sets $S_1, \ldots, S_N$ of elements of some kind. Each set contains at most A elements, and the sets need not be disjoint. The probability of set S_i is α_i, $i=1, \ldots, N$. Given any S_i, there is a conditional probability distribution on its elements. Let $\sum$ be the union of the S_i. To each element of $\sum$ one of the numbers $1, \ldots, B$ is assigned at random (with equal probability), independently of the assignment to the other elements. An "error occurs" (at an element of a set) if another element in the same S_i is assigned the same number as the first element.

Conclusion. The expected probability of an error is $<\frac{A}{B}$.

Proof. Consider an element σ of any set, say S_i. The conditional expected value of the error that occurs whenever another element in S_i has the same number as σ is not greater than $\frac{A-1}{B}$. This proves the desired result.

The above is the essence of the S-W theorem with Cover's proof ([1], Theorem 12.1.1). The theorem is usually applied with A and B functions of n such that $\frac{A}{B} \to 0$ as $n \to \infty$.

Let us now consider the problem of complete information, $g_n(Y_n) \equiv Y_n$. Let V be a chance variable which takes the values $1, \ldots, (a+1)$, $V \to X \to Y$, $E\delta(V, Y) \leqq z$, and $\bar{I}(V, X \mid Y)$ is minimal among all chance variables V which satisfy the above conditions. Let $\pi(Y)$ be the distribution of Y, and let y_n be a $\pi(Y)$-sequence in the space of Y_n. Consider the sequences generated $Y \to X$ by y_n, and for any sequence x_n among the latter, the sequences v_n generated $(Y, X) \to V$ by (y_n, x_n). Holding y_n, we apply Lemma 11.2.1 of [1], and obtain a code $V \to X$ for the fixed y_n. The length of this code is less than

$$\exp_2 \{n\bar{I}(V, X \mid Y) + K\sqrt{n}\}. \tag{7.1}$$

Each y_n designates a set S_i as just above, the elements of S_i being the message sequences of the code just constructed. Thus the quantity (7.1) is an upper bound on A. Now let

$$B = \exp_2 \{n\bar{I}(V, X \mid Y) + 2K\sqrt{n}\}. \tag{7.2}$$

Then $\frac{A}{B} \to 0$ as $n \to \infty$. Applying the S-W theorem in the above form, we obtain that (7.2) is an upper bound on $\|f_n\|$. This is the result of W-Z ([2]) cited in Sect. 3.

In the general case $g_n(Y_n)$ of partial information, the argument is the same, except that Y is now replaced by U. We have $V \to X \to Y \to U$, U takes the values $1, \ldots, (b+2)$, $E\delta(V, U) \leqq z$, and $\bar{I}(U, Y) \leqq R_0$. The upper bound on $\|f_n\|$ is now

$$\exp_2 \{n\bar{I}(V, X \mid U) + 2K\sqrt{n}\}, \tag{7.3}$$

which is the result of [3].

References

1. Wolfowitz, J.: Coding theorems of information theory. Third edition. Berlin-Heidelberg-New York: Springer 1978
2. Wyner, A.D., Ziv, J.: The rate distortion function for source coding with side information at the decoder. IEEE Trans. Information Theory, IT-22, 1–10 (1976)
3. Berger, T., Housewright, K., Omura, J., Tung, S.-Y., Wolfowitz, J.: An upper bound on the rate distortion function for source coding with partial side information at the decoder. [To appear in the IEEE Trans. Information Theory]
4. Eggleston, H.G.: Convexity. Cambridge and New York: Cambridge University Press 1958
5. Shannon, C.E.: Coding theorems for a discrete source with a fidelity criterion in Information and decision processes. R.E. Machol, editor. New York: McGraw-Hill 1960

Received June 1, 1979; in revised form July 27, 1979

Note Added in Proof. Professor R. Ahlswede has kindly informed the author that the S-W Theorem of Sect. 7 appeared in his paper "Coloring hypergraphs etc.", Jour. Combinatorics, Information and System Sciences, **4**, No. 1 (1979) 76–115.

Bibliography of the Publications of Jacob Wolfowitz

1939

[1] Confidence limits for continuous distribution functions, (with A. Wald), *Annals of Math. Stat.*, June, 1939. 10, 105–118, Note on same, *Annals of Math. Stat.*, 1941, 12, 118.

1940

[2] On a test whether two samples are from the same population, (with A. Wald), *Annals of Math. Stat.*, June, 1940, 11, 147–162.

1942

[3] Additive partition functions and a class of statistical hypotheses, *Annals of Math. Stat.*, September, 1942, 13, 247–279.

1943

[4] On the theory of runs with some applications to quality control, *Annals of Math. Stat.*, September, 1943, 14, 280–288.

[5] An exact test for randomness in the non-parametric case, (with A. Wald), *Annals of Math. Stat.*, December, 1943, 14, 378–388.

1944

[6] The covariance matrix of runs up and down, (with H. Levene) *Annals of Math. Stat.*, March, 1944, 15, 58–69.

[7] Note on runs of consecutive elements, *Annals of Math. Stat.*, March, 1944, 15, 97–98.

[8] Asymptotic distribution of runs up and down, *Annals of Math. Stat.*, June, 1944, 15, 163–172.

[9] Statistical tests based on permutations of the observations, (with A. Wald), *Annals of Math. Stat.*, December, 1944. 15, 358–370.

1945

[10] Sampling inspection plans for continuous production (with A. Wald) *Annals of Math. Stat.*, March, 1945, 16, 30–49.

1946

[11] Tolerance limits for a normal distribution, (with A. Wald) *Annals of Math. Stat.*, June, 1946, 17, 208–215.

[12] Confidence limits for the fraction of a normal population which lies between two given limits, *Annals of Math. Stat.*, December, 1946, 17, 483–488.

[13] On sequential binomial estimation, *Annals of Math. Stat.*, December, 1946. 17, 489–493.

1947

[14] Consistency of sequential binomial estimates, *Annals of Math. Stat.*, March, 1947, 18, 131–135.

[15] The efficiency of sequential estimates, *Annals of Math. Stat.*, June 1947. 18, 215–230.

1948

[16] Optimum character of the sequential probability ratio test, (with A. Wald), *Annals of Math. Stat.*, September, 1948. 19, 326–339.

1949

[17] Non-parametric statistical inference, *Proceedings Berkeley Symposium*, University of California Press, 1949.

[18] Bayes solutions of sequential decision problems, (with A. Wald) *Proceedings National Academy of Science*, U.S.A., February, 1949.

[19] The distribution of plane angles of contact, *Quarterly of Applied Math.*, April, 1949.

[20] Remarks on the notion of recurrence, *Bulletin of the American Math. Society*, 1949, 55, 394–5.

[21] The power of the classical tests associated with the normal distribution, *Annals of Math. Stat.*, December, 1949. 20, 540–551.

[22] On Wald's proof of the consistency of the maximum likelihood estimate, *Annals of Math. Stat.*, December, 1949.

1950

[23] Bayes solutions for sequential decision problems, (with A. Wald), *Annals of Math. Stat.*, March, 1950. 21, 82–99.

[24] Elimination of randomization in certain problems of statistics and the theory of games, (with A. Dvoretzky and A. Wald), *Proc. Nat. Acad. Sci., U.S.A.*, April, 1950.

[25] Minimax estimate of the mean of a normal distribution with known variance, *Annals of Math. Stat.*, June 1950. 21, 218–230.

1951

[26] Relations among certain ranges of vector measures, (with A. Dvoretzky and A. Wald), *Pacific J. of Math.*, March, 1951, Vol. 1. 59–74.

[27] Elimination of randomization in certain statistical decision procedures and

zerosum two person games, (with A. Dvoretzky and A. Wald), *Annals of Math. Stat.*, March, 1951, Vol. 22, 1–21.

[28] Two methods of randomization in statistics and the theory of games, (with A. Wald), *Annals of Math. Stat.*, 1951. 53, 581–6.

[29] Characterization of the minimal complete class of decision functions when the number of distributions and decisions is finite, (with A. Wald), *Proc. 2nd Berkeley Symp. on Prob. and Stat.*, University of California Press. 1951.

[30] Sums of random integers reduced modulo m, (with A. Dvoretzky), *Duke Math. J.*, 1951. 18, 501–507.

[31] On ϵ-complete classes of decision functions, *Annals of Math. Stat.*, September, 1951, 22, 461–464.

1952

[32] On a limit theorem in renewal theory, (with K.-L. Chung), *Annals of Math.*, January, 1952.

[33] The inventory problem I, (with A. Dvoretzky and J. Kiefer), *Econometrica*, April, 1952.

[34] The inventory problem II, (with A. Dvoretzky and J. Kiefer), *Econometrica*, July, 1952.

[35] On the stochastic approximation method of Robbins and Monro, *Ann. Math. Stat.*, 1952. 23, 457–461

[36] Stochastic estimation of the maximum of a regression function, (with J. Kiefer), *Ann. Math. Stat.*, 1952. 23, 462–466.

[37] Consistent estimators of the parameters of a linear structural relation, *Skandinavisk Aktuarietidskrift*, pp. 132–151, 1952.

1953

[38] The method of maximum likelihood and the Wald theory of decision functions, *Indagationes Matematicae*, Vol. 15, No. 2, 1953. 114–119.

[39] Sequential decision problems for processes with continuous time parameter. Testing hypotheses, (with A. Dvoretzky and J. Kiefer), *Ann. Math. Stat.*, June, 1953, pp. 254–264.

[40] Sequential decision problems with continuous time parameter. Problems of Estimation, (with A. Dvoretzky and J. Kiefer), *Ann. Math. Stat.*, June, 1953, pp. 403–415.

[41] On the optimal character of the (s,S) policy in inventory theory, (with A. Dvoretzky and J. Kiefer), *Econometrica*, October, 1953, pp. 586–596.

[42] Estimation by the minimum distance method, *Ann. of the Inst. of Stat. Math.*, Tokyo, Vol. 5(1953), No. 1, pp. 9–23.

1954

[43] Generalization of the theoreum of Glivenko-Cantelli, *Ann. Math. Stat.*, March, 1954, pp. 131–138.

[44] Estimation by the minimum distance method in non-parametric stochastic difference equations, *Ann. Math. Stat.*, June, 1954, pp. 203–217.

[45] Estimation of the components of stochastic structures, *Proc. Nat. Ac. Sc., U.S.A.*, Vol. 40, No. 7, July, 1954, pp. 602–606.

1955

[46] On the theory of queues with many servers, (with J. Kiefer), *Trans. Amer. Math. Soc.*, Vol. 78, No. 1, (1955), pp. 1–18.

[47] On tests of normality and other tests of goodness of fit based on the minimum distance method, (with M. Kac and J. Kiefer), *Ann. Math. Stat.*, June, 1955, pp. 189–211.

1956

[48] On the characteristics of the general queueing process, with applications to random walk, (with J. Kiefer), *Ann. Math. Stat.*, March, 1956, pp. 147–161.

[49] Asymptotic minimax character of the sample distribution function and of the classical multinomial estimator, (with A. Dvoretzky and J. Kiefer), *Ann. Math. Stat.*, September, 1956, pp. 642–669.

[50] Consistency of the maximum likelihood estimator in the presence of infinitely many incidental parameters, (with J. Kiefer), *Ann. Math. Stat.*, December, 1956, pp. 887–906.

[51]. On stochastic approximation methods, *Ann. Math. Stat.*, December, 1956, pp. 1151–1155.

[52] Sequential tests of hypotheses about the mean occurrence time of a continuous parameter Poisson process, (with J. Kiefer), *Naval Research Logistics Quarterly*, Vol. 3, No. 3, 1956, pp. 205–219.

1957

[53] The minimum distance method, *Ann. Math. Stat.*, March, 1957, pp. 75–88.

[54] The coding of messages subject to chance errors, *Ill. Jour. Math.*, Vol. 1, No. 4, December, 1957, pp. 591–606.

1958

[55] On the deviations of the empiric distribution function of vector chance variables, (with J. Kiefer), *Trans. Amer. Math. Soc.*, Vol. 87, No. 1, January, 1958, pp. 173–186.

[56] An upper bound on the rate of transmission of messages, *Ill. Jour. Math.*, Vol. 2, No. 1, March, 1958, pp. 137–141.

[57] The maximum achievable length of an error correcting code, *Ill. Jour. Math.*, Vol. 2, No. 3, September, 1958, pp. 454–458.

[58] Information theory for mathematicians, *Ann. Math. Stat.*, Vol. 29, No. 2, June, 1958, pp. 351–356.

[59] Distinguishability of sets of distributions, (with W. Hoeffding), *Ann. Math. Stat.*, Vol. 29, No. 3, 1958, pp. 700–718.

1959

[60] Optimum designs in regression problems, (with J. Kiefer), *Ann. Math. Stat.*, Vol. 30, No. 2, June, 1959, pp. 271–294.

[61] Asymptotic minimax character of the sample distribution function for vector

chance variables, (with J. Kiefer), *Ann. Math. Stat.*, Vol. 30, No. 2, June, 1959, pp. 463–489.

[62] Strong converse of the coding theorem for semi-continuous channels, *Ill. Jour. Math.*, December, 1959, Vol. 3, No. 4, pp. 477–489.

1960

[63] Convergence of the empiric distribution function on half-spaces, *Contributions to Probability and Statistics in honor of Harold Hotelling*, Stanford University Press, 1960.

[64] The equivalence of two extremum problems, (with J. Kiefer), *Canad. Jour. Math.*, 1960, pp. 363–366.

[65] Simultaneous channels, *Archive for rational mechanics and analysis, Vol. 4, No. 4, 1960, pp. 371*–386.

[66] A note on the strong converse of the coding theorem for the general discrete finite-memory channel, *Information and Control,* Vol. 3, No. 1, (1960), pp. 89–93.

[67] On channels in which the distribution of error is known only to the receiver or only to the sender, in *Information and Decision Processes,* edited by Robert E. Machol, New York, 1960, McGraw-Hill Book Co.

[68] Contributions to information theory, *Proc. Nat. Acad. Sci., U.S.A.*, Vol. 46, No. 4, pp. 557–561, April, 1960.

[69] On coding theorems for general simultaneous channels, *IRE Trans. On Circuit Theory,* Vol. CT-7, No. 4, December, 1960.

1961

[70] A channel with infinite memory, *Proc. Fourth Berkeley Symposium.* Berkeley and Los Angeles University of California Press, 1961.

[71] Coding theorems of information theory, Berlin. Springer-Verlag, First edition, 1961, Second edition, 1964. (Book)

1962

[72] Channels with arbitrarily varying channel probability functions, (with J. Kiefer), *Information and Control,* Vol. 5, No. 1, March, 1962. pp. 44–54.

[73] Bayesian inference and axioms of consistent decision, *Econometrica,* Vol. 30, No. 3, July, 1962, pp. 470–479.

1963

[74] Products of indecomposable, aperiodic, stochastic matrices, *Proc. Amer. Math. Soc.,* Vol. 14, No. 5, October, 1963, pp. 733–37.

[75] The capacity of an indecomposable channel, *Sankhya,* Series A. Vol. 25, Part I, June, 1963, pp. 101–108.

[76] On channels without a capacity, *Information and Control,* Vol. 6, No. 1, March, 1963, 49–54.

1964

[77] Optimum extrapolation and interpolation designs I (with J. Kiefer), *Ann. of the Inst. of Stat. Math.* (Japan), XVI, (1964), 79–108.

[78] Optimum extrapolation and interpolation designs II, (with J. Kiefer), *Ann. of the Inst. of Stat. Math.* (Japan), XVI, 1964, 295–303.

1965

[79] On a problem connected with the Vandermonde determinant, (with J. Kiefer), *Proc. Amer. Math. Soc.*, 16, No. 5, October, 1965, 1092–5.

[80] Asymptotic efficiency of the maximum likelihood estimator, *Teoriya Vyeroyatnostey*, 10, No. 2, 267–281, June, 1965.

[81] On a theorem of Hoel and Levine, (with J. Kiefer), *Ann. Math. Stat.*, 36, No. 6, December, 1965, 1627–1655.

[82] Approximation with a fidelity criterion, *Proc. Fifth Berkeley Symposium on Prob. and Math. Stat.*, 1965, I, 565–573.

1966

[83] Generalized maximum likelihood estimators, (with L. Weiss), *Teoriya Vyeroyatnostey*, 11, No. 1, 1966, 68–93.

[84] Existence of optimal stopping rules for linear and quadratic rewards, (with H. Teicher), *Zeitschrift fuer Wahrscheinlichkeitstheorie*, 5(1966), 361–368.

[85] Remark on the optimum character of the sequential probability ratio test, *Ann. Math. Stat.*, 37, No. 3, June, 1966, pp. 726–727.

1967

[86] The moments of recurrence time, *Proc. Amer. Math. Soc.*, Vol. 18, No. 4, August, 1967, 613–614.

[87] Estimation of a density function at a point, (with L. Weiss), *Zeitschrift fuer Wahrscheinlichkeitstheorie*, 7, 327–335, (1967).

[88] Maximum probability estimators (with L. Weiss), *Ann. Inst. Stat. Math.* (Tokyo), 19, No. 2, (1967), 193–206.

[89] Memory increases capacity, *Information and Control*, 11, 423–428 (1967).

[90] Remarks on the theory of testing hypotheses, *The New York Statistician*, Vol. 18, No. 7, March, 1967, 1–3.

1968

[91] Generalized maximum likelihood estimators in a particular case (with L. Weiss), *Teoriya Vyeroyatnostey*, 13, No. 4, (1968), 657–662.

[92] Note on a general strong converse, *Information and Control*, 12, No. 1, January, 1968, 1–4.

[93] Note on the Gaussian Channel with feedback and a power constraint, *Information and Control*, 12, No. 1, January, 1968, 71–78.

1969

[94] Reflections on the future of mathematical statistics, S. N. Roy Memorial Volume (1969), 739–750.

[95] Maximum probability estimators with a general loss function, (with L. Weiss), Proc. International Symposium on Prob. and Information Theory, held at McMaster University, Hamilton, Ontario, Canada, April 4 and 5, 1968, pp. 232–256. Springer-Verlag, Berlin-Heidelberg-New York, Lecture Notes in Mathematics, #89, 1969.

[96] Asymptotically minimax tests of composite hypotheses, (with L. Weiss), *Zeitschrift fuer Wahrscheinlichkeitstheorie.* 14 (1969), 161–168.

[97] The structure of capacity functions for compound channels, (with R. Ahlswede), Proc. International Symp. on Prob. and Information Theory, held at McMaster University, Hamilton, Ontario, Canada, April 4 and 5, 1968, pp. 12–54. Springer-Verlag, Berlin-Heidelberg-New York, Lecture Notes in Mathematics, #89, 1969.

[98] Correlated decoding for channels with arbitrarily varying channel probability functions, (with R. Ahlswede), *Information and Control,* 14, #5, (1969), 457–473.

1970

[99] The capacity of a channel with arbitrarily varying channel probability functions and binary output alphabet (with R. Ahlswede), *Zeitschrift fuer Wahrscheinlichkeitstheorie,* 15, #3, (1970), 186–194.

[100] Maximum probability estimators and asymptotic sufficiency (with L. Weiss), *Annals Institute Statistical Mathematics,* 22, #2 (1970), 225–244.

[101] Asymptotically efficient non-parametric estimators of location and scale parameters, (with L. Weiss) *Zeitschrift fuer Wahrscheinlichkeitstheorie,* 16, #2, (1970), 134–150.

[102] Asymptotically efficient tests and estimators, address at the 1970 International Congress of Mathematicians at Nice, France. Proc. vol. 3, 259–263.

[103] On systems on channels, *Proceedings of Symposium on Information Measures,* April 10–14, 1970, University of Waterloo, Canada.

[104] Asymptotically efficient estimation of non-parametric regression coefficients, (with L. Weiss), Statistical decision theory and related topics, Proc. Symp. held at Purdue University, November, 1970. Shanti S. Gupta and James Yaeckel, eds., pp. 29–39.

1971

[105] Channels without synchronization, (with R. Ahlswede), *Advances in Applied Probability,* Vol. 3 (1971), 383–403.

1972

[106] Optimal, fixed length, non-parametric, sequential confidence limits for a translation parameter, (with L. Weiss), *Zeitschrift fuer Wahrscheinlichkeitstheorie u. verw. Gebiete.* 24, #3, (1972), 203–209.

[107] An asymptotically efficient, sequential equivalent of the t-test, (with L. Weiss), *Journal of the Royal Statistical Society.* Series B, 34, #3 (1972), 456–400.

1973

[108] Maximum likelihood estimation of a translation parameter of a truncated distribution, (with L. Weiss), Annals of Statistics, 1, #5, (1973), 944–947.

1974

[109] Maximum probability estimators and related topics (with L. Weiss). Lecture Notes in Mathematics #424. Springer-Verlag Berlin-Heidelberg-New York, 1974.

[110] Asymptotically efficient non-parametric estimators of location and scale parameters II. *Zeitschrift fuer Wahrscheinlichkeitstheorie u. verw. Gebiete,* 30, 117–118 (1974).

1975

[111] Maximum probability estimators in the classical case and in the "almost smooth" case, *Teoriya Vyeroyatnostey i y. P.,* 20, No. 2, (1975), 371–379.

[112] Signalling over a Gaussian channel with feedback and autoregressive noise. *Journal of Applied Probability,* 12, No. 4 (1975), 713–723.

1976

[113] Asymptotically minimax estimation of concave and convex distribution functions (with J. Kiefer). *Zeitschrift fuer Wahrscheinlichkeitstheorie u. verw. Gebiete,* 34, (1976), 73–85.

[114] Asymptotically efficient estimators when the densities of the observations have discontinuities. *Ann. Inst. Stat. Math.* (Tokyo), 28, No. 3, (1976), 359–370.

[115] Asymptotically minimax estimation of concave and convex distribution functions II (with J. Kiefer). Proceedings Purdue Symposium on Statistical Decision Theory and Related Topics, held at Purdue University, May 17–19, 1976.

1978

[116] Coding theorems of information theory, Springer-Verlag Berlin-Heidelberg-New York. Third Edition, considerably revised, 1978.

1979

[117] On list codes. Accepted by the *Journal of Combinatorics, Information and System Sciences,* 4, No. 2, 117–122.

[118] An upper bound on the rate distortion function for source coding with partial side information at the decoder. IEEE Transactions on Information Theory, IT-25, No. 6, 664–666.

[119] Codes within codes. *Zeitschrift fuer Wahrscheinlichkeitsrechnung und v. G.* 46 (1979), 307–315.

[120] The rate distortion function for source coding with side information at the decoder. *Zeitschrift fuer Wahrscheinlichkeitstheorie und v. G,* 50, 245–255.

Printed in the United States
By Bookmasters